AF664805

Selected Topics in Mass Spectrometry in the Biomolecular Sciences

NATO ASI Series

Advanced Science Institutes Series

A Series presenting the results of activities sponsored by the NATO Science Committee, which aims at the dissemination of advanced scientific and technological knowledge, with a view to strengthening links between scientific communities.

The Series is published by an international board of publishers in conjunction with the NATO Scientific Affairs Division

A Life Sciences **B Physics**	Plenum Publishing Corporation London and New York
C Mathematical and Physical Sciences **D Behavioural and Social Sciences** **E Applied Sciences**	Kluwer Academic Publishers Dordrecht, Boston and London
F Computer and Systems Sciences **G Ecological Sciences** **H Cell Biology** **I Global Environmental Change**	Springer-Verlag Berlin, Heidelberg, New York, London, Paris and Tokyo

PARTNERSHIP SUB-SERIES

1. Disarmament Technologies	Kluwer Academic Publishers
2. Environment	Springer-Verlag / Kluwer Academic Publishers
3. High Technology	Kluwer Academic Publishers
4. Science and Technology Policy	Kluwer Academic Publishers
5. Computer Networking	Kluwer Academic Publishers

The Partnership Sub-Series incorporates activities undertaken in collaboration with NATO's Cooperation Partners, the countries of the CIS and Central and Eastern Europe, in Priority Areas of concern to those countries.

NATO-PCO-DATA BASE

The electronic index to the NATO ASI Series provides full bibliographical references (with keywords and/or abstracts) to more than 50000 contributions from international scientists published in all sections of the NATO ASI Series.
Access to the NATO-PCO-DATA BASE is possible in two ways:

– via online FILE 128 (NATO-PCO-DATA BASE) hosted by ESRIN, Via Galileo Galilei, I-00044 Frascati, Italy.

– via CD-ROM "NATO-PCO-DATA BASE" with user-friendly retrieval software in English, French and German (© WTV GmbH and DATAWARE Technologies Inc. 1989).

The CD-ROM can be ordered through any member of the Board of Publishers or through NATO-PCO, Overijse, Belgium.

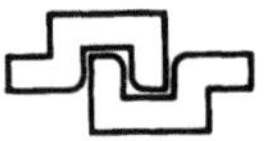

Series C: Mathematical and Physical Sciences – Vol. 504

Selected Topics in Mass Spectrometry in the Biomolecular Sciences

edited by

Richard M. Caprioli
University of Texas,
Medical School,
Houston, Texas, U.S.A.

Antonio Malorni
SESMA-C.N.R.,
Napoli, Italy

and

Giovanni Sindona
Università della Calabria,
Dipartimento di Chimica,
Arcavacata di Rende, Italy

Springer-Science+Business Media, B.V.

Proceedings of the NATO Advanced Study Institute on
Mass Spectrometry in the Biomolecular Sciences
Altavilla-Milicia (PA), Italy
7–18 July 1996

A C.I.P. Catalogue record for this book is available from the Library of Congress.

ISBN 978-94-010-6171-1 ISBN 978-94-011-5165-8 (eBook)
DOI 10.1007/978-94-011-5165-8

Printed on acid-free paper

CONTENTS

Part II : Instrumentation for Mass Analysis and Detection

Part III: Application to Biomolecules

PREFACE

This book is the result of efforts of both young and senior scientists who met together in Altavilla-Milicia (Italy) on July 1996 on the occasion of the NATO advanced study institute (ASI) in *Mass Spectrometry in Biomolecular Sciences*. The recent developments in instrumentation and methodologies of mass spectrometry have been discussed in this beautiful area of the Mediterranean located on the island of Sicily.

The aim of the book is to provide young scientists with accounts of the recent developments of methods and applications in biomolecular sciences. It was decided therefore to present in part I only those ionization methods such as electrospray and MALDI techniques which are currently under development and refinement. Two chapters have been devoted to GC/MS and quantitative analysis and four contributions deal with the physical-chemical aspects of the methodology.

In part II are presented the recent achievements in instrumentation for mass analysis and detection which are especially suited for the detection and characterization of biological molecules.

In part III of the book applications of methods and instrumentation to biomolecules are presented. An extensive coverage of the impact of mass spectrometric methods in protein analysis on pharmaceutical industry is given. Three more contributions explore the possibility of obtaining structural information from intact proteins with reference to their secondary structure as well. The protein/peptide subject also includes two chapters related to the contribution of the methodology to cancer therapy. A unique tool in DNA studies is provided by mass spectrometry which can be used in the analysis of a variety of compounds, from small nucleosides up to large oligonucleotides. In addition it also plays a role in the evaluation of DNA adducts in human carcinogenesis, the object of three of the chapters in part III. The importance of conventional and desorption ionization methods in the characterization of polysaccharides, lipids and plant metabolites is also examined, together with the role that mass spectrometry plays in the control of the environment and in the preservation of work of arts from pollution.

The editors acknowledge the special effort of the lecturers who have contributed to this book. Special thanks go to the NATO scientific affairs division, the Italian National Research Council (Rome), The University of Palermo, the University of Calabria and the manufacturers Micromass (UK), Thermo (US) , Perkin-Elmer-Sciex (US), Bruker (GER) Perseptives (US) who provided most of the funds for the institute. The editors also thank the French Society for Mass Spectrometry (SFSM), the Portuguese JNCT, Turkish TUBITAK, Greek Ministry of Industry, Energy and Technology and the U.S. National Science Foundation for providing travel grants to some junior participants at the ASI.

R. M. Caprioli
A. Malorni
G. Sindona
June, 1997

CONTRIBUTING AUTHORS

P. Agozzino	Dipartimento di Chimica e Tecnologie Farmaceutiche, Università di Palermo, Via Archirafi 32, 90123 Palermo, Italy
O. Altungoz	Department of Chemistry, Izmir Institute of Technology, Gaziosmanpasa Bulvari 16, Cankaya Izmir, Turkey
J. Amster	Department of Chemistry, University of Georgia, Athens Georgia 30602-2556, Usa
P. Arpino	Laboratoire de Chimie Analytique, Institut National Agronomique, 16 rue Claude Bernard, 75231 Paris-05, France
U. Bahr	Institute for Physical and Theoretial Chemistry, Department of Instrumental Analytical Chemistry, University of Frankfurt, Theodor-Stern-Kai 7, 60590 Frankfurt, Germany
H. Bartsch	German Cancer Research Center, Im Neuenheimer Feld 280, D-69120 Heidelberg, Germany
M. S. Bolgar	Michael Barber Centre for Mass Spectrometry, Department of Chemistry, UMIST, Manchester M60 1QD, UK
R. M. Caprioli	Analytical Chemistry Center, Department of Biochemistry and Molecular Biology, University of Texas Medical School, P.O. Box 20708, Houston, Tx 77225
V. Carbone	Servizio di Spettrometria di Massa del CNR c/o Facoltà di Medicina dell'Università di Napoli Federico II, Via Pansini 5, I - 80131 Napoli, Italy
B. A. Cerda	Department of Chemistry, University of Akron, Akron, OH 44325-3601, Usa
G. Chen	Department of Chemistry, Purdue University, West Lafayette IN 47907, Usa
R. G. Cooks	Department of Chemistry, Purdue University, West Lafayette IN 47907, Usa
M. Coulot	Central Research Services, Ciba-Geigy Ltd, Ch-4002 Basel, Switzerland
A. De Nino	Dipartimento di Chimica, Università della Calabria, I-87030 Arcavacata di Rende (CS), Italy

M. W. Duncan	Biomedical Mass Spectrometry, Unit University of New South Wales, Sydney, Australia
M. L. Easterling	Department of Chemistry, University of Georgia, Athens Georgia 30602-2556, Usa
T. B. Farmer	Analytical Chemistry Center, Department of Biochemistry and Molecular Biology, University of Texas Medical School, P.O. Box 20708, Houston, Tx 77225
P. Ferranti	Servizio di Spettrometria di Massa del CNR c/o Facoltà di Medicina dell'Università di Napoli Federico II, Via Pansini 5, I - 80131 Napoli, Italy
G. Forte	Servizio di Spettrometria di Massa del CNR c/o Facoltà di Medicina dell'Università di Napoli Federico II, Via Pansini 5, I - 80131 Napoli, Italy
S. Foti	Dipartimento di Scienze Chimiche, Università di Catania, Viale A. Doria 6, 95125 Catania, Italy
D. Garozzo	Istituto per la Chimica e la Tecnologia dei Materiali Polimerici, Viale A. Doria 6, 95125 Catania, Italy
S. J. Gaskell	Michael Barber Centre for Mass Spectrometry, Department of Chemistry, UMIST, Manchester M60 1QD, UK
J. K. Gooden	Department of Chemistry, Washington University, One Brookings Drive, St. Louis, Missouri 63130, Usa
M. L. Gross	Department of Chemistry, Washington University, One Brookings Drive, St. Louis, Missouri 63130, Usa
H. Grossenbacher	Core Drug Discovery Technologies, Ciba-Geigy Ltd, CH-4002 Basel, Switzerland
M. Karas	Institute for Physical and Theoretial Chemistry, Department of Instrumental Analytical Chemistry, University of Frankfurt, Theodor-Stern-Kai 7, 60590 Frankfurt, Germany
P. Kofel	Institute of Organic Chemistry, University of Bern Freiestr. 3, CH-3012 Bern, Switzerland
A. Napoli	Dipartimento di Chimica, Università della Calabria, I-87030 Arcavacata di Rende (CS), Italy

W. Maerki	Core Drug Discovery Technologies, Ciba-Geigy Ltd, CH-4002 Basel, Switzerland
A. Malorni	Servizio di Spettrometria di Massa del CNR c/o Facoltà di Medicina dell'Università di Napoli Federico II, Via Pansini 5, I - 80131 Napoli, Italy
T. Marino	Dipartimento di Chimica, Università della Calabria, I-87030 Arcavacata di Rende (CS), Italy
R. J. Maguire	National Research Institute, Department of the Environment, 867 Lakeshore Road, Burlington, Ontarioo L7R 4A6, Canada
G. Mamone	Servizio Spettrometria di Massa, CNR-Università di Napoli, Via Pansini 5, 80131 Napoli, Italy
G. Marino	Servizio Spettrometria di Massa, CNR-Università di Napoli, Via Pansini 5, 80131 Napoli, Italy
D. Muller	Central Research Services, Ciba-Geigy Ltd, Ch-4002 Basel, Switzerland
O. V. Nemirovskiy	Department of Chemistry, Washington University, One Brookings Drive, St. Louis, Missouri 63130, Usa
R. P. Newton	Biochemistry Group, School of Biological Sciences, University of Wales Swansea, Singleton Park, Swansea SA2 8PP, United Kingdom
M. J. Nold	Department of Chemistry, University of Akron, Akron, OH 44325-3601, Usa
F. I. Onuska	National Research Institute, Department of the Environment, 867 Lakeshore Road, Burlington, Ontarioo L7R 4A6, Canada
I. T. Özgen	Department of Chemistry, Izmir Institute of Technology, Gaziosmanpasa Bulvari 16, Cankaya Izmir, Turkey
G. Pocsfalvi	Servizio Spettrometria di Massa, CNR-Università di Napoli, Via Pansini 5, 80131 Napoli, Italy
P. Pucci	Servizio Spettrometria di Massa, CNR-Università di Napoli, Via Pansini 5, 80131 Napoli, Italy
A. Raffaelli	Centro di Studio del CNR per le Macromolecole Stereordinate ed Otticamente Attive, Dipartimento di Chimica e Chimica Industriale, Università di Pisa, Via Risorgimento 35, 56126 Pisa, Italy

R. Ramanathan	Department of Chemistry, Washington University, One Brookings Drive, St. Louis, Missouri 63130, Usa
I. Riba	Michael Barber Centre for Mass Spectrometry, Department of Chemistry, UMIST, Manchester M60 1QD, UK
W. J. Richter	Central Research Services, Ciba-Geigy Ltd, Ch-4002 Basel, Switzerland
M. Ruoppolo	Servizio Spettrometria di Massa, CNR-Università di Napoli, Via Pansini 5, 80131 Napoli, Italy
N. Russo	Dipartimento di Chimica, Università della Calabria, I-87030 Arcavacata di Rende (CS), Italy
B. Salih	Department of Chemistry, Hacettepe University, Beytepe Campus, Ankara, Turkey
N. Sannolo	Dipartimento di Biochimica e Biofisica, II Università, 80131 Napoli, Italy
P. Schindler	Central Research Services, Ciba-Geigy Ltd, Ch-4002 Basel, Switzerland
U. P. Schlunegger	Institute of Organic Chemistry, University of Bern Freiestr. 3, CH-3012 Bern, Switzerland
E. Sicilia	Dipartimento di Chimica, Università della Calabria, I-87030 Arcavacata di Rende (CS), Italy
G. Sindona	Dipartimento di Chimica, Università della Calabria, I-87030 Arcavacata di Rende (CS), Italy
B. Spengler	Institute of Laser Medicine, Heinrich-Heine-University of Duesseldorf, P.O. Box 101007, D-40001 Duesseldorf, Germany
S. G. Summerfield	Michael Barber Centre for Mass Spectrometry, Department of Chemistry, UMIST, Manchester M60 1QD, UK
M. J. F. Suter	Department of Chemistry, Swiss Federal Institute for Environmental Science and Technology EAWAG, CH-8600 Duebendorf, Switzerland
F. Talamo	Servizio Spettrometria di Massa, CNR-Università di Napoli, Via Pansini 5, 80131 Napoli, Italy
B. G. M. Vandeginste	Unilever Research Vlaardingen, P.O. Box 114, 3130 AC Vlaardingen, The Netherlands

K. Vékey — Central Research Institute for Chemistry, H-1025 Budapest, Pusztaszeri ùt 59-67, Hungary

C. Weil — II Physikalisches Institut, Justig Liebig Universitat, Giessen, Germany

PARTICIPANTS IN THE INSTITUTE

AUSTRALIA

Mark DUNCAN
Biomedical Mass Spectrometry Unit
University of New South Wales
Sydney, New South Wales 2052

CANADA

Joseph H. BANOUB
Department of Fisheries and
Oceans Science Branch
Environmental Sciences Division
P.O. Box 55667
St. John's, NF A1C 5X1

Francis I. ONUSKA
Aquatic Ecosystem Protection Branch
National Water Research Institute
Canada Centre for Inland Waters
867 Lakeshore Road
Burlington, Ontario

DENMARK

Tina HENRIKSEN
Department of Chemistry
University of Copenhagen
H.C. Orsted Institutet
Univrsitetsparken 5
2100 Copenhagen

Thomas JOERGENSEN
Department of Chemistry
Odense University
Odense

Thomas KROGH
Department of Molecular Biology
Odense University
Campusvej 55
5230 Odense M

Martin R. LARSEN
Department of Molecular Biology
Odense University, ampusvej 55
5230 Odense M

Ekatharina MIRGORODSKAYA
Department of Molecular Biology
Odense University
Campusvej 55
5230 Odense M

Malene MOHR
Department of Chemistry
University of Copenhagen
H.C. Orsted Institutet
Universitetsparken 5
2100 Copenhagen

Peter ROEPSTORFF
Department of Molecular Biology
Odense University
Campusvej 55
5230 Odense M

FRANCE

Badia AMEKRAN
Laboratoire de Chimie Organique Structurale
Universitè Pierre et Marie Curie
Boite 45
4 Place Jussieu
75252 Paris Cedex 05

Denis GIRAUD
Institut National Agronomique
16, rue Claude Bernard
75231 Paris Cedex 05

Patrick J. ARPINO
Institut National Agronomique
Paris-Grignon
16, rue Claude Bernard
75231 Paris Cedex 05

Sébastien GREGOIRE
10 rue du A. Manceau
77400 Lagny sur Marne

Laure CAPRON
Ecole Normale Supèrieure
24, rue Lhomond
75231 Paris cedex 05

Frèdèric HALGAND
Institut d e Biologie Structurale
CEA/CNRS
41 Avenue des Martyrs
38027 Grenoble Cedex

Myriam FERRO
LPTF-CNRS
205 route de Narbonne
31077 Toulose

Laurent SERANI
Centre Nat. de la Recherche Scientifique
Inst. de Chimie des Substances Naturelles
Lab. de Spectromètrie de Masse
1, avenue de la Terrase
91198 Gif sur Yvette cedex

FINLAND

Antti KOIVUNIEMI
Dept. of Pharmacy
University of Kuopio
P.O. Box 1627
Sf-70211 Kuopio

Marja RIMPILAINEN
Dept. of Medical Microbiology
University of Turku
Kiinamyllynkatu 13
Fin-20520 Turku

GERMANY

Ute BAHR
Chemie fur Mediziner
JW Goethe-Universitat
Theodor-Stern-Kai 7
60590 frankfurt am Main

Jurgen SCHAFER
Division for Instrumental Analytical Chemistry
Chemistry for Medical Students
JW Goethe-University
Theodor-Stern-Kai 7
60590 Frankfurt

Frank LUETZENKIRCHEN
Corneliustrasse 93
40215 Dusseldorf

Andrea SCHMIDT
Division for Instrumental Analytical Chemistry
Chemistry for Medical Students
JW Goethe-University
Theodor-Stern-Kasi 7
60590 Frankfurt

Michael KARAS
Chemie fur Mediziner
J.W. Goethe - Universitat
Theodor-Stern-Kai 7
60590 Frankfurt am Main

Bernhard SPENGLER
Institute of Laser Medicine
University of Duesseldorf
P.O. Box 101007
40001 Duesseldorf

Sabine METZGER
Institute of Laser Medicine
University of Dusseldorf
P. O. Box 101007
40215 Dusseldorf

GREECE

Costantinos ATHANASSOPOULOS
Dept. of Chemistry
University of Patras
Patras

Nestor RODIOS
Dept. of Chemistry
University of Thessaloniki
Thessaloniki

Nikolaos KARAMANOS
Dept. of Chemistry
University of Patras
Patras

HUNGARY

Krisztina LUDANYI
Central Research Institute for Chemistry
Pusztaszeri ùt 59 - 67
1025 Budapest II

Karoly VEKEY
Central Research Institute for Chemistry
Pusztaszeri ùt 59 - 67
1025 Budapest II

ITALY

Pasquale AGOZZINO
Dip. di C. T. F.
Università di Palermo
Via Archirafi, 32
90123 Palermo

Angela AMORESANO
SESMA/CNR
II Facoltà di Medicina
Via S. Pansini, 5
80131 Napoli

Giuseppe ATTORRESE
Dip. di Chim. Ing.Chim. e dei Materiali
Università di L'Aquila
Via Vetoio - Coppito II
67100 L'Aquila

Giuseppe BARONE
SESMA/CNR
II Facoltà di Medicina
Via S. Pansini, 5
80131 Napoli

Davide BONGIORNO
Dip. di C. T. F
Università di Palermo
Via Archirafi, 32
90123 Palermo

Virginia CARBONE
SESMA/CNR
II Facoltà di Medicina
Via S. Pansini, 5
80131 Napoli

Anna CASELLI
Dept. of Biochemical Sciences
Viale Morgagni, 50
50134 Firenze

Fulvio MAGNI
IRRCS San Raffaele
Via Olgettina, 60
20132 Milano

Antonio MALORNI
SESMA/CNR
II Facoltà di Medicina
Via S. Pansini, 5
80131 Napoli

Gianfranco MAMONE
SESMA/CNR
II Facoltà di Medicina
Via S. Pansini, 5
80131 Napoli

Gennaro MARINO
Dept. of Organic and Biological Chemistry
Università di Napoli
Via Mezzocannone
Napoli

Bruno NERI
Ist. Zooprofilattico Sperimentale
delle Regioni Lazio e Toscana
Via Appia Nuova 1411
00178 Roma Capannelle

Stefania ORRU'
SESMA/CNR
II Facoltà di Medicina
Via S. Pansini, 5
80131 Napoli

Piero PUCCI
Dept. of Organic and Biological Chemistry
Università di Napoli
Via Mezzocannone
80131 Napoli

Bruno CASETTA
Perkin Elmer Sciex
Via Tiepolo 2
Monza

Leopoldo CERAULO
Dip. di C. T. F.
Università di Palermo
Via Archirafi, 32
90123 Palermo

Antonio DE NINO
Dip. di Chimica
Università della Calabria
87030 Arcavacata di Rende (CS)

Vita DI STEFANO
Dip. di C. T. F.
Università di Palermo
Via Archirafi, 32
90123 Palermo

Pasquale FERRANTI
SESMA/CNR
II Facoltà di Medicina
Via S. Pansini, 5
80131 Napoli

Salvatore FOTI
Dip. Scienze Chimiche
Viale A. Doria
Università di Catania
90125 Catania

Antonio FUSCO
Centro Tec. Militare Chimico
Fisico Biologico
Civitavecchia

Domenico GAROZZO
Ist. per la Chimica e la Tecnologia
dei Materiali Polimerici
Università di Catania
Viale A. Doria
95125 Catania

Roberto RABEZZANA
Dip. di Chimica Gen. ed Organica Applicata
Università degli Studi di Torino
Corso M. D'Azeglio, 48
10125 Torino

Andrea RAFFAELLI
Centro di Studio del CNR
c/o Dip. di Chim. e Chim. Industr.
Università di Pisa,Via Risorgimento, 35
56126 Pisa

Nino RUSSO
Dip. di Chimica
Università della Calabria
87030 Arcavacata di Rende

Rosaria SALETTI
Dip. Scienze Chimiche
Università di Catania
Viale A. Doria
90125 Catania

Stefania SEGRETO
Dip. di C. T. F.
Università di Palermo
Via Archirafi, 32
90132 Palermo

Nicola SERPIERI
Dept. of Pharmacology
Viale Morgagni, 65
50134 Firenze

Giovanni SINDONA
Dip. di Chimica
Università della Calabria
87030 Arcavacata di Rende (CS)

Antonio TRIOLO
Menarini Ricerche
Via Settesanti, 3
50131 Firenze

Emanuele MAGI
Dip. di Chimica e Chimica Industriale
Sez. di Chimica Analitica e Ambientale
Università di Genova
Via Dodecaneso, 31
16146 Genova

THE NETHERLANDS

Connie JIMENEZ
Vrije Universteit Amsterdam
Faculty of Biology
1081 HV Amsterdam

NORWAY

Ann Margot WHYATT
Department of Chemistry
University of Bergen
5007 Bergen

POLAND

Wanda GRADOWSKA
The Children's Memorial
Health Institute
Al. Dzieci Polskich
04-736 Warszawa

PORTUGAL

Pedro Miguel DIMAS DAS NEVES
Universidade de Aveiro
Aveiro

Paula ESPERANCA
Facultade de Ciencias
Universidade de Lisboa
Lisboa

Maria Conceicao OLIVEIRA
Centro de Espectrometria de Massa
Complexo I
Avenida Rovisco Pais
1096 Lisboa Codex

SLOVAK REPUBLIC

Kenneth D. ONUSKA
Faculty of Medicine
Comenius University
Department of Medical Chemistry
Biochemistry and Clinical Biochemistry
Sasinkova 2, 811 08 Bratislava

SPAIN

Imma FERRER
Dept. de Quìmica Ambiental
CID-CSIC
Jordi-Girona 18-26
08304 Barcelona

Rosa Elena MONSERRAT
Dept. de Macroestructura y Procesos
CID-CSIC
Jordi-Girona 18-26
08304 Barcelona

Encarnaciòn MOYANO
Facultad de Quìmica
Universidad de Barcelona
Barcelona

SWITZERLAND

Wilhelm RICHTER
Ciba Geigy
K - 127.508
CH - 4002 Basel

Urs SCHLUNEGGER
Department of Organic Chemistry
University of Bern
Freiestrasse 3
CH - 3012 Bern

Marc SUTER
Dept. of Chemistry
Swiss Federal Institute for Environmental
Science and Technology
EAWAG
Ueberlandstrasse 133
CH-8600 Duebendorf

TURKEY

Oya ALTUNGOZ
Ismir Institute of Technology (IZTECH)
Gaziosmanpasa Bulvari, 16
CanKaya, Izmir

Isik CEPNI
Chemmical Eng. Dept.
Istambul Technical Univ.
80626 Maslak - Istambul

Ebru APAK
Chemical Eng. Dept.
Istambul Technical Univ.
80626 Maslak-Istambul

Tamerkan OZGEN
Izmir Institute of Technology
Gaziosmanpasa Bulv. n. 16
Cankaya-Izmir

UKRAINE

Marina Vadimovna KOSEVICH
Department of Molecular Biophysics
B. Verkin Institute for low Temperature
Physics and Engineering of the
National Academy of Science of Ukraine
47, Lenin Ave
310164 Kharkov

UNITED KINGDOM

Louisa Maria MARQUES PEREIRA
Biochemistry Group
School of Biological Sciences
University of Wales Swansea
Singleton Park
Swansea SA2 8PP

Russell P. NEWTON
Biochemistry Group
School of Biological Sciences
University of Wales Swansea
Singleton Park
Swansea SA2 8PP

Martin McCULLAGH
Biochemistry Group
School of Biological Sciences
University of Wales Swansea
Singleton Park
Swansea SA2 8PP

Isabel RIBA
Department of Chemistry
Institute of Science and Technology
University of Manchester
P.O. Box 88
Manchester M60 1QD

Mark Mc DOWALL
Micromass UK Limited
3 Tudor Road
Altrincham Cheshire WA14 5RZ

Scott SUMMERFIELD
Department of Chemistry
Institute of Science and Technology
University of Manchester
P.O. Box 88
Manchester M60 1QD

Simon GASKELL
Department of Chemistry
Institute of Science and Technology
University of Manchester
P.O. Box 88
Manchester M60 1QD

USA

Jon AMSTER
Dept. of Chemistry
University of Georgia
Athens, GA 30602-2556

R. M. CAPRIOLI
Medical School
University of Texas
P:O: Box 2078
Houston TX 77030

Guodong CHEN
Dept. of Chemistry
Purdue University
West Lafayette, IN 47907-1393

Graham COOKS
Department of Chemistry
Purdue University
1393 Herbert C.
Brown Laboratory Building
47907-1393West Lafayette IN

Terry B. FARMER
Analytical Chemistry Center
University of Texas
Medical School
6431 Fannin Street
MSB 6130, Houston, TX 77030

Robert C. MURPHY
Pediatrics Department K929
National Jewish Center for
Immunology & Respiratory Medicine
1400 Jackson Street
80206 Denver CO

Olga NEMIROVSKAYA
Dept. of Chemistry
Washington University
Campus Box 1134
One Brookings Drive
St. Louis, Missouri 63130 - 4899

Lenore Marie POLO
Chemistry Department
Lousiana State University
232 Choppin Hall
Baton Rouge, LA 70803

Ragalan RAMANATHAN
Dept. of Chemistry
Washington University
Campus Box 1134
One Brookings Drive
St. Louis, Missouri 63130 - 4899

Michael S. STORY
Thermo Instrument Systems Inc.
15745 Wood Acre Road
Los Gatos CA 95030

Jonathan GOODEN
Dept. of Chemistry
Washington University
Campus Box 1134
One Brookings Drive
St. Louis, Missouri 63130 - 4899

Michael GROSS
Department of Chemistry
Washington University
Campus Box 1134
One Brookings Drive
63130-4899 St. Louis MO

Scott A. MILLER
Dept. of Chemistry
Purdue University
West Lafayette, IN 47907

Marvin VESTAL
PerSeptive Biosystems
500 Old Connecticut Path
01701 Framingham MA

Russell J. WAUGH
Division of Basic Sciences
national Jewish Center for
Immunology and Respiratory Medicine
1400 Jackson Street
Denver, CO 80206

Chrys WESDEMIOTIS
Dept. of Chemistry
University of Akron
Akron, OH 44325-360

Part I
Methods in Mass Spectrometry

ELECTROSPRAY IONIZATION: THEORY AND APPLICATION

SIMON J. GASKELL, MARK S. BOLGAR, ISABEL RIBA and
SCOTT G. SUMMERFIELD
Michael Barber Centre for Mass Spectrometry
Department of Chemistry
UMIST
Manchester M60 1QD, UK

1. Introduction

A great deal of chemistry and most of biochemistry concerns reactions taking place in solution. The development and application of an introduction and ionization technique which allows direct sampling from solution has therefore been a prominent objective of analytical mass spectrometry for many years. Furthermore, the removal of the requirement for the volatilization of neutral analyte prior to ionization has long been recognized as the *sine qua non* for the analysis of many molecules of biological importance, including biopolymers such as peptides and proteins, oligosaccharides and oligonucleotides. The development by Barber and his colleagues [1,2] of fast atom bombardment (FAB) significantly extended the range of mass spectrometric analyses to less volatile compounds than were amenable to traditional techniques such as electron and chemical ionization. The elaboration of the FAB method to allow ionization from a surface continuously replenished by a flowing stream of analyte solution [3,4] introduced new capabilities (albeit rudimentary by current standards) for direct mass spectrometric analyses of solutions. Thermospray [5], expressly developed for the coupling of mass spectrometry and condensed-phase separation methods such as high-performance liquid chromatography (HPLC), addressed the need for sampling from solution but was limited with respect to the volatility and polarity of the analytes amenable to the technique.

The electrospray technique is unique in providing simultaneously a logical coupling of solution introduction of analytes for mass spectrometry and the facility for ionization of highly polar and extremely involatile compounds. Early work on the development of electrospray was performed by Dole and co-workers and was summarized in a review in 1984 [6]. This work was, however, hampered by the use of an inadequate ion analysis system. In the 1980s, Fenn and his colleagues built on Dole's ideas to develop electrospray as a sample introduction and ionization technique for mass spectrometry [7]. Subsequent contributions from numerous research groups have established electrospray as a standard technique in mass spectrometry

R. M. Caprioli et al. (eds.), Selected Topics and Mass Spectrometry in the Biomolecular Sciences, 3–16.

laboratories concerned with the analysis of many compounds of biological significance and in those involved with many other applications.

The purpose of this brief review is to outline the fundamental concepts underlying the electrospray technique and to exemplify applications. Reference is made both to the original literature and to review articles which explore aspects of the subject in greater depth.

2. Essential Features of the Electrospray Process

Figure 1 shows a highly simplified schematic of an electrospray interface sufficient to explain the essential features. A solution of the analyte is introduced through a capillary at a flow rate typically (though not necessarily; see below) in the range 3-10 μl/min. Most commonly, the solvent consists of a mixture of water and an organic such as methanol or acetonitrile, with the pH adjusted to optimize ion yield. The capillary is held at a potential of several kilovolts relative to ground and to a counter electrode, termed a skimmer in Figure 1.

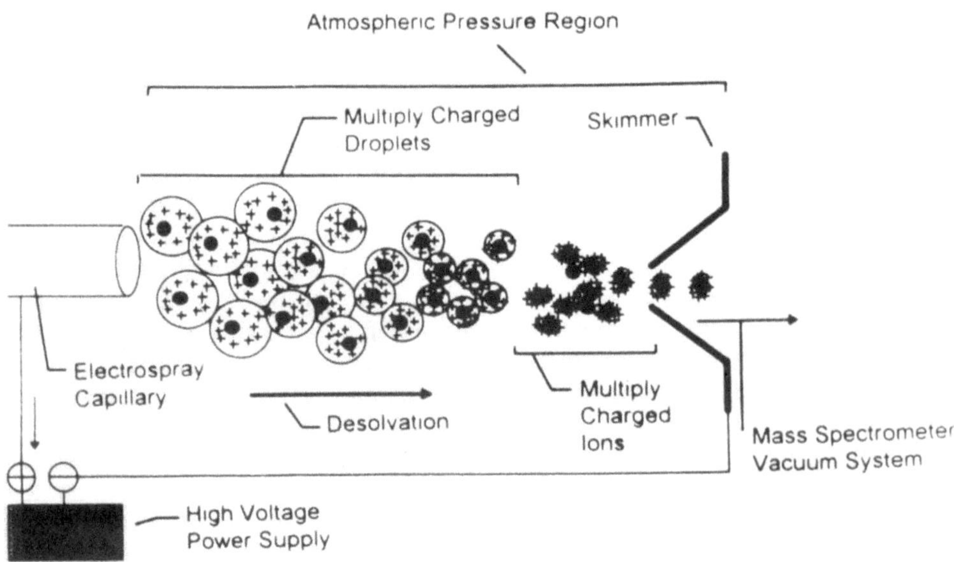

Figure 1. Simplified schematic of an electrospray interface.

The mechanism(s) of ion transfer to the gas phase will not be discussed in detail here; the reader is referred to several reviews [7-10] for thorough discussions of the process (which remains imperfectly understood). In outline, however, the initial formation of a mist of droplets is achieved by electrostatic dispersion as the solution is expelled from the capillary. The nebulization process may be assisted by an annular flow of nitrogen. Henion and coworkers, who pioneered this elaboration of the electrospray technique [11], coined the separate term "ion spray". Aspects of the ion spray technique are discussed elsewhere in this volume by Raffaelli; readers should, however, be aware that many current designs of electrospray sources incorporate the facility for a nebulizer gas and that its use has little effect on the ionization chemistry. Initial droplet formation is followed by solvent evaporation, which may be promoted by the use of a drying gas. When the surface charge density on a droplet exceeds the so-called "Rayleigh limit", Coulombic explosion gives rise to smaller droplets. This process may continue until each droplet contains a single ion. Alternatively or additionally, there is a contribution from ion evaporation, which is expected to take place when surface charge densities are sufficiently high.

Gas-phase ions formed in the electrospray source traverse a region of intermediate pressure where their mean-free path is sufficiently long that energetic collisions can take place with residual neutral molecules. This can result in ion activation sufficient to promote significant fragmentation; the extent to which this occurs can be influenced by the setting of the potential gradients in this region. The use of this technique (too often given unnecessarily obscure names such as "cone voltage fragmentation") can be valuable in increasing the yield of diagnostic fragment ions without the use of a tandem mass spectrometer.

In sum, the electrospray source effects the transfer of ions from the condensed phase to the gas phase, with the charge state modified by the transfer process and both charge state and ion integrity modified by activating collisions.

3. The Properties of Electrosprayed Ions

3.1 CHARGE STATE

Gas-phase ions generated by the electrospray process are commonly multiply charged; the charge state in the gas phase is undoubtedly influenced by the charge state in solution but (as noted above) is likely modified by ion-molecule interactions in the electrospray interface. In the analysis of peptides by electrospray, a good correlation is observed between the maximum observed charge state and the number of basic sites in the structure. Figure 2 shows the electrospray mass spectrum of the 14-residue oligopeptide, Leu-Phe-Thr-Gly-His-Pro-Glu-Thr-Leu-Glu-Lys-Phe-Asp-Lys. The most abundant ion species is the tetra-protonated molecule, reflecting the presence of three amino acid residues with basic side-chains (two lysines and a histidine) and the primary amino group at the N-terminus.

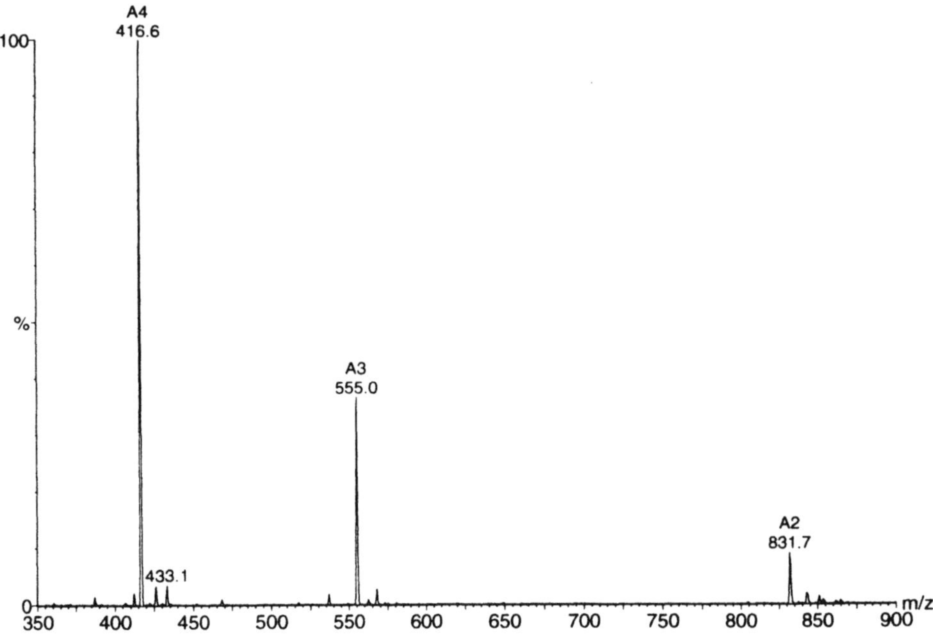

Figure 2. Electrospray mass spectrum of the 14-residue oligopeptide, Leu-Phe-Thr-Gly-His-Pro-Glu-Thr-Leu-Glu-Lys-Phe-Asp-Lys.

In this instance, the relationship between the m/z-ratio values of the ion species is immediately apparent, so that independent determination of mass and charge is readily made. The data may be less readily interpretable if a single charge state predominates to the effective exclusion of others or if interpretation is complicated by the presence of multiple molecular components. If the mass spectrometric resolution is sufficiently high, the separation of the components of the isotope cluster allows determination of the charge state; thus, the separation in m/z-ratio units is 1/n, where n is the charge state. Figure 3 illustrates this approach in the interpretation of data obtained by electrospray MS analysis of the unfractionated peptide mixture resulting from tryptic digestion of myoglobin (M_r 19952); the Figure shows a portion of the spectrum including signals derived from two tryptic fragments. Alternatively, charge state determination can be based on the concomitant observation of other cationized species. Thus, the presence of ion pairs separated by 22/n m/z-units suggests both protonated species and ions in which one proton is replaced by a sodium cation (Figure 4).

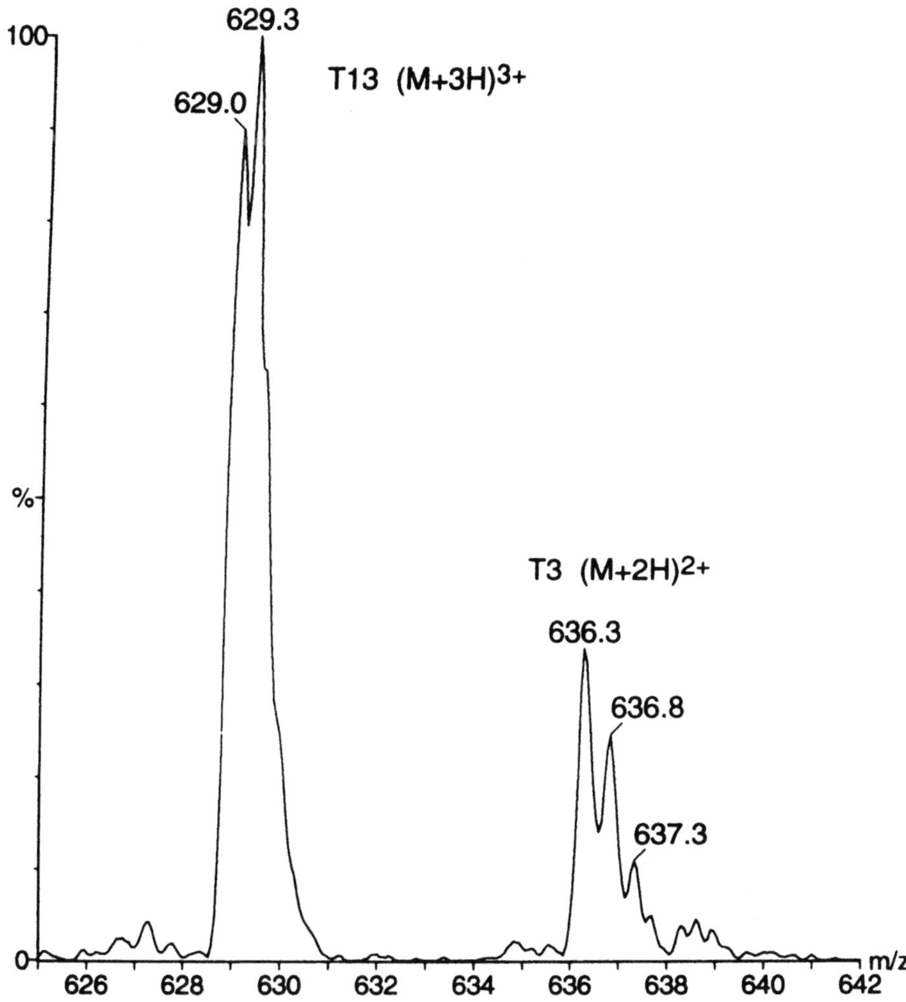

Figure 3. Partial electrospray spectrum of the unfractionated mixture resulting from the tryptic digest of myoglobin, illustrating the determination of charge state based on m/z-ratio separation of components of the isotopic cluster.

Cunniff and Vouros [12] have described an analogous approach to charge state determination based on the observation of 1:1 non-covalent complexes between peptides and a crown ether (18-crown-6, of molecular weight 264) during electrospray analysis. The crown ether is added to the analyte solution or is introduced in a liquid sheath at the electrospray capillary tip. The determination

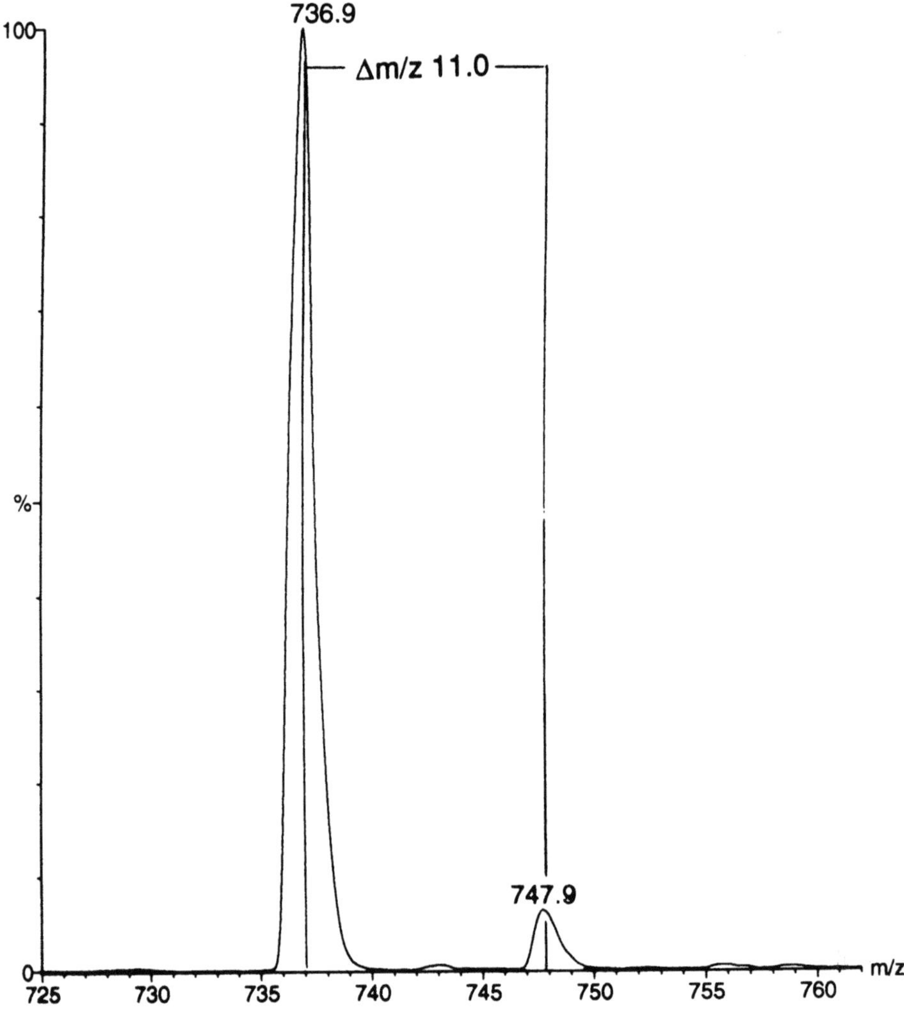

Figure 4. Partial electrospray spectrum of a peptide with a molecular weight of 1471.8, showing the determination of charge state (+2 in this instance) based on the m/z-ratio separation of protonated species and ions in which a proton is replaced by a sodium cation.

of the charge state of individual components of the spectrum of a complex mixture is facilitated by constant neutral loss scanning during tandem MS analyses [12]; the neutral loss of the crown ether is evident in scans detecting the loss of 264/n m/z-units.

During the analysis of high molecular weight compounds, such as proteins and other biopolymers, the formation of multiple charge states gives rise to an apparently complex spectrum for each analyte. Conveniently, in the electrospray of proteins, the number of attached protons usually averages approximately one per kDa (determined by the statistical frequency of basic residues), so that the required m/z ratio range of the mass spectrometer is commonly less than 2000. The deprotonation of oligonucleotides to form negative ion distributions leads to a similar m/z range requirement in most applications. Figure 5a shows the electrospray spectrum of rho protein (M_r 47005) which includes charge (protonation) states in the range of +40 to +70. The assumption that adjacent members of the ion series differ by a single proton in mass and charge allows mass assignment. Thus, the charge state of an ion of the general form, $[M+nH]^{n+}$, is given by

$$n = [x_1 - 1] / [x_2 - x_1] \qquad (1)$$

where x_2 is the mass/charge ratio of the ion of charge, n, and x_1 is the mass/charge ratio of the ion of charge, n+1. Such transformation of m/z-ratio data to a "true" mass spectrum is commonly performed by the mass spectrometer data system, achieving an estimate of molecular weight which is typically within 0.01% of the true value. More sophisticated approaches to data deconvolution, such as the maximum entropy algorithm [13], may provide the further benefit of effectively enhancing the resolution of mixture components. Maximum entropy processing is a probabilistic method which functions by repeatedly processing trial spectra of different mass and charge and comparing with the experimental data. The net result is a candidate mass spectrum, generally showing enhanced resolution, the validity of which may be assessed, in part, by comparing the experimental spectrum with that predicted from the final deconvoluted spectrum. Figure 5b shows the result of maximum entropy processing of data from the analysis of rho protein. The facility with which electrospray data transformation and processing may be initiated by the analyst should not disguise the assumptions on which the treatment is based. Thus, for example, studies of hydrogen/deuterium exchange using deuterated solvents (an increasingly common approach to the study of protein conformation) result in the formation of multiply *deuterated* ions.

The charge state distributions observed during electrospray MS of proteins may provide direct evidence for conformation changes in solution. An early example of such observations was the report by Chait and coworkers [14] of the variation in the charge state distribution of ions derived from the protein cytochrome *c*. Changes in the pH of the electrospray solvent resulted in the observation of three distinct distributions, in agreement with earlier observations (based on a number of alternative techniques) of the existence of three conformations in solution. In broad terms, the observation of differing extents of protonation are rationalized by varying accessibilities of basic sites; further studies are required, however, to establish more precise explanations.

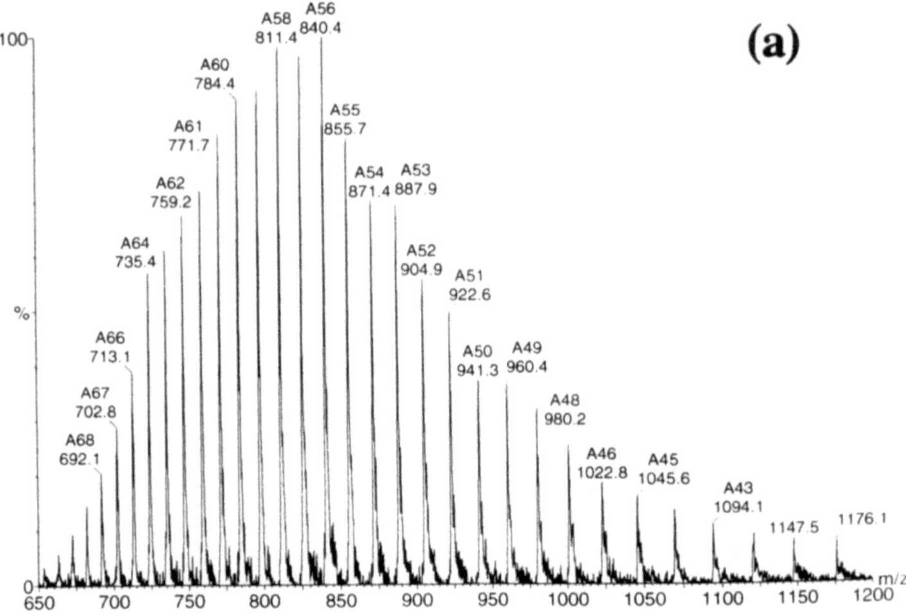

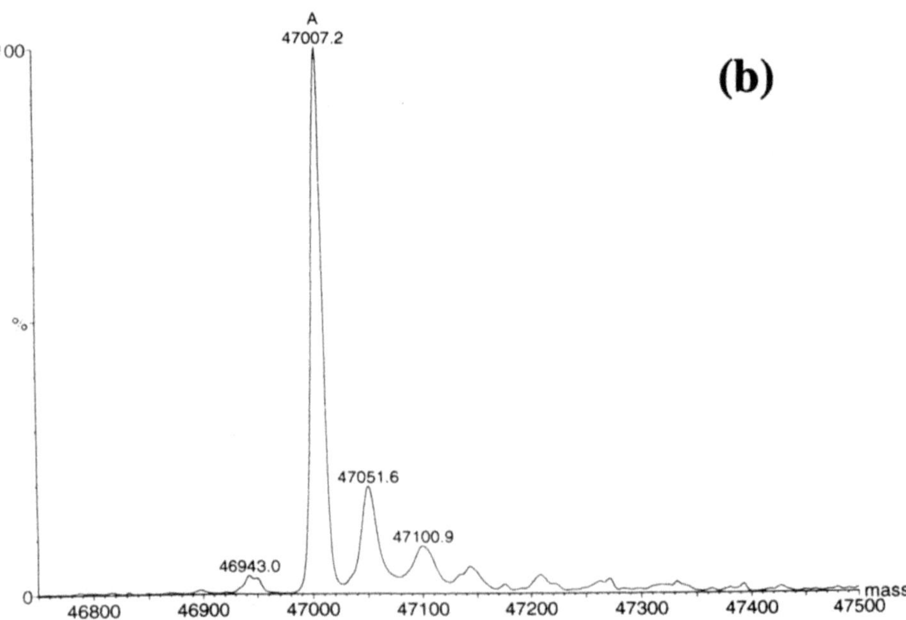

Figure 5. Electrospray MS analysis of rho protein (M_r 47005). (a) Conventional mass/charge ratio spectrum. (b) Mass spectrum obtained by maximum entropy processing of the same data.

3.2 FRAGMENTATIONS OF MULTIPLY CHARGED IONS

The formation of multiply charged ions in the electrospray process has important implications for ion stability and, in consequence, for the facility to induce fragmentation following collisional activation. Early studies of electrospray tandem MS analyses of peptides derived from tryptic digestion of proteins noted a remarkable increase in the product ion yield following low energy collisional activation of doubly protonated ions, in comparison with their singly protonated counterparts [15]. The comparison is particularly striking for peptides incorporating a C-terminal arginine (rather than a C-terminal lysine, which is the other trypsin cleavage site). It is now understood that this observation results directly from the generalization that low energy fragmentation processes are predominantly directed by the site of charge [16-18]. Populations of arginine-containing peptide ions incorporating a single proton are expected to be highly homogeneous with respect to the site of charge: the guanidino group of the arginine side-chain is by far the most basic site. In consequence, the extent and diversity of fragmentations observed following low energy collisional activation are both poor. When the same peptide is converted to a doubly protonated species, as commonly observed in electrospray ionization, the second proton is "mobile" [18] or, equivalently, the ion population is heterogeneous with respect to the (second) site of charge [16,17]. In consequence, efficient and diverse fragmentation is observed.

These ideas are not limited to the example of doubly charged peptide ions. An illustration is the example of the synthetic nonapeptide, Arg-Leu-Cys-Ile-Phe-Ser-Cys-Phe-Arg. In this instance, the presence of *two* highly basic sites (the two arginine side-chains) ensures that the population of *doubly* protonated peptide ions is homogeneous. The product ion spectrum obtained following low energy collisional activation (Figure 6a) shows a very poor yield of fragment ions and a consequent paucity of structural information. In contrast, the triply protonated analogue incorporates a mobile proton which confers heterogeneity of charge location and promotes a diversity of fragmentation pathways (Figure 6b).

4. Electrospray at low flow rates

It was noted above that sample introduction for electrospray ionization is most commonly achieved using flow rates of 3 - 10 μl. Some effort has been made, particularly in Henion's laboratory [19] to adapt electrospray to higher flow rates to enable coupling to conventional "analytical" scale HPLC. The benefits of *reduced* flow rates, however, were recognized by several laboratories [8,20,21] and "micro-electrospray" or "nanospray" now represents an active area of technique development. The principal advantages are high sensitivity and stability, and the facility for electrospraying aqueous solutions without the need for a sheath flow of organic solvent.

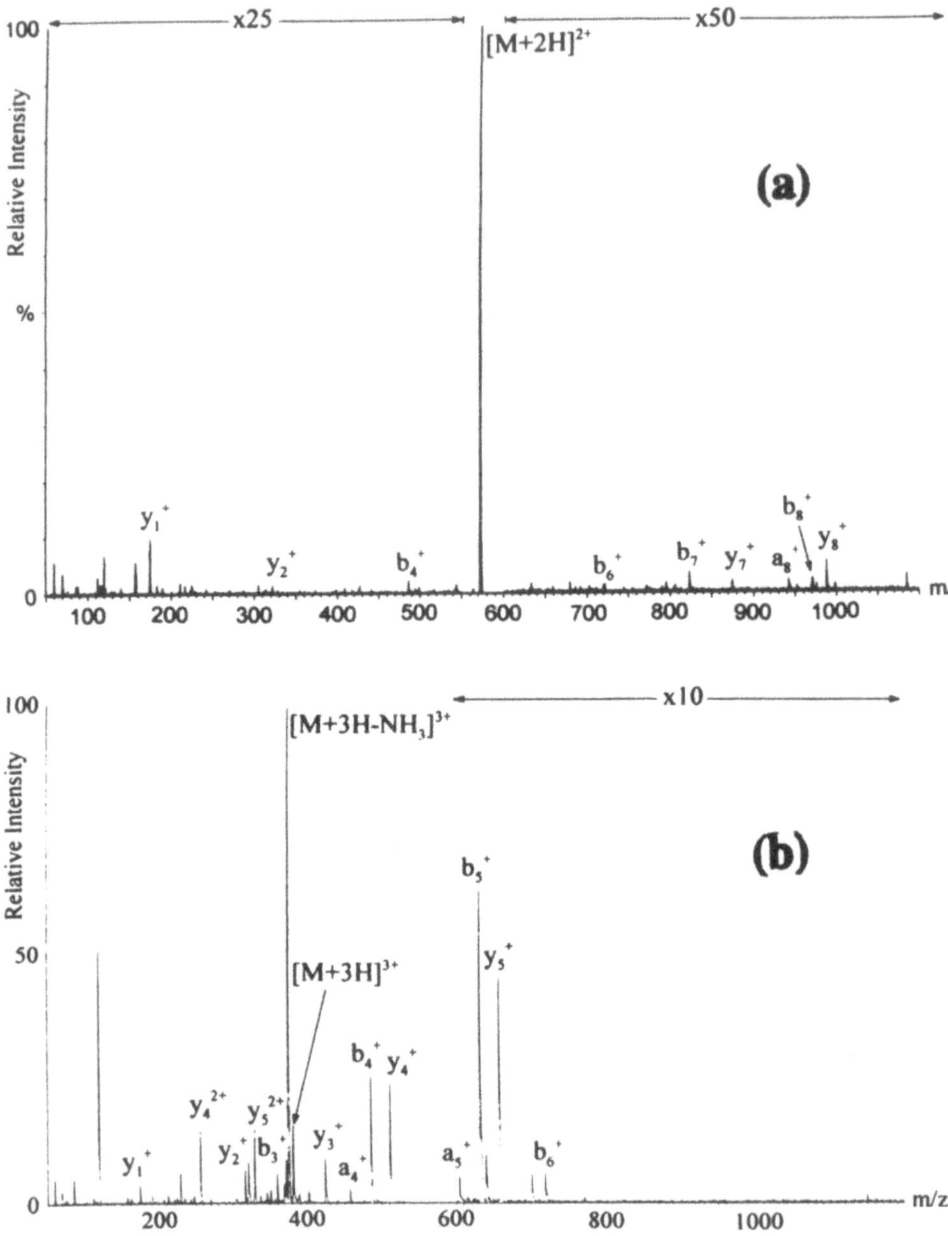

Figure 6. Product ion spectra recorded during the electrospray tandem MS analysis of the nonapeptide, Arg-Leu-Cys-Ile-Phe-Ser-Cys-Phe-Arg. $[M+2H]^{2+}$ and $[M+3H]^{3+}$ precursor ions were subjected to identical conditions of low energy collisional activation in a tandem quadrupole instrument. (a) Products of $[M+2H]^{2+}$ precursors. (b) Products of $[M+3H]^{3+}$ precursors.

Caprioli and coworkers [21] have adopted an approach in which sub-microlitre/minute flows are pumped to the electrospray capillary. The benefit of reduced sample consumption is illustrated in Figure 7 which shows electrospray mass spectra recorded using two flow rates of the same solution of three authentic analytes routinely used as m/z-calibrants in this laboratory. Similar signal intensities were observed in the two analyses (using similar acquisition times) despite the near thirty-fold difference in sample consumption.

The low-flow electrospray arrangement used by Wilm and Mann [8] incorporates a short drawn-out capillary which serves as a sample reservoir with a total volume of a few microlitres. Droplets emitted from the capillary, of 1-3 μm internal diameter, are predicted [8] to contain on average only one analyte molecule when the concentration is 1 pmol/μl. With an estimated flow rate of about 20 nl/min using such a device, Wilm and Mann estimate an electrospray efficiency improved by two orders of magnitude in comparison with "conventional" electrospray on the same instrument.

5. Future developments

The development of electrospray ionization has transformed the application of mass spectrometry to the biomolecular sciences. Several areas of future research and development can be predicted with some confidence:

a) Electrospray ionization, hitherto most commonly installed on quadrupole mass spectrometers, will be increasingly used in conjunction with ion trappping instruments (both Paul ion traps and ion cyclotron resonance instruments) and time-of-flight analyzers. An advantage of the latter is that the extended mass/charge ratio range will enable the analysis of macromolecules with relatively low charge states, permitting the use of electrospray solvents that may promote the retention of non-covalent interactions in the transition from condensed phase to gas phase.

b) Systematic study is required of the generation by electrospray of ions incorporating non-covalent interactions. Successful application of the electrospray process requires the efficient *destruction* of non-covalent bonds (of solvation) so that the observation of non-covalently bound adducts requires fine-tuning of the interface conditions. The present situation is that positive evidence for the observation of non-covalent adducts is probably meaningful (when analyses are made from dilute solution) but negative evidence is equivocal.

c) A continuing emphasis on the use of low flow rates and the achievement of high sensitivities of detection will encourage attention to the development of analyte extraction and processing stages on-line to the electrospray introduction capillary, thereby minimizing sample handling and transfer steps.

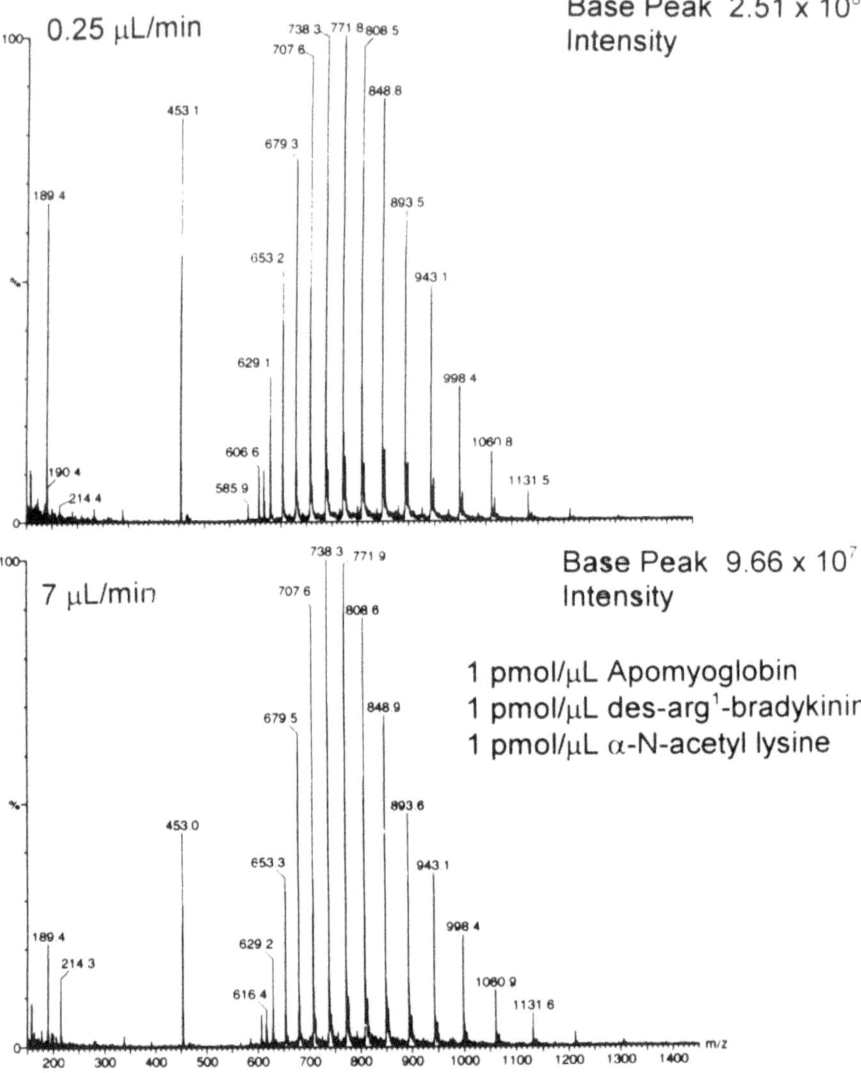

Figure 7. Electrospray MS analyses of a solution (1 pmol/μl of each component) of three authentic analytes used as an m/z calibrant: apomyoglobin, des-Arg[1]-bradykinin, and α-N-acetyl lysine. a) flow rate 0.25 μl/min; b) flow rate 7 μl/min. Both analyses were performed using a Micromass Quattro instrument; spectra were accumulated for 1 min in each case. The electrospray capillary was stainless steel, with an internal diameter of 75 μm. For the low flow rate analysis, the high voltage lens was removed from the electrospray interface and the capillary was placed within approximately 2 mm of the sampling cone.

d) Continuing study of the properties of multiply charged ions will further enhance the analytical utility of electrospray in conjunction with tandem MS.

e) The substantial variation in response factors for analytes of similar structure (such as peptides of differing sequence) merits further study. Potentially significant contributors include differential adsorptive losses in the electrospray interface, varying ionization efficiencies, and different extents of ion survival in passage through the pressure gradient preceding the mass spectrometer analyzer.

6. Acknowledgements

We are grateful to Dr. Kathleen Cox and Holger Hermann for their contributions to this review. Research at UMIST is funded by the UK Engineering and Physical Sciences Research Council (grant no. GR/K18658); further support has been received from the Royal Society and from the Bristol-Myers Squibb Pharmaceutical Research Institute.

7. References

1. Barber, M., Bordoli, R.S., Sedgwick, R.D., and Tyler, A.N. (1981) *J. Chem. Soc. Chem. Comm.* 325-327.
2. Barber, M., Bordoli, R.S., Garner, D.V., Gordon, D.B., Sedgwick, R.D., Tetler, L.W., and Tyler, A.N. (1981) *Biochem. J.* **197,** 401-404.
3. Ito, Y., Takeuchi, T., Ishi, D., and Goto, M. (1985) *J. Chromatog.* **346,** 161-170.
4. Caprioli, R.M., Fan, T., and Cottrell, J.S. (1986) *Anal. Chem.* **58,** 2949-2954.
5. Blakley, C.R. and Vestal, M.L. (1983) *Anal. Cm.* **55,** 750-754.
6. Gieniec, J., Mack, L.L., Nakamae, K., Gupta, C., Kumar, V., and Dole, M. (1984) *Biomed. Mass Spectrom.* **11,** 259-268.
7. Fenn, J.B., Mann, M., Chin Kai Meng, Shek Fu Wong, and Whitehouse, C.M. (1990) *Mass Spectrom. Rev.* **9,** 37-70.
8. Wilm, M.S. and Mann, M. (1994) *Int. J. Mass Spectrom. Ion Processes* **136,** 167-180.
9. Kebarle, P. and Tang, L. (1993) *Anal. Chem.* **65,** 972A-986A.
10. Fenn, J.B. (1993) *J. Am. Soc. Mass Spectrom.* **4,** 524-535.
11. Bruins, A.P., Covey, T.R., and Henion, J.D. (1987) *Anal. Chem.* **59,** 2642-2646.
12. Cunniff, J.B. and Vouros, P. (1995) *J. Am. Soc. Mass Spectrom.* **6,** 1175-1182.
13. Ferrige, A.G., Seddon, M.J., Green, B.N., Jarvis, S.A., and Skilling, J. (1992) *Rapid Commun. Mass Spectrom.* **6,** 707-711.

14. Chowdhury, S.K., Katta, V., and Chait, B.T. (1990) *J. Am. Chem. Soc.* **112,** 9012-9013.

15. Tang, X.-.. and Boyd, R.K. (1992) *Rapid Commun. Mass Spectrom.* **6,** 651-657.

16. Burlet, O., Yang, C.-Y., and Gaskell, S.J. .(1992) *J. Am Soc. Mass Spectrom.* **3,** 337-344.

17. Cox, K.A., Gaskell, S.J., Morris, M., and Whiting, A. (1996) *J. Am. Soc. M Spectrom.* **7,** 522-531.

18. Dongre, A.R., Jones, J.L., Somogyi, A., and Wysocki, V.H. (1996) *J. Am. Chem. Soc.* **118,** 8365-8374.

19. Hopfgartner, G., Wachs, T., Bean, K., and Henion, J. (1993) *Anal. Chem.* **65,** 439-446.

20. Gale, D.C. and Smith, R.D. (1993) *Rapid Commun. Mass Spectrom.* **7,** 1017-1021.

21. Emmett, M.R. and Caprioli, R.M. (1994) *J. Am. Soc. Mass Spectrom.* **5,** 605-613.

ATMOSPHERIC PRESSURE IONIZATION (ISI AND APCI)

Theory and Application

A. RAFFAELLI
Centro di Studio del CNR per le Macromolecole Stereordinate ed Otticamente Attive, Dipartimento di Chimica e Chimica Industriale, Università di Pisa, Via Risorgimento, 35, 56126 PISA, Italy.

1. Introduction

Chemical ionization (CI) is based on the transfer of charged species from a reagent ion to an analyte molecule. This process produces an analyte ion that can be analysed by a mass spectrometric analyser. The ions formed in positive mode are usually adducts of the analyte molecule with small cations such as H^+, NH_4^+, Na^+ and, in negative mode, result from loss of protons, electron capture, adduct formation with small anions such as Cl^-, $HCOO^-$, CH_3COO^- and so on[1]. CI spectra are usually simple with little fragmentation and the technique can be very sensitive, as the ion current is concentrated in a few species. CI requires a higher pressure with respect to that found in the common Electron Ionization (EI) sources and in the mass analyser. It is very efficient in producing ions, as the rate constants for the ion-molecule reactions are generally the collision rate of the reagents.
On the basis of the last observation, we can postulate that CI performed at atmospheric pressure would have the highest efficiency: the creation of ions at atmospheric pressure is quite straightforward and the major difficulty is transferring the ions produced into the high vacuum (10^{-5}-10^{-6} Torr) region of the mass analyser.
The above considerations induced the development of mass spectrometry systems with an ion source operating at atmospheric pressure and led to the development of three ionization methods: Atmospheric Pressure Chemical Ionization (APCI), Electrospray (ESI) and Ionspray (ISI). We will discuss briefly ISI first, in order to illustrate the differences with respect to ESI, elsewhere described in this book (see Part I, Chapter 1), and then we will focus mainly on APCI.

2. Ionspray Ionization

2.1 INTRODUCTION

In 1986 Andries Bruins, Tom Covey and Jack Henion[2] at Cornell University developed a new ionization technique moving from the Ion Evaporation ionization of Iribarne and

R. M. Caprioli et al. (eds.), Selected Topics and Mass Spectrometry in the Biomolecular Sciences, 17–31.

Thompson[3]. As it was also similar to ESI[4], because of a high voltage directly applied to the sprayer, the name "Ionspray" was chosen as a combination of "Ion Evaporation" and "Electrospray". The main difference between ESI and ISI is that ESI, in its original configuration, use only a strong electric field for the production of a spray of charged particles, whereas ISI produces the spray by a pneumatic nebulizer, like Ion Evaporation, but also uses a strong electric field, like ESI. Indeed ISI is also known as "Pneumatic Assisted Electrospray". The term "Ionspray" is a trademark of Perkin Elmer Sciex, but nowadays in practice all ESI systems use some kind of pneumatic assistance to nebulization. The use of a nebulizing gas (usually air or nitrogen) allows a more stable ion current, a higher percentage of water (up to 100%) in the mobile phase, a less critical positioning of the nebulizer with respect to the ion entrance orifice of the mass spectrometer and a higher flow rate (classic ESI requires flow rates lower than 5 μl/min, whereas ISI can work still nicely at 200 μl/min). At present the term "ESI" is used also to indicate ISI applications, and the two terms are used almost indifferently. Therefore we will focus here only on the different aspects of ISI with respect to ESI, while the general features of ESI can be found in Part I, Chapter 1.

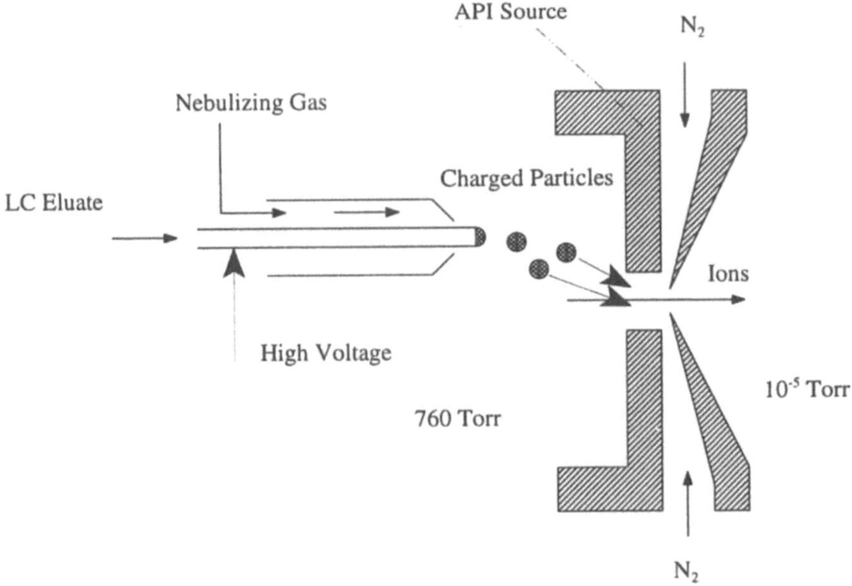

Figure 3.1. Schematic representation of an Atmospheric Pressure Ionization source with Ionspray interface.

2.2 PRINCIPLES AND APPARATUS

A schematic representation of an Atmospheric Pressure Ionization (API) source with an Ionspray interface is shown in Figure 3-1. The sample solution flows through a capillary inserted in a coaxial pneumatic nebulizer. The nebulizer is kept at high voltage (3-6 kV). Solvent evaporation and ion declustering is assisted by a curtain gas of dry nitrogen. The ions are introduced into the mass spectrometer through a 100 μm orifice driven by electric fields. The ionization occurs by the "ion evaporation" mechanism in the same fashion as in ESI, schematized in Figure 3-2.

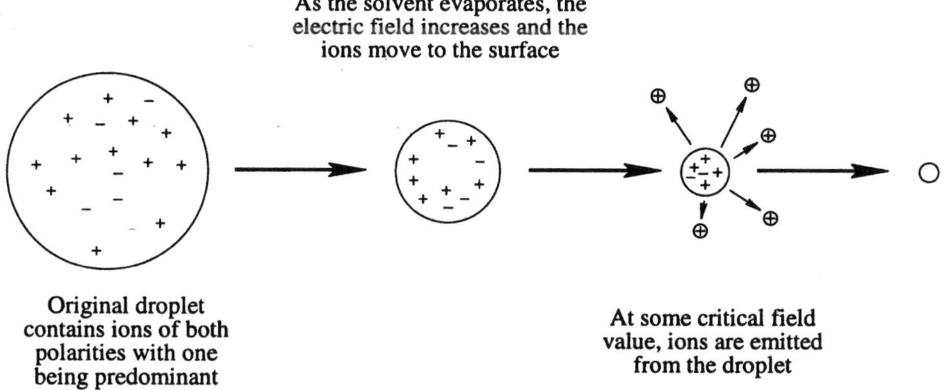

Figure 3-2. The Ion Evaporation mechanism.

3. Atmospheric Pressure Chemical Ionization

3.1 INTRODUCTION

Atmospheric Pressure Chemical Ionization (APCI) is older than ESI or ISI, but it was used only for gas analysis up to 1974. A lot of literature reports from the late 60s to the late 70s[5] deal with the development of this ionization technique using different ways to get the primary ions and its application to Ion Mobility Mass Spectrometry. The main primary ionization techniques used were a Ni^{63} β-particle emitter and needle-to-plane corona discharge ion sources, but this last was later used more widely. The use with standard mass analysers working under high vacuum conditions came in 1973 by Horning[6], and shortly afterwards the same author illustrated the combination of APCI-MS with liquid chromatography for the first time[7]. This work also introduced the acronym "API" (for Atmospheric Pressure Ionization) which became widely used. Nowadays the term "APCI" is preferred since it suggests the use of chemical ionization techniques at very high pressures (up to atmospheric or even slightly above) rather than other high-pressure ion sources such as plasma discharge and so on. Moreover at present the term "API" denotes an ion source working at atmospheric pressure, which can accommodate different interfaces, such as ESI/ISI, APCI, Townsend, etc., whereas the term "APCI" denotes just one of the interfaces used in an API source.

The first application of tandem mass spectrometry coupled with APCI was performed in 1982 by Henion et al[8]. Further developments to the interface used by Henion led to the Sciex Heated Nebulizer LC-MS Inlet, the first commercially available LC-APCI-MS interface, which provided the capability for detecting a wide variety of polar, semi-volatile compounds and quasi-molecular ions with minimal thermal degradation.

3.2 PRINCIPLES AND APPARATUS

The coupling with HPLC created at the beginning a large interest in the scientific community, as it was, under several aspects, the ideal HPLC-MS interface: "The best interface is 'no interface at all'". Indeed APCI is the closest approach to this situation, also because it can arrange flow rates up to 1 or even 2 ml/min, so that a direct coupling to 4.6 I.D. analytical HPLC columns is possible without any splitting.

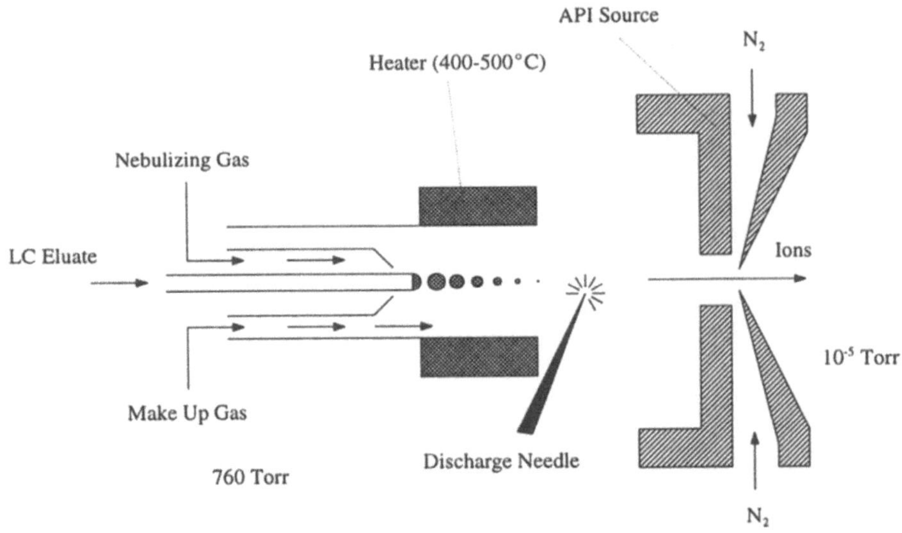

Figure 3-3. Schematic representation of an Atmospheric Pressure Ionization source with a Heated Nebulizer interface for Atmospheric Pressure Chemical Ionization.

A schematic representation of an API source with a Heated Nebulizer Interface for APCI is reported in Figure 3-3. The sample solution flows through a capillary inserted in a coaxial pneumatic nebulizer. A make up gas (usually air) is added to assist the ionization process. The mixture of air and nebulized solution passes through a heated zone that assist the solvent evaporation. Despite the high temperature of the heater (400-500°C), minimal degradation of the sample occurs, since the heat provided is used just for the vaporization of the solvent and the sample temperature usually do not exceed 100-120°C. A corona discharge needle kept at high voltage (5-6 KV) is positioned between the exit of the heated nebulizer and the source interface plate. The needle generates a discharge current of ca. 2-3 μA, which ionizes air producing primary ions (mainly N_2^+, O_2^+, H_2O^+ and NO^+ in positive mode, O_2^-, O^-, NO_2^-, NO_3^-, O_3^- and CO_3^-). The primary ions react very rapidly (within 10^{-6} sec) transferring their charge to solvent molecules to produce the CI reactant ions that finally produce analyte quasi-molecular ions by charge or proton transfer reactions. The total reaction time in the source is about 5×10^{-4} sec. Overall we can observe $[M+H]^+$ ions in positive mode and $[M-H]^-$ ions in negative mode. Adduct ions are not readily formed in APCI. Even polyoxygenated molecules that normally produce mainly ammonium and sodium adducts under Ionspray ionization conditions, such as non-ionic surfactants, form predominant protonated molecules when analysed by APCI-MS (figure 3-4).

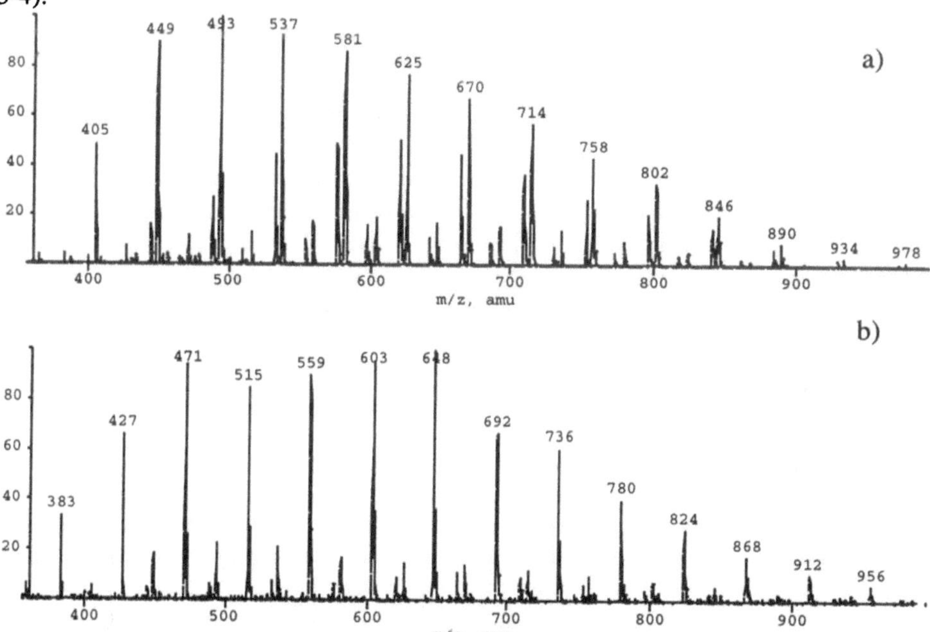

Figure 3-4. ISI-MS (a) and APCI-MS (b) spectra of a non-ionic surfactant (fatty alcohol ethoxylated).

Polyethoxylated non-ionic surfactants are oligomeric mixtures giving rise to family of ions equally spaced by 44 mass units (the -CH_2-CH_2-O- subunit). Under Ionspray conditions (Figure 3-4a) we can observe ammonium (M+18) and sodium (M+23) adducts as the predominant ions, whereas the protonated molecules give quite weak signals, if any. The same sample, analysed by APCI-MS (Figure 3-4b), shows the protonated molecules as the most abundant ions, whereas the ammonium and sodium adducts are below 20%[9]

As soon as the analyte ions are formed they are driven into the mass analyser through a 100 μm orifice by the electric field generated between the discharge needle, the interface plate, the orifice plate and the rest of the mass spectrometer. A curtain of dry nitrogen prevents neutral particles from entering the vacuum region. At the same time this curtain gas assists the declustering process to obtain free ions from their solvated forms.

All together, the main advantages of APCI mass spectrometry can be summarized as follows:

- ☺ CI-like spectra, with molecular weight information;
- ☺ Suitable for volatile or semi-volatile analytes;
- ☺ Relatively easy to install and "friendly" for the user;
- ☺ Very rugged by an analytical point of view;
- ☺ Flow rates up to 1-2 ml/min can be used without any problem;
- ☺ Very good sensitivity;

☺ It is enough "universal";
☺ Good chromatographic fidelity.

The main disadvantages, on the other hand, are not very heavy:

☹ No fragmentation, and hence no structural information;
☹ Some thermal degradation can occur;
☹ Inorganic buffers can create problems if too concentrated (above 5-10 mM).

The ideal mass spectrometric system for the implementation of APCI is a tandem mass spectrometer such as a triple quadrupole, so that the disadvantage of the lack of fragmentation can be overcome using MS-MS. The use of Selected Reaction Monitoring (SRM), moreover, allows to reach very good detection limits for quantitation with a good specificity as well, much better than in Selected Ion Monitoring (SIM) available on single analyser instrumentation.

4. Applications

4.1 INTRODUCTION

ISI and APCI can use the same apparatus, as the ion source is practically the same, and the switching from one technique to the other is a matter of changing the interface, which operation takes about 10-15 min. As most instruments providing an API source it is available with both interfaces, it is quite common to have both the techniques in the same laboratory. It can be useful first to give here a brief summary of the characteristics of both systems.

4.1.1 *Summary of Ionspray features*

☞ Ionspray is suitable for those compounds that are present as ions in solution (acids and bases);
☞ Response to basic compounds can sometimes be enhanced by reducing pH of the solution;
☞ The normal flow rate used in ISI is 2 to 200 μl/min. A typical operating flow rate can be 50 μl/min;
☞ The typical Ionspray nebulizer voltage is + 4000 to 6000 V for positive ion mode, - 3000 to -4500 V for negative ion mode.
☞ ISI can generate multiply charged ions if multiple basic or acidic groups are present on the molecule (as in the case of biopolymers such as peptides, proteins and oligonucleotides).

On the basis of these considerations, the classes of molecules suitable to ISI-MS can be listed as follows:

☞ Ionic molecules, salts (for instance quaternary ammonium salts);

- Zwitterion compounds (for instance amminoacids);
- Polar molecules containing heteroatoms (for instance carbohydrates, esters, etc.);
- Organic acids and bases (for instance carboxylic acids and amines);
- Proteins, peptides and oligonucleotides generating multiply charged ions;
- Ionic organometallic compounds;
- Covalent organometallic compounds bearing an acidic or basic site or capable to accept or release one electron.

4.1.2 *Summary of APCI features*

- APCI is suitable for not so polar compounds, which may be charged in solution, but typically are not strong acids or bases;
- pH does not have a strong influence on the response;
- The normal flow rate used in APCI is 0.5 to 2 ml/min. A typical operating flow rate can be 1 ml/min;
- Usually under APCI conditions the parameter to set is the discharge current, typically limited to 2-3 μA. This induces a needle voltage comparable with the values normally set for ISI.
- APCI cannot generate multiply charged ions in any case.

On the basis of the above considerations, compounds ranging from low to mid polarity, volatile or semivolatile are suitable for APCI-MS analysis. Sometimes APCI can provide a better sensitivity with respect to ISI even with molecules amenable to ISI, for instance if a phosphate buffer is used, as ISI is quite sensitive to inorganic salts present in solution, as APCI can tolerate buffers more easily.

4.2 IONSPRAY APPLICATIONS

The fields of application of ISI mass spectrometry are quite the same as ESI mass spectrometry that is protein characterization, drugs characterization, metabolism studies, food contaminants, environmental control, etc. As ESI and its applications have been described elsewhere in this book, we present here only an application in the field of characterization of natural products of pharmaceutical interest.

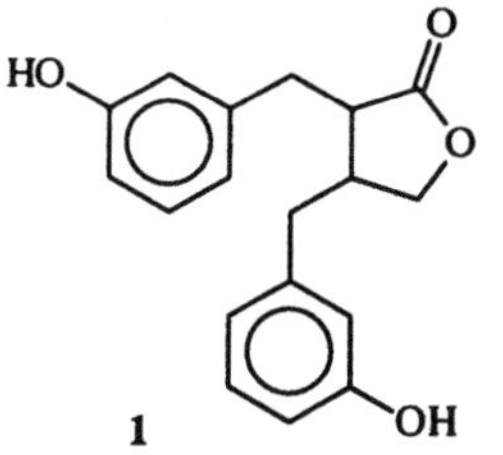

HO, O, O, OH — 1
HO, OH, OH, OH — 2 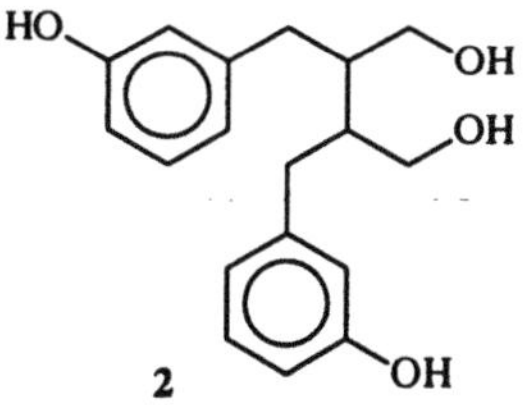
HO, OGlucose, OGlucose, OH — 3 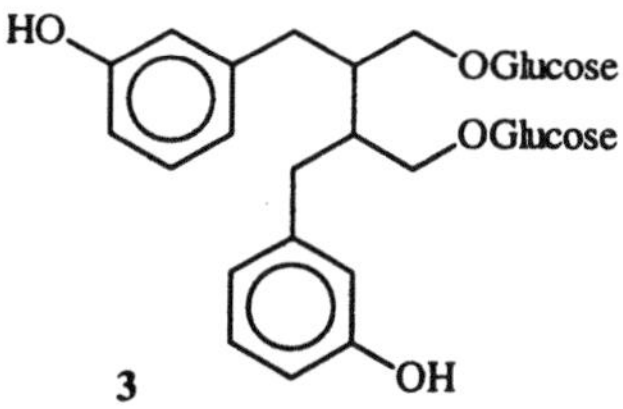

A strong research effort has been devoted in recent years to the mammalian lignans enterolactone (**1**) and enterodiol (**2**) because of their claimed protective effects against breast and colon cancer. Some plant lignans appear to be the precursors of these molecules. One of these precursors is the 2,3-bis(3-methoxy-4-hydroxybenzil)butane-1,4-diglucoside (**3**), commonly known as secoisolariciresinol diglucoside (SDG). This compound was discovered in the early 50s by Klostermann and co-workers[10] but it was not previously characterized by modern spectroscopic methods. Direct analysis of the mixture coming from alkaline methanolysis of polar solvent extractable of fat-free flax seed meal by ISI-MS both in positive and negative ion modes provided a final proof of the molecular weight of this substance (Figure 3-5)[11].

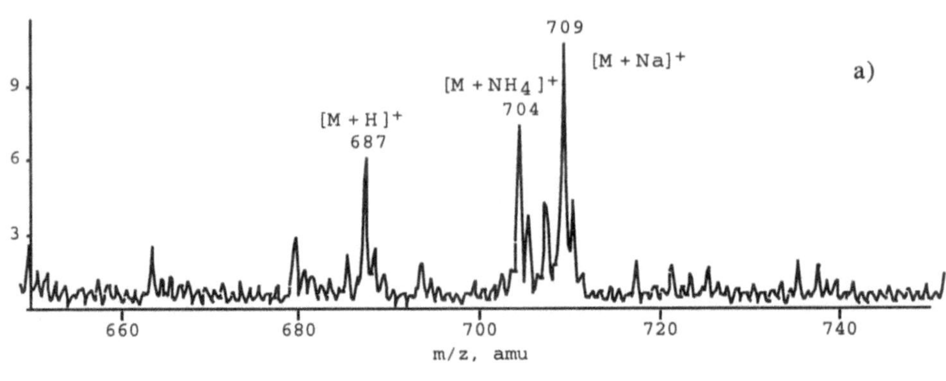

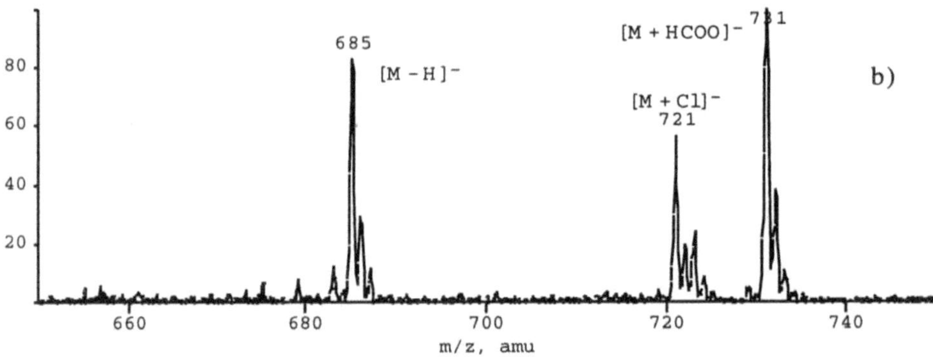

Figure 3-5. Positive ion (a) and negative ion (b) spectra of SDG in a sample of crude methanolyzed extracts from flax seeds.

The presence of ions at m/z 687 ($[M+H]^+$), 704 ($[M+NH_4]^+$) and 709 ($[M+Na]^+$) in the positive ion spectrum (Fig. 3-5a) and at m/z 685 ($[M-H]^-$), 721 ($[M+Cl]^-$) and 731 ($[M+HCOO]^-$) strongly support the presence of a substance having MW 686. MS-MS measurements by CID fragmentation of the ions at m/z 687 in positive ion mode and 685 in negative ion mode allowed also to get a good indication of the correct structure. The MS-MS spectra reported in figure 3-6 are in a very good agreement with the structure of the compound **3**.

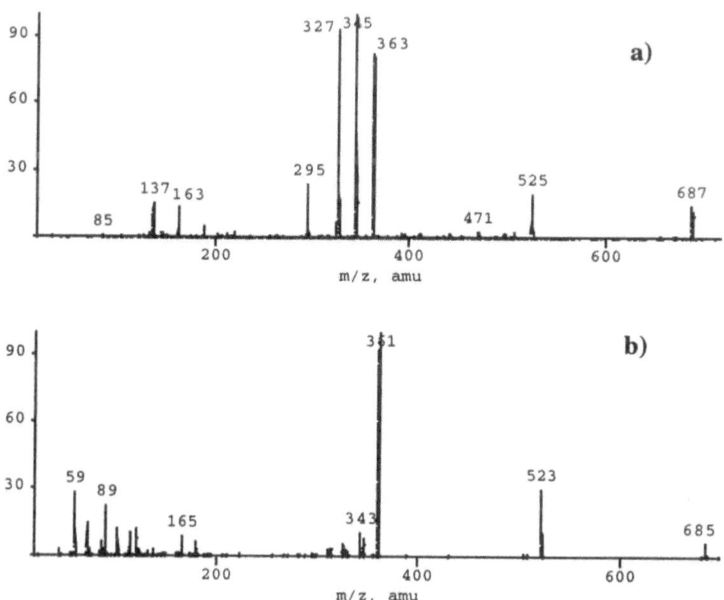

Figure 3-6. Positive ion (a) and negative ion (b) MS-MS spectra of the $[M+H]^+$ (a) and $[M-H]^-$ (b) ions of SDG.

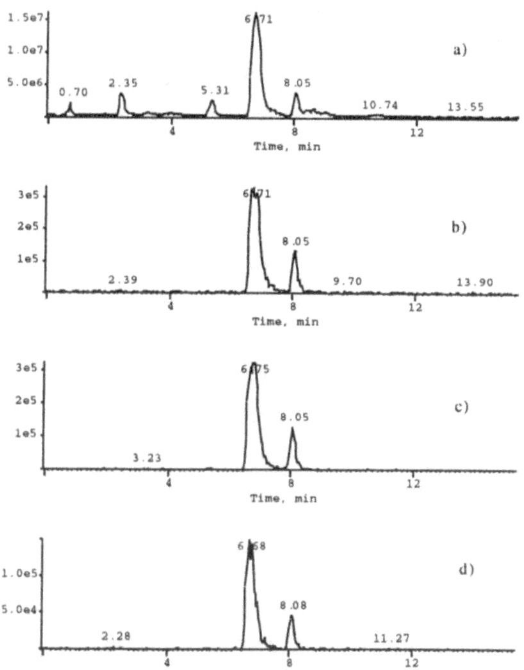

Figure 3-7. TIC chromatogram (a) and XIC of the ions at m/z 687 (b), 704 (c) and 709 (d) of the mixture containing SDG.

In order to set up a separation and purification method for SDG, the same mixture was examined by HPLC-MS on a C18 column.

This investigation allowed to identify two isomeric forms of SDG, as shown in Figure 3-7[12]. The CID MS-MS spectra of the two components are completely superimposable, indicating that they are very likely two stereoisomeric forms due to the stereogenic centers at carbon atoms 2 and 3.

In this application the use of ISI mass spectrometry both in single MS mode and in MS-MS allowed to perform a good characterization of SDG for the first time, without any need of isolation and purification. More applications can be found in literature by searching the keywords "Ionspray" or "Electrospray".

4.3 APCI APPLICATIONS

APCI HPLC-MS is a very powerful and rugged technique for routine analysis in different fields, such as environmental control, pharmacokinetic studies, food analysis and so on. Its capability to provide only one single ion from most analytes (protonated molecule in positive ion mode, deprotonated molecule in negative mode) makes it ideal for quantitation based on Selected Ion Monitoring (SIM). Also, the MS-MS sensitivity is good as well, allowing the large specificity of Selected Reaction Monitoring (SRM). We present here two applications in the fields of environmental control and food analysis.

4.3.1 *Determination of Thiourea in Waste Water by FIA-APCI-MS-MS*

Thiourea is one of the most toxic substances for nitrification bacteria, with a toxicity limit of 76 ppb. Normally it is not present in waste water, but it can be a serious problem for industrial waste water treatment plants, in particular if the waste water comes from tannery factories. Thiourea or some possible precursor can be used in some stage of the tanning process. Therefore, a fast and reliable method for its determination is strongly needed. The approach to this problem must face four main problems:

1 The samples of industrial waste water to be analysed contain a large number of different substances, some of which with a relatively high concentration.
2. Thiourea is very soluble in water and practically insoluble in organic solvents such as ethyl ether and methylene chloride so that its extraction with organic solvents is almost impossible. Solid Phase Extraction fails as well, with a recovery below 20-30 % in the best cases.
3. Thiourea is relatively unvolatile and its GC analysis is not very easy, at least using common apparatus.
4. Preliminary HPLC measurements caused the irreversible degradation of the HPLC column after a few injections due to the build up of some components of the matrix onto the stationary phase.

APCI tandem mass spectrometry in SRM allowed a brilliant solution of this problem[13], with detection limits down to 1 ppb and minimal sample pre-treatment. Fig. 3-8 shows the APCI-MS-MS spectrum of thiourea[14].

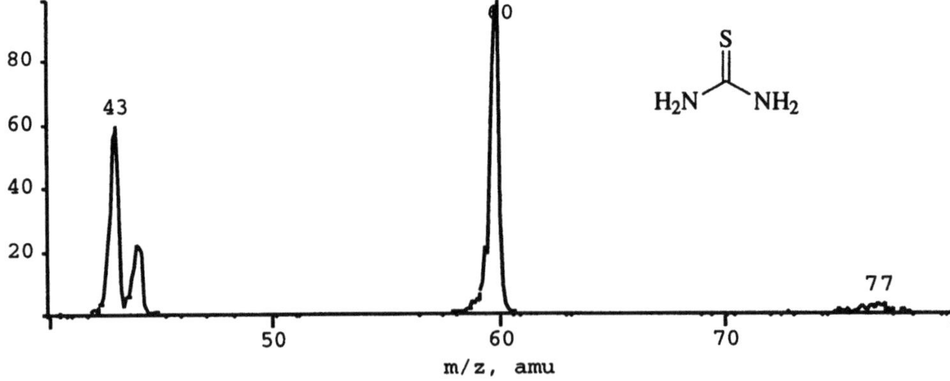

Figure 3-8. APCI-MS-MS spectrum of thiourea.

The possibility to monitor the two fragmentations 77→60 and 77→43 on a triple quadrupole mass spectrometer allows to quantitate thiourea directly by Flow Injection Analysis (FIA) in the waste sample with a high sensitivity and specificity. The figure 3-9 shows the chromatogram related to the calibration curve, where standard water solutions of thiourea with concentrations of 10, 50, 100, 500, 1000 and 5000 ppb.

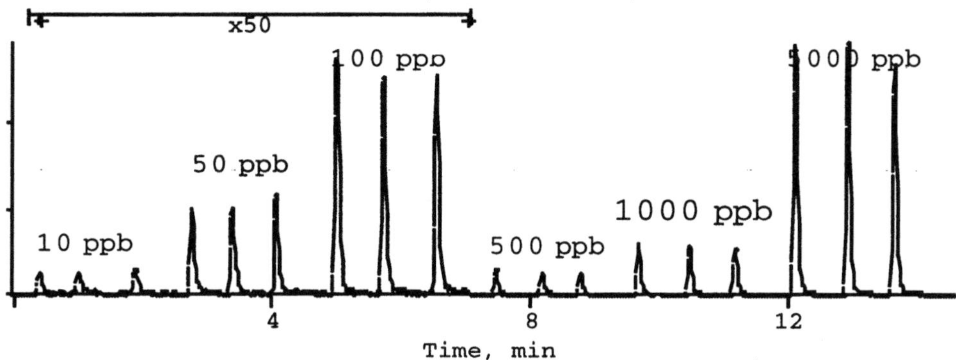

Figure 3-9. FIA-APCI-SRM-MS-MS TIC chromatogram of standard solutions of thiourea.

The calibration curve obtained showed a quite good linearity (correlation factor R=0.999931), with a detection limit around 1-2 ppb and a variation coefficient below 6 %. The reliability of this analytical procedure was tested by the analysis of "spiked" samples and it was very good. It is also worthy to note that we never found thiourea levels above 50 ppb when the plant was working properly, and we never found thiourea levels below 100 ppb when something was wrong and there was an ammonia release from the plant. This means that the experimental amounts found were in a very good agreement with the toxicity limit known of 76 ppb. Moreover, we found values between 50 and 100 ppb when the plant was working not too bad, but a slight release of ammonia was still present.
Moreover, the specificity of the procedure is also very high. The monitoring of two fragmentations minimizes the possibility of any interference. Some samples (all coming from the same tanning factory) showed a partial interference: there was a compound in the matrix giving the 77→43 fragmentation, but not the 77→60 fragmentation, so that the signal observed in the TIC chromatogram could not be attributed to thiourea.

4.3.2 *Analysis of Hydrosoluble Vitamins by APCI Mass Spectrometry*
Particle Beam HPLC-MS gives very good results with lipophilic vitamins, whereas it fails for the analysis of hydrosoluble vitamins. We analysed a group of these components both by ISI-MS and APCI-MS, but APCI gave the best results due to the use of a 20 mM ammonium formate buffer. The study has been performed on vitamin B1 (thiamine), vitamin B2 (riboflavin), B6 group vitamins (pyridoxine, pyridoxal, pyridoxamine), vitamin PP (nicotinamide, nicotinic acid), vitamin C (ascorbic and dehydroascorbic acids), vitamin H (biotin) and vitamin B5 (pantothenic acid)[15]. Single MS measurements of these compounds gave the expected molecular weight information, with low fragmentation. Structural information can be easily accessed by MS-MS. As known, MS-MS can discriminate

molecules having very similar molecular weight and structure. This is the case, for instance, of the three vitamers of the B6 group. Their APCI-MS-MS spectra are reported in the Figure 3-10.

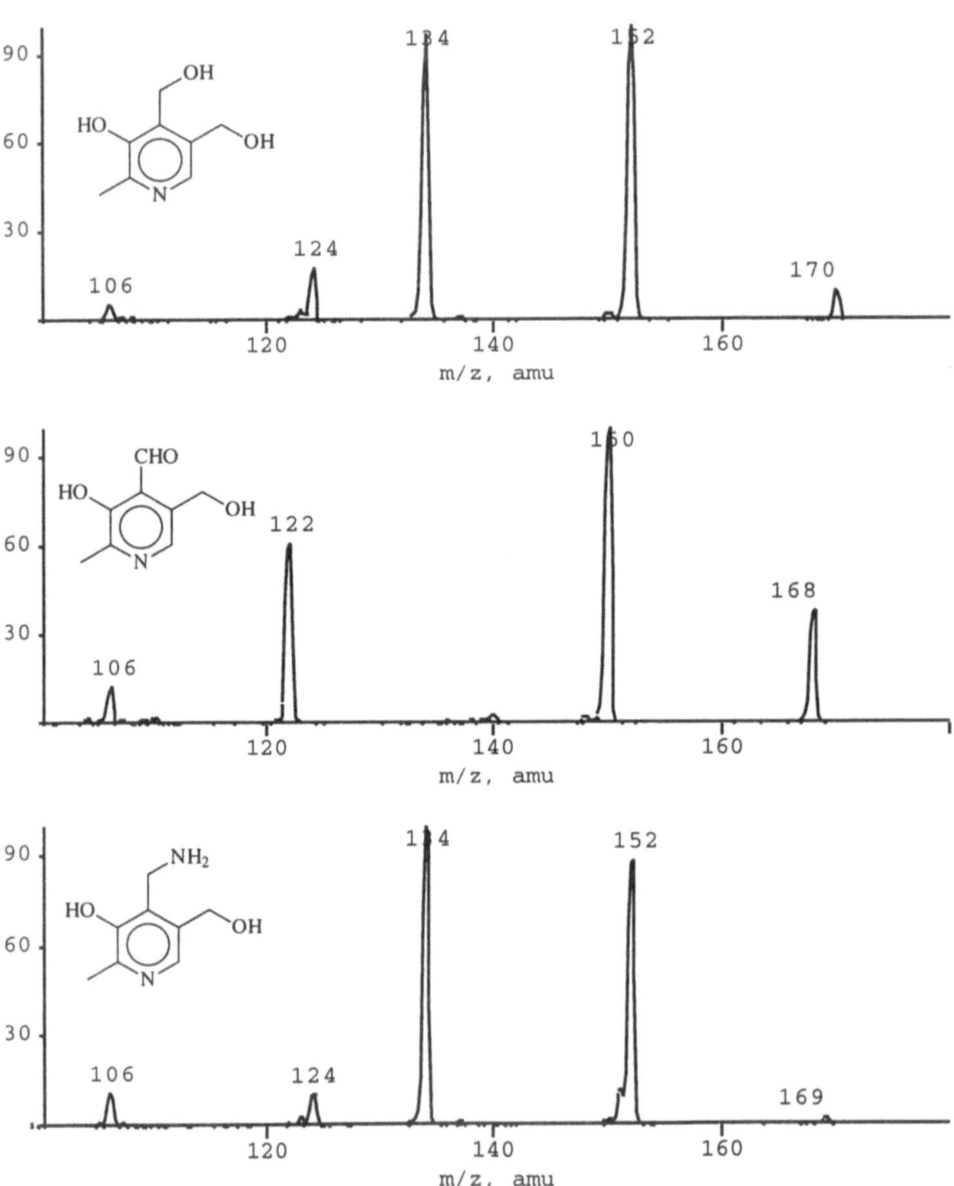

Figure 3-10. APCI-MS-MS spectra of pyridoxine, pyridoxal and pyridoxamine.

It is worthy to note the consecutive losses of two water molecules from pyridoxine, the loss of a single water molecule from pyridoxal and the consecutive losses of ammonia first and then water from pyridoxamine.
Based on MS-MS investigations, we developed also an HPLC-SRM-MS-MS method for the analysis of these components in food products such as milk, beans, peanuts, fruit juices and so on. Figure 3-11 shows the chromatograms of a mixture of standards.

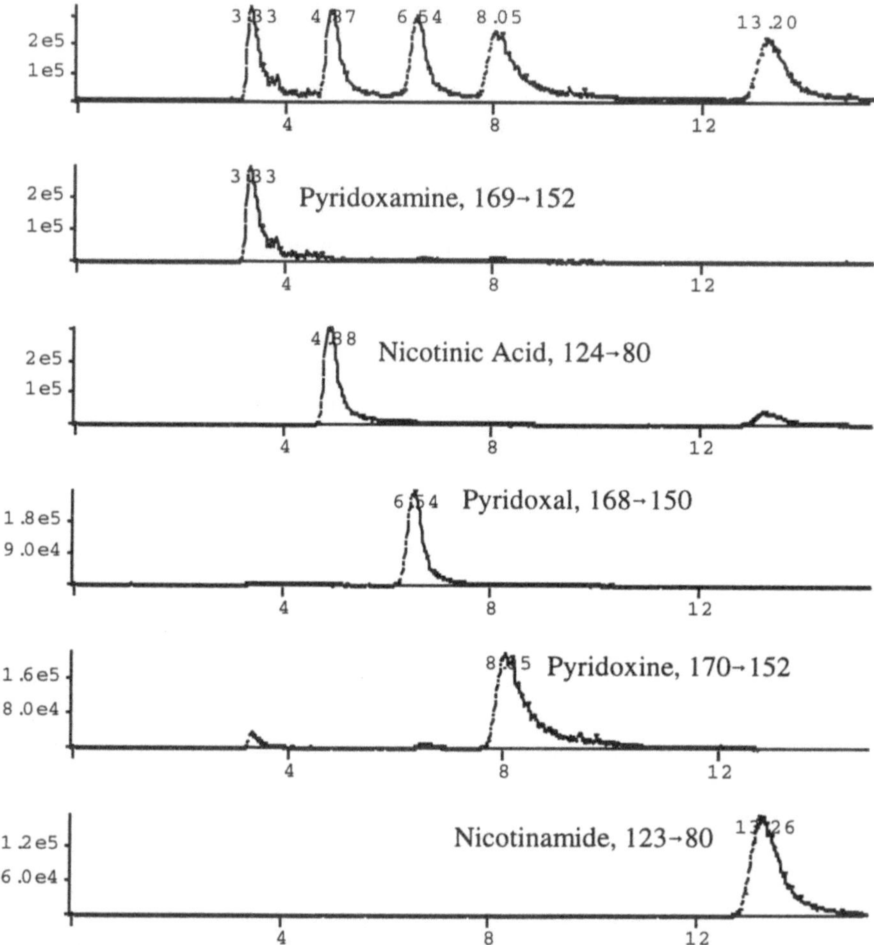

Figure 3-11. TIC and XIC chromatograms relative to a HPLC-SRM-MS-MS analysis of a standard mixture of 5 vitamins.

The analysis of reals samples gave very good results, with good detection limits, specificity and reproducibility. It is worthy to note that the analysis could be performed even without the aid of HPLC, due to the characteristic fragmentations chosen. Quantitation was not very reliable here due to a consistent tailing observed in Flow Injection Analysis (actually some tailing was observed also in HPLC, as shown in figure 3-11), but a qualitative analysis can be performed very quickly (1 min for sample) using this approach.

5. Conclusions

In conclusion, API mass spectrometry, both with ISI and APCI interfaces, can be used in a wide range of applications. The two techniques are somewhat complementary, as ISI is suitable for very polar and/or ionic compounds and low concentrations of salts or buffers, whereas APCI is tailored for medium polarity samples and tolerates relatively high concentration of salts (up to 20 mM).

Under normal operation conditions, that is using 4.6 mm I.D. columns at 1 ml/min flow rate, APCI can be directly interfaced, whereas ISI needs to split the flow so that only 40-50 μl are directed to the source. This does not decrease the sensitivity, as ISI interface is a concentration-sensitive detector, in contrast with APCI, that is a mass flow-sensitive detector. This different behaviour must be taken into account when quantitative analysis is performed. The split, necessary for ISI, can even be an advantage, as about 95% of the sample can be recovered, for instance using a fraction collector. On the other hand, APCI is not suitable for use with narrow bore (2.1 mm I.D.), microbore (1 mm I.D.) and capillary columns, as in these cases the flow rate is too low for a good ionization efficiency. Anyway, the combination of these two ionization techniques represents nowadays the best HPLC-MS interface and allows to approach and solve most problems in biological, pharmaceutical and environmental fields.

References

1. Perkin Elmer Sciex, *The API Book*, Perkin Elmer Sciex Instruments, Toronto, Canada.(1992)

2. A.P. Bruins, T.R. Covey and J.D. Henion, *Anal. Chem.*, **59**, 2642 (1987).

3. B.A. Thomson. and J.V. Iribarne, *J. Chem. Phys.*, **71**, 4451 (1979).

4. C.M. Whitehouse, R.N. Dreyer, M. Yamashita and J.B. Fenn, *Anal. Chem.*, **57**, 675 (1985).

5. D.I. Carroll, I. Dzidic, R.N. Stillwell and E.C. Horning, *Anal. Chem.*, **47**, 1956 (1975); G.W. Griffin, I. Dzidic, D.I. Carroll, R.N. Stillwell and E.C. Horning, *Anal. Chem.*, **45**, 1204 (1973); F.W. Karasek, *Res./Dev.*, December, 25 (1970); F.W. Karasek, *Anal. Chem.*, **46**, 710A (1974); T.W. Carr, *Anal. Chem.*, **49**, 828 (1977).

6. E.C. Horning, M.G. Horning, D.I. Carroll, I. Dzidic and R.N. Stillwell, *Anal. Chem.*, **45**, 936 (1973).

7. E.C. Horning, D.I. Carroll, I. Dzidic, K.D. Haegele, M.G. Horning, and R.N. Stillwell, *J. Chromatog. Sci.*, **12**, 725 (1974).

8. J.D. Henion, B.A. Thompson and P.H. Dawson, *Anal. Chem.*, **54**, 451 (1982)

9. A. Raffaelli, S. Pucci, L. Campolmi and P. Salvadori, *Proceedings of the 44th ASMS Conference on Mass Spectrometry and Allied Topics, Portland, Oregon, U.S.A., 12-16 Maggio 1996*, MPE 115.

10. H.J. Klostermann and J. Smith, *J. Amer. Chem. Soc.*, **76**, 1229 (1954).

11. M. Bambagiotti-Alberti, S.A. Coran, C. Ghiara, V. Giannellini and A. Raffaelli, *Rapid. Commun. Mass. Spectrom.*, **8**, 595 (1994).

12. M. Bambagiotti-Alberti, S.A. Coran, C. Ghiara, G. Moneti and A. Raffaelli, *Rapid Commun. Mass. Spectrom.*, **8**, 929 (1994).

13. A. Raffaelli, S. Pucci, R. Lazzaroni and P. Salvadori, *Proceedings of the 44th ASMS Conference on Mass Spectrometry and Allied Topics, Portland, Oregon, U.S.A., 12-16 Maggio 1996*, WPD 91.

14. A. Raffaelli, S. Pucci, R. Lazzaroni and P. Salvadori, *Anal. Chem.*, Submitted for publication.

15. M. Careri, R. Cilloni, A. Mangia, P. Manini, and A. Raffaelli, *Atti del XII Congresso Nazionale di Chimica Analitica, Firenze, 20-22 Settembre 1995*, p. 70.

MATRIX-ASSISTED LASER DESORPTION-IONIZATION (MALDI) MASS SPECTROMETRY: PRINCIPLES AND APPLICATIONS

M.KARAS, U.BAHR
Institute for Physical and Theoretial Chemistry
Department of Instrumental Analytical Chemistry
University of Frankfurt
Theodor-Stern-Kai 7
60590 Frankfurt
F.R. Germany

ABSTRACT. Matrix-assisted laser desorption/ionization mass spectrometry (MALDI-MS) is an analytical technique for fast and precise mass determination of biological molecules. Intact molecular ions are produced by short pulsed laser irradiation of the biomolecules which are enbedded in a matrix consisting of small highly absorbing organic molecules. Mass analysis is most often carried out in a linear or reflector time-of-flight mass spectrometer. The accessible mass range is > 500 000 Da, a mass accuracy of < 50 ppm for peptides < 2000 Da can be reached, the sample amounts required are 1 pmol or less. Proteins, glycoproteins, oligonucleotides and oligosacharides can be analyzed.

1. Introduction

The development of new ionization techniques during the two past decades has made possible important achievements in bioorganic mass spectrometry for the analysis of biopolymers such as proteins and carbohydrates. The general problem to be overcome is to convert the polar, thermally labile biomolecules from the condensed phase into intact isolated ions in the gas phase. The so-called desorption/ionization techniques use different physical approaches to achieve this goal; field desorption [1] applies a high electric field to the sample; in fast atom bombardment [2] and ^{252}Cf plasma desorption [3] the sample is bombarded by highly energetic ions or atoms; thermospray ionization [4] and electrospray ionization [5] form ions directly from small, charged droplets. Laser desorption [6,7] and the newly developed version of this method, matrix-assisted laser desorption/ionization (MALDI) [8] make use of short, intense pulses of laser light to induce the formation of intact gaseous ions.

2. Development of MALDI-MS

Already in the 1970s, first results of using laser light as mass spectrometric ionization

R. M. Caprioli et al. (eds.), Selected Topics and Mass Spectrometry in the Biomolecular Sciences, 33–53.

method for organic molecules have been reported [9-13]. CO_2-lasers with a wavelength of 10.6 μm in the infrared (IR) and lasers emitting in the far ultraviolet (UV) proved to be most useful. In order to transfer the energy within a very short time and to avoid thermal decomposition of labile organic molecules lasers with pulse widths on the nanosecond timescale, such as Q-swiched Nd-YAG, excimer or TEA-CO_2-laser were used. The pulsed desorption of ions favours the combination of the laser desorption ion source with time-of-flight (TOF) mass analyzers [14,15] or fourier transform ion cyclotron resonance (FT-ICR) mass analyzers [16,17], both make it possible to record complete mass spectra for each laser shot.

The initial experiments on laser desorption of organic ions all showed a restriction of the analysis to biocompounds with molecular masses below ca. 2000 Da. For UV-laser desorption a "soft" desorption of molecular ions was only found for compounds exhibiting a strong resonance absorption at the incident laser wavelength. For larger molecules, even if they contain a chromophoric group, the obtained energy per volume was believed to be insufficient at moderate irradiances (laser power/area, W/cm^2), higher irradiances result in fragmentation (photodissociation) rather than in the desorption of intact molecule ions. Samples which cannot be resonantly excited at the laser wavelength need very high irradiances for ion production which inevitably destroy large organic molecules. These limitations, which prohibited a general use of laser desorption for organic mass spectrometry, promoted the idea to use a matrix for the analysis of non-absorbing and/or high-molecular-mass biopolymers [18]. The matrix consists of a strongly UV-absorbing organic compound known to show a soft laser desorption for the neat compound and the high mass molecules are finely dispersed therein. Thereby, the latter can be desorbed and ionized irrespective of their individual absorption characteristics [19,20].

3. Ion Formation in Matrix-Assisted Laser Desorption/Ionization

The matrix was introduced to serve the following functions :

- Its strong absorption at the wavelength of the incident laser light makes it possible to transfer the energy of the laser light to the solid sample in a controllable way. For the adequate irradiances used in the range of 10^6 to 10^7 W/cm^2, laser pulse lenghts of some ns and a matrix absorptivity of some thousand $l\ mol^{-1}\ cm^{-1}$ typically one photon per matrix molecule in the uppermost layer is absorded. The calculation is based on linear absorption and validity of Beer's Law. This also results in an exponentially decaying in-depth excitation with typical penetration depths ($I=I_o/e$) of ca. 100 nm. This energy is sufficient to induce the ablation of a sample volume setting free intact matrix and analyte neutrals and ions.
- The analyte is diluted in the high-molar excess of matrix forming a „solid solven“. This prevents analyte molecules from aggregation and reduces strong inter-molecular interaction.
- Analyte ionization is promoted by photochemical reactions. Proton transfer from active matrix species, initiated by photoionization resulting in both intact molecule and fragment ions, is supposed to form the major ionization mechanism for analytes which are able to form stable (de)protonated ions.

An increasing number of research investigations is available today illuminating the desorption ionization process [21,22]. Neutral and ion desorption exhibit a strong dependance on the applied laser irradiance [18-20,23]. Once a distinct value is reached (threshold irradiance), ion yield is strongly increasing with the irradiance, typically with the fifth or sixth power. Best results are obtained in a small irradiance range at and slightly above the threshold value. Matrix and analyte ions show a strongly forwarded emission (normal to the target surface) and have supersonic initial velocities irrespective of their mass [24,25].

Despite their high initial velocities, ion accelerated into the mass spectrometer show deficit (too low) energies increasing with increasing irradiance; this can be rationalized by collisions of ions with neutrals in the initially very dense material cloud and/or by ion formation by chemical reactions taking place in the acceleration region [26]. The general notion that the ions formed and observed in the mass spectra are essentially intact molecule ions has to be replaced by a more refined picture. Depending on the matrix and physical conditions in the ion source molecule ions may show strong metastable or even promt dissociation [27,28]. Whereas small peptide ions undergo backbone cleavages resulting in structurally indicative fragments ions, larger proteins show mainly the loss of small neutrals such as ammonium, water and carbon dioxide [29] or of labile functional groups such as sialic acids [30].

Incorporation of analyte molecules in matrix crystals formed upon evaporation of the solvent seems to be an essential prerequisite for a successful MALDI analysis. Formation of a „solid solution" of analyte in matrix crystals has been proven for two matrices, namely 2,5-dihydroxybenzoic acid [31] and sinapic acid [32]. In the latter case a preferential built-in of a test protein onto a hydrophobic crystal surface was documented. This incorporation is also the basis for the observed high tolerance of some high-quality matrices against some ubiquitous contaminants such as inorganic salts and buffers by preferential incorporation of proteins and exclusion of salts.

There is good evidence that the matrix plays an active role in the ionization of analyte ions. The inspection of the LD mass spectra of the matrix compounds themselves revealed some very unexpected features such as high intensity radical cat- and anions accompanying even-electron molecule and fragment ions. This was taken as an indication that a good matrix is not able to form stable ions and that matrix ions function as gas phase acids or bases resulting in (de)protonated analyte ion species [33].

4. Matrices and Sample Preparation

4.1 MATRICES

Even though considerable progress has been obtained in the understanding of the MALDI process, it is not possible yet to select a matrix based on theoretical considerations alone. The required absorption at the UV-laser wavelength is unfortunately not a very restrictive criterion and that is why all of the matrices in use today have been found empirically just by trial and error.

Compatibility of solvents required for matrix and analyte, absence of chemical reactivity and vacuum stability are further obvious properties to be fulfilled by a matrix. Even though many compounds work as matrix in principle, only very few can be regarded as high-quality matrices for UV-MALDI. These are :

- Sinapic acid and some related cinnamic acid derivatives [34,35]

- 2,5-dihydroxybenzoic acid (DHB) [31] often with an admixture of 10% 2-hydroxy,5-methoxybenzoic acid (DHBs) [36].
- α-cyano-4-hydroxycinnamic acid (4-HCCA) [37]
- 3-hydroxypicolinic acid (HPA) [38]
- di- and tri-acetophenones [39].

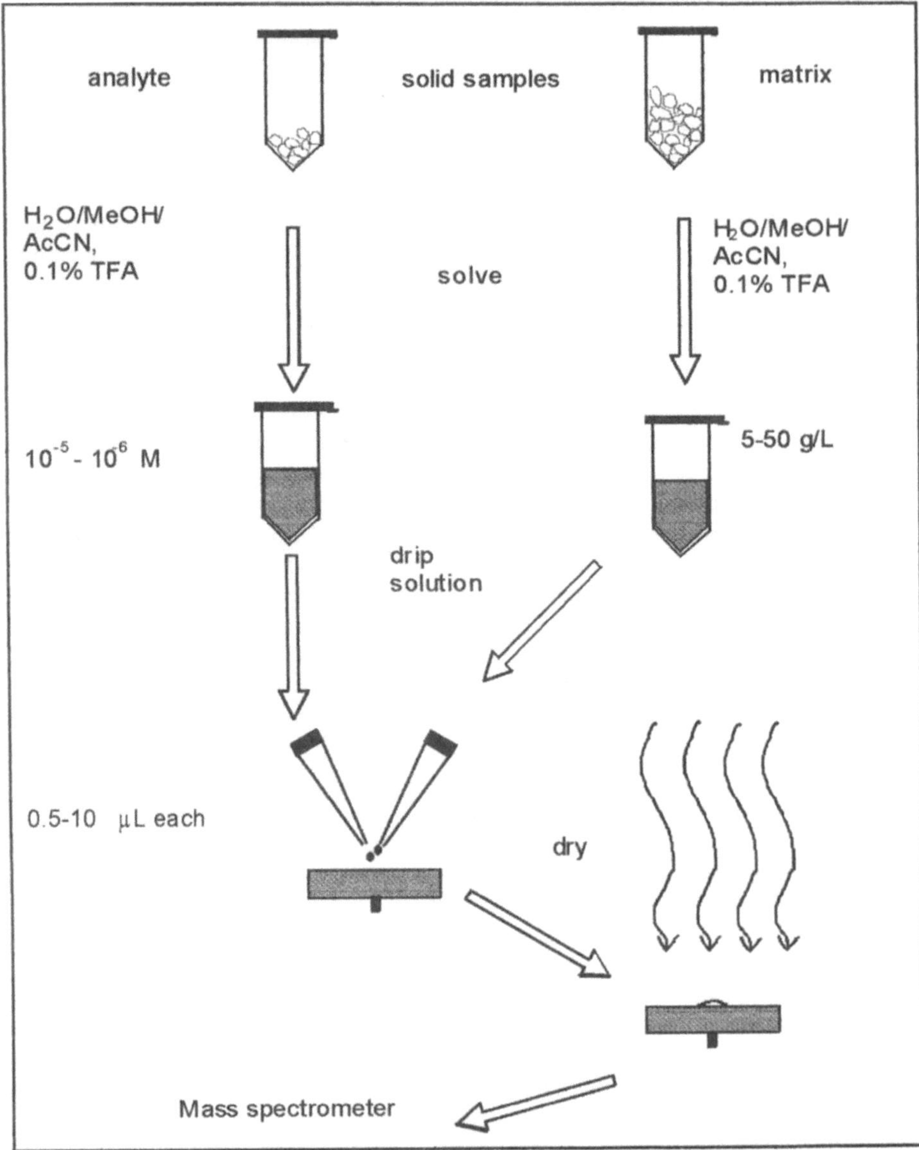

Fig 1: „Normal“ or „dried drop“ sample preparation technique for MALDI mass spectrometry

These widely-used matrices all fulfill some additional criteria. They allow for a fast, easy and reproducible preparation; they show a pronounced tolerance towards several ubiquitous contaminants and reagents; they only yield a low-intensity adduct ion signal, i.e. an ion signal formed by the addition of usually a matrix fragment to an analyte molecule. 3-Hydroxypicolinic acid does not have the latter property for proteins as analytes, but has enabled progress for oligonucleotide MALDI; this indicates that different matrices for different classes of analytes may be required.
For some matrix compounds additives or comatrices have been introduced, resulting in a considerable improvement of the practical performance [40-42].

4.2 SAMPLE PREPARATION

In general, two preparation techniques are used for MALDI analysis. Both are quite simple and fast. For the earliest, the „standard“ or „dried drop“-preparation technique (Fig. 1) a 5-50 g/L solution of the matrix material is prepared in either pure water or a mixture of water and organic solvent (acetonitrile, ethanol). For protein analysis a mixture of water acidified by trifluoroacetic acid (0.1%) and acetonitrile (2:1) is a well-suited solvent. 10^{-5} - 10^{-7} M solutions of the analyte are prepared in the same solvent as the matrix. Small amounts of both solutions (between 0.5 -10 µl) are then mixed together on the metal sample support (usually stainless steel) to give a final analyte concentration of 0.005 - 0.05 µg/µL in the mixture. The solvent is evaporated and the sample tranferred to the vacuum chamber of the mass spectrometer. Depending on the matrix used finely dispersed small crystallites or extended crystalline areas at the rim of the droplet can be seen by a microscopic observation. The most intense ion signals are usually associated with these well-developed crystalline regions of the sample [31,32] whereas the microcristalline or amorphous regions of the droplet contain impurities like salts.
The second preparation technique was reported recently [43], it is called „surface“ preparation technique (Fig. 2). The matrix is solved in an organic solvent like acetone and by applying 0.5-1 µL on the target the solution spreads out very quickly and by instanteneous drying it forms a homogenous thin layer of matrix. On top of this layer 0.5-1µL of analyte, solved in mostly aqueous solvent is deposited and allowed to dry slowly on room temperature. This procedure is only successful when the matrix is only slightly soluble in water like sinapinic acid or α -cyano-4-hydroxy-cinnamic acid. To remove salt impurities the sample-matrix spot can be washed once or twice by applying some µL of water, waiting for some seconds and than blowing or shaking the solvent away.
This preparation technique has some advantages compared to the standard preparation tehnique. Because the sample-matrix layer is very homogenous the spot to spot reproducibility of the mass spectra is enhanced and often a better resolution and a better mass accuracy is obtained. A mass accuracy of < 50 ppm for peptides < 2000 Da is reported [44]. Because the sample is only incorporated into the uppermost and thus accessible matrix layers the sensitivity is enhanced, spectra from attomole amounts of sample solutions can be obtained [43]. The disadvantage of this preparation method is that it works with only few of the known matrices.

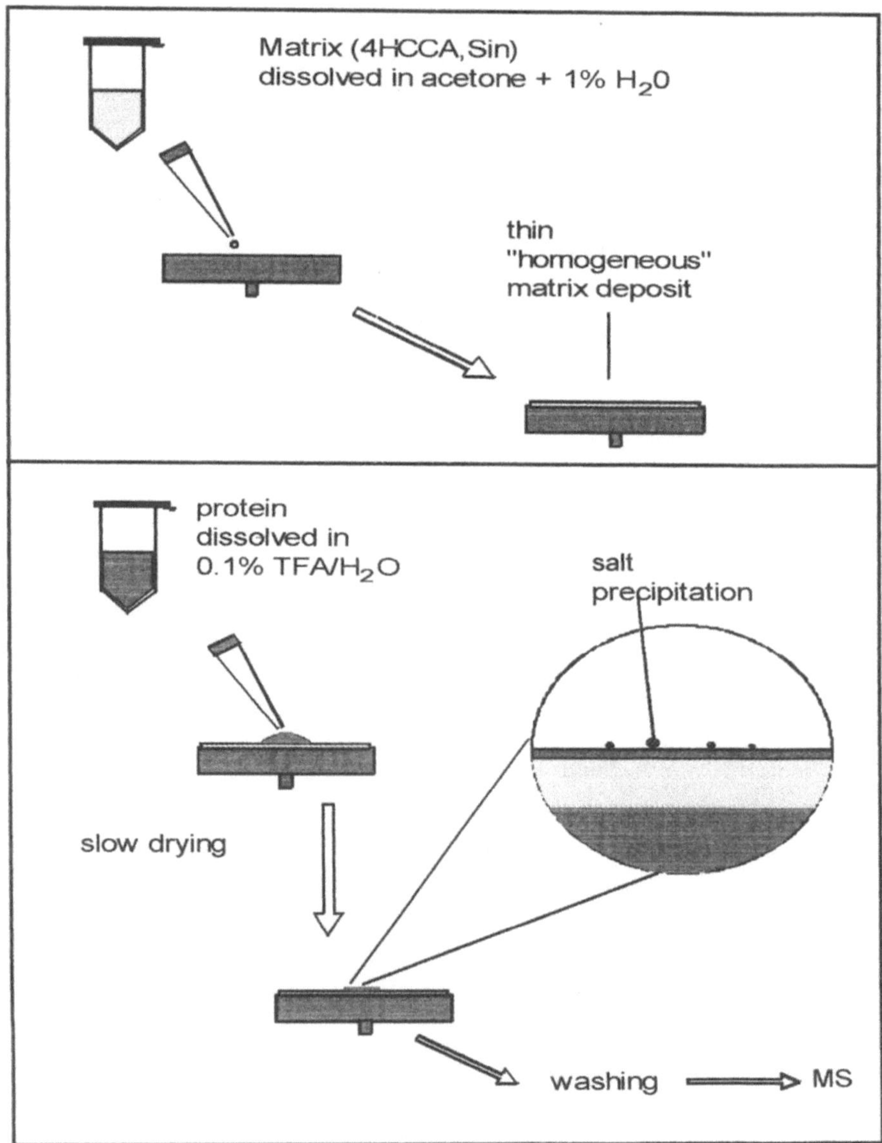

Fig. 2: Surface preparation technique for MALDI mass analysis

5. Instrumentation

5.1. LASER DESORPTION ION SOURCE

Ion sources for MALDI [45-48] use pulsed lasers with pulse durations of 1 - 200 ns. Most commonly used are UV-lasers such as nitrogen (N_2)-lasers emitting at 337 nm or

Nd-YAG lasers, whose emission wavelength of 1064 nm has been transferred to 355 or 266 nm by frequency tripling or quadrupling using nonlinear optical crystals. Moreover, excimer lasers (193, 248, 308 and 351 nm), frequency doubled, excimer pumped dye lasers (220 - 300 nm) have been used. In the visible wavelength region dye lasers and in the infrared TEA-CO_2 at 10.6 μm and Er-YAG at 2.94 μm have been used. For the required irradiances in the range of 10^6 to 10^7 W/cm^2 the laser beams are focussed to spot sizes between 30 and 500 μm by suitable optical lenses. The angle of incidence of the laser beam on the sample surface varies between 15 - 70°. The irradiance at the sample surface is a critical parameter, the minimum (threshold) irradiance to produce ions is well defined and best results are obtained for a laser irradiance no more than ca 20% above the threshold. Thus the intensity of the laser beam on the sample has to be carefully adjusted either by neutral density filters, angle-

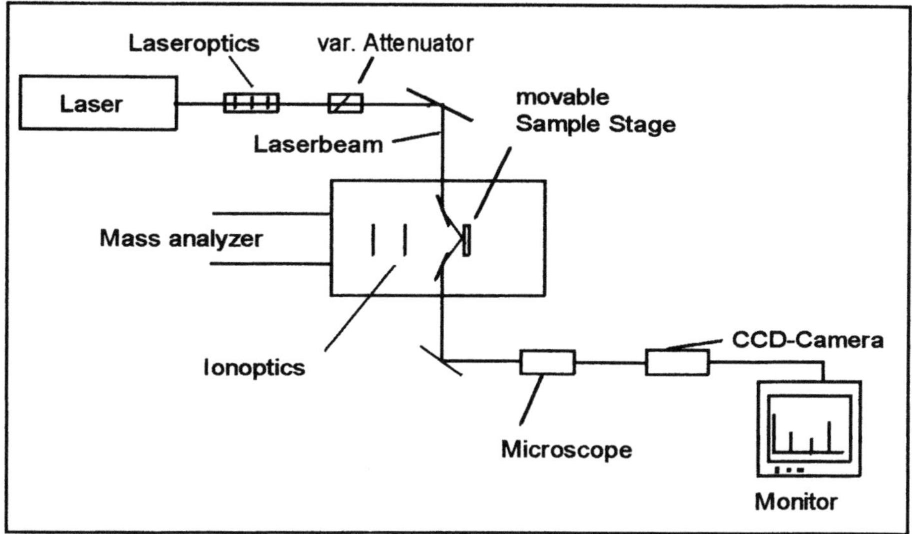

Fig 3: Schematic drawing of a MALDI ion source

dependant reflection attenuaters or polarizers. The position of the laser focus on the sample surface can be changed either by moving the sample towards a fixed laser spot or by steering the laser beam. Optical control of the sample is a valuable means to yield optimal MALDI results. In Fig. 3 laser ion source with movable sample stage and a video-microscope for observation is shown schematically.

5.2. MASS ANALYZER

MALDI of large molecules is usually coupled to a TOF mass analyser, although several applications have been performed on FT-ICR [49-53] or magnetic sector analyzers [29,54-56], results with quadrupole ion trap mass spectrometer have also been reported [57-60]. In TOF analyzers the mass to charge ratio of an ion is determined by measuring their flight time. After acceleration of the ions in the ion source to a fixed kinetic energy, they pass a field free drift tube with a velocity proportional to

$(m_i/z_i)^{-1/2}$ (m_i/z_i is the mass-to-charge ratio of a particular ion species). Due to their mass-dependant velocities, ions are separated during their flight. A detector at the end of the flight tube produces a signal for each ion species. Typical flight times are between a few microseconds and several 100 µs. Fig. 4 shows schematically ion

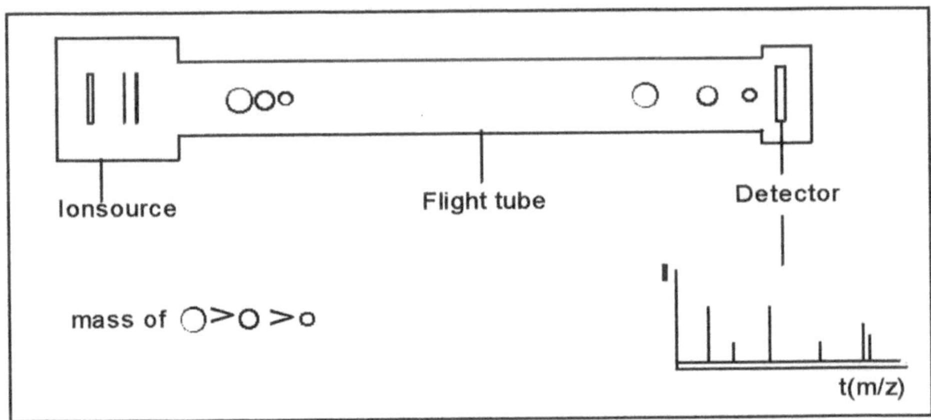

Fig 4: Schematic drawing of ion separation in a linear time-of-flight instrument

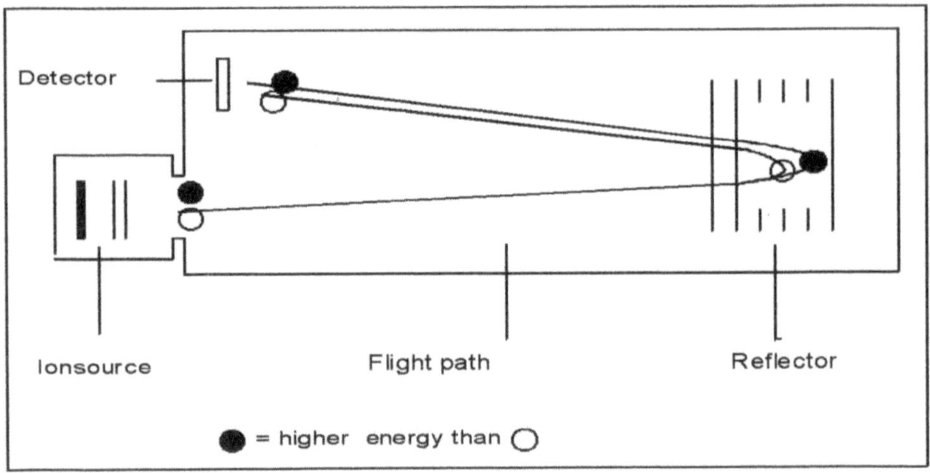

Fig 5: Schematic drawing of ion separation in a reflector time-of-flight instrument

separation in linear time-of-flight instrument. The flight path lengths are typically between 0.5 and 3 m, the acceleration voltage between 10 - 30 kV. The mass resolution m/δm in linear TOF instruments is limited to 200 - 800 (hwfm) due to the initial energy distribution ions receive during desorption. To enhance mass resolution a reflectron TOF mass analyzer can be used (see Fig. 5). The ions are decelerated in the ion mirror and turn around at different locations in the reflecting electric potential gradient. Ions of higher kinetic energy spend a longer time in the ion mirror than ions of lower kinetic energy. The arrival time spread can thus largely corrected for at the detector and an increase in mass resolution up to 6000 can be obtained.

5.3 DELAYED ION EXTRACTION

A delay time between ion formation and extraction results in a dramatical enhancement of mass resolution [61-64]. The ions are produced in a field-free region or a weak electrical field and subsequently extracted by application of a high voltage pulse after a mass dependant time delay. Linear as well as reflector instruments can be used with this technique. Depending on the length of the analyzer mass resolutions between 4000 for angiotensin in a 1.2 m linear TOF and 8600 in a reflector TOF (6.6 m) can be reached [64].

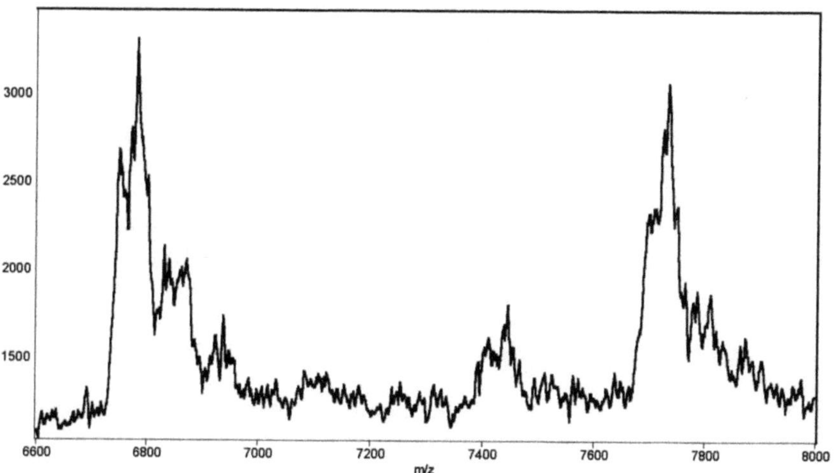

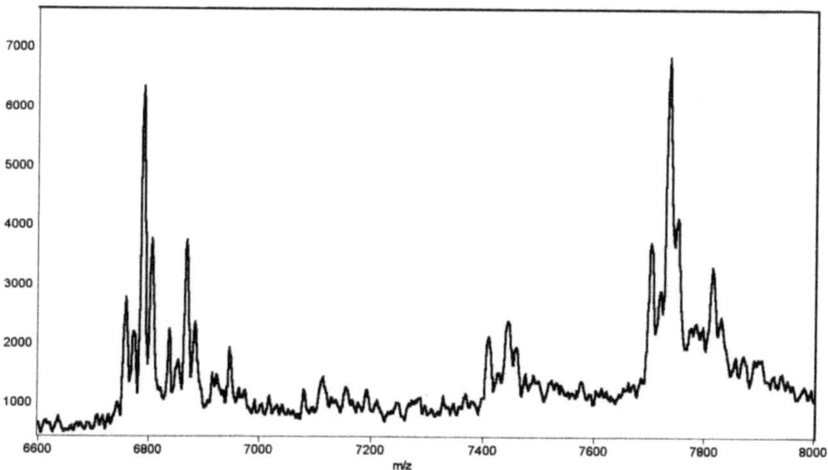

Fig. 6: Section of a MALDI spectra of a highly glycosylated and phosphorylated peptide. Top: without delayed extraction. Botton: with delayed extraction.

Fig. 6 shows a comparison between spectra of a glycosylated and phosphorylated labile peptide which could not be analyzed in a reflector mass spectrometer obtained without delayed extraction (upper spectrum) and with delayed extraction (lower spectrum) in a linear reflector TOF instrument of 1.3 m length. With this technique the mass resolution is limited in short TOF-analyzers for small peptides by the limitation in time measurement devices and in general for higher mass ions. In addition to improved mass resolution, delayed extraction yields a lower level of chemical noise and a reduction of matrix background is observed. The critical dependance of the spectra quality of laser irradiance is reduced. The higher mass resolution also leads to an enhancement in mass accuracy. Without internal calibration a mass accuracy of > 10 ppm has been reported in reflectron instruments [44].

5.4. ION DETECTION AND DATA COLLECTION

Ion detection in MALDI-MS is done by either convential secondary electron multipliers or microchannelplate detectors. The high-mass ions produce either electrons or low-mass ions at the conversion dynode of the multiplier. These particles are than used to start the multiplication cascade in an electron multilier. In most MALDI instruments where the ions are accelerated to a high kinetic energy double channelplate detectors are used. In MALDI instruments using a low ion acceleration energy postacceleration of high mass ions is needed in order to compensate for the lower detection efficiency at lower ion velocities. This is achieved by a separated dynode held at a potential of typically 20 kV placed in front of the multplier. However, this detection mode uses both secondary electrons and ions which - even though beneficial for high-mass ion intensity- deterioate signal quality in the low and medium mass range.

The signal of the detector is either amplified with a fast linear amplifier or directly digitized by a digital oscilloscope (transient recorder) with a sampling rate of 100 Msample/s or higher. Unsing delayed ion extraction the peak widths have considerably decreased and thus time measurement requirements have increased; the ongoing progress in electronics makes possible today to collect data with a sampling rate of a quarter of a nanosecond. The data are then transferred to a PC for spectrum evaluation. For an accurate mass determination, typically 10 - 50 spectra, each from a single laser shot, are summed to improve the signal-to noise ratio.

6. MALDI-mass analysis of biomolecules

Most applications on MALDI-MS have been done for proteins and no limitations of the application caused by primary, secondary or tertiary structure has yet been discovered. Proteins with different solution-phase properties, including proteins that are insoluble in aqueous solutions can be analyzed as well as glycoproteins with large amount of carbohydrate. Moreover MALDI has been widely applied to carbohydrate and nucleic acid analysis.

6.1. CHARACTERISTICS OF MALDI MASS ANALYSIS

Fig. 7 shows as typical MALDI spectrum the spectrum of the protease Subtilisin Carlsberg with a molecular weigth of 27288. The molecular ion peak usually is the most prominent signal accompanied by doubly and triply charged molecular ion signals as well as a dimer. The relative abundances of the ion signals depend on the matrix and the concentration of the analyt. In the mass range below 500 Da ion signals from the matrix appear. Both positive and negative ion spectra with comparable signal intensities can be obtained. Positive ions appear as protonated, negative ions as deprotonated species.

The technique is relatively immune to many non-proteinaceous impurities. Salts or buffers in concentrations normally used for biochemical procedures often do not have to be removed before analysis. With sinapinic acid as matrix inorganic salts like NaCl and denaturation agents like urea or guanidinium hydrochloride up to 2 mM in the protein solution and buffering agents like ammonium bicarbonate, ammonium acetate, citrate, glycine, hepes, tris up to 200 mM do not strongly affect the analysis [34,65]. Up to 10% SDS (sodium dodecyl sulfate) in the protein solution is tolerable with DHB as matrix [31]. This enables the use of the technique in direct combination with biochemical reactions, such as enzymatic or chemical digests. Complex mixtures of different peptides and proteins can be measured.

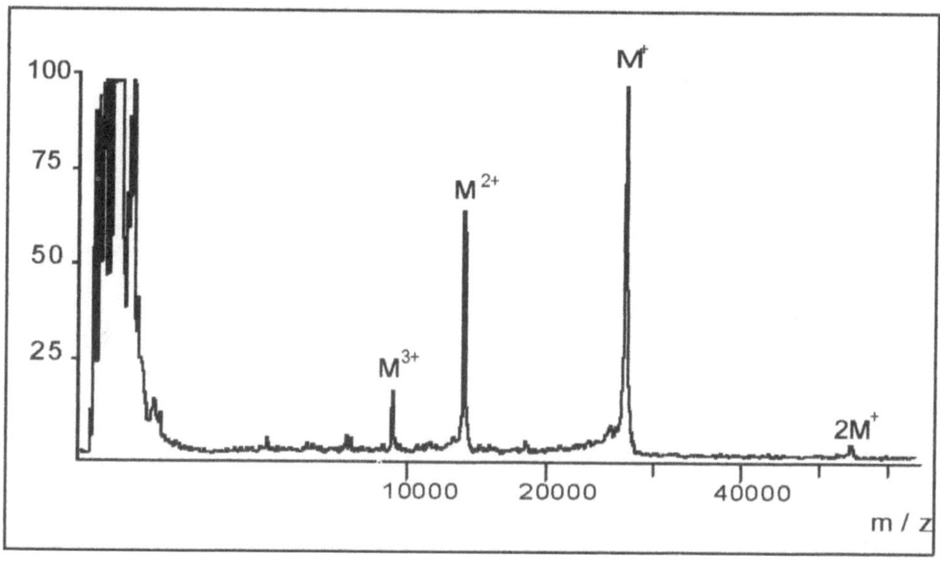

Fig 7:MALDI spectrum of Subtilisin Carlsberg

Mass determination accuracy as high as 0.01% for proteins below 30 000 Da can be achieved [65,66]. For this purpose mass calibration has to be done using well-defined reference compounds internally by adding them to the analyte-matrix mixture rather than externally by measuring their masses in a separate analysis. Above 30 000 Da a decrease in mass accuracy is observed which is supposed to be due to non-resolved adducts between analyte and matrix.

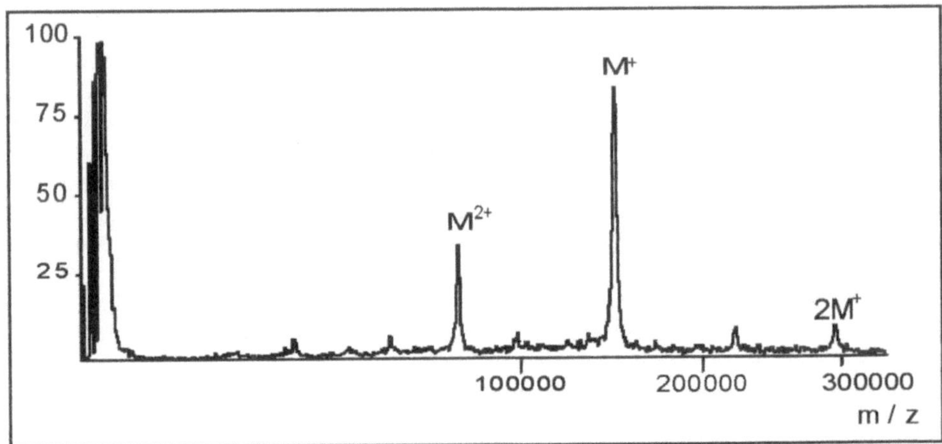

Fig 8: MALDI mass spectrum of a monoclonal antibody

Typical sample amounts used for MALDI analysis are about 1 pmol, in several cases 1-10 fmol have been used for analysis [31,67], with the surface preparation technique even lower sample amounts have been applied [43].

The mass range accessible for TOF mass analysis is in principle unlimited. Fig. 8 shows the MALDI spectrum of a monoclonal antibody with a decent signal for the dimer at a mass of nearly 300000 Da. Among the highest masses detected so far in MALDI analysis are the doubly charged molecular ions of human IgM with molecular masses above 900 000 Da [68] and the distributions of synthetic polymers with molecular weights between 200 000 and 1.5 million Da [69-71].
The time for sample preparation and for data acquisition is only some minutes, more time-comsuming are the calibrations generally needed to obtain high mass accuracies.

6.2 ION FRAGMENTATION AND POST-SOURCE-DECAY (PSD)

Although MALDI is regarded to be a soft ionization technique often metastable fragmentation of ions can occur in dependance of the analyte, the matrix, the composition and pressure of the residual gas [28] or the strength of the acceleration field [72,21]. Whereas in linear TOFs the fragments from metastable decay appear at the same time than the intact ions at the detector, most of the fragments will either be sorted out by the reflector or they appear at lower flight times in the spectrum eventually leading to a brodening of the mass peaks [21,30]. A comparison of linear and reflector mass spectrum gives immediately information over the „softness" of a matrix. Fig. 9 shows spectra of ubiquitine with sinapinic acid and DHB as matrices in a linear and reflector TOF. The laser irradiance was kept constant for measurement in the linear and reflector mode. Whereas the linear-mode spectra look identical with the different matrices the reflector spectra show that with sinapinic acid the fragmentation is drastically enhanced. The metastable decay is strongly dependant on the kind and mass of the analyte. Several carbohydrate moieties, e.g. sialic acids and fucoses in glycoproteins show a high tendency towards fragmentation [30] as well as oligonucleotides and nucleic acids [73]. Table 1 lists the typical application modes for

TABLE 1. Matrices which can be used for compounds in a linear and/or reflector TOF

Matrix	lin TOF	refl TOF
DHB(s)	Proteins	Proteins
Sinapinic acid	Proteins	Peptides
α-cyano-4-hydroxycinnamic	Proteins	Peptides
3-HPA	Nucleic acids, Glycoproteins	Nucleic acids, Glycoproteins

the most commonly used matrices. Metastable fragmentation can, however, be used for structure analysis by tuning the refletor potential according to the kinetic energy of the ions [27,74]. In this way a series of spectrum segments can be obtained from which a complete fragment ion spectrum can be reconstructed. This method, called „post source decay (PSD)" analysis has been used for analysis of small peptides showing the complete series of sequence fragments in the mass spectrum, in favourable cases obtained in several minutes. The method is described in detail elsewhere in this book.

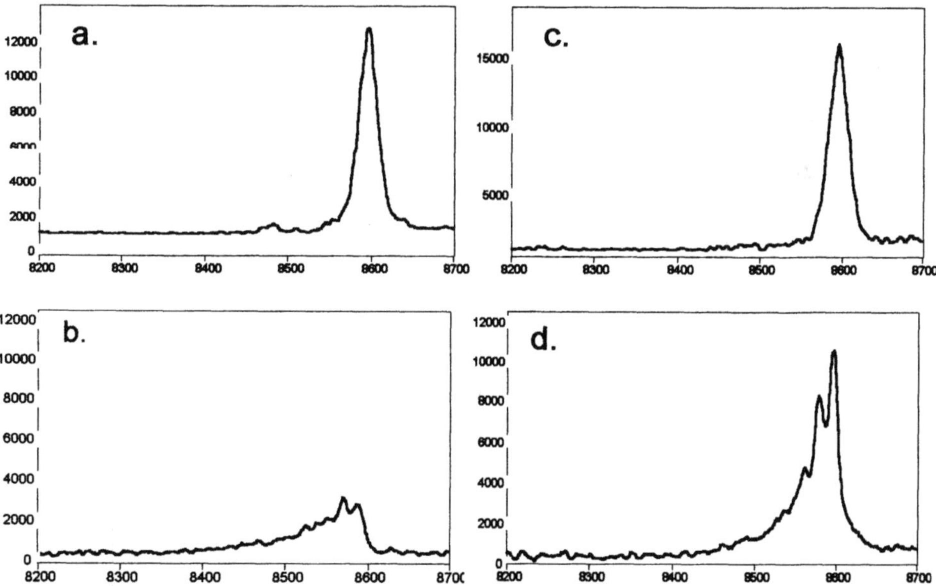

Fig 9: Spectra of ubiquitine: a) with sinapinic acid in the linear mode and b) in the reflector mode.c) With dhb in the linear and d) in the reflector mode

6.3. APPLICATIONS IN PROTEIN ANALYSIS

6.3.1. *Characterization of intact proteins.*

Because of its speed and sensitivity MALDI-MS is used for characterization of intact proteins via determination of the molecular weight where a quick control of integrity and purity of a protein preparation is needed. A MALDI mass spectrum can give a first indication to posttranslational modification like phosphorylation or glycosilation [75-78].

6.3.2. *Characterization of proteins by inspection of cleavage peptides.*

Characterization of proteins by MALDI-MS can also be done via molecular weight determination of peptide fragments [42]. Peptide fragments produced by enzymatic or chemical degradation of a protein can either be analyzed in the unfractionated mixture or after separation with HPLC or gel electrophoresis [79].

Another approach of protein sequence analysis by MALDI MS consists of two steps, a controlled generation of sequence-defining peptide fragments, each differing from the next by one amino acid and the subsequent MALDI-MS analysis of this fragment mixture [80]. Each amino acid in the sequence is identified from the mass difference between adjecent peaks. Peptides with molecular weights up to 3500 Da have been sequenced in this manner with high accuracy. This method was also used to directly locate a phosphoserine residue in a phosphopeptide [80]. It, however, appears that this protein ladder sequencing approach, even though convincing as an idea, suffers from lack of general applicability and only some examples have been reported.

6.3.3. *Identification of proteins.*

A rapid and very sufficient approach for identifiying proteins is the comparison of masses determined by MALDI-MS from protein digests (fingerprint) and calculated masses in a database [81,82]. Using this method it has been shown that often as few as three or four experimentally determined peptide masses are sufficient to uniquely identify the sample when screened agained a fragment data base [82,83].

6.4. COMBINATIONS OF MALDI-MS WITH SDS-GEL-ELECTROPHORESIS

Because of the high speed and accuracy of MALDI mass analysis the method is ideally suited for combination with biochemical methods which have a high potential in separation of highly complex mixtures.

Sodium dodecyl sulfate-polyacrylamide gel electrophoresis (SDS-PAGE) and high-resolution two-dimensional polyacrylamide electrophoresis (2-DE) provide rapid and reproducible methods for separation of proteins from complex mixtures. Unfortunately, the mass accuracy achieved with these methods is only 5-10%. For more precise mass determination, the separated proteins can be removed from the gel and measured by MALDI-MS [84,85]. MALDI-MS direcly from the gel has not yet become a viable method. For this reason the proteins are electroblotted onto membranes such as polyvinylidene diflouride (PVDF) or polyamide membranes, a method often used for further protein chemical analysis like sequencing or amino acid composition analysis. The piece of membrane containing the protein of interest is cut out, soaked in matrix solution and transferred to the mass spectrometer for infrared-MALDI analysis using an Er-Yag laser (2.94 μm) [86]. Fig.10 shows a spectrum of Carbonicanhydrase

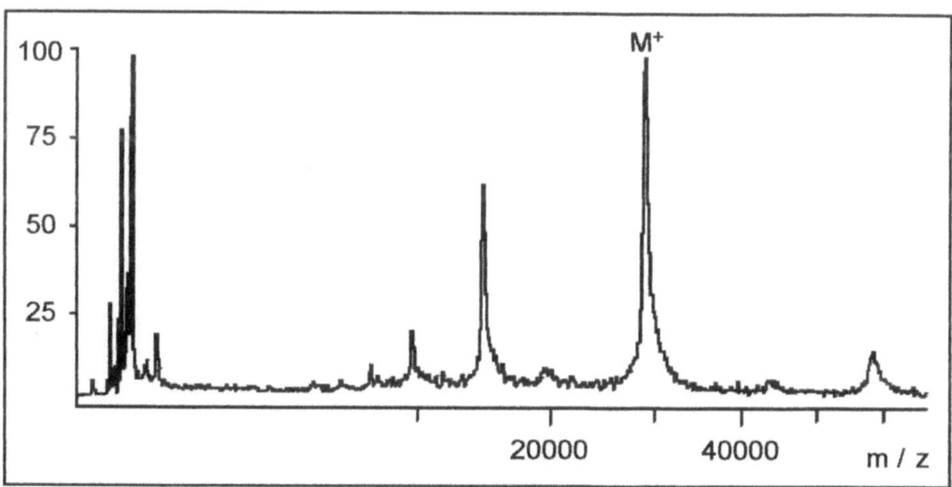

Fig 10: MALDI spectrum of carbonic anhydrase directly from a PVDF membrane

gained in this manner with succinic acid as matrix.. The spectrum has nearly the same quality than a MALDI spectrum from a normal preparation. A problem arises from staining of the proteins, as multiple attachment of dye molecules to the analyte can lead to peak broadening. Staining by colloidal pigments is well-suited for MALDI analysis. Scanning of complete lanes of the blotting membrane to yield mass contour images is possible as the lateral resolution of the electrophoretic separation is preserved after incubation of the matrix solution [87,88]. First results with UV-MALDI of electroblotted proteins are encouraging although the quality of the spectra is not yet comparable to IR-MALDI [89,90].

7. MALDI Mass Analysis of Nucleic Acids

Compared to protein analysis the analysis of nucleic acids is more difficult because this class of compounds is more labile and tend to multiple cation attachment. With the introduction of suitable matrices, 3-hydroxypicolinic acid is the most widely used one, numerous applications have been reported, ranging from natural and modified, single and double-stranded oligo(deoxy)ribonucleotides and mixtures thereof [91-96,73] to high mass RNA [97] and DNA samples [98]. Moreover, sequencing of olignonucleotides can be performed [99,100]. A more detailed description of MALDI-MS of oligonucleotides and nucleic acids is given in a separate chapter in this book.

8. MALDI Mass Analysis of Carbohydrates

Underivatized oligosaccharides have been successfully analyzed by MALDI-MS with detection limits down to the low pmol range [101-104]. Only 10 fmol of sample were needed to obtain spectra from native and permethylated glycoshingolipids and gangliosides [105,106]. In contrast to proteins, carbohydrates are ionized in MALDI

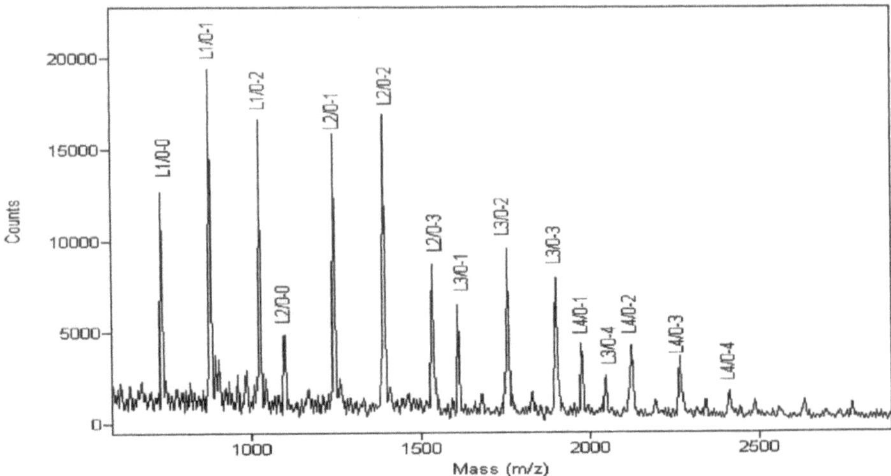

Fig .11: Mixture of neutral oligosaccharides extracted from human milk.
Nomenclature: Lx/y-z, L = lactose, x = number of N-acetyllactosamins, y: number of fucoses, z: number of sialic acids
(sample kindly provided by Milupa Research, Friedrichsdorf, Germany)

by cation-attachment. Fig. 11 shows the MALDI spectrum of a mixture of neutral oligosaccharides extracted from gynolactose. All mass peaks are $[M+Na]^+$-species. Besides DHB and DHBs some other matrices or matrix mixtures are used such as 3-aminoquinoline [107], a mixture of DHB and 1-hydroxy isoquinoline 3:1 [108] or a mixture of 2,4,6-trihydroxyacetophenone with nicotinic acid and di-ammoniumhydrogencitrate [109]. It is much more difficult to analyze high-molecular weight polysaccarides compared to large proteins. One reason is the broad molecular weight distribution of natural oligosaccharides where the total ion intensity is distributed over a large number of ion species. MALDI results from polysaccharides with narrow molecular weight distrubution are in good aggreement with those obtained with classical methods [102]. Fig. 12 shows the MALDI spectrum of a standard glucan with a declared M_n of 44 000 which is in good agreement with the value calculated from the spectrum. MALDI spectra from native and permethylated dextran standards with large polydispersity show a large discrepancy between the determined molecular weight distribution and the expected [110]. A fractionation of the dextrans with SEC on Sephadex columns and subsequent MALDI analysis of the fractions showed that it is possible to obtain mass spectra of polysaccarides up to 100 000 Da when the polydispersity is < 1. Thus it seeems that not unefficient ionization of carbohydrates in the high mass range but suppression of desorption and ionization of

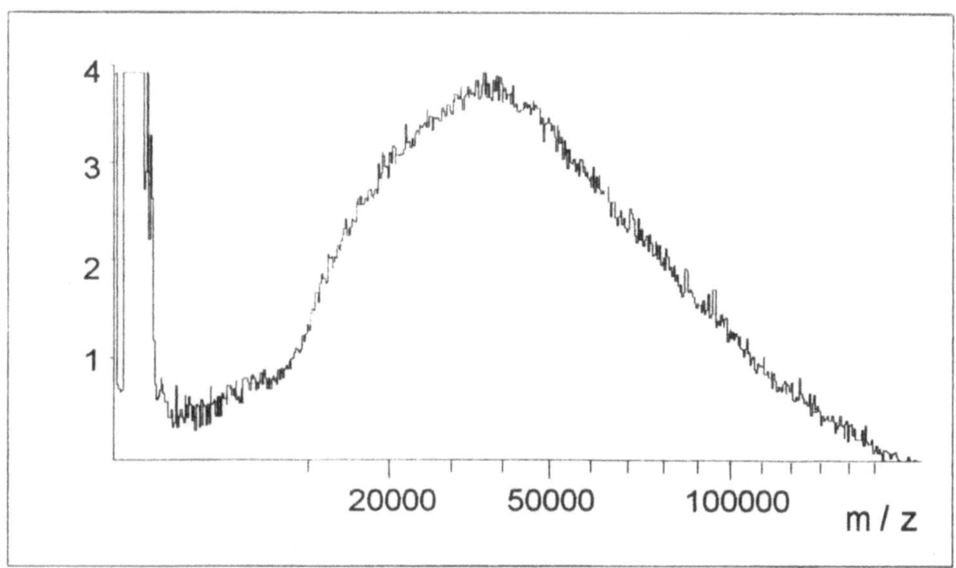

Fig. 12: MALDI spectrum of a standard glucan with an average molecular weight of 44 000 Da.

larger molecules by the lighter molecules is the problem in MALDI of polysaccharides.

For structure analysis of carbohydrates MALDI mass analysis of metastable ions (PSD analysis) is a valuable method [111,72].

9. References

1. H.D. Beckey *Principles of field desorption mass spectrometry*. Pergamon Press, Oxford (1977).
2. M. Barber, R.S. Bordoli, R.S.Sedgwick, A.N. Tyler, *Chem. Soc Chem. Commun.* **325** (1981).
3. R.D. Macfarlane, F.D. Torgerson, *Science* **109**, 920 (1976).
4. C.R. Blakely, J.J. Carmody, M.L. Vestal, *Anal. Chem.* **52**, 1636 (1980).
5. C.M. Whitehouse, R.N. Dreyer, M. Yamashita, J.B. Fenn,*Anal. Chem.* **57**, 675 (1985).
6. F. Hillenkamp, H. Ehring, In: Gross ML (ed) *Mass spectrometry in the biological sciences: a tutorial*. Kluwer Academic Publishers, Dordrecht, Boston, London. pp 165-179 (1992).
7. J.R. Cotter, *Anal. Chim. Acta* **195**, 45 (1987).
8. F. Hillenkamp, M. Karas, R.C. Beavis, B.T. Chait, *Anal. Chem.* **63**, 1196A (1991).
9. F.J. Vastola, R.O. Mumma, A.J. Pirone, ***Org. Mass Spectrom.*** **3**, 101(1970).
10. M.A. Posthumus, P.G. Kistemaker, H.L.C. Meuzelaar, M.C. deBrauw,*Anal.Chem.* **50**, 985 (1978).
11. R. Stoll, F.W. Röllgen, *Org. Mass Spectrom.* **14**, 642 (1979).
12. R.J. Cotter, *Anal. Chem.* **53**, 1306 (1981).
13. E.D. Hardin, M.L. Vestal, *Anal. Chem.* **53**, 1492 (1981).

14. F. Hillenkamp, E. Unsöld, R. Kaufmann, R. Nitsche, *Appl. Phys.* **8**, 341 (1975).
15. R.B. van Breemen, M. Snow, R.J. Cotter,*Int. J. Mass Spectrom. Ion Phys.***49**, 35 (1983).
16. R.R. Weller, J.J. MacMahon, B.S. Freiser, In: D. Lubman (Ed.): *Lasers and Mass Spectrometry*, Oxford University Press, New York (1990), pp 249-270.
17. C. Köster, M. S. Kahr, J. A. Castoro, C.L. Wilkins, *Mass Spectrom. Reviews* **11**, 495 (1992).
18. M. Karas, D. Bachmann, U. Bahr, F. Hillenkamp,*Int. J. Mass Spectrom. Ion Proc.* **78**, 53 (1987).
19. M. Karas, F. Hillenkamp, *Anal Chem* **60**, 2299 (1988).
20. M. Karas, U. Bahr, F. Hillenkamp, *Int. J. Mass Spectrom. Ion Proc.* **92**, 231 (1989).
21. M. Karas, U. Bahr, J. Stahl-Zeng in: *Large Ions: Their Vaporization, Detection and Structural Analysis*, T. Bear, C.Y. Ng, I. Powis (Eds), Wiley and Sons, Chichester (1996), pp 27-47.
22. K. Dreisewert, M. Schürenberg, M. Karas, F. Hillenkamp, *Int. J. Mass Spectrom. Ion Proc.* **141**, 127 (1995).
23. W. Ens, Y. Mao, F. Mayer, K.G. Standing, *Rapid Commun. Mass Spectrom.* **5**, 117 (1991).
24. R.C. Beavis, B.T. Chait, *Chem. Phys. Lett.* **181**, 479 (1991).
25. T. Huth-Fehre, C.H. Becker, *Rapid Commun. Mass Spectrom.* **5**, 198 (1991).
26. J. Zhao, W. Ens, K.G. Standing, A. Verentchikov, *Rapid Commun. Mass Spectrom* **6**, 671 (1992).
27. B. Spengler, D. Kirsch, R. Kaufmann, E. Jaeger, *Rapid Commun. Mass Spectrom* **6**, 105 (1992).
28. B. Spengler, R. Kaufmann, *Analusis* **20**, 91 (1992).
29. R.S. Annan, H.J. Köchling, J.A. Hill, K. Biemann, *Rapid Commun Mass. Spectrom* **6**, 298 (1992).
30. M. Karas, U. Bahr, K. Strupat, F. Hillenkamp, A. Tsarbopoulos, B.N.Pramanik, *Anal. Chem.* **67**, 675 (1995).
31. K. Strupat, M. Karas, F. Hillenkamp, *Int. J. Mass Spectrom. Ion Proc.* **111**, 89 (1991).
32. R.C. Beavis, J.N. Bridson, J Phys D, *Appl. Phys.* **26**, 442 (1992).
33. H. Ehring, M. Karas,F. Hillenkamp, *Org. Mass Spectrom.* **27**, 472 (1992).
34. R.C. Beavis, B.T. Chait, *Rapid Commun. Mass Spectrom.* **3**, 436 (1989).
35. R.C. Beavis, B.T. Chait, *Rapid Commun. Mass Spectrom.* **3**, 432 (1989).
36. M. Karas, H. Ehring, E. Nordhoff, B. Stahl, K. Strupat, F. Hillenkamp, M.Grehl, B. Krebs, *Org. Mass Spectrom.* **28**, 1476 (1993).
37. R.C. Beavis, T. Chaudhary, B.T. Chait, *Org. Mass Spectrom.* **27**, 156 (1992).
38. K.J. Wu, A. Steding, C.H. Becker, *Rapid Commun Mass Spectrom* **7**, 142 (1993).
39. U. Pieles, W. Zürcher, M. Schär, H. Moser, *Nucl. Acid Res.* **21**, 3191 (1993).
40. K. Köster, J.A. Castoro, C.L. Wilkins, *J. Am. Chem. Soc.* **114**, 7572 (1992).
41. T.M. Billeci, J.T. Stults, *Anal. Chem.* **65**, 1709 (1993).
42. A. Tsarbopoulos, M. Karas, K. Strupat, B.N. Pramanik , T.L. Nagabhushan,F. Hillenkamp, *Anal. Chem.* **66**, 2062 (1994).
43. O. Vorm, P. Roepstorff, M. Mann, *Anal. Chem.* **66**, 3281 (1994).

44. O. Vorm, M. Mann, *J. Am. Soc. Mass Spectrom.* **5**, 955 (1994).
45. P. Feigl, B. Schueler, F. Hillenkamp, *Int. J. Mass Spectrom. Ion Phys.* **47**, 15 (1983)
46. R.C. Beavis, B.T. Chait, *Rapid Commun. Mass Spectrom.* **3**, 233 (1989).
47. M. Salehpour, I. Perera, J. Kjellberg, A. Hedin, M.A. Islamian, P. Hakansson, B.U.R. Sundquist, *Rapid Commun. Mass Spectrom.* **3**, 259 (1989).
48. B. Spengler, R.J. Cotter, *Anal. Chem.* **62**, 7932 (1990).
49. M.V. Buchanan, R.L. Hettich, *Anal. Chem.* **65**, 245A (1993).
50. J.A. Castoro, C. Köster, C.L. Wilkins, *Anal. Chem.* **65**, 784 (1993).
51. T. Robert, J. McIver, Y. Li, R.L. Hunter, *Proc. Natl. Acad. Sci. USA* **91**, 4801 (1994).
52. L. Pasa-Tolic, Y. Huang, S. Guan, H.S. Kim, A.G. Marshall, *Mass Spectrom.* **30**, 825 (1995).
53. J. Wu, S.T. Fannin, M.A. Franklin, T.F. Molinski, C.B. Lebrilla, *Anal. Chem.* **67**, 3788 (1995).
54. J.A. Hill, R.S. Annan, K. Biemann, *Rapid Commun Mass Spectrom* **5**, 395 (1991).
55. R.S. Bardoli, K. Howes, R.G. Vickers, R.H. Bateman, D.V. Harvey *Rapid Commun. Mass Spectrom.* **8**, 585 (1994).
56. V.S.K. Kolli, R. Orlando, *Rapid Commun. Mass Spectrom.* **10**, 923 (1996).
57. J.C. Schwartz, M.E. Bier, *Rapid Commun. Mass Spectrom.* **7**, 27 (1993).
58. D.M. Chambers, D.E. Goeringer, S.A. McLuckey, G.L. Glish, *Anal. Chem.* **65**, 14 (1993).
59. K. Jonscher, G. Currie, A.L. McCormack, J.R. Yates, *Rapid Commun. Mass Spectrom* .**7**, 20 (1993).
60. V.M. Doroshenko, R.J Cotter, *Anal. Chem.* **67**, 2180 (1995).
61. R.S. Brown, J.J. Lennon, *Anal. Chem.* **67**, 1998 (1995).
62. S.M. Colby, T.B. King, J.R. Reilly, *Rapid Commun. Mass Spectrom.* **8**, 865 (1994).
63. R.M. Whittal, L. Li , *Anal. Chem.* **67**, 1950 (1995).
64. M.L. Vestal, P. Juhasz, S.A. Martin, *Rapid Commun. Mass Spectrom.* **9**, 1044 (1995).
65. R.C. Beavis, B.T. Chait, *Proc. Natl. Acad. Sci. USA* **87**, 6873 (1990).
66. R.C. Beavis, B.T. Chait, *Anal. Chem.* **62**, 1836 (1990).
67. M. Karas, A. Ingendoh, U. Bahr, F. Hillenkamp, *Biomed. Environm. Mass Spectrom.* **18**, 841 (1989).
68. R.W. Nelson, D. Dogruel, P. Williams, *Rapid Commun. Mass Spectrom.* **8**, 627 (1994).
69. P.O. Danis, D.E. Karr, *Org. Mass Spectrom.* **28**, 923, (1993).
70. P.O. Danis, D.E. Karr, A. HolleA, F. Mayer-Posner, C. Watson, Proc 41th ASMS Conference on Mass Spetrom and Allied Topics, San Francisco, p 1093 (1993).
71. D.C. Schriemer, L. Li, *Anal. Chem.* **68**, 2721 (1996).
72. B. Spengler, D. Kirsch, R. Kaufmann, J. Lemoine, *Mass Spectrom.* **30**, 782 (1995).
73. E. NordhoffE, A. Ingendoh, R. Cramer, A. Overberg, B. Stahl, M. Karas, F. Hillenkamp, P.F. Crain, *Rapid Commun. Mass Spectrom.* **6**, 771 (1992).

74. R. Kaufmann, B. Spengler, F. Lützenkirchen, *Rapid Commun. Mass Spectrom.* **7**, 902 (1993).
75. H.C. Hedrich, K. Isobe, B. Stahl, K. Nokihara, M. Kordel, R.D. Schmid, M. Karas, F. Hillenkamp, F. Spener, *Anal. Biochem.* **211**, 288 (1993).
76. M.C. Huberty, J.E. Vath, W. Yu, S.A. Martin, *Anal. Chem.* **65**, 2791(1993).
77. M.A. Baldwin, A.L. Burlingame, S.B. Prusiner, *Trends in Analytical Chemistry* **12**, 239 (1993).
78. D.W. Ownby, Z. Hao, K. Schneider, R.C. Beavis, B.T. Chait, A.F. Riggs, *Biol. Chem.* **268**, 13539 (1993).
79. M. Mann, G. Talbo, *Current Opinion Biotechnol.* **7**, 11 (1996).
80. B.T. Chait, R. Wang, R.C. Beavis, S.B.H. Kent, *Science* **262**, 89 (1993).
81. W.J. Henzel, T.M. Billeci, J.T. Stults, S.C. Wong, C. Grimley, C. Watanabe, *Proc. Natl. Acad. Sci. USA* **90**, 5011 (1993).
82. M. Mann, P. Hojrup, P. Roepstorff, *Biol. Mass Spectrom.* **22**, 338(1993).
83. D.J.C. Pappin, P. Hojrup, A.J. Bleasby , *Current Biol.* **3**, 327 (1993).
84. Y.K. Wang, P.-C. Liao, J. Allison, D.A. Gage, P.C. Andrews, D.M. Lubman, S.M. Hanash, J.R. Strahler, *J. Biol. Chem.* **268**, 14269 (1993).
85. B.T. Chait, S.B.H. Kent, *Science* **275**, 1885 (1992).
86. C. Eckerskorn, K. Strupat, M. Karas, F. Hillenkam, F. Lottspeich, *Electrophoresis* **13**, 664-665 (1992).
87. K. Strupat, M. Karas, F. Hillenkamp, C. Eckerskorn, F. Lottspeich, *Anal. Chem.* **66**, 464 (1994).
88. K. Strupat, C. Eckerskorn, M. Karas, F. Hillenkamp, in: *Mass Spectrometry in the Biological Sciences,* A.L. Burlingame, S.A. Carr (Eds.), Humana Press, Totowa (1996), pp. 203-216.
89. M. Schreiner, K. Strupat, F. Lottspeich, C. Eckerskorn, *Electrophoresis* **17**, 954 (1996).
90. J.C. Blais, P. Nagnan-Le-Meillour, G. Bolbach, J.C. Tabet, *Rapid Commun. Mass Spectrom.* **10**, 1 (1996).
91. M. Karas, U. Bahr, U. Giessmann, *Mass Spectrom. Rev.* **10**, 335 (1991).
92. K.O. Börnsen, M. Schaer, H.M. Widmer, *Chimica* **44**, 412 (1990).
93. G.R. Parr, M.C. Fitzgerald, L.M. Smith, *Rapid Commun. Mass Spectrom.* **6**, 369 (1992).
94. E. Nordhoff, R. Cramer, M. Karas, F. Hillenkamp, F. Kirpekar, Kristiansen, P. Roepstorff, *Nucl. Acid Res.* **15**, 3347 (1993).
95. P. Lecchi, L.K. Pannell, *J. Am. Soc. Mass Spectrom.* **6**, 972 (1996).
96. W.H. Benner, D. Horn, J. Katz, J. Jaklevic, *Rapid Commun. Mass Spectrom.* **9**, 537 (1995).
97. F. Kirpekar, E. Nordhoff, K. Kristiansen, P. Roepsdorff, A. Lezius, S. Hahner, M. Karas, F. Hillenkamp, *Nucl. Acid Res.* **22**, 3866 (1994).
98. K. Tang, N.I. Taranenko, S.L. Allman, L.Y. Chang, C.H. Chen, *Rapid Commun. Mass Spectrom.* **8**, 727 (1994).
99. A.S. Cohen, D.L. Smisek , B.W. Wang , in: P.R. Brown, E. Grushka (Eds) *Advances in Chromatography* **36**, 127 (1996).
100. P. Juhasz, M.T. Roskey, I.P. Smirnov, L.: Haff, M.L. Vestal, S.A. Martin, *Anal. Chem.* **68**, 941 (1996).

101. K.J. Wu, A. Steding, C.H. Becker, *Rapid Commun Mass Spectrom* 7, 142(1993).
102. F. Hillenkamp, M. Karas, A. Ingendoh, B. Stahl in: A.L. Burlingame AL, J.A. Mc Closkey JA (eds) *Biological mass Spectrometry*, Elsevier, Amsterdam (1990), pp49-60.
103. H. Egge, J. Peter-Katalinic, M. Karas, B. Stahl, *Pure Appl. Chem.* **63**, 491(1991).
104. B. Stahl, S. Thurl, J. Zeng, M. Karas, F. Hillenkamp, M. Steup, G.Sawatzki, *Anal. Biochem.* **223**, 218 (1994).
105. K.K. Mock, M. Davey, J.S. Cotrell , *Biochem. Biophys. Res. Commun.* **177**, 644 (1991).
106. B. Stahl, M. Steup, M. Karas, F. Hillenkamp, *Anal. Chem.* **63**, 1463 (1991).
107. J.O. Metzger, R. Woisch, W. Tuszynski, R. Angermann, *Fresenius J. Anal. Chem.* **349**, 473(1994).
108. M.D. Mohr, K.O. Börnsen, H.M. Widmer, *Rapid Commun. Mass Spectrom.* **9**, 809 (1995).
109. K.O. Börnsen, E. Gassmann, H.M. Widmer, Proc. 13th Int. Mass Sectrom. Conf. Budapest, p. 174 (1994).
110. D. Garrozzo, G. Impallomeni, E. Spina, L. Sturiale, F. Zanetti, *Rapid Commun. Mass Spectrom.* **9**, 937 (1995).
111. D.J. Harvey, T.J. Naven, B. Küster, R.H. Bateman, M.R. Green, G. Critchley, *Rapid Commun. Mass Spectrom.* **9**, 1556 (1995).

HIGH SENSITIVITY ESI MS AND MALDI MS

TERRY B. FARMER and RICHARD M. CAPRIOLI
University of Texas Medical School, Department of Biochemistry and Molecular Biolory, and Analytical Chemistry Center
6431 Fannin Street, Houston, TX 77030

Abstract: The increase in the use of electrospray ionization mass spectrometry (ESI MS) and matrix-assisted laser desorption ionization mass spectrometry (MALDI MS) in the biomedical sciences has fueled an intense interest in analyzing smaller and smaller amounts of sample. Increases in sensitivity have allowed the analysis of samples in the low femtomole to subattomole range. This has been accomplished primarily by changes in the capillary size for ESI MS and in sample deposition methods for MALDI MS.

1. Introduction

The introduction of ESI MS and MALDI MS has greatly facilitated the analysis of biomolecules. In particular, high molecular weight proteins can be analyzed by these two methods in order to accurately determine the mass of biopolymers, particularly proteins or peptides. The sensitivity for both ESI and MALDI early in the development of these techniques was found to be in the picomole range for proteins in the 20-40 kDa range.

Electrospray ionization sources spray liquid electrostatically out of a capillary (usually 100-150 µm i.d.) with a flow rate of 1-10 µl/min. Detection limits reported previously were from 200 femtomol/µl to 1 picomol/µl for proteins with constant infusion of the sample solution. When ESI was coupled to liquid chromatography (LC) and capillary electrophoresis (CE), the typical sensitivity achieved for peptides was in the range of 100-500 femtomole. However the need for higher sensitivity and lower flow rates gave rise to several new ESI sources termed micro-electrospray [1], nano-electrospray [2], and pico-electrospray [3]. The sources utilize nanoliter/min and picoliter/min flow rates and employ nozzle capillary orifices with inner diameters of less than 10 μm.

R. M. Caprioli et al. (eds.), Selected Topics and Mass Spectrometry in the Biomolecular Sciences, 55–84.

MALDI MS, also widely used for the analysis of peptides and proteins, requires that the analyte be deposited on a target in a mixture with a matrix compound that absorbs energy at the output wavelength of the laser. The commonly used sample preparation procedure utilizes the co-deposition of analyte and matrix solutions by the "dried droplet" method where a saturated matrix solution is placed onto the target and the analyte is deposited on the matrix solution and allowed to dry. Analyte concentrations in the 10-100 picomol/μl range were not uncommon initially. Recent advances in sample preparation such as the rapid evaporation method [4], small sample spotting [5,6], picoliter vials [7], and specially prepared membrane surface targets [8], have resulted in increased sensitivity for the detection of attomole concentrations of peptides and proteins. In addition, improvements in time-of-flight analyzer technology have further aided MALDI MS in achieving high sensitivities. Post-source decay techniques also provide a high level of confidence for the sequence specific analysis of these molecules.

The purpose of this chapter is to review advances in ESI and MALDI MS for achieving high sensitivity in the last few years, examining not only instrumental developments, but also applications and methodologies.

2. ESI MS

Two different low-flow rate electrospray sources were developed at approximately the same time, the micro-electrospray source [1] and the nano-electrospray source [2]. Both employed a smaller capillary bore to decrease the flow rate to nanoliters/minute, thereby decreasing the volume of sample needed as well as decreasing the concentration necessary for analysis of peptides and proteins.

2.1 MICRO-ELECTROSPRAY

The micro-ESI source design [1], a modification of a commercial ESI source (Vestec, Inc.) is diagrammatically shown in Figure 1. In order to achieve maximum sensitivity, the flow rate, various voltages, and physical dimensions were optimized so that nearly all of the eluant was sprayed into the analyzer from a fused silica capillary needle. The source was mounted on a Finnigan TSQ70 upgraded to TSQ700 performance. The spray needle, typically a 22.5 cm length of 50 μm i.d. fused silica capillary containing a 10 μm i.d. orifice was placed to within less than 1 mm of the flat plate nozzle (containing a 400 μm orifice) leading into the analyzer. The needle voltage was typically 3.5 kV with a nozzle voltage of 200-250 V and a repeller voltage of 10 V. Tandem mass spectra were obtained using Ar gas in the collision cell with a pressure of 1.6 mTorr.

Samples could be introduced into the source by constant infusion or direct injection via a valve-loop system using flow rates of 200-800 nL/min. Ultrapure N_2 gas was employed as a drying gas and to assist transport of the droplets to the

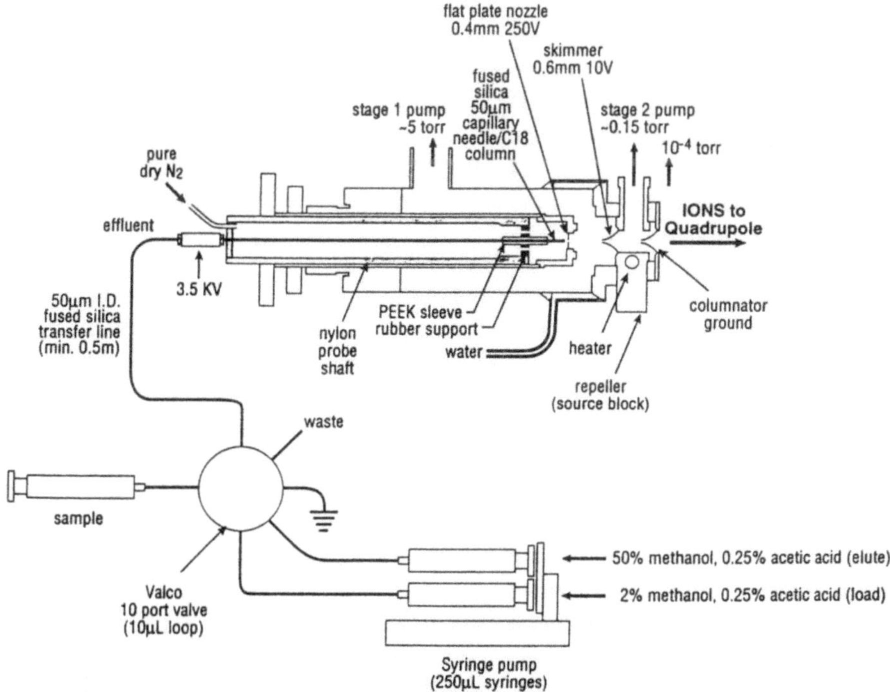

Figure 1. Schematic drawing of micro-electrospray source coupled to a 10-port injection valve (10 µl loop). (Reproduced from reference 1)

converging nozzle. To optimize spray conditions, needles of various diameters were tested with a constant infusion of 100 femtomol/µl methionine-enkephalin ([Met]^5-enkephalin) in 50% methanol containing 0.25% acetic acid. The results of these experiments showed that as the inner diameter of the capillary needle decreased, the flow rate decreased, the minimum high voltage needed to establish a steady spray increased, and the sensitivity increased. A stable spray could not be routinely established for capillaries with less than 10 µm i.d. because of problems of plugging, although acceptable spray conditions were obtained intermittently. The outer diameter of the capillary did not significantly affect the spray stability. Capillaries that were etched to the minimum possible outer diameter (wall thicknesses were approximately 10-20 µm at the tip) typically gave performance similar to those that were unetched.

To obtain on-column concentration and desalting, the fused silica capillary needles were packed with 10 µm C18 particles, ODS-AQ 120A S10, (YMC Corp.). To keep the packing in place, 5 µm spherical silica (SIL 120A-S5, YMC Corp.) was sintered into the tip to a length of 100-200 nm prior to packing. Samples that were introduced via direct injection utilized a Rheodyne injector valve with a 5 μl loop and a syringe pump providing solvent flow at 500-800 nl/min.

Direct injection of 10 μl of a solution containing 5 attomol/μl of [Met]^5-enkephalin produced an $[M+H]^+$ ion at m/z 574.2 with a S/N of greater than 5:1 obtained over a scan range of m/z 550-600. Sensitivity in the MS/MS mode was also high. Sequence specific ions, e.g. the A_4 cleavage ion at m/z 397.5 were scanned in the MS/MS mode after injection of 10 μl of a 50 attomol/μl solution of [Met]^5-enkephalin onto a packed needle column and subsequent elution into the ion source.

Neurotensin was used as the compound to test the sensitivity and overall performance of the micro-ESI source [9]. The sensitivity for neurotensin was sufficient to give a S/N of 5:1 for the injection of 10 μl of 320 zeptomol/μl solution of neurotensin, a total injection of 3.2 attomoles (Figure 2a). Continuous infusion of 320 zeptomol/μl in 50% methanol containing 0.25% acetic acid with a flow rate of 0.8 μl/min was performed to show the amount of sample consumed in the time period required to produce a particular spectrum. The molecular $[M+3H]^{+3}$ ion of neurotensin, shown in Figure 2b, was obtained without C-18 packing in the needle. Approximately 35 zeptomoles of neurotensin were consumed during the acquisition of this spectrum obtained from an average of 16 scans of 0.5 seconds each.

MS/MS capabilities are also important in measuring endogenous peptides because of the structure specific nature of the analysis [9]. For example, a narrow m/z range MS/MS scan for the $[M+3H]^{+3}$ ion of neurotensin (m/z 558.6) with collision decomposition shows a doubly-charged Y_5 fragment ion at m/z 578.9. A spectrum was obtained of the product ion at m/z 578.92 with S/N > 20:1 from the continuous infusion of 32 attomol/μl at 0.8 μl/min with 3.5 attomoles of neurotensin consumed using an average of 16 scans of 0.5 sec each.

It should be noted that these results represent the highest sensitivity achieved with this ion source in that a peptide was used which gives an abundant signal by ESI in aqueous solutions and by scanning relatively narrow mass ranges. For physiological samples, for example, one could expect a lower overall detection limit for neurotensin (i.e. about 30 attomoles total injected). The ultimate sensitivity achievable in a given experiment is dependent on the nature of the peptide of interest, and not all peptides will give the same results as neurotensin. For example, some other peptides tested, such as [Met]^5-enkephalin and substance P, gave over a 10 fold less intense signal than neurotensin for equivalent sample amounts.

2.2 NANOSPRAY ESI

A "micro-electrospray" source, later termed nano-electrospray (nano-ESI) was developed, based on the original Taylor Cone theory [2,10]. An extension to the dynamic spraying mode resulted in the relation between the size of the zone at the tip of the Taylor Cone which emits droplets and the flow rate. The nano-ESI source was

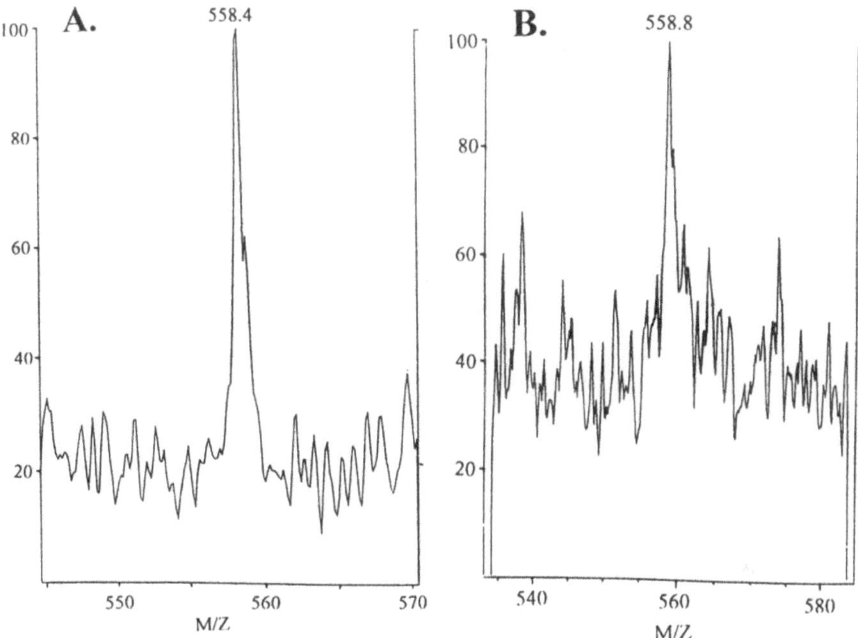

Figure 2. A.) Direct injection of 10 µl of a solution of 320 zeptomol/µl of neurotensin onto a C-18 packed micro-electrospray needle with subsequent elution by 50% methanol/0.25% acetic acid. The spectrum is an average of scans of 0.5 s each taken at the apex of an LC peak for about 5 s total width. B.) Continuous infusion of a solution of neurotensin containing 320 zeptomol/µl into the micro-electrospray source. This spectrum is an average of 16 scans of 0.5 s each, consuming a total of 35 zeptomoles of peptide. (Reproduced from reference 9)

mounted on a Sciex API III triple quadrupole mass spectrometer as depicted in Figure 3 [2]. The liquid was sprayed from the tip of a gold coated borosilicate glass capillary with an orifice of 1-3 µm formed using a micropipette puller. The capillary, centered 1-2 mm from an inner orifice, had a length of 2-3 cm. A counter flow of heated N_2 gas (0.2-0.6 l/min, 58° C) was used to assist in desolvation. The source droplet size was ≤200 nm. A voltage of 600 V was put on the needle with 100 V on the interface plate and 80 V to the inner orifice. This is lower than other similar sources where needle voltages of 3-5 kV are generally used, and is the result of the sharpness of the needle and the small distance of the needle to the orifice, allowing a reduction in the applied voltage while maintaining the same field strength.

Aqueous solutions can be sprayed in both the positive and negative ion mode and the source can tolerate up to 200 mM NaCl. A typical flow rate is 25 nl/min. As a test sample, a synthetic peptide (mass 1875.1 u) was dissolved in 1:1 methanol:water with 1% formic acid. For peptide concentrations of 0.5 picomol/µl and a volume of 1 μL,

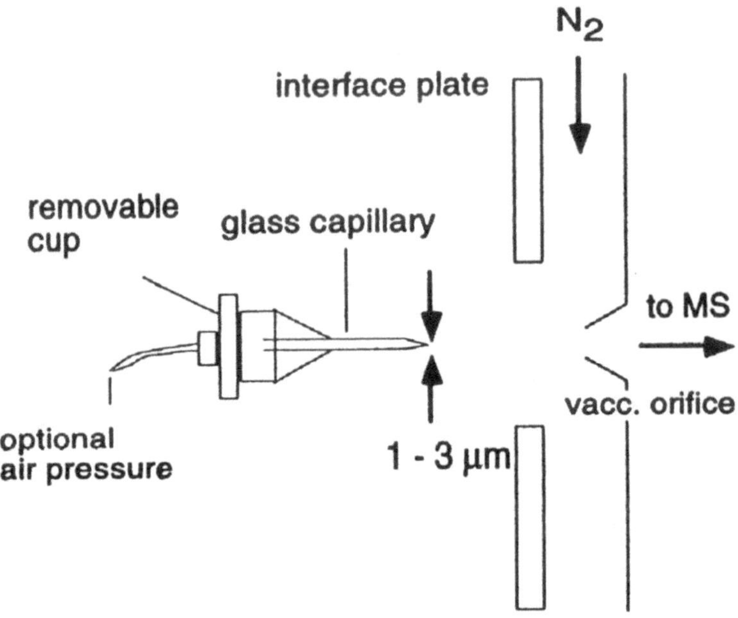

Figure 3. Schematic of the Nano-ES setup [2].
(Reproduced with permission)

the spray lasted about 45 minutes (Figure 4) [2]. Overall efficiency was calculated by measuring the doubly-charged ions ($[M+2H]^{+2}$) at the detector until no fluid remained in the capillary. The total number of detected ions in comparison to the number of peptide molecules injected into the capillary was used to calculate total transmission by dividing the total number of ions detected by the number of peptide molecules sprayed. The efficiency calculated was $8x10^{-4}$ which meant that 1 out of 1300 peptide molecules were detected. A loss of a factor of 100 was assumed due to the detection system and transmission of the quadrupole. Therefore, the desorption/ionization efficiency and transfer into the vacuum system was estimated to have an efficiency of about 10%.

2.3 PICO-ELECTROSPRAY

Electrospray ion sources operating at picoliter/min flow rated were constructed by using needles with smaller sample volumes and smaller i.d. tips to minimize sample

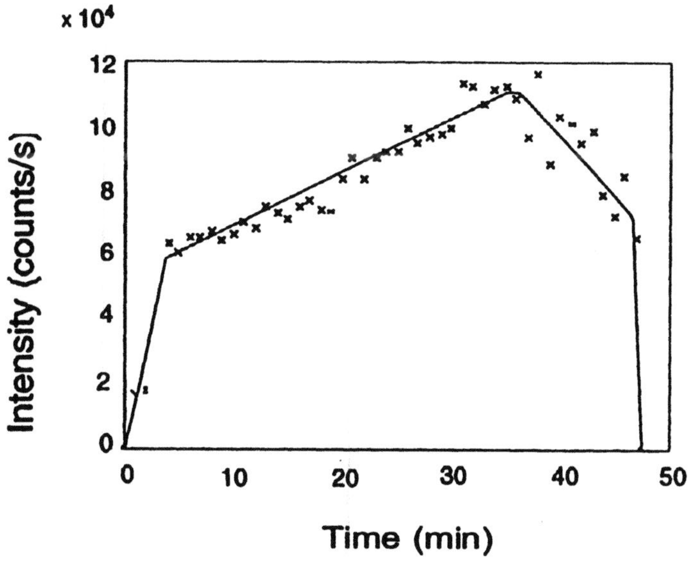

Figure 4. Spraying time for a 1 microliter sample volume monitoring the $(M+2H)^{+2}$ peptide ion [2]. (Reproduced with permission)

volume and droplet size [3]. These needles were fabricated by pulling down fused-silica capillary tubing to get picoliter/min solvent flow. Figure 5 depicts this pico-ESI source and needle set-up.

Fused silica capillaries with 5, 10, and 20 μm i.d. were pulled in a modified micropipet puller that utilized a 25 W CO_2 laser (Figure 6). The inner diameter of 50-100 nm was determined by electron microscopy and the tubing was cut to 1-4 cm lengths. The tips were immersed in 49% HF for 30-60 seconds and then flushed with water for 20-30 minutes to obtain a tip i.d. of 1-5 μm. A gold film with a thickness of 25-150 nm was deposited on the needle using a 20 mA sputter current. Needle lifetimes are approximately 20-60 minutes. Clogging of the needle, occurring from particle impurities or precipitation of the analyte, and loss of electrical contact, due to degradation of the gold coating on the tip of the needle, contributes to the low life time of the needle. A fused silica needle was glued to the end of the glass/copper mount. Electrical contact between the gold-coated portion of the needle tip and the copper plate was made with silver paint. The tip mount is attached to an X, Y, Z stage for fine position with respect to the heated capillary inlet (100° C). The pico-spray source was placed on a 6 T Extrel-Waters FTMS-2000 instrument with quadrupolar transport through five stages of differential pumping into a cylindrical trapped ion cell held at 10^{-9} Torr. The optimum distance for the tip to inlet distance is 0.25-1.5 mm. A positive or negative tip-inlet potential of 600-1300 V is typically applied. Decreasing the

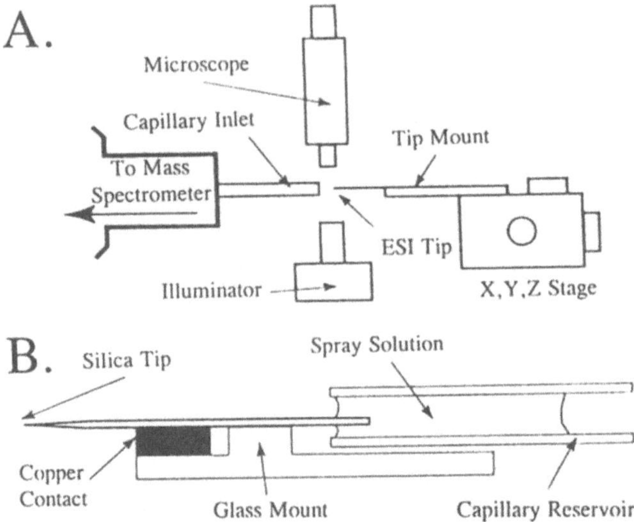

Figure 5. Schematic drawing of A.) Electrospray inlet system B.) Detail of the needle mount [3]. (Reproduced with permission)

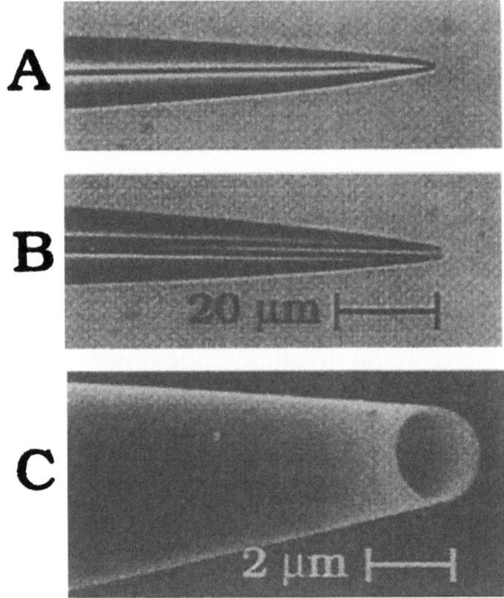

Figure 6. A. Light photomicrograph of a needle from 5 µm i.d. tubing after pulling and B.) After etching tip i.d. ~2 µm. C.) SEM photomicrograph of a 2.3 µm i.d. tip [3]. (Reproduced with permission)

distance increases the flow rate at a fixed potential between the tip and inlet capillary. In practice, the flow rate and total spray current is determined by applied voltage. The flow rate was measured by the duration of the spray current from a completely filled needle without the reservoir.

As a test example, cytochrome C was dissolved in 2% acetic acid and mass spectra were acquired with 5 μm i.d. capillary for 3 seconds. At a flow rate of 0.5 nl/min, 150 attomoles of cytochrome C was consumed giving a spectrum with a S/N of 400:1, high resolving power (>10^5), and excellent mass accuracy (12,359.3 Da theoretical, 12,359.2 Da experimental) (Figure 7A and B). A flow rate of 0.23 nl/min with approximately 3 s acquisition for a total of 0.22 femtomoles loaded onto the instrument yielded a spectrum with S/N of about 60:1 with an experimentally determined molecular weight of 12,359.1 Da (Figure 7C).

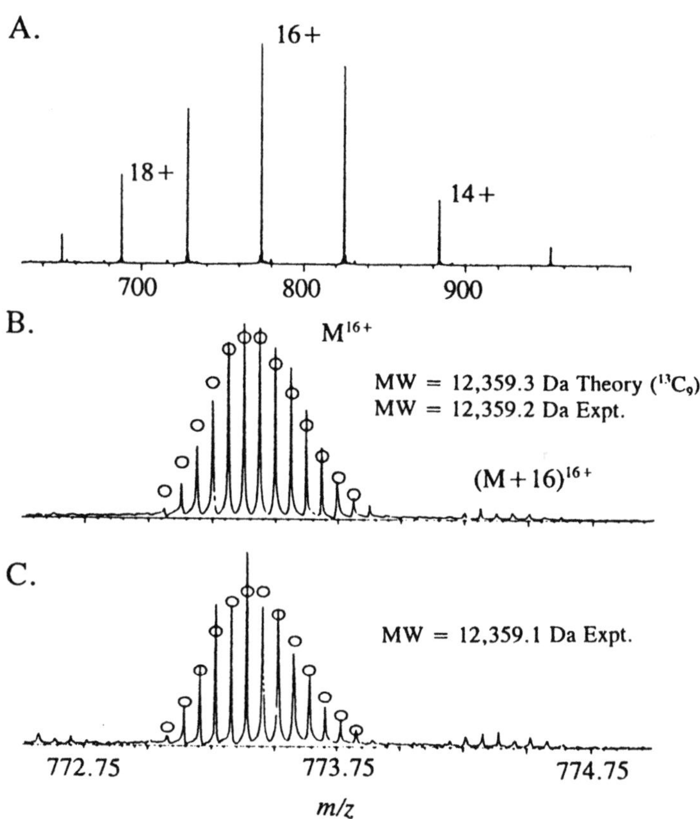

Figure 7. Mass spectra (3 s data acquisition) of equine cytochrome C using an ESI needle with 2 μm i.d. tip. (A) and (B) 0.86 nL of 10 μM solution (8.6 fmol) at 0.3 nL/min (C) 0.22 nL of 1 μM solution (0.22 fmol) at 0.23 nL/min. (o indicates theoretical isotopic peak abundances) [3]. (Reproduced with permission)

3. Peptide and Protein Analysis

3.1 SAMPLE PRECONCENTRATION

To provide preconcentration and desalting for the analysis of samples containing physiological fluids, the micro-ESI source contains a spray capillary needle packed with C-18 particles whereby the eluant is sprayed directly from the end of the column into the analyzer to eliminate dead volumes. $[Met]^5$-enkephalin, oxytocin, neurotensin, and substance P were examined using this set-up [1]. The limit of detection for $[Met]^5$-enkephalin in artificial CSF (5 mM KCl, 120 mM NaCl, 1.2 mM $MgCl_2$, 1.8 mM $CaCl_2$, 0.15% phosphate buffered saline, pH 7.4) was 1 femtomole in a total volume of 10 μl injected onto the column giving a S/N of 3:1. Similar results were obtained for the other peptides tested with neurotensin being somewhat more sensitive (<30 attomole detection limit) while oxytocin and substance P were slightly less sensitive. MS/MS analysis of 500 attomole of $[Met]^5$-enkephalin injected in 10 μl of 50 attomol/μl concentration showed the A_4 product ion (m/z 397) recorded over a 15 mass unit range.

Analyses of proteins were performed using the micro-ESI source using both constant infusion and direct injection [1]. The mass spectrum obtained from horse skeletal apomyoglobin with constant infusion of a solution containing 480 attomol/μl gave rise to a molecular weight peak having a S/N ratio of 4:1 with a centroid of 16,955.1 (theoretical molecular weight is 16,951.49). Direct injection of 2.5 μl of 10 femtomol/μl bovine serum albumin (theoretical molecular weight 66,430.3) produced a spectrum that showed the bovine serum albumin sample was a mixture of three species and gave an experimental mass of 66,433.3 (<0.005% accuracy).

Optimization of the microelectrospray source for neurotensin and $[Met]^5$-enkephalin gave higher sensitivities using a C-18 packed capillary needle [9]. Injection of 10 μl of 320 zeptomol/μl neurotensin gave an $[M+3H]^{+3}$ ion at m/z 558.4 with S/N ratio of >8:1 and a total of 35 zeptomoles of peptide were consumed.

Other similar approaches have also been reported for sample concentration and desalting [11]. Approximately 100 nL of POROS R2 sorbent (PerSeptive Biosystems) was placed in the tip of a pulled capillary. The chromatographic material was not packed and no frit or other micro LC assembly was used with a new capillary and new resin used for each analysis. The sample adsorbed to the POROS resin was washed extensively (several microliters were used) with 5% formic and 5% methanol in water. After aligning the capillary in the instrument, the sample was then eluted using approximately 1 μl of 60% methanol/5% formic acid/35% water. Direct analysis of a peptide mixture was performed after extracting a protein that had been isolated by gel electrophoresis and then digested in the gel with trypsin. A total of 4 picomoles of protein appeared to be present in the gel as determined by Coomassie blue staining. Mapping and identification of a putative protein, importin, was accomplished after electrospray analysis and a database search. Other peaks were mapped to another protein, NF 90, present in a 1:3 ratio with the importin.

3.2 APPLICATIONS

3.2.1. *In Vivo Sampling by Microdialysis*

Endogenous neuropeptides have been implicated in numerous neurological disorders including schizophrenia, depression, drug addiction, movement disorders, and chronic pain. In order to study the effects of these compounds in such disorders and to integrate neuropeptide metabolism with that of other neurotransmitters, it is essential to analyze these endogenous peptides and their metabolites *in vivo.* Microdialysis is a sampling technique for obtaining neuropeptides and their metabolites present at low concentrations in brain tissue. Micro-ESI/tandem mass spectrometry was coupled with *in vivo* microdialysis to study neurochemical metabolism [12]. Tandem mass spectrometry plays a vital role in increasing sensitivity by lowering background signals and greatly increases the confidence in the identification of the molecule being analyzed. For example, such an assay for neurotensin will record ions of m/z 579.2, the C-terminal Y_5 fragment of neurotensin, only if this fragment originates from a molecule of m/z 558.6 ($[3M+H]^{+3}$ ion of neurotensin). The same C-terminal fragment, m/z 579.2 produced from other peptides or proteins, will not be recorded even if they contain this sequence fragment if they have a different molecular weight. Thus, the assay not only measures the molecular weight of the molecule, but also verifies the structure by identifying a given structure-specific fragment that originates from the precursor molecule.

In one experiment, for example, samples were collected by microdialysis from unstressed, unanesthetized, freely-moving animals to avoid any inhibition of neuropeptide release by anesthetics. A microdialysis probe was positioned in the globus pallidus/ventral pallidum or the hypothalamus, through a cannula implanted 5 days earlier under anesthesia. Artificial CSF was perfused through the probe for a 2 hour equilibration period followed by a 1 hour sample collection for baseline analysis. Release of the $[Met]^5$-enkephalin from the globus pallidus/ventral pallidum was accomplished by presenting a 1 µl bolus of artificial CSF supplemented with 100 mM KCl, to the tissue through the microdialysis probe. The flow rate was 436 nL/min and microdialysate samples were collected at 30 minute intervals after KCl administration. The mass spectrum of the release of $[Met]^5$-enkephalin from the freely moving rat by direct neuronal depolarization caused by KCl injection shows the m/z 397.5 product ion which corresponds to the A_4 product ion from endogenous $[Met]^5$-enkephalin (Figure 8). The m/z 397.5 ion is structurally specific, i.e., it is derived from a protonated molecule of m/z 574.7 which loses mass 177.2 to give a product ion of m/z 397.5. Other peptides that might contain the $[Met]^5$-enkephalin sequence have different molecular masses and therefore are not recorded in this assay. An estimate of the amount of this neuropeptide, released based on an external standard curve, showed that approximately 6.5±2.2 femtomoles of $[Met]^5$-enkephalin (n=4 rats) was contained in the 10 µl of dialysate collected following KCl administration.

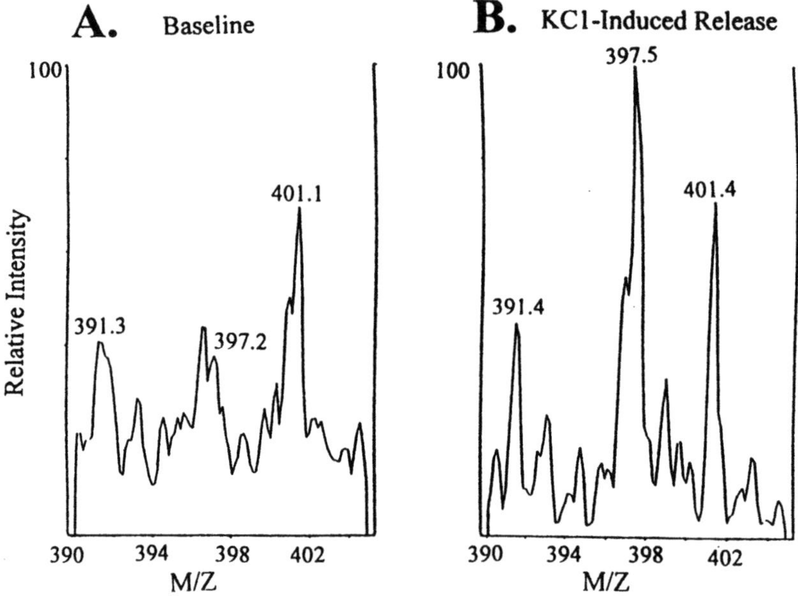

Figure 8. Endogenous release of [Met]^5-enkephalin from the globus pallidus/ventral pallidum region of rat brain. Panel A is the control from perfusion of the microdialysis probe with artificial CSF, and Panel B by perfusion with 100 mM KCL. (Reproduced from reference 12)

A similar experiment was performed to measure the release of neurotensin from the hypothalamus of a freely moving rat initiated by neuronal depolarization after KCl injection. A two hour dialysate was collected and analyzed by micro-ESI/MS/MS immediately following depolarization. The sequence-specific doubly-charged ion at m/z 579.2 representing the C-terminal fragment containing residues 5-13, was recorded. An estimate of the amount of neurotensin released in this experiment, based on an external standard curve, was 499±192 attomoles/10 µl of dialysate (n=6 rats) collected over a 30-minute period.

3.2.2. *Sequencing of proteins from polyacrylamide gels*

Proteins isolated by polyacrylamide gel electrophoresis were sequenced using nano-electrospray and tandem MS analysis [13]. To test the method, 800 femtomoles of BSA were loaded onto a gel and subjected to electrophoresis. The band containing BSA was removed and washed with 100 mM NH_4HCO_3 and acetonitrile. The gel pieces were then allowed to swell in a buffer containing 50 mM NH_4HCO_3, 5 mM $CaCl_2$, and 12.5 ng/µl trypsin. After 45 minutes, the supernatant was aspirated off and replaced with the same buffer without trypsin and stored overnight at 37° C. The hydrolyzed peptides were extracted from the gel by three volumes of 5% acetic acid and acetonitrile, and these subsequently evaporated to dryness. The tryptic peptides were redissolved in 5% formic acid and passed through a capillary containing a small volume of perfusion sorbent (Poros R2). The adsorbed peptides were washed extensively and eluted in 60% methanol/5% formic acid in a volume of 1 µl which was then directly placed into the source. Eleven BSA peptides were sequenced by MS/MS techniques from 800 femtomoles loaded onto the gel. Unambiguous sequence information was obtained from 4 peptides when 80 femtomoles (5 ng) of BSA was loaded onto the gel.

This methodology has also been applied to determine the sequence of protein-RNA interaction partners [13]. Approximately 100 ng of immunoaffinity-purified proteins interacting with RNA binding proteins were isolated by PAGE techniques. The band, corresponding to a protein with M_r of 160 kDa, was removed and digested by trypsin. The mass spectrum of the tryptic peptides extracted from the gel are shown in Figure 9A. Peaks that were not present in the blank were sequenced by tandem MS. For example, peak C at m/z 749 gave the spectrum and sequence shown in Figure 9B.

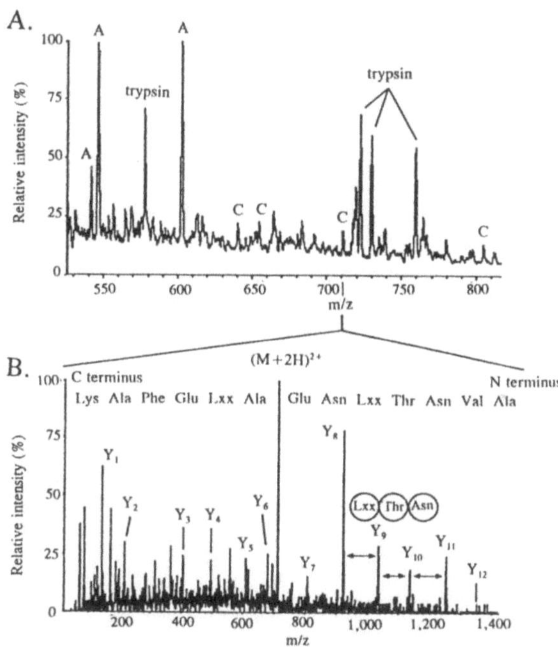

Figure 9. Identification of protein interaction partners. A.) Part of the peptide ion spectrum of a tryptic digest of a band of 160K protein excised from a polyacrylamide gel. Peptides not present in the blank were sequenced. The major peaks are tryptic peptides from rabbit antibody light chain (A). B.) Tandem mass spectrometry of one of the other peptide ions shown in A.) The masses of the ion and of four fragment ions were assembled by Peptide Search into a peptide sequence tag that was used to retrieve carbamoyl phosphate synthetase from a nonredundant, inclusive sequence data base (NRDB, maintained by C. Sander, EMBL) As shown in B.) The complete series of Y" ions confirms the peptide sequence is correct [13]. (Reproduced with permission)

Using a protein search algorithm, PeptideSearch, a peptide sequence contained in carbamoyl phosphate synthetase was retrieved. Mass spectra of three other ions also identified the same protein. The small amount of sample, the excess antibody protein over sample protein and the large size of the protein would have made use of traditional sequencing methods very difficult.

4. MALDI MS

MALDI MS can also be used quite effectively at high sensitivity for molecular weight analysis for samples obtained from biological sources. Care in sample preparation is critical for achieving high sensitivity. Sample loading, where sensitivity is not an issue, is normally accomplished by spotting 0.5-1 μl of a matrix solution onto the target followed by the immediate addition of 0.5-1 μl of the analyte solution and the mixed droplet allowed to dry slowly ("dried droplet" method). This usually yields detection limits in the range of 0.1-1 picomole of protein, although the amount of sample consumed in the analysis is much less. The consumption of sample per spectrum has been estimated to be ≥10 attomoles (10^{-18} mol) [14]. Using the "dried droplet" method, sample typically covers an area on the target of several cubic millimeters.

Another sample preparation method has been shown to achieve higher sensitivity and utilizes the matrix dissolved in acetone with or without 1-2% water [4]. Approximately 0.5 μl of the dissolved matrix solution was quickly applied to the smooth, polished stainless steel probe tip. The solvent evaporation took only a few seconds and the resulting surface appeared homogeneous when checked by a light microscope when compared with that of the dried droplet method (Figure 10). The

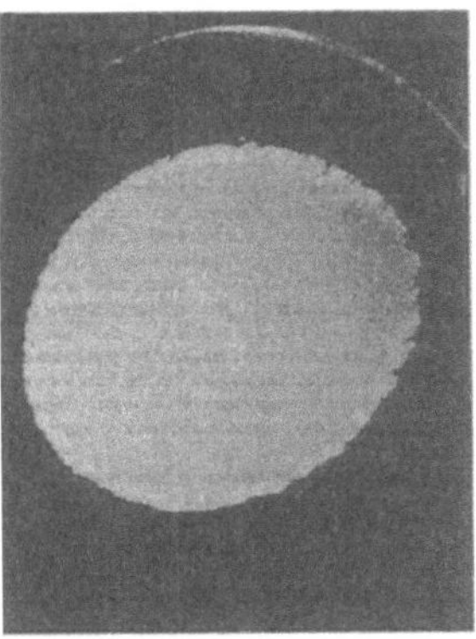

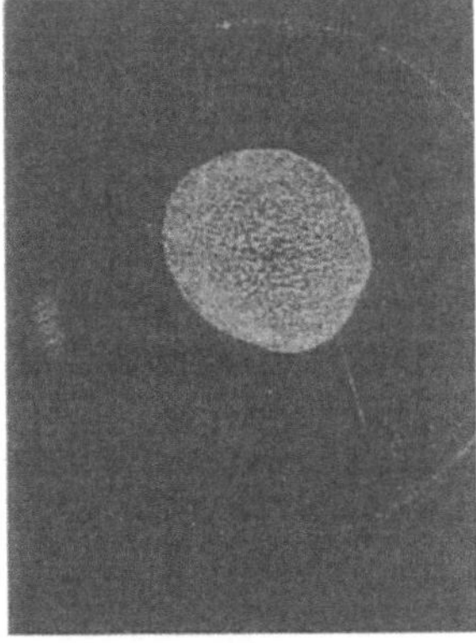

Figure 10. Photographs of 0.5 μl of αCHCA solution evaporated on a stainless steel probe. The probe tip is polished with a diameter of 8 mm. A.) Rapid evaporation matrix preparation using matrix dissolved in 1-2% acetone. B.) Crystals formed from a saturated solution of matrix in 33% aqueous acetonitrile [4]. (Reproduced with permission)

volume placed on the tip determined the area of the dried material; a 0.2 μl volume spread to a circle of only a few millimeters in diameter while a 1 μl volume spread to form a circle with a 10 mm diameter. For probe tips that were less than 1.3 mm in diameter, a matrix solution ten times less concentrated was applied in the same volume to the tip. The analyte, dissolved in any solvent that did not completely redissolve the matrix crystals, was placed on top of the matrix surface and dried. Analyte concentrations tested ranged from 10 attomol/μl to 200 picomol/μl and α-cyano-4-hydroxy cinnamic acid (α-CHCA), ferulic, and sinapinic acid were successfully used as matrices. In general, this procedure increases sensitivity one or two orders of magnitude over the "dried droplet" method. Excellent results were obtained from 100 attomoles of chicken egg white lysozyme loaded onto a probe tip. The limits of detection of approximately 5 attomoles were determined for substance P using α-CHCA as shown in Figure 11. The practical lower limit for samples was generally in the high attomole range because of contamination by other peptides or detergents.

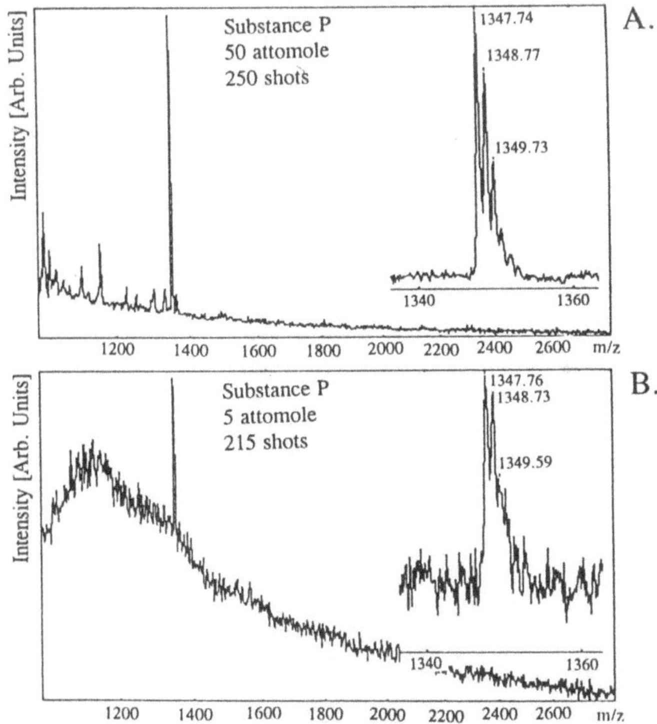

Figure 11. Reflector mass spectra of substance P (monoisotopic M_r 1347.736) using αCHCA as the matrix. Volumes of 0.5 μl of substance P solution (10% acetonitrile/0.1% aqueous TFA were added. A.) Substance P concentration was 100 attomol/μl. B.) Substance P concentration was 10 attomol/μl both obtained from serial dilution [4]. (Reproduced with permission)

Biological samples were generally diluted to 25 femtomol/μl and applied in volumes of 0.3-1 μl. At high sample levels, if greater than 100 picomoles were applied to the target, a significant decrease in resolution was seen. This loading method was tested on ten different instruments in eight different laboratories and, as expected, the absolute limits of sensitivity depend on the individual instruments used.

4.1 REDUCTION OF SAMPLE VOLUME

Several investigators have shown low absolute detection limits by reducing the sample volume, thereby reducing the sample area on the target. In one report a monocrystalline silicon wafer with a specific surface orientation was micromachined to give arrays of 100x100 μm wells or vials with a 200 μm space between each vial (Figure 12) [7]. The volume of each vial was 250 pL and the size of the laser spot was

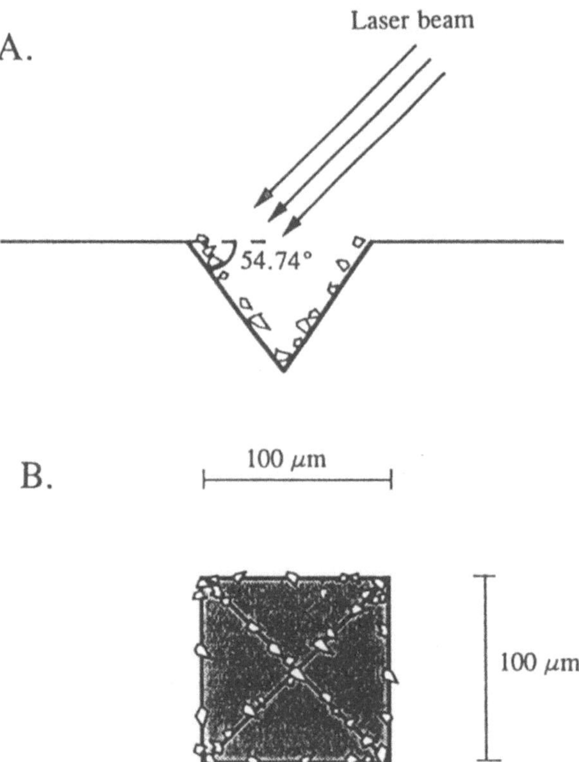

Figure 12. Schematic drawing of a 250 pL vial showing the position of the microcrystals in A.) cross-sectional view of the vial and B.) top view of the vial [7]. (Reproduced with permission)

approximately 100 mm^2 diameter. The silicon wafer was covered with a sputtered film of chromium and gold to establish an electrical contact between the sample holder and the surface of the wafer. The picoliter vials were filled with sample and matrix (in this case 2,5-dihydroxybenzoic acid) and the solvent was allowed to evaporate. A series of dilutions for bradykinin with a total of 100, 10, 5, or 2.5 attomoles was analyzed and the resulting spectra are shown in Figure 13. Signal decreased with decreasing

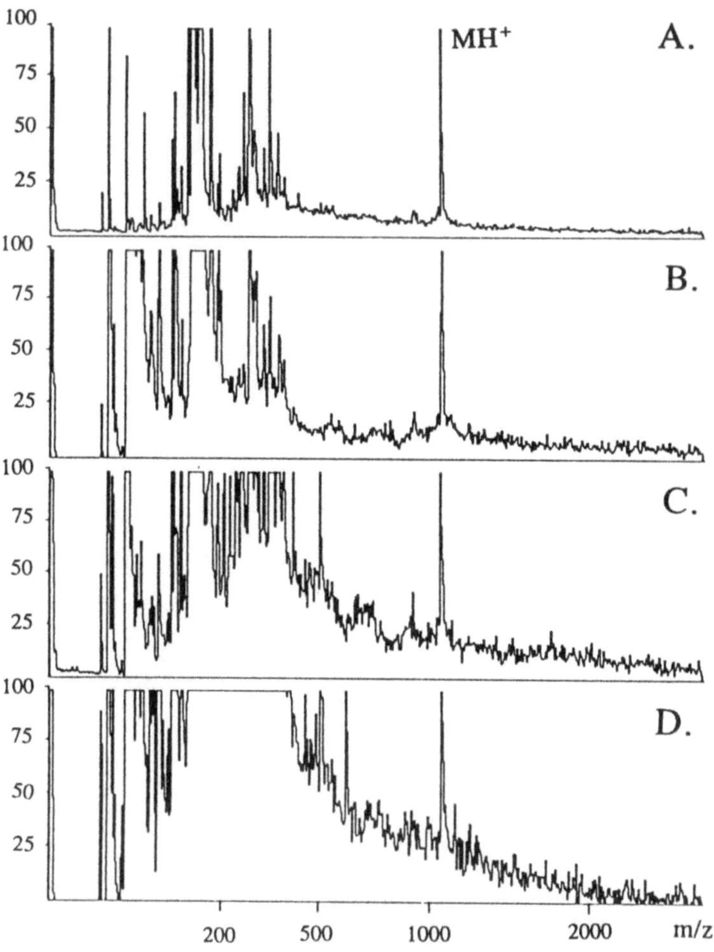

Figure 13. MALDI MS spectra of bradykinin obtained from the 250 pL vials containing absolute amounts of A.) 100 attomoles B.) 10 attomoles C.) 5 attomoles D.) 2.5 attomoles [7]. (Reproduced with permission)

amounts of sample and threshold irradiance was also seen to decrease with decreasing amounts of sample. However, a signal for bradykinin was still recorded at 2.5 attomoles. Similar results were seen for cytochrome C with the lower limit of detection being 25 attomoles. Approximately ten spectra could be obtained from each vial and so the calculated consumption of sample appears to be 250 zeptomoles for bradykinin and 2.5 attomoles for cytochrome C for each spectrum.

One of the difficulties in placing sample on the probe tip for MALDI MS analysis is sample delivery. A method of delivering very small volumes (0.2-130 nL) to the target reliably and reproducibly was described utilizing calibrated capillaries with different inner diameters (Figure 14) [5]. The sample solution was delivered onto a

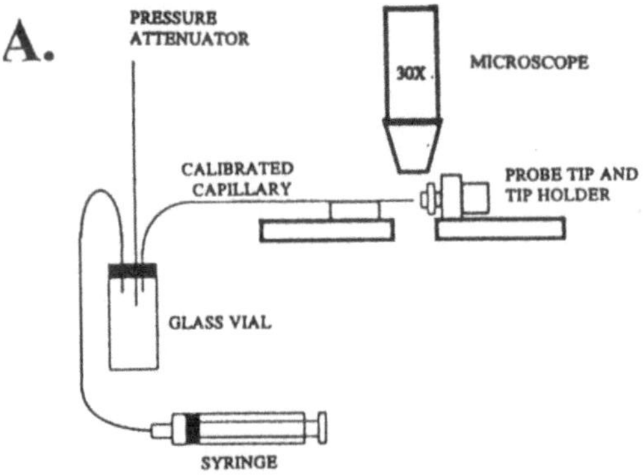

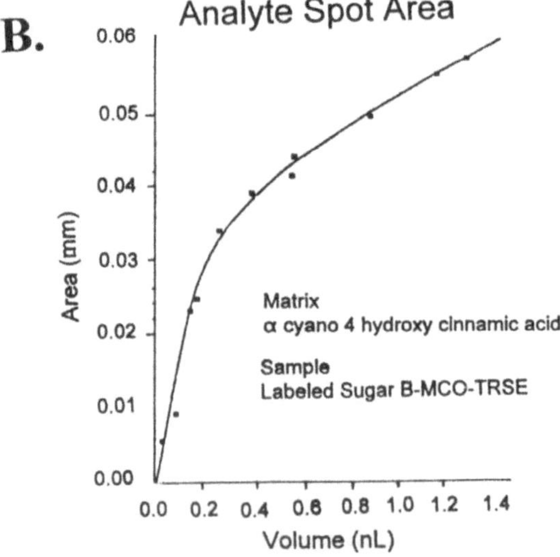

Figure 14. A.) Schematic drawing of nanopipette for spotting small volume samples onto a MALDI probe tip. B.) Size of analyte spot area compared to volume of sample loaded [5]. (Reproduced with permission)

solid matrix layer of small crystals (0.5-2 μm diameter) formed using the technique of Vorm, et al. [4]. A 1 μM solution of bradykinin with a volume of 0.15 nL was loaded onto the probe tip giving a spot with a diameter of approximately 177 μm^2. The total amount of protein loaded was 150 attomoles and the spectrum is shown in Figure 15.

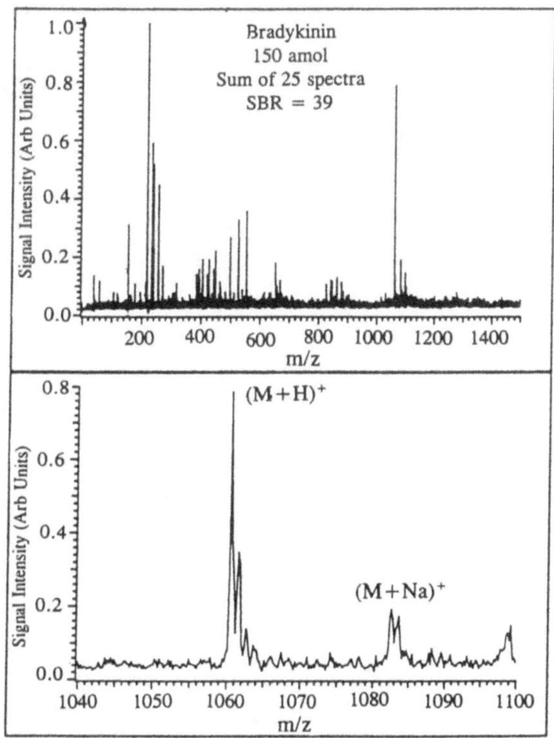

Figure 15. Average spectrum for a 0.15 nL volume of 1 μM bradykinin. A total of 150 attomoles was loaded onto the probe tip [5]. (Reproduced with permission)

The S/N ratio is well above the detection limit indicating lower initial concentrations could be analyzed successfully.

Another approach to small sample volume preparation combined microspotting with multiple remeasurement of ions from a single laser shot [6]. A sample probe tip was modified to allow deposition of a 1 μl aliquot of dilute sample solution by microsyringe onto the sample spot indentation (Figure 16). Fourier transform ion cyclotron resonance (FT-ICR) experiments were carried out in the source side of a dual cubic ion trap of an instrument with a 3T superconducting magnet. A total of 8 attomoles of substance P (M_r=1347.6) was applied to the sample probe. To ensure the indentation containing the analyte was aligned with the laser beam, scanning electron micrograph pictures were taken of the laser spot. The sample was well

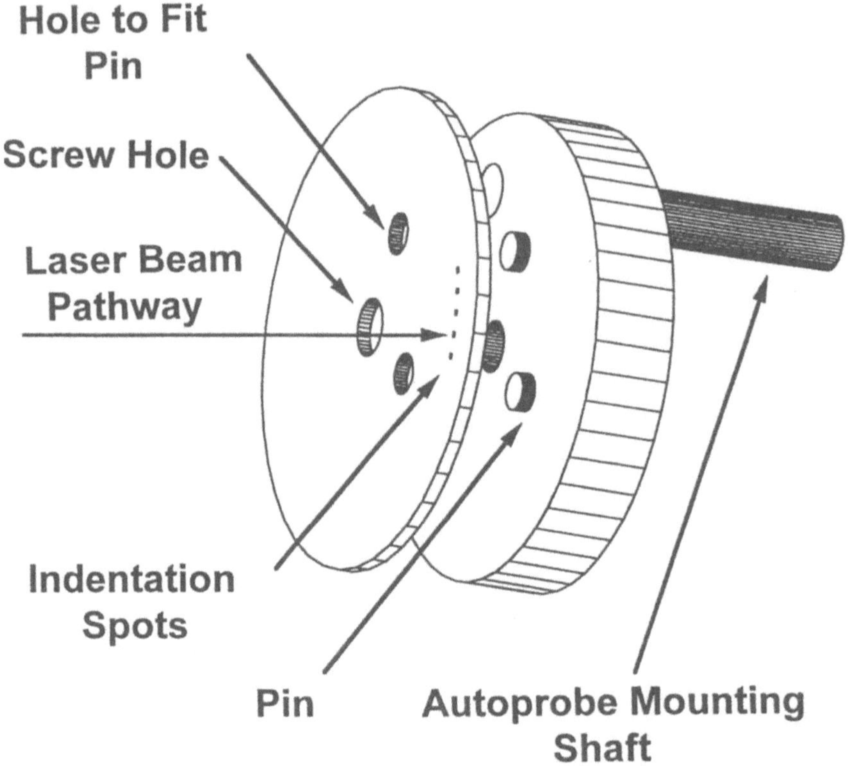

Figure 16. Modifications to a standard Extrel sample probe tip. The indention spots are located at position 1-5 and are 5° apart. The two guide pins and the corresponding holes on the indented plate provide alignment of the laser beam with the sample placed in the indentations [6]. (Reproduced with permission)

within the boundaries of the laser spot as shown in Figure 17. The present limitation of this mode of deposition is the inability to carefully control the size of the droplet. Positive ion mass spectra for 8 attomoles of substance P with both single measurement (Figure 18A) and the sum of 40 remeasured time domain data sets (Figure 18B) were obtained. Although substance P was visible in the single measurement spectrum, remeasurement greatly improved the S/N ratio and several structurally significant fragment ions could be detected upon remeasurement. At low concentration levels, adduct species (e.g. $(M+Na)^+$, $(M+K)^+$) also increased.

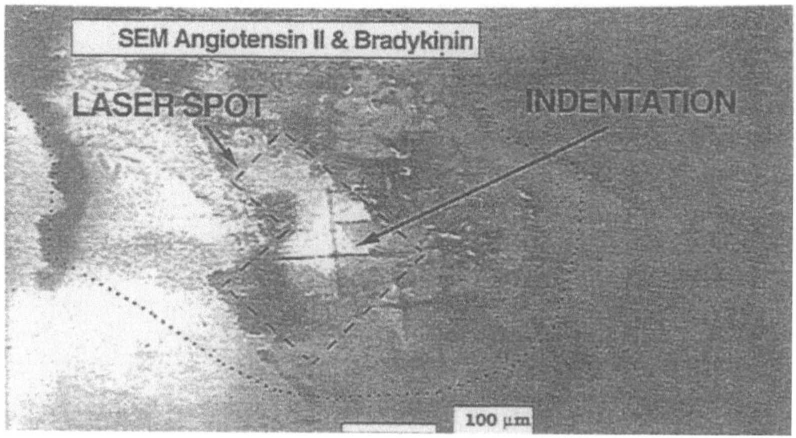

Figure 17. Scanning electron microscopy of stainless steel sample probe tip after 16 laser shots on same spot. Sample applied consisted of 2,5-dihydroxy benzoic acid, fructose, angiotensin II, and bradykinin with mole ratios of 10^8:10^8:1:1 [6]. The indentation and the laser spot are denoted by arrows. The laser spot is outlined by dashes and the area over which sample was deposited is outlined by dotted lines. (Reproduced with permission)

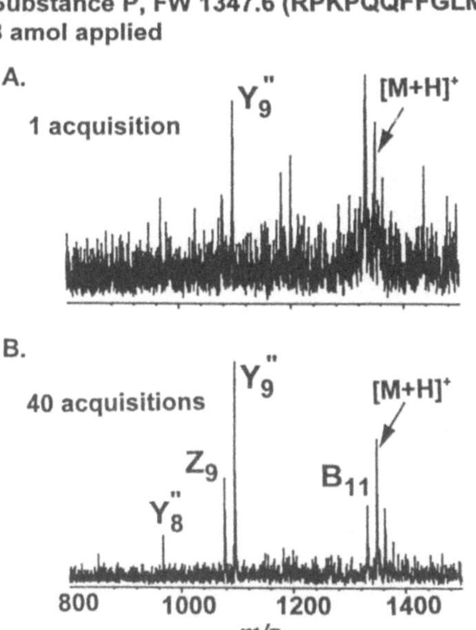

Figure 18. Mass spectra of substance P (M_r 1347.6) from a total of 8 attomoles applied. A.) Single measurement B.) Remeasurement of 40 times [6]. (Reproduced with permission)

Another approach to decreasing spot size and increasing sensitivity has been the use of membranes where sample is transferred to the membrane. In one report [8] a regenerated cellulose dialysis membrane was utilized to obtain mass spectra of high quality from attomole amounts of peptides and proteins [8]. The membrane was precoated with matrix solution (e.g. 10 mg/ml α-CHCA in 80% acetonitrile with 2% acetic acid) and allowed to dry. No crystals were visible on the surface. Samples in aqueous solutions were deposited in 3-5 nL volumes using a 50 μm i.d. fused silica capillary. The diameter of the laser spot impinging on the membrane was approximately 50 μm. Samples containing 200 attomoles of neurotensin ($[M+H]^+$= 1674) and 1 femtomole horse heart cytochrome C ($[M+H]^+$=12,362) were dissolved in 50 mM phosphate buffer (pH 2.5) and deposited on the membrane in 4 nL volumes. The results are shown in Figure 19. Several peptides with a concentration of approximately 40 attomoles and several proteins (m/z >20,000 Da) with a concentration of 200 attomoles were examined and the resulting spectra gave a S/N ratio >3:1.

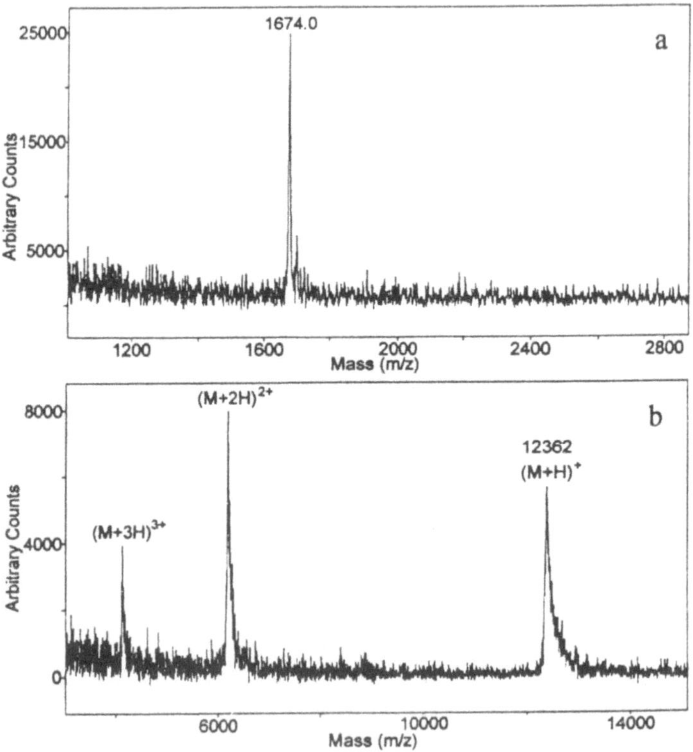

Figure 19. MALDI mass spectra produced from a matrix precoated cellulose membrane with A.) 200 attomoles neurotensin and B.) 1 femtomole of horse cytochrome C. In both cases, samples were spotted in 4 nL of 50 mM phosphate buffer (pH 2.5). (Reproduced from reference 8)

4.2 PRECONCENTRATION AND DESALTING

Preconcentration and desalting of a sample on a C-18 microcapillary column allows analysis of samples containing contaminants that adversely affect MALDI MS analysis [8, 15]. For example, a 50 μm i.d. capillary column was employed with 1 cm of C-18 packing held in place by a constriction (~20 μm i.d.) formed by heating the capillary and pulling it gently. A schematic drawing of the microconcentration set-up is shown in Figure 20. Sample was injected onto the capillary column and washed with 2% acetic acid. Elution was achieved by using a solvent containing matrix in 75% acetonitrile (1 part saturated α-CHCA in 50% acetonitrile and 2% acetic acid and 1 part acetonitrile with 2% acetic acid) at a flow rate of 0.1 μl/min. Approximately 5 nL was spotted onto the individual wells on a multi-well sample target, giving a spot area of less than 0.3 mm^2. A S/N ratio better than 3:1 was achieved for a total of

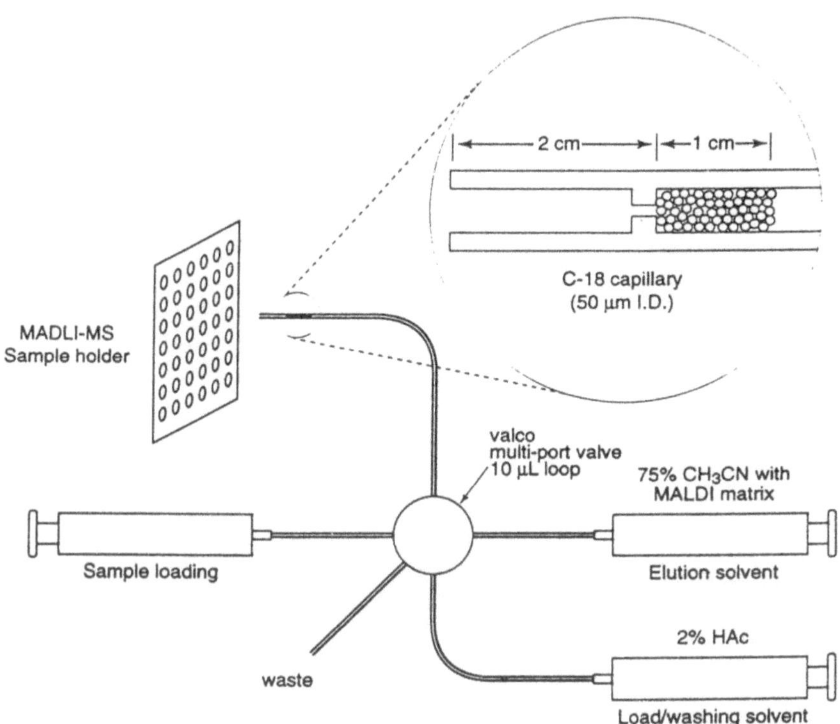

Figure 20. Schematic diagram of the MALDI sample micro-concentration/salting/matri addition device. Loading and washing solvent was 2% acetic acid with the elution matrix being αCHCA in 75% acetonitrile with 2% acetic acid. The elution rate was 0.1 μl/min and the sample volume spotted on the target was ~5 nL per fraction. (Reproduced from reference 15)

200 attomoles neurotensin spotted from a 10 μl injection of 20 attomol/μl solution (Figure 21A). Samples analyzed without preconcentration and washing did not give a signal that could be distinguished from noise below 100 femtomoles loaded (Figure 21B). The dried droplet method, loading 0.5 μl of matrix and sample gave a spot size of 9 mm^2 and 1 femtomole of neurotensin was required for a S/N ratio of 4:1.

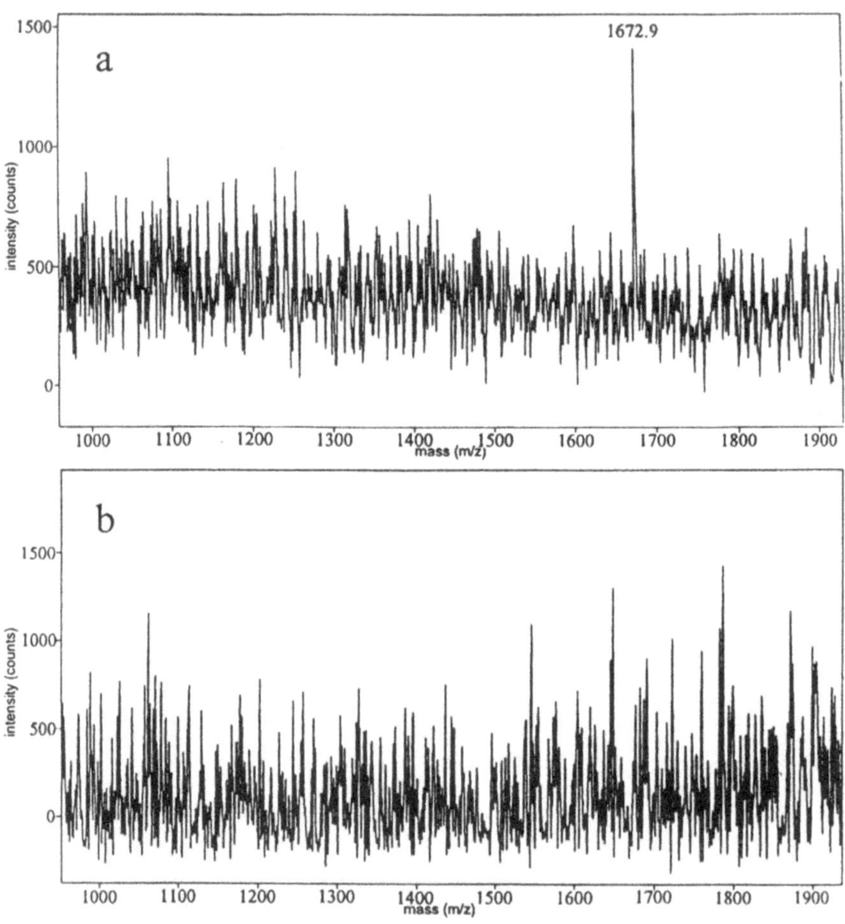

Figure 21. MALDI mass spectra obtained A.) using the C18 desalting/preconcentration procedure for a solution of 200 attomoles of neurotensin loaded in 10 μl from a 20 attomol/μl solution containing 0.2 M phosphate buffer (pH 7.0) and B.) from direct spotting of 100 femtomoles of neurotensin from a 1:1 mixture of 40 picomol/μl neurotensin in 0.2 M phosphate buffer and MALDI matrix. In both spectra 5 nL volumes were spotted. (Reproduced from reference 15).

4.3 APPLICATIONS

4.3.1 *On-Target Sequencing of Bioactive Peptides*

MALDI MS has been combined with on-target exopeptidase digestion to identify and directly sequence a model tumor-specific peptide antigen [16]. Cells were transfected with vaccinia virus, V69, which contains a gene encoding the influenza virus nucleoprotein (NP). NP 147-155 peptide was isolated from 1.6 x 10^9 cells by immunoprecipitation and RP-HPLC. Before on-target enzymatic digestion, the major ion recorded was m/z 1108 (Figure 22A). Carboxypeptidase P was added to the target and allowed to digest for 9, 15, 30, 60, 90, and 600 s. After 9 s, ions of m/z 1108, 1009, 896, 825, 669, and 568 were visible in the spectrum corresponding to the full length peptide and the loss of the carboxy-terminal V, L, A, R, and T, respectively (Figure 22B). The mass spectrum after aminopeptidase M digestion showed peaks with m/z 1108, 1007, 844, and 716, corresponding to the full length peptide and the loss of amino terminal T, Y, and Q (Figure 22C). Subtraction of the five amino acids from carboxypeptidase P digestion and the three amino acids from aminopeptidase M digestion, gives a value of 156 which corresponds to arginine, the fourth amino acid from the N-terminus. On-target digestion was repeated for influenza NP peptide isolated from retrovirally transduced tumor cells to examine the feasibility of isolating and sequencing peptides from a gene integrated into the host genome. The same results were obtained as those seen for the vector infected cell line. The authors also found that addition of 0.3 µl saturated ammonium sulfate solution to 0.3 µl analyte sample followed by 0.6 µl of matrix solution aided detection of the peptide.

The limits of detection for this method were evaluated using a synthetic nonameric peptide TYQRTRALV (M_r= 1107.4) representing amino acids 147-155 of NP. Dilutions (3.3, 1, 0.1, 0.067, 0.033, 0.001 picomol/µl) of the synthetic NP 147-155 were prepared in pooled HPLC fractions of whole cell lysates of the tumor cells. A volume of 0.3 µl was deposited onto the target and 0.3 µl of α-CHCA matrix was added. The solutions were analyzed with and without the addition of ammonium sulfate. Without ammonium sulfate the 3.3 and 1 picomol/µl dilutions had a prominent peak at m/z 1108 with a smaller peak at m/z 973. For concentration of 0.1 picomol/µl, m/z 973 was the prominent peak with a small, distinct peak at m/z 1108 that disappeared at the more dilute concentrations. However, upon addition of ammonium sulfate, m/z 1108 was the predominant peak in the spectrum with a small peak at m/z 973 for concentrations of 3.3, 1, 0.1, 0.067, 0.033, and 0.001 picomol/µl. At concentrations less than 0.001 picomol/µl no peak was recorded at m/z 1108. NP 147-155 is a fairly hydrophobic peptide and addition of ammonium sulfate appeared to precipitate the more hydrophobic peptides, thereby assisting in MALDI MS detection. One-fifth of the total cell preparation was used for mass spectral analysis, therefore the limit of detection of a known peptide in an RP-HPLC fraction obtained from 10^9 tumor cells is 10-30 femtomoles of NP peptide.

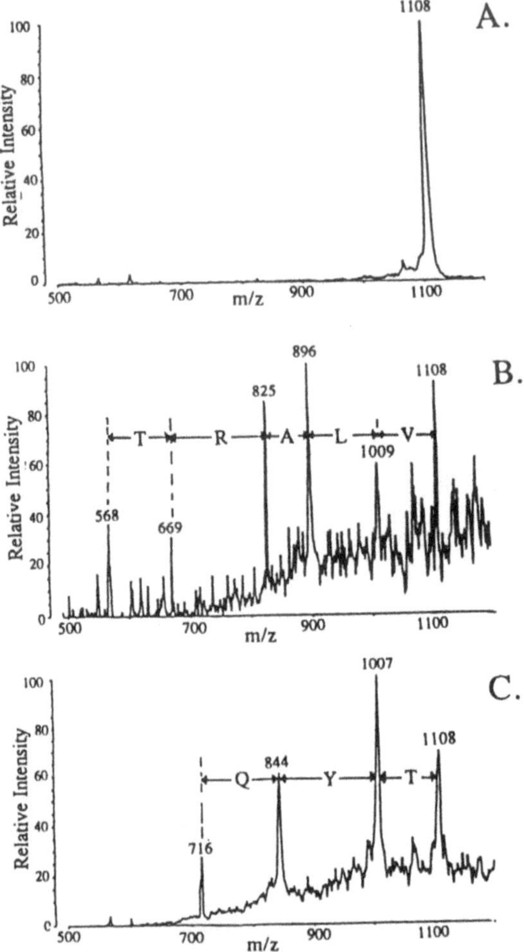

Figure 22. MALDI spectrum of the H-2K^d-restricted NP147-155 peptide isolated from CT26 tumor cells infected with the recombinant vaccinia virus V69. A.) NP147-155 peptide. B.) Carboxypeptidase P digest (9 s at room temperature) showing $(M+H)^+$ values corresponding to the consecutive loss of the C-terminal V, L, A, R, and T. C.) Aminopeptidase M digest (120 s at room temperature) showing $(M+H)^+$ values corresponding to the consecutive loss of the N-terminal T, Y, and Q [13]. (Reproduced with permission)

4.3.2 *In vivo Analyses*

In order to assess the suitability of microconcentration/desalting for the analysis of complex biological mixtures containing physiological levels of salts, a sample from an in vivo microdialysis study of the metabolism of neurotensin was analyzed [8]. In this experiment, a microdialysis probe was stereotaxically implanted in the striatum of an anesthetized rat. The probe was used both to deliver the peptide into the brain and to collect the degradation products. An infusion of 100 femtomol/μl of neurotensin in artificial CSF was administered at 300 nL/min for 30 minutes for 2 hours following the neurotensin infusion and samples were taken every 30 minutes. The dialysate should therefore contain metabolized products of neurotensin in addition to unmetabolized (full length) neurotensin. As shown in Figure 23A, several peptides were identified from their m/z values corresponding to neurotensin fragments 9-13, 1-8, 1-10, 1-11, and 1-12. In contrast, no neurotensin fragments could be identified or detected from the same microdialysate if the microconcentration/desalting preparation procedure was omitted (Figure 23B).

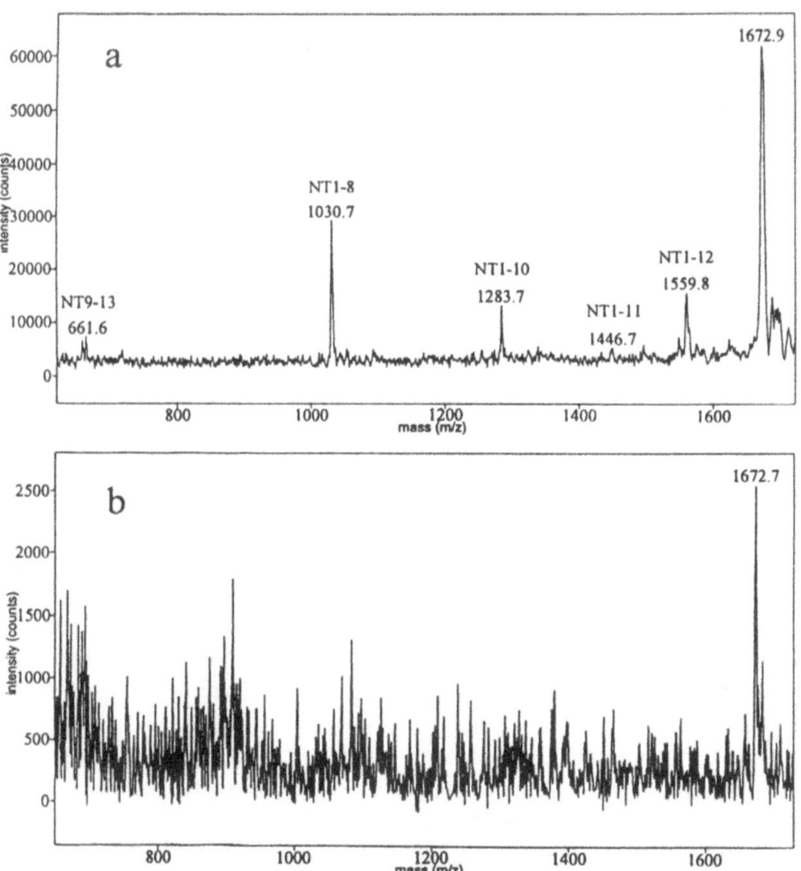

Figure 23. MALDI mass spectra of in vivo microdialysate containing artificial cerebrospinal fluid obtained from the metabolism of exogenous neurotensin in rat brain. A.) C18 desalting/preconcentration procedure with a 7 μl loading of dialysate. B.) Direct MALDI analysis of 1 μl of a 1:1 mixture of the dialysate sample and MALDI matrix without preconcentration and desalting. (Reproduced from reference 15)

4.3.3 *Combining CE and MALDI*

Continuous deposition of effluent from a capillary electrophoresis (CE) capillary onto a target membrane can produce a 'track' containing the CE separated compounds [17]. This technique utilizes thin matrix films on the target to achieve high sensitivity. A MALDI target plate with a matrix-coated strip of Spectra/Por 2 dialysis membrane (Spectrum Medical Industries) is mounted onto movable stage (Figure 24). The sample solution was loaded electrokinetically onto a polyethylene glycol coated

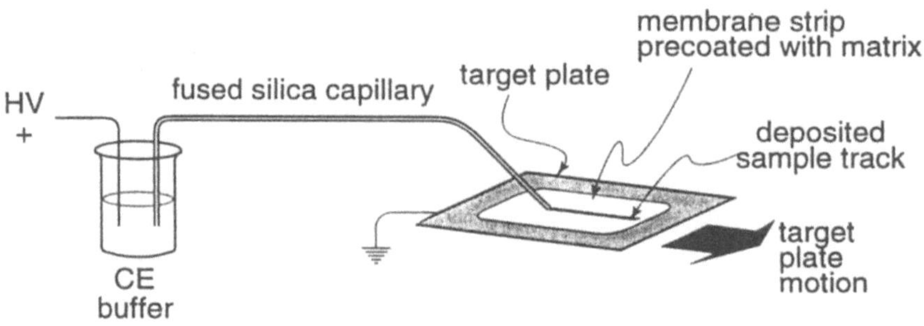

Figure 24. Schematic diagram of the membrane sample deposition apparatus for off-line coupling of CE and MALDI MS. (Reproduced from reference 17)

capillary (to avoid wall adsorption) at 5 kV for 4 ns and separated electrophoretically using 30 kV and 20 mM phosphate buffer (pH 2.8). Movement of the target plate was usually begun 90-120 s after the beginning of CE analysis and terminated 200 s later (50 mm travel). A composite selected ion eletropherogram is shown in Figure 25A for the $[M+H]^+$ ion for each protein in a four protein mixture while Figure 25B shows a similar electropherogram obtained using a UV detector after loading 600 femtomoles of the sample. Representative MALDI mass spectra along the CE track are shown in Figures 25C and 25D for a loading of approximately 2 femtomoles for each protein in this CE analysis.

The use of the precoated membranes provides means for the effective off-line coupling of CE and MALDI, for several reasons. First, the water insoluble matrix can be solubilized on the surface of the gel-like membrane, forming a homogeneous thin film of matrix material. Second, the membrane-matrix combination provides an aqueous/organic media that interacts with the analyte. Third, the aqueous solution containing the analyte solubilizes the thin matrix film and forms microcrystals on drying. The resulting technique is thus compatable with conditions necessary to achieve high resolution CE separations and those necessary to achieve high sensitivity MALDI MS.

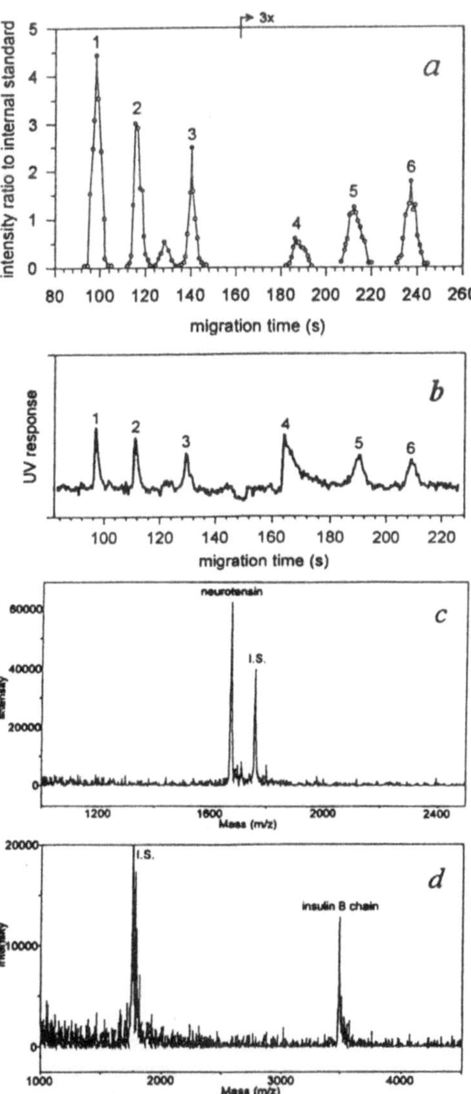

Figure 25. A.) Composite selected ion electropherogram obtained using a α-CHCA pre-coatedregenerated cellulose membrane for the separation of (1) horse cytochrome C, (2) chicken egg white lysozyme, (3) bovine lactoglobulin A, and (4) bovine ribonuclease A. total amount protein injected was 2 femtomoles each from a sample solution containing 500 femtomol/μl of each protein in 10 mM sodium phosphate (pH 2.8). B.) Electropherogram obtained using an on-line UV detector for a similar separation except that a total of 80 femtomoles of each protein was injected from a solution with a concentration of 20 picomol/μl of each protein. C.) MALDI spectrum taken at 516.8 s in panel A.D.) MALDI spectrum taken at 633.6 s from panel A. (Reproduced from reference 17)

5. Conclusions

Both ESI and MALDI mass spectrometry offer the research investigator very powerful analytical tools with which to probe the chemical dynamics of biological systems. ESI has many advantages and provides an excellent and very effective on-line device for flow-injection analysis of aqueous samples, in addition to providing CE/MS and LC/MS capabilities. MALDI is an excellent off-line analytical device for the high sensitivity and high mass detection of biomolecules, and is especially useful in applications in peptide and protein chemistry. Both methods allow analysis of peptides, in the low attomole range, and analysis of proteins, in the low femtomole to high attomole range. These two mass spectrometric techniques are complementary and research laboratories engaged in analysis at the molecular level would find it advantageous to have both techniques readily available.

6. References

1. M. R. Emmett and R. M. Caprioli, *J. Am. Soc. Mass Spec.* **5**, 605-613 (1994)
2. M. Wilm and M. Mann, *Int. J. Mass Spectrom. Ion Processes* **136**, 167-180 (1994)
3. G. A. Valaskovic, N. L. Kelleher, D. P. Little, D. J. Aaserud and F. W. McLafferty, *Anal. Chem.* **67**, 3802-3805 (1995)
4. O. Vorm, P. Roepstorff and M. Mann, *Anal. Chem.* **66**, 3281-3287 (1994)
5. R. Golding, R. M. Whittal and L. Li, *Proc. of the 43rd ASMS Conference on Mass Spectrometry*, Atlanta, 1223 (1995)
6. T. Solouki, J. A. Marto, F. M. White, S. Guan and A. Marshall, *Anal. Chem.* **67**, 4139-4144 (1995)
7. S. Jesperson, W. M. A. Niessen, U. R. Tjaden, J. van der Greef, E. Litborn, U. Lindberg and J. Roeraade, *Rap. Commun. Mass Spectrom.* **8**, 581-584 (1994)
8. H. Zhang and R. M. Caprioli, *J. Mass Spectrom.* **31**, 690-692 (1996)
9. P. E. Andren, M. R. Emmett and R. M. Caprioli, *J. Am. Soc. Mass Spec.* **5**, 867-869 (1994)
10. G. Taylor, *Proc. R. Society London Ser. A*, **280**, 383 (1964)
11. M. Wilm and M. Mann, *Anal. Chem.* **68**, 1-8 (1996)
12. M. R. Emmett, P. E. Andren and R. M. Caprioli, *J. Neuroscience Methods* **62**, 141-147 (1995)
13. M. Wilm, A. Schevchenko, T. Houthaeve, S. Breit, L. Schweigerer, T. Fotsis and, M. Mann, *Nature* **379**, 466-479 (1996)
14. M. Karas, U. Bahr and F. Hillenkamp, *Int. J. Mass Spectrom. Ion Processes* **92**, 231-242 (1989)
15. H. Zhang, P. E. Andren and R. M. Caprioli, *J. Mass Spectrom.* **30**, 1768-1771 (1995)
16. A. Woods, A. Y. C. Huang, R. J. Cotter, G. R. Pasternack, D. M. Pardoll and E. M. Jaffee, *Anal. Biochem.* **226**, 15-25 (1995)
17. H. Zhang and R. M. Caprioli, *J. Mass Spectrom.* (1996) (In Press)

Optimised GC/MS interfacing systems and operating conditions

P. J. ARPINO
Institut National Agronomique,
Laboratoire de Chimie Analytique
16, rue Claude Bernard, F-75231 Paris Cedex 05, France

ABSTRACT. The different methods for directly coupling a capillary column to a mass spectrometer in a compact GC/MS system are reviewed.

1. Introduction

Combined GC/MS is an old and well established technique. Hundreds of instruments are sold each month and are installed in laboratories around the world. In a review published in 1996 [1] , from the 1300 GC papers published over a period of two years, 58 % used mass spectrometry as the detection method. The dominant use of capillary columns has eliminated the need for intensive research into complex interface methodologies. However, there are still some misunderstanding concerning the optimisation of the chromatographic parameters in a GC/MS instrument. As early as in 1974, after a decade of GC/MS development, Nau and Biemann had already noted that « the overemphasis of the mass spectrometer as the sole parameter for identification has relegated the gas chromatograph to the secondary role of a separation device » [2]. Two decades later, the situation is about the same, and it is too frequently estimated that perfect chromatographic resolution is not essential to identify and quantitate compounds, as much of it can be left to the mass spectrometer. To correct this misconception, some fundamental of gas chromatography pertinent to GC/MS operation is recalled.

2. Theory

A brief discussion of the basic situation in a GC/MS system is presented below. A more complete study was published by Cramers, Leclercq and co-workers [3, 4], and by Davies [5].

Mobile phases used in GC can be assimilated to perfect gases. From the injector of the GC to the MS vacuum system, the mass flow rate of gas, Q, is constant, thus :

$$Q = PF = P_i F_i = P_o F_o = P_{atm} F_{atm} \quad (1)$$

$$P_i u_i = P_o u_o = P_{atm} u_{atm}$$

R. M. Caprioli et al. (eds.), Selected Topics and Mass Spectrometry in the Biomolecular Sciences, 85–101.

P is the absolute gas pressure, $F = \frac{dV}{dt}$, the volume flow rate, u is the gas linear velocity ; i and o refer to variables at the inlet or outlet of the capillary column, atm when the outlet is at atmospheric pressure.

The viscous hydrodynamic regime established within the capillary column follows the Poiseuille law :

$$Q = \frac{\pi d_c^4 \left(P_i^2 - P_o^2\right)}{256 \eta L} \tag{2}$$

$$u_o = \frac{\pi d_c^2 \left(P_i^2 - P_o^2\right)}{64 \eta L P_o} \tag{3}$$

η is the gas dynamic viscosity, d_c the column inner diameter and L is the column length.

In a chromatographic separation, the resolution R between two consecutive peaks 1 and 2 is defined by the relation :

$$R = 1{,}18 \left(\frac{t_{r2} - t_{r1}}{\omega_{h1} + \omega_{h2}} \right) \tag{4}$$

t_r is the retention time, ω_h is the chromatographic peak width at half-height.

Optimum resolution is achieved when the following equation is maximised :

$$R = \frac{\sqrt{N}}{4} \left(\frac{k'}{1 + k'} \right) \left(\frac{\alpha - 1}{\alpha} \right) \tag{5}$$

The analyst optimises the separation by selecting the most appropriate stationary phase, which sets the proper phase selectivity α, then the column temperature, t_c, which affects both the retention factor k' and α. The last parameter to optimise is the number of theoretical plates N. As a rough approximation, N is mainly affected by the gas average linear velocity $\bar{u}$ and the column diameter d_c. However, once the column is installed in the GC/MS system, the only factor which is under direct control of the analyst is the tuning of $\bar{u}$.

A parameter related to N is the height equivalent to a theoretical plate h :

$$h = \frac{L}{N} \tag{6}$$

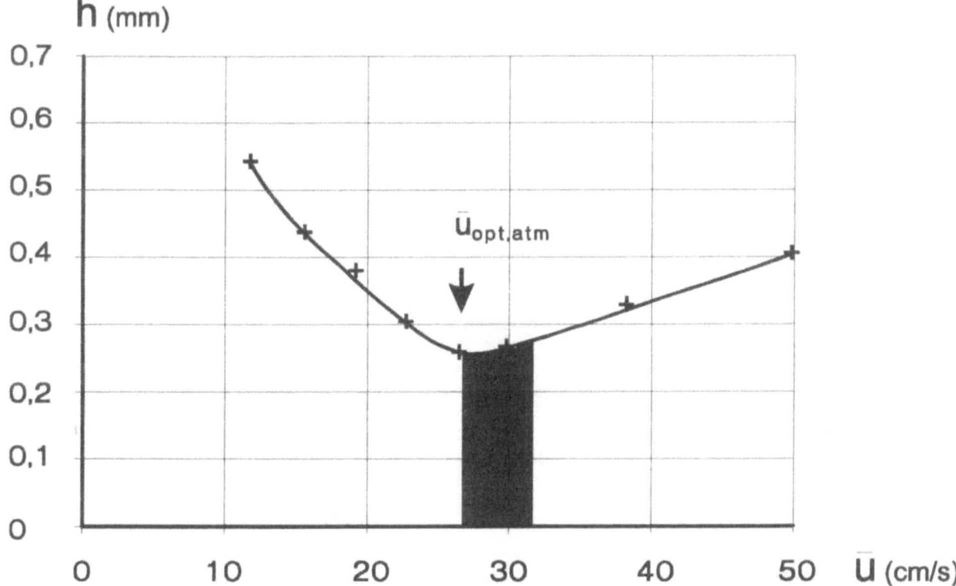

Figure 1. Experimental Van Deemter curve. 25 m x 0.22 mm id column coated with 0.25 μ of methylsilicone phase. The solute is cumene eluted by helium at 80°C (k'=1.79).

and maximum separation power is obtained when h is minimum. A typical $h=f(\bar{u})$ curve, or Van Deemter curve, is shown in Fig. 1.

Although maximum efficiency is obtained at a given linear velocity $\bar{u}_{opt}$ (for example at ca. 26 cm/s in Fig. 1), working at a slightly higher linear velocity is still acceptable as the analysis time is reduced for a minimal loss of plates. On the other hand, lower velocities result in both long and inefficient analyses and these conditions should be totally avoided.

Taking into account the dimension of common commercially available columns, and the nature of the most commonly used gases for GC/MS - helium or hydrogen - good separating conditions are obtained when the average linear velocity is close to the values listed in table 1 ; e_f is the average film thickness of the stationary phase. When using thicker films, compared to those listed in Table 1, the corresponding gas velocity should be slightly reduced, but we shall neglect the correction in this text. These values apply when the outlet of the column is at atmospheric pressure. Higher speeds are required when the column outlet is under a vacuum (see below).

TABLE 1. Practical values of the average carrier gas velocity for a GC separation with the column outlet at atmospheric pressure.

d_c (mm) (e_f μ)	helium (cm/s)	hydrogen (cm/s)
0.53 (1.0)	21	28
0.32 (0.5)	26	34
0.22 (0.2)	28	36
0.15 (0.1)	30	38

The average linear velocity can be experimentally measured either from the retention time t_o of an inert solute, e.g., the air peak in a GC/MS system, or from the gas velocity at the column outlet, u_o.

$$\bar{u} = \frac{L}{t_0} \tag{7}$$

$$\bar{u} = j u_o \tag{8}$$

j is the compression factor of James and Martin :

$$j = \frac{3}{2} \frac{P_o \left(P_i^2 - P_o^2\right)}{\left(P_i^3 - P_o^3\right)} \tag{9}$$

From combination of equations 3,8,9, it can be derived that :

$$\bar{u} = \frac{3 d_c^2}{128 \eta L} \frac{\left(P_i^2 - P_o^2\right)^2}{\left(P_i^3 - P_o^3\right)} \tag{10}$$

The gas viscosity increases with the column temperature, according to the equation :

$$\eta_2 = \eta_1 \left(\frac{T_2}{T_1}\right)^x \tag{11}$$

T is the absolute column temperature, and x is a constant close to 0.7. The gas viscosity of hydrogen or helium at a specific temperature can be obtained in the literature, for example in ref. [6].

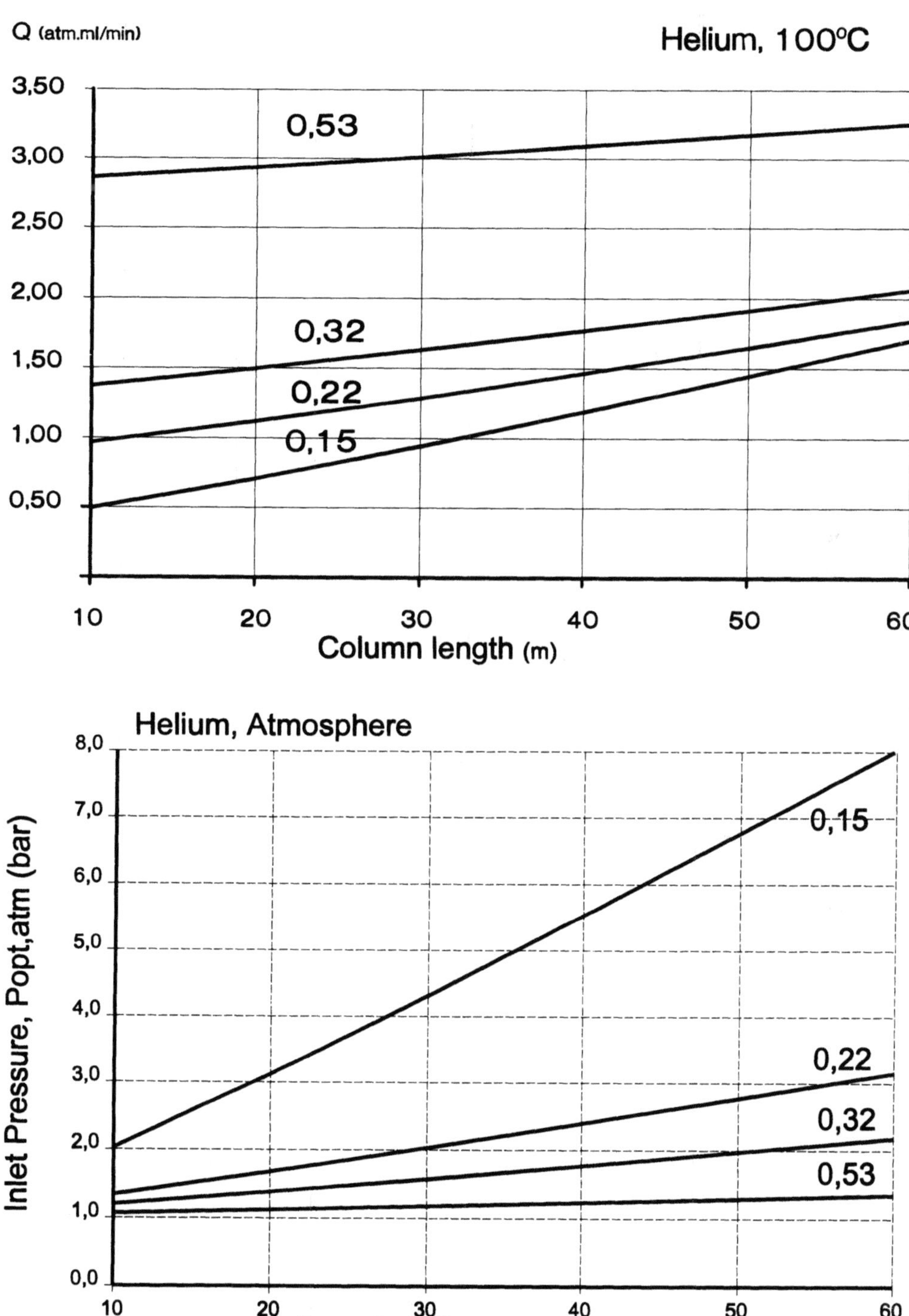

Figure 2. Optimum gas flow rate and column inlet pressure for the column dimensions and optimum gas velocities in Table 1, for helium at 100°C.

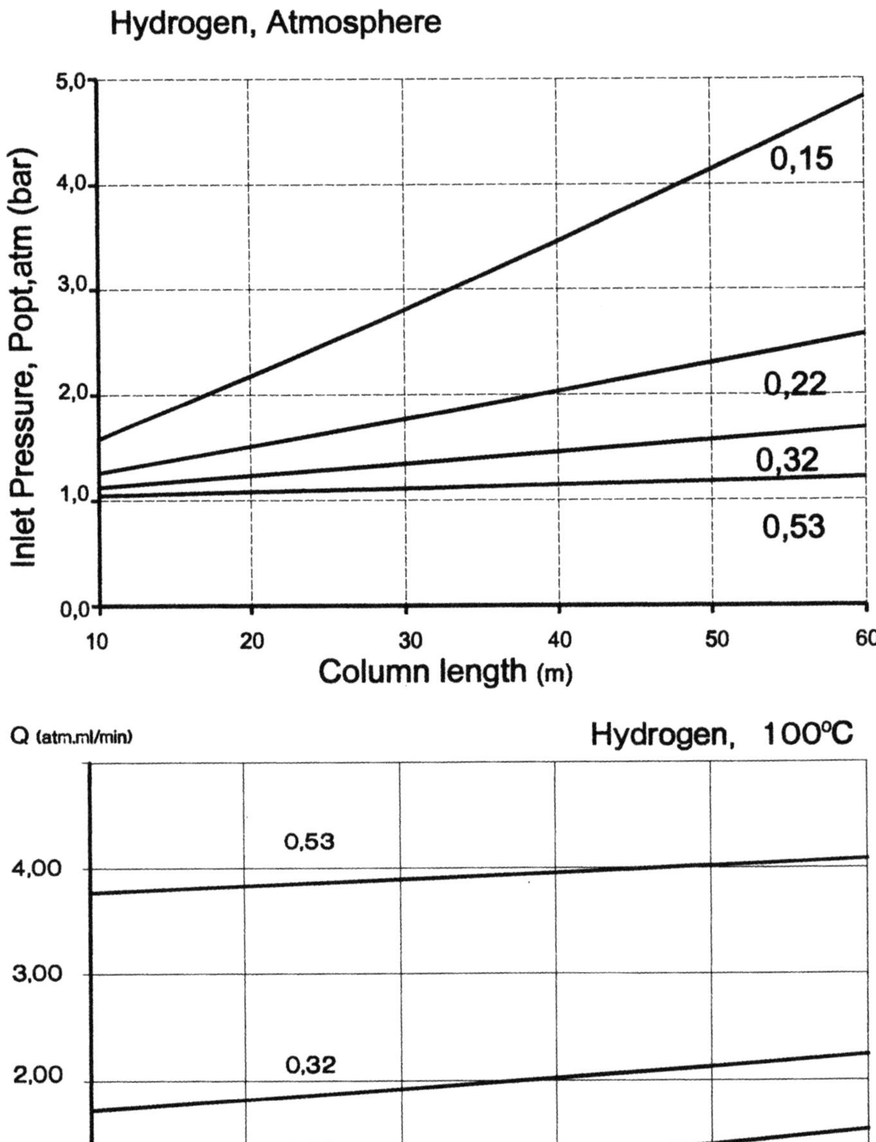

Figure 3. Optimum gas flow rate and column inlet pressure for the column dimensions and optimum gas velocities in Table 1, for hydrogen at 100°C.

At the moment, most GC/MS instruments are still fitted with a conventional gas chromatograph equipped with a mechanical head pressure regulator (see sect. 5 for a short discussion on modern GC with computerised controlled pneumatic systems). Thus the operator must manually set the GC inlet pressure P_i. The mass flow rate of gas at the GC column outlet, Q_{GC}, results from this pressure setting. It may or may not match the acceptable mass flow rate of gas acceptable by the mass spectrometer Q_{MS}, which is not adjustable and is specific to a particular GC/MS instrument.

Equations 2 and 10 can be solved for the unknowns P_i and Q, as a function of the column length L, choosing the column diameter d_c and optimal average gas velocities $\bar{u}$ listed in Table 1, a given column temperature, for example 100°C for the computation of the gas viscosity, and taking P_o=1 atm. The results can be represented graphically and are shown for the cases of helium (Fig. 2) or hydrogen (Fig.3) at 100°C. The optimum absolute inlet pressure when the column outlet is at atmospheric pressure is $P_{optm,atm}$.

The choice of a particular column length and inner diameter depends on the type of analytical problem. Long and narrow columns with inner diameter less than 0.25 mm have excellent separating properties and are well suited to the analysis of complex mixtures. On the other hand, sample injection is difficult, and the column is rapidly overloaded. Larger columns are often preferable for quantitative analyses or for mixtures containing samples over a wide range of individual concentrations. Short columns can provide fast separations for quantitative measurements. At any rate, the analyst should be able to select the right column, depending on the separation to be accomplished. However, the larger columns require high optimum gas flow rates that may be too large for simple GC/MS instruments, even for those that are presented as a general GC detector.

Frequently, when Q_{GC} becomes too large and exceeds the tolerance of the mass spectrometer, the analyst tends to reduce the inlet pressure P_i. The consequence is an increase of the solute retention times, but much more detrimental to the analysis, a loss of separating power. A better approach is to reinvestigate the GC/MS connection, in order to always maintain optimum chromatographic performance.

3. Main GC/MS configurations

There are roughly three basic types of GC/MS instruments. They mainly differ by the performance of the vacuum equipment attached to the mass spectrometer.

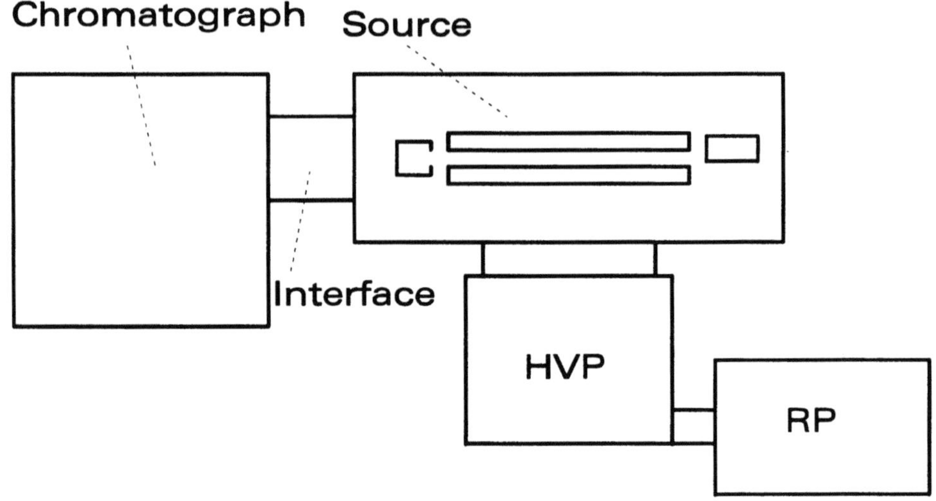

Figure 4 . Simple quadrupole GC/MS system with one main line of vacuum pumping.

Type 1 instruments are the most simple machines. They have only one line of vacuum pumping (Fig. 4). The high vacuum pump, HVP, is sometime an oil diffusion pump, but more and more machines are fitted with a turbomolecular pump. The high vacuum pump is backed by a conventional rotary pump, RP. This simple arrangement does not permit the introduction of more than 2 atm.ml/min of gas, and in practice, it is safer to limit the gas load to half this value, as both the sensitivity and the mass resolution of the mass spectrometer tend to degrade at higher gas loads.

The more costly type 2 GC/MS systems are also designed for mass spectrometry under chemical ionisation mode. They have two separate lines of vacuum pumping (Fig. 5), and accept the direct introduction of up to 10 atm.ml/min of mobile phase. The high vacuum pumps can also be oil diffusion pumps or turbomolecular pumps. A single large roughing pump is sometime used instead of two separated pumps RP1 and RP2.

Figures 4 and 5 represent GC/MS systems fitted with linear quadrupole analysers, but the discussion is comparable when the mass analyser is a magnetic sector, and the tolerable gas flow rates for singly pumped or differentially pumped machines are of the same order of magnitude.

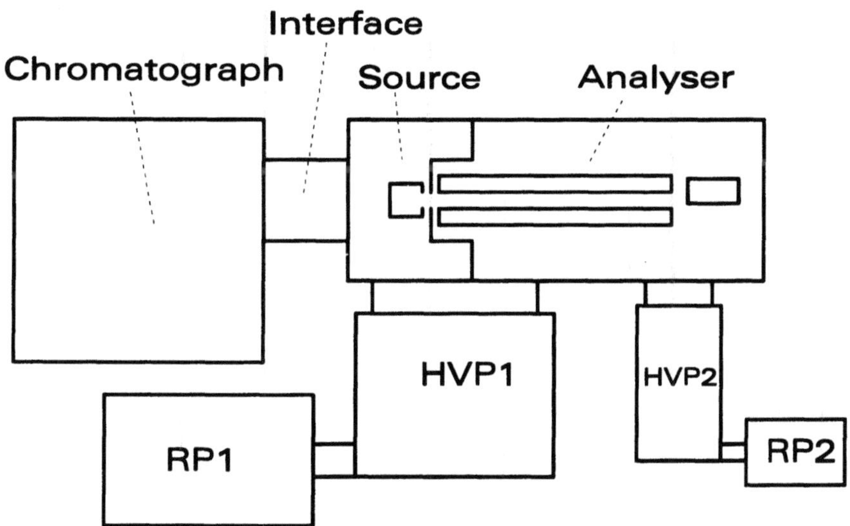

Figure 5. Quadrupole GC/MS system with differential vacuum pumping.

The third type of common GC/MS systems (Fig. 6) is fitted with an ion trap analyser. It can be made very compact in size. The maximum tolerable gas load is also less than 2 atm.ml/min

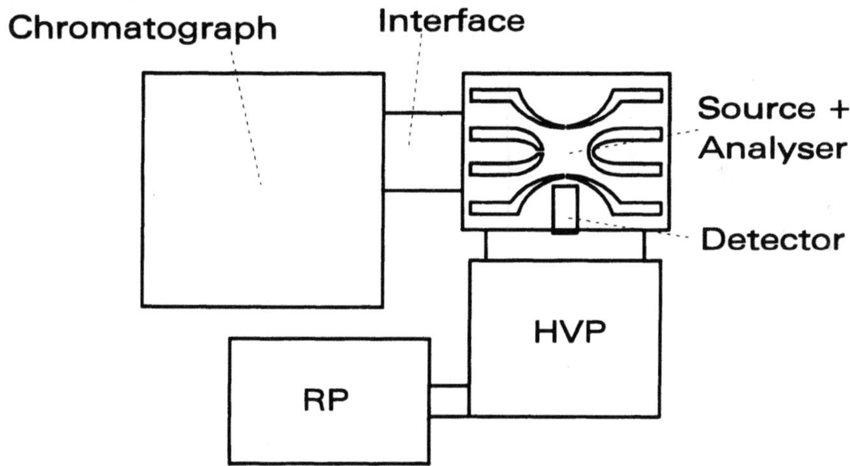

Figure 6. GC/MS system with an ion trap analyser.

4. Methods for connecting the GC column to the mass spectrometer

4.1 DIRECT COUPLING WITH TOTAL SAMPLE TRANSFER.

This is the most straightforward method (Fig. 7), and is possible with any type of GC/MS instrument, as long as it may accept the direct introduction of up to 2 atm.ml/min. This is true for GC columns with d_c less than 0.25 mm, and lengths less than 40 m (Fig. 2 and 3). Larger columns are only possible to use with type 2 instruments.

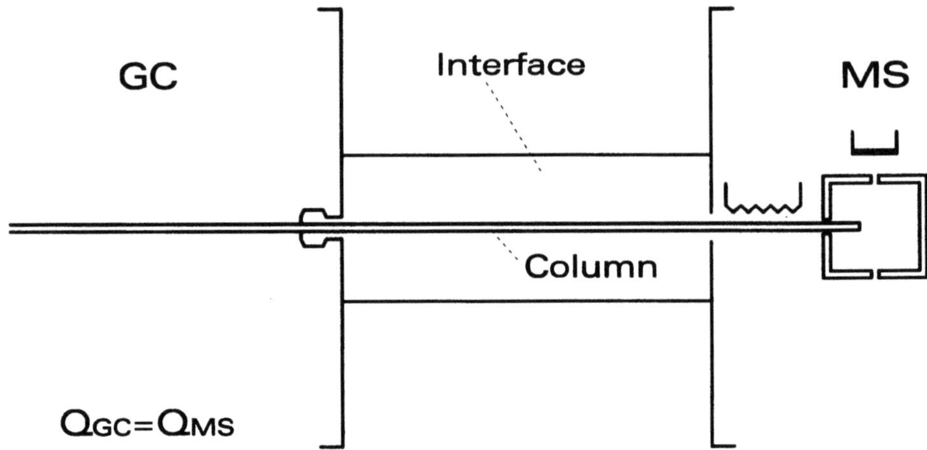

Figure 7. Direct coupling of the capillary column to the MS ion source

It is important that the capillary column be uniformly heated along the last centimetres, with no cold spots. The connection of the GC column to the interface must be leak free. On the other hand, tight connection to the MS ion source block is not essential : the column end should not be in direct line of sight with the ionising electron beam, but generally a few millimetres backward. The exact position is a function of the particular instrument being used.

Operating a GC column with vacuum at the outlet has no great detrimental effect on the column performance, as long as the average linear velocity $\bar{u}$ is increased (Fig. 8). The plate loss should not exceed 7% in a well-designed GC/MS system, which is hardly visible on a chromatogram. Thus similar resolution can be obtained in a GC or in a directly coupled GC/MS system, but with faster analysis times with the later instrument. Note that, faster retention times can be considered as a curse or a blessing : fast analysis allows the determination of a larger number of samples, specially when only a few mass peaks are monitored. On the other hand short retention times also mean narrow chromatographic peaks. Under some circumstances, quadrupole analysers are too slow to record 5-10 full scans across the resulting narrow GC peaks, and faster MS recording conditions must be investigated. The use of a time of flight instrument [7] or of a supersonic molecular beam analyser [8] has been recommended.

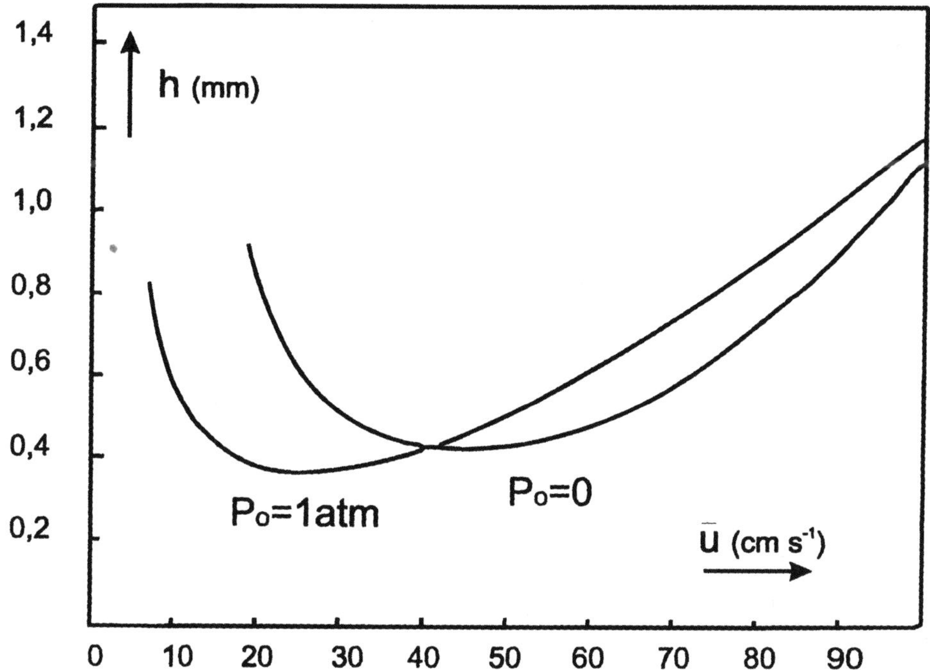

Figure 8. Experimental Van Deemter curves for the same column successively connected to atmospheric pressure and vacuum. 34 m x 0.4 mm i.d. column, coated with 1 μ of methylsilicone phase, for n-dodecane at 150°C (k'=2). Reproduced with permission from Ref. 3.

The new exact value of $\bar{u}$ with vacuum outlet operation does not need to be computed exactly. Cramers et al. [3, 4] have shown that optimum conditions for either atmospheric pressure outlet or vacuum outlet correspond to the same mass flow rate Q of eluting mobile phase. Thus the new inlet pressure $P_{opt,vac}$ can be obtained from Eq. 2, taking $P_o = 0$. Although the corresponding average linear velocity is now higher, by a factor of 1.2 to 3, depending on the column dimensions, the new inlet pressure $P_{opt,vac}$ is generally inferior to that under atmospheric pressure conditions, $P_{opt,atm}$ (Fig. 8). If the separation was previously optimised in a stand-alone GC, the same efficient separation by GC/MS using the same column requires a lower inlet pressure, although the solutes retention times are diminished.

The reduction of the inlet pressure may be such that it becomes below 1 atm, that would mean sub-ambient pressure conditions. This must be avoided, as air would enter the GC column, when using conventional syringe injection through a septum injector. Air leaks are detrimental to the column stationary phase and to several components of the mass spectrometer. Besides, the inlet pressure should be at least of 1.5 bar (absolute) when injecting dilute solutions by the so-called « split/splitless injection technique ». Thus a 0.32 mm i.d. column should be at least 40 to 60 m long, depending on the nature of the carrier gas, and 0.53 mm i.d. columns cannot be directly connected.

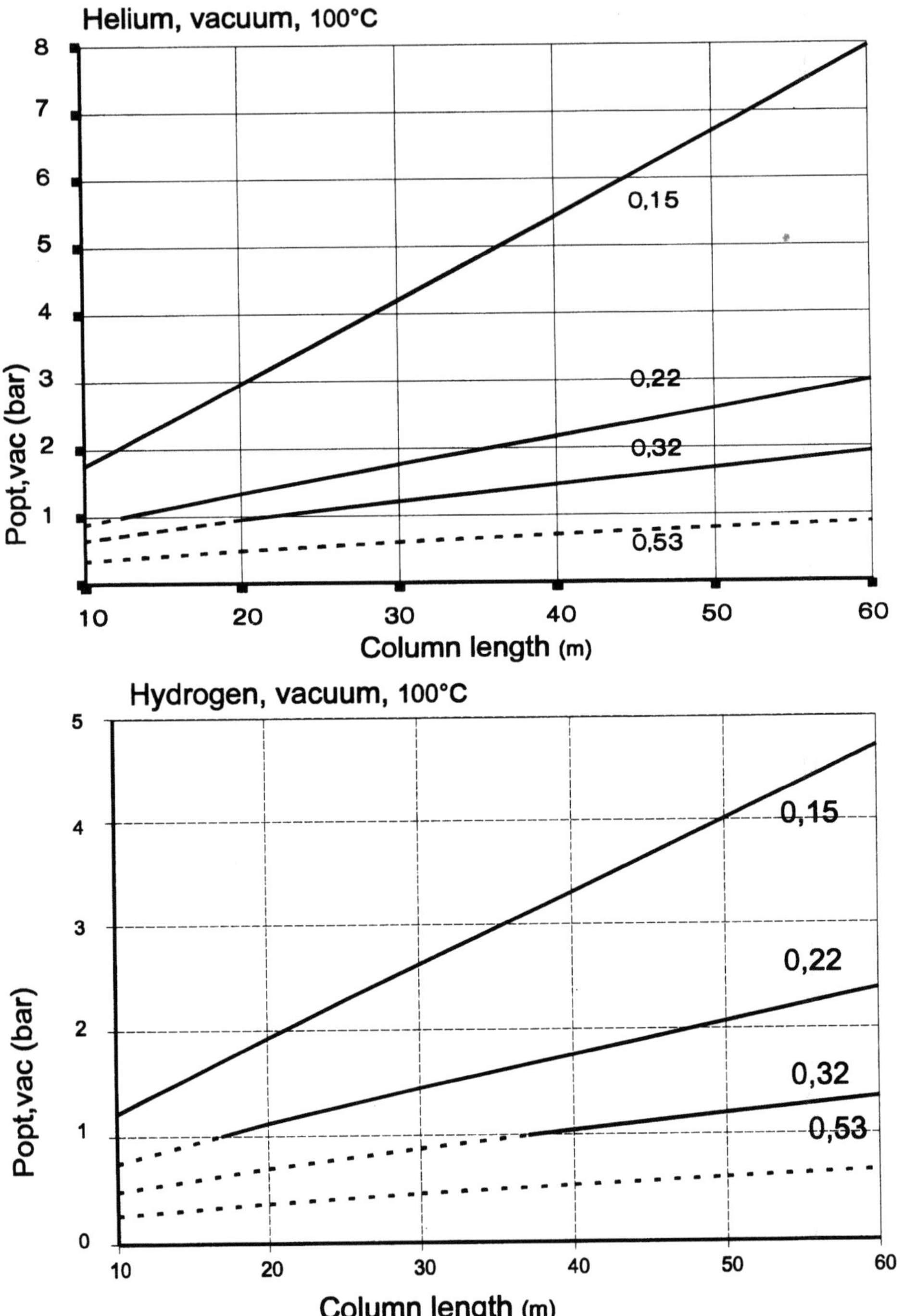

Figure 9. Injector inlet pressures corresponding to optimum GC conditions, for helium or hydrogen, and for the column dimensions listed in Table 1.The dashed-lines correspond to unrealistic sub-ambient conditions.

When chromatographic conditions require to use 0.32 or 0.53 mm columns, direct coupling can still be achieved by inserting 15-30 cm of a narrow capillary tube (ca. 0.1 mm i.d.) between the GC column end and the MS ion source (Fig. 10). The added pressure drop across the capillary tube increases the injector inlet pressure. Dimensions of the capillary restrictor can be such that the GC column end is at near atmospheric pressure, thus the GC retention times are the same as in a normal GC. Column exchange is also facilitated as the connection is made in the GC part of the GC/MS system.

On the other hand, the advantage of increased speed of analysis is lost, and the capillary restrictor should be changed periodically to avoid plugging or loss of polar samples because of active sites. Such arrangement does not allow the direct connection of a 0.53 mm column using a type 1 or type 3 GC/MS instrument. A suitable capillary could be found that would increase the inlet pressure above the limit of 1.5 bars. Nevertheless, the optimum Q values are still above 5 atm.ml/min, thus if the flow rate is reduced to ca. 2 atm.ml/min, the linear velocity is very low and corresponds to very inefficient chromatographic conditions.

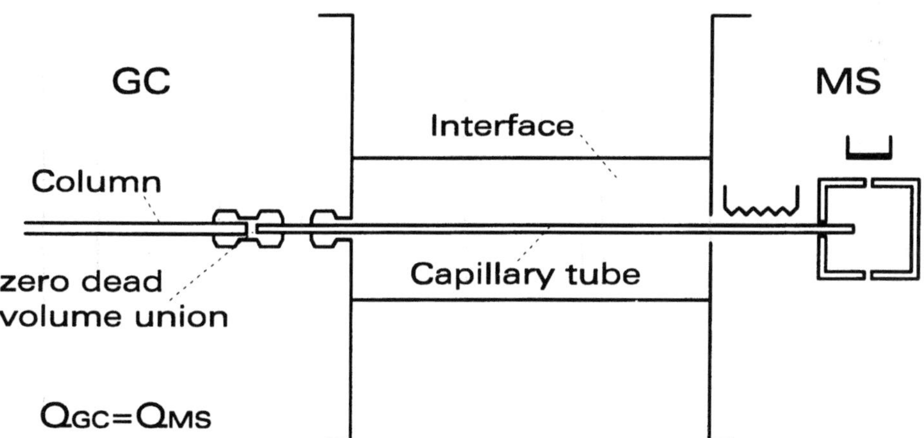

Figure 10. Direct connection to the MS spectrometer via a fixed restriction of much smaller inner diameter than the GC column. Column diameter is typically 0.25-0.32 mm i.d. and restriction diameter of 0.1-0.15 mm i.d.

4.2 SPLITTED GC/MS COUPLING.

4.2.1 Conventional coupling

Splitting the effluent at the outlet of the GC column (Fig. 11) is done easily using a mini-T-connector and a capillary restrictor. The system leaves the column end at atmospheric pressure, thus GC retention times under GC or GC/MS are identical, and the head pressure P_i is the same under both conditions. The dimension of the restrictor is such that, by applying Eq.2, the mass flow rate Q does not exceed the tolerable input flow to the mass spectrometer.

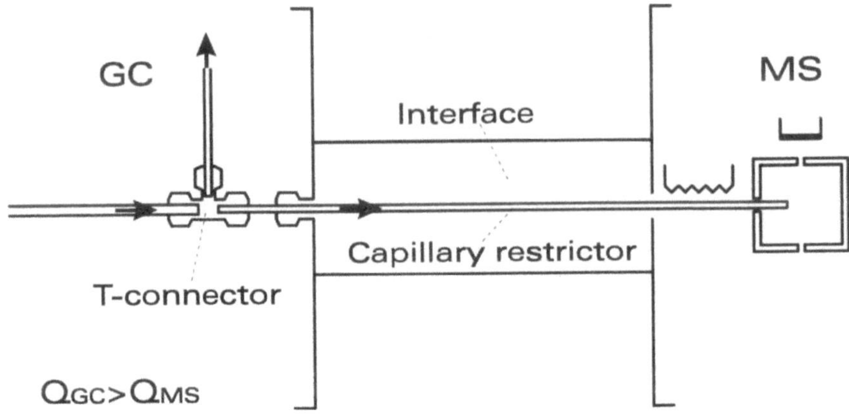

Figure 11. Splitting of the GC effluent using a fixed restriction and a T connector.

There are very few situations where such set-up is advantageous. Parallel recording using a mass spectrometer and another detector may sometime be justified, but not when the second detector is a flame ionisation detector. Connections to sample traps or FT/IR cells have sometime been used. When using capillary columns under column temperature and constant injector pressure conditions, the split ratio changes continuously as the flow rate of sample entering the mass spectrometer is constant, whereas the flow rate delivered by the GC column decreases regularly with increasing column temperature. The only realistic situation is for the connection of 0.53 mm i.d. columns - the so-called « macrobore, megabore, ultrabore ... capillary columns » -, to type 1 or type 3 GC/MS instruments. These large bore GC columns require Q values of 5-8 atm.ml/min, but they may be operated under constant flow rate conditions. The restrictor in the interface can be of such dimension that 1-2 atm.ml/min enter the mass spectrometer, and the rest is vented. This results in a sample loss for the mass spectrometer of more than 50%. The situation is rather strange as large bore columns have fewer plates than narrow capillaries, but they are recommended for their capacity to accept higher sampling loads. Thus what is gained at the column entrance is lost at the outlet.

A modification of the set-up is to add an isolation valve at the end of the venting line, and to connect it to a vacuum pump. Normally, the valve is closed, and the system behaves as in Fig.10. When the valve is open, most of the GC flow rate is diverted. This can be useful to avoid large solvent peaks to enter the mass spectrometer : it prevents contamination of the ion source block and damage to the electron emitting filament.

4.2.2 *Open-split coupling.*

This simple device took its name from the work of Henneberg et al. [9, 10], but a somewhat similar device was already used by Grob and Jaeggi [11]. The connection between the GC column outlet and the capillary restrictor (Fig. 12) is at atmospheric pressure and continuously flushed by a stream of helium. The dimension of the restrictor is such that a fixed mass flow rate of gas enters the mass spectrometer (typically 1 atm.ml/min). For a 0.15 mm i.d. tube, the necessary restrictor length is 1.2 m (Eq. 2). When the GC column flow is higher than this value, the difference is lost ; when it is inferior, it is diluted by helium up to 1 atm.ml/min.

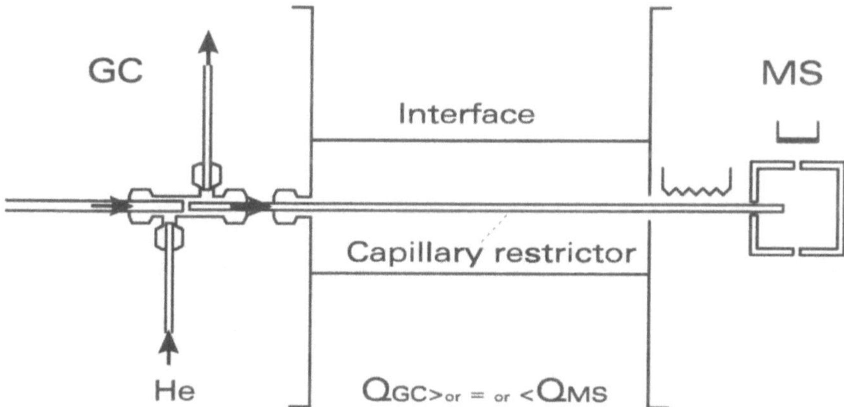

Figure 12. Open-split coupling

For some time, the system was frequently used with ion trap instruments (type 3 GC/MS). An ion trap exhibits better sensitivity and mass resolution at a relatively high ion source pressure (ca. 100 Pa), unlike a quadrupole mass spectrometer that favours lower ion source pressures (ca. 10 Pa or less). On the other hand, the gas pressure within the ion trap should remain constant. The open-split interface achieves this purpose as it always sample the same flow of mobile phase, independently of the GC conditions.

Despite this advantage, the system has been progressively abandoned because of the anaesthetic length of the transfer line. When using narrow GC columns with $d_c \leq$ 0.25 mm, the addition of a constant make-up of helium of ca. 1.5 ml/min, at the column outlet using a set-up similar to Fig. 11, may also provide a roughly constant Q_{MS} flow. The development of new capillary GC that can be operated under constant flow rate conditions (see below) is another reason to avoid the use of an open split interface with ion trap analysers.

5. Computerised pneumatic control.

In most GC/MS systems, the GC inlet pressure P_i is maintained constant. When the column temperature is programmed, as very frequently done, the carrier gas viscosity is changed (Eq. 11), resulting in a change in column flow Q_{GC} and linear velocity. With a viscous gas such as helium, the effect can be severe, and for a temperature increase from 100°C to 300°C, the flow and the velocity can be reduced by a factor of 1.5. This effect is often disregarded by analysts using GC/MS equipped with a linear quadrupole with direct vacuum tight column coupling, as the performance of the mass spectrometer improves when the flow of gas is reduced (the situation is the reverse with an ion trap analyser). The loss of chromatographic resolution as a consequence of the reduced linear velocity is too frequently neglected. Some advised operators initially adjust the pneumatic conditions with the column temperature set at mid-interval, around 200°C.

Modern GC systems fitted with computer-controlled pneumatic systems can eliminate these changes in gas flow and carrier gas velocity [12]. The heart of the system is a variable pressure regulator whose set point is controlled by a micro-processor. The operator only needs to enter the column dimension, the nature of the carrier gas, and the desired mobile phase flow rate or the carrier gas velocity, into the system parameters, then the pressure regulator is automatically adjusted to maintain the flow rate constant. A mathematical model of the viscosity change with temperature is also included so that the system also adjusts to temperature column variations.

A detailed discussion is beyond the topic of this paper, but it can be reasonably predicted that these systems will equip most GC/MS systems in the future. At the moment, computer controlled pressure regulators are still expensive. As the competition between GC/MS manufacturers is severe, price is often a strong marketing argument, thus costly accessories are often removed from the basic system configuration.

The availability of computer-controlled pneumatic system does not eliminate the need to understand the importance of the mobile phase flow rate on the GC performance. The system is still unable to choose the right flow as a function of the chromatographic resolution. Selection of the right column and operating parameters will still be a function of the separation to be obtained, and not of the GC/MS instrument to be used.

6. Conclusion

As the cost of mass spectrometers decreases and their reliability increases, an ever greater number of mass spectrometers will be used as gas chromatographic detectors in the future. The chromatographic column remains the main separation device of the system. Column selection and correct setting of the various chromatographic parameters are among the few variables that are still left under direct control of the GC/MS operator, thus it is still important to understand their significance and to be able to tune them correctly.

7. Abbreviations

α	Stationary phase selectivity
η	Dynamic gas viscosity
ω_h	Chromatographic peak width at half-height
d_c	Capillary column inner diameter
e_f	Stationary phase film thickness
F	Volumetric flow rate of gas
h	Height equivalent to a theoretical plate
HVP	High vacuum pump
j	Compression factor of James and Martin
k'	Retention factor
L	Column length
N	Number of theoretical plates
P	Absolute gas pressure (P_i, P_o, P_{atm})
Q	Mass flow rate of gas (Q_{GC}, Q_{MS})
RP	Roughing pump
t_r	Absolute chromatographic retention time
T	Absolute GC column temperature
t_c	Column temperature (°C)
u	Local linear gas velocity (u_i, u_o, u_{atm})
$\bar{u}$	Average gas linear velocity

8. References

1. Eiceman, G.A., Hill, H.H., Davani, B. and Torresday, J.G. (1996) *Anal. Chem.* **68**, 291R.
2. Nau, H. and Biemann, K. (1974) *Anal. Chem.* **46**, 426.
3. Cramers, C.A., Scherpenzeel, G.J. and Leclercq, P.A. (1981) *J. Chromatogr.* **203**, 207.
4. Leclercq, P.A., Scherpenzeel, G.J., Vermeer, E.A.A. and Cramers, C.A. (1982) *J. Chromatogr.* **241**, 61.
5. Davies, N.W. (1985) *J. Chromatogr.* **325**, 23.
6. Lide, D.R.(1992) *Handbook of Chemistry and Physics, 73rd Edition.*
7. Wollnik, H., Becker, R., Götz, H., Kroft, A., Jung, H., Chen, C. C, Van Ysacker, P. G., Janssen, H. G., Snijders, H. M. J., Leclercq, P. A. and Cramers, C. A. (1994) *Int. J. Mass Spectrom. Ion Processes*, **130**, L7.
8. Dagan, S. and Amirav, A., (1996) *J. Am. Soc. Mass Spectrom.*,**7**, 737.
9. Henneberg, D., Henrichs, U. and Schomburg, G. (1975) *J. Chromatogr.*, **112**, 343.
10. Henneberg, D., Heinrichs U and Schomburg, G. (1975) *Chromatographia.*, **8**, 449.
11. Grob, K. and Jaeggi. J. (1973) *Anal. Chem.* **45**, 1788..
12. Hinshaw, J.V. (1995) *LC-GC Int.*, **8**, 631.

QUANTITATIVE ANALYSIS BY MASS SPECTROMETRY: SOME IMPORTANT CONSIDERATIONS

M. W. DUNCAN
Biomedical Mass Spectrometry Unit
University of New South Wales
Sydney, Australia

1. Outline

Quantitative analysis represents one of the major growth areas in mass spectrometry and all indicators are that this trend will continue unabated for some time. Despite increased interest and activity in this area, there remains a paucity of published information detailing the key aspects of quantitative mass spectrometry. This chapter outlines some of the basic principles, but space limitations preclude a detailed coverage; instead, several of the issues specific to biological mass spectrometry are covered and references to relevant literature are provided.

In the context of this chapter, the term 'mass spectrometry' is used to refer to analytical methods based on ion generation, separation and detection. Such methods invariably incorporate either gas chromatography (GC) or high performance liquid chromatography (LC) as on-line separation techniques, but chromatographic methods are not discussed in any detail here.

2. Defining the Analytical Task

Qualitative analysis aims to provide a definitive answer to a question: what is this compound? There is no grey-zone, either the structure is defined, or it is not. Quantitative analysis aims to provide the best *estimate* of the concentration of a substance (analyte), often when it is contained in a complex biological matrix. For example, a detailed and labor-intensive study might be necessary to generate

R. M. Caprioli et al. (eds.), Selected Topics and Mass Spectrometry in the Biomolecular Sciences, 103–119.

quantitative data with high accuracy and precision for a clinical trial, or alternatively, a semi-quantitative determination might be all that is required to give a "ballpark" approximation of the concentration of an analyte in a drug overdose case. In other words, it is necessary to tailor the approach to the question, and the objective should always be to provide the most *appropriate* answer for the situation at hand. High accuracy, precision and sensitivity are achievable, but they come at a price and are not necessarily universal requirements.*

Some representative bioanalytical tasks are shown in Table 1, and the major considerations in each instance are indicated. For example, accuracy and precision are not major concerns when testing for a drug overdose. Here, to identify the class of drug, and to estimate the dose is all that is necessary to instigate appropriate medical treatment. (Whether the patient has overdosed on 10 or 20 times the prescribed dose is rarely of clinical significance.) In this situation the emphasis is necessarily on providing an answer as quickly as possible. By contrast, regulatory authorities, such as those responsible for drug registration, will insist on the highest standards of accuracy and precision and these objectives must be obtained at the expense of considerations such as cost and time.

Table 1

Analytical Requirements for a Variety of Selected Tasks

Purpose of Analysis	Selectivity	Sensitivity	Precision	Accuracy	Speed
Forensic Toxicology	low-medium	low	low	low	low
Overdose	medium	low	low	low	high
Drug Monitoring	medium	medium	medium	medium	high
Pharmaco-kinetics	high	high	high	high	medium
Phase II Clinical Trial	high	high	high	high	high

This table is based on a similar version appearing in: Chamberlain, Analysis of Drugs in Biological Fluids, CRC Press, Boca Raton (1985).

* An appendix provides basic definitions for some of the terms used in this chapter.

3. When to Use Mass Spectrometry for Quantitative Analysis

The great strengths of the approach are the high precision and accuracy that is afforded by selective (mass) detection and the application of stable isotope labeled internal standards (isotopomers). Additionally, mass spectrometry offers a wide dynamic range and a good limit-of-quantification. Where precise and accurate determinations of trace components in a complex sample matrix are critical, this is likely to be the best analytical approach. One of the most important applications of mass spectrometry has been its application as a reference, or definitive approach to quantitative analysis.[1] Formerly, quantification by mass spectrometry was considered complex and too expensive for most applications, and it therefore became the last analytical resort, only employed when no other option was appropriate. However, the advent of affordable benchtop systems means that the equipment is now commonplace and quantification based on mass spectrometry may be considered a viable option sooner, rather than later.

However, there are certainly instances where mass spectrometry may not be appropriate. Establishing and validating a mass spectrometric method can be complex and time-consuming, so sample number and time-cost benefits must enter into consideration. The effort is only justified when sample throughput is likely to be substantial. Mass spectrometry is not an appropriate option when other (simpler) methods exist that provide adequate analytical precision and accuracy. For example, a mass spectrometric method could be developed to monitor blood glucose levels in diabetic patients, but the effort would not be justified; clinically relevant data are readily available using a low-cost hand-held glucose monitor.

4. The Principles of Quantitative Analysis

Regardless of the separation and detection devices used, quantitative analysis employs essentially the same approach. To a sample(s) containing the analyte of interest is added a fixed amount of a suitable internal standard. Following extraction and on-line separation steps, signals characteristic of both the analyte and the internal standard are monitored. The ratio of the peak intensities for the analyte and the internal standard is determined, and this value is converted to a quantitative measure by reference to a calibration curve. The calibration curve is generated from a series of samples prepared and analyzed at the same time as the test samples, and containing the internal standard (at the same fixed amount), and various dilutions (of known concentration) of a reference sample of the analyte of interest. The steps in a typical quantitative analysis are represented in diagrammatic form in figure 1.

In the sections that follow, some of the steps are discussed in detail. Special emphasis is given to the issues that specifically relate to mass spectrometric detection.

Figure 1
Steps in a Typical Quantitative Analysis

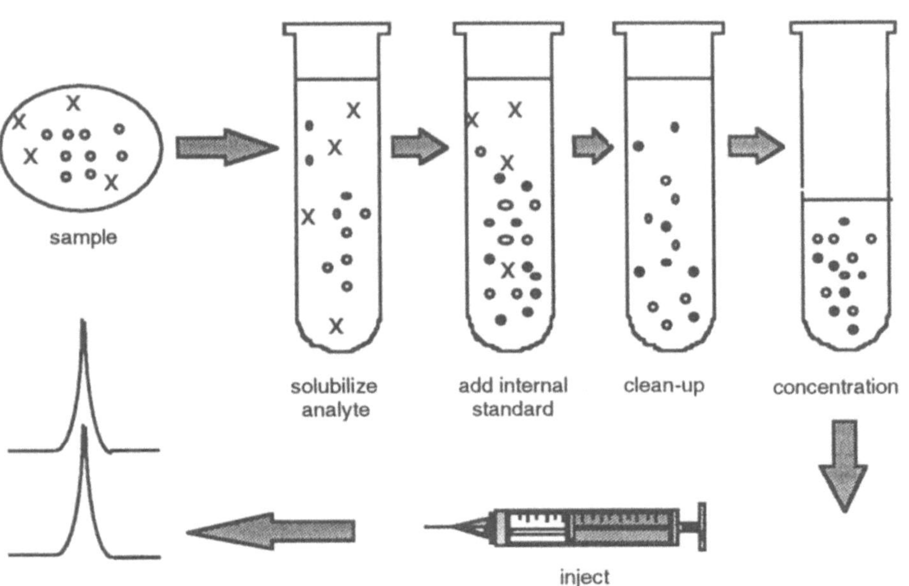

acquire data, calculate values from calibration curve and report results

Where: O = analyte; ● = internal standard and X = other sample constituents.

Although the steps will vary depending on the analyte and the matrix, this scheme is typical of that used for the majority of quantitative determinations based on mass spectrometry. Internal standards are routinely used, and a signal is monitored for the analyte and for the internal standard. Absolute concentration is determined by reference to a calibration curve prepared at the time of the analysis.

5. The Analyte

In large part, the analyte dictates which mass spectrometric method is most suitable for a given quantitative task. Properties such as molecular weight, boiling point, functional groups, stability and fragmentation characteristics will be major factors in determining the best analytical option. In this section it is assumed that on-line chromatography - either GC or LC - is a necessary requirement for the analysis, and with the maturation of electrospray ionization (ESI) and atmospheric pressure chemical ionization (APCI) there are now several matched separation/ionization options appropriate for a wide range of compounds. In this section some of the possible scenarios for different analyte types are examined.

5.1 ANALYSIS OF VOLATILE ANALYTES

Most volatile analytes will be amenable to an approach based on combined GC-MS analysis. Typically, these analytes will have a MW of less than *ca.* 500 Da. (i.e., within the mass range of low-cost commercial instruments), and in the case where the electron ionization (EI) spectrum is characterized by intense, diagnostic, high mass ions, then EI is a practical option. If EI delivers a spectrum dominated by non-specific low mass ions (e.g., *m/z* 29, 43, 57, 71 ...), then chemical ionization (CI) mass spectrometry may yield more diagnostic high mass ions and should therefore be considered. While direct GC-MS analysis is an attractive option, it is worth noting that it is estimated that only about 20% of known organic compounds can be examined by GC without chemical modification.[2,3]

5.2 NONVOLATILE, OR THERMALLY LABILE, LOW MASS ANALYTES

Two separate approaches can be adopted here: GC-MS (following chemical derivatization), or LC-MS (either APCI or ESI). Often only GC-MS is available, so it is important to consider the issues related to this approach.

Intractable low mass analytes owe their high boiling points to strong intermolecular interactions between polar functional groups; however, these functionalities can be chemically modified (or derivatized) to impart volatility to the molecule. For example, the replacement of polar hydrogen atoms associated with -NH, -OH & -SH groups (e.g., by alkylation, acylation or silylation) dramatically reduces hydrogen-bonding and significantly decreases the boiling point. Modifications based on the formation of silyl and perfluoro derivatives are particularly advantageous in this regard.[4]

The aim is to prepare a single, stable product of the analyte that is more suited to the subsequent analysis. Chemical derivatization is widely adopted, and there is an extensive literature - even an industry - associated with reagents and approaches to chemical modification.[4-6] Derivatization can also be used to stabilize a labile compound, improve chromatographic performance, increase the MW of low mass analytes, modify mass spectral properties or even, in some instances, to reduce the volatility of low boiling point compounds.

While both silyl and perfluoro derivatives are frequently employed, quantitative mass spectrometry is generally simplified if fluorinated derivatives are used.[7] Fluorine has no naturally occurring isotopes, so the integrity of the isotope pattern associated with major ions is maintained. By contrast, there is a significant complication when silyl derivatives are used: silicon is an A+2 element,[8] and the incorporation of one or more silicon atoms complicates the isotope pattern. Of particular concern is the potential for spill-over to occur between analyte and internal standard ion channels when isotopomers are combined with silyl derivatives. Perfluorinated derivatives do not contribute to isotopic spillover and can also offer the additional advantage that they facilitate electron-capture ionization, an approach capable of affording a significant enhancement in the limit-of-detection.[7]

Although chemical derivatization is undoubtedly a valuable tool, it is not a panacea. If the low volatility of the analyte is due to its sheer mass, then the product of the derivatization process, if not the parent compound, may well have a mass outside the range of the instrument. (This is not a consideration when GC is used with something other than a mass spectrometer as the detector.) Further, derivatization may yield a product with a mass spectrum that is inappropriate for mass spectral analysis, dominated by derivative-specific ions and devoid of useful analyte-specific (diagnostic) ions. If this is the case with EI, CI may provide higher mass, analyte-specific ions.

Until relatively recently, and the advent of practical LC-MS options, chemical derivatization followed by GC-MS was the only option for quantitative mass spectrometry of many compounds, and as a consequence, there is an extensive literature on derivatization schemes and operating parameters for a wide range of analytes. Some of these tasks might now be better addressed by LC-MS methods. (See below.)

5.3 COMPOUNDS NOT AMENABLE TO GC INTRODUCTION: NON-VOLATILE, THERMALLY LABILE, OR HIGH MASS COMPOUNDS

Compounds with a MW of greater than *ca.* 500-600 Da. are rarely amenable to GC introduction, even following chemical derivatization, and in these instances other analytical options need to be considered. Quantitative LC-MS, based on either ESI or APCI, is now a practical option in many of these instances. Sample introduction based on LC is not bound by the limitations of GC, and additionally, the complementary techniques of APCI and ESI effectively ionize a diverse array of compounds.

Although ionization methods such as ESI and MALDI have markedly extended the available mass range for qualitative applications, quantitative mass spectrometry of high molecular weight analytes is restricted for other practical reasons. As a guide, the upper mass limit for quantification should be considered to be *ca.* 1-2,000 Da; above this value it is probably the exception, rather than the rule, that mass spectrometry offers advantages.

One limiting factor is the natural isotopic distribution of high mass ions. As the molecular weight of an analyte increases, the total ion current becomes split across multiple species. (See figure 2.) As a result, the ion current for any given *m/z* value is markedly diminished. In the example shown, the base peak (*m/z* 2136) carried only about 30% of the total ion current of the molecular ion cluster. (By contrast, the molecular ion for methane carried 99% of the ion current.) If unit resolution is maintained, then the selected ion current (*m/z* 2136) will represent only 30% of the total signal and sensitivity will be compromised; if the resolution is reduced to allow a greater proportion of the total ion current to be recorded, then the specificity of selected mass detection will be lost.

ESI may confound this situation even further because multiply charged ions are formed and further dissipate the ion current. At a mass of 1,000-2000 Da, the ion current is likely to be distributed between $[M+H]^+$, $[M+2H]^{2+}$ and $[M+3H]^{3+}$ and this will contribute to a further reduction in the specific ion current for any one of these species. While the isotopic distribution about a given mass is predictable, the distribution of the charge across multiply charged species is dependent on many factors, including the sample matrix. The total ion current and the current for a selected ion may therefore vary significantly and across a range of samples, independent of analyte concentration, and this makes quantification impossible unless an appropriate internal standard is employed. The effect becomes increasingly important as the molecular weight (and charge distribution) increases, so selection of the internal standard becomes critical, as will be discussed later.

Figure 2

Natural Isotopic Distribution for a Peptide of Mass ca. 2,000 Daltons

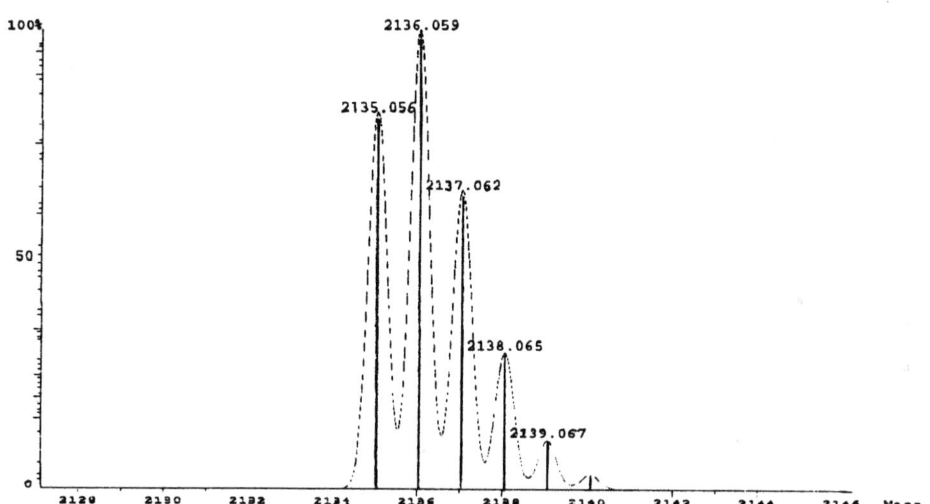

The theoretical isotope distribution for the a peptide of monoisotopic mass 2135.05 is shown above. (Sequence: WEELHDAKYGFLSAAELI, $C_{98}H_{144}N_{25}O_{29}$). Note that at a resolution of 2,000, (dotted line), the total ion current for the molecular ion is distributed over 6 separate species.

Some recent work has indicated that matrix assisted laser desorption ionization (MALDI) combined with time-of-flight (tof) mass spectrometry holds promise as an approach to biopolymer quantification,[9,10] but as yet there are few practical applications. MALDI offers the advantage that predominantly singly charged ions are formed. Another significant advantage of MALDI is its remarkably sensitivity. However, MALDI is a solid state technique that does not lend itself to efficient, direct LC coupling. Elaborate off-line clean-up is therefore required, and the selection of an appropriate internal standard still remains an issue.

In summary, above a mass of 1-2,000 Da, other analytical options must be given careful consideration. Specific mass detection may seem an attractive option, but immunometric techniques may prove more practical and offer high sensitivity with acceptable accuracy and precision.

6. The Internal Standard

The internal standard acts as a mimic of the analyte of interest and is added early in the sample-handling process to adjust for losses throughout the analysis (figure 1). It serves to improve both accuracy and precision. The ideal internal standard is one that has physical and chemical properties as close as possible to the analyte of interest. Suitable mimics (or internal standards) come in several forms and include: structural analogues, homologues, isomers and isotopomers.

The great strength of mass spectrometric detection is that it allows the use of isotopomers as internal standards. Stable isotope labeled internal standards must be of high isotopic purity, but their chemical purity is less of a concern. When selecting an isotopomer standard it is vital to ensure that the labeled atoms do not exchange at any point during the analysis, and that the ions monitored incorporate the labeled atoms. Ideally, several labeled atoms (> 2 if possible) should be incorporated into the standard to avoid "natural" isotopic spill-over between analyte ions and the corresponding ions of the internal standard. Although deuterium labeled standards are likely to be the easiest to acquire, by synthesis, exchange or from commercial sources, there is a significant isotope effect that may serve to outweigh some of the advantages associated with a stable labeled standard. Notably, when multiple deuterium atoms are incorporated into a molecule a difference between the retention time for the labeled and unlabeled material may result. (With the deuterium labeled form eluting first.) This phenomenon is not restricted to GC separations; for example, several separate deuterium-labeled forms of 3,4-dihydroxyphenylalanine have been separated by LC, with baseline resolution obtained between the native and [$^2H_7$]-labeled forms.[11] Differences in retention times can be significant because this means that the analyte and standard are not present in the ion source at the same time, and therefore the isotopomer does not adequately compensate for ion source events (e.g., fluctuations in source pressure that affect ion abundance). A much better option is to employ ^{13}C, ^{15}N, and/or ^{18}O labeled standards whenever possible. Standard synthetic methods can be used to incorporate these labeled atoms, or in some instances labeled isotopomers are available from commercial sources. In this context, an elegant exchange-labeling technique for the incorporation of ^{18}O has also been reported.[12,13]

In summary, when employing a mass spectrometer as the detector it is a significant compromise to use anything other than an isotopomer as the internal standard. In fact, if a suitable isotopically labeled standard is not available, or cannot be prepared, it might be appropriate to consider the merits of alternative analytical techniques.

7. The Sample Handling, Clean-up, Concentration and Injection Steps

The objective - to derive signals specific for the analyte and internal standard - is usually achieved by both incorporating clean-up steps, and by optimizing the selectivity of the detection process: in very rare instances clean-up can be avoided altogether. For example, a vast excess of inorganic salts will invariably compromise the quantification of a trace organic component in crude samples, regardless of the ionization mode. In addition, multiple components in the ion source at the same time will invariably suppress the ion current arising from the compound of interest, and the build-up of salts and other components will progressively degrade source performance.

In practice, for any given quantitative task, the nature and number of clean-up steps must be determined empirically. However, as a general rule, the complexity of the clean-up should be minimized to the point where the limit-of-detection and selectivity are adequate and where stable performance of the instrument can be maintained.

Although there are no special considerations that relate specifically to quantitative mass spectrometry, the importance of these steps justifies some discussion of the key concerns. The objectives of these first few steps in the analytical process are: to efficiently solubilize the analyte; to remove potentially interfering substances; and to concentrate the analyte into a practical volume of solvent. The published literature often provides valuable information relating to sample handling. If a quantitative assay for the compound of interest is not available, then work on structural analogues may be reported and relevant. For analyses that will be employed to analyze large numbers of samples, special consideration should be given to the cost and ease of automation: sample handling is one of the most labor-intensive and expensive components of the analytical process.

Where information on analyte extraction is limited, it is worth considering the synthesis of a suitable radiolabeled form to employ during the development phase of the assay. Tritiated or ^{14}C-labeled forms can prove invaluable aids when determining suitable solvent and extraction systems. The radiolabeled form of the analyte must be of high chemical purity, but specific activity is a lesser concern: it need only be adjusted so that extraction studies can be undertaken on an appropriate scale. The radiolabeled form allows many of the issues relating to sample work-up and extraction to be addressed quickly and efficiently without recourse to a mass spectrometer, and this can facilitate assay development.

While liquids (e.g., urine, cell culture medium) can be extracted directly, solid samples (e.g., tissue) must be disrupted in the presence of the solvent to allow complete dissolution. This step can be undertaken by any one of several approaches: e.g., mortar & pestle (i.e., sample grinding), blades or blenders, sonication or enzymes/chemicals. The most appropriate approach will be determined by the nature of the sample and the analyte of interest. However, throughout the process of analyte extraction, careful consideration must be given to: analyte stability (e.g., to enzymes, heat and chemicals), artifactual formation, foaming, and the consistency and efficiency of the process. Of special concern, especially with ultra-trace analysis, is the issue of cross-contamination of samples.

There is an extensive array of extraction/clean-up options and discussion of these is well beyond the scope of this chapter. Among the most commonly employed are: solvent extraction, solid phase extraction, chemical modification followed by extraction (e.g., extractive alkylation) and immunoaffinity clean-up. The more selective the clean-up step(s), the less likely it will be that interferences will compromise subsequent steps in the analytical procedure, but it is worth noting that mass spectrometry offers far greater specificity than alternative detection methods and often work-up steps can be streamlined.

8. Instrumental Considerations

8.1 CHROMATOGRAPHY

The objective with any chromatographic system coupled to a mass spectrometer is to adjust temperatures (GC) or mobile phase polarity (LC) to provide adequate chromatographic resolution, while minimizing the retention time of the analyte of interest. This minimizes peak broadening, and thereby provides the best possible signal-to-noise ratio for the analyte of interest. Additionally, this approach will ensure that the analyte and isotopomer elute at or near the same time.

Sample introduction into the mass spectrometer is also an important step. The lowest possible volume consistent with efficient transfer to the system will give the best limit-of-detection. For example, GC injection volumes are typically 1-2 μL, but final sample volumes in most instances are much greater (e.g., 20-1,000 μL). Because only a fraction of the total analyte is introduced, the practical limit of detection may be compromised significantly by this step alone. Recent developments in high volume GC injectors help to overcome this problem, but these are not universally available. By contrast, LC injection volumes are generally much higher than those possible with a standard GC injector, and therefore a greater proportion of the final sample can be introduced into the system. While the absolute limit-of-detection may be less with this approach, in practice, a better limit-of-quantification might well be achieved with a LC-MS system in some instances.

8.2 MASS SPECTROMETRY

As discussed, analytical specificity is obtained by a combination of selective extraction and selective detection. In almost all practical quantitative analyses, on-line chromatography will also be employed and help improve specificity by eliminating interferences.

As the analyte elutes from the chromatographic column into the mass spectrometer, quantitative measurements can be made by repetitive scanning across a large mass range, scanning a limited mass range, or by monitoring only selected ions (i.e., selected ion monitoring, or SIM). In the scan mode the mass spectrometer spends very little time monitoring any given *m/z* value, so the number of ions constituting a peak can be very small. Additionally, most of the ions of interest do not reach the detector because the vast majority of the acquisition time is spend monitoring masses of little or no relevance to the quantitative task. For example, in a quadrupole mass spectrometer scanning, in 1 sec, the range *m/z* 50-550 at 0.1 unit intervals, there are 5,000 separate data points in the mass spectrum. If a GC peak is of 5 secs duration, then a maximum of 5 separate data points define the GC peak, and only 1/5000 of the scan time has been employed to monitor any single analyte-specific ion. (That this to say that only 1/5,000 of all of the ions formed at any given *m/z* value are detected.)

If, however, selected ions characteristic of the analyte of interest and the internal standard alone are monitored, the sensivity is greatly enhanced. This arises because: a much higher proportion of the target ions is detected; ion statistics improve dramatically, and now many more data points define the GC peak. The magnitude of the improvement can be calculated, and is a function of several variables including the GC peak width, the mass range and scan function used in the scan mode, the number of ions and their dwell times, and the system overhead time (or switching time) in SIM. An improvement in the limit-of detection of several orders of magnitude is routine, and arises because the ion statistics are improved dramatically in the SIM mode. Hence, increased sensitivity and precision result, but at the expense of the qualitative information inherent in full-scan analysis.

Quantitative analysis based on SIM should monitor appropriate analyte-specific ions. By employing high abundance, high mass ions, the sensitivity and selectivity of the analysis is optimized. Monitoring "noisy" ions (e.g., those arising from GC column bleed), or ions characteristic only of the derivatizing reagent, can compromise specificity and/or sensitivity. Typically, one or more ions arising from the analyte and its isotopomer are monitored. If only one ion is monitored, the limit-of-detection is optimized, but specificity is reduced. In this instance the appearance of the ion at the appropriate retention time alone is considered sufficient to establish the presence of the analyte in the sample. However, retention times can shift, and additionally, co-eluting sample components can sometimes display the target (SIM) ion.

The first of these problems can be countered by using an isotopomer as the internal standard. Shifts in the retention time of the analyte will result in a similar shift in the retention time of internal standard. In the ideal situation where both standard and analyte co-elute, then modest shifts in retention time are of little consequence. Co-eluting species, however, still present a problem, and this can only be addressed by increasing the specificity of the sample work-up and/or the analysis steps.

The simplest and most common approach to dealing with this issue is to monitor two or more ions characteristic of the analyte and this affords several advantages (figure 3). Under these conditions the specificity is enhanced greatly because two parameters serve to increase analytical certainty: the retention time and the relative intensities of the analyte ions. In fact, in this instance, two independent determinations of the analyte concentration can be made and compared, and under these conditions the limit of detection will be determined by the ion of least relative abundance. Although a compromise in the limit-of-detection results, in many instances the increase in specificity is considered to be of more significance.

Figure 3
Selected Ion Monitoring with On-Line Chromatography

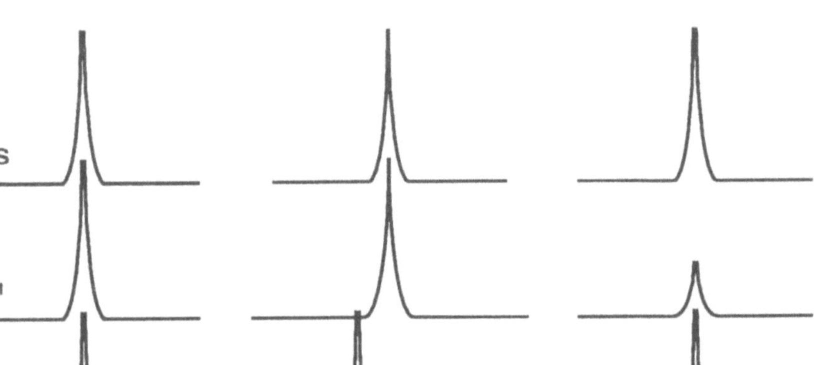

In this example, two analyte ions (A1 & A2) are monitored along with one ion derived from the internal standard (IS). (a) When both analyte ions occur at the appropriate retention time (indicated by the arrow), and where they have the correct relative intensities, quantification can be undertaken with a high degree of certainty and two independent values for the analyte can be determined. (b) When only one of the two diagnostic ions occurs at the correct retention time, the analyte is deemed to be absent from the sample. (c) If the two ions occur at the correct retention time, but with the wrong relative intensities, accurate quantification is impossible. Interferences are compromising the specificity of the analysis.

One key parameter in SIM analysis is the dwell time for each of the ions (i.e., the time spent monitoring a single *m/z* value before moving to the next selected *m/z* value). GC peaks eluting from a capillary column may only be a few seconds in duration, so it is important to select ion dwell times that provide sufficient data points per GC peak (i.e., to allow good chromatography), while at the same time offering good ion statistics. In instances where several ions from the one compound are being monitored, and where their relative abundances are very different, optimal specificity and sensitivity may require that unique dwell times are set for each separate *m/z* value.

9. Approaches to Increasing Analytical Specificity

The most common approaches to improving analytical specificity include: increasing the number of ions monitored; adding sample work-up steps; improving the chromatography; or even changing the ionization mode. However, there may be instances where incorporating sample clean-up steps may not be an option and these other approaches fail to deliver the requisite specificity. In these instances, mass spectrometry can offer additional specificity when operated in alternative scanning/detection modes: e.g., SIM at high(er) resolution, or selected reaction monitoring (SRM). For example, in a study aimed at determining trace levels of the plasticizer *n*-butylbenzenesulfonamide (NBBS) in biological and environmental samples, sample-handling had to be kept to a minimum to avoid contamination.[14] In this instance, minimal sample work-up was combined with capillary GC and SIM operated at increased (i.e., 5,000) resolution to provide adequate analytical specificity.

Because these options afford improved selectivity, it may be possible to detect the analyte of interest despite the presence of components that might otherwise interfere with the determination. Increasing the specificity of the detection may actually reduce the absolute signal intensity, but at the same time dramatically improve the signal-to-noise ratio, and thereby improve the limit of detection. Beyond a certain point, the benefits of increased specificity will be lost, and the signal-to-noise will no longer improve; more to the point, the limit-of-detection will be compromised.

High(er) resolution SIM has served well as an approach to increased instrumental specificity for many years, but selected reaction monitoring (SRM) now provides additional specificity and is more readily available. The major issues relating to these alternative scanning options have been discussed in detail by Chapman.[15]

10. Standards, The Calibration Curve & Statistical Considerations

Quantification requires the construction of a calibration curve, prepared from reference material, and all concentrations are determined by reference to this. If the reference material is not pure a systematic error will be introduced into all determinations (i.e., precise but inaccurate results are obtained). Analytes that are deliquescent, or that decompose on storage, are not uncommon. Reference material should therefore either be certified as such, or the investigator should establish the integrity of the sample by elemental analysis. Although mass spectrometry can provide information about the purity of analyte, results can be misleading. For example, GC-MS analysis will not provide evidence for the presence of involatile contaminants in the standard, nor for the contribution of water-of-crystallization in a hydrated sample.

The calibration curve should include a minimum of 6 points, and must cover the entire analytical range. Wherever possible, standards should be prepared in (analyte-free) sample matrix. The analysis should include samples containing: analyte only, internal standard only and a solvent blank. A control sample (sample matrix and a known concentration of the analyte) should be available and assayed in each run. This sample should return the same value for each assay. The amount of internal standard added should be similar to the amount of analyte contained in the samples because the errors in quantification are reduced when the relative responses for analyte and internal standard are comparable (i.e., *ca.* equal).[16]

The calibration curve relates the ratio of the response of the analyte and the response of the internal standard to the quantities of analyte and internal standard. (Although quantities are often expressed as concentrations, mass or mole ratio are sometimes used). The ion abundance ratios for samples (or unknowns) spiked with a fixed amount of internal standard are determined, and these are compared with the calibration curve to determine the quantity of analyte in the unknowns.

If we use an isotopomer as our internal standard, the analysis is simplified if the analyte does not contribute to the ion abundance for the internal standard, and the standard does not interfere with the mass for the analyte (i.e., there is no cross-talk between ion channels). For this reason the ideal isotopic standard contains three or more labeled atoms and is therefore separated from the native isotope peaks (i.e., at M+1 & M+2) of the analyte itself. Under these conditions the calibration curve is reduced to a straight line passing through the origin. This simplifies the analysis considerably and the calibration curve can be prepared by least squares linear regression analysis.[16] In this situation the regression line should be forced through the origin to minimize errors associated with determinations at the low end of the calibration curve. The correlation coefficient for the calibration curve of > 0.999 is generally regarded as evidence of an acceptable fit of the data to the regression line.[17] Further, if the analysis is to extend over a wide concentration range (i.e., several orders of magnitude), then weighting the data or the regression should be considered.[16,18,19] This approach is helpful in ensuring that the regression line is less sensitive to the large absolute errors at the upper extreme of the calibration curve.

If the internal standard contributes to the analyte ion abundance, the calibration curve remains linear, but has a non-zero (positive) intercept on the y-axis. This situation is common and arises when the isotopomer contains (or has been contaminated with) unlabeled material.

If the analyte ion abundance contributes to the ion abundance for the internal standard, with or without the standard contributing to that of the analyte, this cross-talk means that the calibration curve is hyperbolic, and the analysis is necessarily more complicated. Under these conditions several approaches can be employed: the impact can be minimized by adding large amounts of internal standard and discarding the non-linear portion of the calibration curve; the data can be linearized by transformation (e.g.,[20], or the data can be fitted to a non-linear equation (e.g.,[21]).

Some of these issues relating to calibration curves and instrumental conditions have been discussed in detail elsewhere.[15,22-25]

11. Assay Validation

This important step in the development of a quantitative analysis is often compromised, and data of variable and uncertainly quality are then generated. Little of the validation process is unique to mass spectrometry so in this area there is an extensive literature that can be consulted. A concise guide to aspects of analytical method validation has recently been published,[17] and other useful references are available.[26-28] Parameters such as: specificity, linearity, range, accuracy, precision and the limit of quantification must be determined. Additionally, the ruggedness of the procedure should be assessed. Wherever possible, comparisons with existing methods should be made.

However, one unique feature of quantification by mass spectrometry is that several separate (independent) determinations can be made at the one time, simply by monitoring several analyte-specific ions. When 2 or more ions are monitored, the analyst can simultaneously calculate two (or more) separate values for the analyte, and these can be correlated for all unknowns. If the correlation is acceptable, then the mean value can be reported; if not, the analyst must discard both values. This exercise provides valuable information about the appropriateness of the procedure, and indicates possible interferences when the assay is applied routinely. If the correlation is consistently poor, the specificity of the extraction and/or analysis phase(s) must be increased. If the correlation holds up across a wide range of samples, then specificity is adequate. This powerful approach is not generally available in quantitative analysis.

In a research setting, investigators may be tempted to short-cut some of these steps, and to use an assay without rigorous validation. Compromises may lead to invalid data and erroneous conclusions and thorough validation should be encouraged in all instances.

Acknowledgments

I thank Ian Gregor, Phil Harris, Ray Lidgard, George Smythe and Al Yergey for their input into this chapter.

References

1. L. Siekmann, *in* "Mass Spectrometry" (A. M. Lawson, ed.), Vol. 1, p. 645. Walter de Gruyter, Berlin, 1989.
2. V. R. Meyer, "Practical High-Performance Liquid Chromatography," p. 376. Wiley, Chichester, 1994.
3. L. R. Snyder and J. J. Kirklan, "Introduction to Modern Liquid Chromatography," p. 863. Wiley, New York, 1979.
4. D. R. Knapp, "Handbook of Analytical Derivatization Reactions," p. 741. Wiley, New York, 1979.
5. A. E. Pierce, "Silylation of Organic Compounds,". Pierce Chemical Co., Rockford, IL, 1968.
6. K. Blau, (ed.), "Handbook of Derivatives for Chromatography." Second Edition, Wiley, Chichester, 1993.
7. K. F. Faull and J. D. Barchas, *in* "Methods of Biochemical Analysis" (D. Slick, ed.), Vol. 29, p. 325. Wiley, New York, 1983.
8. F. W. McLafferty and F. Turecek, "Interpretation of Mass Spectra," p. 371. University Science Books, Mill Valley, 1993.
9. R. W. Nelson, M. A. McLean, and T. W. Hutchens, *Anal. Chem.* **66**, 1408 (1994).
10. R. W. Nelson, J. R. Krone, A. L. Bieber, and P. Williams, *Anal. Chem.* **67**, 1153 (1995).
11. C. F. Masters, M. W. Duncan, S. P. Markey, and I. N. Mefford, *Anal. Chem.* **60**, 2131 (1988).
12. R. C. Murphy and K. L. Clay, *Biomed. Mass Spectrom.* **6**, 309 (1979).
13. R. C. Murphy and K. L. Clay, *Mass Spectrometry* **193**, Chapter 17 (1990).
14. P. Duffield, D. Bourne, K. Tan, R. M. Garruto, and M. W. Duncan, *J. Anal. Toxicol.* **18**, 361 (1994).
15. J. R. Chapman, "Practical Organic Mass Spectrometry: A Guide for Chemical and Biochemical Analysis," p. 330. Wiley & Sons, Chichester, 1993.
16. B. J. Millard, "Quantitative Mass Spectrometry," p. 171. Heyden, London, 1978.
17. J. M. Green, *Anal. Chem.* , 305A (1996).
18. D. A. Schoeller, *Biomed. Mass Spectrom.* **3**, 265 (1976).
19. M. Claeys, S. P. Markey, and W. Maenhaut, *Biomed. Mass Spectrom.* **4**, 122 (1977).
20. M. W. Duncan, S. P. Markey, B. G. Weick, P. G. Pearson, H. Ziffer, Y. Hu, and I. J. Kopin, *Neurobiol. Aging* **13**, 333 (1992).

21. G. C. Thorne, S. J. Gaskell, and P. A. Payne, *Biomed. Mass Spectrom.* **11**, 415 (1984).
22. F. C. Falkner, *Biomed. Mass Spectrom.* **8**, 43 (1981).
23. S. P. Markey, *Biomed. Mass Spectrom.* **8**, 426 (1981).
24. A. P. De Leenheer and L. M. Thienpont, *Int. J. Mass Spectrom. Ion Processes* **118/119**, 723 (1992).
25. A. P. De Leenheer, M. F. Lefevere, W. E. Lambert, and E. S. Colinet, *Adv. Clin. Chem.* **24**, 111 (1985).
26. W. Wegscheider, *in* "Accreditation and Quality Assurance in Analytical Chemistry" (H. Gunzler, ed.), p. 135. Springer-Verlag, Berlin, 1994.
27. L. A. Currie, *Pure & Appl. Chem.* **67**, 1699 (1995).
28. R. K. Boyd, J. D. Henion, M. Alexander, W. L. Budde, J. D. Gilbert, S. M. Musser, C. Palmer, and E. K. Zurek, *J. Am. Soc. Mass Spectrom.* **7** (1996).

Appendix: A Glossary of Key Terms

The reader is also referred to an extensive discussion of the nomenclature of quantification [27].

Precision: The scatter in results that occurs when repeat determinations are made.

Accuracy: The closeness of the measured result to the true value.

Specificity (or selectivity): The ability to measure the response arising from one analyte independent of any other sample component(s).

Linearity: The range over which test results are proportional to the concentration of the analyte.

Range: The interval over which acceptable accuracy, linearity and precision are achieved.

Limit of Detection: The lowest possible amount of analyte that gives a response detectable above the noise level (defined by the signal-to-noise ratio) and can therefore be reliably detected.

Limit of Quantification: The lowest possible amount of analyte that gives a response detectable above the noise level and can be quantified with acceptable precision and accuracy.

OPTIMIZATION OF MASS SPECTROMETRIC IONISATION EFFICIENCY DATA

İ. T. ÖZGEN, O. ALTUNGÖZ
Department of Chemistry, İzmir Institute of Technology, Gaziosmanpaşa Bulvari, No 16, Çankaya, İzmir, TURKEY

and

B. SALİH
Department of Chemistry, Hacettepe University, Beytepe Campus, Ankara, TURKEY

and

B. G. M. Vandeginste
Unilever Research Vlaardingen,P.O. Box 114, 3130 AC Vlaardingen, The NETHERLANDS

ABSTRACT. A new method is developed and tested to find out the excited electronic and vibrational energy levels (fine structure) of molecules from mass spectrometric ionisation efficiency data which were obtained by a conventional mass spectrometer ion source having normal electron energy distribution.

Electrons emitted in a conventional mass spectrometer ion source are not monoenergetic, therefore, evaluation of Ionisation Potentials (IP) and Appearance Potentials (AP) from experimental data creates some problems. This is even worse in the evaluation of Fine Structure (excited electronic and vibrational energy levels) from ionisation efficiency data [1] . This is overcome either by using specially designed monoenergetic ion sources which have their own problems (manufacture of special design, difficulty of operation, too much decrease of ion current), or by eliminating disturbing effects in the ionisation efficiency data by some deconvolution techniques [2,4,5,6,7,9].

In this study attempts were made to eliminate the deteriorating effects (mainly arising from electron energy distribution and noise) in the ionisation efficiency data by a deconvolution technique.

The technique was applied to the ionisation efficiency data of molecular nitrogen,oxygen and methylammine and its fragment and interesting results were obtained.

1. Introduction

Molecular and fragment ions, excited to various electronic levels may be formed with excess energy , and if the transferred energy is appropriate, excitation to different vibrational (and translational) energy levels is also possible. Difference between these energy levels can be detected by appropriate techniques [3]. Electrons emitted from a conventional mass spectrometer ion source obey the Maxwell energy distribution and have an energy spread. However it is possible to decrease this energy spread by some deconvolution techniques. Some of these deconvolution techniques are The First

R. M. Caprioli et al. (eds.), Selected Topics and Mass Spectrometry in the Biomolecular Sciences, 121–127.

Differential of the ionisation efficiency data [3,4], The Second Differential of the ionisation efficiency data [5]; Energy Distribution Difference Technique (EDD) [4]; Double EDD [6] ; and The Inverse Convolution Technique [7] . The Inverse Convolution Technique proposed by Morrison gives very satisfactory results for ionisation efficiency data. This technique uses Fourier Transformation and has gained importance after the improvement of Fast Fourier transformation (FFT) [8]. Smoothing Technique of Savitzky and Golay [9] was also used by some experimenters to eliminate the effect of noise from experimental data. However, smoothing spreads the deteriorating effect of a bad data to a certain data region therefore prevents precise and satisfactory results [10] .

In this study, spoiling effects of electron energy spread and noise on the mass spectrometric ionisation efficiency data were eliminated by using filters and FFT deconvolution.

The deconvolution technique used to eliminate a distorting effect from an experimental data can be explained as follows. The function of the distorting effect is formed. Inverse Fast Fourier Transform of both this function and the experimental data are obtained. Then Fast Fourier Transform of the experimental data are divided by the Fast Fourier Transform of the function of the distorting effect. Inverse Fast Fourier Transform of this function gives the signal data which is stripped from the distorting effect (figure 1).

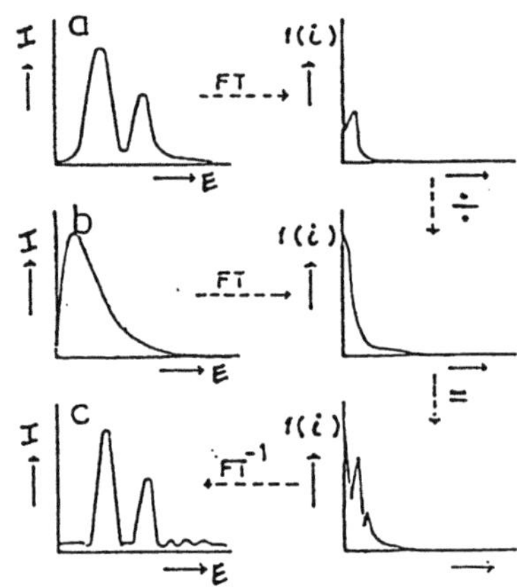

Figure 1. Improvement of experimental data by using the deconvolution technique. a)deteriorated signal b) deteriorating signal, c) improved signal.

2.Experimental

The experiments were carried out by a CH7 model VARIAN MAT mass spectrometer connected to a PC-XT computer. The mass spectrometer has a conventional electron impact ion source, a magnetic sector analyser and a faraday cage detector. External vacuum measuring / control system was connected to the mass spectrometer. Also an isolated external electron energy control module which itself could be controlled by the computer was connected between the mass spectrometer and the computer, and electron energy could be increased by small (0.01V) steps. The collector of the mass spectrometer was also connected to the computer so that data could be stored and processed afterwards.

The computer contained a 14 bit Analog to Digital (AD) and Digital to Analog (DA) cards both operating within the ranges -8.5 V / + 8.5 V.

Measurements were started approximately 1.5 - 2 volts below the treshold values (I.P. or A.P.) which were roughly determined in the first runs. Ion currents (collector signals) were collected 2 seconds after every voltage increase so that the filament was given enough time to stabilise. Six to ten volt ranges were scanned for every ionisation efficiency data.

3. Studies on Molecular Fine Structure

Thermal energy spread of electrons emitted from a filament is explained by Honig [11]. According to Honig the number of electrons emitted from a filament at a unit time is given by equation 1.

$$d\,Ne(U)/dU = (4\,P\,mA/h^3)\,EXP\,[-\,(F+U)/kT] \qquad (1)$$

In the equation [d Ne(U)/dU] represents the number of electrons emitted from a filament in a unit time, [m] the mass of the electron, [A] the surface area of the filament, [h] the Plancks constant, [F] the work function of the filament, [U] the energy applied to the filament, [k] the Boltzman constant and [T] the temperature of the filament. To eliminate the distorting effects of the electron energy distribution (and other effects) from experimental data three computer software were developed. The computer software which were prepared are:

3.1. BANDGEN (For forming energy distribution data)

With this program electron energy distribution data, similar to formed in the mass spectrometer ion source were generated for a suitable energy range. In the calculations the surface area of the filament was taken as A = 0.157 cm^2 (calculated by measurements), the work function of the tungsten filament as F=4.5 eV and temperature of the filament as 1800 K (temperature generally referred in the literature). Calculations for other temperatures also gave similar values. Since this distribution extends to infinity, a certain range of this data had to be used in the calculations. In our calculations 3 eV range of this data was used because it covered most of the energy distribution.

3.2. CONVOLUT (a convolution program to convolve data of disturbing effects (such as energy distribution and noise) with the data of an ideal experiment.

According to Knewstubb [13] an ideal ionisation efficiency curve should be in the form of straight lines which intersect with each other (fig.2). The value of each intersection point on the energy scale corresponds to the energy level of a different process (excited energy level). With disturbing effects (such as electron energy distribution and electronic noise) these intersection points are smoothed out and experimental data give a smooth curve and some scatter at low energy region.

To test whether ideal ionization efficiency data can be obtained from experimental data by deconvolution, an artificial data set representing an ideal ionisation efficiency curve for nitrogen molecule was formed using literature data for excited states of nitrogen molecule (figure 3). A noise data set was also developed taking limits of a standard deviation of a blank mass spectrometric run data (figure 4). Then these electron energy distribution and electronic noise data (formed by previously explained BANDGEN program) were convoluted to the ideal data to obtain a data set which represented an experimental data (figure 5). As is seen in figure 5 the ideal ionisation efficiency curve giving breaks is smoothed into a continuous curve after convolution. This generated data set was used to test our deconvolution program (Huntfil Program) .

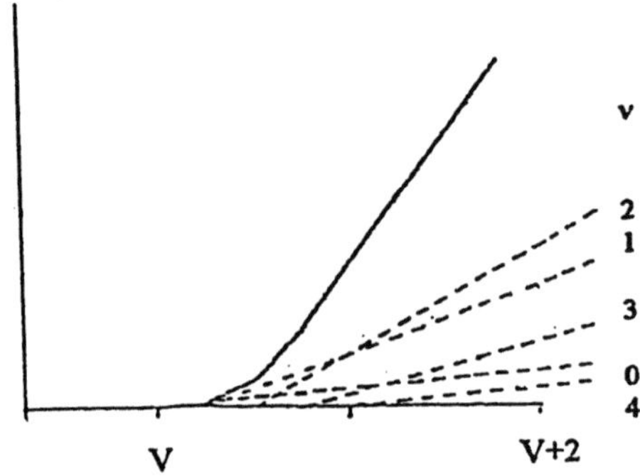

Figure 2. Ionisation curve and separate vibrational levels of a molecule [13].

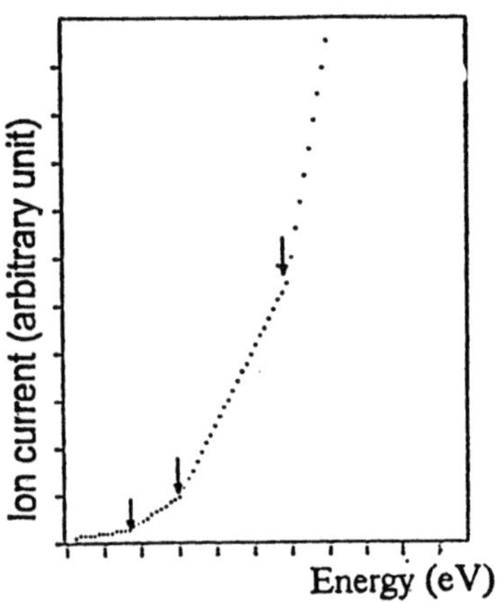

Figure 3. Ideal ionisation efficiency curve for nitrogen generated using literature data.

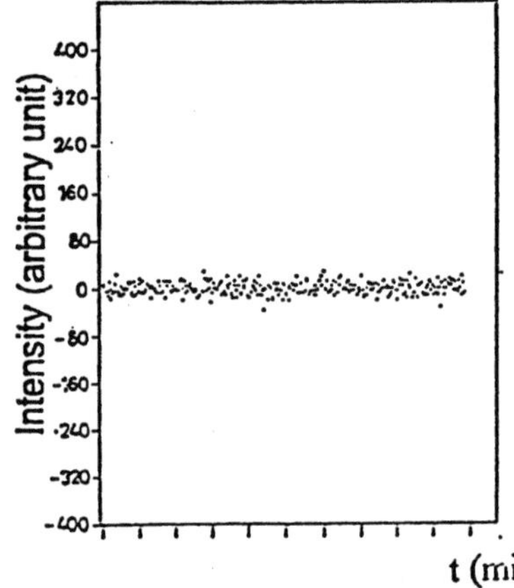

Figure 4. Graph of the noise of the mass spectrometer blank run standard deviation taken as limits.

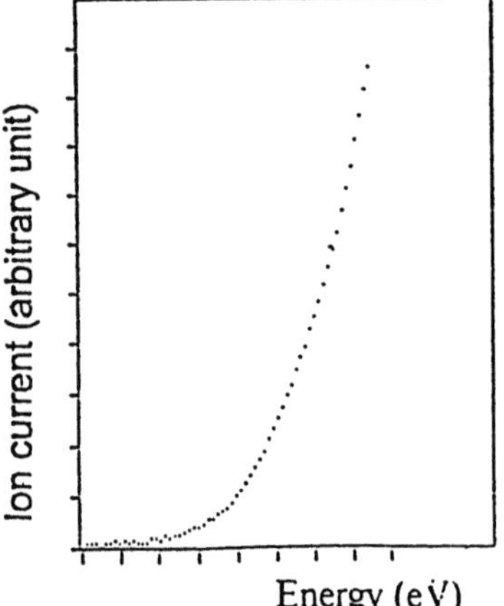

Figure 5. Distorted graph of ideal ionisation efficiency curve for nitrogen after adding noise and electron energy distribution.

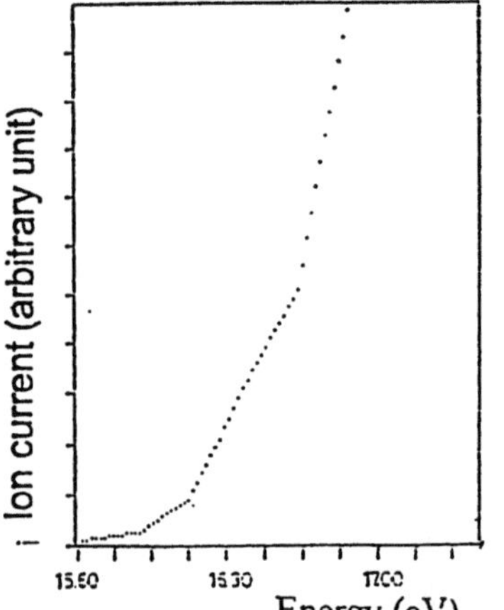

Figure 6. Graph of the resulting data obtained after applying deconvolution technique to the distorted ideal ionisation efficiency data of nitrogen (fig 5).

3.3. HUNTFIL (Deconvolution program)

This program was developed to eliminate the deteriorating effects (such as electron energy distribution and electronic noise) from the experimental data. Figure 6 shows the resulting ionisation efficiency curve after application of this program to the data of figure 5. The fine structure of the nitrogen molecule (breaks in the curve as in figure 3) is recovered again.

This deconvolution program first generates the Fast Fourier Transform of experimental data and energy distribution and noise data. Then the Fast Fourier Transform of experimental data is divided to the Fast Fourier Transform of energy distribution and noise data. Inverse Fast Fourier Transform generates the resulting ionisation efficiency data.

4. Results

Our deconvolution program (HUNTFIL) was tried on the experimental data of methylammine and fragment ions. We found that intersection points of the lines correspond to excited vibration levels, whereas the jumps correspond to the excited electronic levels. Figure 7 shows the experimental and improved data curves for methyl ammine molecular ion. Processing of experimental data measured at different dates for the same ions show the similar jumps at similar energy intervals. Repetitions in the curves show the success of the method.

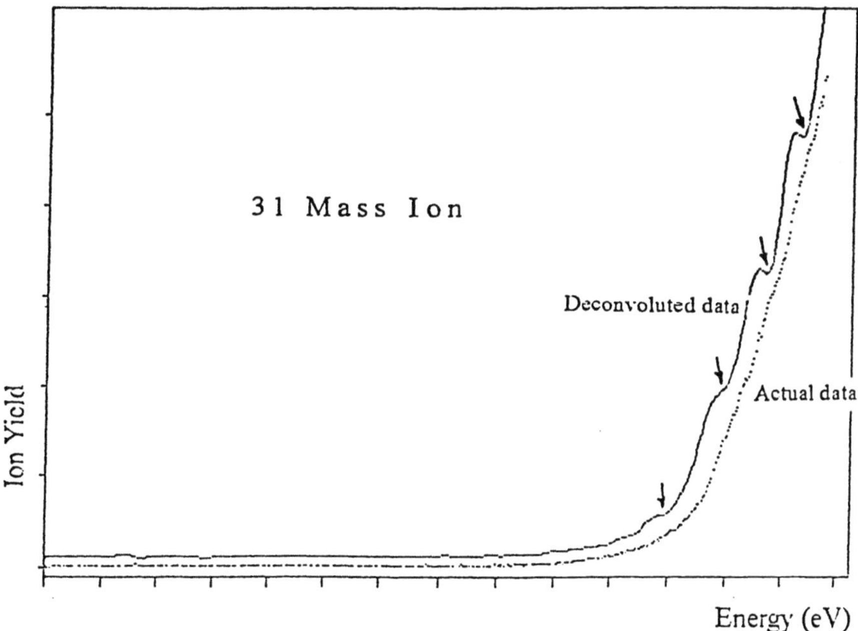

Figure 7. Differences in experimental (actual) and deconvoluted data curves showing the effect of elimination of noise and energy spread from experimental data.

5. Conclusion

With this method conventional mass spectrometric ionisation efficiency data can be used to get good and detailed information on the excited states (electronic and vibrational energy levels) of molecular and fragment ions .

Results for methyl ammine molecular ion (m/e = 31) and fragment ion (m/e = 30) are given in Tables 1 and 2. The results in the tables are the average of at least 10 independent runs. Values are given with their standard deviations.

Table 1. Electronic and vibronic energy levels of molecular ion of methyl amine .

Electronic energy levels after threshold (Literature values) (a)	Excited states after threshold (This work)
3.50	3.50 ±0.02
	3.69 ±0.05
	3.91 ±0.06
	4.09 ±0.03
	4.32 ±0.04
4.80	4.67 ±0.04
	4.95 ±0.04
	5.18 ±0.04
	5.39 ±0.02
	5.59 ±0.04
	5.75 ±0.02
5.90	6.00 ±0.05
7.40	—

a. See Reference 12.

Table 2. Electronic and vibronic energy levels of methyl amine fragment m/e=30.

Electronic energy levels after threshold (Literature values)	Excited states after threshold (This work)
	1.98 ±0.02
	2.21 ±0.02
	2.39 ±0.01
	2.59 ±0.02
	2.77 ±0.01
	2.92 ±0.01
	3.07 ±0.01
	3.32 ±0.02
	3.50 ±0.01
	3.68 ±0.03
	3.96 ±0.03

6. Acknowledgements

This collaboration research study was supported by NATO with project no 347/84.

7. References

1. Field, F.H. and Franklin, J.L., 1970, Electron Impact Phenomena and the Properties of Gaseous Ions, Academic Press, New York, p.17.

2. Collins, J.H., Winters, R.E. and Engerholm, G.G., J. Chem. Phys., 49, 2469 (1968).

3. Franklin,J.L., Dillard,J.G., Rosenstock,H.M., Herron,J.T., and Draxl,K., "Ionisation Potentials, Appearance Potentials and Heats of Formation of Gaseous Positive Ions", U.S. Department of Commerce, NSRDS-NBS, U.S. Government Printing Office, Washington D.C. (1969).

4. Winters, R.E., Collins, J.H. and Courchene, W.L., J. Chem. Phys., 45, 1931 (1966)

5. Morrison, J.D., J. Chem. Phys., 21, 1767 (1953).

6. Johnstone, R.A. and McMaster, B.N., Advances in Mass Spectrometry, 6, 451 (1974).

7. Morrison J.D., J. Chem. Phys., 39, 200 (1963).

8. Cooley, J.W. and Tukey, J.W., Math. Computing, 9, 297 (1965)

9 Savitzky, A. and Golay, M.J.E., Anal. Chem., 36, 1627 (1964).

10. Raznikov, V.V., Dodonov, A.F. and Zelenov, V.V., Intr. J. Mass Spec. and Ion Proc., 71, 1 (1986).

11. Honig, R. E., J. Chem. Phys. 16, 105 (1948).

12. A. B. Cornford, D. C. Frost, F. G. Herring, and C. A. McDowell, Can. J. Chem. 49, 1135 (1971).

13. Knewstubb, P.F., 1969, Mass Spectrometry and ion-molecule reactions, Cambridge University Press, p.101.

ROLE OF INTERNAL ENERGY IN MASS SPECTROMETRIC FRAGMENTATION

KÁROLY VÉKEY
Central Research Institute for Chemistry
H-1025 Budapest, Pusztaszeri út 59-67, Hungary

ABSTRACT. The internal energy of an ion has a very large effect on the appearance of the mass spectrum. The first part of the paper describes the theoretical background: the reaction rate theory and the calculation of ion abundances from reaction rates. The effect of the reaction timescale, the degree-of-freedom effect and the breakdown curve is discussed. Finally, typical internal energy distributions in electron impact, metastable ions, collision induced fragmentations are shown.

1. Introduction

Ions formed in a mass spectrometer are excited to a degree and, depending on their internal energy, they fragment. The internal energy (E_{int}) of an ion determines the available fragmentation channels, and the corresponding reaction rates. Highly excited ions will produce a spectrum of a wide variety of abundant fragments. If the internal energy is low, the ions will not decompose, or will produce only few fragments of low abundance. This effect is well illustrated by high and low energy electron impact spectra, but has a large influence in most other areas of mass spectrometry as well.

The purpose of the present paper is to give a clear picture on the effect of internal energy on fragmentation processes observed in mass spectrometry. The fundamentals will be described in some detail: Dependance of the reaction rate on the internal energy; connection of reaction rates and fragment ion abundances; of the internal energy and fragment abundances (the breakdown diagram); and of the internal energy distribution and the mass spectrum. From these relationships important consequences can be derived regarding degree-of-freedom effects, the effective internal energy and effective temperature and the influence of the observation time window. Particular emphasis will be placed on how internal energy effects relate to the study of large (bio)molecules. Most of these topics are discussed in detail in books like that of Cooks, Beynon, Caprioli & Lester [1], Levsen [2], Howe, Williams & Bowen [3], and in a recent review of Vékey [4].

R. M. Caprioli et al. (eds.), Selected Topics and Mass Spectrometry in the Biomolecular Sciences, 129–142.

2. Theoretical Background

2.1. REACTION RATE THEORY

Ions produced in a mass spectrometer are typically energized in a single event (e.g. ionization or collision) and are not in equilibrium or in energy-exchange with their environment. To apply statistical rate theories to mass spectrometry modifications (relating to non-thermal energization) are required, these are incorporated in the so-called Quasi-Equilibrium Theory (QET), developed by Rosenstock et. al. [5]. A good description of statistical rate theories can be found in the book of Robinson and Holbrook [6], and a more qualitative discussion, giving a number of semi-quantitative applications relating to mass spectrometry, is published in the treatment on metastable ions [1].

The fundamental assumptions of the QET are the following: (1) the time required for dissociation is long compared to the time of interaction leading to excitation; (2) the rate of dissociation is slow relative to the rate of distribution of internal energy over all degrees of freedom; (3) an ion in the mass spectrometer represent an isolated system in the state of internal equilibrium. ("Isolated" means that there is no energy exchange with the environment, "internal equilibrium" means that the energy within the ion is distributed statistically over the internal oscillators.) (4) Fragmentation products are formed by a series of competing and consecutive unimolecular reactions.

These assumptions are usually satisfied in mass spectrometric experiments. There are some exceptions, however. Assumption (2) is probably not true for very large molecules (consisting of ~1000 or more atoms). In some experiments (e.g. Fourier-transform (FT-MS) or ion trap instruments) ions are excited by sequential collisions and thermal equilibrium may be approached, so assumption (3) will not be valid. In the majority of cases, however, QET was found to be a good approximation, even when the conditions above are not completely satisfied.

The early version of QET used the classical approximation of the absolute rate theory (the oscillators are not quantized, the Rice-Rampsberger-Kassel (RRK) theory). The reaction rate constant, k(E), is expressed as a function of internal energy, (E) :

$$k(E) = \nu\,(1 - E_0/E)^{s-1} \qquad (1)$$

ν is the "frequency factor", s is the number of oscillators in the molecule (sometimes also called the "degree of freedom"). E_o is the critical energy of the reaction: difference of the heat of formation of the activated complex and that of the ground state. It is the minimum amount of internal energy of the precursor ion to overcome the energy barrier of the reaction. The critical energy is similar to the more familiar activation energy, but the latter can properly be used in thermal reactions only.

The use of this expression is often very helpful for qualitative purposes, e.g. Equation 1. clearly indicates that the reaction rate increases exponentially with the internal energy. Quantitatively the expression is, however, grossly inaccurate, due to a bad mathematical treatment (using non-quantized oscillators). As a compensation the concept of "effective" or "active" oscillators has been introduced, decreasing the number of oscillators (value of s in Equation 1.) typically by 3-5 times. *It is important to emphasize, that "effective oscillators" do not have a physical significance, their only purpose is to compensate for a bad mathematical treatment!* Even using a reduced number of oscillators, this form of the theory should not be used for quantitative or semi-quantitative purposes.

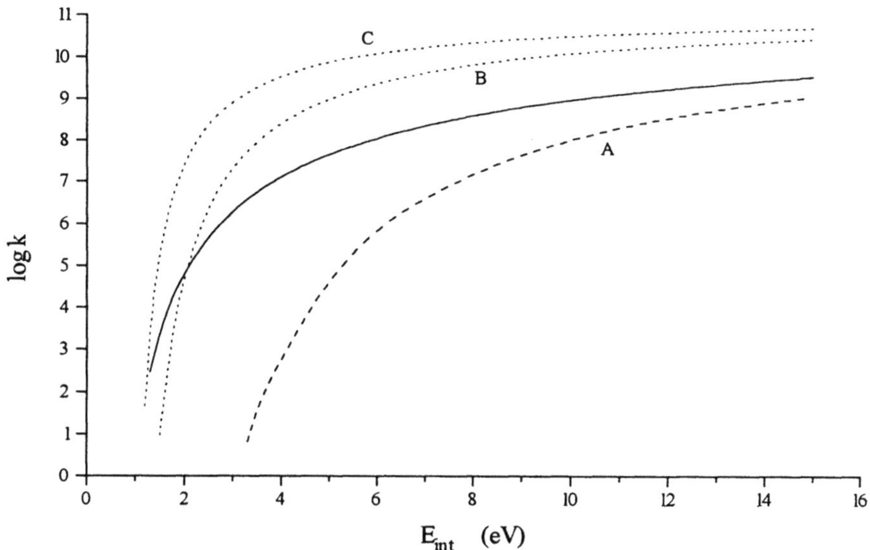

Figure 1. Reaction rate as a function of internal energy calculated by the proper RRKM expression (2), using a direct counting of states (solid curve) [8,9]. The reaction rate was also calculated using the classical RRK approximation (Equation (1); dashed (A) curve). Using the classical approximation but assuming that only 1/3 or 1/5 of the oscillators are "active" a better approximation was obtained, as shown by the dotted curves B and C, respectively.

The mathematically correct form of the reaction rate, called the Rice-Rampsberger-Kassel-Marcus (RRKM) theory, takes into account the quantized nature of vibrations and rotations, therefore there is no need to assume "effective" oscillators. The rate constant takes the form:

$$k(E) = \frac{\sigma}{h} \frac{G^*(E - E_0)}{\rho(E)} \tag{2}$$

where σ is the reaction path degeneracy, h is the Planck's constant, $G^*(E\text{-}E_o)$ is the number of states in the energy range $E\text{-}E_o$; ρ(E) is the density of states in the reactant ion of E internal energy. The main drawback of this formalism that it is not a straighforward analytical expression, and evaluation requires a sophisticated computer program (determination of $G^*(E-E_0)$ and $\rho(E)$). Evaluation requires a knowledge of vibrational frequencies in the reactant (typically known) and in the transition state (typically unknown). The latter (or both) are often estimated - which procedure is analogous to the estimation of the frequency factor (ν) in Equation (1). Many applications are not sensitive to the choice of vibrational frequencies, in such cases quantitative studies can be performed without detailed knowledge of the transition state [4,7,8].

The RRK and RRKM expressions of QET (Equation (1) and (2), respectively) can be compared numerically in the case of a model system, propene loss from buthylbenzene (Fig. 1). This is a rearrangement reaction with 1.0 eV critical energy and 10^{11} s^{-1} frequency factor (or, in the case of RRKM, a frequency model resulting in that frequency factor) [4]. In the RRK model we have calculated the reaction rate assuming all, 1/3 or 1/5 of the oscillators being "active". Fig. 1 shows that the RRK theory, taking into account all oscillators, very badly underestimates the reaction rate compared to the proper RRKM result. Assuming "active" oscillators increases the reaction rate, but the k(E) curves so calculated never overlap with that of RRKM. The number of "effective" oscillators properly describing the reation rate strongly depends on the internal energy. Deviations are most significant at relatively low internal energy - close to the fragmentation threshold.

2.2. REACTION RATES AND ION ABUNDANCES

A molecular or parent ion may dissociate by a series of competing and consecutive reactions. The reaction rates of individual reactions define the product ion abundances in a straightforward manner - i.e. from the reaction rates the product ion abundances can be calculated. The calculations are further simplified in mass spectrometry, where predominantly unimolecular reactions are observed. Collisional activation can be regarded as instantaneous process followed by a unimolecular decomposition. High accuracy is usually not attempted (errors connected to the calculation of reaction rates are typically far larger), so complications due to the loss of fragments during the flight time are neglected. In the simplest case, when the molecular ion has one unimolecular decomposition reaction only, with rate constant k, the abundance of M at time τ is given by:

$$M^+ \rightarrow A^+ \tag{3}$$

$$[M^+] = e^{-k\tau} \tag{4}$$

τ is measured from the formation (excitation) of the molecular (parent) ion, the initial abundance of M (corresponding to 0 time) is normalized to unity. In a mass spectra this normalization means that the sum of all ion abundances is unity. The abundance of A will, therefore, be equal to ($1- e^{-k\tau}$). This expression also implies that the average lifetime of a precursor ion ($\tau_{1/2}$, the time in which 50% of the precursor ions fragment) is the inverse of the reaction rate constant ($\tau_{1/2}= \ln 2 / k = 0.69 / k$).

Competitive reactions are of high interest in mass spectrometry. These reactions and ion abundances can be expressed as:

$$M^+ \rightarrow A^+ \quad \text{and} \quad M^+ \rightarrow B^+ \tag{5}$$

$$[M^+] = e^{-\Sigma k\tau}; \quad [A^+] = (k_A/\Sigma k)\,(1- e^{-\Sigma k\tau}); \qquad [B^+] = (k_2/\Sigma k)\,(1- e^{-\Sigma k\tau}) \tag{6}$$

here k_A and k_B are the individual rate constants, while Σk is their sum, which determines the molecular ion abundance. The ratio of the fragments ions is determined by the ratio of the respective rate constants ($[A^+]/[B^+] = k_A/k_B$). Further competitive (parallel) reactions (e.g. $M^+ \rightarrow C^+$) can be regarded in a similar manner.

Consecutive reactions can also be described, approximatively, with simple analytical expressions:

$$M^+ \rightarrow A^+ \rightarrow B^+ \tag{7}$$

$$[M^+] = e^{-k_A\tau}; \quad [A^+] = (1-e^{-k_A\tau})\,e^{-k_B\tau}; \quad [B] = (1-e^{-k_A\tau})\,(1-e^{-k_B\tau}) \tag{8}$$

Further consecutive steps (e.g. $B^+ \rightarrow C^+$) can be regarded in a similar manner. Expressions (4), (6), (8) indicate that beside reaction rates the timescale of the reaction is also important - this is mainly determined by the type of the instrument or experiment.

3. Rate Calculations: Consequences

3.1. DIRECT CLEAVAGES AND REARRANGEMENTS

Based on the frequency factor, fragmentation reactions are usually divided into two broad categories. *Direct bond cleavages* are characterized by a "loose" transition state and a corresponding high frequency factor (ν in Eq. 1), usually ca. 10^{13}-10^{15} s^{-1}. The reaction usually involves stretching of a bond along the reaction coordinate, a few other vibrational frequencies may decrease or these may become internal rotations. The structure of the transition state is close to that of the products. These reactions usually have relatively high critical energies. The reaction rate as a function of internal energy of such a reaction (propyl radical loss from buthylbenzene [4]) is shown in Fig. 2 (broken curve, data taken from).

Rearrangement reactions are characterized by a "tight" transition state and a corresponding low frequency factor, usually ca. 10^6-10^{12} s^{-1}. Rearrangements usually proceed through a cyclic transition state, where some internal rotations "freeze" (become vibrations), and several other vibrations change as well. To compete efficiently with direct bond cleavages, rearrangements have low activation energies (otherwise only direct cleavages would be observed). The reaction rate of a typical rearrangement reaction (propene loss from buthylbenzene [4]) is shown in Fig. 2 (continuous line).

In typical cases direct bond cleavages and rearrangements compete efficiently: one having a high critical e energy prevails. A above this internal energy propyl radical loss, proceeding with direct bond cleavage and so having a high frequency factor, is more probable. If the internal nergy and high frequency factor, the other a low critical energy and low frequency factor. The two curves, in the example shown in Fig. 2, cross at 5.8 eV internal energy. Below this internal energy propene elimination, requiring low critical energy prevails. A above this internal energy propyl radical loss, proceeding with direct

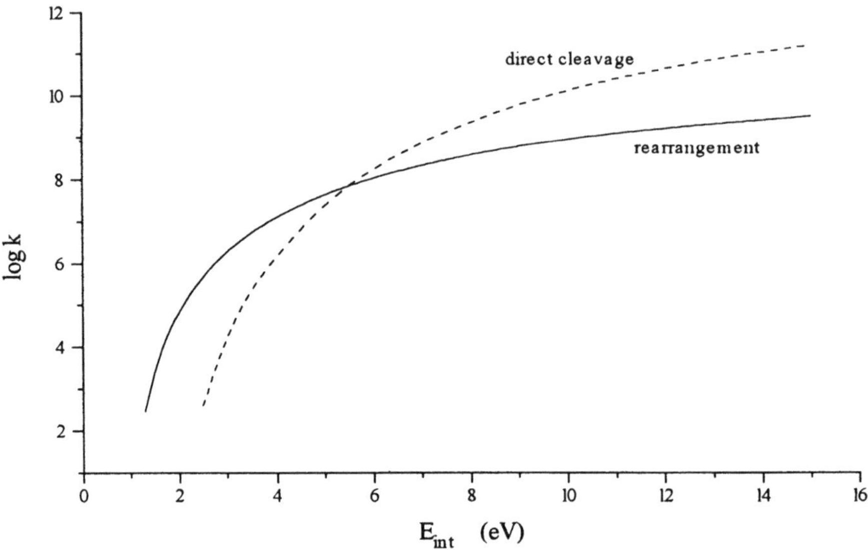

Figure 2. Reaction rates of a direct bond cleavage and a rearrangement reaction as a function of internal energy calculated by the RRKM theory. The model calculation relates to propyl radical ($\nu=10^{14}$ s^{-1}, $E_0=1.7$ eV) and propene eliminations ($\nu=10^{11}$ s^{-1}, $E_0=1.0$ eV) from buthylbenzene, respectively. Data were taken from [4].

bond cleavage and so having a high frequency factor, is more probable. If the internal energy of a particular precursor ion is increased (e.g. by collisional excitation) products formed by direct bond cleavages are likely to become more abundant.

3.2. REACTION TIMESCALE

Equations (3), (5) and (6) show that beside the reaction rate constants the timescale of the mass spectrometric experiment is also important determining fragment ion abundances. This may vary from 10^{-12} s to several s, depending on the experiment. The residence times in various parts of the instrument do vary somewhat according to the experimental conditions. The typical values listed below refer to medium mass molecules, large ones (in excess of 1000 dalton) may have significantly longer residence times. Fast reactions (10^{-12}-10^{-9} s lifetime) can be studied by coincidence techniques or by field ionization kinetics. Fast and moderately fast reactions (larger than rate constant $\sim 10^{6}$ s^{-1}) are observed in most conventional instruments as fragments formed in the ion source (the residence time in an ion source is $\sim 10^{-6}$ s). Metastable ions observed in sector instruments have a lifetime $\sim 10^{-6}$-10^{-4} s. Precursor ions in collision induced decomposition (CID) processes in sector or quadrupole-type instruments has a lifetime over $\sim10^{-5}$-10^{-4} s when reaching the collision cell. The internal energy content is than increased by collision, consequently the fragmentation rate will increase as well. To observe the ion fragmenting within the collision cell, its lifetime (after collision) should be less than $\sim10^{-7}$-10^{-6} s. Ions with lifetime longer than $\sim10^{-4}$ s are observed in sector and quadrupole type mass spectrometers as undecomposed molecular ions. Ions have up to minute lifetimes in ion trap or FT-MS instruments.

The time window of an experiment is important, as it defines the observable ion lifetimes, which are, in turn inversely proportional to the corresponding unimolecular reaction rates. As shown in Fig. 2, high reaction rates (short ion lifetimes) are possible only if the internal energy of the precursor ion is high, and in such cases direct bond cleavages are typical reactions. Long lifetimes are possible only if the internal energy of the precursor ion is low, and in such a case rearrangement reactions predominate. These also mean that by selecting the experiment one is also selecting the ion lifetime, and this typically determines the range of internal energies and reaction types observed.

Large (bio)molecules require a very large amount of internal energy to fragment in conventional (sector, quadrupole) instruments. It is very difficult, if not impossible, to impart 50-100 eV or more into an ion by present techniques. If the observation time is long, like in FT-MS, a much smaller amount of energy would be sufficient to induce fragmentation - this is part of the explanàtion of the success of MS/MS by this technique. The long ion lifetimes also mean that in FT-MS predominantly rearrangement reactions are observed.

3.3. DEGREE-OF-FREEDOM EFFECT

Increasing the molecular size decreases the fragmentation rates drastically, if the frequency factor, the critical energy and the internal energy are the same. (In mass spectrometry excitation energis are typically independent of the molecular size.) This dependence is obvious from Eq. 1, large molecules fragment less, or require higher internal energies to fragment. Calculating the reaction rate using Eq. 2. yields a similar result, but can be studied only numerically. The result of such a calculation is shown in Fig. 3. - showing the reaction rate of the same process ($\nu=10^{14}$ s^{-1}, $E_0=1.6$ eV, typical values for peptides) for three different molecules size - 24 (buthylbenzene), 78 (Leucine Enkephalin) and 194 atoms (Substance P) [4].

In practice a fragment ion to be observable in a spectrum it should have an intensity ~1 % of the molecular ion, to be called an abundant peak it should have a relative abundance at least ~10 %. Residence time in the ion source can be assumed (as before) to be ~1 μs. Using these values rate constants should be 10^4 to give an observable fragment in ion source fragmentations in sector, quadrupole and time-of-flight instruments. This requires over 10 eV internal energy for a medium sized peptide (Fig. 3). To observe fragments in an FT-MS or ion trap instument, were residence times of ca. 1 s can be obtained, 20-50% smaller internal energies are sufficient, but these are still quite large. The high internal energies necessary to fragment large molecules are especially striking if compared to the critical energy (1.6 eV). The difference is often called the "kinetic shift": the internal energy above the critical energy, necessary to drive the reaction fast enough to observe fragmentation inside the mass spectrometer. To observe abundant fragments, even higher internal energy is required.

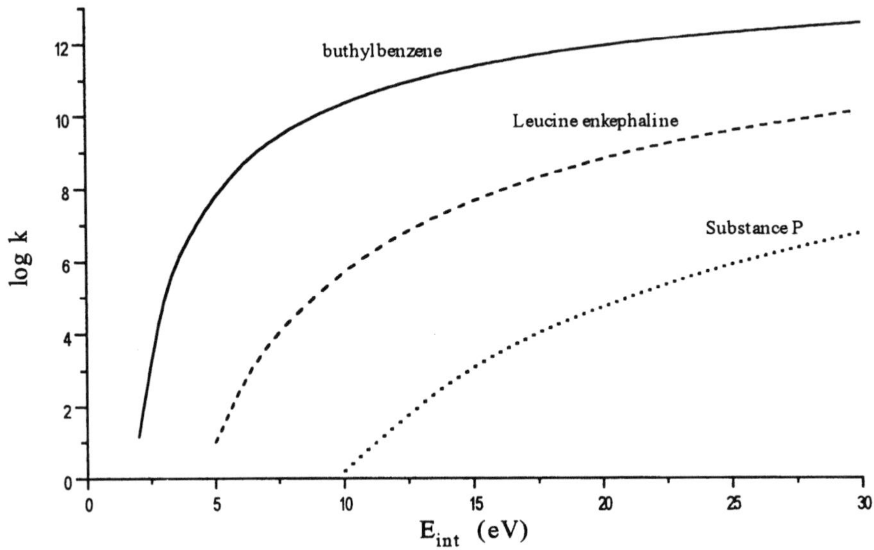

Figure 3. Reaction rates as a function of internal energy modelled in the case of molecules of different size. Activation energies and frequency factors were the same in all three cases. Calculations were performed by the RRKM theory, data were taken from [4].

As a rule of thumb, the internal energy necessary for fragmentation increases nearly linearly with the degrees of freedom (i.e. the size) of the molecule. This can be qualitatively explained as to induce fragmentation, the internal energy per oscillator (degree of freedom) should be roughly the same in various molecules. The effect of fragmentation rates decreasing with the is size of the molecule is therefore often called the "degree of freedom" effect.

3.4. THE BREAKDOWN CURVE

From reaction rate constants and residence times ion ratios can be determined as a function of the internal energy of the ion. This correlation is usually called the breakdown curve, or to be more precise, it is the breakdown curve as a function of internal energy. From the rate constants shown in Fig. 2 using 1 μs residence time in the source, and assuming that all ions leaving the source will reach the detector, fragment ion ratios were determined, the result is shown in Fig. 4. There are some complications making precise calculations difficult, but these usually do not change the results substantially. It is important to note, that the residence time does have a significant effect on the breakdown curve.

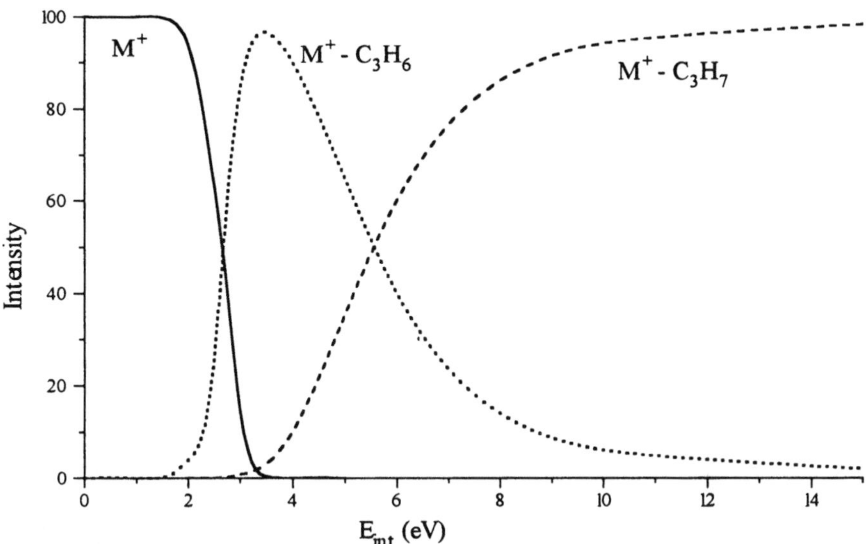

Figure 4. Breakdown diagram (as a function of internal energy) of the buthylbenzene molecular ion, calculated from the rate constants shown in Fig. 2, using 1 μs residence time in the ion source and assuming that all ions leaving the source will be detected. Data were taken from [4].

Breakdown curves (as functions of internal energy) can be determined experimentally. It is, however, fairly difficult, and these are known only in a few cases. There are related correlations, which are often also (loosely and confusingly) defined as breakdown curves. Foremost among these is the correlation between ion ratios and collision energy, but this should properly be called the breakdown diagram as a function of (laboratory frame) collision energy. The two are related, as higher collision energies typically result in higher internal energy products. The ratio between collision energy and internal energy is, however, not straightforward, so transformation of the two curves into each other is difficult (at present usually not possible). Ion ratios can be measured as functions of other variables, related to the internal energy. In chemical ionization the difference between the gas phase basicity of the sample and the reagent gas defines the energy balance of protonation. This way the breakdown diagram as a function of the energy balance of the reaction can be determined.

4. Internal Energy Distribution

Typically ions in a mass spectrometer are formed with a wide range of internal energies. The mass spectrum, in such a case, is the convolution of the breakdown diagram and the internal energy distribution (usually indicated as P(E)). In other words the ion ratios defined by the breakdown graph will be weighed by the probability of the ion having that particular internal energy, than summed over all possible internal energies.

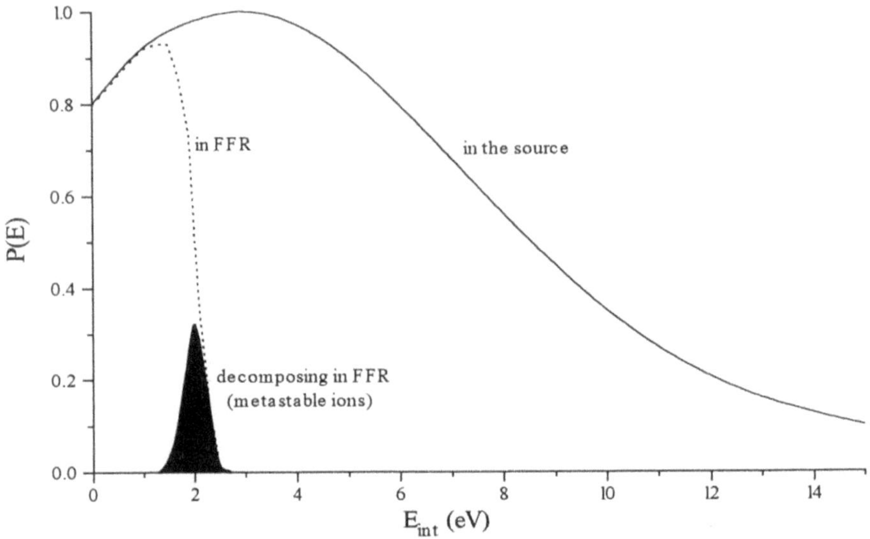

Figure 5. Typical internal energy distribution of the molecular ion in 70 eV electron impact ionization (solid line), and of those, which reach the FFR undecomposed (dotted curve). Internal energy distribution of molecular ions fragmenting in the FFR (metastable ions) are shown by the shaded area. Data were taken from [4].

The internal energy content of molecular ions derives from two different sources: the thermal energy of the molecule before ionization and the amount of energy deposited in the ionization process. Typically (organic molecules below mass 500, at ion source temperature) the contribution of the thermal energy to the internal energy of an ion can be neglected. There are two exceptions, in the case of large molecules (above 500-1000 dalton), where the thermal energy may exceed energy deposition by ionization even at room temperature and in the case of ionization of very labile molecules, which decompose easily (e.g. some alcohols). [4, 7, 8].

Electron impact is by far the most thoroughly studied and the best understood ionization technique. It is generally assumed that at 70 eV ionizing electron energy the internal energy distribution of molecular ions is similar to that shown in Fig. 5 (continuous line): a wide distribution having a long tail to high internal energies. Increasing the ionizing electron energy above this value (to 100 or 1000 eV) practically does not change the internal energy distribution or the mass spectra. Low energy EI is assumed to have a similar distribution, only the high energy side of the curve is cut off at the value corresponding to the difference of the ionizing electron energy and the IE of the compound (this "cut off" is never sharp, may be ~2 eV wide). The precise internal energy distribution of a particular compound, however, may be different. There may be structure in the P(E) curve, and the average and the maximum value may not be exactly like that shown on Fig. 5. For semiquantitative purposes the curve shown on Fig. 5 can be, however, taken as a general guide.

The expressions "hard" and "soft" ionization are connected to internal energy content. The distinction between them is not rigid, and often depends on the comparison intended. "Hard" ionization indicates a method resulting in high average internal energy molecular ions. The result is a spectrum with low intensity molecular ion and abundant fragments, like in 70 eV EI spectra. "Soft" ionization is the opposite expression, means low average internal energy, where the molecular ion is of high abundance (typically the base peak in the spectrum), while fragments are absent or are of low abundance. In soft ionization not only the fragment ions have low abundance, but also the number of reaction channels (fragment ion types) is small. Typical low electron energy EI is a typical example of soft ionization.

Internal energy distribution in fast atom bombardment (FAB) ionization is similar to that observed in electron impact (Fig. 5, solid line). Fragmentation is, however, often masked by the matrix, the relatively high abundance of the molecular ion is mainly due to the fact that large molecules require very high internal energy to fragment (Fig. 3). Chemical ionization is typically a soft ionization technique, where the protonated molecular ion ($MH^{\oplus}$) is usually the base peak in the spectrum. Electrospray and MALDI are also soft ionization techniques, producing little fragmentation. Like in FAB this, in part, may be due to the slow reaction rates of large molecules.

4.1. INTERNAL ENERGY IN TANDEM MASS SPECTROMETRY

The internal energy of the precursor ion in tandem mass spectrometry is determined by three effects: (1) internal energy distribution following ion formation (a distribution like that shown in Fig. 5. is often a good approximation), (2) "loss" of the high energy ions by fragmentation and (3) internal energy deposition by excitation.

Ions formed with high internal energy are likely to fragment before being selected for MS/MS. This means that precursor ions selected for MS/MS will have a lower average internal energy than the same ions studied in the source (they will lack the high energy tail of the internal energy distribution). This can be illustrated in the example of buthylbenzene in a sector instrument: The molecule ion needs 10 μs to reach the field free region (FFR, calculated from the instrument geometry, accelerating voltage and ion mass). Those, which have a lower lifetime, will decompose beforehand and are lost for subsequent analysis. The internal energy distribution of undecomposed molecule ions reaching the FFR can be calculated using the internal energy distribution following ionization (Fig. 5, solid curve) and the rate constants shown in Fig. 2. The result is shown by the dotted curve in Fig. 5. Using the same data and 5 μs residence time in the FFR, the internal energy distribution of metastable ions (those, which ions fragment spontaneously in the FFR) can be determined (shown by the shaded area in Fig. 5).

Internal energy deposition by secondary excitation is of profound interest in MS/MS. The resulting internal energy (distribution) will be the sum of the internal energy before excitation (as discussed above) and the energy deposition. By far the most commonly used secondary excitation is collisional activation (CA), and the resulting fragmentation is called collision induced (or collisionally activated) decomposition (CID or CAD). In this process accelerated ions collide with a neutral (collision or target) gas. CID processes are divided into high (better defined as keV) and low energy (1-100 eV) collisions. This expression refers to the kinetic energy of the collision partners, not the internal energy transfer! Typically the laboratory frame collision energy are discussed; the center-of-mass (c.o.m.) collision energy is typically much lower. The correlation between collision energy and internal energy deposition is not very well understood.

In keV (high) collision energy CID (used on sector-type instruments) typically one or only few collisions occur (at 70-90% main beam transmission predominantly one, at 30% on average two). The internal energy deposition does not depend significantly on the collision energy. Average internal energy deposition is usually estimated to be a few (1-3) eV. The distribution of internal energy deposition in keV CID is wide and characterized by a high energy tail (energy deposition of 10 eV or more also occurs). This means that in keV energy CID (a) the undecomposed molecular ion will have a high abundance and (b) fragmentation processes requiring large critical energy are also observed. The presence of this high internal energy tail can be very important in some applications. Peptide sequencing is one such example [10]: internal fragments and immonium ions (which contain significant information on the molecular structure) are frequently observed in keV CID, while not in low energy CID.

Low kinetic energy (usually 1-100 eV eV in the laboratory frame) collisions are observed in quadrupole, ion trap or Fourier-transform type mass spectrometers. In this energy range the internal energy deposition increases with the collision energy, although not necessarily linearly. The *maximum* amount of energy, which may be converted to internal energy in one collision, is equal to the c.o.m. collision energy. Usually, however, only a fraction, maybe a few %, of this is energy will become internal energy of the incoming ion. In experiments utilizing low (kinetic) energy CID the precursor ion typically collides many times, so the pressure of the collision gas will also have a large effect. The average internal energy deposition is estimated to be between 1-10 eV, depending on the experimental conditions. The distribution of internal energy deposition is, however, much narrower than in high kinetic energy collisions. A significant limitation is that high internal energies can not be reached (ions obtaining a relatively high internal energy will fragment before they could be further excited). The overall result is that (a) the molecular ion is often absent or of low abundance but (b) fragmentation processes requiring large critical energy are frequently not observed.

Colliding an ion beam with a surface leads to surface induced decomposition (SID, [11]). In this case usually a low kinetic energy (10-100 eV) single collision occurs, the laboratory frame and the c.o.m. collision energy being equal. The kinetic to internal energy transfer is relatively well described: depending on the surface and the precursor ion, 15 - 25% of the collision energy is converted to internal energy. The average value of energy conversion depends linearly on the collision energy and the resulting internal energy distribution is narrow relative to CID.

5. Conclusions

The internal energy content of an ion is a very important parameter in mass spectrometry. It has a similar effect, as the temperature in thermal reactions. The internal energy, and its distribution, determines the relative abundance of molecular and fragment ions, and also the variety of fragmentation processes. The internal energy of an ion is rarely characterized by a well defined single value, but typically by a wide distribution. The internal energy effects are often masked by other changes. Conversely, various effects may be masked by a concommittant change in internal energy distribution. An example is the comparison of metastable and collision induced fragmentations, where the internal energy distribution of the precursor ions is markedly different. The two techniques, however, sample a different section of the precursor ions: those giving metastable fragments initially had more internal energy, so may have isomerized to a different structure than the rest, which fragments by CID.

Theoretical modeling of internal energy effects is often very helpful - partly to reveal internal energy effects, partly for a better understanding of fragmentation processes, ion structures, energetics.

6. References

[1] R.G. Cooks, J.H.Beynon, R.M.Caprioli, G.R. Lester, *Metastable Ions*, Elsevier, Amsterdam, 1973

[2] K. Levsen, *Fundamental Aspects of Organic Mass Spectrometry*, Verlag Chemie, Weinheim, 1978

[3] I. Howe, D.H. Williams, R.D.Bowen, *Mass Spectrometry, Principles and Applications*, McGraw-Hill, New York, 1981

[4] K. Vékey, *J. Mass Spectrom.*, **31**, 445 (1996)

[5] H.M. Rosenstock, M.B. Wallenstein, A.L. Wahrhafig, H. Eyring, *Proc. Nat. Acad. Sci.*, **38,** 667 (1952)

[6] P.J. Robinson, K.A. Holbrook, *Unimolecular reactions*, Wiley, London, 1972

[7] K. Vékey, Á.Somogyi, V. Wysocki, *Rapid Comm. Mass Spectrom.*, **10,** 911 (1996)

[8] P.J. Derrick, P.M.Loyd, J.R. Christie in *Advances in Mass Spectrometry* vol. 13., Eds. I. Cornides, Gy. Horváth, K. Vékey, Wiley, Chichester, 1995

[9] The use of the "RRKM Large" program, kindly provided by J.R. Christie is gratefully acknowledged.

[10] K.F. Medzihradszky, A.L.Burlingame, *A Companion to Methods in Enzymology.*, **6**, 284 (1994)

[11] R.G. Cooks, T. Ast, T. Pradep, V.H. Wysocki, *Acc. Chem. Res.*, **27**, 316 (1994)

INTERACTION OF METAL IONS AND PROTONS WITH SIMPLE BIOMOLECULES. STRUCTURE INFORMATION FROM BOND ENERGIES AND ENTROPY EFFECTS

B. A. CERDA, M. J. NOLD, AND C. WESDEMIOTIS
Department of Chemistry
The University of Akron
Akron, OH 44325-3601, USA

ABSTRACT: A modified version of the kinetic method that determines bond energies when entropy effects are not negligible is presented. The enthalpic *and* entropic data derived this way provide unique information on the structures of cationized biomolecules, as demonstrated for alkali ion adducts of nucleobases. The proton affinities of N- and C-terminal breakup products of protonated peptides are also evaluated by the kinetic method, in an effort to gain new insight on the decomposition mechanism of peptide ions.

1. Introduction

Metal ions and protons play a pivotal role in biology both through direct involvement in biochemical reactions [1] and through the mediation of molecular recognition [2]. The latter process helps specific reactants and reagents find each other, mainly via hydrogen bonding and electrostatic metal ion interactions. In vivo, metal ions are stored and transported by biomolecules that selectively recognize and bind them [3]. Metalation and protonation also are encountered in analytical mass spectrometry, where they represent important modes of ionization for biomolecules [4]. The structure and function in vivo of a metalated or protonated biomolecule as well as its reactivity in the mass spectrometer are significantly influenced by its thermochemical properties. For this reason, knowledge of the proton and metal ion affinities of the different classes of biomolecules is of fundamental interest.

R. M. Caprioli et al. (eds.), Selected Topics and Mass Spectrometry in the Biomolecular Sciences, 143–161.

The affinity of a biomolecule (B) to a cation Cat^+ (e.g., a proton or a metal ion) is defined as the energy needed to break the B−Cat^+ bond (eq 1).

$$B\text{-}Cat^+ \rightarrow B + Cat^+ \qquad \Delta H^o_{rxn} = \Delta H^o_{Cat} \qquad (1)$$

$$\Delta G^o_{Cat} = \Delta H^o_{Cat} - T\Delta S^o_{Cat} \qquad (2)$$

In ion B−Cat^+, B represents a nucleophile or ligand (i.e. a Lewis base) and Cat^+ an electrophile (i.e. a Lewis acid). Cation - Lewis base bonds can be measured by several methods [5,6], including guided ion beam experiments involving endothermic reactions [7], equilibrium [8] and bracketing [9] approaches, photodissociation [10], analysis of the kinetic energy release distributions resulting upon B−Cat^+ decomposition [11], or from the kinetics of dissociation of Cat^+-bound dimers (Cooks kinetic method) [12]. Affinities are bond enthalpies (eq 1); however, the majority of the mentioned techniques actually supplies free energies, ΔG^o_{Cat} (also called basicities), which are related to the desired affinities by eq 2. Experimentally acquired ΔG^o_{Cat} values depend on both the B−Cat^+ bond energy (ΔH^o_{Cat}) and the B−Cat^+ bond entropy (ΔS^o_{Cat}). Here, we present a modified version of the Cooks kinetic method that deconvolutes enthalpy from entropy, thereby leading to cation affinities as well as information on bond entropies [13]. It is shown that the enthalpic and entropic data obtained this way provide valuable insight on the structure of the cationized biomolecule.

2. Experimental Approach

The kinetic method assesses an unknown B−Cat^+ bond energy by comparing it to a known B_i−Cat^+ bond energy in the Cat^+-bound heterodimer B−Cat^+−B_i. This is accomplished by measuring the kinetics of dissociation of B−Cat^+−B_i to each individual cationized monomer (eq 3), from which the relative cation affinity (i.e. affinity difference) between B and B_i can be derived.

$$B\text{-}Cat^+\text{-}B_i \xrightarrow{k} B\text{-}Cat^+ + B_i \qquad (3a)$$

$$B\text{-}Cat^+\text{-}B_i \xrightarrow{k_i} B + Cat^+\text{-}B_i \qquad (3b)$$

According to transition state theory [14], the rate constant ratio k/k_i, which is equivalent to the ratio of the abundances of $B-Cat^+$ and Cat^+-B_i in a decomposition spectrum of dimer $B-Cat^+-B_i$ is given by eq 4. In this equation, Q^* and Q_i^* are the partition functions for the activated complexes in reactions 3a and 3b, respectively, ε^o and ε_i^o are the corresponding activation energies, and T_{eff} is the effective temperature [15] of the dissociating dimer ion $B-Cat^+-B_i$. T_{eff} is a measure of the internal energy of $B-Cat^+-B_i$ and depends on the lifetime of this ion as well as instrumental parameters [12]. If the reverse activation energies of fragmentations 3a and 3b are negligible, then $\varepsilon_i^o - \varepsilon^o$ can be approximated by $\Delta H^o_{Cat}(B) - \Delta H^o_{Cat}(B_i) = \Delta(\Delta H^o_{Cat})$, i.e. by the relative cation affinity of the two bases being compared. If further B and B_i are chemically similar species and thus bind Cat^+ in an analogous fashion, the partition functions Q^* and Q_i^*, which describe entropy, are quite similar and $\ln(Q^*/Q_i^*) \approx 0$ [12]. Under these conditions, the experimentally measurable $\ln(k/k_i)$ ratio becomes directly proportional to the difference in affinities between B and B_i, viz. eq 5.

$$\ln(k/k_i) = \ln(Q^*/Q_i^*) - (\varepsilon^o - \varepsilon_i^o)/RT_{eff} \qquad (4)$$

$$\ln(k/k_i) \approx [\Delta H^o_{Cat}(B) - \Delta H^o_{Cat}(B_i)]/RT_{eff} = \Delta(\Delta H^o_{Cat})/RT_{eff} \qquad (5)$$

With a series of reference bases B_i, a plot of $\ln(k/k_i)$ versus $\Delta H^o_{Cat}(B_i)$ supplies T_{eff} and the desired affinity $\Delta H^o_{Cat}(B)$.

The simple relationship of eq 5 has been applied extensively to determine unknown bond energies [12], such as proton affinities [16], gas phase acidities [17], metal ion affinities [18], and electron affinities [19]. Deviations have however been reported [12,18,20,21], indicating that the above mentioned assumptions are not always valid. Reverse activation energies can be probed by measuring the kinetic energy releases associated with the competitive dissociations 3a and 3b [22,23]. For singly charged heterodimers, we have found that the kinetic energy releases are small (typically <50 meV irrespective of Cat^+), consistent with no significant reverse activation energies. Discrepancies can still arise from entropic effects. Entropy changes can not be neglected if the heterodimer $B-Cat^+-B_i$ undergoes rearrangement upon dissociation [12,24] or, most commonly, when $B-Cat^+-B_i$ is prepared with molecules from different chemical classes because chemically similar reference bases do not exist (as is true for most biomolecules under study) [12,18,21]. When entropy effects are not negligible, eq 5 must include the $\ln(Q^*/Q_i^*)$ term, cf. eq 6. In the absence of reverse activation energies, $\ln(Q^*/Q_i^*)$ corresponds to the difference in entropy change between reactions 3a and 3b, cf. eq 7. Combination of equations 6 and 7 leads to relationship 8, which shows that the kinetic method provides free energies (i.e. basicities) if the entropy changes of reactions 3a and 3b do not cancel.

$$\ln(k/k_i) = \Delta(\Delta H^o_{Cat})/RT_{eff} + \ln(Q^*/Q_i^*) \qquad (6)$$

$$\ln(Q^*/Q_i^*) = -\Delta(\Delta S^o_{Cat})/R = -[\Delta S^o_{Cat}(B) - \Delta S^o_{Cat}(B_i)]/R \qquad (7)$$

$$\ln(k/k_i) = \Delta(\Delta H^o_{Cat})/RT_{eff} - \Delta(\Delta S^o_{Cat})/R = \Delta(\Delta G^o_{Cat})/RT_{eff} \qquad (8)$$

$$\ln(k/k_i) = [\Delta H^o_{Cat}(B)/RT_{eff} - \Delta(\Delta S^o_{Cat})/R] - \Delta H^o_{Cat}(B_i)/RT_{eff} \qquad (9)$$

Expanding the enthalpic term in eq 8 yields eq 9. If the reference bases used in the dimers $B{-}Cat^+{-}B_i$ are kept chemically similar to each other, the net entropy change $\Delta(\Delta S^o_{Cat})/R$ should remain constant [21]. With a set of $B{-}Cat^+{-}B_i$ dimers, a plot of $\ln(k/k_i)$ versus $\Delta H^o_{Cat}(B_i)$ now provides T_{eff} and the term $\Delta H^o_{Cat}(B)/RT_{eff} - \Delta(\Delta S^o_{Cat})/R$ from which an apparent cation affinity, viz. $\Delta G^{app}_{Cat}(B)$, can be calculated (eq 10). Since the latter quantity contains enthalpic and entropic contributors, it is designated a basicity acronym.

$$\Delta G^{app}_{Cat}(B) = \Delta H^o_{Cat}(B) - T_{eff}\,\Delta(\Delta S^o_{Cat}) \qquad (10)$$

$$\Delta(\Delta S^o_{Cat}) = [\Delta G^{app}_{Cat}(B)_{MI} - \Delta G^{app}_{Cat}(B)_{CAD}]/[T_{CAD} - T_{MI}] \qquad (11)$$

By investigating the $B{-}Cat^+{-}B_i$ set at two effective temperatures, e.g., under metastable ion (MI) and collisionally activated dissociation (CAD) conditions, it is possible to determine the entropic term $\Delta(\Delta S^o_{Cat})$ based on eq 11; then the cation affinity of B, viz. $\Delta H^o_{Cat}(B)$, can be deconvoluted using eq 10 [13]. As will be shown in this chapter, both these thermochemical quantities provide important information on the binding of Cat^+ to the biomolecule under study. Moreover, the given procedure allows one to obtain true bond energies (affinities), which are preferred over basicities in quantitative correlations of intrinsic structural properties. Experiments at one effective temperature, which yield basicities if entropy effects do not cancel (eqs 8-9), are still useful for a qualitative analysis of the binding of simple cations to biomolecules.

3. Alkali Metal Ion Affinities of Purine and Pyrimidine Nucleobases

The modified kinetic method was applied to determine the Li^+, Na^+, and K^+ affinities of the DNA and RNA nucleobases depicted in Figure 1. The reference bases B_i utilized for

Li^+ and Na^+ were the structurally similar aminoacids glycine, alanine, and valine [25], while the B_i set for K^+ consisted of the unidentate amines pyridine, *n*-propylamine, and aniline [26] (Table 1).

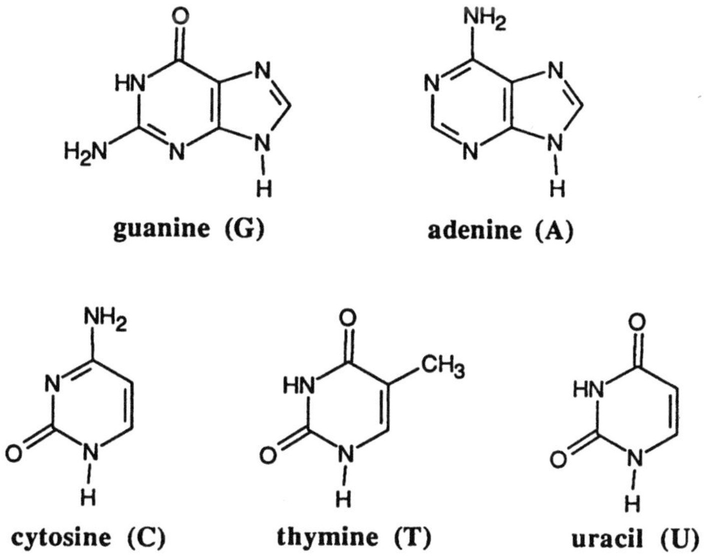

guanine (G) adenine (A)

cytosine (C) thymine (T) uracil (U)

Figure 1 DNA (G, A, C, T) and RNA (G, A, C, U) nucleobases studied.

TABLE 1. Alkali metal ion affinities of references bases in kJ mol^{-1}.

B_i	$\Delta H^o{}_{Li}$	$\Delta H^o{}_{Na}$	$\Delta H^o{}_{K}$	*Ref.*
glycine (Gly)	213	159		25
alanine (Ala)	220	165		25
valine (Val)	229	172		25
pyridine (Py)			86.6	26
n-propylamine (Pr)			91.2	26
aniline (An)			95.4	26

For each nucleobase (NB) and metal ion (M^+), the corresponding [NB + B_i] M^+ dimers were generated by FAB ionization in a glycerol matrix, and their MI and CAD (He) spectra were acquired with a sector tandem mass spectrometer following established procedures [13,18,27]. Two representative MS/MS spectra are depicted in Figure 2 and belong to the K^+-bound dimer of cytosine (C) and pyridine (Py), viz. [C + Py]K^+. The predominant fragments in these spectra are the metalated monomers, in support of the connectivity NB–M^+–B_i. Further, the peaks in the MI spectra are of gaussian shape and their widths correspond to kinetic energy releases ($T_{0.5}$) of 20 meV or less, pointing out that the dimer ions decompose without appreciable reverse activation energies [22,28], as required by the kinetic method.

The relative abundances of the metalated monomers in Figure 2 indicate that cytosine binds K^+ stronger than pyridine. A quantitative assessment of the C–K^+ bond energy is obtained by application of eq 9 to the entire [C + B_i]K^+ set (B_i = Py, Pr, An, cf. Table 1). The regression line resulting from the MI spectra (Figure 3) yields T_{MI} = 620 K (from the slope) and $\Delta G^{app}{}_K(C)_{MI}$ = 106 kJ mol^{-1} (from the intercept). Similarly, the CAD data (Figure 3) give rise to T_{CAD} = 1258 K and $\Delta G^{app}{}_K(C)_{CAD}$ = 102 kJ mol^{-1}. Based on this information, eqs 10 and 11 lead to $\Delta H^o{}_K(C)$ = 110 kJ mol^{-1} and $\Delta(\Delta S^o{}_K)$ = +6.3 J mol^{-1} K^{-1}, respectively. The enthalpic value represents the potassium ion affinity of cytosine, i.e. the C–K^+ bond energy, while the entropic value is the net entropy change upon dissociation of the C–K^+–B_i dimers, i.e. the difference in bond entropies between the C–K^+ and B_i–K^+ bonds. Application of this procedure to all other nucleobases and alkali metal ions furnishes the enthalpic and entropic data summarized in Table 2. The error limits in these values are ±<4 kJ mol^{-1} and ±<1.5 J mol^{-1} K^{-1}, respectively, and mainly reflect the uncertainty in $\Delta H^o{}_M(B_i)$. Relative alkali ion affinities, i.e. $\Delta(\Delta H^o{}_M)$ between individual nucleobases, are more accurate (±<2 kJ mol^{-1}).

For the three metal ions studied, the affinities are found to increase in the order uracil (U) < thymine (T) < adenine (A) < cytosine (C) < guanine (G); guanine always forms the strongest and uracil the weakest NB–M^+ bond. Li^+, Na^+, and K^+ are spherical ions and are considered as hard Lewis acids [29]; hence, their common behavior towards the nucleobases is not surprising. With a given nucleobase, $\Delta H^o{}_M$ increases from K^+ to Na^+ to Li^+. Evidently, the NB–M^+ bond becomes weaker as the radius of the Lewis acid increases, as expected [25,30].

The entropy term $\Delta(\Delta S^o{}_M)$, which is also listed in Table 2, corresponds to the entropy difference upon breaking the NB–M^+ vs. the M^+–B_i bond in NB–M^+–B_i.

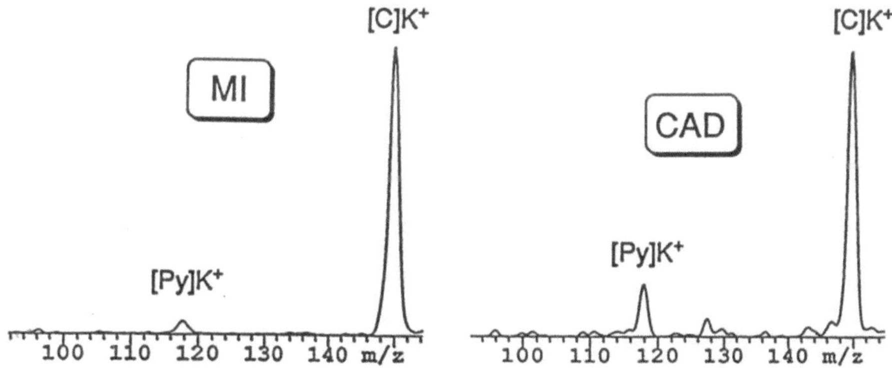

Figure 2. (a) MI and (b) CAD spectra of the K^+-bound dimer of cytosine (C) and pyridine (Py), viz. $[C + Py]K^+$. He (90% transmittance) served as the CAD target.

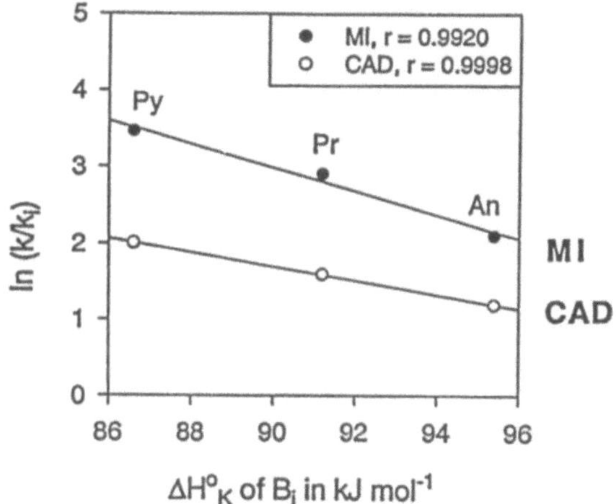

Figure 3. Plot of $\ln(k/k_i)$ versus $\Delta H^o{}_K(B_i)$ at two effective temperatures for cytosine dimers $[C + B_i]K^+$. The ratio k/k_i is calculated by dividing the abundances of $[C]K^+$ and $[B_i]K^+$ in the corresponding MI (filled circles) or CAD spectra (open circles). Reproduced with permission from ref. [13].

TABLE 2. Alkali ion affinities of the nucleobases (NB) and net entropy changes in the decomposition of $[NB + B_i]M^+$ dimers. The values given are in kJ mol^{-1} for $\Delta H^o{}_M$ and in J mol^{-1} K^{-1} for $\Delta(\Delta S^o{}_M)$.

Nucleobase	$\Delta H^o{}_{Li}$	$\Delta(\Delta S^o{}_{Li})$	$\Delta H^o{}_{Na}$	$\Delta(\Delta S^o{}_{Na})$	$\Delta H^o{}_K$	$\Delta(\Delta S^o{}_K)$
guanine (G)	239	−4.3	182	−6.7	117	+6.9
cytosine (C)	232	−3.9	177	−7.3	110	+6.3
adenine (A)	226	−5.1	172	−8.1	106	+6.9
thymine (T)	215	−10.8	144	−13.8	102	+2.3
uracil (U)	211	−13.9	141	−11.9	101	+1.9

Thus, it is expected to depend on the nature of B_i. Indeed, the $\Delta(\Delta S^o{}_M)$ values resulting from the aminoacid B_i set (used for Li^+ and Na^+) differ substantially in sign and magnitude from those resulting from the monoamine B_i set (used for K^+). The trends observed can adequately be accounted for by attributing $\Delta(\Delta S^o{}_M)$ to the rotational entropy change occurring upon dissociation of the $[NB + B_i]M^+$ dimers [13,26,31].

Figure 4 shows the dissociation of an "idealized" K^+-bound dimer between adenine and pyridine, in which both ligands bind K^+ in a similar, unidentate fashion ("chemically similar" ligands). Rotational degrees of freedom are not affected in the metalated dimer or its dissociation products and $\Delta(\Delta S^o{}_K)$ should be zero; in such a case the simple form of the kinetic method (eq 5) could have been used. However, the actual $\Delta(\Delta S^o{}_K)$ measured (+6.9 J mol^{-1} K^{-1}) does not support this mechanism.

Theoretical calculations by Rodgers and Armentrout [32] predict that the most stable structures of metalated nucleobases have the metal ion chelated (when possible), as shown in Figure 5 for the K^+ adduct of adenine. Such coordination freezes the rotations of the amino group, which in turn causes entropy changes upon dissociation. From the dimer $[Py + A]K^+$ (Figure 5), the formation of $[Py]K^+ + A$ is associated with a more positive entropy change than the formation of $Py + [A]K^+$, because adenine recovers the NH_2 rotations once it is released from the dimer; overall, entropy is gained and $\Delta(\Delta S^o{}_K)$ is positive.

Figure 4. Dissociations of the "idealized" [Py + A]K^+ dimer, in which both monomers bind K^+ in a unidentate fashion. $\Delta(\Delta S^o_K)$ should be zero. Reproduced with permission from ref. [13].

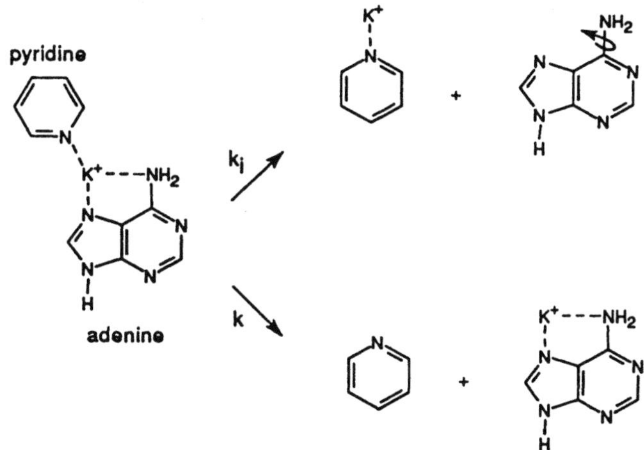

Figure 5. Dissociations of [Py + A]K^+ with involvement of the amino group in the coordination of K^+. Structures justify $\Delta(\Delta S^o_K) = +6.9$ J mol^{-1} K^{-1}. Reproduced with permission from ref. [13].

Inspection of Table 2 shows that the K^+-bridged dimers of cytosine and guanine lead to similar net entropy changes. Evidently, their amino groups are also involved in the binding of the metal ion (vide infra). On the other hand, thymine and uracil, which lack amino substituents, show the smallest $\Delta(\Delta S^o_K)$ values.

The positive values for $\Delta(\Delta S^o_K)$ are due to the fact that the B_i set used for K^+ contains no rotors that can be restricted in the metalated species. The situation changes dramatically with the B_i set used for Li^+ and Na^+, viz. Gly, Ala, and Val. These aminoacids complex the metal ion between the amino and carbonyl functionalities [30], as exemplified for Gly–Li^+–A in Figure 6. Breakup of the Li^+–A bond frees one rotation but breakup of the Gly–Li^+ bond frees two, making $\Delta(\Delta S^o_{Li})$ negative (Table 2); the same is true for $\Delta(\Delta S^o_{Na})$. The $\Delta(\Delta S^o_{Li/Na})$ values of cytosine and guanine are quite similar to those of adenine, consistent with the presence of an amino group in these nucleobases. In contrast, thymine and uracil show much larger net entropy losses, for both Li^+ and Na^+ (Table 2). With these nucleobases, which possess no free rotor that can participate in the binding of the metal ion, practically none of the rotational entropy gained upon fragmentation to B_i + $[NB]M^+$ is counter-balanced upon fragmentation to $[B_i]M^+$ + NB, thus causing $\Delta(\Delta S^o_{Li/Na})$ to be overall more negative.

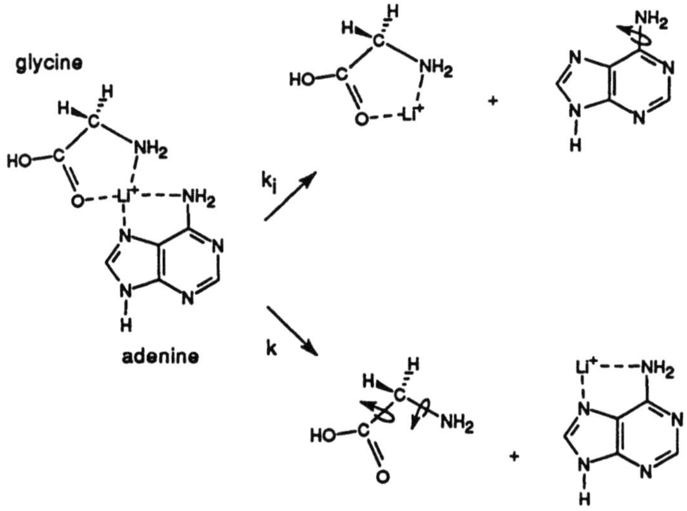

glycine

adenine

k_i

k

Figure 6. Dissociations of [Gly + A]Li^+. The reference base recovers more rotations, justifying $\Delta(\Delta S^o_{Li}) = -5.1$ J mol^{-1} K^{-1}. Reproduced with permission from ref. [13].

The entropy trends in Table 2 can be summarized as follows: the net entropy change $\Delta(\Delta S^{o}_{M})$ has the same sign and comparable values for guanine, cytosine, and adenine, all of which contain amino rotors. Similarly, thymine and uracil, which lack amino rotors, show parallel entropic behavior. Confirmatory characteristics are observed when the nucleobases are compared against each other in $[NB + NB_i]M^+$ dimers, where one nucleobase (NB) acts as the molecule of unknown affinity and two to three of the others (NB_i), which must be chemically similar, as the reference base set [13]. For example, $[G + NB_i]M^+$ dimers with NB_i = C and A, in which both the unknown (guanine) and the reference set (cytosine and adenine) include NH_2 groups, yield $\Delta(\Delta S^{o}_{M}) \approx 0$. In contrast, $[G + NB_i]M^+$ dimers with NB_i = T and U, which contain no rotors, lead to overall entropy gains ($\Delta(\Delta S^{o}_{M}) \approx +10$ J mol^{-1} K^{-1}). Expectedly, the reverse arrangement, namely $[T + NB_i]M^+$ or $[U + NB_i]M^+$ with NB_i = G, C, and A, brings upon a net entropy loss ($\Delta(\Delta S^{o}_{M}) \approx -10$ J mol^{-1} K^{-1}). In the $[NB + NB_i]M^+$ dimers, the sign and magnitude of $\Delta(\Delta S^{o}_{M})$ primarily depend on the nature of NB and NB_i and much less on the metal ion, corroborating that Li^+, Na^+, and K^+ bind similarly to the nucleobases (vide supra).

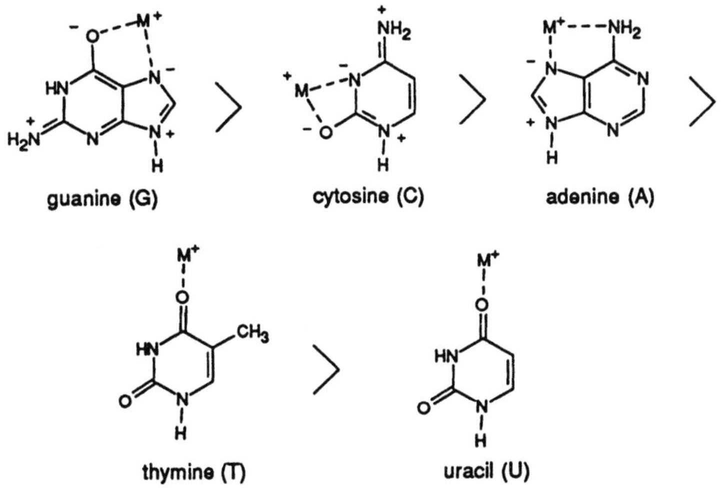

Figure 7. Binding modes of M^+ (Li^+, Na^+, or K^+) to DNA and RNA nucleobases. The $[NB]M^+$ ions are shown in an order of decreasing alkali metal ion affinities. Adapted from ref. [13] and reproduced with permission.

The entropy data unequivocally prove that NH_2 substituents, when present, affect the bonding of the metal ion. The three nucleobases that carry NH_2 groups, viz. guanine, cytosine, and adenine, show superior alkali ion affinities. The experimentally derived ΔH^o_M order and $\Delta(\Delta S^o_M)$ values strongly support the binding modes displayed in Figure 7 [13]. These M^+ attachment sites create the thermodynamically most stable $[NB]M^+$ structures according to the theoretical studies of Rodgers and Armentrout [32]. Moreover, similar binding sites have been suggested in solution [33], indicating that the stability of the resulting complexes is intrinsic and not solvent-mediated. Note that the structures of $[G]M^+$ and $[C]M^+$ involve the amino group through resonance, not directly. Such resonance leads to additional stabilization via the formation of salt bridges (ion pairs) [34]; it also restricts the rotational flexibility of the amino rotor, in line with the experimental observations regarding $\Delta(\Delta S^o_M)$.

The M^+ affinity order of the nucleobases can be rationalized by the attachment sites available in the nucleobases and their capability of developing salt bridges. The higher M^+ affinities of guanine, cytosine, and adenine are ascribed to their ability to complex the metal ion in a bidentate arrangement promoting the formation of salt bridges (Figure 7). Alkali metal ions are oxyphilic. Hence, the lower ranking of adenine vis à vis guanine and cytosine could result from the absence of a carbonyl group in it; additionally, the adenine conjugate system allows only one salt bridge whereas guanine and cytosine permit the formation of salt bridges at both sites attaching to M^+. Further, guanine can coordinate M^+ in a five-membered ring arrangement, while only a four-membered ring is possible with cytosine, explaining $\Delta H^o_M(G) > \Delta H^o_M(C)$. The inferior affinities of thymine and uracil are most probably due to the unidentate bonding of the metal ion. Finally, the slightly higher ranking of thymine vs. uracil must be the consequence of inductive effects.

4. Proton Affinities of Peptide Segments

Among the most important sequence ions arising upon the decomposition of protonated peptides are the N-terminal b_n and the C-terminal y_n series, which result from breakup of the peptide (i.e. amide) bonds in the backbone [35]. The structures of b_n and y_n and of the complementary neutral losses have been established by combined labeling, MS^n, and neutral fragment-reionization studies [36,37,38]. The current status of knowledge is illustrated in Figure 8 for the b_2 and y_2 fragments from protonated tetraalanine. The C-terminus is cleaved as a smaller peptide, either neutral (accompanying b_2) or protonated (y_2); on the other hand, the N-terminus is cleaved either as a protonated oxazolone (b_2) or as a diketopiperazine (accompanying y_2). The mechanism of these reactions could be dependent on the proton affinity (PA) of the products involved. In an effort to shed more light on this issue, the PAs of the peptide, oxazolone, and diketopiperazine structures have been evaluated. Here, we followed a more qualitative approach, conducting experiments at only one effective temperature, viz. T_{MI}. In such cases, the kinetic method leads to apparent basicities, $\Delta G^{app}{}_{H}(B) = PA(B) - T_{eff}\,\Delta(\Delta S^{o}{}_{H})$, cf. eq 10. How much the apparent basicity differs from the proton affinity depends on $\Delta(\Delta S^{o}{}_{H})$, whose value is in turn influenced by the reference base set used (vide supra). The contribution of the term $T_{eff}\,\Delta(\Delta S^{o}{}_{H})$ is minimized at the lowest effective temperature possible, which corresponds to metastable dimer ions.

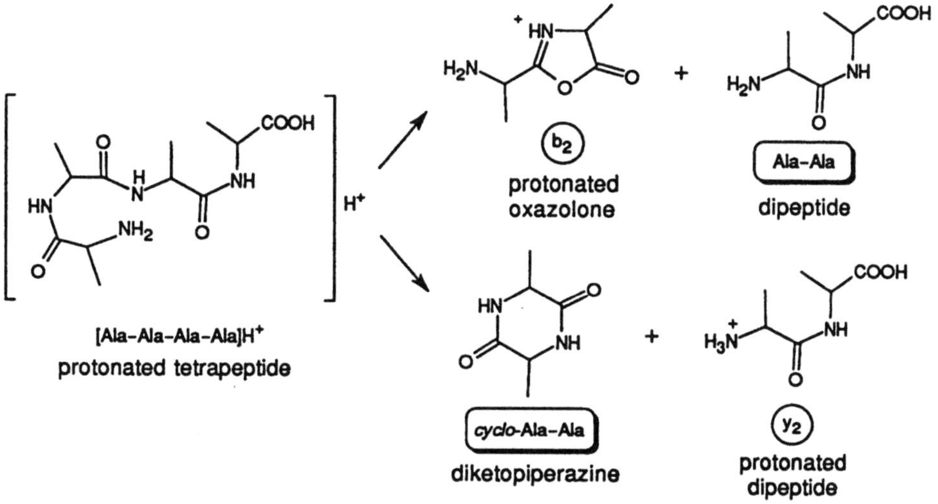

Figure 8. N- and C-terminal sequence ions and their complementary neutral losses formed upon cleavage of the central peptide bond in protonated tetraalanine.

The biomolecules (B) investigated are shown in Figure 9 and encompass 2-phenyl-5-oxazolone, the cyclic dipeptide *cyclo*-Gly–Leu (a diketopiperazine), and the isomeric linear dipeptides Gly–Leu and Leu–Gly. The dipeptides consist of glycine and leucine residues. An oxazolone with the same aminoacid constituents is difficult to synthesize, as would be any other oxazolone with *unprotected* aminoacid residues (e.g., that shown in Figure 8), presumably due to a high *inter*molecular reactivity. In contrast, 2-phenyl-5-oxazolone can be prepared easily (by dehydration of hippuric acid) and is quite stable in the typical FAB matrices (e.g., glycerol). Notice that the oxazolone moiety generated from a protonated peptide contains two basic nitrogens (Figure 8), whereas the selected oxazolone contains only one (Figure 9). For this reason, 2-phenyl-5-oxazolone most likely is less basic than oxazolones emerging from decomposing protonated peptides [39], and its gas phase basicity should be viewed as a lower limit of the basicities expected for "real" oxazolones.

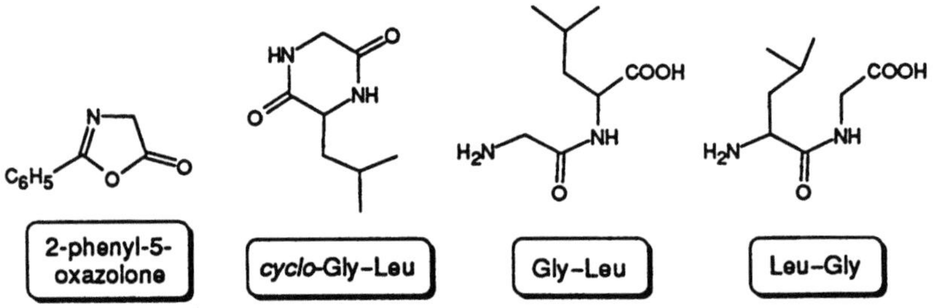

2-phenyl-5-oxazolone | *cyclo*-Gly–Leu | Gly–Leu | Leu–Gly

Figure 9. Biomolecules (B) whose neutral or protonated forms are found in N- or C-terminal peptide fragments.

Table 3 lists the reference bases (B_i) used for each of the molecules in Figure 9 along with reported PA(B_i) values. Proton-bound dimers $[B + B_i]H^+$ were generated by FAB in a suitable matrix (glycerol or thioglycerol), and their competitive dissociations to $[B]H^+$ and $[B_i]H^+$ were assessed from metastable ion spectra. Application of eq 9 leads to the apparent basicities included in Table 3.

As mentioned above, the experimentally determined quantity is the apparent basicity $\Delta G^{app}_H(B) = PA(B) - T_{eff}\,\Delta(\Delta S^o_H)$. Depending on $\Delta(\Delta S^o_H)$, the apparent basicity could be closer to PA(B) (if $\Delta(\Delta S^o_H) \approx 0$) or closer to the true gas phase basicity $GB(B) = PA(B) - T_{eff}\,\Delta S^o_H$ (if $\Delta(\Delta S^o_H) \approx \Delta S^o_H(B)$ i.e. if most net entropy change results from B and not the reference base set B_i in the dissociation of B_i–H^+–B).

TABLE 3. Basicities of the N- and C-terminal peptide segments shown in Figure 9. All thermochemical data given are in kJ mol^{-1}.

B	B_i	$PA(B_i)$ [ref.]	$\Delta G^{app}{}_H(B)$ (from eq 9)
2-phenyl-5-oxazolone	alanine	899 [39]	
	serine	909 [40]	
	proline	931 [40]	**914**
cyclo-Gly–Leu	alanine	899 [39]	
	serine	909 [40]	
	phenylalanine	920 [40]	**874**
Gly–Leu	ethylamine	908 [41]	
	diethylamine	945 [41]	
	di-*n*-propylamine	952 [41]	**933**
Leu–Gly	the same as for Gly–Leu		**938**

Based on the data in Table 3, the oxazolone system is substantially more basic than the diketopiperazine system. It is reminded at this point that N-terminal sequence ions are protonated oxazolones while N-terminal neutral losses are diketopiperazines (vide supra and Figure 8). Thus, the N-terminus of a protonated peptide is cleaved with the structure of higher basicity as a charged particle, but with the structure of lower basicity as a neutral particle. Presumably, the barriers to these (instead of alternative) combinations are most favorable.

The $\Delta G^{app}{}_H(B)$ values for *cyclo*-Gly–Leu, Gly–Leu, and Leu–Gly further support the following conclusions: (i) with the linear dipeptides, H^+ forms the stronger bond when the more basic aminoacid (Leu) is located at the N-terminus [16b]. This result agrees well with protonation at the N-terminus, as has been proposed in previous studies [9d]. (ii) The much lower basicity of the cyclic peptide, which only contains amide bonds, indicates that free termini bring upon a higher basicity. This could be the result of a higher inherent proton affinity for the amine group than for the amide group [39] as well as of a higher stabilization of the proton in the linear structures through additional H-bonding [9d].

5. Conclusions

A modified version of the kinetic method has been presented that accounts for entropy effects, to thereby obtain the metal ion (or other) affinities of biomolecules for which no chemically similar reference bases exist. The approach is demonstrated with the derivation of the Li^+, Na^+, and K^+ affinities of DNA and RNA nucleobases. Deconvolution of entropy is essential for obtaining the correct order of the quite similar M^+–NB bond energies. The entropic quantities measured also provide valuable insight on the attachment site of the metal ion. On the other hand, entropic correction may not be necessary for a qualitative assessment of bond energies that differ significantly from each other, as do the proton affinities of the peptide segments studied.

Structure and sequence are found to greatly affect the bonding between cations (guests) and biomolecules (hosts). This means that thermochemical values gained by the kinetic method could develop into a powerful tool for probing the attachment sites in guest-host complexes and molecular recognition. As shown for the nucleobases, entropy data shed light on the rotational flexibility of $[B + Cat]^+$ and, thus, could be useful in studies of biomolecular folding. Current studies in the Akron laboratory focus on protonated and metalated oligopeptides and on metalated saccharides for a better understanding on how these cations are coordinated by flexible (and hence foldable) vs. more rigid structures.

6. Acknowledgements

We thank Dr. William G. Kofron for the synthesis of 2-phenyl-5-oxazolone and Michael J. Polce for helpful and stimulating discussions. This work was supported by the Ohio Board of Regents, the National Institutes of Health, and the University of Akron.

7. References

1. Zubay, J. (1988) *Biochemistry*, 2nd. ed., MacMillan Publishing Company, New York.
2. Ball, P. (1994) *Designing the Molecular World: Chemistry at the Frontier*, Princeton University Press, Princeton.
3. Lippard, S. J. and Berg, J. M. (1994) *Principles of Bioinorganic Chemistry*, University Science Books, Mill Valley.
4. Siuzdak, G. (1996) *Mass Spectrometry for Biotechnology*, Academic Press, San Diego.
5. Martinho Simões, J. A. and Beauchamp, J. L. (1990) *Chem. Rev.* **90**, 629-688.
6. Eller, K. and Schwarz, H. (1991) *Chem. Rev.* **91**, 1121-1177.
7. Armentrout, P. B. (1989) In D. H. Russell (ed.), *Gas Phase Inorganic Chemistry*, Plenum Press, New York, pp. 1-42.
8. Aue, D. H. and Bowers, M. T. (1979) In M. T. Bowers (ed.), *Gas Phase Ion Chemistry*, Academic Press, New York, vol. 2, pp. 1-51.
9. (a) Locke, M. J. and McIver, R. T. (1983) *J. Am. Chem. Soc.* **105**, 4226-4232. (b) Uppal, K., Lebrilla, C. B.; Drewello, T., and Schwarz, H. (1988) *J. Am. Chem. Soc.* **110**, 3068-3071. (c) Gorman, G. S., Speir J. P., Turner, C. A., and Amster, J. A. (1992) *J. Am. Chem. Soc.* **114**, 3986-3988. (d) Zhang, K., Zimmerman, D. M., Chung-Phillips, A., and Cassady, C. J. (1993) *J. Am. Chem. Soc.* **115**, 10812-10822.
10. Cassady, C. J. and Freiser, B. S. (1984) *J. Am. Chem. Soc.* **106**, 6176-6179.
11. Van Koppen, P. A. M., Jacobson, D. B., Illies, A., Bowers, M. T., Hanratty, M, and Beauchamp, J. L. (1989) *J. Am. Chem. Soc.* **111**, 1991-2001.
12. Cooks, R. G., Patrick, J. S., Kotiaho, T., and McLuckey, S. A. (1994) *Mass Spectrom. Rev.* **13**, 287-339.
13. Cerda, B. A. and Wesdemiotis, C. (1996) *J. Am. Chem. Soc.*, submitted.
14. Robinson, P. J. and Holbrook, K. A. (1972) *Unimolecular Reactions*, Wiley-Interscience, London.
15. Vékey, K. (1996) *J. Mass Spectrom.* **31**, 445-463.
16. (a) Wu, Z. and Fenselau, C. (1992) *J. Am. Soc. Mass Spectrom.* **3**, 863-866. (b) Wu, Z. and Fenselau, C. (1993) *Tetrahedron* **49**, 9197-9206.
17. Graul, S. T., Schnute, M. E., and Squires, R. R. (1990) *Int. J. Mass Spectrom.*

Ion Processes **96**, 181-198.

18. Cerda, B. A. and Wesdemiotis, C. (1995) *J. Am. Chem. Soc.* **117**, 9734-9739.
19. Burinsky, D., Fukuda, E. K., and Campana, J. E. (1984) *J. Am. Chem. Soc.* **106**, 2770-2771.
20. Bliznyuk, A. A., Schaefer, H. H., III, and Amster, I. J. (1993) *J. Am. Chem. Soc.* **115**, 5149-5154.
21. Cheng, X., Wu, Z., and Fenselau, C. (1993) *J. Am. Chem. Soc.* **115**, 4844-4848.
22. Cooks, R. G., Beynon, J. H., Caprioli, R. D., and Lester, G. R. (1973) *Metastable Ions*, Elsevier, Amsterdam.
23. Busch, K. L., Glish, G. L., and McLuckey, S. A. (1988) *Mass Spectrometry/Mass Spectrometry*, VCH Publishers, New York.
24. Grützmacher, H. -F and Caltapanides, A. (1994) *J. Am. Soc. Mass Spectrom.* **5**, 826-836.
25. Bojesen, G., Breindahl, T., and Andersen, U. (1993) *Org. Mass Spectrom.* **28**, 1448-1452.
26. Davidson, W. R. and Kebarle, P. (1976) *J. Am. Chem. Soc.* **98**, 6133-6138.
27. Polce, M. J., Cordero, M. M., Wesdemiotis, C., and Bott, P. A. (1992) *Int. J. Mass Spectrom. Ion Processes* **113**, 35-58.
28. Levsen, K. (1978) *Fundamental Aspects of Organic Mass Spectrometry*, Verlag Chemie, Weinheim.
29. Flemming, I. (1976) *Frontier Orbitals and Organic Chemical Reactions*, John Wiley & Sons, Chichester, pp. 34-40.
30. Bouchonnet, S. and Hoppilliard, Y. (1992) *Org. Mass Spectrom.* **27**, 71-76.
31. Woodin, R. L. and Beauchamp, J. L. (1978) *J. Am. Chem. Soc.* **100**, 501-508.
32. Rodgers, M. T. and Armentrout, P. B. (1996) Poster MPC 32 presented at the 44th ASMS Conference, Portland, Oregon, May 12-16.
33. Marzilli, L. G., Kistenmacher, T. J., and Eichhorn, G. L. (1980) In T. G. Spiro (ed.) *Nucleic Acid - Metal Ion Interactions*, John Wiley & Sons, New York, pp. 179-250.
34. Campbell, S., Rodgers, M. T., Marzluff, E. M., and Beauchamp, J. L. (1995) *J. Am. Chem. Soc.* **117**, 12840-12854.
35. Biemann, K. (1990) *Meth. Enzym.* **193**, 351-360 and 455-479.
36. Mueller, D. R., Eckersley, M., and Richter, W. (1988) *Org. Mass Spectrom.* **23**, 217-221.

37. Cordero, M. M., Houser, J. J., and Wesdemiotis, C. (1993) *Anal Chem.* **65**, 1594-1601.
38. Yalcin, T., Khouw, I. G., Csizmadia, M. R., and Harrison, A. G. (1995) *J. Am. Soc. Mass Spectrom.* **6**, 1165-1174.
39. Lias, S. G., Bartmess, J. E., Liebman, J. F., Holmes, J. L., Levin, R. D., and Mallard, W. G. (1988) *J. Phys. Chem. Ref. Data* **17**, Suppl. No. 1.
40. Bojesen, G. and Breindahl, T. (1994) *J. Chem. Soc., Perkin Trans. 2* 1029-1037.
41. Lias, S. G., Liebman, J. F., and Levin, R. D. (1984) *J. Phys. Chem. Ref. Data* **13**, 695-808.

A PROMISING MARRIAGE BETWEEN THEORY AND EXPERIMENT: DENSITY FUNCTIONAL METHOD VERSUS MASS SPECTROMETRY.

T. MARINO, N. RUSSO, E. SICILIA
and M. TOSCANO

*Dipartimento di Chimica, Universita' della Calabria,
I-87030 Arcavacata di Rende (CS), Italy*

ABSTRACT. The reliability of gaussian density functional method in reproducing and predicting a number of gas-phase thermochemical properties, that are currently obtained experimentally with modern mass spectrometry techniques, is shown through some molecular computations on systems including simple diatomics, amino acids and nucleic acid bases. The gas-phase thermochemical determinations concern:

- Proton affinities
- Affinities for metal ions
- Absolute basicities and acidities

Structural characterizations include:

- Ion structures and potential energy surfaces
- Preferred attach sites for proton and metal ions.

Comparisons with experimental data reveal a good agreement and open a new perspective for density functional based methods that can be used closely with modern mass spectrometry methodologies in order to complete gas-phase thermochemical data and characterize better simple and complex chemical systems.

1. Introduction

In recent years the applications of density functional techniques to molecular systems of chemical and physical interest has grown enormously. The interest is essentially due to the:

- powerful and practical scheme proposed by Kohn and Sham for the many-body problem in a single-particle framework;
- progress in the methodology development;

R. M. Caprioli et al. (eds.), Selected Topics and Mass Spectrometry in the Biomolecular Sciences, 163–179.

- improving and expanding of the availability of density functional based computer codes;
- accuracy in reproducing many properties (i.e. conformational equilibria, spectroscopic constants, binding energies and thermochemical parameters) for a wide range of chemical systems;
- reasonable computational requests that are smaller than those necessary for the more sophisticated post-Hartree-Fock approaches (i.e. multireference configuration interaction and coupled cluster);

The main conceptual characteristic of density functional theory (DFT) is to consider the electron density of a system as the principal variable rather than the wave function as in the traditional quantum chemistry. DFT has its origin in the works of Thomas [1] and Fermi [2] in the 1920s, but it became a well found theory with the papers of Hohenberg and Kohn [3] and Kohn and Sham [4] in the early 1960s. Practical applications to systems of chemical interest have been allowed by the introduction of the so-called $X\alpha$ approximation of Slater [5] and after the definition of reliable exchange-correlation potentials in the framework of the local density approximation [6-8]. Recently, with the development of nonlocal gradient-corrected, functionals for the exchange-correlation terms, in inhomogeneous systems [9-12], or with the hybrid ones [13], the applications of DFT in many fields of modern chemistry have increased considerably [14-27]. In fact, accurate DF studies have indeed been reported for compounds (i.e. organic, inorganic, organometallic molecules, clusters, crystalline solids) and for many properties (i.e. ground-state geometries, activation and dissociation energies, spectroscopic parameters, electrostatic potentials, proton affinities, chemisorption energies, band structures, etc).

In this work we will show the capabilities of the gaussian density functional method in the computation of different thermochemical properties. The importance of this information in chemistry and physics is so big that does not need justifications.

At experimental level the gas-phase thermochemical constants can be studied with a powerful technique, such is the mass spectrometry, essentially in two different ways: equilibrium [28] and kinetic [29] methods. A survey of recent literature shows that nowadays, by modern mass spectrometry equipment, it is possible to know with high accuracy proton and metal affinities, gas-phase basicities and acidities, ion isomer distinction, regioselectivity, isotope and steric effects. Notwithstanding these possibilities, sometimes experimental data are difficult to interpret, in other cases some significant information (i.e. the preferred protonation sites, complete potential energy surfaces, structural parameters) cannot be obtained by these methods. Furthermore, many systems can be degraded during the vaporisation process and their investigation is not possible. In this situation, reliable theoretical methods can be useful to guide the experiment, to help the data interpretation or to predict properties and parameters for molecules that cannot be investigated experimentally. Selected examples help us to show that density functional theoretical methods are powerful tools for gas-phase thermochemistry studies and can be coupled to mass spectrometry techniques for characterizing well the chemical information.

The studied properties include gas-phase proton and metal affinities, basicities and acidities for a series of organic, inorganic and biological systems.

2. Density functional theory background and computational details

In DFT [14,15] the total energy of a many-particle system can be expressed as a functional of the single particle density ρ(**r**)

$$E[\rho] = \int d\mathbf{r} v(\mathbf{r})\rho(\mathbf{r}) + F[\rho]$$

the energy functional, for a fixed external potential v(**r**), assumes a minimum value for the true density. Minimizing this equation with respect to the density and under the normalization constrain, for a N-electron system, one obtains the Euler equation

$$\mu = \frac{\partial E[\rho]}{\partial \rho(\mathbf{r})} = v(\mathbf{r}) + \frac{\partial F[\rho]}{\partial \rho(\mathbf{r})}$$

$$\int d\mathbf{r}\rho(\mathbf{r}) = N$$

where μ is the Lagrange multiplier representing the chemical potential.

The exact form of the universal potential F[ρ] is not known, and approximations can be used for practical computations.

In the Kohn and Sham approach [4], two components of F[ρ], the noninteracting kinetic energy functional and the classical electrostatic energy, are evaluated exactly while the remaining exchange-correlation component can be obtained within the local spin density approximation (LSD) [6-8] with some nonlocal corrections.

Using the density defined as

$$\rho_{i\sigma}(\mathbf{r}) = \sum_{occ} \left| \psi_{i\sigma}(\mathbf{r}) \right|^2$$

the energy minimization leads to the Kohn-Sham equations for the spin orbitals

$$[-\tfrac{1}{2}\nabla^2 + v_{\mathrm{eff}}^{\sigma}(\mathbf{r},[\rho_\sigma])]\, \Psi_{i\sigma}(\mathbf{r}) = \varepsilon_{i\sigma}\Psi_{i\sigma}(\mathbf{r})$$

where

$$\rho_\sigma(\mathbf{r}) = \sum_i n_{i\sigma}\rho_{i\sigma}(\mathbf{r})$$

and the occupations number $n_{i\sigma}$ must satisfy the constrain

$$\sum_i n_{i\sigma} = N_\sigma$$

for $\sigma = \alpha$ or β spin.

The spin-dependent Khon-Sham effective potential is given by

$$v_{\mathrm{eff}}^{\sigma}(\mathbf{r}) = v(\mathbf{r}) + \int \frac{d\mathbf{r}'\rho(\mathbf{r}')}{|\mathbf{r}-\mathbf{r}'|} + \mu_{xc}^{\sigma}(\mathbf{r};[\rho])$$

in which μ denotes the exchange-correlation potential.

The total energy is obtained from the following expression:

$$E = \sum_\sigma \sum_i \varepsilon_{i\sigma} - \frac{1}{2}\int\int \frac{d\mathbf{r}d(\mathbf{r}')}{|\mathbf{r}-\mathbf{r}'|} + E_{xc}[\rho_\sigma] -$$

$$\sum_\sigma d\mathbf{r}\mu_{xc}^{\sigma}[\rho_\sigma]\rho_\sigma(\mathbf{r})$$

In the deMon code [30], used for our computations, the orbitals are given as linear combination of gaussian orbitals and the local spin density approximation is employed

$$E_{XC}[\rho] = \int \rho(\mathbf{r})\, \varepsilon_{XC}[\rho(\mathbf{r})]\mathbf{dr}$$

where the $\varepsilon_{XC}[\rho]$ is the exchange-correlation energy per electron in an interacting electrons system of constant density ρ.

Different LSD functionals have been proposed and applied in finite electron systems . The local functional used in this study is that proposed by Vosko, Wilk and Nusair (VWN) [8].

Generally the local approximation overestimates the exchange-correlation contributions and consequently the energetic parameters. The introduction of the so-called nonlocal corrections improves strongly the results.

Many nonlocal correction schemes have been implemented in the Kohn-Sham method. One of the most used is the gradient expansion approximation (GEA) [31]. The GEA consists in truncating the asymptotic Taylor series of E_{XC} around ρ (r) at the lowest finite order beyond the LSD:

$$E_{XC}[\rho] = E_{XC}^{LDA} + \int \mathbf{dr}\, C_{XC}(\rho(\mathbf{r}))[\nabla(\mathbf{r})]^2$$

where the coefficient C_{XC} is initially assumed independent on the density gradient itself.

In all our computations the exchange and correlation potentials proposed by Perdew and Wang [9] and Perdew [10] respectively have been used.

Full energy optimizations have been performed by using the VWN+ gradient corrected exchange-correlation potentials (hereafter indicated as NLSD) for all the presented systems.

The employed orbital basis sets are of triple-zeta plus polarization (TZVP) quality [32].

The high quality of the used orbital basis set exhibits a small basis set superposition error since the core orbitals are represented well.

Proton affinity (PA) or metal ion affinity (MIA) is assumed to be the negative of the enthalpy for the process:

$$B + X^+ \dashrightarrow BX^+ \qquad (X{=}H^+, Li^+, Na^+)$$

and has been calculated as:

$$\Delta H = E(BX^+) - E(B) - E(X^+) + \Delta(PV)$$

$$PA\ (MIA) = -[E_{SCF}(BX^+) - E_{SCF}(B) + (E_{vib}(BX^+) - E_{vib}(B))] + 5/2\ RT$$

where E_{SCF} are obtained from the SCF calculations, E_{vib} includes the zero point energy and temperature corrections to the vibrational enthalpy and 5/2RT includes the translational energy of proton and $\Delta(PV)$ term.
The absolute gas basicity (GB) is calculated as the negative of the standard free energy ΔG :

$$\Delta G = \Delta H - T\Delta S$$

the entropy contribution to the ΔG to be given by

$$- T\Delta S = -T\ [S(BH^+) - S\ (B) - S(H^+)]$$

At 298 K the $TS(H^+)$ term has a value of 7.76 kcal/mol [33].
The BH^+ and B entropies have been obtained from thermochemical calculations by using the VWN and P-PW theoretical harmonic frequencies and equilibrium geometries.

3. Applications

The applications concern different properties and parameters for representative organic, inorganic and biological molecules with great significance in the various fields of modern chemistry and give indications on a broad spectrum of potentialities of DF methods.

3.1. ION STRUCTURES AND POTENTIAL ENERGY SURFACES

The reactivity and the fragmentation mechanisms of ions are studied largely by mass spectrometry because of their importance in fundamental and applied chemistry. Although the thermochemical data that these experiment give are precious, structures and potential energy surfaces remain unknown using these techniques. In this field, theoretical methods are very useful and complete the information coming from the experiments.

Recently the O_3H^+ cation has been detected experimentally and the proton affinity of this elusive specie has been measured [34].

Because of its role in catalysis, the knowledge of the structure and the potential energy surface of this ion becomes significant. As it is shown in figure 1, this system can assume different isomeric forms and different transition states for the interconvertsion between them are possible. Our study shows that the planar trans structure is the absolute minimum, but the planar cis and planar C2v as well as the cyclic form are minima in the PES and lie, using the P-PW exchange-correlation potential, at 3.7, 58.9 and 56.2 kcal/mol respectively above the absolute one. Results are in good agreement with previous high level CISD+Q correlated study [35].

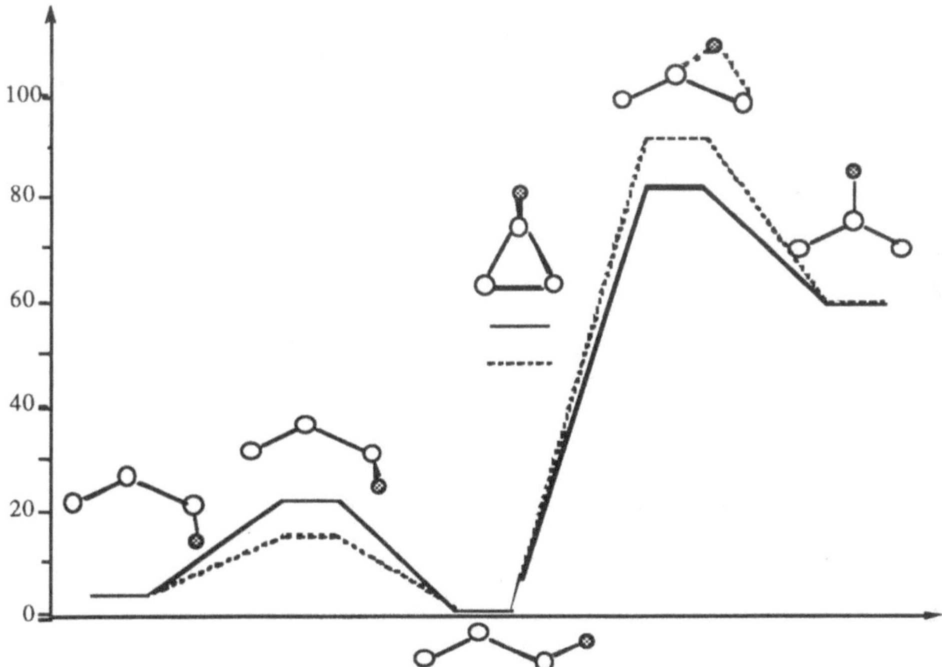

Figure 1. NLSD potential energy surface of O_3H^+. Continuum line shows the NLSD path while dashed line the CISD one. Values are in kcal/mol.

An indirect way to ascertain the reliability of the result is the calculation of proton affinity considering the minimum structures of O_3H^+ ion and neutral ozone. The obtained value of 145.5 kcal/mol is in excellent agreement with the experimental counterpart (148.0 $\pm$ 3 kcal/mol) [34].

The presence of N_2H^+, P_2H^+ and other valence isoelectronic species has been revealed for some time in interstellar space and theoretical and experimental investigations for these protonated species have been made extensively [36]. Recently their proton affinity have been determined to be 118.8 (N_2H^+) [37] and 162.0 (P_2H^+) [38] kcal/mol. Which is the structure of these two cations?

Also in this case the answer cannot come from mass spectrometry measurements and the question whether these protonated species are linear or bent is of considerable interest.

NLSD computations indicate that N_2H^+ prefers the linear structure but the P_2H^+ the bent one. In particular we found that linear N_2H^+ is the only minimum in the potential energy surface while in the case of P_2H^+ the linear form lies at 9.2 kcal/mole above the bent one. Results are in agreement with previous theoretical calculations performed at CISD [39] and MP2 [40] levels of theory. In these studies it is shown that protonation of N_2 favours the linear structure while of P_2 prefers the bridged one.

Calculation of proton affinities for these two cations confirms further the reliability of our structural study (see next section).

3.2. PREFERRED ATTACH SITE

Another important question in the gas-phase study of ions is the preferred site on which proton or other ions can attach the system. For its determination simple chemical arguments as well as mass spectra are sometimes insufficient. As illustrative examples we report here those of the aniline and the CH_2SO species. Four and three possible sites of protonation are present in these molecules respectively. In the latter example proton can bond in principle to carbon, to sulphur or terminal oxygen giving rise to three different cations. Our NLSD computations show that the most stable product is that in which the proton is bond to the oxygen followed by that protonated at the CH_2 group that lies at 25.1 kcal/mole. Finally the less stable isomer is the cation in which the attach of H^+ occurs on the sulphur atom. Also for this kind of problem our results are in very good agreement with previous and more time expansive Hartree-Fock correlated studies [41].

The determination of preferred attach sites for proton or other ions is very important in the case of molecules with biological significance as are the amino acids or nucleic acid bases. Our study on glycine and alanine shows that proton links preferentially to nitrogen and that protonation on oxygen is much less favoured. The oxygen protonated forms lie at about 30 kcal/mol above the nitrogen protonated ones for glycine and alanine respectively.

It has been also discovered theoretically that proton transfer between oxygen and nitrogen, in these two amino acids, occurs without energy barriers [42].

In a recent study we have determined the preferred sites of protonation for the four DNA nucleic bases [43]. In the case of thymine the protonation occurs preferentially on the O4 site being that on the O2 unfavoured by 8.9 kcal/mol. For cytosine the energy difference between protonation at N3 and O2 is of only 0.8 kcal/mol. A similar small energy difference (2.2 kcal/mol) is found when the N1 and N3 sites of adenine are considered. The protonation at N7 position of adenine yields a tautomer that lies at 7.6 kcal/mol with respect to the N1 protonated tautomer. For guanine we found that the preferred protonation site is the N7 with the O6 and N3 at 6.5 and 17.7 kcal/mol respectively. The rationalization of these results is not simple and requires to consider the molecular electrostatic potential, charge distribution, the analysis of the HOMO composition, the charge transfer, polarization and structural changes that occur upon protonation.

For cations other than proton the complexation sites in biomolecules is of great interest. Our results for the metal (Li and Na) affinity on glycine and alanine show that both Li^+ and Na^+ prefer to bond to nitrogen rather than to oxygen. The energy difference in the case of Li^+ on glycine is of 16.6 kcal/mol.

We have considered also the possibility that Li^+ can bond with nitrogen as well as to the carbonyl oxygen in cytosine (see scheme 1). Nitrogen site is favoured with respect to that of oxygen by only 1.5 kcal/mol.

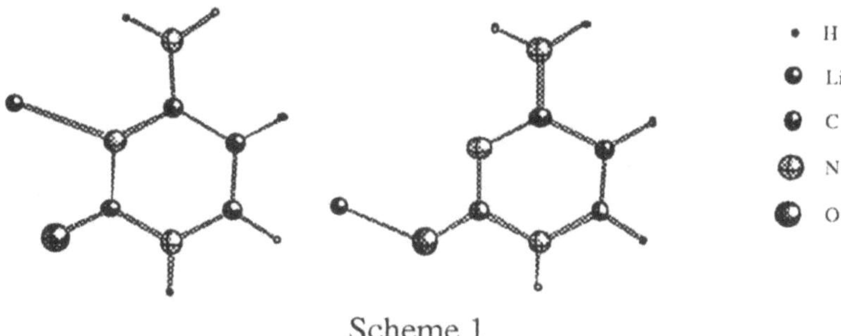

Scheme 1

3.3. PROTON AFFINITY AND GAS PHASE BASICITY

The determination of PAs of different systems (i.e. alkylamines, alcohols, carboxylic acids, amino acids, nucleic acid bases) has been the subject of accurate experimental studies [29, 37,44-53]. From a theoretical point of view, this matter is considered a challenge for modern quantum chemical methods [54-57]. It is established well that reliable PAs can be obtained only if a considerable amount of correlation effects is taken into account. G2 procedure, applied for a series of small size molecules, gives results that are close to the experimental counterparts but requires considerable computational time [58]. Its application to large systems such us molecules with biological importance is difficult. An alternative way is the use of DF methods. Test computations on molecules with different functional groups and protonation sites have shown already their reliability in the PAs determinations [20,59,60].

In Table 1 we report the PAs value for several different compounds determined by using the NLSD approximation, together with experimental and previous theoretical results.

Table 1. Proton affinities (in kcal/mol) for different compounds from different levels of theory and experiment.

System	NLSD	HF or CI	EXP.
N_2	118.2	118.1 (G2)[a]	118.8[g]
P_2	152.4	145.8 (MP2)[b]	162.0h
CH_3CN	188.2	186.6 (G2)[a]	188.1[i], 185.0[l]
C_5NH_5	222.9	221.9 (G2)[a]	220.8[i], 224.0[l]
O_3	145.5	149.5 (CISD)[c]	148.0[m]
H_2CO	166.0	170.2 (G2)[a]	171.7[i]
HCOOH	173.1	177.7(G2)[a]	178.8[i]
CH_3CHO	183.0	184.3(G2)[a]	186.6[i]
Glycine	214.2	213.6(MP4)[d]	213.0[n], 211.7[o], 216.1[p]
Alanine	216.0	225.8 (HF)[e]	215.8[n], 222.1[o], 219.2[p], 213.9[q]
Cytosine	226.8	227.4(MP4)[f]	225.9[r], 223.8[s]

a) ref. 58; b) ref. 40; c) ref. 35; d) ref. 54; e) ref. 55; f) ref. 56; g) ref. 37; h) ref. 38; i) ref. 44; l) ref. 45; m) ref. 34; n) ref. 46; o) ref. 47; p) ref. 48; q) ref. 49; r) ref. 50; s) ref. 37.

For many of these compounds our PA values are in excellent agreement with the experimental ones and very close to the data coming from high level ab initio correlated methods. A considerable discrepancy is found for P_2 protonation for which our value differs by about 10 kcal/mol with respect to the experimental result. A possible explanation is the use of the nonlocal gradient corrected functional used in the computations. As demonstrated recently by Chandra and Goursot [61], the use of different functionals for the exchange-correlation energy improves the results in the PAs value of oxygen containing systems. In fact, using the correlation potential proposed by Proynov et al. [62] and the exchange potential of Becke [11], the PA of H_2CO becomes 171.3 kcal/mol and those of HCOOH and CH_3CHO are found to be 179 and 181.7 kcal/mol respectively [61].

As is shown in Table 1 and from previous works [20,42,43,59,60], PAs of nitrogen containing systems are reproduced well by using Becke exchange and Perdew correlation functionals or by Perdew and Wang exchange and Perdew correlation ones.

It is interesting to note the good agreement between NLSD and experimental data for glycine, alanine and cytosine. This agreement exists also for the PAs of the most common amino acids [63] and for the nucleic acid bases [43].

3.4. GAS-PHASE ACIDITIES (ΔH_{acid})

We have considered of interest to see if gaussian density functional method is able to reproduce the gas-phase acidity. For this purpose the ΔH_{acid} of a series of carboxylic acids, alcohols and amino acids have been calculated. Gas-phase acidity has been computed considering the following process

$$AH \text{------>} A^- + H^+$$

NLSD results are collected in Table 2 together with the experimental values [64-67]. The structures of neutral systems and corresponding anions are shown in Figure 2.

From Table 2 it is evident that the ΔH_{acid} of glycine and alanine is reproduced very well, while, as in the case of PAs, the absolute gas-phase acidities obtained for the systems in which the deprotonation involves oxygen is less satisfactory.

Table 2. Absolute gas-phase acidities (ΔH_{acid}) (kcal/mol) at 298 K for a series of acids, alcohols and amino acids from NLSD computations and from mass spectrometric measurements.

System	NLSD	EXP.
HCOOH	343.0	345.0[a]
CH_3COOH	346.5	348.5[a]
CH_3CH_2COOH	345.9	340.3[b]
$CH_3CH_2CH_2OH$	371.1	368.1[b]
$CH_3CH_2CH_2CH_2OH$	370.5	367.1[b]
$CH_3CH_2CH_2CH_2CH_2OH$	369.1	366.2[b]
Glycine	341.8	342.0[c]
Alanine	339.7	341.0[c]

a) ref. 64; b) ref. 65; c) ref.66.

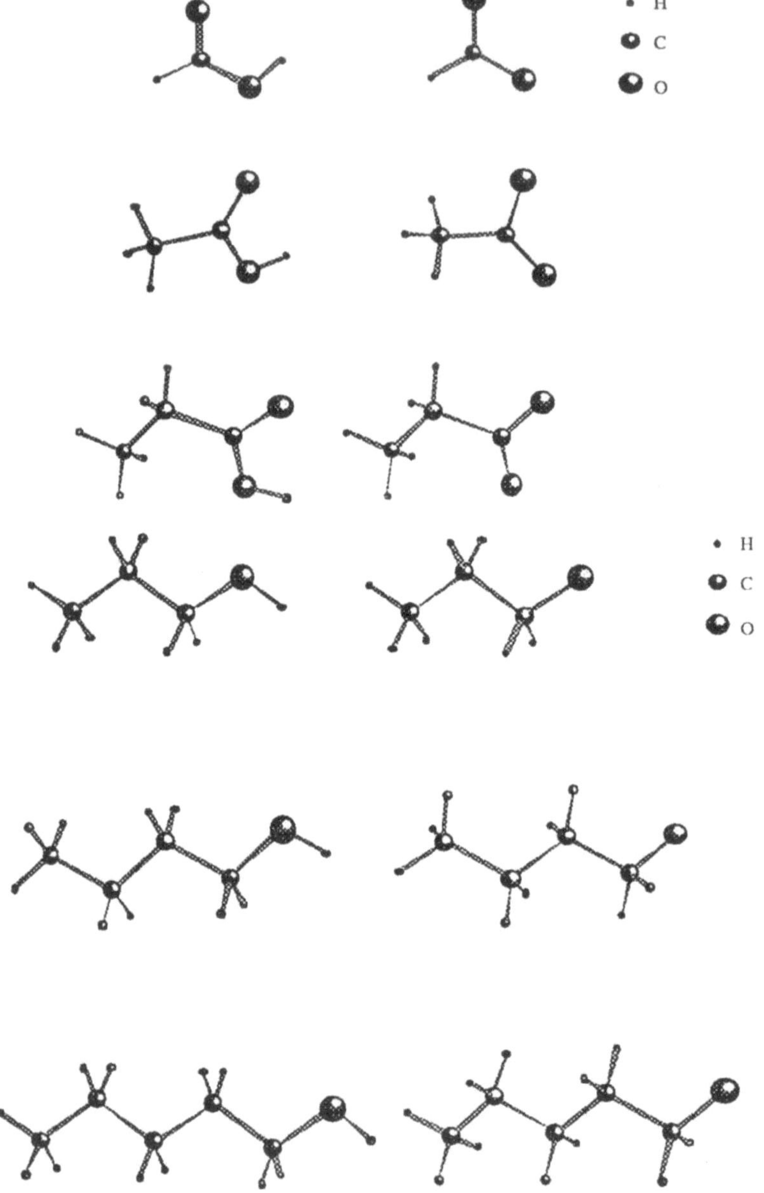

Figure 2. NLSD structures of the neutral and corresponding anions for carboxylic acids and alcohols.

In any case the average error for this class of compounds is of about 3 kcal/mol. This discrepancy can be ascribed to the used vibrational corrections. In fact, because of the high required computational time, we have computed only the corrections relative to HCOOH ------> $HCOO^- + H^+$ process and we have used them for all other similar deprotonation reactions. The use of different nonlocal functionals for the exchange-correlation energy can improve the agreement with the experiment. Work in this direction is in progress.

For glycine and alanine comparisons are possible also with previous theoretical studies [66] performed with Hartree-Fock method and employing the 6-31+G* orbital basis set on geometries optimized at 3-21+G* level. The agreement between our NLSD and HF values is good both for glycine (342.0 versus 344.6 kcal/mol) and alanine (339.7 versus 344.2 kcal/mol). Our computations reproduce better the absolute values as well as the difference between acidity of glycine and alanine ($\Delta\Delta H_{acid}$). This quantity is calculated to be 2.1 kcal/mol at NLSD level while the corresponding experimental value is 1 kcal/mol [66] using the kinetic method and 1.8 kcal/mol ($\Delta\Delta G_{acid}$) (29, 44, 64) using the equilibrium method.

3.5. GAS-PHASE METAL ION AFFINITY (MIA)

The increasing interest for the knowledge of the gas-phase affinity for ions other than proton (i.e. noble, alkali, transition metal ions) [67-69] is due to the demand for thermochemical information on Lewis acids and bases as well as on biochemical systems in gas phase. It is well known that the free and protein-bound forms of the alkali metal ions participate to several processes of nucleic acid metabolism. In particular, synthesis, replication, structural integrity and cleavage of DNA and RNA are affected by the presence of these ions. We have computed, for the first time, the MIA for a series of organic systems and in particular we are working to the determination of MIA in amino acids and nucleic acid bases [70]. In Table 3 the performance of the gaussian density functional method in reproducing this significant thermochemical parameter is shown for glycine and cytosine.

Table 3. Gas-phase metal ion affinities at the nitrogen site (kcal/mol) at 298 K for glycine and cytosine from NLSD computations and from mass spectrometric measurements.

System	NLSD	EXP.
Glycine-Li^+	48.8	51.0[a], 51.9[b]
Glycine-Na^+	36.2	39.4[a], 38.0[b]
Cytosine-Li^+	58.1	55.4[c]

a) ref. 67 ; b) ref. 68; c) ref. 69.

The MIA for lithium of glycine is 48.8 kcal/mol and the corresponding experimental value determined with the kinetic method is 51.0 kcal/mol. In the case of Na^+ our MIA value is underestimated by only 3.2 kcal/mol with respect to the experimental one [67]. Good agreement between our NLSD value (58.1 kcal/mol) and that obtained recently from mass spectrometry by Cerda and Wesdemiotis (55.4 kcal/mol) [69] is found also for cytosine.

The structure of the more stable isomer (see Figure 1) suggests that the Li cation is bond to nitrogen with a distance of 2.083 Å. The corresponding Li-O distance is 2.041 Å. The structure involving the attach of Li^+ on oxygen lies at only 1.5 kcal/mol and has an equilibrium Li-O distance of 1.818 Å. Due to the small energy difference between the two cations it is possible that in gas-phase also the less stable minimum can be populated.

NLSD determination of Ni^+ and Co^+ MIA for organic nitriles and alkylamines is in progress.

4. Conclusions

Gaussian density functional method employing the gradient corrected approximation has been used for thermochemical gas-phase determinations.

Results show that:

• it is possible to obtain reliable information on structures, preferred attach sites and potential energy surfaces for a large series of systems with chemical, physical and biological significance at relatively low computational cost;

• proton affinities and gas-phase basicities are in agreement with the experimental determinations and fall in the range of accuracy of other more expensive high level theoretical methods;

• preliminary results on gas-phase acidities are very encouraging and allow the study of systems of medium-high size;

• the method is able to predict correctly the gas-phase metal ion affinity opening a new way to study these properties of complexes.

The good agreement between calculated and experimental determinations for all these parameters seems to encourage the marriage between theory and experiment.

Acknowledgements

This investigation was supported by CNR, MURST and CINECA. We thank dr. Chris Wesdemiotis for a preprint of its work on metal affinity of DNA and RNA nucleic bases.

References

1. L. H. Thomas, Proc. Cambridge Phil. Soc. 23 (1926) 542.
2. E. Fermi, Z. Phys. 48 (1926) 542.
3. P. Hohenberg and W. Kohn, Phys. Rev. 136, (1964) B864.
4. W. Kohn and L. J. Sham, Phys. Rev. A140, (1965) 1133.
5. J. C. Slater, Phys. Rev. 81 (1951)385.
6. O. Gunnarson, M. Jonson and B. I. Lundqvist, Phys. Rev. B20, (1979) 3136.
7. J. P. Perdew and A. Zunger, Phys. Rev. B23, (1981) 5058.
8. S. H. Vosko, L. Wilk and M. Nusair, Can. J. Phys. 58, (1980) 1200.
9. J. P. Perdew and Y. Wang, Phys. Rev. B33, (1986) 8800.
10. J. P. Perdew, Phys. Rev. B33, (1986) 8822.
11. A. D. Becke, Phys. Rev. A38, (1988) 3098 ; J. Chem. Phys. 88, (1988) 2547.
12. J. P. Perdew and Y. Wang, Phys. Rev. B33, (1986) 8800.
13. C. Lee, W. Yang and R. G. Parr, Phys. Rev. B45 (1992) 13244.
14. R. Parr, and W. Yang , *Density-Functional Theory of Atoms and Molecules* (Oxford University Press, NY, 1989).
15. N. March, *Electron Density Theory of Atoms and Molecules*; (Academic Press, New York, 1991).
16. R. O. Jones and O. Gunnarson, Rev. Mod. Physics 61, (1990) 1280.
17. T. Ziegler, Chem. Rev. 91, (1990) 651.
18. J. Labanowski and J. Andzelm (eds.) *Density Functional Methods n Chemistry* (Springer Verlag: NY, 1991).
19. D. R. Salahub and N. Russo (eds.) *Metal Ligand Interaction. From Atoms, to Clusters, to Surfaces* (Kluwer: Dordrecht, 1992).
20. J. M. Seminario and P. Politzer (eds.) *Density Functional Theory: a Tool for Chemistry* (Elsevier, New York, 1995).
21. N. Russo and D. R. Salahub (eds.) *Metal Ligand Interaction: Structure and Reactivity* (Kluwer: Dordrecht, 1995).
22. D. P. Chong (ed) *Recent Advances in Density Functional Methods*. Vol 1 (World Scientific, Singapore, 1995).
23. V. Malkin, O. L. Malkina, M. E. Casida and D. R. Salahub, J. Am. Chem. Soc. 116 (1994) 5898.
24. A. M. Koster, P. Calaminici and N. Russo, Phys. Rev. **A 53** (1996) 3865.
25. T. Mineva, N. Russo and M. Toscano, Int. J. Quantum Chem. 56, (1995) 663.
26. N. Russo, E. Sicilia and M. Toscano, J. Chem. Phys. 97, (1992) 5031.
27. G. De Luca, N. Russo, E. Sicilia and M. Toscano. J. Chem. Phys.105 (1996) 3206.

28. P. Kebarle, Ann. Rev. Phys. Chem. 28 (1979) 445 and refereces cited therein.
29. R. G. Cooks, J. S. Patrick, T. Kotiaho and S. A. McLuckey, Mass Spectrom. Rev. 13 (1994) 287 and references cited therein.
30. A. St. Amant, PhD Thesis, Université de Montreal, 1991.
31. D. C. Langreth and M. J. Mehl, Phys. Rev. Lett. 47 (1981) 446; Phys. Rev. B28 (1983) 1809; erratum B29, 2310.
32. N. Godbout, D. R. Salahub, J. Andzelm and E. Wimmer, Can. J. Chem. 70, (1992) 560.
33. S. W. Benson, Thermochemical Kinetics, (John Wiley & Sons, New York, 1976).
34. F. Cacace and M. Speranza, Science 265 (1994) 208.
35. C. Meredith, G. E. Quelch and H. F. Schaefer III, J. Am. Chem. Soc. 113 (1991) 1187.
36. G. Winnewisser and E. Herbst, Rep. Prog. Phys. 56 (1993) 1209.
37. S.G. Lias, J.F. Liebman and R.D. Levine, J. Phys. Chem. Ref. Data 13 (1984) 695.
38. R.A.J. O'Hair, M. Krempp, R. Damrauer, C.H. DePuy Inorg. Chem. 1992, 31, 2092.
39. R. Glaser, C. J. Horan and R. D. Levine, J. Phys. Chem. 97 (1993) 1835.
40. Y-N. Su, M-S. Tsai and S-Y. Chu, Int. J. Quantum Chem. 59 (1996) 487.
41. P. J. A. Ruttink, P. C. Burgers and J. K. Terlouw, Chem. Phys. Lett. 229 (1994) 495.
42. I. A. Topol, S. K. Burt, N. Russo and M. Toscano, to be published.
43. N. Russo, M. Toscano, A. Grand and F. Jolibois, submitted.
44. S.G. Lias, J.E. Bartmess, J.F. Liebman, J.L. Holmes, R.D. Levin and W.G. Mollard, J. Phys. Chem. Ref. Data suppl. 1, 17 1988.
45. M. Meot-Ner and W. Sieck, J. Am. Chem. Soc. 113 (1991) 4448.
46. M. J. Locke and R. T. Mc Iver, J. Am. Chem. Soc. 105 (1983) 4226.
47. G. S. Gorman, J. P. Speir, C. A. Turner and I. J. Amster, J. Am. Chem. Soc. 114 (1992) 3986.
48. X. Li and A. G. Harrison, Org. Mass Spectrometry 28 (1993) 366.
49. M. Meot-Ner, E. P. Hunter and F. H. Field, J. Am. Chem. Soc. 101 (1979) 686.
50. A. Liguori, F. Greco, G. Sindona and N. Uccella, J. Am. Chem. Soc. 112 (1990) 9092.
51. G. Bojesen and T. Breindahl, J. Chem. Soc. Perkins Trans. 2 (1994) 1029.
52. K. Isa, T. Omote and M. Amaya, Org. Mass Spectrom. 25 (1990) 620.

53. Z. Wu and C. Fenselau, Rapid Commun. Mass Spectrom. 6 (1992) 403; ibidem J. Am. Soc. Mass Spectrom. 3 (1993) 863.
54. K. Zhang, D. M. Zimmerman, A. Chung-Phillips and C. J. Cassady, J. Am. Chem. Soc. 115 (1993) 10812.
55. L. R. Wright and R. F. Borkman, J. Am. Chem. Soc. 102, (1980) 6207.
56. C. Colominas, F. J. Luque and M. Orozco, J. Am. Chem. Soc. 118 (1996) 6811.
57. A. M. Schmiedekamp, I. A. Topol, S. K. Burt, H. Razafinjanahary, H. Chermette, T. Pfaltzgraff and C. J. Michejda, J. Computat. Chem. 15 (1994) 875.
58. B.J. Smith and L. Radom, J. Phys. Chem. 99 (1995) 6468.
59. E. Sicilia, N. Russo and J. Rubio, Int. J. Quantum Chem. in press.
60. M. Toscano, to be published.
61. A. K. Chandra and A. Goursot, J. Phys. Chem. 100 (1996) 11596.
62. E. I. Proynov, E. Ruiz, A. Vela and D. R. Salahub, Int. J. Quantum Chem., Quantum Chem. Symp. 29 (1995) 61.
63. I. A. Topol, S. K. Burt, N. Russo and M. Toscano, submitted.
64. J. E. Bartmess and R. T. McGiver in *Gas Phase Ion Chemistry* M. T. Bowers Ed.(Academic Press: New York, 1992, Vol. 2).
65. G. Boand, R. Houriet and T. J. Gaumann, J. Am. Chem. Soc. 105 (1983) 2203
66. R. A. J. O'Hair, J. H. Bowie and S. Gronert, Int. J. Mass Spectrom. Ion Processes, 117 (1992) 23.
67. B. A. Cerda and C. Wesdemiotis, 42nd ASMS Conference on Mass Spectrometry and Allied Topics, Chicago, IL, June 1994.
68. G. Bojesen, T. Breindhal and U. Andersen, Org. Mass Spectrom. 28 (1993) 1448.
69. B. A. Cerda and C. Wesdemiotis, J. Am. Chem. Soc. , submitted.
70. A. Grand, N. Russo and M. Toscano, to be published.

Part II
Instrumentation for Mass Analysis and Detection

RECENT ADVANCES IN MAGNET SECTOR MASS SPECTROMETRY

O. V. NEMIROVSKIY, J. K. GOODEN, R. RAMANATHAN and M. L. GROSS*
Department of Chemistry, Washington University
One Brookings Drive
St. Louis, Missouri, 63130 USA

ABSTRACT. The use of double focussing and tandem sector instruments has decreased during the 1990s, owing to the difficulty of interfacing electrospray and matrix-assisted laser desorption ionizations with them. Nevertheless, their high dynamic range, resolving power in the 10,000 range, exact-mass-measurement and high-energy-collisional-activation capabilities continue to recommend their use and improvement. This chapter is a review of analyses for which sectors are well, or even uniquely, qualified and of instrumentation advances that incorporate array detectors and other mass analyzers to give a new generation of hybrid instruments.

1. Introduction

The concept that a combination of an electrostatic analyzer (ESA) and a magnetic sector can achieve direction and energy focussing was realized in the early work of Aston [1]. Following his seminal work, considerable effort has gone into the design of double focussing mass spectrometers. The motivation has been to achieve high resolving power and exact-mass measurement capability. The use of double focussing instruments for exact mass measurements (accuracy of a few parts per million) was a unique capability until recently. Discussion of the design and principles of double focussing mass spectrometers can be found in numerous reviews and book chapters, a few of which are mentioned here for the interested reader [2-4].

Advances that set the stage for new applications of double focussing mass spectrometers began with metastable-ion studies [5] and culminated with the reverse geometry instrument of Beynon and co-workers [6]. Early reverse-geometry instruments were used for ion-structure studies and for analysis of mixtures. They incorporated the necessary scanning functions and collision cells to investigate both metastable-ion and collisionally activated decompositions (CAD). The latter was first demonstrated by Jennings [7], who was awarded the American Society for Mass Spectrometry Award in 1995 for his contribution.

The discovery of fast atom bombardment (FAB) in the early 1980s gave new impetus to the development of reverse-geometry instruments and stimulated new designs of tandem mass spectrometers. FAB made possible the ionization of large and low-volatility molecules, but as ions these species gave fewer fragmentations than

R. M. Caprioli et al. (eds.), Selected Topics and Mass Spectrometry in the Biomolecular Sciences, 183–211.

radical cations produced by conventional electron ionization or by certain forms of chemical ionization. Furthermore, FAB produced large numbers of matrix ions that were usually of little interest to the mass spectrometrist. Tandem mass spectrometers equipped with collisional activation capabilities induced the needed fragmentations, blinded the experiments to the presence of most matrix ions, and permitted simple mixtures to be analyzed.

Two approaches have been taken in developing tandem sector mass spectrometers. One approach involved addition of more sectors in a variety of combinations to existing design double focussing spectrometers, leading to the development of three, four, and five-sector mass spectrometers. The designs and applications of these instruments were reviewed by Gross [8], Burlingame [9] and Trainor and Derrick [10]. A second approach involved the development of hybrid tandem mass spectrometers, which were combinations of two or more different types of mass analyzers. Early interest in hybrid mass spectrometers came about from the desire to access both low- and high-energy CAD spectra by using relatively simple and inexpensive mass spectrometers [11]. Among the hybrid mass spectrometers, those that involve magnetic sector, electric sector and quadrupoles gained popularity because these are components of continuous beam instruments, which are less difficult to concatenate than are instruments whose operating principles are more disparate. Their designs and applications were reviewed by Gaskell [12], Gaskell and Ballard [13] and Yost and Boyd [14], and these hybrids will not be discussed further because there is little current interest in them.

The advent of electrospray ionization (ESI) and matrix-assisted laser desorption ionization (MALDI) has given mass spectrometry a significant opportunity to play a pivotal role in solving problems in the areas of biochemical and biomedical sciences. Minimal structural information is available when biomolecules are ionized by these methods because little fragmentation occurs. This limitation has increased the demands of tandem mass spectrometry to an even higher level, but has also elevated the opportunities. In addition to the requirements for speed and high-information content of tandem mass spectrometry, there continues to be a need for improvements in both detection limit and mass accuracy. Another area that requires improvement is on-line combination of separation methods and tandem mass spectrometry. The availability of appropriate instruments has become more important as sample size has become smaller and smaller, as occurs for samples from in-vitro and in-vivo experiments.

In this chapter, we focus on how sector mass spectrometry has changed during the 1990s to meet the new and demanding role of MS in biomedical and biochemical sciences. Unfortunately the changes have been insufficient to sustain the former high-level interest, and the fraction of spectrometrists who use sectors has steadily decreased during the first half of the 1990s. One reason is the advent of electrospray and MALDI, which are difficult to mate with sectors. Nevertheless, the superior mass resolving power and associated accurate mass measurements, the capacity for high-energy CAD, and the compatibility with array detectors still recommend further developments. For example, making electrospray more compatible with sectors and improving the efficiency of high-energy CAD would promote a comeback. We will consider some opportunities and unique applications of sectors in this review chapter.

2. Advances in Double Focussing Mass Spectrometers

Double focussing mass spectrometers, the most common type of sector mass spectrometer, are composed of an ESA, for energy focussing, and a magnetic sector, for directional focussing of an ion beam. Conventional forward geometry double-focussing instruments have the ESA prior to the magnetic sector (EB configuration). Reversing the positions of the ESA and magnetic sector gives a reverse-geometry sector mass spectrometer (BE configuration), which was widely used because it has the ability to select a precursor ion and study its various decompositions. Tandem mass spectrometry experiments in reverse geometry instruments involved activating a precursor ion after selection by the magnetic sector. Subsequently, the product ions were mass analyzed by scanning the ESA. Because the precursor and product ions were not double focussed, the mass resolving power was not high.

Both forward and reverse geometry sector mass spectrometers can be configured in a C-shape configuration to give a Nier-Johnson arrangement or in an S-shape to give a Mattauch-Herzog arrangement. In the Nier-Johnson arrangement, direction and energy focussing were achieved by deflecting the ions in the same sense in both sectors. The sectors in the Nier-Johnson arrangement could function independently of each other because source-produced ions were focussed directionally after the first sector but double focussed after both sectors where the directional and energy focussing curves intersected. A detector, usually a point detector, was placed at the image focal point and detected one ion at-a-time.

In the Mattauch-Herzog arrangement, directional and energy focussing of the beam were achieved by deflecting the beam in opposite directions. Ions of different mass-to-charge ratios were double focussed along a focal plane, allowing simultaneous detection of a range of different mass ions without scanning the sectors. During the early days of mass spectrometry, a photoplate was used for simultaneous detection of the range of ions, but this type of detection became less popular than electronic detection owing to poor sensitivity and inconvenient operation. Introduction of electronic array detectors, which can also register a wide mass range of ions without scanning, improved the sensitivity and may pave the way to wider usage of the Mattauch-Herzog arrangement. Another advantage of the Mattauch-Herzog arrangement is its compactness, which promoted its use on board the Viking spacecraft [15].

2.1. ADVANCES IN DETECTORS

There are three types of simultaneous (array) detection methods that are in use with contemporary mass spectrometers. The first one is spatial-array detection and is commonly employed with magnetic sector mass spectrometers in an effort to improve detection limits. The second one is time-array detection, and its applicability lies with time-of-flight (TOF) mass spectrometers. This inherent advantage of TOF mass spectrometers has allowed it to gain popularity as one of the choices for coupling with sector mass spectrometers to achieve improved detection limits in tandem mass spectrometry experiments (see **Section 4.1.**). The third type of simultaneous detection is frequency-array detection, and it is inherent to Fourier transform mass spectrometers. The compatibility of sector instruments with spatial-array detection in principle puts double focussing and tandem sectors at an advantage with respect to quadrupole analyzers for which simultaneous recording methods are not apparent.

2.1.1. *Photodiode Array*

Griffin et al. [16] in 1974 developed the microchannel plate (MCP) photodiode array detection for mass spectrometry. Later, a number of groups used MCP-based array detectors for achieving improved sensitivity over single-point detectors. For example, groups involved in stratosphere research used the MCP-based photodiode array (PDA) detectors for simultaneous detection of trace gases [17]. Instrumentalists added array detectors to single focussing mass spectrometers [18] and forward-geometry, double focussing mass spectrometers of Mattauch-Herzog design [19]. In 1986, Cottrell and Evans [20] coupled a PDA detector with a commercial double focussing, Nier-Johnson instrument (Kratos MS50).

The principles governing the array's operation begin with an ion impinging on the front of the MCP (two straight geometry MCPs are coupled in a chevron configuration to improve gain and to suppress ion feedback) and causing emission of a pulse of electrons. The electrons accelerate and multiply as they travel down the channels in the MCP, and finally they come to focus on to a conductive phosphor layer deposited on a fiber-optic bundle. The fiber-optic bundle transfers the light that is produced by impact of electrons to a PDA detector. The PDA is placed after the fiber-optic bundle, often outside the vacuum system, to make provision for more ready cooling and temperature regulation of the detector. In the Cottrell and Evans [20] design, the PDA consisted of a row of 1024 photodiode sensors across a 2.54-cm length.

Using the 2.54-cm PDA detector, Cottrell and Evans [20] showed that the dynamic range for detecting the molecular ion of Substance P in a single exposure to be well over 1000. At larger masses, around the molecular ion region of insulin, 4% of the mass range was detectable with a single exposure. The result was improved dynamic range, sensitivity, and mass accuracy. The ion statistical noise of a mass spectrum obtained by using 1 pmol of bovine insulin improved by eight fold when the photodiode array was employed rather than the single-channel, post-acceleration detector. Another advantage for the array detector is the reduction in acquisition time; for example, the time for recording the molecular ion region of bovine insulin was 7 s and gave equivalent signal-to-noise to when single-point detection was used for 10 min over the same mass range. These encouraging improvements motivated several other research groups to incorporate PDA detectors into two, three, and four-sector and hybrid tandem instruments. Some new developments and applications of array detectors for tandem MS are discussed in **Section 3.1**.

Although PDA detector systems offer improved sensitivity, mass accuracy and dynamic range over single point detectors, the rates of acquisition, however, are not adequate for on-line monitoring and tandem mass spectrometry of samples eluting following high-resolution chromatographic separation. Furthermore, large PDAs are expensive. Among the other detectors that have been considered for magnetic sector mass spectrometers, array detectors based on charge-coupled devices (CCD) and position and time-resolved ion counting (PATRIC) have also been promising (some applications are discussed in **Section 3.3.**).

2.1.2. *CCD Array*

CCD array detectors also consist of a combination of MCPs, phosphor-coated fiber-optic surfaces, and flexible fiber-optic bundles. The use of a CCD readout instead of a PDA reduces cost and improves performance. Hoffman et al. [21] first demonstrated this type of array detector on a commercial double focussing instrument (Kratos Concept). The acquisition of a significant portion of the mass spectrum was accomplished following tryptic digest ion of β-casein. Recently, Evans et al. [22] used a CCD-based scanning array detection system with a tandem mass spectrometer and demonstrated structural characterization of 200 femtomoles of a glycoalkaloid in 10 s. Similar information from an array that used stepped scanning would have required 1-2 minutes.

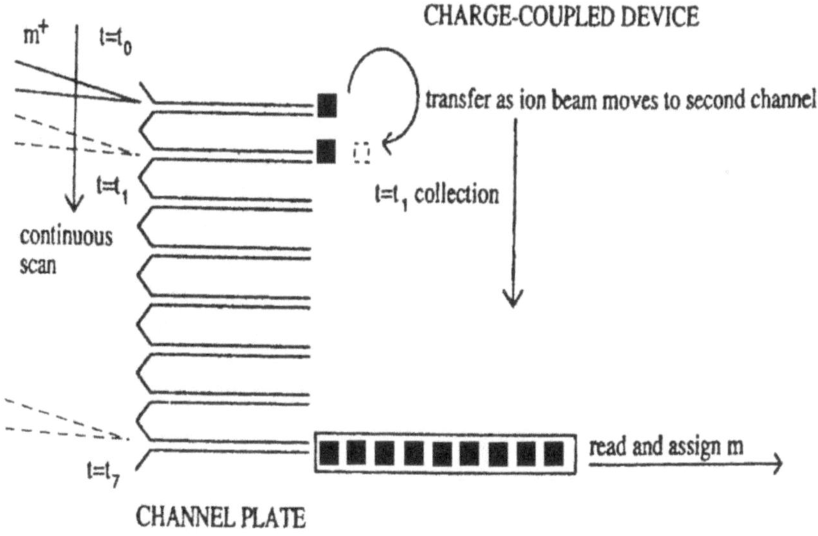

Figure 1. Schematic diagram of the CCD array detector [Ref. 9]

With a scanning CCD array detector (Figure 1), a beam of mass-resolved ions sweeps across the 1024 channels by rapid continuous magnetic scanning. The ion current of each particular *m/z* was integrated and transferred to successive channels in synchrony with the magnet scan function. As the ion of the highest *m/z* dropped off the last channel, the ion current was read out and *m/z* values were assigned.

Musselman and others [23] demonstrated a similar 2.54-cm, CCD-based scanning array detector coupled to a JEOL SX 1024 double focussing mass spectrometer; the sensitivity for detecting lysozyme was 50 to 100 times greater than with the single point detector.

2.1.3. PATRIC Array

Another type of array detector that overcomes some of the limitations with the PDA is the PATRIC scanning array detector. Pesch and others [24] first described the commercial development of the PATRIC detector for a Finnigan MAT 900 double focussing mass spectrometer. Evans [25] also discussed the PATRIC detector in a review chapter. As shown in Figure 2, events in PATRIC detector also started with ions striking surface of a pair of MCPs. Electrons emitted from the MCPs were allowed to expand the and then collected by a plate consisting of 54 conductive strips.

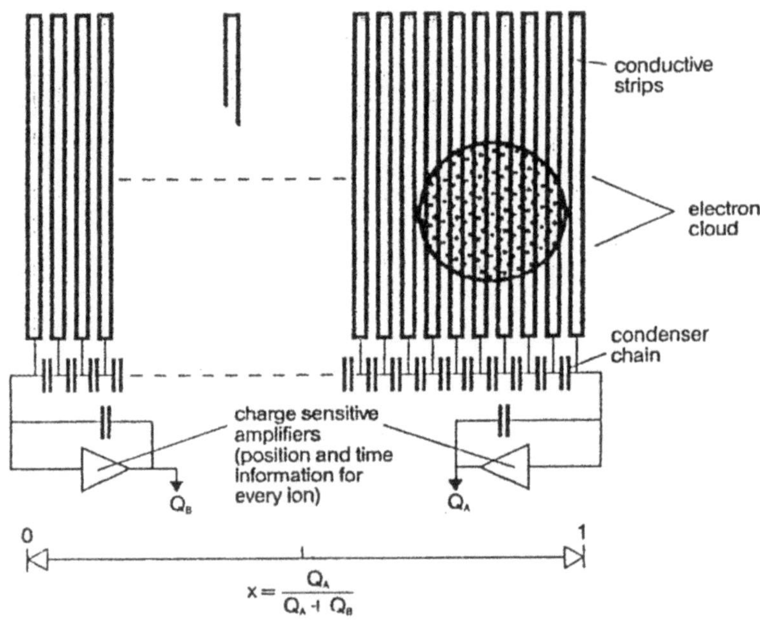

Figure 2. Schematic diagram of the PATRIC array detector [Ref. 25]

The strips were connected to each other via identical capacitors. A charge-sensitive amplifier positioned at each end of the collector plate detected the ratio of currents flowing to the ends of the collector plate. Comparison of the fractions of current shared between the amplifiers allowed one to determine the location of the ion impact.

Because the response time of the detector was 5 μsec, comparative flow of electrons was done rapidly for each and every ion arriving at the detector to produce a complete mass spectrum. The scanning PATRIC array detector used by Pesch et al. [24] covered a mass range of 8% of the mass-to-charge ion centered on the detector. Mass resolution in excess of 5000 was possible with this type of detector. The limitation was dynamic range; that is, every single ion impact must be registered before a second ion strikes the detector. Thus, large ion currents must be avoided.

In sector mass spectrometry, the ability to achieve ultimate detection limits is related to various factors including the ratio between the ion currents from the analyte and the chemical noise, the detector sensitivity, and the mass spectrometer transmission/resolution ratio. We just discussed detection improvements. The ratio between the ion currents from the analyte and the chemical noise improves with the introduction of continuous flow-FAB, MALDI and ESI, as we discuss in **Section 2.3**. The third area, the ratio between the mass spectrometer transmission and resolution has also been improved, and we discuss that in the next section.

2.2 ADVANCES IN DESIGN OF DOUBLE FOCUSSING INSTRUMENTS

We can increase the ratio of transmission and resolving power for conventional forward or reverse-geometry sector mass spectrometers by increasing the overall size of the mass spectrometer or by adding lenses to provide higher order focussing. Similar to any lens, the focussing properties of a sector field are seldom perfect. Focussing aberrations are due to field imperfections, distorted field shape, and fringing fields. To overcome these limitations and to achieve high ion-transmission efficiency, one was required to invoke improved ion-optical designs. Designers at VG (which became Micromass in 1996) implemented one such design in the AutoSpec series of mass spectrometers.

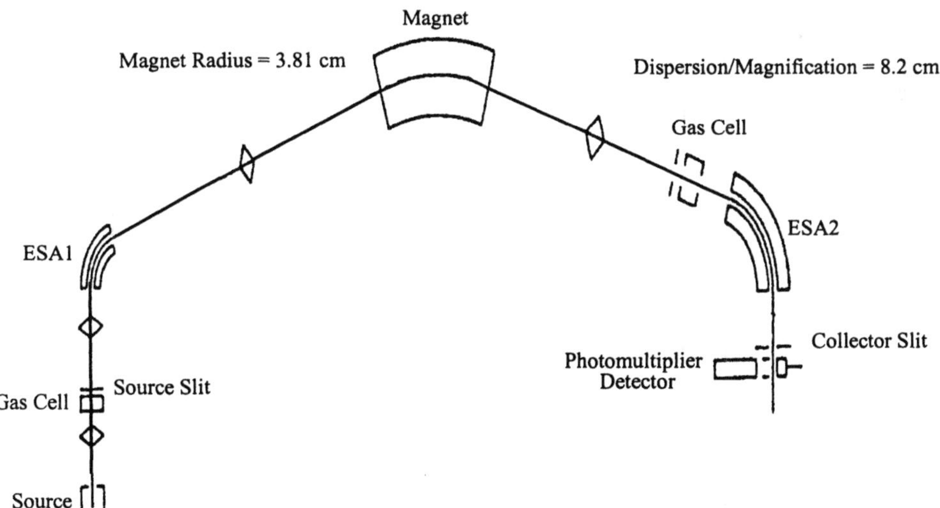

Figure 3. Schematic diagram of the VG AutoSpec double focussing mass spectrometer [Ref. 29]

2.2.1. *AutoSpec Series Instruments*

As shown in Figure 3, the AutoSpec series instruments were based on a design consisting of a magnetic sector and a split ESA arranged in an E_1BE_2 configuration. This design gave a compact instrument with a transmission/resolution product twice as large as other instruments of comparable size. Transmission improved with the addition of z-focussing lens elements before and after the magnetic sector. A fully laminated magnet with a radius of 38.1 cm, a sector angle of 35° and a split ESA with a sector angle of 72.5° were positioned to give a total turn angle of approximately 180° and a reasonably sized footprint of 1.5 m x 0.5 m for the tube unit. The magnet had a large pole gap to accommodate a flight tube with a large z dimension, which enabled good ion transmission with little z-axis loss. The radii of the first and second ESAs were 6.35 and 12.7 cm, respectively.

The AutoSpec configuration benefited from the combination of forward and reverse geometries in a single mass spectrometer. For example, one could use the collision cell in the first field-free region for B/E linked scan experiments [26], and the second and third collision cells situated in the third field-free region could be exploited for MIKES [27] and neutralization/reionization MS [28] experiments. The instrument had an upper mass limit of *m/z* 3000 at full accelerating potential (8 kV). Mass resolving power of 5000 and 60,000 were achievable following single focussing (after the magnet) and double focussing (after the second half of the ESA), respectively. In addition to an improvement in ion transmission and a reduction in instrument size, full-scan mass spectra were free of metastable peaks because the second half of the ESA discriminated against ions that decomposed in the flight tube. This advantage was demonstrated by Bertrand et al. [29], who also demonstrated that the AutoSpec design can provide mass resolving power in excess of 80,000.

Hybrid mass spectrometers such as AutoSpec-Q, AutoSpec-FPD and AutoSpec-OA-TOF evolved from the basic AutoSpec design. To permit both low and high-energy tandem mass spectrometry experiments, Micromass personnel designed the AutoSpec-Q with an EBEqQ configuration, where q is RF-only quadrupole and Q is a quadrupole mass filter. Because hybrid tandem mass spectrometers that combined magnetic sectors and quadrupoles were recently reviewed [30], no further discussion will be given; the interested reader may wish to check recent applications in biomedical problem solving [31].

In MALDI, ions are produced in a pulsed manner, making the technique incompatible with conventional scanning instruments. To overcome this limitation, the Micromass scientists developed the AutoSpec-FPD, where FPD refers to a focal-point detector. In addition to using a magnet and a split ESA, Micromass personnel positioned an inhomogeneous field ESA between the magnet and the second half of the ESA (E_2) to focus the beam onto a 5.0-cm PDA. The first use of the inhomogeneous ESA was for the tandem, four-sector mass spectrometer at Washington University [8]; see **Section 3.3**. Using the inhomogeneous ESA, one was able to obtain a well focussed beam covering the mass range from 1.2:1 to 1.075:1, with mass resolving power higher than 2000. When the array detector was not in use, the inhomogeneous ESA plates were grounded to direct the beam through the second half of the ESA and then to the point detector. For a description of the AutoSpec-FPD, see Bordoli et al. [32].

To select precursor ions at the high resolving power of a double focussing instrument and to exploit the high transmission of a time-of-flight mass spectrometer were motivations for the AutoSpec-OA-TOF series. Two combinations have been available: (i) AutoSpec-OA-TOF-FBD and (ii) AutoSpec-OA-TOF (see **Section 4.1** for a more complete description).

2.2.2 *Benchtop Double-Focussing Instrument*

The combination of mass spectrometry and on-line separation techniques is the most sensitive hypernated method for detecting and characterizing biological samples. Although GC/MS is now a routine experiment and is available in most research laboratories, LC/MS still need improving. The availability of ESI has stimulated a number of groups to seek the improvement. Most laboratories have been involved in coupling LC with ion traps and quadrupoles, which have been usually benchtops and have smaller footprints than conventional magnetic sector mass spectrometers [33].

The coupling of ESI with magnetic sectors came later than with quadrupoles and has not met with high success. The instrument had as motivation the use of high mass-resolving power. To take advantage of high resolving power and its attendant mass measurement capability during LC/MS experiments, Fujimaki et al. [34] developed a benchtop, double-focussing, magnetic sector instrument with LC capabilities. Several combinations and designs of magnetic and electric sectors have been tried to reduce the scale of double focussing mass spectrometers [35]. To make the LC/MS instrument compact, the designers adopted a QQQBQEQ geometry. A magnet, with a radius of 15 cm and a sector angle of 27.8° and an ESA with a sector angle of 52.2° and a radius of 6.0 cm shrunk the footprint of the instrument to only 0.85 x 0.95 m. The compact design and efficient focussing by the quadrupoles gave an image magnification of 0.18 and allowed the source-slit opening to be five times wider than that in a conventional sector mass spectrometer of same size. This instrument may also be a low-cost alternative to larger sector instruments for making exact-mass measurements of low molecular weight (< 700 Da) synthetic organic, organometallic, and natural-product molecules.

2.3. APPLICATIONS OF DOUBLE FOCUSSING INSTRUMENTS AND ARRAYS

The early work using mass spectrometers with array detectors involved FAB ionization. Although the major advantage of array detectors was to increase the sensitivity by two- or three-orders of magnitude, the improvement pertained to the analyte as well as to interferences (chemical noise) from the matrix. The improvement in sensitivity was more apparent when the arrays were used for tandem mass spectrometry or in conjunction with MALDI or ESI because, with these methods, the amounts of interferences were reduced.

2.3.1. *ESI*

Initially, Cody et al. [30] coupled an ESI source with a reverse-geometry, double focussing mass spectrometer (JEOL SX-102) and demonstrated that voltage and narrow-magnet scans could be used to determine accurate masses (to within 12 ppm) of compounds ranging from small organic molecules to proteins. For example, one could use the ESI mass spectrum of lysozyme acquired at mass resolving power of 10,000 to determine the charge state of $[M + 9H]^{9+}$ by measuring the $^{12}C/^{13}C$ isotope-peak separation. Later, the same group used a forward-geometry double-focussing mass spectrometer (JMS-HX110A) equipped with an ESI source and a 5.08-cm PDA, similar to the one described by Hill et al. [31], to demonstrate a ten-fold improvement in the detection limits for proteins when using the PDA instead of the point detector [32]. Although they demonstrated that the minimum detectable quantity of hen egg-white lysozyme with a signal-to-noise ratio of 2:1 is 500 amol/μL, the practical sample concentration for proteins is in the range of 10-100 fmol/μL.

Loo et al. [33] used a Finnigan MAT 900 forward-geometry mass spectrometer equipped with a PATRIC array detector to demonstrate that the higher m/z range available with this instrument was an advantage for studies of non-covalent adducts of proteins. In addition, the same group evaluated B/E link scanning and in source-CID for obtaining structural information on several peptides, and they also demonstrated that the detection limit improves because the detector discriminates against the lower charge (lower energy) and lower molecular weight ions [40]. Multiply charged ions have higher kinetic energy compared to singly charged ions when accelerated in the same electric field. When they hit the MCP, they generate many more secondary electrons than singly charged ions. Thus, the PATRIC electronics will consider these events as normal events. To detect the multiply charged ions, the voltage across the MCP is reduced. At reduced MCP voltages, singly charged ions are not detected. Using a 3 fmol/μL solution of bovine ubiquitin and 850 V across the MCP resulted in very poor sensitivity, but lowering the MCP to 690 V allowed the detection of 90 amol of bovine ubiquitin.

2.3.2. *MALDI*

MALDI has been successful with time-of-flight mass spectrometry because both are pulsed techniques. This combination, however, suffers from poor mass resolving power. A magnet sector mass spectrometer with an array detector allows one to overcome some of the limitations and couple MALDI with a scanning mass spectrometer. Another advantage of coupling MALDI with a sector is the access to higher mass-to-charge range to detect the predominantly singly charged ions produced by MALDI. Biemann et al. [41] used the MS-2 of the four sector mass spectrometer described in **Section 3.2** to detect MALDI ions produced in the region between MS-1 and MS-2. A detection limit of 11 fmol and a mass resolving power of 4500 were demonstrated while using substance P. Analysis of a tryptic digest of sperm-whale apomyoglobin showed that MALDI on magnetic sectors can be used to detect peptides present in complex mixtures.

Both Cody et al. [42] and Bateman et al. [43] reported later the coupling of MALDI with the JMS-HX110 forward-geometry double focussing and AutoSpec-FPD mass spectrometers, respectively. These two groups suggested that the MALDI/magnetic sector combination is useful when mass accuracy and high mass resolving power are required for molecules having molecular masses less than 15,000 Da. Recently, Bordoli et al. [32] compared the results from an AutoSpce-FPD with those obtained using a MALDI-TOF. The improved mass resolving power available with the magnetic sector resolved ambiguities in the MALDI-TOF data.

3. Advances in Tandem Sector Mass Spectrometers

3.1. INCORPORATION OF ARRAY DETECTORS ON TANDEM INSTRUMENTS

The search for high quality spectra from small amounts of material promotes new developments and improvements of mass spectrometers including four-sector tandem mass spectrometers. There are a number of ways that ionization methods, ion optics, and detectors can be arranged to achieve better performance with tandem sector mass

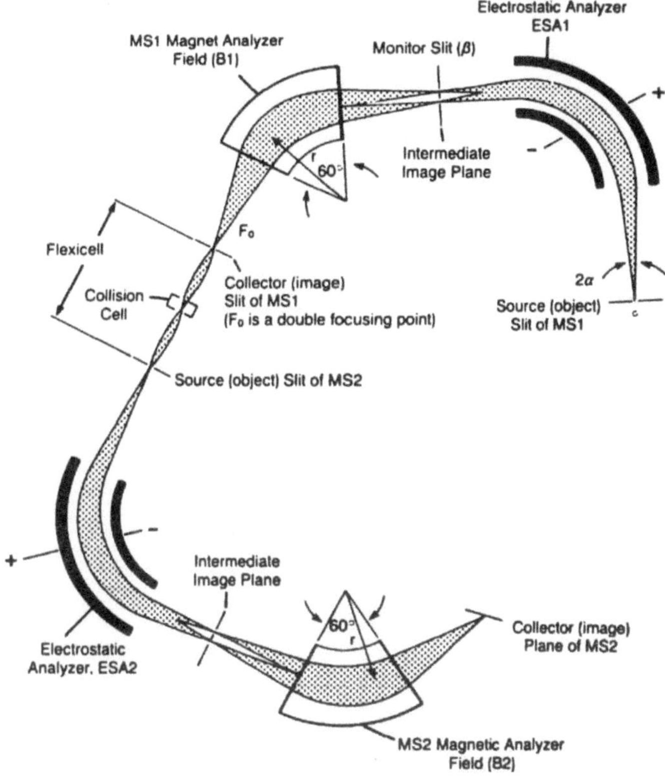

Figure 4. Schematic diagram of the Kratos Concept tandem four-sector mass spectrometer [Ref. 8]

spectrometers. Addition of an array detector (see **Section 2.1**) is one. Three types of tandem four-sector instruments have been modified to operate with array detectors. The first design is the simplest and involves adding a small, 2.5-cm array at the focal point of a standard four-sector. The third design is the most innovative and involves a redesign of the second stage to incorporate a Mattauch-Herzog type instrument that is configured to accommodate a large, 15.2-cm array. The second design is intermediate and represents a clever modification of an existing instrument.

The first commercial tandem four-sector mass spectrometer with an array detector [20] on the second stage was a Kratos Concept IIHH (Figure 4), which was designed for the NIH mass spectrometry facility at the University of California-San Francisco. We discussed the PDA used for this instrument in **Section 2.1.1**.

In 1993, Kratos upgraded the four-sector with a charge-coupled detector (CCD--see **Section 2.1.2**) [9,44]. The CCD permitted rapid data acquisition for continuous scanning of high-energy CAD product-ion spectra. Employing this fast scanning mode, Burlingame and co-workers [45] demonstrated acquisition of CAD spectra of all the peptides from a protein digest that was separated on line with the tandem mass spectrometer. They obtained CAD spectra of peptides eluting at low femtomole levels for a 50-pmol digest of recombinant HIV-2 protease.

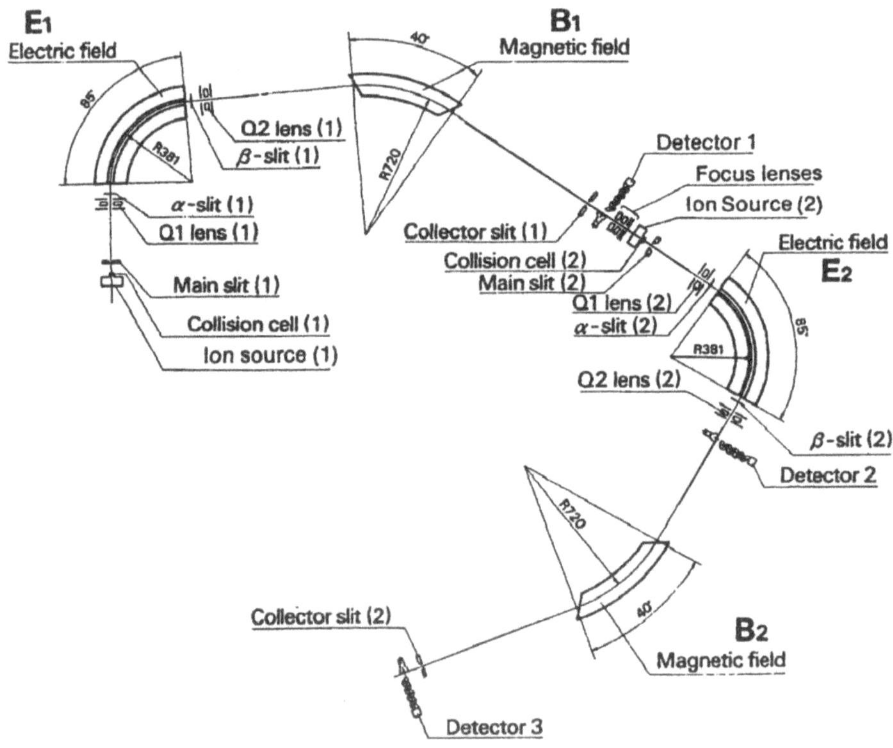

Figure 5. Schematic diagram of the JEOL HX 110/HX110 tandem four-sector mass spectrometer [Ref. 8]

The second design of an array detector for a four-sector mass spectrometer was by Biemann and co-workers [46] for the JEOL HX110/HX110. This $E_1B_1E_2B_2$ design permitted the simultaneous recording of a 6.6% mass range of product ions and allowed one to vary the ratio of low to high *m/z* from 1:1.04 to 1:1.13 [47-48]. A pair of quadrupole lenses and one octapole lens after the second magnetic field of the instrument dispersed the ions across the array (Figure 5); the dispersion range was governed by the ion optics (i.e., the degree of coincidence of the α and β focal planes and their flatness).

One advantage of this 2048-pixel, 5.08-cm array detector was an increase in the mass range covered by a single exposure compared to that of a 2.54-cm array; fewer individual segments needed to be recorded and assembled for a complete product-ion spectrum. The speed and resolving power are of particular importance in MS/MS when obtaining CAD spectra of eluents from an HPLC that is coupled to the mass spectrometer. For example, the CAD spectrum of a tryptic fragment from 50 pmol of the β-chain of human hemoglobin was recorded in 15.5 s [49]. This was adequate for the peak widths (~ 30 s) in conventional HPLC. Furthermore, owing to the integrating properties of the detector, one could employ a pulsed laser to accomplish photodissociation.

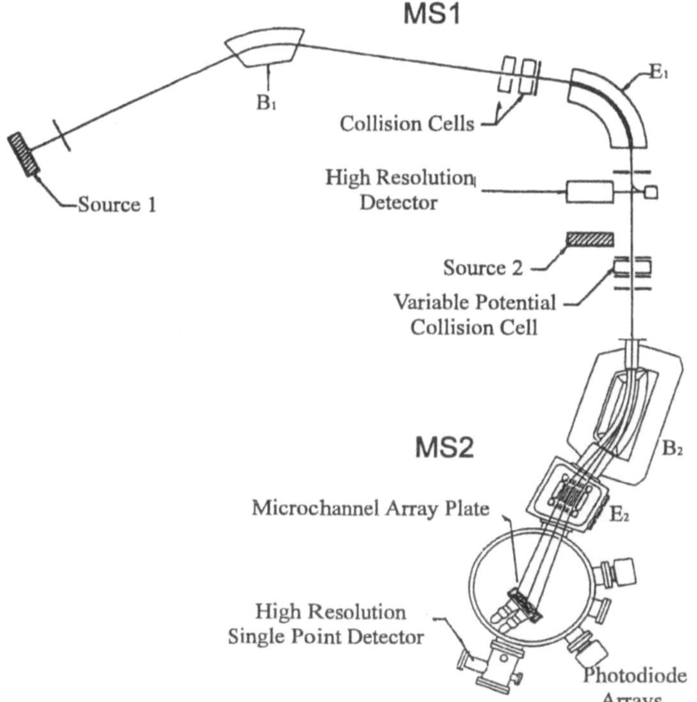

Figure 6. Schematic diagram of the VG-ZAB-T tandem four-sector mass spectrometer [Ref. 8]

The third design of an array-detector instrument resulted in the extended 6144-pixel photodiode array, originally part of a ZAB-T four sector that was built for Gross by VG Instruments (Figure 6). Incorporation of a 15.2-cm array after the second stage (a reverse-geometry Mattauch-Herzog-type MS2) [9, 50] allowed simultaneous recording of 1.6:1 mass range of product ions while maintaining a collision-cell voltage at 50% of the acceleration potential. When the collision cell was at ground potential, one could detect simultaneously product ions over nearly a 1.4:1 mass range.

This tandem instrument had a second ESA with an inhomogeneous electric field. The novel design of the second electric sector allowed a wide mass dispersion of ions to be transmitted and double focussed (to first order) on the large array detector. One achieved the inhomogeneous electric field by applying voltages to a set of shim electrodes, between which the ion beam was traveling. Therefore, this ESA served not only as an ion deflector but as a double-focussing device.

Taking advantage of the design of this instrument, Gross and co-workers [51] successfully coupled tandem MS and capillary electrophoresis by using a coaxial continuous-flow FAB interface and obtained structural analysis of mixtures of small peptides at low-femtomole levels in electrophoretic real-time. They also obtained significant improvement in the S/N of tandem mass spectra or in the speed of acquisition when they compared the results with those of similar experiments carried out with a four sector equipped with a single-point detector.

3.2. NEW COLLISION CELL DESIGN

Fenselau and others [52] designed a new interface system for collision cells; the new design consisted of an ion decelerator, a floating collision-cell, chemical-ionization ion source, and an ion extractor for a JEOL HX110/HX110 four sector. This design permitted high transmission of ions through the collision cell at low to intermediate ion kinetic energies (3 eV to 1 keV) and made use of cylindrical deceleration lenses and an extraction-lens assembly. The ion-retarding system had three cylindrical-shaped lenses and permitted deceleration through an energy ratio in excess of 330:1. Incorporation of the dual-purpose collision cell facilitated CAD over a range of collision energies from 3 eV to 10 keV, ion-molecule reactions of mass-selected ions, higher order tandem MS experiments (e.g., MS^4), and neutralization-chemical reionization of analyte ion beams.

The length and position of the collision cell affects the resolving power for ions produced in collisionally activated processes in sector instruments. Brenton et al. [53] showed that collisional broadening could be reduced to less than 0.001 eV for 1-keV collisional energy if both the length and position of the central collision cell were chosen carefully. Under their experimental conditions, only the size of the resolving slits and broadening, due to the thermal motion of the target gas, determined the energy resolution.

3.3. APPLICATIONS OF FOUR-SECTOR TANDEMS

3.3.1. *Charge-remote fragmentations of fatty acids.*

Charge-remote fragmentations were discovered while using sector mass spectrometers [54-55], and now many problems associated with structural studies of fatty acids, fatty alcohols, prostaglandins, lipids, and steroids can be solved. The advantage of charge-remote fragmentations is that processes such as double-bond migration and hydrogen scrambling, that are competitive with charge-driven decompositions, do not compete for charge-remote processes. Thus, charge-remote fragmentations effectively reveal the site of functional groups such as double bonds, multiple double bonds, branch points, cyclopropyl and cyclopropenyl rings, epoxides, and hydroxyl groups. A key requirement is that the charge must be tightly localized. The charge for a gas-phase $[M - H]^-$ of a fatty acid resides on the carboxyl group, for example, yet most of the fragmentation reactions take place along the hydrocarbon chain. For a saturated, unbranched fatty acid, the resulting CAD spectrum contains a series of product ions separated by 14 u. The regularity of the CAD spectrum is disturbed if unsaturation, branch points or other functional groups are present in the molecule.

The originally proposed mechanism for the fragmentations along an hydrocarbon side chain was a 1,4-elimination of H_2 and a normal olefin. Wesdemoitis showed with neutralization-reionization mass spectrometry (NRMS) [56] on a sector instrument that the major neutral losses are alkenes, not alkanes or alkyl radicals, which is consistent with the original mechanism.

$$\text{R}\text{—(cyclic H,H transition state)—}(CH_2)_nCO_2^- \longrightarrow \text{R—CH=CH}_2 + H_2 + CH_2\text{=CH—}(CH_2)_nCO_2^- \quad (1)$$

3.3.2. *High energy CAD of peptides and glycopeptides.*

The application of tandem sector mass spectrometry to problems in structure elucidation has been one of its important uses; peptide sequencing is a specific application [57-70]. Collisional activation is at high kinetic energy, and some fragmentations are unique, reflecting the high center-of-mass energies and the short time for decomposition. A distinguishing feature is the prominence of ions formed by cleavages of the peptide backbone and certain amino-acid side chains giving d, w, and v sequence ions. These product ions provide valuable information for complete peptide sequence determination. Peptides that contain basic amino acids at the N-terminus usually fragment to give N-terminal product ions, whereas positioning the basic residue at the C-terminus results in the formation of fragment ions that retain charge at or near the C-terminus. Although the mechanisms of formation of some fragment ions are not understood, many may be charge-remote.

Biemann et al. [58] studied the requirements for the formation of c_n ions in high-energy CAD of protonated peptides. Because one commonly sees c_n ions in FAB mass spectra but seldom in CAD spectra, the authors investigated whether c-ion formation depends more on sequence than that of other backbone fragment ions. The propensity to form c_n ions is determined by the nature of the amino acid on the C-terminal side of the cleavage site and is large when that amino acid is threonine, and to lesser extent tryptophan, lysine, and serine. There is a little influence of the amino acid on the N-terminal of the cleavage site. The proposed mechanism for the favored formation of c_n ions involves hydrogen transfer from a β-carbon to the charged N-terminal part of the molecule, producing a neutral containing a double bond (see Eqn 2), which can be stabilized through conjugation or induction by appropriate c-terminal-side amino acids.

$$\sim\mathrm{CH(R_n)}-\mathrm{C(=O)}-\mathrm{NH}-\mathrm{CH}[\mathrm{C(OH)(CH_3)H}]-\mathrm{C(=O)}\sim \;\longrightarrow\; \sim\mathrm{CH(R_n)}-\mathrm{C(=O)}-\mathrm{NH_2} + \mathrm{CH}[=\mathrm{C(OH)CH_3}]-\mathrm{C(=O)}\sim \quad (2)$$

Biemann and co-workers [59] also studied the effect of charge state and charge localization on the CAD of protonated peptides under high-energy collision conditions. They compared CAD spectra of singly and doubly charged precursor ions (generated by FAB and ESI, respectively) for peptides with similar primary structures. Multiply charged ions have higher kinetic energies than their singly charged counterparts when formed under same acceleration potential. Therefore, more energy is available for electronic and vibrational excitation when these ions collide with target gas. Furthermore, coulombic repulsion also facilitates most decompositions of multiply charged ions. Results show that the decomposition pathways of singly and doubly charged ions are similar and give product ions that are strongly influenced by the position of basic residues. The abundance of peptide ions containing two charges is still favored when both charges reside in close proximity. These results challenge the idea that, in multiply charged ions, the charges must be maximally separated to minimize coulombic repulsion.

Kolli and Orlando [60] investigated by four-sector tandem MS the decompositions of multiply protonated peptides in the 4-5 kDa range. They achieved sequence confirmation, including leucine-isoleucine distinctions, for multiply protonated peptides. The mass resolving power permitted the determination of the charge state of the fragment ions by allowing the spacing between isotope peaks to be measured. The high-mass accuracy assured identification of amino acids that differed by one mass.

Burlingame and co-workers [66] compared the high energy CAD spectra of synthetic peptides with MALDI-PSD and MALDI-CAD spectra obtained with an EBE-oa-TOF tandem (see **Section 4.1.1**). The CADs for the MALDI-produced ions were characteristic for high-energy activation. Abundant d, w, and v fragment ions formed although the collision energy was approximately six times greater than that used to promote CAD of FAB-produced ions on other tandem four sectors. Processes that typically take place as PSD, such as sequential neutral losses from Arg-containing ions, also occurred in the CAD of the MALDI-produced $[M + H]^+$ ions of the antigenic peptide, APRTLVLLL .

Martin et al. [68] compared the types, extent, and distribution of peptide fragmentations produced in PSD of MALDI-produced ions with those occurring for FAB-produced ions that fragment as metastable ions and upon low-energy collisions in a tandem four sector. The results suggest that the nature of MALDI-PSD and low-energy CAD of FAB-produced ions is remarkably similar for des-Arg^1 and des-Arg^9 bradykinin Both methods presumably deposited comparable amounts of internal energies into the $[M + H]^+$ precursor ions. The distribution of N-terminal, C-terminal, immonium, and internal ions from MALDI-PSD did not change appreciably with respect to the position of Arg residue unlike for metastable-ion and low and high-energy CAD of FAB-produced ions. This indicates that the internal energy deposited in MALDI and the 14 ms of flight time conspire to promote steady-state decomposition kinetics.

Summerfield and Jennings [69] showed that the small neutral losses that occur in charge-remote fragmentations of $[M + H]^+$ and $[M + Met]^+$ adducts (where Met is alkali metal ion) involved cleavages of the amino-acids side chains. The neutral losses varied according to the structure of the amino-acid side chain and, in some instances, provided a rapid means of verifying the presence of an amino acid in a peptide. For the TyrPheMetArgPhe/Li^+ complex, for example, the neutral losses occurring remote to the charged metal center are characteristic of the constituent amino acids: 48 (CH_2S, H_2), 62 (CH_2S, CH_4), and 75 ($(CH_2)_2SCH_3$) for Met; 44 (NHCNH, H_2), 59 (NHCNH, NH_3), 73 ($CH_2NHCHNHNH_2$), 87 ($CH_2CH_2NHCHNHNH_2$); and 100 ($(CH_2)_3NHCHNHNH_2$) for Arg; and 98 (C_6H_7OH, H_2) for Tyr.

Structural information for glycopeptides could also be obtained from high energy CAD experiments on tandem sector mass spectrometers. Burlingame and co-workers [70] established the peptide identity and glycosylation site(s) for O-linked glycopeptides from a tryptic digest of bovine fetuin and from synthetic N-linked glycopeptides. One may also establish oligosaccharide composition and sequence from the presence of oxonium ions and carbohydrate losses from the molecular ion. The analysis of the CAD of O-linked glycopeptides was difficult, however, owing to the ready fragmentation of the glycosidic bond, giving low-abundance fragment ions containing the oligosaccharide. In contrast, N-linked glycopeptides, at least those with a single N-acetyl-glycosamine, have a relatively stable linkage between the carbohydrate unit and an Asn residue.

3.3.3. *High energy CAD of glycoalkaloids, oligosaccharides, and glycosides.*

Four-sector tandem mass spectrometry with CCD array detection was used in the structural characterization of a glycoalkaloid [71]. α-Tomatine, which was isolated at the 200-fmol level from tomato fruit, was determined by high energy CAD in a 10-s scan [22].

High-energy CAD spectra of $[M + H]^+$ ions of glycoalkaloids and aglycons contain a wealth of information [22, 71-72] on the carbohydrate sequence and linkage sites, on the mass of the aglycons, and on structural features of the aglycon. The observed fragment ions resulted from charge-remote fragmentations, multiple cleavages, or complex charge-driven rearrangements.

Series of linear and branched synthetic mannosyl oligosaccharides containing phosphate substituents on either or both the non-reducing termini or on penultimate residues were investigated by means of high energy CAD [73]. These naturally occurring phosphorylated oligosaccharides were usually available only at low pmol levels. The presence of the phosphate group allowed formation of an abundant fragment ion that arises from single cleavages at the glycosidic bond with charge retention on the phosphorylated fragments. The CADs of linear structures containing non-reducing terminal Man-6-phosphate residue were dominated by Y-type ions. On the other hand, branched oligosaccharides that contained one or two terminal Man-6-PO_4 residues, displayed double cleavages with charge retention on the phosphate-containing fragment. The location of the phosphorylated residue can be determined from presence or absence of specific B and Y ions.

Gross and co-workers in collaboration with workers at Schering Plough [74] examined glycosidic antibiotics that were introduced by FAB into a four-sector tandem mass spectrometer as $[M + H]^+$ or $[M + Met]^+$, where Met is Na or K. Remarkably different collisionally induced processes occurred for the $[M + H]^+$ compared to the $[M + Met]^+$ ions. These results underscore the effectiveness of using the simple expedient of introducing an $[M + Met]^+$ and using high-energy MS/MS to obtain complementary structural information. The same group characterized unknown cardenolide glycosides, isolated as natural products, by means of tandem MS [75]. The CADs of FAB-produced ions gave information to identify the aglycons, the sugar moieties, and their sequence. NMR provided the stereochemistry of the linkages.

3.3.4. *Endohedral complexes of fullerenes.*

A unique application of high-energy collisions is the insertion of noble gases into the cavity of fullerene radical cations to form "endohedral complexes" [76-81]. The first $Ar@C_{60}^{+\cdot}$ and an analogous series with C_{70}. These endohedral complexes then underwent sequential elimination of C_2 from principally the endohedral complex, and not the original fullerence radical cation, giving a series of endohedral product ions. The extent of C_2 elimination depended on the nature of fullerenes (C_{60} or C_{70}) and the size of the trapped gas. The end point, or smallest mass product, for C_2 loss from $He@C_{60}^{+\cdot}$ was $He@C_{44}^{+\cdot}$, whereas for $Ne@C_{60}^{+\cdot}$ the end point was $Ne@C_{46}^{+\cdot}$. Surprisingly, the fragmentations of $Ar@C_{60}^{+\cdot}$ gave rise to all odd-carbon species by expelling C_5, C_7, C_9, and C_{11} neutrals.

There is a significant kinetic-energy shift of the product ions when a precursor ion captures the target gas; the energy in the center-of-mass frame is lost to translation The energy shift increases with the mass of the target gas. The largest shift for the endohedral complexes seen by us occurred for the formation of the argon-containing complexes from $Ar@C_{60}^{+\cdot}$. The wide energy bandpass of the ESA on MS2 of the

ZAB-T (see **Section 3.1**) was sufficient to transmit the product ions, which had less translational ions than those formed by processes in which the target gas was not captured.

The instrument was used more recently to detect the capture of He by $He@C_{60}^{+\cdot}$, $Ne@C_{60}^{+\cdot}$, $Ar@C_{60}^{+\cdot}$ and $Kr@C_{60}^{+\cdot}$ [80-81]. The results show that it is possible to make endohedral complexes containing two noble gases. As example is in Figure 7, which gives evidence for the capture of He by $Ar@C_{70}^{+\cdot}$

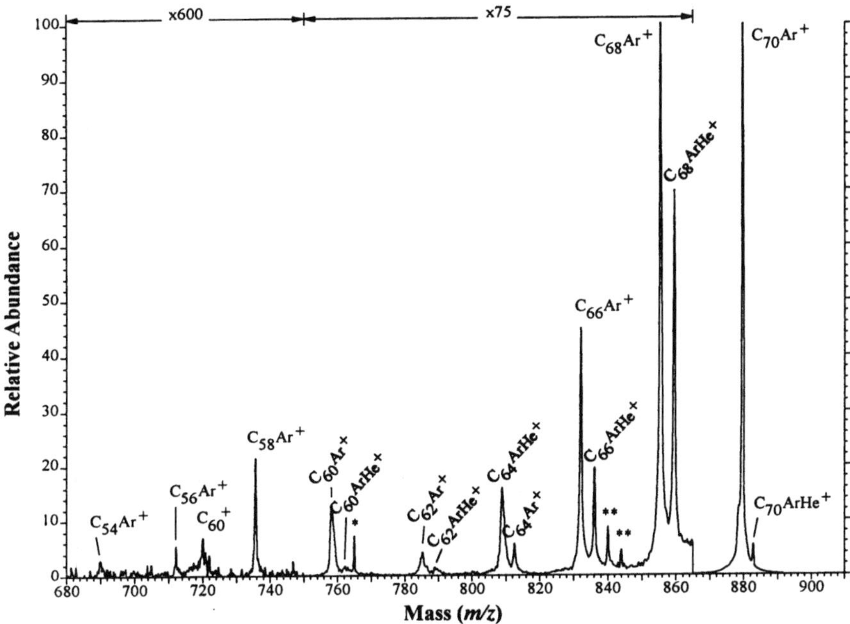

Figure 7. CAD of $Ar@C_{70}^{\ +}$ by He using the VG-ZAB-T tandem four-sector.

4. Five-Sector Mass Spectrometers

An evolution of tandem four-sectors is the five-sector instruments in which the fifth sector is an ESA. These instruments have been generally built to investigate the details of product-ion fragmentations. The basis is mass-analyzed ion kinetic energy spectrometry (MIKES) experiments, which have been carried out on reverse geometry (BE) mass spectrometers for over 20 years. The shapes of the product ion peaks from metastable-ion decompositions can be used to determine kinetic-energy-release distributions of the products of metastable-ion reactions. Kinetic-energy release can be used to retrieve details about fragmentation mechanisms such as whether the fragment ions are formed from strongly or loosely bound complexes and the extent of charge separation on doubly charged molecules. For the measured kinetic energy release to be accurate, however, the kinetic energy width of the precursor ion beam must be measured correctly. In a 1995 study, Adams and co-workers [82] described the correct procedure for measuring the kinetic-energy width of the precursor ion beam. Although they used a five-sector instrument (JEOL JMS-SX102/SX102A/E) of $B_1E_1/B_2E_2/E_3$ geometry, their results are general.

A five-sector VG Autospec-T of E_1B_1 /E_2B_2/E_3 geometry was employed in 1994 by Lange and co-workers [83] to obtain sequential product-ion spectra (MS^3). The technique involved reaction-intermediate scanning with array detection; the sample was the cyclopeptide E-destruxin. The ESA, E_3, on this instrument was an inhomogeneous electric field that allows ions with a range of kinetic-energy values to be focused simultaneously onto a planar array detector (see **Section 3.1**). The first-generation product-ions were selected and subjected to collision-induced dissociation in a variable-potential collision cell in the fourth field-free region; argon was the collision gas. These first-generation product ions were then analyzed with the B_2E_3 portion of the instrument (MS2).

5. Advances in Hybrid Tandem Mass Spectrometers

There are several reasons why one would want to incorporate a mass analyzer other than a magnetic/electrostatic sector as MS2 on a tandem instrument. Four-sector instruments are large, expensive, not very sensitive (even with array detection), and collisional activation is not efficient when compared to that on other commonly used mass analyzers. Increased sensitivity and flexibility are paramount when trying to solve complex, modern-day biological research problems.

5.1. INCORPORATION OF TIME-OF-FLIGHT

5.1.1. *Axial or parallel TOF*

Time-of-flight mass spectrometers will become, in our opinion, the most common second mass analyzer on a tandem sector and tandem quadruple instruments; this hybrid will replace the sector/quadrupole hybrid that was popular in the 1980s. Time-of-flight mass spectrometers are relatively simple and are a much cheaper alternative to double focussing sectors. Time-of-flight analyzers also offer a multichannel advantage (for simultaneous ion detection) and have inherently high ion transmissions, which translates into increased sensitivity. These features are especially attractive when performing CAD experiments on samples at low levels. There are, however, some inherent difficulties in the incorporation of time-of-flights to magnetic sector instruments. A notable and obvious one is the need to pulse the ion beam. Different research groups have proposed different means to do this. The following is a review of some tandem instruments incorporating time-of-flight mass spectrometers as the second stage.

The first example was reported in 1983 by Enke and co-workers [84]. The authors measured the flight time of ions with a LKB-9000, using a modified source to provide brief pulses of ions, and a time-resolved ion detector. The entire instrument was used as the time-of-flight mass spectrometer. The authors used a continuous ion source and pulsed portions of the ion beam by applying a square-wave modulation to beam-focussing plates. Ions entered MS1 only on the rising and falling edges of the square wave. One can use the instrument to make unambiguous assignments of the origin of the product ions.

Russell and co-workers [85] incorporated a time-of-flight equipped with a reflectron on line with the precursor ion beam from a Kratos MS-50 double focussing mass spectrometer. The authors used MALDI to produce pulsed peptide ions and studied their CADs. This constitutes, to our knowledge, the first results of a MALDI/tandem mass spectrometry experiment. The authors selected a peptide ion beam with MS1, decelerated it from 8 keV to 1 keV into a collision cell and then reaccelerated the product ions to 8 keV into the TOF. Partial sequence was obtained from 10 picomoles of peptide loaded on the probe.

Russell and co-workers then used the instrument to do neutral-ion correlation measurements for the acquisition of tandem mass spectra [86]. These measurements can be made using pulsed as well as continuous ionization methods. Neutral-ion correlation measurements, which are made by using two detectors, one behind, and one in front of the reflectron, allow one to determine the product ions from a number of precursors in one experiment.

In 1994, Cotter and co-workers [87] proposed to use an axially mounted, curved-field reflectron TOF after a sector mass spectrometer of EB design. The pulsed ion beam would enter the collision cell after MS1 at 10 keV and exit at 4-5 keV into the TOF. The curved field reflectron would provide nearly complete fragmentation information simultaneously without the traditional stepping of the reflectron's retarding electric field.

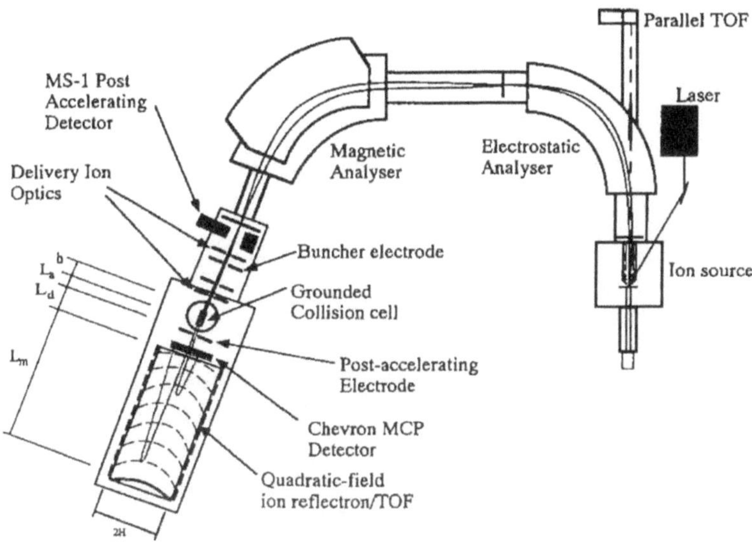

Figure 8. Schematic diagram of the magnetic sector/TOF mass spectrometer [Ref. 88]

Derrick and co-workers [88] proposed in 1995 to use a time-of-flight with a quadratic ion reflectron as the second stage of a tandem mass spectrometer (Figure 8). Their goal was to carry out CAD with subsequent analysis by TOF. The instrument combined a reflectrons time focus ions independent of their energy. Ions were produced by a MALDI ion source, and the precursor ions selected by MS1. Precursor-ion time spread, stemming from both ion formation and the passage through the electric and magnetic sectors, was compensated by an ion buncher. The buncher consisted of two electrodes and was located in front of the collision cell at the exit of the double focussing MS1; it compressed the length of the parent ion packet, which was 50-100 ns depending on the mass. Correct application of the ion buncher resulted in a ion packet of 1-5 ns. This focussing technique worked in a similar way as delayed extraction on a linear time-of-flight mass spectrometer. High energy CADs of renin substrate provided full-sequence information despite the high metastable-ion noise.

5.1.2. *Orthogonal TOF*

In a 1992 communication, Bateman [89] proposed that many of the incompatibilities of sectors and TOFs can be circumvented by incorporating an orthogonal TOF as the second mass analyzer of a tandem instrument. In a 1996 article, Bateman, Burlingame, and co-workers [66] reported MALDI-CAD data acquired on a VG AutoSpec 5000 orthogonal acceleration (oa)-TOF tandem mass spectrometer (see **Section 2.2.1.**). The instrument used a two-stage deceleration electrostatic lens to focus the ions into a nearly collimated beam before they enter the

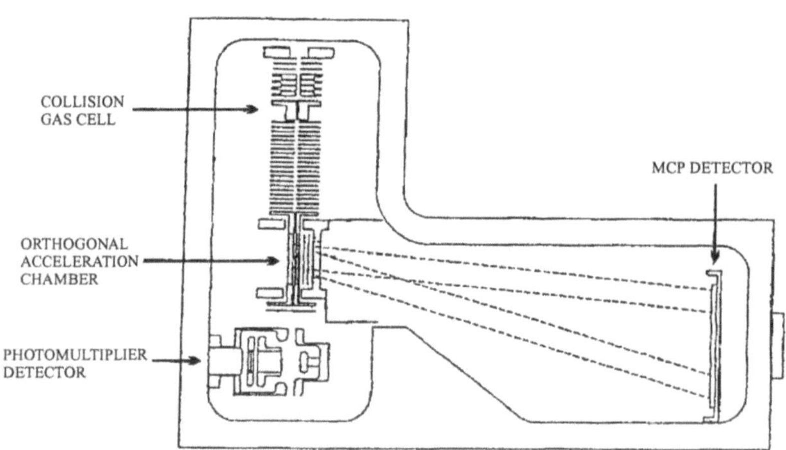

Figure 9. Collision cell and the orthogonal acceleration region of VG AutoSpec-TOF mass spectrometer [Ref. 66]

gas collision cell. After the collision cell, the precursor and product ions passed through the orthogonal acceleration chamber (Figure 9) in which an electric field orthogonal to the direction of travel was established periodically by the sudden application of a voltage to a "push-out" electrode. If orthogonal acceleration were not used, the ions passed through the chamber and registered on a detector, which was at one end of the instrument. The timing of this extraction field was calculated and controlled by the data system.

The ions in the extraction field were accelerated to ground potential through a two-stage voltage gradient into a linear TOF mass analyzer, which had an effective pathlength of approximately 0.5 m. The extraction pulse also served to start the time-of-flight measurement. When the ionization source was pulsed, as with MALDI, the sampling efficiency was 100%. With a continuous ion source, the duty cycle was competitive with a four sector with extended array detection, with a source voltage of 8 kV and a collision energy of 800 eV, the mass range of the TOF was 7200 Da.

Certain models of the AutoSpec allowed one to do traditional time-of-flight; all ions produced by the sample were passed to the push-out electrode. This was accomplished by bypassing the magnet and allowing the MALDI ions to traverse both electrostatic sectors before reaching the push-out electrode [90]. This permitted one to determine the *m/z* of precursor ions in a preliminary experiment before doing MS/MS. MALDI-CAD data that were acquired on this instrument were compared to CAD spectra of FAB-produced ions on a tandem four sector (this comparison is discussed in **Section 3.3.2.**). The authors demonstrated the high sensitivity of this setup by obtaining the MALDI-CAD spectrum of 1 pmol (loaded) of substance P when using DHB as the solid matrix.

5.2. INCORPORATION OF QUADRUPOLE ION TRAPS

Another type of hybrid is a tandem sector coupled with a quadrupole ion trap. Because the ion trap is capable of MS^n studies alone, these instruments offer a high degree of experimental flexibility. As early as 1984, March and co-workers [91] reported an improved hybrid mass spectrometer for CAD. The instrument was of EB geometry to which a quistor and a quadrupole mass filter were attached. The quistor was mounted in the detector housing of a Kratos MS-30 (a now obsolete, medium performance Nier-Johnson double focussing instrument) and oriented to allow the ion beam to pass diagonally through the ion trap. A pulsed gas valve introduced pressure pulses of collision gas. Ions that underwent collisions sustained a loss of momentum, which permitted the ions to be stored in the quistor.

In 1990, Cooks and co-workers [92] published a study in which they used an instrument of BE/trap configuration. The authors used the ion trap not only for ion containment and reaction, but as an independent mass analyzer as well. Recently, Schroder and co-workers [93-94] built a commercial tandem mass spectrometer that included an ion-trap as MS2 (the overall configuration is EB-ion trap). The ion optics from the sector instruments into the ion trap consisted of an electrostatic deceleration lens system and a dual-octapole assembly to facilitate the effective transfer of ions into the ion trap, which was floated at the acceleration potential of MS1 to slow down the ions as they enter the trap (see Figure 10). One benefit of using the hybrid instrument over the ion trap alone for MS/MS was the selectivity to fill the ion trap with precursor ions only. This feature allowed better detection limits and a larger dynamic range. Using this instrument, the authors obtained the complete sequence of bombesin by MS/MS and MS^3 experiments. By fragmenting the doubly charged molecular ion $[M + 2H]^{2+}$ at *m/z* 810.4 by MS/MS, and following with MS^3 experiments revealed an early complete series of b ions starting from the C-terminus. MS/MS analysis of the $[M + Na]^+$ ion at *m/z* 1641.9 showed that N-terminal sequence ions (y ions) can be read through eight amino acids.

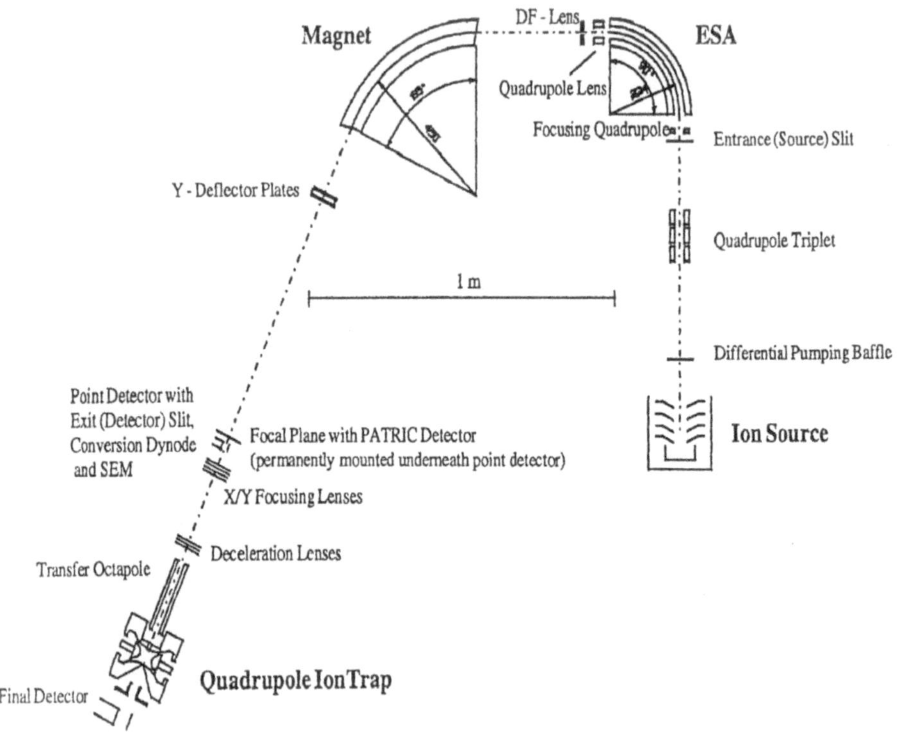

Figure 10. Schematic diagram of the MAT 900 S-trap mass spectrometer [Ref. 93,94]

5.3. INCORPORATION OF WIEN FILTERS

Wollnik and co-workers [95] described another type of hybrid tandem mass spectrometer in 1992. The first stage was a double focussing sector instrument of BE configuration. The second stage incorporated a Wien velocity filter and a sector magnet as the double focussing combination. The Wien filter-sector magnet combination allowed double focussing and simultaneous detection of a wide range of masses. This analyzer was chosen because, for high-energy collisional activation, the product ions have variable energy but constant velocity. Therefore, a velocity analyzer (a Wien filter) was chosen just as an energy analyzer (an ESA) was chosen for a double focussing instrument that is operated at constant energy.

A Wien filter consists of a magnetic field and an electric field that are perpendicular to each other and to the optical axis. Fragment ions leave the collision cell with approximately the velocity of the incoming molecular ions and with kinetic energies that are proportional to their masses. The use of a conventional cylindrical ESA was difficult because a wide-energy bandpass analyzer would have been needed, such as the inhomogeneous planar ESA described in **Section 3.1.** The mass-to-energy ratio of the fragment ions equals approximately the mass-to-energy ratio of the precursor ions because they all have nearly equal velocities. Ions of low energy, for example, can be lost to the electrodes of the ESA. The use of the Wien filter overcomes this disadvantage of the ESA, providing wide-energy acceptance. With the Wien filter, ions of similar velocities move in the Wien filter along neighboring trajectories. Therefore, ions of widely different masses pass through a Wien filter of constant E_o/B_o ratios provided the ion velocities are similar.

6. Conclusions

At this writing, it appears that double focussing and tandem sector instruments will play a minor role in future mass spectrometry. They are too complicated, too costly to build and maintain, and not readily compatible with ESI and MALDI. There will always be a need for exact-mass measurement, but this can also be fulfilled by using FTMS. Moreover, the mass accuracy of TOF instruments is approaching the 10 part-per million range. It is too early, however, to predict that scientists who need high resolving power and exact-mass measuring capability will make an FT mass spectrometer their first choice for a high-performance instrument purchase.

The invention of ESI and its successful coupling with FTMS has made unnecessary grand-scale mass spectrometers that were built to explore high-mass, high-resolving-power applications. Even the large, commercially available sector instruments (e.g.,VG ZAB, Kratos Concept, JEOL HX110), which were built through the late 1980s to provide mass measurement to *m/z* 15,000, will no longer be assembled for future applications.

A diminished role in the future, however, does not preclude any role. Double focussing instruments now have smaller footprints than previously, are more manageable, and operate well at resolving powers of 10,000. Their dynamic range recommends them for EI, for mixture analysis, and for high-resolving-power analyses of target compounds. They have played an important role in capillary column GC/high resolution MS for dioxins and other materials that need to be determined with high certainty at the sub parts-per-trillion range. This role may expand to target compound analyses by LC/MS, a role not easily played by low dynamic-range instruments such as ion traps and FT mass spectrometers.

Tandem four-sector instruments are being replaced by a new generation of hybrids. Particularly promising are those using TOFs as the second stage. These instruments will be capable of high-energy CA, a currently unique application of tandems with a magnet-sector component. High-energy CA is important for studying endohedral complexes of fullerenes and charge-remote fragmentations and is complementary to low energy CA in biomolecule-structure studies. The future for hybrids will brighten considerably when someone discovers more efficient ways to couple ESI and MALDI with sectors and to make high-energy CA more efficient.

7. Acknowledgments

Preparation of this review was supported by the NIH (Grants 2P41RR00954 and 2PO1CA49210). The authors also thank NATO for support, and JG thanks NSF for a travel grant.

8. References

1. Aston, F. W. (1924) *Isotopes*, Edward Arnold & Co., London.
2. Futrell, J. H. (1996) *Proceedings of the 44th ASMS Conference on Mass Spectrometry and Allied Topics*, Portland, Oregon, 34.
3. Nier, A. O. C. *Int. J. Mass Spectrom. Ion Processes* **100**, 1-13.
4. Svec, H. J. (1985) *Int. J. Mass Spectrom. Ion Processes* **66**, 3-29.
5. Cooks, R. G., Beynon, J. H., and Caprioli, R. M. (1973) *Metastable Ions*, Elsevier, Amsterdam, The Netherlands.
6. Beynon, J. H., Cooks, R. G., Amy, J. W., Baitinger, W. E. and Ridley, T. Y. (1973) *Anal. Chem.* **45**, 1023A-1031A.
7. Jennings, K. R. (1968) *Int. J. Mass Spectrom. Ion Phys.* **1**, 227-235.
8. Gross, M. L. (1990) *Methods in Enzymology: Mass spectrometry*, Vol. 193, Academic Press, Inc., San Diego, California, 131-153.
9. Burlingame, A. L. (1994) *Biological mass spectrometry: Present and future*, John Wiley & Sons Ltd., West Sussex, England.
10. Trainor, J. R. and Derrick, P. J.(1992) *Mass spectrometry in the biological sciences: A tutorial*, Kluwer Academic Publishers, Dordrecht, The Netherlands.
11. Reilly, M. H. and Gaskell, S. J. (1988) *Proceedings of the 36th ASMS Conference on Mass Spectrometry and Allied Topics*, San Francisco, California, 137-138.
12. Gaskel, S. J. (1996) *Mass spectrometry in biomolecular sciences*, Kluwer Academic Publishers, Dordrecht, The Netherlands.
13. Gaskell, S. J. and Ballard, K. D. (1992) *Mass spectrometry in the biological sciences: A tutorial*, Kluwer Academic Publishers, Dordrecht, The Netherlands.
14. Yost, R. A. and Boyd, R. K. (1990) *Methods in Enzymology: Mass spectrometry*, Vol. 193, Academic Press, Inc., San Diego, California, 154-200.
15. Nier, A. O. and Schlutter, D. J. (1985) *Rev. Sci. Instrum.* **56**, 214-219.
16. Giffin, C. E., Boettger, H. G. and Norris, D. D. (1974) Int. J. Mass Spectrom. Ion Phys. 15, 437.
17. Murphy, D. and Mauersberger, K. (1987) *Int. J. Mass Spectrom. Ion Proc.* 76, 85-93.
18. Hedfjall, B. and Ryhage, R. (1979) *Anal. Chem.* **51**, 1687-1690.
19. Murphy, D. and Mauersberger, K. (1985) *Rev. Sci. Instrum.* **56**, 220-226.
20. Cottrell, J. S. and Evans, S. (1987) Anal. Chem. **59**, 1990-1995.
21. Hoffman, A. D., Ryan, P. A. and Moore, C. (1992) *Proceedings of the 40th ASMS Conference on Mass Spectrometry and Allied Topics*, Washington, DC, 519-520.

22. Evans, S., Buchanan, R., Hoffman, A., Mellon, F. A., Price, K. R., Hall, S., Walls, F. C., Burlingame, A. L., Chen, S. and Derrick, P. J. (1993) *Org. Mass Spectrom.*, **28**, 289-290.
23. Finch, J. W., Musselman, B. D., Sims, G. R. and Kunihiro, F. (1995) *Proceedings of the 43rd ASMS Conference on Mass Spectrometry and Allied Topics*, Atlanta, Georgia, 1074.
24. Pesch, R., Jung, G., Rost, K., and Tietje, K. H. (1989) *Proceedings of the 37th ASMS Conference on Mass Spectrometry and Allied Topics*, Miami Beach, Florida, 1079-1080.
25. Evans, S. (1990) *Methods in Enzymology: Mass spectrometry*, Vol. 193, Academic Press, Inc., San Diego, California, 61-86.
26. Griffiths, W. J., Yang, Y., Zhang, J., Lindgren, J. A. and Sjovall, J. (1995) *Proceedings of the 43rd ASMS Conference on Mass Spectrometry and Allied Topics*, Atlanta, Georgia, 1160.
27. Maquestiau, A., Meyrant, P. and Flammang, R. (1982) *Org. Mass Spectrom.* **17**, 96-108.
28. Polce, M. J., Cordero, M. M., Wesdemiotis, C. and Bott, P. A. (1992) *Int. J. Mass Spectrom. Ion. Processes* **113**, 35-58.
29. Bertrand, M. J., Bott, P. A., Porter, C. J. and Bateman, R. H. (1989) *Proceedings of the 37th ASMS Conference on Mass Spectrometry and Allied Topics*, Miami Beach, Florida, 1019-1020.
30. Boyd, R. (1994) *Mass Spec. Rev.*, **13**, 359-410.
31. D'Agostino, P. A., Hancock, J. R. and Provost, L. R. (1996) *Proceedings of the 44th ASMS Conference on Mass Spectrometry and Allied Topics,* Portland, Oregon, 301.
32. Bordoli, R. S., Howes, K., Vickers, R. G., Bateman, R. H. and Harvey, D. J. (1994) *Rapid Commun. Mass Spectrom.* **8**, 585-589.
33. Hopfgartner, G., Wachs, T., Bean, K. and Henion, J. (1993) *Anal. Chem.* **65**, 439-446.
34. Fujimaki, S., Ishihara, M., Kunihiro, F., Finch, J. W., Cody, R. B., Kammei, Y. and Kubota, E. (1996) *Proceedings of the 44th ASMS Conference on Mass Spectrometry and Allied Topics*, Portland, Oregon, 1146.
35. Chambers, D. M., Gregg, H. R. and Anderson B. D. (1993) *Proceedings of the 41st ASMS Conference on Mass Spectrometry and Allied Topics*, San Francisco, California, 934a-934b.
36. Cody, R. B., Tamura, J. and Musselman, B. D. (1992) *Anal. Chem.* **64**, 1561-1570.
37. Hill, J. A., Biller, J. E., Martin, S. A. and Bieman, K. (1989) *Int. J. Mass Spectrom. Ion Proc.* **92**, 211-230.
38. Cody, R. B., Tamura, J., Finch, J. W. and Musselman, B. D. (1993) *J. Am. Soc. Mass Spectrom.* **5**, 194-200.
39. Loo, J. A., Ogorzalek, R. R. and Andrews, P. C. (1993) *Org. Mass Spectrom.* **28**, 1640-1649.
40. Loo, J. A. and Pesch, R. (1994) *Anal. Chem.* **66**, 3659-3663.
41. Annan, R. S., Kochling, H. J., Hill, J. A. and Bieman, K. (1992) *Rapid Commun. Mass Spectrom.* **6**, 298-302.
42. Bateman, R. H., Birchenough, A. D., Bordoli, R. S., Phillips, A. M., Vickers, R. G., Woolfit, A. R. and Harvey, D. J. (1993) *Proceedings of the 41st ASMS Conference on Mass Spectrometry and Allied Topics*, San Francisco, California, 422a-422b.
43. Cody, R. B., Finch, J. W., Tamura, J. and Musselman, B. D. (1993) *Proceedings of the 41st ASMS Conference on Mass Spectrometry and Allied Topics*, San Francisco, California, 426a-426b.
44. Burlingame, A. L., et al. (1993) *Presented at the Pittsburgh Conference*, Atlanta, GA, Abstract 1222.
45. Hall, S. C., Smith, D. M., Masiarz, F. R., Soo, V. W., Tran, H. M., Epstein, L. B. and Burlingame, A. (1993) *Proc. Natl. Acad. Sci. USA* **90**, 1927-1931.
46. Hill, J. A., Martin, S. A., Biller, J. E. and Biemann, K. (1988) *Biomed. Environ. Mass Spectrom.* **17**, 147-151.
47. Hill, J. A., Biller, J. E. and Biemann, K. (1991) *Int. J. Mass Spectrom. Ion. Processes* **111**, 1-25.
48. Hill, J. A., Annan, R. S. and Biemann, K. (1991) *Rapid Commun. Mass Spectrom.* **5**, 395-399.
49. Cappiello, A., Palma, P., Papayonnopoulos, I. A. and Biemann, K. (1990) *Chromat.* **30**, 477.
50. Cerny, R. and Gross, M. L. (1992) *Mass spectrometry in the biological sciences: A tutorial*, Kluwer Academic Publishers, Dordrecht, The Netherlands.
51. Deterding, L. J., Wellemans, J. M. Y., Cerny, R. L., Gross, M. L. and Tomer, K. B.manuscript in preparation.
52. Cheng, X., Wu, Z., Fenselau, C., Ishihara, M. and Musselman, B. D. (1995) *J. Am. Soc. Mass Spectrom.* **6**, 175-186.

53. Leather, C., Wilkins, C. R. and Brenton, A. G. (1995) *J. Am. Soc. Mass Spectrom.* **6**, 212-219.
54. Gross, M. L. (1992) *Int. J. Mass Spectrom. Ion. Processes* **118/119**, 137-165.
55. Gross, M. L. (1994) *Accounts of Chem Research.* **27**, 361-369.
56. Cordero, M. M. and Wesdemiotis, C. (1994) *Anal. Chem.* **66**, 861-866.
57. Jensen, N. J., Tomer, K. B. and Gross, M. L. (1985) *J. Am. Chem. Soc.* **107**, 1863-1868.
58. Downard, K. M. and Bieman, K. (1993) *J. Am. Soc. Mass Spectrom.* **4**, 874-881.
59. Downard, K. M. and Bieman, K. (1994) *J. Am. Soc. Mass Spectrom.* **5**, 966-975.
60. Kolli, V. S.and Orlando, R. (1995) *J. Am. Soc. Mass Spectrom. 6*, 234-241.
61. Zaia, J. and Bieman, K. (1995) *J. Am. Soc. Mass Spectrom.* **6**, 428-436.
62. Downard, K. M.,Papayannopoulos, I. A., Costello, C. E. and Bieman, K. (1993) *Proceedings of the 41st ASMS Conference on Mass Spectrometry and Allied Topics*, San Francisco, California, 360a.
63. Falick, A. M., Hines, W. M., Medzihradszky, K. F., Baldwin, M. A. and Gibson, B. W. (1993) *J. Am. Soc. Mass Spectrom.* **4**, 882-893.
64. Burlingame, A. L., Boyd, R. K. and Gaskell, S. J. (1994) *Anal. Chem.* **66**, 634R-683R.
65. Medzihradszky, K. F. and Burlingame, A. L. (1994) *Methods: A Companion to Methods in Enzymology* **6**, 284.
66. Medzihradszky, K. F., Adams, G. W., Burlingame, A. L., Bateman, R. M. and Green, M. (1996) *J. Am. Soc. Mass Spec.* **7**, 1-10.
67. Medzihradszky, K. F., Yu, Z., Qiu, Y., Hall, S. C., Chen, Y. D., Maltby, D. D. and Burlingame, A. L. (1996) *Proceedings of the 44th ASMS Conference on Mass Spectrometry and Allied Topics*, Portland, Oregon, 709.
68. Rouse, J. C., Yu, W. and Martin, S. A. (1995) *J. Am. Soc. Mass Spectrom.* **6**, 822-835.
69. Summerfield, S. and Jennings, K. (1995) *Eur. Mass Spectrom.* **1**, 183-194.
70. Medzihradszky, K. F., Gillece-Castro, B. L., Townsend, R. R., Burlingame, A. L. and Hardy, M. R. (1996) *J. Am. Soc. Mass Spectrom.* **7**, 319-328.
71. Chen, S., Derrick, P. J., Mellon, F. A. and Price, K. R. (1994) *Anal. Biochem.* **218**, 57-169.
72. Claeys, M., Van den Heuvel, H., Chen, S., Derrick, P. J., Mellon, F. A. and Price, K. R. (1996) *J. Am. Soc. Mass Spectrom.* **7**, 173-181.
73. Lipniunas, P. H., Townsend, R. R., Burlingame, A. L. and Hindsgaul, O. (1996) *J. Am. Soc. Mass Spectrom.* **7**, 182-188.
74. Cerny, R. L., MacMillan, D. K., Gross, M. L., Mallams, A. K. and Pramanik, B. N. (1994) *J. Am. Soc. Mass Spectrom.* **5**, 151-158.
75. Cabrera, G. M., Deluca, M. E., Seldes, A. M., Gros, E. G., Oberti, J. C., Crockett, J. and Gross, M. L. (1993) *Phytochem.* **32**, 1253-1259.
76. Wieske, T., Bohme, D. K., Hrusak, J. and H. Schwarz, H. (1991) *Chem. Phys. Lett.* **186(4-5)**, 459-62.
77. Caldwell, K. A., Giblin, D. E., Hsu, C. S., Cox, D. and Gross, M. L. (1991) *J. Am. Chem. Soc.* **113**, 8520-8521.
78. Caldwell, K. A., Giblin, D. E. and Gross, M. L. (1991) *J. Am. Chem. Soc.* **113**, 3744-3756.
79. Saunders, M., Jimenez-Vazquez, A., Gross, R. J., Mroczkowski, S., Gross, M. L., Giblin, D. E. and Poreda, R. (1994) *J. Am. Chem. Soc.* **116**, 2193-2194.
80. Weiske, T., Schwarz, H., Giblin, D. E. and Gross, M. L. (1994) *Chem. Physics Letters* **227**, 87-90.
81. Giblin, D., Gross, M. L., Saunders, M., Vazquez, J., Cross, R. J., Mroczkowski, S. and Poreda, R. J. (1994) *Proceedings of the 42nd ASMS Conference on Mass Spectrometry and Allied Topics*, Chicago, Illinois, 584.
82. Strobel, F. and Adams, J. (1995) *J. Am. Soc. Mass Spectrom.* **6**, 1232-1242.
83. Lange, C., Loutelier, C. and Bordoli, R. S. (1994) *Rapid Commun. Mass Spectrom.* **8**, 960-66.
84. Enke, C. G., Stults, J. T., Holland, J. F., Pinkston, J. D. and Allison, J. (1983) *Int. J. Mass Spectrom. Ion Phys.* **46**, 229-232.
85. Strobel, H., Solouki, T., White, M. A. and Russell, D. H. (1991) *J. Am. Soc. Mass Spectrom* **2**, 91-94.
86. Strobel, F., Preston, L., Washburn, K. and Russell, D. (1992) *Anal. Chem.*, **64**, 754-762.
87. Cotter, R. and Cornish, T. (1991) *Rapid. Commun. Mass Spectrom.* 6, 339-340.

88. Makarov, A., Raptakis, E. and Derrick, P. (1995) *Int. J. Mass Spectrom. Ion Proc.* **146/147**, 165-182.
89. Clayton, E. and Bateman, R. (1992) *Rapid Commun. Mass Spectrom.* **6**, 719-720.
90. Bateman, R. H., Green, M. R. and Scott, G. (1995) *Proceedings of the 43rd ASMS Conference on Mass Spectrometry and Allied Topics*, Atlanta, Georgia, 1074.
91. Ho, M., Hughes, R. J., Kazdan, E.M., Matthews, P.J.,Young, A.B. and March, R.E. (1987) *Proceedings of the 35th ASMS Conference on Mass Spectrometry and Allied Topics* , Denver, Colorado, 237.
92. Schwartz, J.C., Kaiser, R.E. Jr., Cooks, R.G. and Savickas, P.J., (1990) *Int. J. Mass Spectrom. Ion Processes*, **88**, 97-105.
93. Heuls, W., Muenster, H., Pesch, R. and Schroeder, E. (1996) *Proceedings of the 44th ASMS Conference on Mass Spectrometry and Allied Topics,* Portland, Oregon, 1147.
94. Muenster, H., Huels, W. and Schroeder, E. (1996) *Proceedings of the 44th ASMS Conference on Mass Spectrometry and Allied Topics,* Portland, Oregon, 149.
95. Li, G., Duhr, A. and Wollnik, H. (1992) *J. Am. Soc. Mass Spectrom.* **3**, 487-492.

QUADRUPOLE MASS FILTERS and QUADRUPOLE ION TRAPS

R. GRAHAM COOKS, GUODONG CHEN and CARSTEN WEIL [+]
Department of Chemistry, Purdue University, West Lafayette, IN 47907, U.S.A.
[+] Also from II Physikalisches Institut, Justig Liebig Universität, Giessen, Germany

Abstract: Characteristics of quadrupole mass filters and quadrupole ion traps are summarized in comparison with other mass analyzers. The theoretical basis for mass analysis using quadrupole fields is described in conjunction with the various scan methods and operating parameters. The high performance of quadrupole ion traps is illustrated by introducing some new developments of the instrumentation, including mass range extension, high resolution, MS/MS experiments, SWIFT and non-destructive detection. The importance of ion motion in ion traps is also illustrated by the simulation and laser tomography measurements.

1. Introduction

1.1. COMPARISON OF VARIOUS MASS ANALYZERS

There are a number of methods of performing mass analysis which are based on quite different physical principles, employ quite different instrumentation and have different strengths and weaknesses. This situation is a source of some difficulty to students on first approaching mass spectrometry and it is compounded by the array of ionization techniques which can be used with each of the mass analyzer types. Table 1 summarizes the most common of the mass analyzers and lists some characteristics by which they can be compared [1, 2]. Characteristics important in one type of application of mass spectrometry may be less so in others so that the choice of mass analyzer can be dovetailed to the problem in hand.

1.2. SUMMARY OF CHARACTERISTICS OF QUADRUPOLE MASS FILTERS

The quadrupole mass filter [3, 4], or linear quadrupole, is a device in which forces are exerted in a plane normal to the direction of motion (z-direction) of ions moving through a linear array of four concentric rods (Figure 1). The forces are produced by the application of rf and dc potentials to opposite pairs of rods of hyperbolic cross-section which run parallel to the direction of the electrical fields. Ions respond to these forces by oscillating in the x,y-plane with frequencies which depend on their mass/charge ratios and with excursions which depend on the amplitudes of the applied potentials. Mass analysis

R. M. Caprioli et al. (eds.), Selected Topics and Mass Spectrometry in the Biomolecular Sciences, 213–238.

TABLE 1. Characteristics of Different Mass Analyzers

Method	Quantity measured	Mass/Charge range (Thomson)[b]	Resolution at 1000 (Thomson)	Dynamic range
Sector magnet	momentum/ charge	10^4	10^5	10^7
Time of flight	flight time	10^6	10^3 - 10^4	10^4
Cyclotron resonance	cyclotron frequency	10^5	10^6	10^4
Quadrupole ion trap	frequency	10^4	10^4	10^4
Quadrupole	filters for *m/z*	10^3-10^4	10^3- 10^4	10^5

[a] Adapted with permission from Reference 1. The values chosen in the table are those which are currently available using commercial instruments but are not the best-demonstrated values.
[b] Thomson= atomic mass units/unit positive atomic charge [2].

achieved by setting the conditions so that ions of a given m/z ratio (or group of m/z ratios) pass through the array while those of higher or lower mass are ejected from the device, hitting or passing between the rods. The basis for this selection, including the ability to select a group of ions of any desired range of mass/charge ratios, is described in the next section.

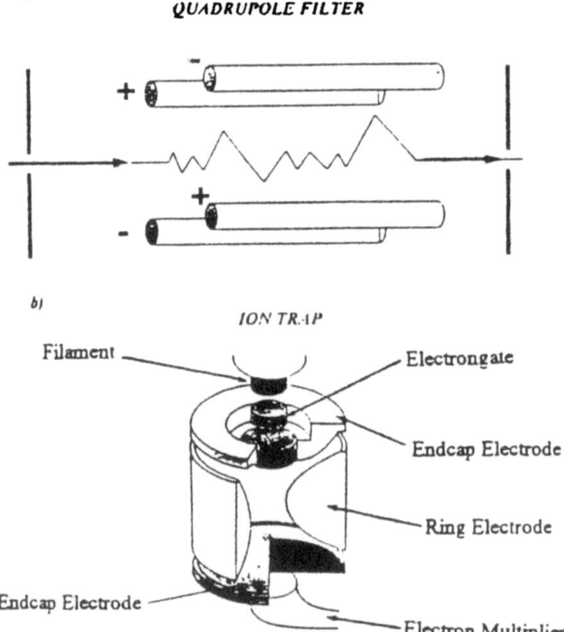

Figure 1. Schematic diagram showing the operation of (a) the quadrupole mass filter and (b) the quadrupole ion trap. [adapted with permission from Reference 1; for convenience round rods are shown.]

Quadrupole mass filters are analyzers which give only modest performance, as is evident from Table 1. Their very wide use is the result of their relatively low cost, the ease with which their resolution can be electronically controlled and the simplicity with which they can be used in multi-analyzer instruments, especially tandem arrangements of quadrupoles for MS/MS experiments [5]. [Note that in spite of the importance of triple quadrupoles, lack of space prevents their being covered in this paper.]

One can summarize the attractive characteristics of quadrupole mass filters as follows:

- o They are compact, relatively inexpensive devices
- o Resolution can be controlled by electronic adjustment of operating voltages
- o They can be operated in a broadband transmission device or as a mass-analyzer with unit (or somewhat better) resolution
- o Quadrupoles are readily coupled with other types of analyzers, including other quadrupoles in tandem arrangements for MS/MS experiments
- o The ion beam which is examined need not have a well-defined kinetic energy
- o The quadrupole serves as an excellent collision cell, which allows the products of collision-induced dissociation and ion/molecule reactions to be transmitted with high efficiency to a subsequent mass filter.

In addition to these advantageous characteristics, the device has significant limitations, especially in mass range and resolution, as indicated in Table 1.

1.3. SUMMARY OF CHARACTERISTICS OF QUADRUPOLE ION TRAPS

The quadrupole ion trap is the three-dimensional analog of the linear quadrupole mass filter [6-8]. Ions are again subjected to forces based on their mass/charge ratios but these are applied in three-dimensions and stable motion results in trapping within a small volume confined by a system of three electrodes - a ring electrode and two end-cap electrodes. As with quadrupole mass filters, these lens elements have cross-sections of hyperbolic shape. Once again, the device is small and relatively simple but the performance achieved is significantly improved over that found for the two-dimensional analog. The principal advantages of quadrupole ion traps can be summarized as follows:

- o Extremely high sensitivity
- o Compact and very simple device which is nevertheless capable of high performance experiments
- o Tandem mass spectrometry experiments available using a sequence of operations in time; good dissociation efficiency in MS/MS
- o Ion/molecule reactions can be studied for mass-selected ions
- o High resolution ($>10^6$ at m/z values $> 10^3$ Th) is accessible through slow scans, but mass measurement accuracy is relatively poor
- o High mass/charge range is accessible using the resonant ejection experiment
- o Non-destructive detection available through fourier transform measurements.

2. Mass Analysis Using Quadrupole Fields

2.1. EQUATIONS OF MOTION IN 1-, 2- AND 3-DIMENSIONAL QUADRUPOLE FIELDS

To understand the operation of quadrupoles and ion traps, it is useful to consider ion motion in fields in which the force exerted on an ion is proportional to its displacement from the center of the device. Such a field must have a potential which varies quadratically with displacement in order that the gradient, the field strength, vary linearly. Fields with four-fold symmetry produce the necessary forces. It is useful to consider, in turn, the cases of (hypothetical) one-dimensional, two-dimensional and three-dimensional quadrupole fields. The force F on an ion mass m, and charge e, exerted by an electric field, E, is

$$F=ma=\frac{md^2z}{dt^2}=eE \tag{1}$$

where z is the displacement. By definition, a quadrupole field is one in which the field strength varies linearly with displacement,

$$E=E_oz \tag{2}$$

hence,

$$a=\frac{d^2z}{dt^2}=\frac{eE_oz}{m} \tag{3}$$

That is, the acceleration varies linearly with displacement, z, linearly with field strength, E_o, and inversely with m/e. The potential ϕ, must vary quadratically in order that the field strength varies linearly with z. Hence

$$\Phi=f(z^2) \tag{4}$$

and

$$E=-\frac{d\Phi}{dz}=f(z) \tag{5}$$

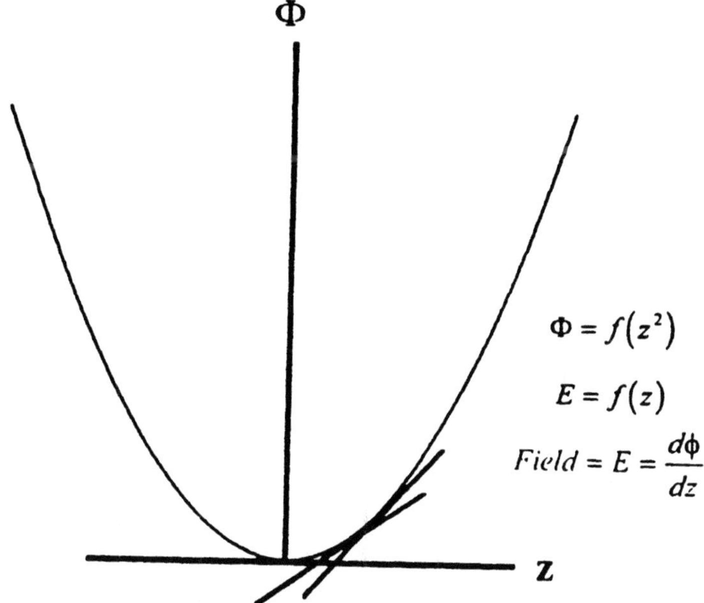

Figure 2. Potential and field strength in a hypothetical one-dimensional quadrupole field.

If it were possible to employ a system of electrodes and construct a one-dimensional quadrupole field, then the potential distribution and the field strength would be as shown in Figure 2. Ions falling an increasing distance from the center are subjected to a force which increases linearly with displacement and which tends either to return the ions to the center of the device (the case shown) or to accelerate them away from it (if the field direction were reversed). Ions could be trapped in such a field provided the direction of the field was reversed at an appropriate frequency. There are no stable solutions to the one-dimensional case since it does not satisfy the conservation law (Laplace equation, see below). [In other words, trapping in a one-dimensional field is not possible, the forces leading to motion towards the origin in the z-direction are accompanied by forces in the orthogonal directions.]

The situation for a two-dimensional device can be developed analogously from that just described and in this case the conservation law can be satisfied and real devices built.

In general, a quadrupolar electric field in which the forces are linear with displacement, is described by a quadrupolar electric potential of the form:

$$\Phi = \alpha x^2 + \beta y^2 + \gamma z^2 \qquad (6)$$

and the field strength is

$$E = E_o(\delta x + \epsilon y + \zeta z) \qquad (7)$$

where E_0 is time dependent and where δ, ϵ and ζ are weighting parameters. The Laplace conservation law $\nabla E=0$ is satisfied when $\delta + \epsilon + \zeta = 0$.

There are two solutions, (i) $\delta + \epsilon = 0$, $\zeta = 0$, which corresponds to the linear quadrupole device in which there are no forces in the z-direction as indicated by the zero value of the weighting factor, ζ, and (ii) $\delta + \epsilon = 2$, $\zeta = -2$ which corresponds to the three-dimensional ion trap.

The field strength in the linear quadrupole is given by

$$E=E_o\delta(x-y) \tag{8}$$

hence, the potential, the negative of the field strength, is given by

$$\Phi=-0.5E_o\delta(x^2-y^2) \tag{9}$$

where $\delta=-r_o^{-2}$, and r_o is the radius. The forces in such a field are uncoupled, and contracting the arguments which can be found in standard references on quadrupoles and ion traps [3, 7], in the x direction the electrical force E_xe balances the acceleration force $m\delta^2x/\delta t^2$,

$$E_xe=m\frac{\delta^2x}{\delta t^2}=-e\frac{\delta\Phi}{\delta x}=\frac{-xe\Phi_o}{r_o^2} \tag{10}$$

where $\Phi_0= E_0$. Hence,

$$\frac{\delta^2x}{\delta t^2}=\frac{-xe\Phi_o}{mr_o^2} \tag{11}$$

as x increases, the acceleration towards the center (x=0) increases, however, simultaneously, as y increases, the ions are accelerated from the center. Such a device, in which the polarities are switched at an appropriate frequency, can operate as an ion trap. Figure 3 illustrates the lines of force in a two-dimensional quadrupole field and the shapes of the electrode surfaces which are required to generate the field.

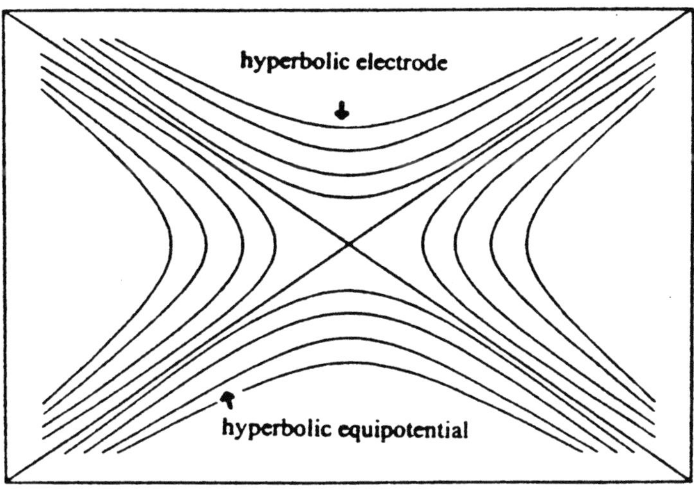

Figure 3. Lines of force in a two-dimensional quadrupole field generated using electrodes of hyperbolic cross-section.

Corresponding equations can be written for the three-dimensional ion trap, for which the field is $E = E_0(x + y - 2z)$, where E_0 is position-independent but time-dependent. The device can be operated by applying a potential $\phi_0 = U + V\cos\omega t$ between the ring and endcap, where U is the DC potential applied to the ring electrode and V is the amplitude (zero to peak) of the RF potential, also applied to the ring electrode, with angular frequency $\omega = 2\pi f$. As shown in detail elsewhere [8], for appropriately shaped hyperbolic electrodes, such a field provides potentials at any point x,y, z or in radial coordinates, r,z ($x^2+y^2=r^2$) given by

$$\Phi_{r,z}=(U+V\cos\omega t)\frac{r^2-2z^2}{2r_o^2}+\frac{U+V\cos\omega t}{2} \tag{12}$$

Solutions to the equations of motion of ions in such a field can be expressed in the form of a second order differential equation of a form first discussed by Mathieu in studies of waves on surfaces. The ion motion is described, independently in each of the independent directions, by defining the Mathieu parameters

$$a_z=-2a_r=-\frac{8eU}{m\omega^2 r_o^2} \tag{13}$$

$$q_z=-2q_r=\frac{4eV}{m\omega^2 r_o^2} \tag{14}$$

in the case of the ion trap and

$$a_x = -a_y = \frac{-4eU}{m\omega^2 r_o^2} \tag{15}$$

$$q_x = -q_y = \frac{2eV}{m\omega^2 r_o^2} \tag{16}$$

for the quadrupole filter. Using these parameters, the ion motion can be described as

$$\frac{d^2u}{d\zeta^2} + (a_u - 2q_u \cos 2\zeta)u = 0 \tag{17}$$

Note the use of the generalized coordinate u, to represent motion in any of the coupled directions, x, y and z (trap) and x and y (quadrupole). Note also that the Mathieu parameters include the m/e values of the ions but the description of ion motion is dependent both on mass and on other experimental conditions, and an ion of any mass having a particular Mathieu parameter will move with the same characteristic frequency. Hence, the behavior of an ion of given m/e value, including its transition from stability to instability, can be controlled by controlling the value of the parameter a_u or q_u. This is conveniently done by increasing the dc voltage in order to increase a_u and increasing the amplitude of the rf in order to increase q_u. Note also the similarity in the Mathieu equations for the ion trap and the mass filter, the small differences relating back to the weighting parameters already discussed.

2.2. SOLUTIONS TO THE MATHIEU EQUATION

The motion of ions in two- and three-dimensional quadrupole fields is described, as just indicated, by equations of the Mathieu form. These equations have finite solutions for certain values of the parameters. It is convenient to express these bounded solutions graphically in terms of the combined parameters a_u and q_u. Notice, again, that the solutions are for motion in each of the independent directions in which the forces operate (two in the case of the mass filter, three in the case of the ion trap). Notice, too, that the Mathieu parameters include experimental variables such as the amplitude of the applied voltages and the choice of mass/charge ratio. Values of a_u and q_u which correspond to stable motion are indicated graphically in Mathieu stability diagrams as illustrated in Figure 4.

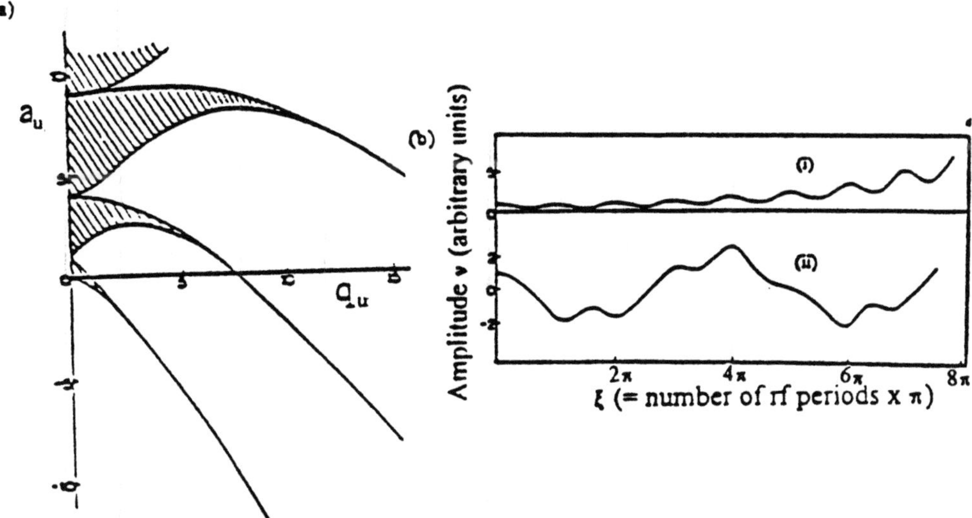

Figure 4. (a) Mathieu stability diagram for the x-component of motion in a two-dimensional trap. (b) Motion in one direction of (i) unstable and (ii) stable ion plotted as a function of time. [Adapted with permission from reference 3]

Ions of a given mass/charge ratio can be placed at any point in the region of stability indicated in Figure 4(a) by appropriate adjustment of the values of U and V. Alternatively, if the U and V values are fixed, ions of various m/z values will take various positions within a, q space and some of these may correspond to stable motion. When ions of different m/z values are placed at a point in a,q space, they will display the identical characteristics. In particular, they will move with a characteristic set of frequencies. The frequencies are related directly, but not simply, to the Mathieu parameters a and q. The major component of motion - in each of the uncoupled directions -has a frequency which is given by a β-value the expression for which is

$$\beta_u^2 = a_u + \cfrac{q_u^2}{(\beta_u+2)^2 - a_u - \cfrac{q_u^2}{(\beta_u+4)^2 - a_u - \cfrac{q_u^2}{(\beta_u+6)^2 - a_u - \ldots}}} + \cfrac{q_u^2}{(\beta_u-2)^2 - a_u - \cfrac{q_u^2}{(\beta_u-4)^2 - a_u - \cfrac{q_u^2}{(\beta_u-6)^2 - a_u - \ldots}}} + \ldots \qquad (18)$$

The principal component of the (angular) frequency $\omega_{secular} = [\omega/2]\beta$ and at low values of a and q, the value of β is approximated as

$$\beta_u = [a_u + \frac{q_u^2}{2}]^{\frac{1}{2}} \tag{19}$$

At these low values, that is when the ion is far from the stability diagram, the motion of the trapped ions is essentially harmonic, and the ion can be thought of as being trapped in a potential well of depth $\mathbf{D = q_u V_{rf}/8}$ i.e, the well depth increases with q_u as the ion moves towards instability. It is possible to illustrate this motion at several different values of the trapping parameters as shown in Figure 5, which displays data for the ion trap. Note that the behavior of a single ion is shown, so that ion/ion interactions are not considered.

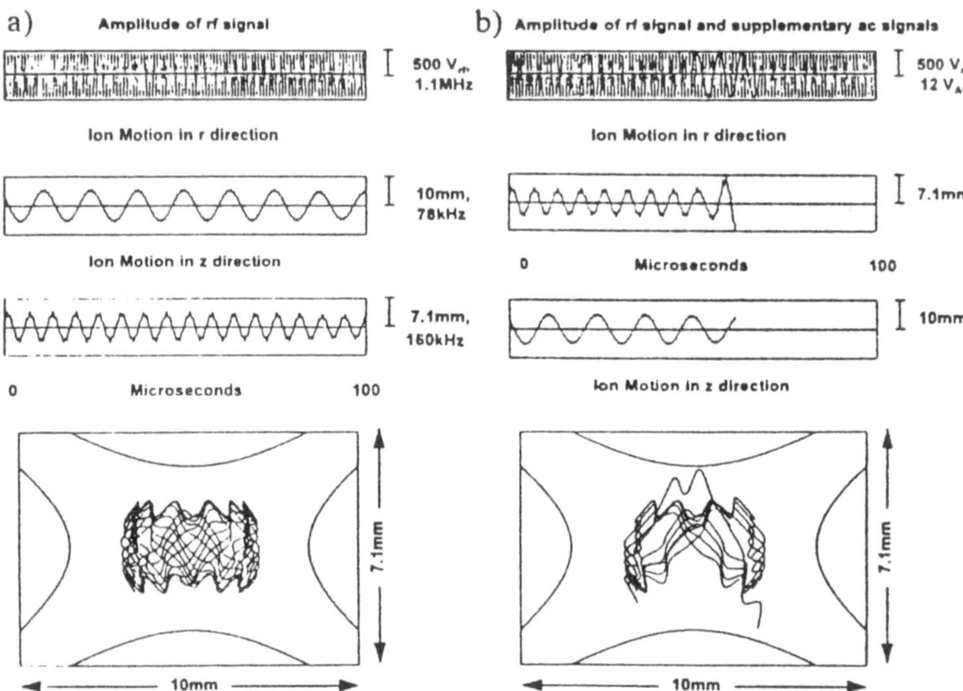

Figure 5. **Illustration of ion motion under different trapping conditions, showing separately the motion in the r- and z-directions in the ion trap. Note the fact that the ions have characteristic frequencies of motion which are different in the independent directions and which are essentially harmonic at low values of the trapping potentials. In part (b) a supplementary resonant signal is used to energize the ions.**

Note too, that ions of lower mass, present under a given set of trapping conditions, will experience a shallower trapping well and the lighter ions which leave the trap first and are trapped nearer the center of the device in the deeper potential well must pass through the heavier, outlying ions, in order to be ejected. The resulting interference effects add to the range of times of ion ejection and decrease the resolution.

Obviously, the motion of ions in the quadrupole filter and ion trap is complex and in fact there are additional complex sets of frequencies which have not been discussed which arise from higher order field components. In addition, the ion/ion interactions mentioned briefly above, are in fact extremely complex.

2.3. SCAN METHODS AND OPERATING PARAMETERS

There are several methods of recording a mass spectrum using either of the quadrupole devices. One is to employ the fact that the ions move with characteristic frequencies and to create a resonance condition in which power absorption by a supplementary field is measured. This was one of the earliest methods and has recently been used in broad-band detection (section 5). The two principal methods of analysis are mass-selective stability and mass-selective instability scans. The former is used almost exclusively on the mass filter and the latter, almost exclusively, with the ion trap.

In the mass-selective stability scan, a population of ions must be continuously remade and remeasured while the operating conditions are adjusted to allow ions of different m/e values to be stable. As this adjustment of the magnitudes of both U and V is made, while keeping their ratio constant, a mass/charge spectrum is recorded. This type of scan can be illustrated by considering the Mathieu stability diagram again where it is represented as a line of constant a/q ratio. (Figure 6). Because of the shape of the stability boundary, this line, which can be chosen to have any desired slope, will define a range of m/z values for which the ions have stable motion. Stable ions can pass through the device (2-D case) or remain in stable motion within it (3-D case). Ions of other m/e values will be lost. Thinking of such a line as a scan line (Figure 6) and varying the values of U and V, one sees that a group of ions, often just a single m/e unit in width (m_2^+ in Fig. 6), is defined and can be passed into the detector in the case of the mass filter. As the values of U and V are increased ions of increasing m/z acquire a and q values which correspond to stable trajectories and pass into the detector. This process results in a mass scan being recorded.

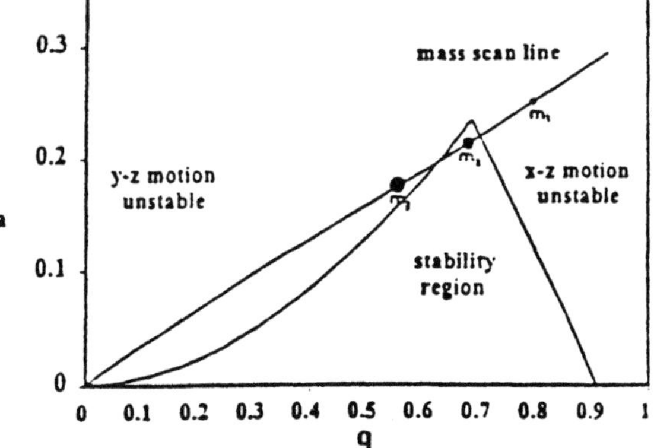

Figure 6. Mass selective stability scan shown as a scan line on the Mathieu stability diagram. The range of m/e values which are in stable motion are selected by the ratio of U/V and this ratio is kept constant while the values of U and V are scanned to record the mass spectrum.

The equation for mass analysis using the mass filter operated in the mass-selective stability mode is simply based on the definition of the Mathieu parameters, so that

$$\frac{m}{e}=\frac{2V}{q\omega^2 r_o^2} \tag{20}$$

or in units of Thomson, cm, MHz, and volts,

$$\frac{m}{e}(Thomson)=\frac{0.07V(Volts)}{f^2(MHz)^2 r_o^2(cm^2)} \tag{21}$$

where we have assumed q= 0.706. Typical values of the parameters are V_{max} = 4000 v, f = $2\pi\ \omega$ = 500 kHz, U/V = 0.17, hence the mass/charge range $(m/z)_{max}$ achieved by maximizing V and minimizing r_0 and f, is typically, for an instrument of 1 cm radius: m/z_{max} = 0.07x 4000 / $(0.5)^2 1^2$ = 1120 (Thomson)

The theoretical mass resolution or resolving power, R= m/Δm, can be obtained by differentiating the mass analysis equation, δm/m= δV/V, but V ∝ m/z, hence R ∝ (m/z)/δV so that if V is held constant, R increases with m/z. This is an interesting characteristic, not shared by most mass analyzers. For example, R decreases with m/e in a Fourier transform icr instrument and is independent of m/e in a magnetic sector instrument. Note that in practice, resolution falls with increasing mass in the mass selective stability scan. The resolution depends on the appropriate choice of a,q values but it also depends on the ions experiencing a sufficiently large number of rf cycles. In the mass selective stability mode of operation, ions must experience a certain number of rf cycles for stability/instability to be established. Resolution therefore depends on the time the ions spend in the field. This means that ions should experience some tens of rf cycles and for common values of operating conditions, ion translational energies of ca. 10 eV are optimum. In practice, it is noted that resolution ∝ $time^2$, i.e. resolution ∝ l^2/v^2 or resolution ∝ l^2 / KE where l is the length of the device and v is the velocity in the direction of motion through the device. For the typical ion energy of 10 eV and at mass 100, this corresponds to a velocity of 4x 10^5 cm/sec, hence these ions pass through a typical 10 cm quadrupole in a time of 25 μseconds and, at a frequency of 500 kHz, they experience 13 rf cycles. The ion frequency is significantly lower than the rf drive frequency (compare Figure 5), so that these conditions are minimum conditions for successful filtering since at least few cycles of ion motion are required for successful discrimination between stable and unstable ions. Ions of higher mass will have lower velocities at a fixed kinetic energy so that mass analysis can be performed over a wide mass/charge range with ion axial kinetic energies of ca. 10 eV.

The second main type of scan, the mass selective instability scan, is performed using the quadrupole ion trap. In this experiment, which does not employ a DC potential, operation is along the line a_u = 0. The equation for mass analysis is

$$q_z=\frac{-8eV}{m\omega^2(r_o^2+2z_o^2)} \tag{22}$$

[note the use of the expanded expression here] or

$$\frac{m}{e}=\frac{4V}{q_{eject}\omega^2 r_o^2} \tag{23}$$

when $r_o^2 = 2z_o^2$. This equation can be expressed in common units as

$$\frac{m}{e}=\frac{0.10748(Volts)}{[f^2(MHz)^2 r_o^2(cm^2)]} \tag{24}$$

where $q_{eject} = 0.908$. The z-direction is chosen because ion detection is performed using an external electron multiplier detector, placed behind one of the endcap electrodes. In this ion trap experiment, a collection of ions is generated and all ions remain in stable motion provided that $q_z < 0.908$. As q_z values for the different trapped ions are increased by increasing V, the amplitude of the applied rf voltage, ions become unstable in order of their m/e ratios, hit the detector and are recorded. A mass spectrum is recorded on a single population of ions by a simple V scan. Clearly, such an experiment has the advantage of a very high duty cycle, no ions are wasted. It is straightforward to estimate the mass/charge range of the instrument. From eq. (24) , and using common values for the various parameters, one can write,

$$\left(\frac{m}{e}\right)_{max}=0.08898V(volts) \tag{25}$$

where $r_o = 1.0$ cm, f= 1.1 MHz. For the usual values of the commercial Finnigan ITMS this corresponds to an upper mass/charge limit of 666 Thomson.

3. High Performance and Biological Applications of Ion Traps

3.1. RESONANCE EJECTION AND MASS RANGE EXTENSION

The mass analysis equation for the ion trap, when operated in the mass selective instability mode (eq. 23), is very similar to that for the mass-selective stability mode of the quadrupole mass filter (eq. 20). The differences are in the parameters which are used to solve the Laplace equation ($\nabla E=0$) and the fact that the experiments use different q_z points of operation viz. 0.908 and 0.706.

The mass range achieved using the mass-selective instability mode of operation can be extended by (i) increasing the maximum operating voltage, V; (ii) decreasing the operating frequency, f_{RF}; (iii) decreasing the size of the trap or (iv) decreasing the q_z-value. The first three options are relatively inaccessible while the last is easily implemented. It simply requires that ions be expelled from the trap at a q_z-value below the normal value and this can be achieved by resonantly exciting the ions using a supplementary frequency source. The frequency will correspond to a certain m/e value at a given value of V and, by scanning V, ions of different m/z values can be brought through the resonance point in succession. The supplementary ac frequency can be applied with a relatively low amplitude and still serve to eject ions efficiently. This method of resonance ejection often uses q_z-values as low as 0.1 or even 0.01 instead of the usual 0.908, and hence gives a mass/charge range extension of a factor of 10 or even 100. Figure 7 shows a spectrum of the test compound, CsI, examined in the mass/charge range up to 21,000 Thomson.

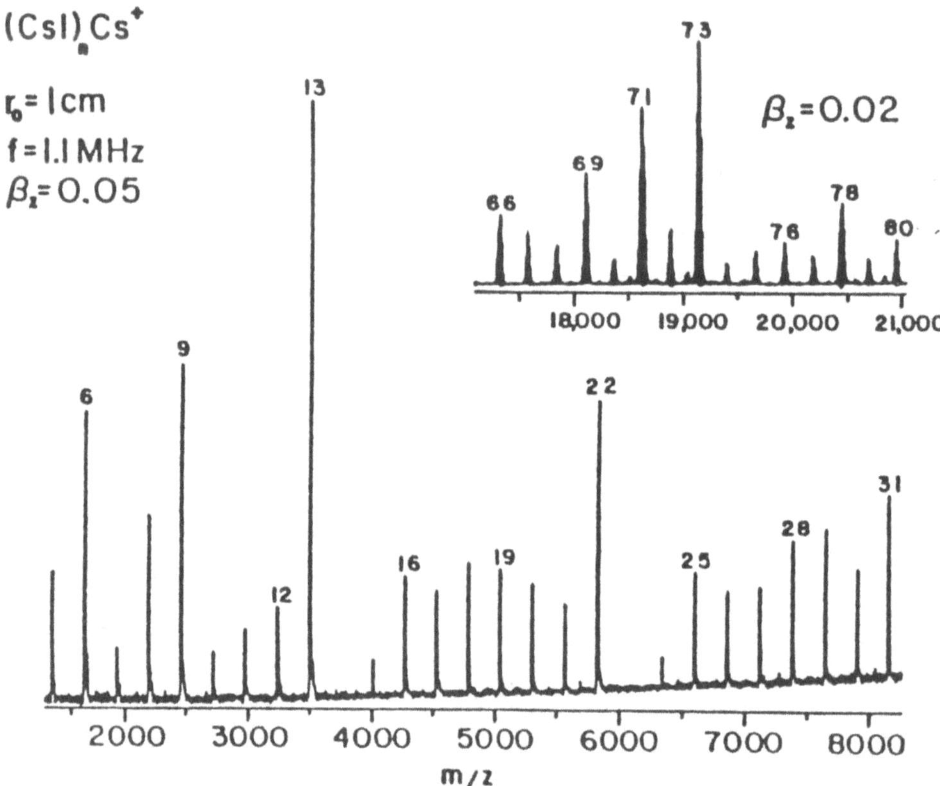

Figure 7. Access to high m/e values by resonant ejection: case of CsI ionized by Cs^+ bombardment with mass analysis using $q_z = 0.027$ which corresponds to a mass range extension of 33 times. [Adapted with permission from reference 9]

Note that the resonant ejection method of mass analysis should not be confused with the related experiment of axial modulation in which ions near the instability point, $q_z = 0.908$, are "tickled" using a supplementary frequency of approximately half the RF frequency and so ejected in a sharper package than would be the case in the usual mass-selective instability scan.

3.2. SLOW SCANS AND HIGH RESOLUTION

The resolution achieved in mass analysis using an ion trap may depend, in practice, on factors such as the stability of the power supplies, but a more fundamental limitation is the time taken for a population of ions to leave the device. This depends on their initial conditions, viz. position and velocity, as determined during internal ionization (or ion injection into the trap) and modified by subsequent collisions occurring in the trap. One of the advantages of resonance ejection is that the ions move coherently, so that the time taken for their ejection (and hence the peak width) is decreased over that for the standard mass-selective instability scan. If one simply scans the voltage, V, more slowly, then the time between peaks separated by (say) $\Delta m = 1$ increases while the peak width remains constant. Figure 8 shows that this principle can be used to obtain high resolution mass spectra in peptide analysis, the individual isotopic forms of the peptide being well separated with a resolution of 10,000 FWHM.

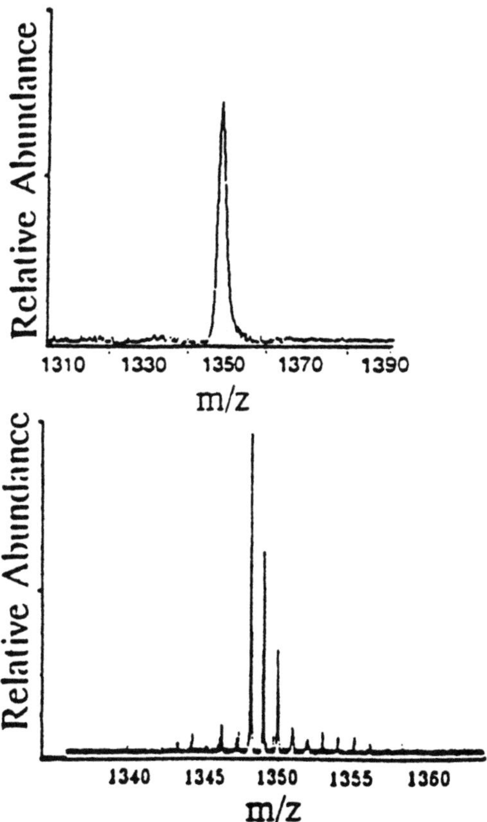

Figure 8. Effect of scan on the resolution achieved for isotopic forms of the protonated molecular ions of substance P. (Lower, scan rate was 312Th/sec; upper, usual scan rate of 12,500 Th/sec). [Adapted with permission from reference 10]

This type of experiment has been shown to produce extremely high resolution: in excess of 10^6 at m/e 3000 Thomson. However, it suffers from the disadvantage that the spacing of the isotopic peaks is not uniform unless the ion densities used are low. It is most appropriately applied in a zoom-scan mode in which only the ions of interest are examined. This procedure also minimizes the scan time required since a full scan taken using an attenuation of a factor of 40 as used in Figure 8 corresponds to a scan rate of 31sec/decade.

3.3. MS/MS EXPERIMENTS

Tandem mass spectrometry in the ion trap depends on three steps: (i) Isolation of the ions of interest; (ii) excitation of these ions and (iii) mass analysis of the product ions. Since ions of a wide range of m/e values are trapped in an ion trap, it is straightforward to isolate individual ions and to carry out experiments on them. The isolation step in MS/MS experiments was first done by a reverse-forward RF scan in which ions above the mass of interest were first ejected from the trap under resonance ejection conditions, then ions of lower mass were removed by the same procedure [11, 12]. This procedure allows ions to be selected within a range of 0.02% Th with little or no loss of signal and ions of a single Th to be selected with little loss of sensitivity up to 1000 Thomson. The key to MS/MS experiments is the fragmentation of the isolated ions and this is readily achieved by a resonance excitation process which is exactly analogous to that used to eject ions but which uses a lower resonant ac amplitude so that the ions of interest are translationally excited but not ejected. Since a bath gas is used to cool ions to the center of the trap and so homogenize their initial positions and velocities, the buffer gas is used as the target gas in the MS/MS experiment. The last of the three stages of the MS/MS experiment is simply a conventional mass scan and, again, this is normally done using resonant ejection in order to extend the mass/charge range and improve performance. It is clear from this description that an MS/MS experiment of this type, a product ion spectrum, is almost as easily performed using an ion trap as is a conventional MS scan. The fact that MS/MS data can be taken using temporal separation of parent and product ions, but using the same device for their mass/charge analysis is evident from the scan timing diagram shown in Figure 9.

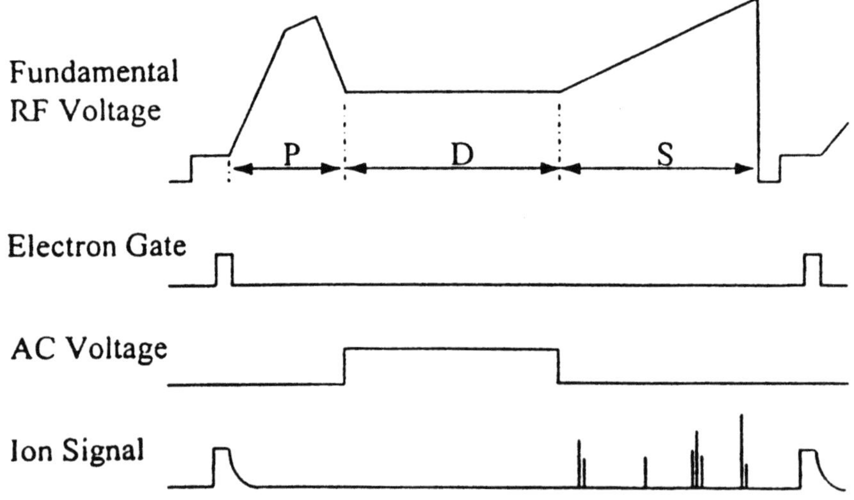

Figure 9. Scan diagram showing the sequence of steps used to produce product ion spectra in an ion trap.

The characteristics of MS/MS experiments in an ion trap are as follows. (i) Internal energy is normally supplied by a series of collisions, each of which transfers a relatively small amount of energy into internal energy. (ii) The efficiency of dissociation of relatively small ions (< 1000 Th) is remarkably high, often approaching 100%. (iii) The fact that ion traps operate with a low mass cut-off (see Figure 4) means that a small portion of the low end of the MS/MS product ion spectrum can not be recorded in a single scan from the parent ion. (iv) It is very easy to extend the experiment to include multiple stages of analysis, fragmentation and detection, viz. to do MS^n experiments. This latter capability is demonstrated in Figure 10 which shows MS^4 data on ions generated by desorption ionization using an external probe.

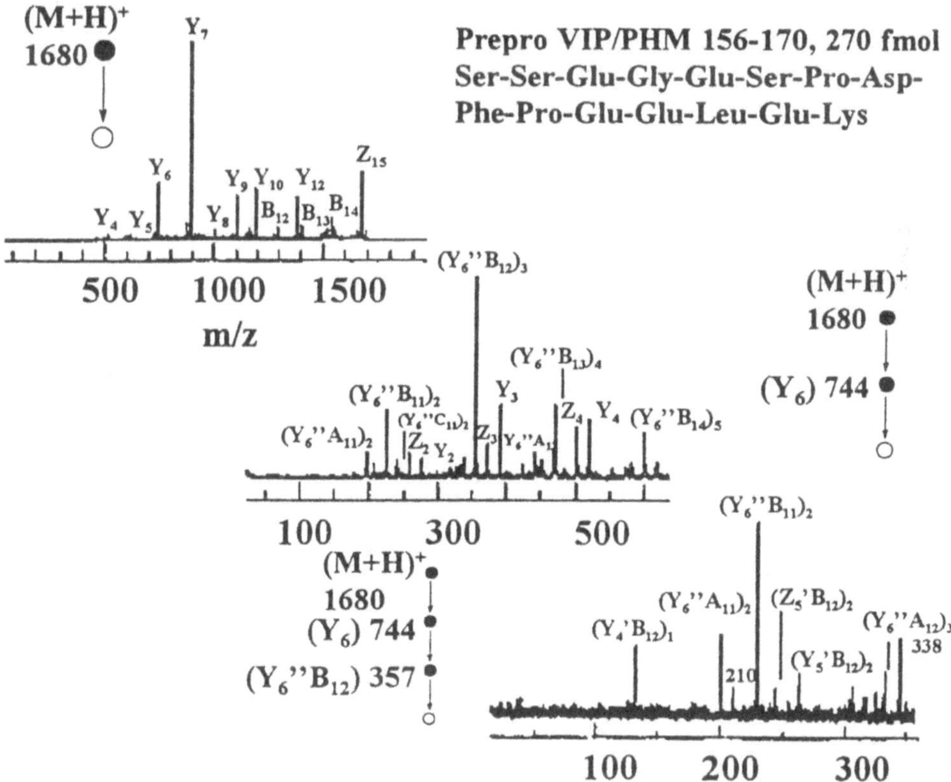

Figure 10. Multiple stage MS^n experiments on a peptide performed in the quadrupole ion trap. [Adapted with permission from reference 11]

Note that it is possible [13] but not a simple matter to perform MS/MS experiments other than the product ion scan using the ion trap. This is a disadvantage of the method as currently developed.

3.4. POPULATION CONTROL BY SWIFT AND RELATED METHODS

It should be evident that it is simple to select ions of a single m/z value for subsequent MS/MS experiments, whether these are ion/molecule reaction products or fragments of collision-induced dissociation. However, there are many reasons to wish to select for (or against) ions of particular m/e values. These reasons include exclusion of matrix-derived ions in experiments such as FAB where there are abundant glycerol cluster ions of a variety of m/e values. It is also often desired to select ions a variety of m/e values for activation or mass-analysis in MS/MS experiments and including selected reaction monitoring experiments. A significant weakness of ion traps is that they suffer from a poor dynamic range. This is because space charge effects - defined as changes in ion motion which result from the interaction of ions with each other - occur in principle as soon as a second ion is introduced into a trap containing a single ion and in practice, in analytical mass spectrometry, when the trap contains 10^5 or more ions. These effects are more easily noticed in high resolution and other demanding experiments.

Space charge effects can be alleviated by removing matrix and other ions not germane to the experiment in hand. Similar reasons for developing a capability for exciting ions of any specified m/e ratios exist in ion cyclotron resonance and led to the development of a capability known as SWIFT [14], for stored waveform inverse fourier transform. Quite naturally, this was the first procedure applied to quadrupole ion traps for the general purpose of ion population control. The value of this experiment in improving signal/noise ratios in a trace analysis example is illustrated in Figure 11 [15].

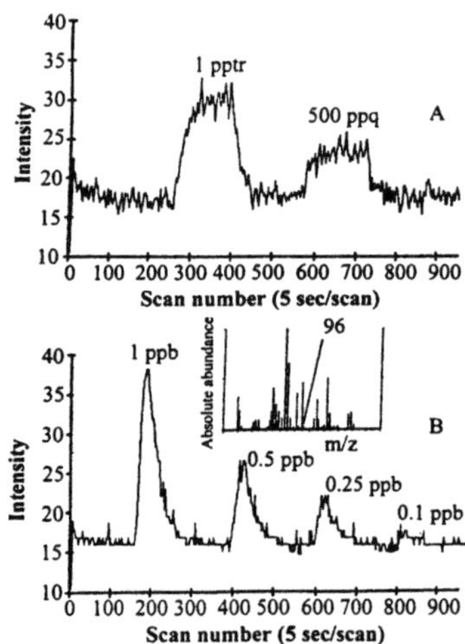

Figure 11. Application of SWIFT to the improvement in the S/N ratios obtainable in on-line trace analysis of water by membrane introduction methods. (a) Ion chromatogram obtained from monitoring m/e 96 Thomson (trans-1,2-dichloroethene) in a mixture of volatile organic compounds (each 1 ppb) in tap water with SWIFT. (b) Ion chromatogram obtained from monitoring m/e 96 Thomson (trans-1,2- dichloroethene) in a mixture of volatile organic compounds (each 1 ppb) in tap water without SWIFT. The inset shows the full mass spectrum recorded at these low levels. [Adapted with permission from reference 15]

4. Ion Motion in Ion Traps

4.1. RELATIONSHIP TO ION TRAP PERFORMANCE

The motion of ions in quadrupole ion traps is complex and intimately related to performance. This makes it important that the motion of ions be accurately understood before strategies for ion manipulation can be successful at improving performance.

An illustration of this can be found in the topic of resonant ion ejection using a supplementary ac field. As already noted, this procedure extends the mass range of the instrument. The resolution achieved depends on the range of times taken for ions to experience excitation to such a degree that they leave the device. Ejection of ions is facilitated when the phases of the ac and the main trapping rf are such that they supplement each other. This can only occur over a minimum number of cycles, if there is an integral number of rf cycles per ac cycle (i.e. their frequencies must be multiples of each other). Simulations of ion motion (see below) first suggested the importance of ac/rf phase and phase angle relationship in enhancing the resolution in resonant ejection.

4.2. LASER TOMOGRAPHY

Given the importance of ion motion in determining ion trap performance, it follows that efforts have been made to characterize the motion of trapped ions experimentally. What is desired in such an experiment is that the spatial and temporal distribution of the ion population be measured. Laser photodissociation measurements made *in situ* answer this requirement, since the positioning of the laser beam provides the required spatial information and the time of laser firing provides the temporal information. It is necessary only to chose for examination, ions which are readily photodissociated at an appropriate photon wavelength and then to translate the laser to cover the spatial range of interest. By firing the laser at particular points in time, in relationship to the trapping rf field, one can map out the temporal behavior of the ion population too.

Figure 12 illustrates some of these points by showing the distribution of trapped ions along the z-axis as a function of time within the ac cycle for a resonant excitation experiment performed using the benzoyl cation, m/e 105 [16]. This particular ion is chosen since it is readily photodissociated using a nitrogen laser to yield a single photoproduct, m/e 77. The data show that the ion cloud is distributed within a relatively narrow range of values along the z-axis and that it moves coherently under the influence of the resonant field. Additional data shows that the frequency of motion of the cloud in the z-direction matches that of the ac frequency. A great deal of further information of this type is acquired by direct experimental examination, in this fashion, of the ions whose mass and MS/MS spectra are being recorded.

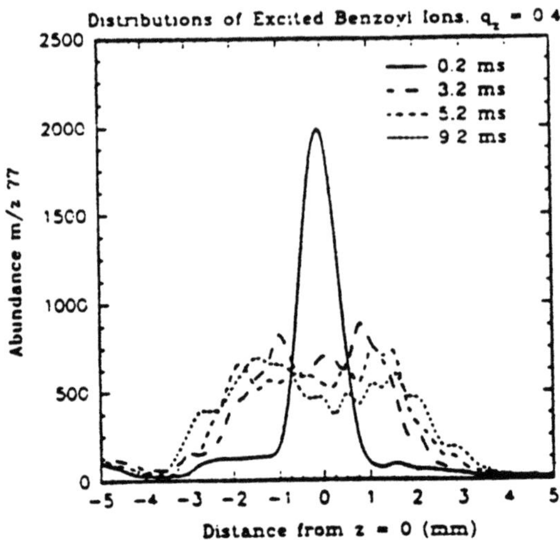

Figure 12. Laser tomography showing the spatial (z-axis) and temporal distribution of benzoyl ions (105 Th) subjected to resonant excitation. The ion cloud was sampled by the laser at 0.2, 3.2, 5.2 and 9.2 ms after the excitation voltage was turned on. The AC voltage was on for 10 ms. The value of q_z equaled to 0.4, and helium bath gas was set at a pressure 2.1×10^{-4} Torr. [Adapted with permission from reference 16.]

4.3. SIMULATIONS USING THE PROGRAM ITSIM

The behavior of ions in the quadrupole ion trap can be characterized in detail by an appropriate simulation program. Ideally, such a program should consider a population of ions and include their interactions with each other as well as the effects of collisions with residual gas and the effects of imperfections in the fields within which the ions are confined. Many programs exist which allow the motion of single ions to be followed, and the program ITSIM, developed at Purdue, allows the motion of large populations of ions to be characterized [17]. Many predictions of this program have been reproduced by experiment and a number of new experiments were first suggested by considering simulations. These features, together with the didactic value of visualizing the motion of ions under the influence of various fields and having access to different methods of visualization of that motion, make the program invaluable. The program allows a sequence of different operations on a population of ions to be performed and so simulates a wide variety of experiments including injection from an external ion source, trapping, ac and dc pulse excitation.

Figure 13 illustrates the value of the simulation program by showing the effect of applying a short dc pulse, to the endcap electrodes, as a means of activating a population of ions. This pulse results in coherent motion of trapped ions of a particular m/e value, the cloud of ions moving with a characteristic frequency. If ions of other m/e values are also present, they will also move coherently but with different frequencies, characteristic of their mass/charge ratios. The motion of the ion clouds is towards and away from the endcap electrode. This result suggests that it should be possible to use the image current generated in a suitable electrode by the moving ion cloud to perform non-destructive detection [18]. It further suggests that fourier transformation of the temporal motion should lead to a distribution of frequencies which represent a mass spectrum: This fourier transform ion trap experiment was first simulated, as just described, before being implemented as described in the next section.

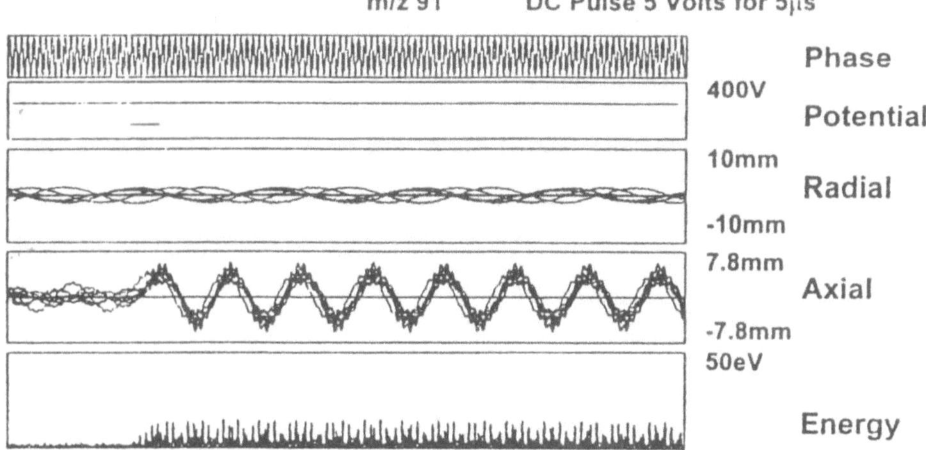

Figure 13. Simulation of the motion of a population of ions under the influence of a dc impulse applied to one of the endcap electrodes. All trapped ions are excited, those of the same m/z value moving coherently with the same frequency.

5. Non-Destructive Detection and Re-Measurement

5.1. RATIONALE AND BACKGROUND

Since the first application of array detectors to mass spectrometry in 1978 [19], it has been clear that the increase in sensitivity achieved through the enhanced duty cycle made available is extremely important. The high efficiency of detection of all the ions created is one of the most important attributes of time-of-flight mass spectrometry, for example [20]. This advantage extends to the ion trap mass spectrometer although the ions are recorded in sequence on a relatively slow time scale, at least when compared to the arrival time distribution of ions in time-of-flight mass spectrometry. There is an alternative procedure which increases the speed of analysis while retaining the high duty cycle and this is fourier transform mass spectrometry. Although a fourier transform TOF experiment has been reported [21], the methodology is essentially confined to ion cyclotron resonance [22-24], where it provides capabilities for examination of all ions in the spectrum without the resolution/sensitivity trade-off which is characteristic of sector and quadrupole mass spectrometry. Indeed, the longer the signal is observed the greater the resolution, which can therefore be traded against observation time. These experiments have a further advantage and that is that the ions can be detected non-destructively [24]. This is a separate and very important advantage of the fourier transform techniques.

For all these reasons, it is obviously of interest to enquire whether non-destructive, fourier transform methods are applicable to the quadrupole ion trap. The answer is yes, and several authors have reported the observation of ions by this method [25-27]. However, the full advantages of non-destructive detection are only available if a wide range of m/e values can be examined simultaneously. The quest for broadband fourier transform mass spectrometry involved simulations of ion motion as described in the previous section, followed by experiments, as described in the succeeding one.

5.2. METHODOLOGY FOR NON-DESTRUCTIVE DETECTION

While a variety of methods of detection can be conceived, the simplest is that in which ions approach an electrode and polarize it so that a current flows towards and away from the electrode through an external conductor in response to the ion motion. This current flow will have a frequency equal to that of the ion motion and will therefore contain frequency components which depend on the mass-to-charge ratios of the ions making up the population of trapped ions. The experiment will only be successful if all ions of a given m/z ratio move coherently and this is readily achieved by application of a dc pulse across the endcap electrodes to create a dipolar field as discussed above. The impulsive excitation step lasts only a few rf cycles, in a typical case 4 microseconds.

There are experimental difficulties in reducing this concept to practice, including the need for sophisticated amplification and discrimination electronics, and most importantly the fact that the high amplitude of the rf trapping field contributes a very large background signal which makes the image current difficult to measure. This background signal can be removed by filtering methods but only for a narrow range of observation frequencies. A method which reduces the contribution of the rf field has recently been developed using a receiver-electrode set in the endcap of the ion trap mass spectrometer (Figure 14) [28]. Reduction of the capacitance of the detection circuit is a key to the enhanced performance achieved.

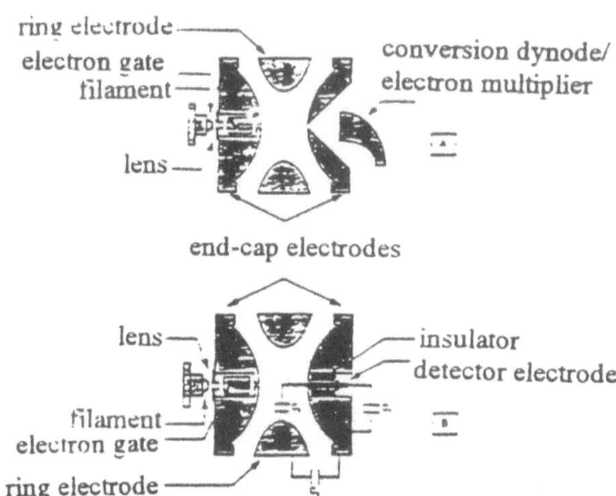

Figure 14. **Ion trap mass spectrometer modified for non-destructive detection by removal of the electron multiplier and addition of a receiver electrode set in but isolated from the endcap electrode (B). A is the standard Finnigan ion trap configuration. [Adapted with permission from reference 28]**

5.3. REMEASUREMENT AND TANDEM MASS SPECTROMETRY

The use of the device just described, together with fourier transformation of the amplified signals, gives mass/charge spectra which are represented by that shown for acetophenone in Figure 15. This broadband spectrum was acquired without ion destruction in a time of ca. 4.5 seconds by adding together some 100 transients recorded in the time domain and fourier transforming the resultant signal.

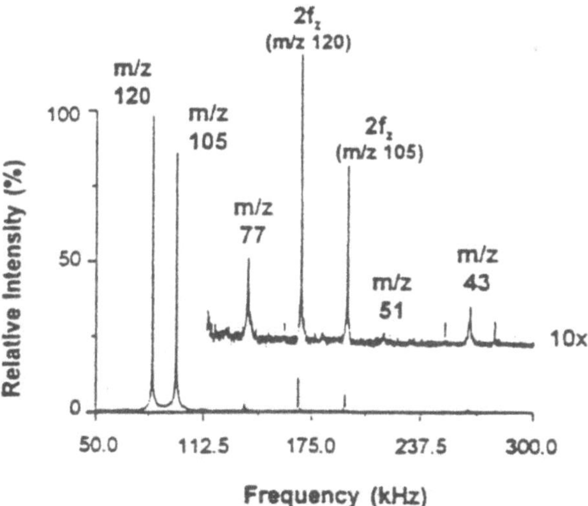

Figure 15. **Mass spectrum of acetophenone recorded by non-destructive detection using fourier transformation of the image current in an ion trap mass spectrometer. [Adapted with permission from reference 28]**

Since ions are not destroyed by the measurement, the *same population of ions* can be subjected to collision-induced dissociation and then the products can be mass analyzed, and again recorded in a broadband by non-destructive detection. This type of experiment is represented by the result shown in Figure 16 , which displays the product ion spectrum of the molecular ion of acetophenone. This ion is first selected from the distribution of ions which are created by electron ionization of the gas phase compound. The mass selection is achieved by the reverse/forward scan described earlier or by the SWIFT method also described earlier.

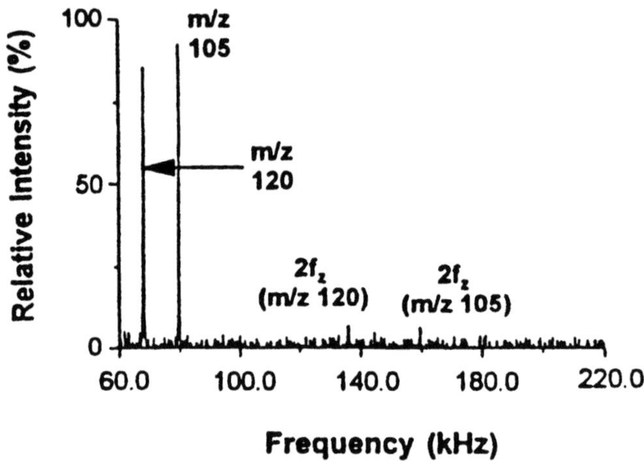

Figure 16. Product ion MS/MS spectrum (m/z 120 parent ions, collisions on helium) recorded by non-destructive detection and fourier transformation in an ion trap mass spectrometer. [Adapted with permission from reference 28]

Although these developments are in the process of being refined, the resolution is excellent and it has been shown that the decay in the coherent signal as a result of collisions with bath gas can be reversed by re-application of the dc impulse and remeasurement of the signal. This feature is illustrated in Figure 17 in a case in which four separate transients were measured for the same population of Ar^+ ions and the results were accumulated. The average remeasurement efficiency is better than 90%.

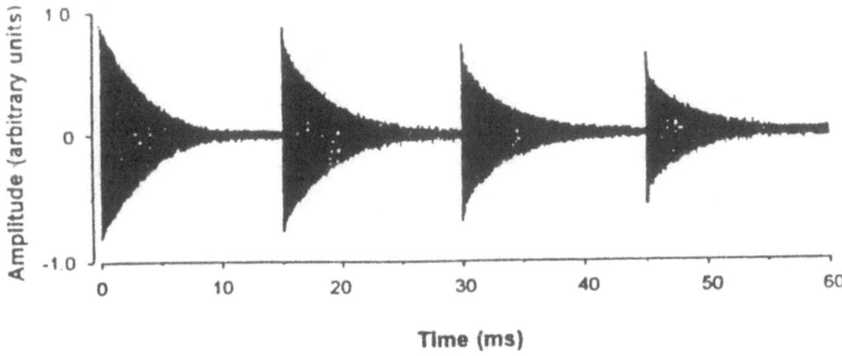

Figure 17. **Remeasurement experiment in fourier transform ion trap mass spectrometry (FT-ITMS). The individual transients decay due to collisions with the bath gas and the entire population is then re-excited by re-application of the dc pulse and then re-measured. [Adapted with permission from reference 28]**

6. Conclusions

The quadrupole mass filter and quadrupole ion trap share common characteristics in terms of their fundamental principles. However, they have developed very differently - the mass filter is very widely used in many branches of science, including the biological sciences, as a rugged, low cost and relatively low performance instrument. By contrast, the ion trap mass spectrometer, although invented at the same time as its two-dimensional counterpart, has recently undergone very rapid development and has become a high performance mass spectrometer which shows signs of becoming one of the leading tools in the discipline. It allows tandem mass spectrometry experiments which are possible only using combinations of multiple quadrupoles (especially the highly successful triple quadrupole instrument). Its extension to high mass/charge measurements, the development of high resolution capabilities and the very recent demonstration of non-destructive, broad-band fourier transform capabilities, all suggest an increased role in the future.

The detailed characteristics of ion motion, including the secular frequencies of the ions, underlie the performance of the quadrupole and the ion trap. These characteristics, although complex, can be calculated with high precision using simulation programs such as ITSIM, and these programs assist in both understanding the devices and in predicting new modes of operation. Theory has consistently been ahead of experiment in developments of ion trap capabilities. Direct measurements of the positions of trapped ions and of their motion as a function of time has been made by laser tomography in which spatial and temporal investigations are carried out using photodissociation. The continued investigation of these instruments is likely to produce even more advanced analytical capabilities.

7. Acknowledgements

The work described in this chapter was supported by the US Department of Energy, Office of Basic Energy Sciences, and by Finnigan Corporation. The contributions to this work of former and present coworkers are acknowledged with appreciation. These individuals include Bobette Nourse, Raymond Kaiser, John Louris, Randy Julian, Peter Reiser, Kenny Morand, Mario Nappi, Carsten Weil, Manish Soni, Vladimir Frankevich, Kathy Cox, Mitch Wells, Steve Lammert, Jon Williams, Curtis Cleven, Huy Anh Bui, Lynn Ann Horn.

8. References

1. J. B. Lambert, H. F. Shurvell, D. A. Lightner and R. G. Cooks, Introduction to Organic Spectroscopy, Macmillan Publishing Company, New York (1987).
2. R. G. Cooks and A. L. Rockwood, *Rapid Commun. Mass Spectrom.* **5**, 93 (1991).
3. P. H. Dawson, Quadrupole Mass Spectrometry and Its Applications, Elsevier, Amsterdam (1976).
4. W. Paul and H. Steinwedel, Apparatus for Separating Charged Particles of Different Specific Charges, German Patent 944, 900, 1956; U. S. Patent 2, 939, 952, 1960.
5. R. A. Yost and C. G. Enke, *J. Am. Chem. Soc.* **100**, 2274 (1978).
6. W. Paul and H. Steinwedel, *Z. Naturforsch* **8a**, 448 (1953).
7. R. E. March and R. J. Hughes, Quadrupole Storage Mass Spectrometry, John Wiley & Sons, New York (1989).
8. R. E. March and J. F. J. Todd, Practical Aspects of Ion Trap Mass Spectrometry, CRC Press, New York (1995).
9. R. E. Kaiser, Jr., J. N. Louris, J. W. Amy and R. G. Cooks, *Rapid Commun. Mass Spectrom.* **3**, 225 (1989).
10. J. D. Williams, K. A. Cox, R. G. Cooks, R. E. Kaiser, Jr. and J. C. Schwartz, *Rapid Commun. Mass Spectrom.* **5**, 327 (1992).
11. R. E. Kaiser, Jr., R. G. Cooks, J. E. P. Syka and G. C. Stafford, Jr., *Rapid Commun. Mass Spectrom.* **4**, 30 (1990).
12. K. A. Cox, J. D. Williams, R. G. Cooks and R. E. Kaiser, Jr., *Biological Mass Spectrom.* **21**, 226 (1992).
13. J. V. Johnson, R. E. Pedder and R. A. Yost, *Int. J. Mass Spectrom. Ion Processes* **106**, 197 (1991).
14. A. G. Marshall, T.-C. L. Wang and T. L. Ricca, *J. Amer. Chem. Soc.* **107**, 7893 (1985).
15. M. Soni, S. Bauer, J. W. Amy, P. Wong and R. G. Cooks, *Anal. Chem.* **34**, 1409 (1995).
16. J. D. Williams, Ph. D. thesis, Purdue University (U. S. A.), 1993.
17. H. P. Reiser, R. K. Julian and R. G. Cooks, *Int. J. Mass Spectrom. Ion Processes* **121**, 49 (1992).
18. R. G. Cooks, C. Cleven, L. Horn, M. Nappi, C. Weil, M. Soni and R. Julian, *Int. J. Mass Spectrom. Ion Processes* **146/147**, 147 (1995).
19. J. H. Beynon, D. O. Jones and R. G. Cooks, *Anal. Chem.* **47**, 1734 (1975).
20. R. J. Cotter, Time-of-flight Mass Spectrometry, American Chemical Society, Washington, D. C. (1994).
21. P. Oksman, *Int. J. Mass Spectrom. Ion Processes* **141**, 67 (1995).
22. A. G. Marshall, *Acc. Chem. Res.* **18**, 316 (1985).
23. M. B. Comisarow and A. G. Marshall, *Chem. Phys. Lett.* **25**, 282 (1974).
24. I. J. Amster, F. W. McLafferty, M. E. Castro, D. H. Russell, R. B. Cody and S. Ghaderi, *Anal. Chem.* **58**, 483 (1986).
25. J. E. P. Syka and W. J. Fies, Jr., U. S. Patent 4, 755, 670, 1988.
26. J. H. Parks, S. Pollack and W. Hill, *J. Chem. Phys.* **101**, 6666 (1994).
27. D. E. Goeringer, R. I. Crutcher and S. A. McLuckey, *Anal. Chem.* **67**, 4164 (1995).
28. M. Soni, V. Frankevich, M. Nappi, R. Santini, J. W. Amy and R. G. Cooks, *Anal. Chem.*, in press (1996).

TIME-OF-FLIGHT MASS SPECTROMETRY

Marvin L. Vestal
PerSeptive Biosystems
Framingham, MA.

1. Introduction

The emphasis in this discussion is on time-of-flight mass spectrometry as it relates to present and potential biological applications. It is not intended to be a comprehensive review of time-of-flight technology. Some excellent reviews have been published recently, and the reader is referred to those for subjects not covered here. [1-3] We begin with a brief history covering some of the seminal developments, proceed to an overview of TOF theory and practice, and then discuss the present state-of-the-art of TOF instruments for biological applications. The discussion concludes with a brief (and perhaps cloudy) gaze into the future.

2. History

Although there are a few earlier papers [4-6], the history of time-of-flight (TOF) mass spectrometry begins for practical purposes with the work of Wiley and McLaren. [7]. They described a two-field pulsed ion source which could produce either space focusing or velocity focusing, but not both simultaneously. The basic geometry is shown in Figure 1. This system was applied to the dominant ionization technique of that time, electron ionization in the gas phase, where both velocity spread and

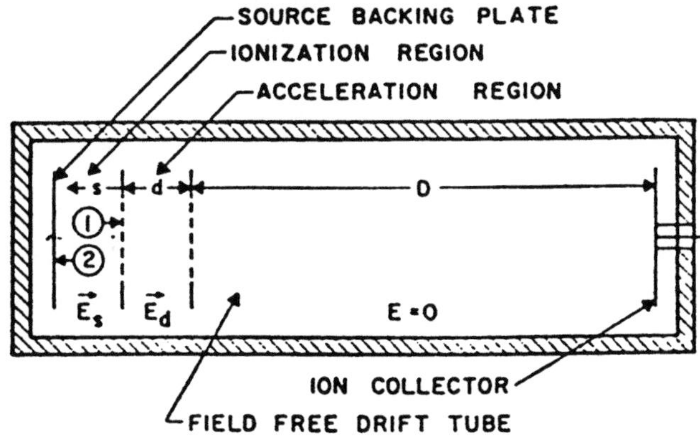

Figure 1. Basic geometry of the TOF mass spectrometer with two-field source [7]

R. M. Caprioli et al. (eds.), Selected Topics and Mass Spectrometry in the Biomolecular Sciences, 239–262.

distribution of initial positions make significant contributions to limiting the mass resolution. Experimental studies were conducted to find a compromise between these two competing effects. They concluded

"Both the best compromise and the quality of the resulting resolution depend on the actual initial space and energy distributions. Lag systems give preferred treatment to energy resolution; and no lag systems to space resolution. Hence, the best method of operating depends on the space and energy distributions encountered. In some cases the use of time lag focusing gives greatly improved resolution. In others it seems to make little difference."

To the aficionados of modern ion optics, this statement is equivalent to the Liouville Theorem which states that phase space is incompressible, and a consequence is that simultaneous space and velocity focusing is impossible. Thus to achieve high resolution in TOF we must either find the optimum compromise for focusing from the initial ion phase space provided by the ionization technique, or arrange the system so that one of the initial distributions is negligible.

A commercial instrument based on the Wiley and McLaren design was developed by the Bendix Corporation. While these instruments never achieved mass resolutions greater than about 500, they were extensively used in applications such as GC-MS [8] and fast kinetic studies[9] where the sensitivity and speed for selected ions were unmatched by scanning instruments.

2.1. THE COINCIDENCE TOF MASS SPECTROMETER

In 1960 Henry Rosenstock [10] proposed a coincidence TOF mass spectrometer in which the time measurement was initiated by detection of the secondary electron produced by a gas phase ionization event, and time measurement terminated by detection of the corresponding positive ion at a second detector. A schematic diagram of the apparatus is shown in Figure 2.[11] The coincidence principle was previously employed in studies of nuclear phenomena and in measuring the time-of-flight of high energy particles, but this was the first application on the molecular scale of single ion counting with time-of-flight measurement using a time-to-digital converter (TDC). The advantages of using coincident detection are that ionization processes with very small cross sections can be observed, and distributions of electron energy/and or angle may be measured and correlated with the simultaneously measured mass spectra. The mass resolution obtained in this instrument was similar to that obtained by Wiley focusing as shown by the spectrum of the xenon isotopes shown in Figure 3. This instrument used constant electric fields and relied on a well-collimated electron or photon beam to produce ions in a very small region, and relatively high accelerating field to overcome the effect of the initial velocity distribution. Miniaturized version of this device were developed under the Apollo program at NASA for monitoring cabin atmospheres, but were never flown. Laboratory applications included x-ray and UV photoionization studies [12] coupled with coincident measurement of secondary electron energy spectra. These techniques have been further developed by a number of laboratories[13-16] and have become important techniques for fundamental studies, but not for analytical applications.

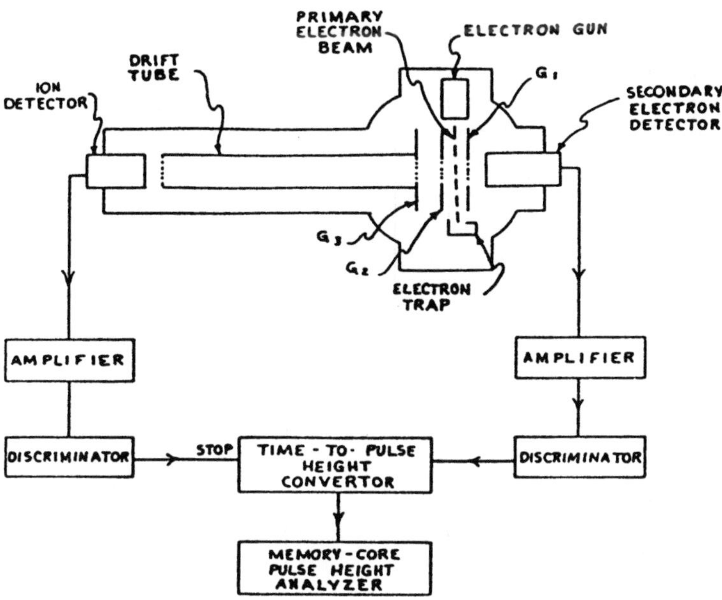

Figure 2. Schematic diagram of the coincidence time-of-flight mass spectrometer .

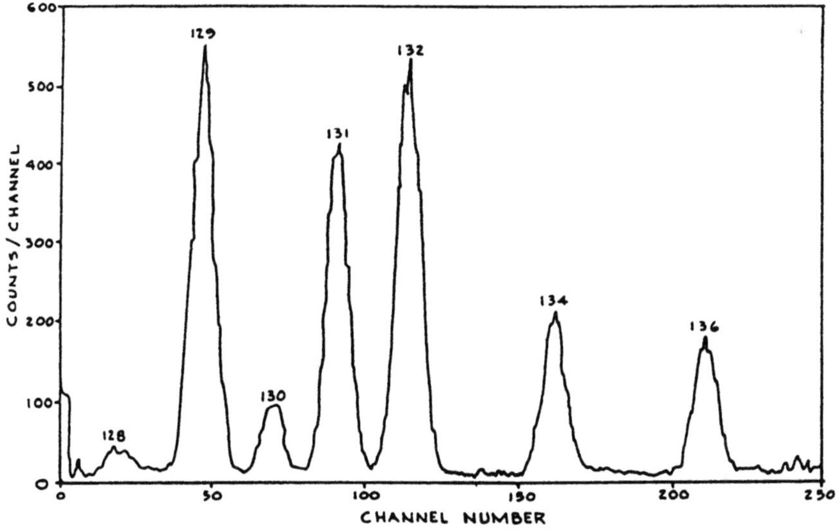

Figure 3. Mass spectrum of xenon isotopes measure in the coincidence mass spectrometer

2.2. THE REFLECTRON

In the decade between 1965 and 1975, TOF mass spectrometry as an analytical technique was virtually dead. However, two key developments during this period were largely responsible for its resurrection. The first of these was the development of the "mass-reflectron" by Mamyrin in 1966[17], although it was nearly a decade later before the significance of this work was widely recognized in the west.[18] Early results give mass resolutions of about 3500, which was nearly an order of magnitude better than that available with conventional TOF. A schematic diagram of the two-stage reflectron developed by Mamyrin is shown in Figure 4. In this device ions with higher energy spend a longer time in the mirror which compensates for their shorter flight time in the field-free region. By using the two-stage reflector shown, both the first and second order dependence on ion energy spread can be removed. Theoretical resolving power for this device is very high, but in general very high resolving powers are not achieved because other limitation may become dominant.

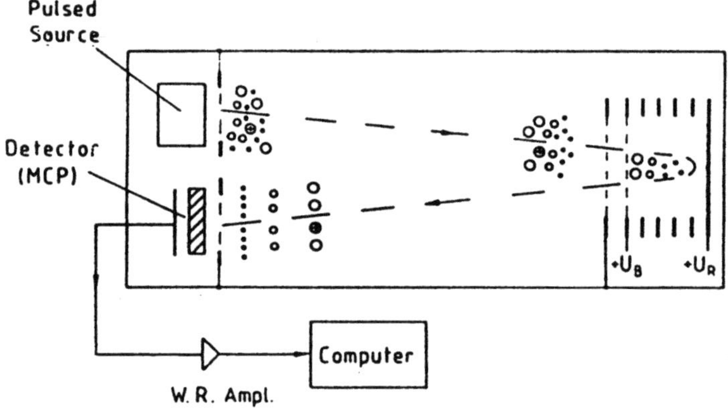

Figure 4. Basic geometry of the TOF mass spectrometer with two-stage reflector

More recently, Standing and coworkers have developed a TOF instrument for secondary ion mass spectrometry (Manitoba TOF II)[19,20], shown schematically in Figure 5. which employs a single-stage reflector. While the theoretical resolution of this geometry is somewhat lower than that for the Mamyrin design, the actual resolution achieved is comparable as indicated by the spectra shown in Figure 6. The resolution for melittin at m/z 2847 is about 5000 FWHM.

2.3. PLASMA DESORPTION

The second key development was the discovery in 1974 by Torgerson, Skowronski, and Macfarlane[21] that intact high molecular weight ions could be desorbed into the gas

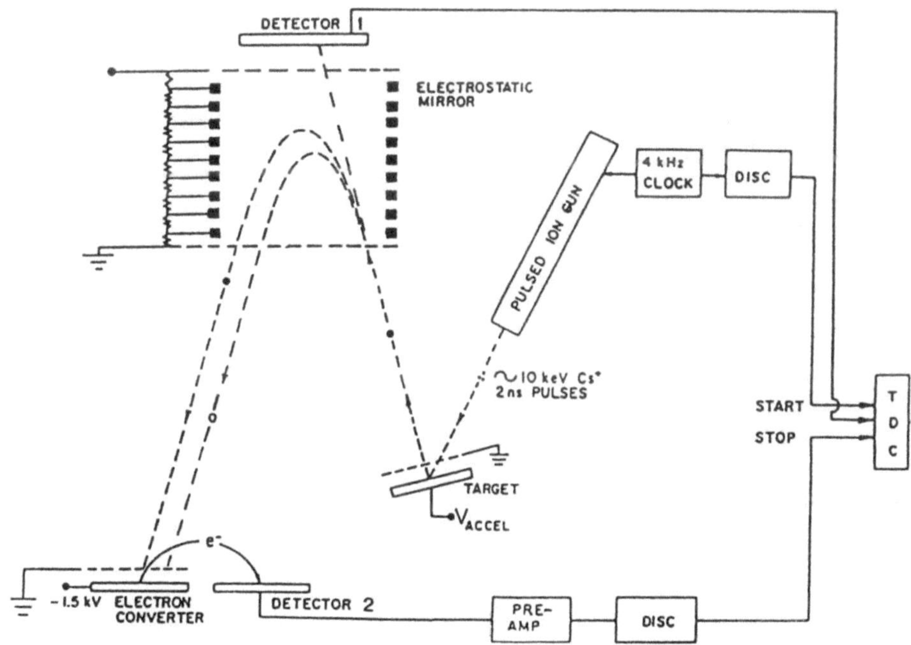

Figure 5. Schematic diagram of the TOF spectrometer with single stage reflector (Manitoba TOFII) showing typical flight paths of a parent ion O and a daughter ion O. The angle between the ion path and the spectrometer axis has been exaggerated for clarity; the actual angle is 1.4 degrees.

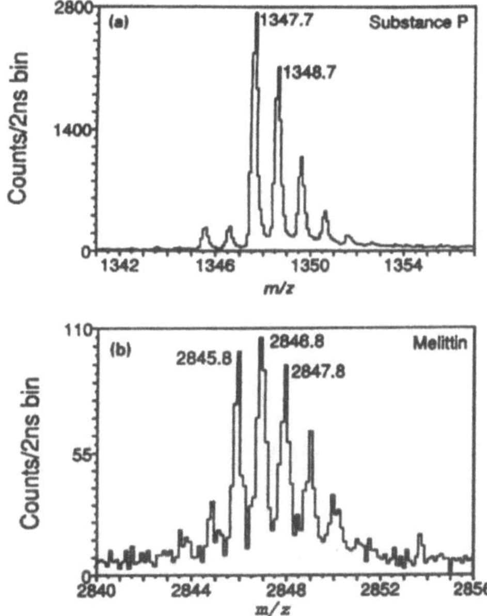

Figure 6. Mass spectra of the MH+ ions from the peptides substance P (Mr=1347) and melittin (Mr=2847) obtained with Manitoba TOFII. [20]

phase by ^{252}Cf plasma desorption. A schematic diagram of the plasma desorption mass spectrometer is shown schematically in Figure 7. The remarkable similarity between this instrument and the earlier coincidence mass spectrometer, shown in Figure 2, reflects the fact that both were the result of applying techniques developed for nuclear physics to molecular applications. The major difference is that in plasma desorption one of the fission fragments triggers the start detector and the other desorbs ions from a solid sample, while in the earlier device ionization occurs in the gas phase and the secondary electron triggers the start detector. During the next decade there was considerable interest in this technique, but adoption was rather slow. A review [22] published in 1986 lists 22 papers from only five different laboratories. There were undoubtedly two major reasons for this relatively slow acceptance. First, and perhaps most important, commercial plasma desorption TOF instruments were not available until about 1985, and secondly, the development of the competing technique, FAB, by Barber and coworkers in 1981[23]. While this later technique was not useful beyond about molecular weights of ca. 5000, it gave excellent results at lower mass, and most importantly it could be readily installed on existing magnetic sector mass spectrometers. Thus FAB quickly became commercially available, and became the dominant technique for analyzing nonvolatile compound in the 80's.

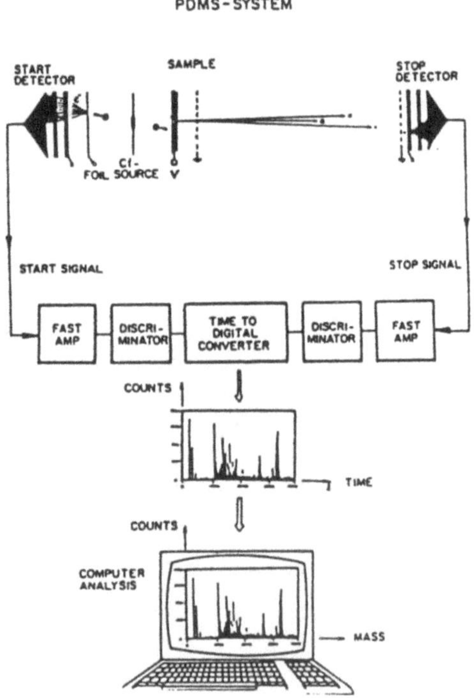

Figure 7. Schematic diagram of the ^{252}Cf Plasma Desorption TOF mass spectrometer system [22]

2.4. MALDI-TOF

Laser ionization mass spectrometry was first reported in about 1963. [24]. The emphasis of the early work was on elemental analysis, but subsequently several groups produced spectra showing intact molecular ions from nonvolatile compounds in the 1000-2000 u mass range.[25-28] However, it was not until 1988 when Karas and Hillenkamp[29] at the Bordeaux International Meeting showed a convincing spectrum of a protein (B-galactosidase) with molecular weight in excess of 100,000 that laser desorption received widespread attention. That spectrum is shown in Figure 8. The key to success of this technique appears to be that isolated molecules of the high molecular weight sample are embedded in crystals made up mostly of a matrix that absorbs strongly at the wavelength of the UV laser. Within a few months several of the laboratories equipped with TOF instruments for PDMS had converted existing instruments or built new instruments for MALDI and were rapidly expanding the applications of this new technique. Karas, Hillenkamp and coworkers have continued studying fundamental aspects of the new technique [30] and have investigated the alternative us of IR lasers.[31] The first commercial instrument was built and delivered in 1990, and the field has exploded since that time.

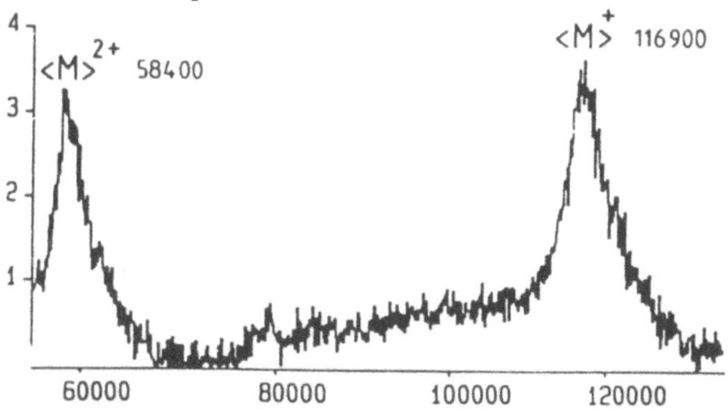

Figure 8. First mass spectrum of a protein with M_r>100,000 u. Karas and Hillenkamp.[29]

The resolution and mass accuracy obtained in the early work with MALDI was not spectacular, but it was somewhat better than that obtained with PDMS and results were obtained much more rapidly, with smaller amounts of sample, and routinely for much larger proteins and oligonucleotides than was possible with PDMS or FAB. With linear analyzers, mass resolutions of 500 could be obtained with careful attention to sample preparation and by keeping the laser fluence near the threshold for ion production. With reflecting analyzers isotopic resolution of peptides in the 1000-2500 u range could be obtained but the resolution at higher mass was not significantly better than with the linear analyzers. With reasonable care the linear analyzers yielded mass accuracy of ca. 0.1% with external calibration and up to 0.01% with internal calibration.[32] With the reflector accuracy for peptides were nearly an order of magnitude better, but for proteins it was often difficult to do as well as with the linear analyzers.[33]

2.5. DELAYED EXTRACTION

In 1994 Lennon and Brown [34] presented the first results from applying time lag velocity focusing as described by Wiley and McLaren [7] to MALDI with a linear analyzer. The showed significant enhancement in resolution for both proteins and peptides as shown in Figure 9. They also observed an extensive series of fragment ions formed during the pulse delay period from a number of proteins which provided easily interpreted structural information.

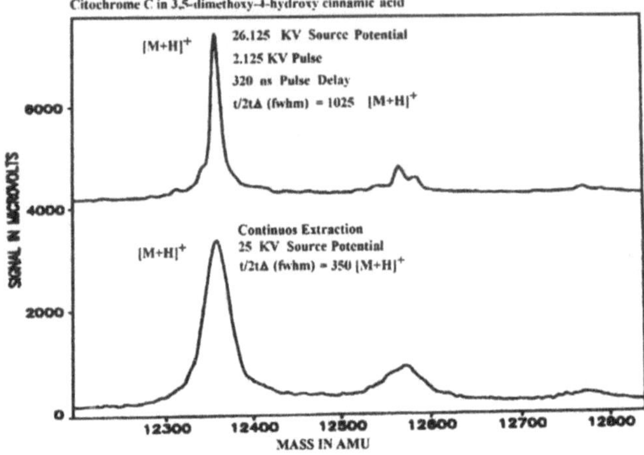

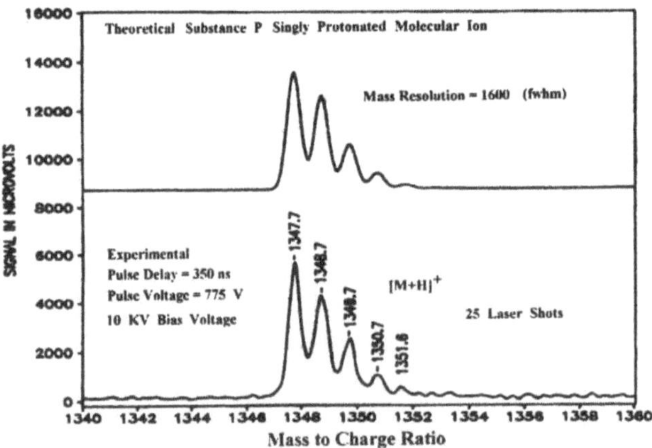

Figure 9. Spectra from Brown and Lennon [34] showing enhanced resolution obtained by delayed extraction with a linear TOF analyzer.

Similar resolution enhancement was reported soon thereafter by two additional laboratories. [35,36] The development of delayed extraction techniques on a series of commercial instrument with both linear and reflecting analyzers was reported in

1995 [37] and this instrumentation was applied to sequencing of DNA oligomers [38], and proteins and peptides [39]. A schematic diagram of the ion source and electronics for delayed extraction is shown in Figure 10. Some examples of the first results are shown in Figures 11 and 12. With the larger reflecting analyzer resolutions in excess of 10,000 were obtained routinely over a broad mass range extending up to at least 6000 u. For the first time the resolution available from TOF was not only clearly superior to that available from quadrupoles, but also is beginning to compete with that available from large magnetic deflection instruments.

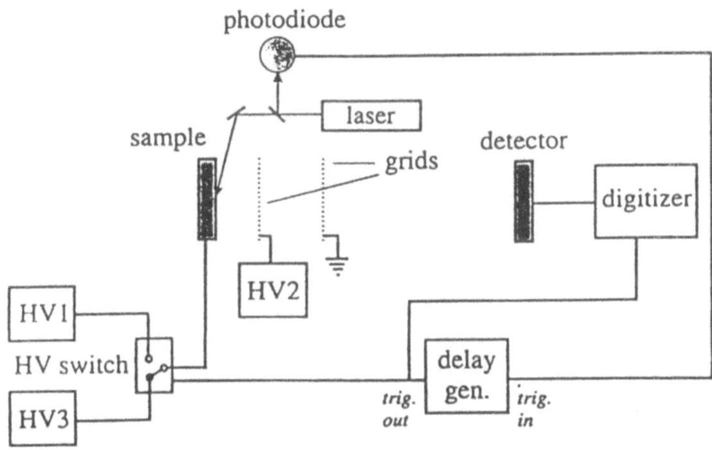

Figure 10. Schematic diagram of the ion source and electronics for delayed extraction MALDI.

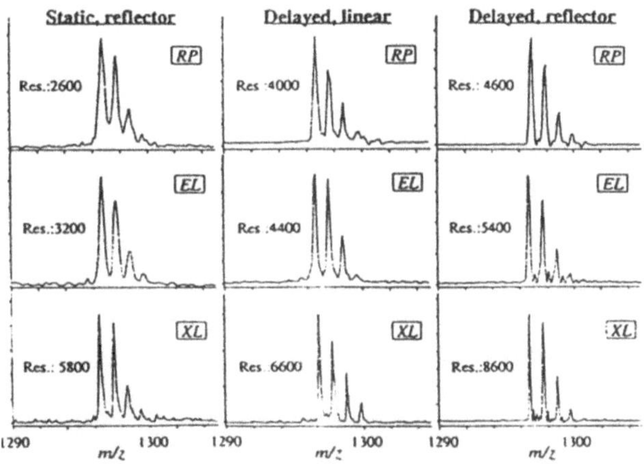

Figure 11. Comparison of resolution obtained for MH+ of angiotensin I (monoisotopic m/z=1296.68) with three different operating modes (static reflector, DE linear, and DE reflector) in three analyzers of different size (RP=1.3m linear, 2.0m ref., EL=2.0m linear, 3.0 m ref., XL=4.2m linear, 6.6m ref.)[37]

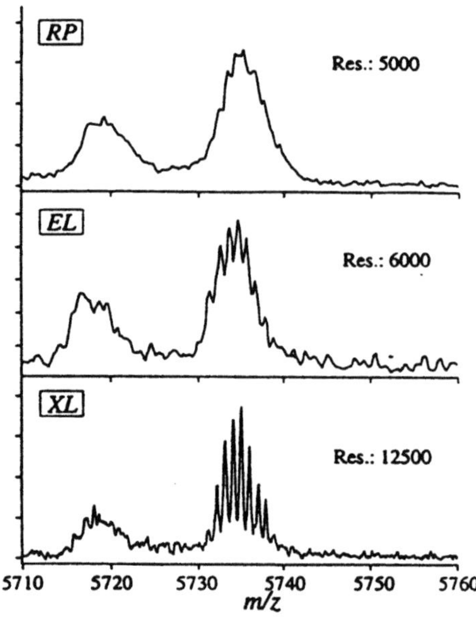

Figure 12. Mass spectra of the molecular ion of bovine insulin (average Mr=5733.5) obtained in reflector mode, illustrating increasing resolution as function of analyzer length.[37]

Orthogonal Injection TOF with Electrospray Ionization

Electrospray as a means of producing molecular ions from large nonvolatile molecules was first proposed by Dole and coworker [40], but while some interesting results were obtained at higher pressure, they were unable to observe ions in the mass spectrometer. This technique was revived by Fenn and coworkers using a quadrupole analyzer [41], but it was not until they showed multiply charged ions from intact proteins in 1988 [42] that the technique received appreciable attention. Applications developed rapidly after the initial discovery partly because adaptation of the technique to existing single and triple quadrupole instruments was easily accomplished, and incorporation in magnetic deflection instruments was also possible, albeit with somewhat more difficulty [43]

Orthogonal injection TOF also has a long, and somewhat checkered history, beginning with some pioneering work in the 50's and 60's.[44] More recently orthogonal injection has been used with a number of different ionization sources including flames [45], liquid SIMS [46], a cluster ion source [47], electron ionization [48,49], atmospheric pressure ionization [50,51], inductively coupled plasma [52], and electrospray [53-56]. The advantages of orthogonal injection are that it provides efficient injection of ions from a continuous beam into a pulsed source, and if the injection beam is well-collimated the velocity spread in the direction of ion acceleration can be kept relatively small. A major recent advance is the incorporation of an RF only quadrupole collision cell between the ion source and the extraction region of the TOF analyzer.[57] By collisional relaxation of the initial ion velocities the ions enter the extraction region with very small velocity components in the direction of ion acceleration. Thus high resolution can be obtained in the TOF analyzer by optimizing the space focusing, since in contrast to MALDI the ions are produced with a relatively broad distribution of initial positions but with a negligible spread in initial velocities. A schematic diagram of this instrument is shown in Figure 13, and examples of the high resolution and high m/z range obtained with this instrument are shown in Figure 14. While the application of TOF to electrospray ionization is relatively new, the advantages over quadrupole analyzers is clearly established. The development and performance of ESI-TOF has recently been reviewed by Chernishevich, Ens, and Standing [58].

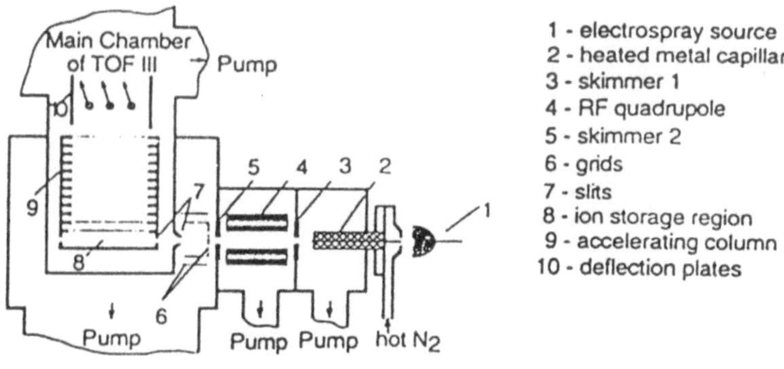

Figure 13. Schematic diagram of the orthogonal TOF mass spectrometer ion source and interface with RF quadrupole.[57]

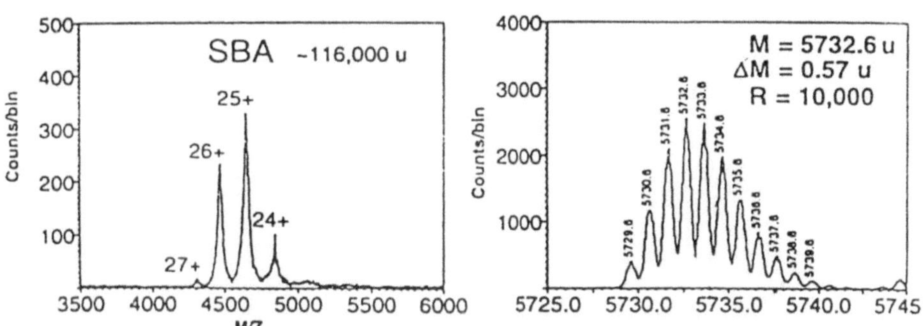

Figure 14. ESI mass spectrum of SBA tetramer (Mr ~116,000) (a) and deconvolution mass spectrum of bovine insulin.(b)[57]

3. The Elements of TOF-MS, Theory and Practice

Two TOF-based techniques have emerged which provide new high performance tools for applying mass spectrometry to a wide variety of analytical problems. The development of these tools is being driven very rapidly by their obvious biological applications, but it should also be recognized that these techniques have great potential for all of the present and future applications of mass spectrometry. The obvious distinction between the two techniques is MALDI versus electrospray, but this is somewhat narrower than necessary. A broader, and perhaps more useful distinction is to consider pulsed desorption from a surface as one approach, and continuous ionization with collisional relaxation and orthogonal injection as the other. Thus MALDI, SIMS, and Plasma Desorption fit neatly into the former, and virtually all ionization techniques can be accommodated by the latter. From an operational view, these may be separated into the following two cases:

Case I. Ions are emitted from a planar surface perpendicular to the axis of the ion optical system in a brief pulse with a distribution of initial velocities.

Case II. Ions are continuously injected into a field-free region in a direction perpendicular to the axis of the ion optical system with a very narrow distribution of initial velocities in the direction parallel to the axis, but with a distribution of initial positions.

In Case I (MALDI) we are concerned only with velocity focusing, and in Case II (ESI) we are concerned only with space focusing. It should be noted that there is a real difference between an initial velocity distribution, and an initial energy distribution resulting from ions originating at different points in an electric field.

In this discussion an attempt is made to discuss in a quantitative way the factors which determine the resolution of a time-of-flight mass spectrometer without becoming totally lost in equations. Rigorous mathematical treatments are given in several excellent reviews.[59-61] We begin with discussions of idealized mass analyzers, including the ion source, for the two cases, consider the deviations from ideality which cause the performance to be less than ideal, and finish with discussions of ion detectors and time digitizers.

3.1. CASE I WITH LINEAR ANALYZER

In the idealized version of this case the ions depart the planar surface all at the same time, and to obtain the best resolution they must all strike the planar detector at the same time. Practical considerations require that the ions be given sufficient velocity that they may be detected efficiently, and the total travel time must be large enough that the resolution of the detector and the timing system are not the primary limits on resolution. The application of the simple linear analyzer with static source to MALDI was first described by Beavis and Chait [62], and a schematic diagram of their analyzer is shown in Figure 15. They used a modified version of this analyzer to determine the initial velocities of the ions produced by MALDI [63] and determined that the average initial velocity was about 750 m/sec, independent of the mass of the ion, and that the full width at half maximum of the velocity distribution was comparable to the average. A typical spectrum obtained with this instrument is shown in Figure 16 where the mass resolution (FWHM) is about 300 at 5734 u. The flight time of an ion in a linear instrument with a two-field ion source is given by

$$t = (v_1 - v_0)/a_1 + (v - v_1)/a_2 + d/v \qquad (1)$$

where v_0 is the initial velocity, v_1 is the velocity exiting the first region of the source in which the ion acceleration is a_1, v is the velocity exiting the second region of the source in which the ion acceleration is a_2, and d is the length of the field-free region between source and detector.

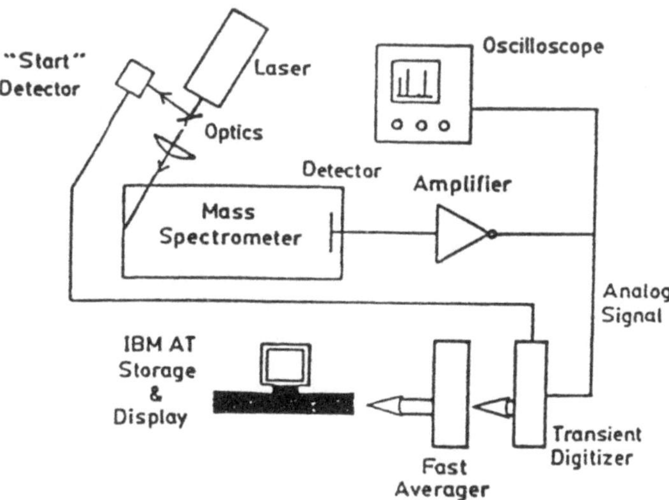

Figure 15. Schematic diagram of MALDI instrument with linear analyzer.[62]

The theoretical resolution of the linear analyzer with continuous extraction can be calculated approximately as the inverse of

$$\Delta m/m = 2\Delta t/t = (\delta v_0)^2 m/2V + 2\delta v_0 (d_1/d)(m/V)^{1/2} + 2\delta d/d + 2\delta t/t \qquad (2)$$

where δv_0 is the width (FWHM) of the initial velocity distribution, V is the total acceleration potential, d_1 is the length of the first source region, δd is the uncertainty in total ion flight path due for example to deviations in flatness or angular orientation of detector or source, and δt is the uncertainty in the measurement of flight times which is approximately equal to the width of the digitizer plus the width of a single ion pulse from the detector. In MALDI the observed resolution may be significantly lower than calculated from this equation, particularly at higher laser fluences, as a result of collisions between ions and neutrals in the laser plume, as has been shown experimentally by Ens and coworkers [64]

With delayed extraction the first order dependence on initial velocity can be removed for a particular mass by proper choice of extraction delay and pulse amplitude. Delayed extraction has at least two beneficial effects. First, by delaying application of the extraction field, the plume of neutral expands sufficiently that loss of translational energy of the ions due to collisions with neutrals is minimized, and by choosing the correct delay, the first order term in the resolution equation is eliminated and the resolution depends only on the distribution of initial velocity relative to the average. Thus the equation for the inverse resolution becomes

$$\Delta m/m = (\delta v_0)^2 m/8V + 2\delta d/d + 2\delta t/t \qquad (4)$$

Calculated resolutions as a function of mass for a linear analyzer with continuous and delayed extraction are shown in Figure 17. The predicted maximum resolution of 6000-7000 in the 2000-6000 u range for delayed extraction has recently been verified experimentally. As an example a spectrum showing isotopic resolution of the insulin is shown in Figure 18 where the measured spectrum is compared with theoretical at a resolution of 7000.

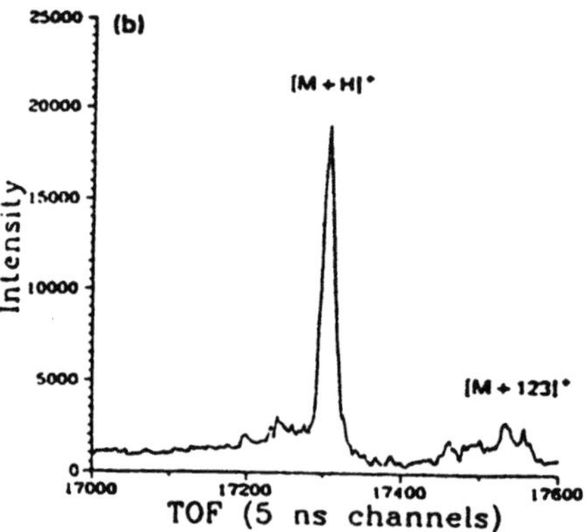

Figure 16. The molecular ion region of the positive-ion LD TOF spectrum of bovine insulin (M_T=5733.5) using vanillic acid matrix with linear analyzer using 10 nsec channels[69].

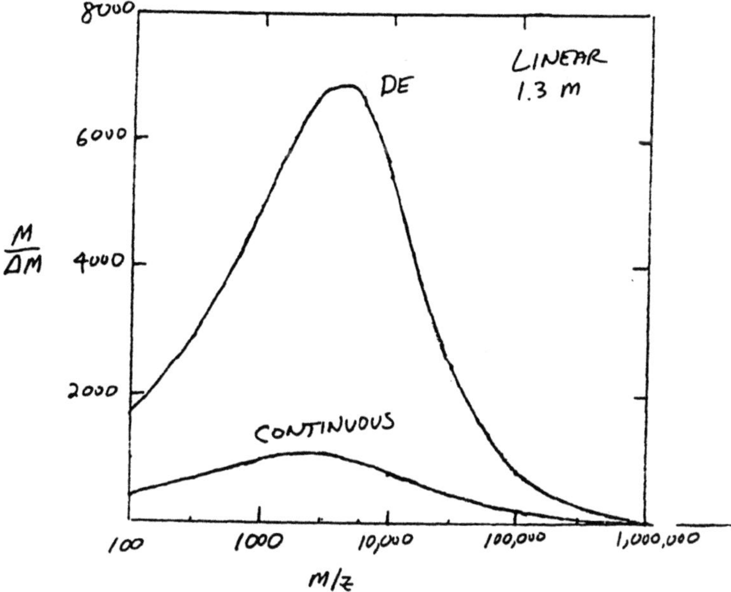

Figure 17. Calculated resolution for a 1.3m linear analyzer in continuous and delayed extraction (DE) modes as a function of mass on a logarithmic scale.

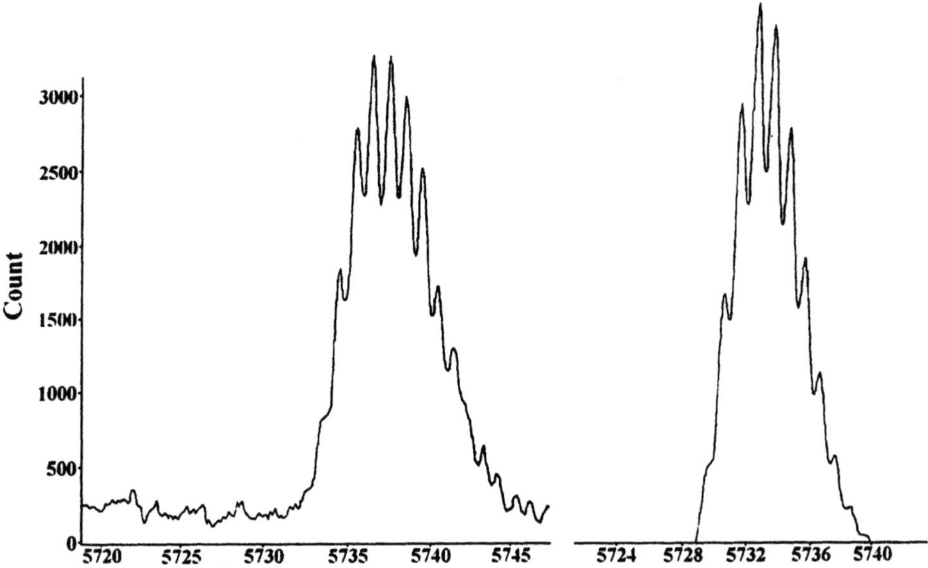

Figure 18. Molecular region of bovine insulin measured in a 1.3m linear analyzer with delayed extraction compared with theoretical at a resolution of 7000. Spectrum was obtained using a transient digitizer (Tektronix 744A) with 0.5 nsec channels.

3.2. CASE II WITH LINEAR ANALYZER

In this case the distances are chosen to provide first order Wiley focusing [7]. For the geometry shown in a simplified schematic in Figure 13, the inverse of the resolution is given by

$$\Delta m/m = (\Delta V/V)(\Delta x/4d_1)^2 + (d_1/d)(V_0 V)^{1/2}/\Delta V + 2\delta d/d + 2\delta t/t \quad (5)$$

where ΔV is the amplitude of the pulse applied to the first region of the source, Δx is the width of the transversely injected ion beam, V_0 is the initial kinetic energy corresponding to motion in the direction of ion extraction, and the other parameters are the same as defined above. Except for the last term which is important only at low mass, the resolution in this case is mass independent.

3.3. CASE I WITH REFLECTING ANALYZER

Adding a reflector to the MALDI system produces a resolution equation of the form

$$\Delta m/m = C_n(\delta v_0)^{2n}(m/2V)^n + 2\delta v_0(d_1/d_e)(m/V)^{1/2} + 2\delta d/d_e + 2\delta t/t (6)$$

where n=2 for a single stage reflector, and n=3 for a two stage reflector. d_e is the effective drift distance which corresponds to the field free distance in a linear analyzer in which the flight times are equal to those in the reflecting system. The coefficient C_2 for a single stage reflector is 1/4, and for a two stage reflector C_3 is between 1/2 and 1 depending on the relative length of the two stages. As discussed by Mamyrin [18], the two stage mirror gives higher energy resolution than the single stage mirror, but this is important only if the first term in equation (6) is dominant. For the initial velocities that are encountered in MALDI this is true only at very high mass or very low accelerating voltage. In general the major effect of adding a reflector is to reduce the first (energy) term to insignificance, and to reduce the other terms as the result of d_e being a factor of 2-3 larger than in a linear analyzer of comparable overall dimensions.

Adding delayed extraction to a reflecting system allows simultaneous cancellation of both the first and second order velocity dependent terms. In this case the resolution equation becomes

$$\Delta m/m = C_d(\delta v_0)^3(m/2V)^{3/2} + 2\delta d/d_e + 2\delta t/t \quad (7)$$

where the coefficient C_d depends on the geometry and the operating point, but it generally sufficiently small that the contribution of the first term can be neglected up to at least 30,000 u.

3.4. CASE II WITH REFLECTING ANALYZER

In this case the resolution is given by

$$\Delta m/m = C_n(\Delta V/V)^n(\Delta x/d_1)^2 + 2(d_1/d)(V_oV)^{1/2}/\Delta V + 2\delta d/d + 2\delta t/t \quad (8)$$

where n=2 for a single stage mirror, and n=3 for a two stage mirror. If the ion beam width Δx is significant compared to the width of the first accelerating region, then the first term may be dominant and use of a two stage mirror may provide higher resolution. On the other hand, unless the transversely injected beam is very well collimated, the second term will be dominant and no gain will be realized by using a higher order mirror.

3.5. ION DETECTORS

At low mass and high accelerating potentials the dominant term in determining resolution is that due to the uncertainty in the time measurement. In many cases the ion detector may limit the attainable resolution. The detector most commonly used in high performance TOF systems is shown schematically in Figure 18. This system consists of dual channel plates oriented in a "chevron" arrangement with a conical anode directly coupled to a standard 50 ohm cable connector. In favorable cases these detectors can produce single ion pulses with width (FWHM) less than 1 nsec. These channel plates are made of semi-conducting lead glass with a large number of parallel cylindrical channels 10-25 microns in diameter set at an angle of ca. 8-12 degrees to the surface of the plate. The thickness of the plate is typically 40-60 time the diameter of the channel. With a potential difference of up to 1000 volts applied, each such plate provides an electron gain in the range 10^3-10^4. Thus two plates in tandem produces a typical output pulse of ca. 10 mV for each incident ion. This is sufficient for single ion detection without additional amplification.

The dual channel plate detector (DCP) is nearly the ideal detector for TOF mass spectrometry in that it is fast, with high gain and low noise, and provides a nearly flat surface over an extended area, but it does have some limitations. For example, only about 60% of the surface contains channels and is active for detection incoming ions. Also, once an individual channel has been fired by multiplication of an ion signal it is inactive for tens of microseconds until the potential difference along the channel has been recharged; this saturation effect may cause discrimination against high mass ions in complex spectra. This detector is very efficient for low mass ions which produced multiple secondary electrons upon impact at high velocity, but is rather inefficient for detecting high mass species which may preferentially produce secondary ions upon impact. Detection systems have been developed which use a conversion dynode to efficiently convert high energy massive ions to secondary ions with acceleration of the

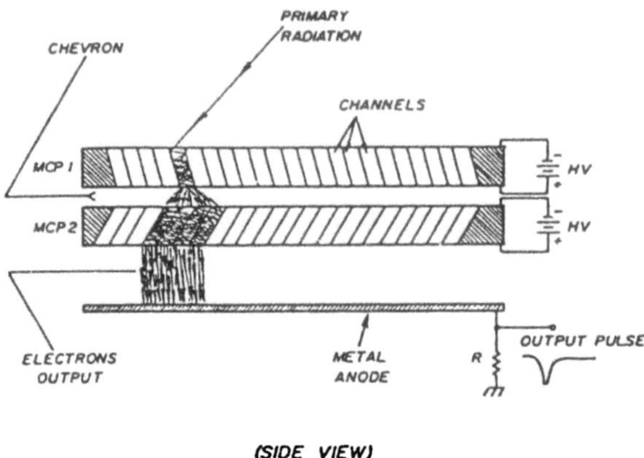

Figure 19. Diagram of the dual channel plate detector in the chevron arrangement with flat metal anode.

secondary ions into a DCP or discrete dynode electron multiplier. These detectors give higher sensitivity at high mass but tend to compromise mass resolution. In high resolution applications the fact that the active surface of the channel plate is not truly flat may be an important limiting factor. For example, with the 25 micron channels at 8 degrees used in the larger diameter detectors, the total distance variation due to ion impact at different points in the channel is about 0.18 mm. With such a detector in a system with 1 m effective flight path the maximum resolution that can be achieved is about 5000 (FWHM) even if all of the other contributions to peak width were negligible. Standing and coworkers [20] have described a detector which uses a flat plate conversion dynode with a deflecting magnet to transport the secondary electrons to a channel plate detector. Other detection schemes have been described which overcome some of the limitations of the DCP detectors, but the perfect detector for all applications of TOF mass spectrometry has not yet appeared.

3.6. TOF MEASUREMENT AND RECORDING OF SPECTRA

There are two highly developed techniques for measuring and recording time-of-flight spectra. These are the digital transient recorder (DTR) such as used in digital oscilloscopes, and the multi-stop time-to-digital converter (TDC) which is the modern version of systems used in the earlier coincidence mass spectrometers. The major distinction between them is that the DTR digitizes the ion current detected for a specified period following a start pulse, and the TDC determines and records the time interval between the start pulse and each detected "stop" pulse corresponding to arrival of an individual ion at the detector. The former can efficiently handle production of a large number of ions per start pulse but is somewhat limited in dynamic range, while the latter must be limited to less than one event per channel per start pulse, but has a dynamic range limited only by the total measurement time. The time resolution is limited only by the number of significant bits in the time measurement since the ion arrival is indicated by the leading edge of the pulse using a constant faction discriminator and the total width of the ion pulse is unimportant. With the DTR the time resolution is determined by the channel width of the digitizer plus the width of the pulse from the detector. In both cases times resolutions of 1 nsec or better are now feasible using commercial products, but most of the TOF instruments presently in use provide somewhat less resolution, 2-5 nsec is more typical. In general the DTR is used with MALDI where high ion currents and low repetition rates are normal, and the TDC is used with electrospray where the ion currents tend to be lower and high repetition rates are desirable to efficiently sample continuous ion production.

4. Practical Instruments

Two types of TOF instruments are now established as practical tools for biological applications. The first of these, MALDI-TOF with Delayed Extraction, is already established and widely applied. The second of these, electrospray ionization with collisional cooling and orthogonal injection TOF (ESI-o-TOF) is not yet commercially available, but recent results on prototype instruments [64,73] clearly show the potential of this technique.

4.1. MALDI-TOF-DE

The addition of DE to MALDI has not only increased the mass resolution in both linear and reflecting analyzers, it has also dramatically improved the sensitivity, reduced the noise, and made the quality of the data much less dependent on precise control of the laser intensity. Most importantly, the mass accuracy routinely possible is significantly improved. Except possibly for some applications of post-source decay (PSD) it is difficult to find any reason to operate the MALDI instrument in continuous extraction mode. A schematic diagram of one version of a practical instrument is shown in Figure 20. A number of details may be different among the various instruments in use, but the high-performance instruments all include the same basic elements. These include (1) a two field ion source with delayed extraction; (2) differentially pumped source and analyzer sections; (3) a collision cell for optionally causing high energy fragmentation; (3) ion lens or beam guide for efficiently transmitting ions; (4) gating system for selecting a specific parent ion for MS/MS; (4) ion reflector with programmable HV supply (5) detector for reflected ions; (6) detector for linear analyzers; (7) transient digitizer, and (8) data system. With the mirror voltage set to zero these system function as linear analyzers; also the mirror voltage may be programmed to selectively focus segments of fragment spectra. For applications not requiring the ultimate in resolution and mass accuracy at lower masses, and where MS/MS capability is not required, items (2) through (5) may be deleted to provide a lower cost system.

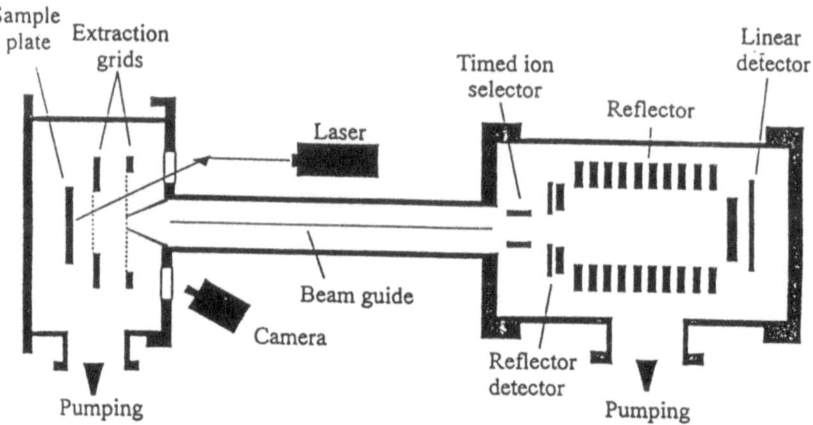

Figure 20. Schematic diagram of a commercial MALDI-TOF instrument with delayed extraction.

4.2. ESI-o-TOF

This system makes the speed, sensitivity, mass range, resolution, and mass accuracy, inherent in modern TOF available for continuous ionization sources. The most obvious ionization source for biological applications is electrospray, but almost any ionization source can be coupled to this mass analyzer. A simplified diagram of the o-TOF mass analyzer is shown in Figure 21, and a schematic of the source region of one version of this instrument is depicted in Figure 12. The essential elements of this system are (1) the electrospray source at atmospheric pressure; (2) nozzle-skimmer system with differential pumping for transmitting ions from atmospheric pressure to vacuum; (3) RF quadrupole in chamber with pressure controlled in the millitorr range; (4) field-free ion storage region with pulsed extraction; (5) field-free drift region; (6) detector for linear analyzer (optional); (7) ion reflector; (8)detector for reflected ions; (9) transient digitizer or TDC; and (10) data system. Similar systems have been described [74] without the reflector, but the cost saving is not large and the reduction in performance, particularly in regard to resolution and mass accuracy is substantial. The cost and performance of this system are largely dictated by the vacuum system, since the vacuum in the analyzer must be maintained below 10^{-6} torr to obtain high performance [65] while efficiently sampling ions from atmospheric pressure.

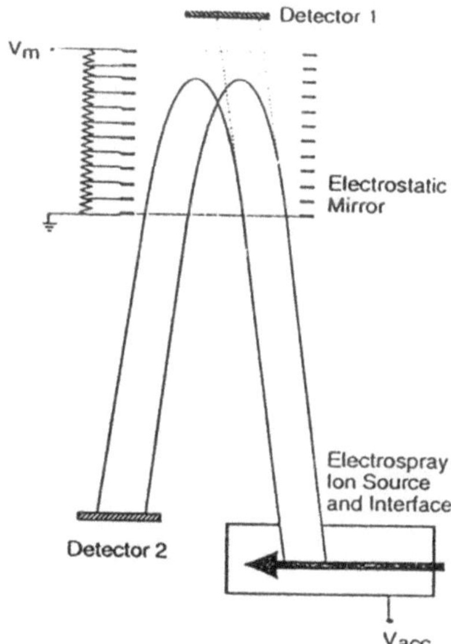

Figure 21. Simplified diagram of ESI-o-TOF system.

5. Present Status and Future Prospects

Dynamic devices such as FT-ICR, quadrupole traps, and time-of-flight are clearly the future of mass spectrometry. Magnetic deflection analyzers are already on the wane, and quadrupole mass filters may not be far behind. The FT instruments with superconducting magnets provide the highest ultimate mass resolution, and the quadrupole trap allows multiple successive fragmentation processes (MS^n) to be studied. While these devices will continue to be used for applications requiring their special properties, it appears that time-of-flight is likely to be the routine work-horse mass analyzer of the future. It provides a unique combination of speed, sensitivity, mass range, mass accuracy, resolution, and dynamic range, and can be made compact and simple without sacrificing performance. Recent developments showing mass resolution in excess of 10,000, mass accuracy at the ppm level, and detection limits in the sub-attomole range are impressive, but even higher performance is expected within the next few years. While further improvement in time-of-flight instruments is expected, it is now rare where the performance of the analyzer is a significant factor in determining the feasibility of a particular application.

For biological molecules the dominant techniques are electrospray and MALDI. Both are now highly developed, and are used in many applications. At present, both suffer some limitations when applied to larger biopolymers. For example, electrospray is a very "soft" technique which can produce intact molecular ions from quite fragile molecules, including some non-covalently bound complexes, but it produces a very broad distribution of charge states which limits its utility for analyzing complex mixtures such as sequencing ladders. MALDI provides a narrower distribution of charges, usually dominated by singly-charged ions, but for larger biopolymers, sufficient excess energy is imparted in the ionization process to cause excessive fragmentation in flight. This causes the observed performance at higher masses (ca.> 30,000 u) to be much poorer than would be expected based on extrapolation from lower mass. These are the major problems limiting the direct applications of mass spectrometry to sequencing of intact proteins and larger DNA fragments.

The performance of mass spectrometry in biological applications is often limited by difficulties in separating the sample of interest from a biological matrix and delivering it in suitable form to the ion source of the mass spectrometer. At the trace levels now accessible these problems are particularly severe. On-line capillary LC and CE systems integrated with mass spectrometric detection will be developed which allow samples to be automatically injected, separated, concentrated, desalted, digested, and analyzed under computer control without operator intervention. Techniques and instrumentation will be developed for characterizing proteins separated on 2-D gels with little or no manual intervention.

Mass spectrometry as a detector for specific chemical or enzymatic reactions is already established as a powerful technique. Strategies for sequencing peptides, proteins, and DNA have been partially developed, but these are not yet widely applied. The keys to future applications of mass spectrometry to biological problems are integration and automation. These will make the power of these new mass spectrometric techniques routinely accessible to the biologist and biochemist.

6. References

1. Schlag, E. W., Ed. (1994) *Time of Flight Mass Spectrometry and its Applications*, Elsevier. Also published in *International J. of Mass Spectrom. and Ion Processes* **131** (1994)
2. Cotter, R. J. (1992) *Anal. Chem.* **64**,1027A-1039A.
3. Price, D. and Milnes, G. J. (1990) *International J. of Mass Spectrom. and Ion Processes* **99**, 1.
4. Cameron, A. E. and Eggers, D. F. Jr.(1948) *Rev. Sci. Instr.* **19**, 605.
5. Wolf, M. M. and Stephens, W. E. (1953) *Rev. Sci. Instr.* **24**, 616.
6. Katzenstein, H. S. and Friedland, S. S. (1955) *Rev. Sci. Instr.* **26**, 324.
7. Wiley, W. C. and McLaren, I. H. (1955) *Rev. Sci. Instr.* **26**, 1150.
8. Gohlke, R. S. (1959) *Anal, Chem.* **31**, 535.
9. Price, D. and Ratajczak, E. (1978) in Dynamic Mass Spectrometry, Vol 5. ed. D. Price and J. F. J. Todd, Heyden, London, p176.
10. Rosenstock, H. M. (1960) U. S. Patent No. 2,999,157.
11. Vestal, M. L., Krause, M., Wahrhaftig, A. L., and Johnston, W. H.(1963) Eleventh Annual Conference on Mass Spectrometry and Allied Topics, ASTM E-14, 358-365.
12. Krause, M. O., Vestal, M. L., and Johnston, W. H. (1964) *Phys. Rev.* **133**, A385.
13. Brehm, B and von Puttkamer, E. (1967) *Z. Naturforsch. Teil A*, **22**, 8.
14. Danby, C. J. and Eland, J. H. D. (1972) *International J. of Mass Spectrom. and Ion Phys* **8**, 153.
15. Stockbauer, R. (1973) *J. Chem. Phys.* **58**, 3800.
16. Riley, J. S. and Baer, T. (1994) *International J. of Mass Spectrom. and Ion Processes* **131**, 295-305.
17. Mamyrin, B. A. (1966), Dokt. Diss. Physico-technical Institute, Academy of Sciences, USSR, Leningrad.
18. Mamyrin, B. A., Karataev, V. I., Shmikk, D. V., and Zagulin, V. A. (1973) Sov. Phys. JETP, 37, 45-48.
19. Tang, X., Beavis, R., Ens, W., Lafortune, F. Schueler, B., and Standing, K. G. (1988) *Int. J. Mass Spectrom. Ion Processes*, **85**, 43.

20. Tang, X., Ens, W., Mayer, F., Standing, K. G., and Westmore, J. B. (1989) Rapid Comm. Mass Spectrom. 3, 443-448.
21. Torgerson, D. F., Skowronski, R. P. and Macfarlane, R. D. (1974) *Biochem. Biophys. Res. Commun.* **60** 616-621.
22. Roepstorff, P. and Sundqvist, B. (1986) *Mass Spectrometry in Biomedical Research*, S. J. Gaskell, Ed., John Wiley & Sons Ltd.
23. Barber, m. Bordoli, R. S., Sedgwick, R. D., and Tyler, A. N. (1981) *Nature* **293**, 270.
24. Honig, R. E.(1963) *Appl. Phys. Letters*, **2**, 138.
25. Posthumus, M. A., Kistemaker, P. G., Meuzelaar, H. L. C., Ten Noever de Brauw, M. C. (1978) *Anal. Chem.* **50**, 985.
26. Zakett, D., Schoen, A. E., Cooks, R. J. (1981) *J. Am. Chem. Soc.* **103**, 1295,.
27. Van Breeman, R. B., Snow, M., Cotter, R. J. (1983) *Int. J. Mass Spectrom. Ion Phys.* **49**, 35.
28. Hardin, E. D., and Vestal, M. L. (1981) *Anal. Chem.* **53**, 1492.
29. Karas, M., and Hillenkamp, F. (1988) Adv. in Mass Spectrometry, Vol 11A. P. Longevialle Ed., Heyden, p354-363 and p.416-417.
30. Ingendoh, Karas, M. Hillenkamp, F., Giessmann, U. (1994) *Int. J. Mass Spectrom. Ion Processes* **131**, 345-354.
31 Overberg, A., Karas, M., Bahr, U., Kaufmann, R., Hillenkamp, F.(1990) *Rapid Comm Mass Spectrom.* **4**, 293; (1991) **5**, 128.
32. Beavis, R. C. and Chait, B. T. (1990) *Anal. Chem.* **62**, 1836-1840.
33. Hillenkamp, F., Karas, M., Beavis, R. C., Chait, B. T. (1991) *Anal. Chem.* **63**, 1193A-1203A.
34. Brown, R. S. and Lennon, J. J. (1995) Anal. Chem. 67, 1998-2003.
35. Colby, S. M. King, T. B., and Reilly, J. P. (1994) *Rapid Comm Mass Spectrom.* **8**, 865-868.
36. Whittal, R. M. and Li, L. (1995) *Anal. Chem.* **67**, 1950-1954.
37. Vestal, M. L., Juhasz, P. and Martin, S. A. (1995) *Rapid Comm Mass Spectrom.* **9**, 1044-1050.
38. Juhasz, P. Roskey, M. T., Smirnov, I. P., Haff, L. A., Vestal, M. L., and Martin, S. A. (1996) *Anal. Chem.* **68**, 941-946.
39. our peptides
40. Dole, M., Mach, L. L., Hines, R. L., Mobley, R. C., Ferguson, L. P., Alice, M. B. (1968) *J. Chem. Phys.* **49**, 2240.
41. Yamashita, M. and Fenn, J. B. (1984) *J. Phys. Chem.* **88**, **4451 and 4471**.
42 Mann, M., Meng, C. K.., and Fenn, J. B. (1988) *Proceeding of the 36th Annual Conference on Mass Spectrometry and Allied Topics*, San Fancisco, ASMS, p 771. Fenn, J. B., Mann, M., Meng, C. K., Wong, S. F., and Whitehouse, C. M. (1989) *Science* **246**, 64-71.
43. Meng, C. K., Larsen, B. S., McEwen, C. N., Whitehouse, C. M., Fenn, J. B. (1989) Second International Symposium on Mass Spectrometry in the Health and Life Sciences, San Francisco. CA.
44. Benson, S. W., Grossman, J. J., Elkin, H. S., (1960) U. S. Patent Nol 2, 938,116.
45. Gerhard, P. H., Loffler, S., Homan, K. H. (1987) *Chem. Phys. Lett.* **137**, 306-310.
46. Olthoff, J. K., Lys, I., Cotter, R. J. (1988) *Rapid Comm Mass Spectrom.* **2**, 171-175.
47. Kennedy, R. A., Kung, C.-Y., Miller, J. P. (1989) in Ion and Cluster Ion Spectroscopy, Maier, J. P. Ed., Elsevier, pp. 213-239.
48. Dawson, J. H. J., and Guilhasu, M. (1989) *Rapid Comm Mass Spectrom.* **3**, 155-159.
49. Coles, J., and Guilhasu, M. (1993) *Trends Anal. Chem.* **12**, 202-213.
50. Dodonov, A. F., Chernushevich, I. V., Laiko, V. V. (1991) *Advances in Mass Spectrometry, Vol 12*
51. Sin, C. H., Lee, E. D., Lee, M. L. (1991) *Anal. Chem.* **63**, 2897-2900.
52. Myers, D. P. Li, G., Mahoney, P. P., Hieftje, G. M. (1995) *J. Am. Soc. Mass Spectrom.* **6**, 400-410 and 411-420.
53. Dodonov, A. F., Chernushevich, I. V., Dodonova, T. F., Raznikov, V. V. , Tal'roze, V. L. (1987) USSR Pat. No. 1681340A1.
54. Boyle, J. G., Whitehouse, C. M. (1992) *Anal. Chem.* **64**, 2084-89.
55. Clayton, E., Bateman, R. H. (1992) *Rapid Comm Mass Spectrom.* **6**, 719-720.
56. Verentchikov, A. N., Ens, W. Standing, K. G., (1994) *Anal. Chem.* **66**, 126-133.
57. Krutchinsky, A. N., Chernushevich, I. V., Spicer, V., Ens, W., Standing, K. G. (1995) *Proceeding of the 36th Annual Conference on Mass Spectrometry and Allied Topics*, p. 126.
58. Chernushevich, I. V., Ens, W., Standing, K. G., in press.
59. Bergmann, T., Martin, T. P. (1994) *International J. of Mass Spectrom. and Ion Processes* **131**, 21-41.
60. Ionnoviciu, D. *International J. of Mass Spectrom. and Ion Processes* **131**, 43-65.
61. Wollnik, H. *International J. of Mass Spectrom. and Ion Processes* **131**, 387-487

62. Beavis, R. C., Chait, B. T. (1989) *Rapid Comm Mass Spectrom.* **3**, 432-435.
63. Beavis, R. C., Chait, B. T. (1991) *Chem. Phys. Lett.* **5**, 479.
64. Zhou, J., Ens, W., Standing, K. G., and Verentchikov, A. (1992) *Rapid Comm Mass Spectrom.* **6**, 671-678.
65. Churnushevich, I. V., Verentchikov, A. N. Ens, W., Standing, K. G. (1996) *J. Am. Soc. Mass Spectrom.* 7,
66. Verentchikov, A. N., Hsieh, F., Tomany, M., Gabeler, S., Vestal, M., Martin, S. (1995) *Proceeding of the 36th Annual Conference on Mass Spectrometry and Allied Topics*, p. 613.

COMBINED ANALYZER TECHNOLOGIES

PETER KOFEL AND URS P. SCHLUNEGGER
Institute of Organic Chemistry, University of Bern
Freiestr. 3, CH-3012 Bern, Switzerland

1. Why are analyzer combinations being used?

There is no one single answer to the question of why different analyzer combinations are being used because the driving force to construct and use such instruments changes with time. Certainly one of the driving forces is the need for further technological development; for example, the need of better sensitivity and higher mass resolution for atomic [1] and molecular [2] mass determination and structure elucidation of natural compounds [3]. This was subsequently achieved by the introduction of double focusing sector instruments (combination of magnetic and electrostatic analyzers) [4,5]. The study of natural compounds, for instance, alkaloids and steroids [6], pushed mass spectrometry to a higher mass range (up to 1 kDa and more). The possibility to analyze larger molecules induced again the development of new ionization methods, thus adding "soft" ionization techniques to the commonly used electron impact ionization (EI). Methods like chemical ionization (CI), field desorption (FD), and fast atom bombardment (FAB) opened a wide research field for mass spectrometry [7-10].

One complication in this development is possible contamination of the ion of interest by isobaric ions in the sample, a situation which is especially important if impure extracts of biological origin have to be analyzed. Usually contamination cannot be excluded by a single step mass spectrometric analysis. Therefore, simultaneously with the development of soft ionization techniques, increased practical mass range and enhanced sensitivity of commercially available instruments, the analysis and control of ion fragmentation became more and more important. When it was shown that the sequence of amino acids in peptides may be analyzed in combined analyzer instruments [11], e.g. by MIKES (mass analyzed ion kinetic energy spectrometry), mass spectrometry entered a new dynamically growing field of research, and actually these methods (tandem mass spectrometry or "MS/MS" [12,13]) have come to be in the toolbox of modern biological sciences. Today, instruments with several analyzers are commercially available and are capable of performing even "multiple" MS/MS (MS^n, n>2) experiments. Some recent applications include carbohydrates [14-17], polymers and dendrimers [18], peptides [19,20], protein folding [21], sequencing of genomes [22], single cells [23], antibiotics [24], drugs [25], molecular biology and biotechnology [26] or even comet analysis [27].

R. M. Caprioli et al. (eds.), Selected Topics and Mass Spectrometry in the Biomolecular Sciences, 263–286.

Sensitivity is always an issue especially in biological research where some applications require very high sensitivity. In a scanning mass spectrometer the time fraction during which one ion of interest is being observed is usually rather short, and all the ions that are not focused on the exit slit are lost. Novel commercial instruments embody analyzer combinations which allow for longer duty cycles, i.e. longer observation times for each ion during spectrum acquisition. These new instruments handle and record many ions simultaneously, thus increasing the duty cycle and consequently increasing sensitivity. Therefore it is not astonishing, that the instruments developed most recently embody quite complex analyzer combinations - such as, e.g., time-of-flight stages (TOF) or ion traps added to magnetic and electric sectors, or to quadrupole analyzers. In addition, these instruments offer different soft and hard ionization methods, as well as ion fragmentation tools.

2. What is essential for analyzer combinations?

Successful analyzer combinations have several common features, although there is no universal mass analyzer that can solve every analytical problem. To recognize the benefits of different analyzer combinations, it is important that the efficiency and versatility of the whole instrument are considered. The efficiency of analyzers or analyzer combinations should not be discussed detached from the rest of a mass spectrometer, but the discussion should also include entrance and exit conditions and restrictions. For an individual analyzer the entrance conditions are defined by the preceding mass spectrometric section (ion source or other analyzer, see Fig. 1) and the exit conditions by the succeeding sections (other analyzer or detector). For a whole analyzer combination the entrance conditions are defined by the ion source or, more

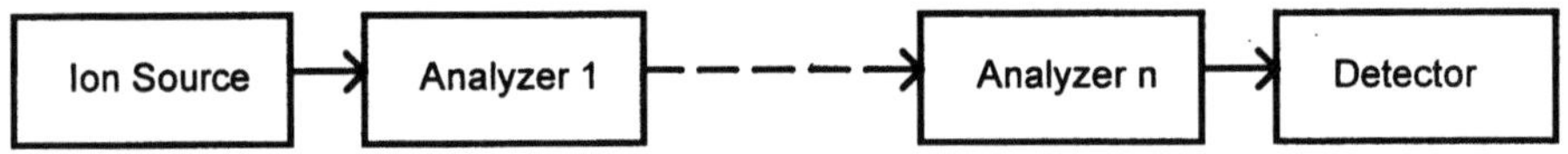

Figure 1 Mass spectrometer sections.

generally, the ionization method, and the exit conditions are defined by the detection method. The important points for analyzer interfacing that have to be addressed are: vacuum compatibility, kinetic energy, beam shape, flight direction, synchronization and efficiency (duty cycle).

An important aspect of the versatility of an analyzer combination is the variety of scan modes [28,29] possible. The success of the "reversed geometry" tandem magnetic sector/electric sector instrument lies in the fact that both sensitive ("doublefocusing") high resolution experiments and MS/MS dissociation experiments are possible on those instruments. Besides the beneficial choice of the sequence of the mass spectrometer section, optimized interfaces between all successive sections are essential. In the paragraphs which follow, after a brief discussion of ion sources and detectors, the different needs and capabilities of the commonly used analyzers are first summarized before the discussion of analyzer interfacing.

2.1 ION SOURCES

Ion sources are discussed in chapters 1 to 3 and elsewhere [30,31]. Only the aspect of continuous versus pulsed ion sources is highlighted here. Continuous sources are ideal for continuous beam analyzers. For pulsed analyzers (traps and time-of-flight), however, pulsed ion sources are needed for analyzer efficiency. Pulsed sources include: direct desorption/ionization sources (e.g. pulsed SIMS and pulsed laser sources); storage sources that collect continuously produced ions and supply pulses of concentrated ion packets to the analyzer [32,33]; and sources with pulsed valve inlets (e.g. pulsed molecular beam sources). These are all truly pulsed sources. In other sources, where neutral flow through the ion source (and into the pump) is continuous, only the ion beam is pulsed and sample is wasted between "ionization" periods.

2.2 DETECTORS

Detectors can be divided into two categories: discharging and non-discharging devices. The discharging detectors [34] neutralize the ions, which are lost at this point. They include the simple ion current sensors (collector plate and Faraday cup) and the multiplying devices (electron multiplier and array detector [35]), which often employ extra charge conversion and/or photoconversion for bridging high voltage differences. The non-discharging detectors (image charge [36], bolometric [37], SQUID [37] and resonance absorption [38]) sense the presence of charges by moving the ions but without having them touch an electrode. When non-destructive detection is used, the ions may be subjected to further experiments (e.g. dissociation) after detection. From the non-discharging techniques, the capacitively coupled broadband image charge detection has found widespread use in Fourier transform ion cyclotron resonance spectrometers [39] to detect the characteristic oscillation frequencies of all m/z values simultaneously.

3. Analyzer characteristics

Table 1 briefly summarizes the characteristics of the most commonly used analyzers. Some recent and early references are included at the bottom of the table. More detailed reviews on the individual analyzer types can be found in the preceding chapters: on magnetic and electric sector analyzers in chapter 4, on quadrupoles and ion traps in chapter 5, on time-of-flight analyzers in chapter 6 and on Fourier transform ion cyclotron resonance traps in chapter 7.

Symbol	B	E	W	Q	q	Trap	ICR	TOF
Analyzer Name	Magnetic Sector	Electric Sector	Wien Filter	Quadrupole	RF Multipole	RF Ion Trap	ICR Trap	Time-of-Flight
Analyzer Type	momentum analyzer	energy analyzer	velocity analyzer	mass analyzer	mass bandpass	mass analyzer	mass analyzer	velocity analyzer
Pressure	$<10^{-6}$	$<10^{-6}$	$<10^{-6}$	$<10^{-4}$	any low mw	$<10^{-3}$ low mw	$<10^{-8}$	$<10^{-6}$
Kinetic Energy	large	large	large	small	any	small	small	large
Ion Path	spatial momentum dispersion	spatial energy dispersion	spatial velocity dispersion	single m/z transmissio n	simultaneous transmission	simultaneous trapping	simultaneous trapping	velocity dispersion in time
Shape	slit	slit	slit	circular	circular	ellipsoidal	ellipsoidal	flat area
Angle	small	small	small	large	large	any	any	small
Synchronization	continuous	continuous	continuous	phase dependent	weakly phase dependent	pulsed, phase dependent	pulsed	very short pulse
Simultaneous Detection	with array detector	with array detector	with array detector	not possible	(limited mass selectivity)	trapped ion full spectrum	truly simultaneous	very fast full spectrum
Application Marks	dynamic range, intensity accuracy	energy filter	dynamic range, sensitivity	dynamic range, sensitivity, small	ion guide, prefilter, beam shape, dissociation, reaction	(high resolution,) Ms^n, reaction, sensitivity, small	high resolution, mass accuracy, Ms^n, reaction, sensitivity	intensity accuracy, sensitivity, high mass
References	[29,40,41]	[29,42]	[43-51]	[52,53]	[54-58]	[59-64]	[39,65-80]	[33,81-90]

Table 1. Analyzer characteristics

4. Interfacing analyzers

4.1 VACUUM

The vacuum interface is usually established by introducing differential pumping devices between different analyzer sections. With the exceptions of the ion trap and rf multipole ion guides, which require a background gas of low molecular weight for best performance [64], all analyzers show better performance at lower pressure. ICR is the most sensitive analyzer towards pressure, where resolution and signal-to-noise of the spectrum depend inversely on pressure [91]. However, ion sources and collision cells for dissociation and reaction usually bring a high gas load and are thus best pumped by separate pumping stages. The analyzer stages are then coupled to the high pressure stages via a differential pumping interface. The purpose of the differential pumping interface is to maximize the ion to neutral ratio in the transport between the two pumping stages. Depending on the type of interface and the pumps used, an increase in the ion to neutral ratio of up to 1000 between two successive stages can be obtained. For very large pumping differentials it is advantageous to use multiple pumping stages between high and low pressure stages. In this way, the ion to neutral ratio of the total pressure differential is the product of the ion to neutral ratios of the individual interfaces between two successive pumping stages [92].

In the molecular flow region, the neutral flow through an aperture is proportional to the area of an orifice or to the cube of the diameter divided by the length of a tubular connection. Ion transmission can be optimized using suitable ion optics and a large kinetic energy for the passage through neutral flow restrictions. A tubular restriction (an "optically open vacuum lock") is only suitable for a high kinetic energy beam.

4.2 KINETIC ENERGY

The kinetic energy interface is made by acceleration or deceleration stages. For many purposes it is advantageous to have high kinetic energies. The main reasons are the decrease in relative energy spread $\Delta E/E$ for obtaining higher resolution, the increase of ion transmission at higher kinetic energies and the higher energy transfer in collisions used for ion fragmentation. However, certain analyzers utilize low energy ions (quadrupoles and traps [92]) and so it is often necessary to accelerate or decelerate the ions for entrance into the next analyzer.

The ultimate goal of the design of both acceleration and deceleration ion optics [92-95] is always to get maximum transmission through a given aperture at the desired ion kinetic energy. Besides desired kinetic energy and aperture size and shape, transmission is determined by the phase space volume occupied by the ions. For an ion beam, the phase space volume is usually characterized by the positional distribution over the beam cross section, the lateral velocity distribution width and the energy distribution width [96]. Taking these parameters into consideration it is possible to accurately model ion beam optics, including acceleration and deceleration ion optics, using simulation programs like SIMION [97].

One example, where a very simple deceleration interface resulted from the preceding simulation was the design of a high transmission injection lens for an rf ion trap [92,98]. The solution that was found was to focus the ion beam at its nominal energy through an orifice placed in front of, but very close to, the ion trap entrance cap and then decelerate the ions over the very short distance between orifice and entrance cap. With such an ion optical element, the ion beam is made strongly diverging, but, due to the short deceleration distance, only after having entered the ion trap. Similar designs later were chosen for injection of a 60 keV ion beam into an ion trap at the isotope separator facility at European Centre for Nuclear Research [99,100].

4.3 BEAM SHAPE

Many interfaces require that the ion beam is reshaped between analyzers, e.g. for optimum passage through collision cells or slits. The shape of the ion beam may be manipulated using dc electric or magnetic multipoles [96]. For higher pressure applications often rf multipole ion guides are used for beam focusing and shaping [54,55].

4.4 FLIGHT DIRECTION

Flight direction seems to be an obvious point. For sector analyzers (E, B, W) entrance and exit points and directions are predefined. But for ion injection into traps or into a TOF analyzer there are different possibilities. For traps [92,93,101,102] entrance and exit may be chosen from and into any direction. For TOF analyzers orthogonal injection [33,103-106] has significant advantages in many cases.

4.5 SYNCHRONIZATION

Synchronization is important for those components of a mass spectrometer which are not continuous. This includes, of course, the synchronization of electrical pulses, rf phases and pulsed ion beams [92]. Also, in a seemingly continuous mass spectrometer there exists a synchronization problem: since the ion beam is never a continuous current but is composed of individual ions, the scan rates of the continuous analyzers have to be slow enough that reasonable statistics are obtained for the ions of interest with lowest abundance.

As an illustrative example, let us consider a mass spectrometer setup consisting of a pulsed ion source, e.g. from laser ionization, a quadrupole analyzer and a detector with a slow analog to digital converter. Without any synchronization, ions will only be efficiently transmitted by the quadrupole, when the quadrupole m/z matches an ion m/z and the laser pulse occurs at the optimum quadrupole phase. Since the detector operates at still another rate, ions will only be detected by chance and each "spectrum" will look very different. A very inefficient way to overcome this problem would be averaging of successive scans.

The most reliably employed technique in synchronization is the derivation of all mass spectrometer frequencies and pulses from a single master clock.

4.6 EFFICIENCY

The efficiency aspect is very important because one needs to have a useful number of detected ions in a given amount of time and from a minimum amount of sample. Numerous efforts have been undertaken in recent years to increase efficiency [34,107]. For a discussion of mass spectrometer efficiency, instruments may be compared using an efficiency factor, defined by the ratio of number of peaks observed in the spectrum to the number of analyte molecules introduced into the mass spectrometer:

$$Efficiency\ factor = \frac{Number\ of\ analyte\ peaks}{Number\ of\ analyte\ molecules}$$

With this definition it is in principal possible that an analyzer has an efficiency of greater than one. This would be the case if the ionization has an efficiency approaching one and the analyzer and detector are capable of gaining information about the molecular weight as well as about fragments from one single ion. Such a situation is imaginable for a highly efficient desorption ionization process combined with nondestructive analysis, and detection (resonance method) combined with remeasurement. Although such mass spectrometers are not yet routine, the potential of analyzer/detection efficiencies of greater than one has already been demonstrated [108,109]. The critical steps for overall efficiency of greater than one remains the ionization process and the interface between ion source and analyzer.

5. Analyzer combinations

In Table 2 most of the analyzer combinations are listed that allow tandem mass spectrometric experiments with MS^n capabilities especially marked. A detailed discussion of the qualities of sector and quadrupole tandem mass spectrometers as well as their applications can be found in Ref. [13]. Excellent discussions of four-sector instruments are found in [51,110], scan modes of BEqQ hybrid instruments in [111], ion trap mass spectrometers in [59], and time-of-flight mass spectrometers in [112]. Instrumentation and applications involving collisions with surfaces is not systematically included in Table 2 and can be found elsewhere [113].

Although not explicitly indicated for all analyzer combinations, the ion trap and ICR trap analyzers always allow MS^n experiments. For these analyzers, the ions can be kept in place and experiments performed in time rather than in space. New capabilities of these analyzers are therefore achieved through new electrode geometries and voltages or new scan modes [114-122]. A particularly interesting finding for high mass structure analysis was the discovery that sustained off-resonance irradiation (SORI) of ions during multiple collisions yields infrared photodissociation like fragmentation [123-125]. Using that technique overcomes the necessity of large collision energies needed for single collision fragmentation. Some further capability enhancements for ICR traps, where actually no additional (external) analyzer was added, are listed in Table 2 as "Combined Analyzers".

	Combination	Purpose	Reference
2 Sectors:			
1	E+B	Resolution, Sensitivity	[1,5]
2	W+B	Resolution	[126]
3	B+E	MS^2, Resolution	[28,127-129]
4	E+Q	MS^2, Kinetic Energy Release	[130]
5	B+B	MS^2	[131]
6	B+Q	MS^2	[132]
7	Q+Q	MS^2	[133]
8	q+Q	Prefilter, Sensitivity	[134,135]
9	Trap+Q	Reaction	[136]
10	Q+Trap	Prefilter	[137]
11	q+Trap	Prefilter, Ion Guide	[138,139]
12	W+ICR	Prefilter, IonGuide	[101]
13	q+ICR	Prefilter, Ion Guide	[140-143]
14	TOF+ICR	Prefilter	[144-147]
15	Trap+ICR	Prefilter, Sensitivity	[148]
16	ICR+ICR	Prefilter, Sensitivity	[148-150]
17	B+TOF	MS^2	[151]
18	E+TOF	MS^2	[152]
19	W+TOF	Low Resolution MS^2	[153]
20	Q+TOF	MS^2, Sensitivity	[103,154,155]
21	q+TOF	Beam Shape, Sensitivity	[103]
22	Trap+TOF	Reaction, Sensitivity	[156-167]
23	ICR+TOF	Resolution	[168-170]
24	TOF+TOF	MS^2, Resolution	[171-177]

Table 2. Analyzer combinations. Analyzer symbols: **B** magnetic sector, **E**: electric sector, **W**: Wien filter, **Q**: quadrupole, **q**: rf multipole, **TOF**: time-of-flight, **Trap**: ion trap, **ICR**: ICR trap

	Combination	Purpose	Reference
3 Sectors:			
25	E+B+E B+E+E	MS^3, Resolution	[178-184]
26	B+E+B	MS^3, Resolution	[13,185,186]
27	Q+q+Q	MS^2, Reaction	[187-191]
28	Q+Trap+Q	MS^2, Reaction	[192-194]
29	Q+E+B	MS^3, Resolution	[195]
30	B+q+Q	MS^2	[132,196]
31	E+B+Trap	MS^n, Resolution, Sensitivity	[197-199]
32	E+B+TOF	MS^2, Resolution, Sensitivity	[200-202]
4 Sectors:			
33	B+E+q+Q	MS^3, Resolution	[203-205]
34	E+B+q+Q	MS^2, Resolution	[206]
35	E+B+Trap+Q	MS^2, Resolution	[207]
36	B+E+Trap+Q	MS^3, Resolution	[208]
37	E+B+E+B	MS^3, Resolution	[51,209-212]
38	B+E+E+B	up to MS^5, Resolution	[213]
5 and more Sectors:			
39	Q+q+Q+q+Q	MS^3, Reaction	[214-216]
40	E+B+E+E+B+E	MS^n, Resolution	[217]
41	E+B+E+Q+E+B+E	MS^n, Resolution	[218]
Combined Analyzers:			
42	ICR/q	High Mass	[109,219-224]
43	ICR/Trap	High Pressure	[225-227]
44	ICR/TOF	Kinetic Energy Analysis	[228,229]

Table 2. Analyzer combinations (continued). Analyzer symbols: **B** magnetic sector, **E**: electric sector, **W**: Wien filter, **Q**: quadrupole, **q**: rf multipole, **TOF**: time-of-flight, **Trap**: ion trap, **ICR**: ICR trap

6. Performance of some modern instruments with combined analyzers

6.1 A QUADRUPOLE, TIME-OF-FLIGHT COMBINATION

If a large molecule such as a peptide is ionized by electrospray, the generated ion is "cooled down" by the evaporation effect, and the excess internal energy - if any remains in the ion - is usually rather small. In addition, a large ion has a huge number of degrees of freedom, thus enabling it to distribute the internal energy in small portions. Therefore very often it is rather difficult to fragment such a large ion, if only a few collisions with low mass partners occur in a high vacuum collision chamber. This is especially true in the case of sector instruments like EB or BE combinations.

The problem may be overcome by using long collision cells working at higher pressures and with higher collision energies (800 eV range), thus inducing multiple collisions. By this method excellent fragmentation of peptides can be achieved, and amino acid sequencing may be done with extremely low sample amounts (attomole range).

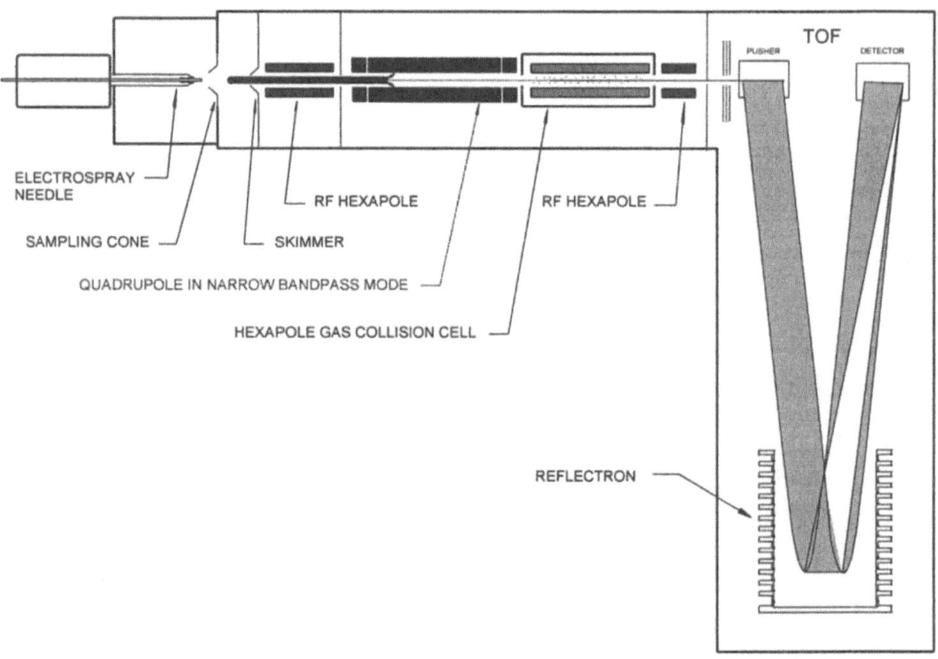

Figure 2. Quadrupole, TOF analyzer combination (courtesy of VG Micromass).

An example of such a dedicated instrument is shown in Fig. 2 [103]. Even in this instrument the analyzer combination is rather complex. After electrospray ionization the ions are collected and focused by a first rf hexapole stage, and introduced into a "normally" operated quadrupole mass filter. One ion type of interest may be selected by this mass filter stage and then introduced into a long rf hexapole collision cell, where the ions fragment. The fragment ions are then collected by a second rf hexapole and focused into a orthogonal reflectron TOF analyzer. This design has the advantage of simultaneously recording all the fragment ions generated in the collision cell, thus enhancing the duty cycle of the instrument. The result is an excellent MS/MS sensitivity, which is shown in the example of Fig. 3.

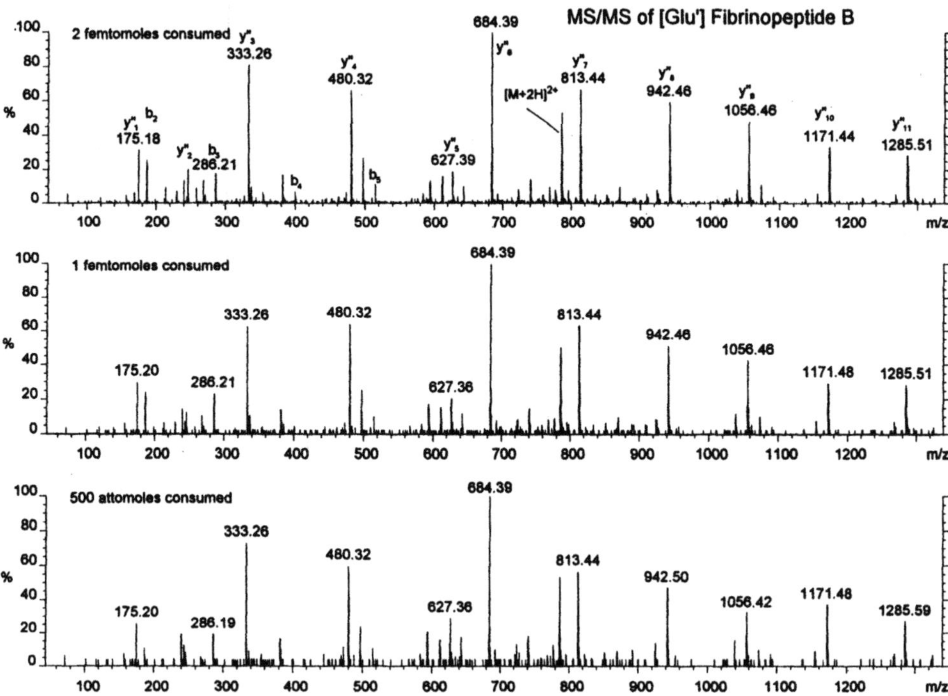

Figure 3. Fragmentation pattern of [Glu']-Fibrinopeptide B recorded by the quadrupole-TOF instrument using a collision energy of ca 800 eV (courtesy of VG Micromass).

6.2 A DOUBLE FOCUSING SECTOR INSTRUMENT WITH AN ION TRAP AT THE BACK END (EB-TRAP)

Many combinations of sector instruments (EB or BE type) with additional sectors or quadrupole mass filters have been realised. Most of those instruments have been designed to do MS^n operations with n = 2 only. The instruments suffer from low sensitivity, particularly at higher mass resolutions of the sector mass spectrometer. In order to overcome these drawbacks, several groups have added an ion trap at the back end of a BE type sector mass spectrometer [197,207,208]. This arrangement potentially offers the possibility to perform multiple MS/MS experiments (n > 2). Additionally it can be expected that collecting ions leaving the sector instrument would bring an enhancement of the sensitivity due to the longer duty cycle of the instrument.

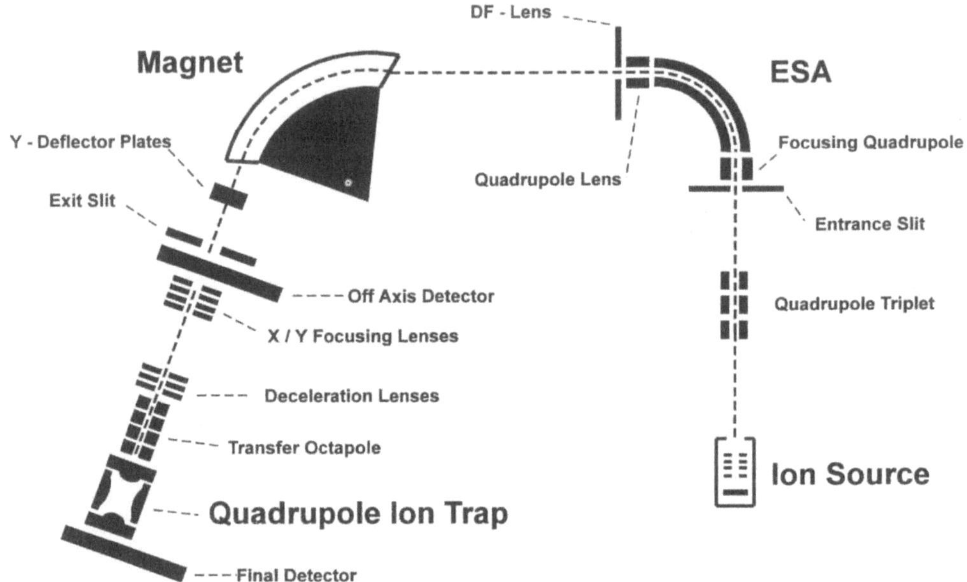

Figure 4. Schematic of the double focusing sector instrument with an ion trap at the back end (Finnigan MAT 900ST).

Such an instrument is now commercially available (Finnigan MAT 900ST, Fig. 4). Ions leaving the ion source of this instrument are collected and focused to the EB entrance slit by a quadrupole triplet. Two more quadrupole systems form the ion beam in front and at the back end, respectively, of the electrostatic analyzer (ESA). After passing the magnetic sector, the ion beam is again reshaped and leaves the sector instrument via the exit slit. At this point an off axis detector system offers the possibility to run the whole system just as a conventional double focusing mass spectrometer.

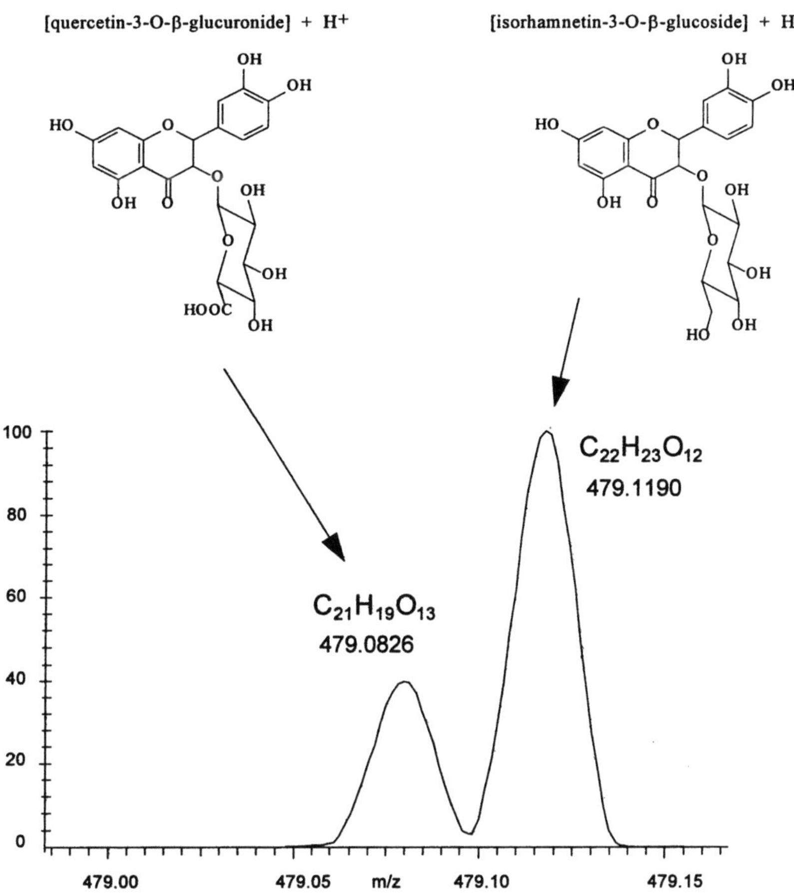

Figure 5. High resolution mass spectrum of mass 479 of the protonated Isorhamnetin-3-O-β-glucoside and Quercetin-3-O-β-glucuronide at a resolution of 14'000. The two peaks are clearly separated (courtesy of Finnigan MAT).

If instead the ions are to be collected and stored in the trap for further fragmentation reactions, the ion beam is again reshaped and focused. Then the ions are decelerated and transferred by an octupole system into the ion trap. Due to the accumulation of the ions in the trap, the sensitivity of the instrument is about 10 times better than on a corresponding instrument with a quadrupole mass filter instead of the trap [198,199].

The performance of this instrument is demonstrated by the analysis of an extract of flavonoid glucosides from Arnica plants. This extract contained Isorhamnetin-3-O-β-glucoside and Quercetin-3-O-β-glucuronide, two very polar and fragile compounds with almost the same molecular weight (Fig. 5). This mixture was ionized by electrospray, yielding the protonated species, and then separated by the double focusing sector instrument, followed by multiple MS/MS experiments within the ion trap. In Fig. 6 the MS^2 mass spectra of the two isobaric ions are shown. They reveal the loss of the glucose and of the glucuronide moieties, respectively.

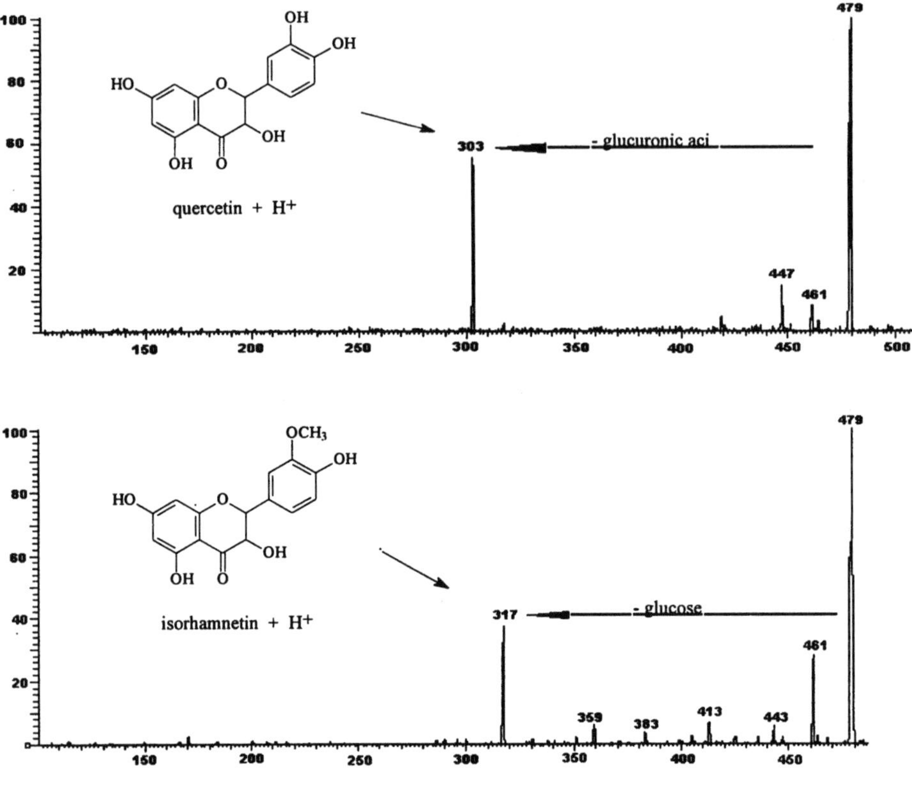

Figure 6. MS^2 mass spectra of the two isobaric ions generated in the ion trap (courtesy of Finnigan MAT).

6.3 A TIME OF FLIGHT ANALYZER COUPLED TO A HIGH RESOLUTION SECTOR INSTRUMENT

The trend of duty cycle enhancement may be seen as well in this commercially available instrument (VG Micromass AutoSpec-TOF [104]). The ion beam of this mass spectrometer (EBE-TOF geometry, see Fig. 7) passes a first electrostatic analyzer, a beam diverter system, the magnetic sector, a second beam diverter system and a second electrostatic analyzer. The beam diverter allows for two possibilities. In the first case the ion beam passes the magnetic sector and is therefore being mass analyzed. In the second case the diverters represent a magnet by-pass arrangement, which may focus the total primary ion beam - now only energy analyzed by the two electrostatic analyzers - directly into the second compartment of the instrument. That part consists of a collision cell, an orthogonal acceleration time-of-flight analyser and an in-line electron multiplier detector system located at the back end of the instrument. The orthogonal TOF analyser itself consists of a high voltage pulse beam deflector, a drift length of 0.5 m orthogonal to the sector ion beam and a micro-channel plate detector. The ions may be recorded directly by the in-line detector or by the orthogonal TOF mass spectrometer, respectively. Of course, the ions may be detected directly or after prior fragmentation in the collision cell.

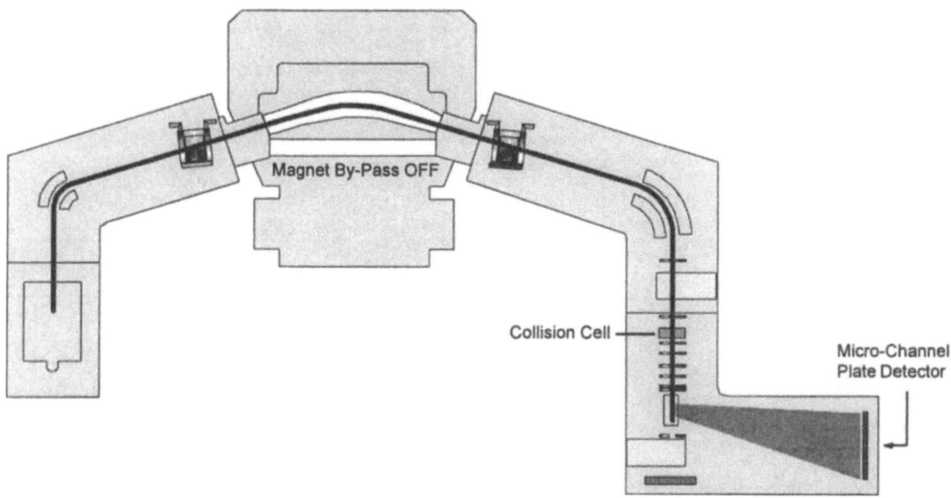

Figure 7. VG Micromass AutoSpec-TOF with magnet by-pass. In the drawing the by-pass system is switched off, thus focusing the ion beam into the magnetic analyzer (courtesy of VG Micromass).

The time-of-flight system offers the simultaneous recording of all ion types present in the moment when the deflecting pulse is driving them onto the micro-channel plate detector. This results in a significant enhancement of the duty cycle of the mass spectrometer, yielding better overall sensitivity. The excellent sensitivity is demonstrated in Fig. 8.

In addition, due to the double focusing EBE instrument accurate mass measurements may be made on this instrument by scanning over a restricted mass range covering the mass of interest and appropriate reference masses. In the case of the [Glu]Fibrinopeptide B shown in Fig, 8, its $(M+2H)^{2+}$ ion at m/z 785 was measured and its exact mass determined [230]. As reference masses the ions of polyethylene glycol at m/z 745.4198 and m/z 789.4460 (PEG + Na^+), respectively, were chosen. Ten individual scans using a mass resolution of 6,300 (10% valley) were acquired and the spectra compared. For the two most abundant isotope peaks the mean m/z values of the 10 scans differed by less than 1 mDa (1 ppm) from the exact m/z values. The RMS

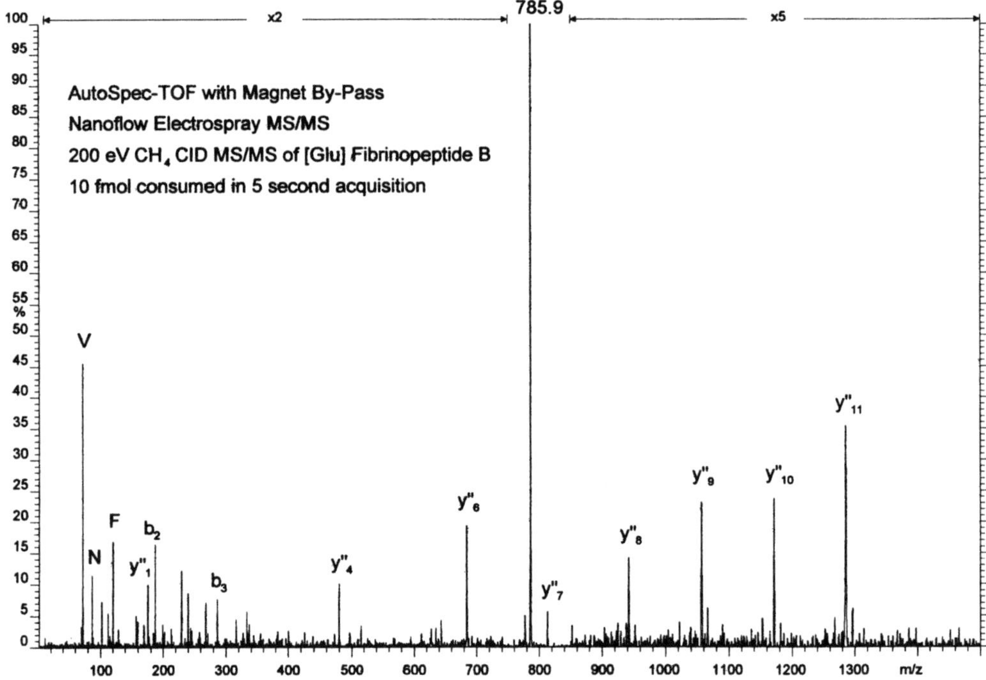

Figure 8. Collision induced MS^2 spectrum of 10 fmol [Glu]Fibrinopeptide B, ionized by nanoflow electrospray. Acquisition time 5 seconds, magnet by-pass off, molecular ion mass analyzed by the magnetic analyzer. Due to the optimized collision energy practically sequence characteristic ions only are generated (courtesy of VG Micromass).

error of the individual measurements was 3.5 mDa (4.5 ppm). Interestingly enough, the sample consumption for each scan was only ~ 10 fmol, demonstrating the excellent sensitivity of this instrumental arrangement.

7. Conclusions

Milestones in the development of analyzer combinations were the double focusing mass spectrometer, classical tandem BE MS/MS instrument, four-sector instrument, "hybrid" BEqQ instrument, combinations with ICR and, more recently, ion trap and time-of-flight analyzers. The trend of the last five years has been clearly towards relatively inexpensive and small instruments with high sensitivity and high mass capabilities, but for more sophisticated applications the high power multiple sector and FT-ICR instruments are still needed. In these instruments the classical magnet analyzer remains the most substantial part.

8. References

1. R. Herzog and J. H. E. Mattauch, *Ann. Phys. (Leipzig)* **19,** 345 (1934).
2. J. H. Beynon, in *Advances in Mass Spectrometry, vol. 2,* ed. by R. M. Elliott. Pergamon, London (1962).
3. K. R. Jennings, *Int. J. Mass Spectrom. Ion Phys.* **1,** 227 (1968).
4. A. O. Nier and T.R. Roberts, *Phys. Rev.* **81,** 507 (1951).
5. W. H. Johnson and A. O. Nier, *Phys. Rev.* **105,** 1014 (1957).
6. A. Maquestiau and R. Flammang, in *Tandem Mass Spectrom., ed. by F. W. McLafferty, p. 401.* Wiley, New York, N.Y. (1983).
7. A. L. Burlingame, R. K. Boyd and S. J. Gaskell, *Anal. Chem.* **68,** 599R (1996).
8. K. R. Jennings and D. Despeyroux, *Adv. Mass Spectrom.* **13,** 3 (1995).
9. C. P. Ball and P. J. Derrick, *Annu. Rep. Prog. Chem., Sect. B: Org. Chem.* **90,** 21 (1994).
10. A. M. Falick, in *Appl. Mass Spectrom. Org. Stereochem.,* ed. by J. S. Splitter and F. Turecek, p. 17. VCH, New York, N. Y. (1994).
11. K. Biemann, in *Methods in Enzymology, vol. 193: Mass Spectrometry,* ed. by J. A. McCloskey, p. 455. Academic Press, New York (1990).
12. E. de Hoffmann, *J. Mass Spectrom.* **31,** 129 (1996).
13. K. L. Busch, G. L. Glish and S.A. McLuckey, *Mass Spectrometry/Mass Spectrometry: Techniques and Applications of Tandem Mass Spectrometry,* VCH, Weinheim, Germany (1988).
14. T. Ii, M. Kubota, T. Hirano, M. Ohashi, K. Yoshida and S. Suzuki, *Glycoconjugate J.* **12,** 282 (1995).
15. V. N. Reinhold, B. B. Reinhold and C. E. Costello, *Anal. Chem.* **67,** 1772 (1995).
16. W. Melaugh, J. J. Engstrom, S. Auriola, N. J. Phillips and B. W. Gibson, in *Mass Spectrom. Biol. Sci., ed. by A. L. Burlingame and S. A. Carr, p. 403.* Humana, Totowa, N. J. (1996).
17. A. L. Burlingame, *Curr. Opin. Biotechnol.* **7,** 4 (1996).
18. C. L. Wilkins and S. Pastor, *Polym. Prepr.* **37,** 284 (1996).
19. M. Mann and G. Talbo, *Curr. Opin. Biotechnol.* **7,** 11 (1996).
20. P. Lecchi and R. M. Caprioli, in *New Methods Pept. Mapp. Charact. Proteins, ed. by W. S. Hancock, p. 219.* CRC, Boca Raton, Fla., p. 219 (1996).
21. A. Miranker, C. V. Robinson, S. E. Radford and C. M. Dobson, *FASEB J.* **10,** 93 (1996).
22. R. C. Beavis and B. T. Chait, U.S. Patent 5,288,644 (1994).
23. S. A. Hofstadler, J. C. Severs, R. D. Smith, F. D. Swanek and A. G. Ewing, *Rapid Commun. Mass Spectrom.* **10,** 919 (1996).
24. A. Triolo and F. Bonelli, *Rapid Commun. Mass Spectrom.* **9,** 707 (1995).
25. K. R. Wehmeyer, P. M. Knight and R. C. Parry, *J. Chromatogr., B: Biomed. Appl.* **676,** 53 (1996).
26. R. Kaufmann, *J. Biotechnol.* **41,** 155 (1995).
27. J. Kissel and F. R. Krueger, *Adv. Space Res.* **15,** 59 (1994).
28. R. K. Boyd, *Mass Spectrom. Rev.* **13,** 359 (1994).
29. K. R. Jennings and G. G. Dolnikowski, in *Methods in Enzymology, vol. 193: Mass Spectrometry,* ed. by J. A. McCloskey, p. 37. Academic Press, New York (1990).
30. T. Matsuo et al. (eds), *Biological Mass Spectrometry: Present and Future,* Wiley, Chichester (1994).
31. M. L. Gross (ed), *Mass Spectrometry in the Biological Sciences: A Tutorial, NATO ASI Series vol. 353,* Kluwer, Dordrecht, The Netherlands (1992).

32. R. Grix, U. Grüner, G. Li, H. Stroh and H. Wollnik, *Int. J. Mass Spectrom. Ion Processes* **93**, 323 (1989).
33. J. Coles and M. Guilhaus, *Trends Anal. Chem.* **12**, 203 (1993).
34. S. Evans, in *Methods in Enzymology, vol. 193: Mass Spectrometry*, ed. by J. A. McCloskey, p. 61. Academic Press, New York (1990).
35. R. Weber, K. Levsen, G. J. Louter and A. J. H. Boerboom, *Anal. Chem.* **54**, 1458 (1982).
36. R. T. McIver Jr. and W. D. Bowers, in *Tandem Mass Spectrom., ed. by F. W. McLafferty, p. 287.* Wiley, New York, N.Y. (1983).
37. F. Vedel and G. Werth, in *Pract. Aspects Ion Trap Mass Spectrom.*, Volume 2, ed. by R. E. March and J. F. J. Todd, p. 237. CRC, Boca Raton, Fla. (1995).
38. G. Rettinghaus, *Z. angew. Phys.* **22**, 321 (1967).
39. F. W. McLafferty, S. C. Beu, J. P. Quinn, M. W. Senko, Y. Shi, D. Suckau, F. M. Wampler III and M.-Y- Zhang, in *Biological Mass Spectrometry: Present and Future*, ed by T. Matsuo et al, p. 199. Wiley, Chichester (1994).
40. H. Matsuda, *Int. J. Mass Spectrom. Ion Processes* **100**, 31 (1990).
41. T. Matsuo, *Mass Spectrom. Rev.* **8**, 203 (1989).
42. J. H. Beynon, J. W. Amy and W. E. Baitinger, *J. Chem. Soc. D* **(13)**, 723 (1969).
43. C. B. Freidhoff and R. M. Young, U.S. Patent 5,536,939 (Cont.-in-part of U.S. 5,386,115) (1996).
44. C. B. Freidhoff, R. M. Young and S. Sriram, PCT Int. Appl., WO 9616430 (1996).
45. C. B. Freidhoff and R. M. Young, PCT Int. Appl., WO 9611492 (1996).
46. C. B. Freidhoff, R. M. Young and S. Sriram, U.S. Patent 5,386,115 (1995).
47. T. Sakurai, M. Toyoda, H. Hayashibara and T. Matsuo, *Int. J. Mass Spectrom. Ion Processes* **146**/147 (1995).
48. M. J. Wouters, *J. Phys. D: Appl. Phys.* **28**, 1547 (1995).
49. K.-H. Homann and W. Wiese, in *Appl. Free-Jet, Mol. Beam, Mass Spectrom. Sampling, Proc.*, ed. by T. Milne, p. 146. National Renewable Energy Laboratory, Golden, Colo. (1994).
50. M. Tanjyo and H. Nakazato, Eur. Pat. Appl., EP 525927 (1993).
51. M. L. Gross, in *Methods in Enzymology, vol. 193: Mass Spectrometry*, ed. by J. A. McCloskey, p. 131. Academic Press, New York (1990).
52. R. E. Finnigan, *Anal. Chem.* **66**, 969A (1994).
53. W. Paul and H. Steinwedel, *Z. Naturforsch.* **8a**, 448 (1953).
54. C. M. Whitehouse and E. Gulcicek, PCT Int. Appl., WO 9523018 (1995).
55. D. J. Douglas, U.S. Patent 5,179,278 (1993).
56. A. E. Schoen and J. E. P. Syka, U.S. Patent 5,089,703 (1992).
57. J. W. Hurst, R. M. Smith and J. M. Horth U.S. Patent 4,990,777 (1991).
58. R. I. Martinez, *Rapid Commun. Mass Spectrom.* **5**, 245 (1991).
59. R. E. March and J. F. J. Todd (eds.), in *Pract. Aspects Ion Trap Mass Spectrom.*, Volumes 1-3, CRC, Boca Raton, Fla. (1995).
60. R. E. March and F. A. Londry, in *Pract. Aspects Ion Trap Mass Spectrom.*, Volume 1, ed. by R. E. March and J. F. J. Todd, p. 25. CRC, Boca Raton, Fla. (1995).
61. J. D. Williams, K. A. Cox, J. C. Schwartz and R. G. Cooks, in *Pract. Aspects Ion Trap Mass Spectrom.*, Volume 2, ed. by R. E. March and J. F. J. Todd, p. 3. CRC, Boca Raton, Fla. (1995).
62. R. G. Cooks and K. A. Cox, in *Biological Mass Spectrometry: Present and Future*, ed. by T. Matsuo et al., p. 179. Wiley, Chichester (1994).
63. J. Franzen, Eur. Pat. Appl., EP 459,602 (1991).

64. G. C. Stafford, Jr., P. E. Kelley, J. E. P. Syka, W. E. Reynolds and J. F. J. Todd, *Int. J. Mass Spectrom. Ion Processes* **60**, 85 (1984).
65. R. D. Smith, J. E. Bruce, Q. Wu, X. Cheng, S. A. Hofstadler, G. A. Anderson, R. Chen, R. Bakhtiar and S. O. Van Orden et al., in *Mass Spectrom. Biol. Sci.*, ed. by A. L. Burlingame and S. A. Carr, p. 25. Humana, Totowa, N. J. (1996).
66. C. L. Holliman, D. L. Rempel and M. L. Gross, *NATO ASI Ser., Ser. C* **475**, 147 (1996).
67. F. W. McLafferty, M. W. Senko, D. P. Little, T. D. Wood, P. B. O'Connor, J. P. Speir, R. A. Chorush and N. L. Kelleher, *Adv. Mass Spectrom.* **13**, 115 (1995).
68. Y. Takigawa, *Reza Kenkyu* **23**, 640 (1995).
69. R. Mazurkiewicz, *Wiad. Chem.* **49**, 265 (1995).
70. J. A. Castoro and C. L. Wilkins *Trends Anal. Chem.* **13**, 229 (1994).
71. F. W. McLafferty, *Acc. Chem. Res.* **27**, 379 (1994).
72. R. L. Hettich and C. Jin *Springer Ser. Mater. Sci.* **28**, 135 (1994).
73. N. M. Nibbering, *Trends Anal. Chem.* **13**, 223 (1994).
74. E. R. Williams, *Trends Anal. Chem.* **13**, 247 (1994).
75. C. L. Holliman, D. L. Rempel and M. L. Gross, *Mass Spectrom. Rev.* **13**, 105 (1994).
76. S. Xu, D. Li, G. Chen, L. Shen and M. Chen, *Hewuli Dongtai* **11**, 42 (1994).
77. T. L. Spell, S. E. DeLong and W. R. Creasy, *Int. J. Mass Spectrom. Ion Processes* **124**, 223 (1993).
78. M. B. Comisarow, *Hyperfine Interact.* **81**, 171 (1993).
79. M. V. Buchanan and R.L. Hettich, *Anal. Chem.* **65**, 245A (1993).
80. M. B. Comisarow and A. G. Marshall, *Chem. Phys. Lett.* **25**, 282 (1974).
81. W. C. Wiley and I. H. McLaren, *Rev Sci. Instrum.* **26**, 1150 (1955).
82. M. Guilhaus, *Adv. Mass Spectrom.* **13**, 213 (1995).
83. M. Guilhaus, *J. Mass Spectrom.* **30**, 1519 (1995).
84. K. O. Bornsen, *Anal. Methods Instrum.* **2**, 202 (1995).
85. M. C. Fitzgerald and L. M. Smith, *Annu. Rev. Biophys. Biomol. Struct.* **24**, 117 (1995).
86. H. Wollnik, *Int. J. Mass Spectrom. Ion Processes* **131**, 387 (1994).
87. A. Woods, R. Wang, M. Chevrier, T. Cornish, C. Wolkow and R. J. Cotter, *Top. Mass Spectrom.* **1**, 199 (1994).
88. R. J. Cotter, in *Time-of-Flight Mass Spectrometry, ACS Symposium Series 549*, ed. by R. J. Cotter, p. 16. American Chemical Society, Washington, DC (1994).
89. D. Price, *ACS Symp. Ser.* **549**, 1 (1994).
90. H. Wollnik, *Mass Spectrom. Rev.* **12**, 89 (1993).
91. C. B. Jacoby, C. L. Holliman and M. L. Gross, in *Mass Spectrometry in the Biological Sciences: A Tutorial, ed. by M. L. Gross, p. 93. NATO ASI Series vol. 353*, Kluwer, Dordrecht, The Netherlands (1992).
92. P. Kofel, in *Pract. Aspects Ion Trap Mass Spectrom., Volume 2*, ed. by R. E. March and J. F. J. Todd, p. 51. CRC, Boca Raton, Fla. (1995).
93. P. Kofel and T. B. McMahon, *Int. J. Mass Spectrom. Ion Processes* **98**, 1 (1990).
94. M. J.-F. Suter, R. M. Stepnowski and U. P. Schlunegger, *Rapid Commun. Mass Spectrom.* **3**, 417 (1989).
95. M. L. Vestal, C. R. Blakley, P. W. Ryan and J. H. Futrell, *Rev. Sci. Instrum.* **47**, 15 (1976).
96. H. Wollnik, *Optics of Charged Particles*, Academic Press, Orlando, Fla (1987).
97. D. A. Dahl, *SIMION 3D, Version 6.0*, Princeton Electronic Systems, Princeton, NJ (1995).

98. P. Kofel, H. Reinhard and U. P. Schlunegger, *Proc. 38th ASMS Conf. Mass Spectrometry and Allied Topics, Tucson, AZ*, p. 886 (1990).

99. G. Rouleau, PhD thesis, McGill University, Montreal, Canada (1992).

100. R. B. Moore and M. D. Lunney, in *Pract. Aspects Ion Trap Mass Spectrom.*, Volume 2, ed. by R. E. March and J. F. J. Todd, p 263. CRC, Boca Raton, Fla. (1995).

101. S. Guan and A. G. Marshall, *Rev. Sci. Instrum.* **67**, 423 (1996).

102. R. T. McIver, Jr., *Int. J. Mass Spectrom. Ion Processes* **98**, 35 (1990).

103. H. R. Morris, T. Paxton, A. Dell, J. Langhorne, M. Berg, R. S. Bordoli, J. Hoyes and R. H. Bateman, *Rapid Commun. Mass Spectrom.* **10**, 889 (1996).

104. E. Clayton and R. H. Bateman, *Rapid Commun. Mass Spectrom.* **6**, 719 (1992).

105. J. H. J. Dawson and M. Guilhaus, *Rapid Commun. Mass Spectrom.* **3**, 155 (1989).

106. A. F. Dodonov, I. V. Chernushevich and V. V. Laiko, in *Time-of-Flight Mass Spectrometry, ed. by R. J. Cotter, p. 108. ACS Symposium Series 549*, American Chemical Society, Washington, DC, (1994).

107. A. L. Burlingame, in *Biological Mass Spectrometry: Present and Future*, ed. by T. Matsuo et al., p. 147. Wiley, Chichester (1994).

108. E. R. Williams, K. D. Henry and F. W. McLafferty, *J. Am. Chem. Soc.* **112**, 6157 (1990).

109. J. P. Speir, G. S. Gorman, C. C. Pitsenberger, C. A. Turner, P. P. Wang and I. J. Amster, *Anal. Chem.* **65**, 1746 (1993).

110. J. R. Trainor and P. J. Derrick, in *Mass Spectrometry in the Biological Sciences: A Tutorial, ed. by M. L. Gross, p. 3. NATO ASI Series vol. 353*, Kluwer, Dordrecht, The Netherlands (1992).

111. S. J. Gaskell and K. D. Ballard, in *Mass Spectrometry in the Biological Sciences: A Tutorial*, ed. by M. L. Gross, p. 29. *NATO ASI Series vol. 353*, Kluwer, Dordrecht, The Netherlands (1992).

112. R. J. Cotter, (ed.), *Time-of-Flight Mass Spectrometry, ACS Symposium Series 549*, American Chemical Society, Washington, DC (1994).

113. R. G. Cooks, T. Ast and M. A. Mabud, *Int. J. Mass Spectrom. Ion Processes* **100**, 209 (1990).

114. V. M. Doroshenko and R. J. Cotter, *Anal. Chem.* **68**, 463 (1996).

115. Y. Wang, *Rapid Commun. Mass Spectrom.* **7**, 920 (1993).

116. J. Franzen, *Int. J. Mass Spectrom. Ion Processes* **125**, 165 (1993).

117. V. H. Vartanian, J. S. Anderson and D. A. Laude, *Mass Spectrom. Rev.* **14**, 1 (1995).

118. S. Guan and A. G. Marshall, *Int. J. Mass Spectrom. Ion Processes* **146/147**, 261 (1995).

119. D. A. J. Laude and S. C. Beu, *Top. Mass Spectrom.* **1**, 153 (1994).

120. J. S. Anderson, H. Vartanian and D. A. Laude, *Trends Anal. Chem.* **13**, 234 (1994).

121. S. Guan, H. S. Kim, A. G. Marshall, M. C. Wahl, T. D. Wood and X. Xiang, *Chem. Rev.* **94**, 2161 (1994).

122. T. Gäumann, *Top. Mass Spectrom.* **1**, 113 (1994).

123. J. W. Gauthier, T. R. Trautman and D. B. Jacobson, *Anal. Chim. Acta* **246**, 211 (1991).

124. K. A. Boering, J. Rolfe and J. I. Brauman, *Int. J. Mass Spectrom. Ion Processes* **117**, 357 (1992).

125. V. C. M. Dale, J. P. Speir, G. H. Kruppa, C. C. Stacey, M. Mann and M. Wilm, *Biochem. Soc. Trans.* **24**, 943 (1996).

126. D. Ioanovicin and C. Cuna, *Rapid Commun. Mass Spectrom.* **9**, 512 (1995).

127. J. H. Beynon, R. M. Caprioli and T. Ast, *Org. Mass Spectrom.* **5**, 229 (1971).

128. J. H. Beynon, R. G. Cooks, J. W. Amy, W. E. Baitinger and T. Y. Ridley, *Anal. Chem.* **45**, 1023A (1973).

129. P. A. D'Agostino, J. R. Hancock, L. R. Provost, P. D. Semchuk and R. S. Hodges, *Rapid Commun. Mass Spectrom.* **9**, 597 (1995).
130. F. M. Harris, G. A. Keenan, P. D. Bolton, S. B. Davies, S. Singh and J. H. Beynon, *Int. J. Mass Spectrom. Ion Processes* **58**, 273 (1984).
131. G. J. Louter, A. J. H. Boerboom, P. F. M. Stalmeier, H. H. Tuithof and J. Kistemaker, *Int. J. Mass Spectrom. Ion Phys.* **33**, 335 (1980).
132. S. A. McLuckey, G. L. Glish and R. G. Cooks, *Int. J. Mass Spectrom. Ion Phys.* **39**, 219 (1981).
133. T.-Y. Yu, T. M. H. Cheng, V. Kempter and F. W. Lampe, *J. Phys. Chem.* **76**, 3321 (1972).
134. S. Hiroki, K. Sakata, N. Sugiyama, S. Muramoto, T. Abe and Y. Murakami, *Vacuum* **46**, 681 (1995).
135. M. Simon, S. Bohatka, C. Trajber and I. Futo, *Rapid Commun. Mass Spectrom.* **9**, 629 (1995).
136. R. F. Bonner, R. E. March and J. Durup, *Int. J. Mass Spectrom. Ion Phys.* **22**, 17 (1976).
137. K. L. Morand, S. R. Horning and R. G. Cooks, *Int. J. Mass Spectrom. Ion Processes* **105**, 13 (1991).
138. A. P. Land, M. Sanders and J. Schwartz, *44th ASMS Conf Mass Spectrom. Allied Topics, Portland, OR*, poster MPN 305 (1996).
139. S. T. Quarmby, R. A. Yost, J. C. Schwartz and J. E. P. Syka, *44th ASMS Conf Mass Spectrom. Allied Topics, Portland, OR*, poster WOD 4:30 (1996).
140. D. F. Hunt, J. Shabanowitz, J. R. I. Yates, P. R. Griffin and N. Z. Zhu *Anal. Chim. Acta* **225**, 1 (1989).
141. R. T. McIver, Jr., R. L. Hunter and W. D. Bowers, *Int. J. Mass Spectrom. Ion Processes* **64**, 67 (1985).
142. D. F. Hunt, J. Shabanowitz, R. T. McIver, Jr., R. L. Hunter and J. E. P. Syka, *Anal. Chem.* **57**, 765 (1985).
143. Y. Huang, S. Guan, H. S. Kim and A. G. Marshall, *Int. J. Mass Spectrom. Ion Processes* **152**, 121 (1996).
144. J. M. Alford, P. E. Williams, D. J. Trevor and R. E. Smalley, *Int. J. Mass Spectrom. Ion Processes* **72**, 33 (1986).
145. P. Kofel, M. Allemann, Hp. Kellerhals and K.-P. Wanczek, *Int. J. Mass Spectrom. Ion Processes* **72**, 53 (1986).
146. Y. Li and R. T. McIver, Jr., *Rapid Commun. Mass Spectrom.* **8**, 743 (1994).
147. L. Van Vaeck, W. Van Roy, H. Struyf, F. Adams and P. Caravatti, *Rapid Commun. Mass Spectrom.* **7**, 323 (1993).
148. P. Kofel, M. Allemann, Hp. Kellerhals and K.-P. Wanczek, *Int. J. Mass Spectrom. Ion Processes* **87**, 237 (1989).
149. R. B. Cody, J. A. Kinsinger, S. Ghaderi, I. J. Amster, F. W. McLafferty and C. E. Brown, *Anal. Chim. Acta* **178**, 43 (1985).
150. S. A. Hofstadler and D. A. J. Laude, *Anal. Chem.* **63**, 2001 (1991).
151. J. T. Stults, C. G. Enke and J. F. Holland, *Anal. Chem.* **55**, 1323 (1983).
152. J. D. Pinkston, M. Rabb, J. T. Watson and J. Allison *Rev. Sci. Instrum.* **57**, 583 (1986).
153. K. Schey, R. G. Cooks, R. Grix and H. Wollnik, *Int. J. Mass Spectrom. Ion Processes* **77**, 49 (1987).
154. G. L. Glish and D. E. Goeringer, *Anal. Chem.* **56**, 2291 (1984).
155. G. L. Glish, S. A. McLuckey and H. S. McKown, *Anal. Instrum.* **16**, 191 (1987).
156. P. Bisling, H. J. Heger, W. Michaelis, C. Weitkamp and H. Zobel, *AIP Conf. Proc.* **329**, 511 (1995).

157. P. Kofel, M. Stoeckli, J. Krause and U. P. Schlunegger, *Rapid Commun. Mass Spectrom.* **10**, 658 (1996).
158. U. Wilhelm, C. Weickhardt and J. Grotemeyer, *Rapid Commun. Mass Spectrom.* **10**, 473 (1996).
159. M. G. Qian, K. Zheng, Y. Chen, C. L. Chang, S. M. Hanash and D. M. Lubman, *Rapid Commun. Mass Spectrom.* **10**, 1079 (1996).
160. R. W. Purves and L. Li, *J. Microcolumn Sep.* **7**, 603 (1995).
161. N. Zhang, Y. Matsuo and M. Takami, *Chem. Phys. Lett.* **244**, 133 (1995).
162. C. Weickhardt and C. Lifshitz, *Eur. Mass Spectrom.* **1**, 223 (1995).
163. M. G. Qian and D. M. Lubman *Anal. Chem.* **67**, 234A (1995).
164. T. L. Grebner and H. J. Neusser, *Int. J. Mass Spectrom. Ion Processes* **137**, L1 (1994).
165. Y. Zerega, J. Andre, G. Brincourt and R. Catella, *Int. J. Mass Spectrom. Ion Processes* **135**, 155 (1994).
166. Y. Zerega, G. Brincourt, J. Andre and R. Catella, *Int. J. Mass Spectrom. Ion Processes* **121**, 77 (1992).
167. S. M. Michael, M. Chien and D. M. Lubman, *Rev. Sci. Instrum.* **63**, 4277 (1992).
168. J. Kern, T. Engel, D. Hagena and G. Werth, *Phys. Scr.* **46**, 575 (1992).
169. S. Becker, G. Bollen, F. Kern, H. J. Kluge, R. B. Moore, G. Savard, L. Schweikhard and H. Stolzenberg, et al., *Int. J. Mass Spectrom. Ion Processes* **99**, 53 (1990).
170. S. Becker, K. Dasgupta, G. Dietrich, H. Kluge, S. Kuznetsov, M. Lindinger, K. Luetzenkirchen, L. Schweikhard and J. Ziegler, *Rev. Sci. Instrum.* **66**, 4902 (1995).
171. B. A. Mamyrin, *Int. J. Mass Spectrom. Ion Processes* **131**, 1 (1994).
172. S. Wei and A. W. J. Castleman, *Int. J. Mass Spectrom. Ion Processes* **131**, 233 (1994).
173. R. C. Beavis, *ACS Symp. Ser.* **549**, 49 (1994).
174. T. J. Cornish and R. J. Cotter, *Rapid Commun. Mass Spectrom.* **8**, 781 (1994).
175. B. A. Mamyrin and D. V. Shmikk, *Zh. Eksp. Teor. Fiz.* **76**, 1500 (1979).
176. B. A. Mamyrin, V. I. Karataev, D. V. Shmikk and V. A. Zagulin, *Zh. Eksp. Teor. Fiz.* **64**, 82 (1973).
177. T. J. Cornish and R. J. Cotter, in *Time-of-Flight Mass Spectrometry,* ed. by R. J. Cotter, p. 95. ACS Symposium Series 549, American Chemical Society, Washington, DC (1994).
178. A. Maquestiau, Y. Van Haverbeke, R. Flammang, M. Abrassart and D. Finet, *Bull. Soc. Chim. Belg.* **87**, 765 (1978).
179. D. H. Russell, D. H. Smith, R. J. Warmack and L. K. Bertram, *Int. J. Mass Spectrom. Ion Phys.* **35**, 381 (1980).
180. M. L. Gross and D. H. Russell, in *Tandem Mass Spectrom.*, ed. by F. W. McLafferty, p. 255. Wiley, New York, N.Y. (1983).
181. V. Vrscaj, V. Kramer, M. Medved, B. Kralj, J. Marsel, J. H. Beynon and T. Ast, *Int. J. Mass Spectrom. Ion Phys.* **33**, 409 (1980).
182. R. S. Bordoli, K. Howes, R. G. Vickers, R. H. Bateman and D. J. Harvey, *Rapid Commun. Mass Spectrom.* **8**, 585 (1994).
183. M. L. Gross, E. K. Chess, P. A. Lyon, F. W. Crow, S. Evans and H. Tudge, *Int. J. Mass Spectrom. Ion Phys.* **42**, 243 (1982).
184. D. J. Burinsky, R. G. Cooks, E. K. Chess and M. L. Gross, *Anal. Chem.* **54**, 295 (1982).
185. J. H. Beynon, F. M. Harris, B. N. Green and R. H. Bateman, *Org. Mass Spectrom.* **17**, 55 (1982).
186. K. Eller and T. Drewello, *Org. Mass Spectrom.* **28**, 1462 (1993).
187. M. L. Vestal and J. H. Futrell, *Chem. Phys. Lett.* **28**, 559 (1974).
188. R. A. Yost and C. G. Enke, *J. Am. Chem. Soc.* **100**, 2274 (1978).

189. R. A. Yost, C. G. Enke, D. C. McGilvery, D. Smith and J. D. Morrison, *Int. J. Mass Spectrom. Ion Phys.* **30**, 127 (1979).

190. R. A. Yost and R. K. Boyd, in *Methods in Enzymology, vol. 193: Mass Spectrometry*, ed. by J. A. McCloskey, p. 154. Academic Press, New York (1990).

191. S. J. Gaskell, in *Biological Mass Spectrometry: Present and Future*, ed. by T. Matsuo, et al., p. 165. Wiley, Chichester (1994).

192. G. G. Dolnikowski, M. J. Kristo, C. G. Enke and J. T. Watson, *Int. J. Mass Spectrom. Ion Processes* **82**, 1 (1988).

193. C. Beaugrand, D. Jaouen, H. Mestdagh and C. Rolando, *Anal. Chem.* **61**, 1447 (1989).

194. P. Kofel, H. Reinhard and U.P. Schlunegger, *Org. Mass Spectrom.* **26**, 463 (1991).

195. G. L. Glish, S. A. McLuckey, E. H. McBay and L. K. Bertram, *Int. J. Mass Spectrom. Ion Processes* **70**, 321 (1986).

196. G. L. Glish, S. A. McLuckey, T. Y. Ridley and R. G. Cooks, *Int. J. Mass Spectrom. Ion Phys.* **41**, 157 (1982).

197. J. C. Schwartz, R. E. Kaiser, Jr., R. G. Cooks and P. J. Savickas, *Int. J. Mass Spectrom. Ion Processes* **98**, 209 (1990).

198. W. Huels, H. Muenster, R. Pesch and E. Schroeder, *44th ASMS Conf Mass Spectrom. Allied Topics, Portland, OR*, poster ThPA 2 (1996).

199. H. Muenster, W. Huels and E. Schroeder, *44th ASMS Conf Mass Spectrom. Allied Topics, Portland, OR*, poster MPD 99 (1996).

200. F. H. Strobel, T. Solouki, M. A. White and D. H. Russell, *J. Am. Soc. Mass Spectrom.* **2**, 91 (1991).

201. S. Schuetz, H. E. Hummel, A. Duhr and H. Wollnik, *J. Chromatogr., A* **683**, 141 (1994).

202. F. H. Strobel and D. H. Russell, in *Time-of-Flight Mass Spectrometry*, ed. by R. J. Cotter, p. 73. ACS Symposium Series 549, American Chemical Society, Washington, DC (1994).

203. A. E. Schoen, J. W. Amy, J. D. Ciupek, R. G. Cooks, P. Dobberstein and G. Jung, *Int. J. Mass Spectrom. Ion Processes* **65**, 125 (1985).

204. A. G. Harrison, R. S. Mercer, E. J. Reiner, A. B. Young, R. K. Boyd, R. E. March and C.J. Porter, *Int. J. Mass Spectrom. Ion Processes* **74**, 13 (1986).

205. C. D. Bradley, J. M. Curtis, P. J. Derrick and B. Wright, *Anal. Chem.* **64**, 2628 (1992).

206. S. W. Lemire and K. L. Busch, *J. Mass Spectrom.* **31**, 280 (1996).

207. M. Ho, R. J. Hughes, E. Kazdan, P. J. Matthews, A. B. Young and R. E. March, *Proc. 32nd ASMS Conf. Mass Spectrometry and Allied Topics*, San Antonio, TX, p. 513 (1984).

208. M. J. F. Suter, H. Gfeller and U. P. Schlunegger, *Rapid Commun. Mass Spectrom.* **3**, 62 (1989).

209. F. W. McLafferty, P. J. Todd, D. C. McGilvery and M. A. Baldwin, *J. Am. Chem. Soc.* **102**, 3360 (1980).

210. P. J. Todd, D. C. McGilvery, M. A. Baldwin and F. W. McLafferty, in *Tandem Mass Spectrometry*, ed. by F.W. McLafferty, p. 271. Wiley, New York (1983).

211. T. Takao, J. Gonzalez, K. Yoshidome, K. Sato, T. Asada, Y. Kammei and Y. Shimonishi, *Anal. Chem.* **65**, 2394 (1993).

212. C. C. Fenselau, in *Biological Mass Spectrometry: Present and Future*, ed. by T. Matsuo, et al., p. 129. Wiley, Chichester (1994).

213. K. B. Tomer, C. R. Guenat and L. J. Deterding, *Anal. Chem.* **60**, 2232 (1988).

214. S. S. Yang, P. Wong, S. Ma and R. G. Cooks, *J. Am. Soc. Mass Spectrom.* **7**, 198 (1996).

215. V. F. Juliano, F. C. Gozzo, M. N. Eberlin, C. Kascheres and C. L. do Lago, *Anal. Chem.* **68**, 1328 (1996).

216. M. N. Eberlin, N. H. Morgon, S. S. Yang, B. J. Shay and R. G. Cooks, *J. Am. Soc. Mass Spectrom.* **6**, 1 (1995).

217. R. H. Bateman, J. Brown, M. Lefevere, R. Flammang and Y. Van Haverbeke, *Int. J. Mass Spectrom. Ion Processes* **115**, 205 (1992).

218. R. Flammang, H. Y. Van C. Braybrook and J. Brown, *Rapid Commun. Mass Spectrom.* **9**, 795 (1995).

219. L. Schweikhard, S. Guan and A. G. Marshall, *Int. J. Mass Spectrom. Ion Processes* **120**, 71 (1992).

220. S. Guan and A. G. Marshall, *J. Chem. Phys.* **98**, 4486 (1993).

221. S. Guan, M. C. Wahl, T. D. Wood and A. G. Marshall, *Anal. Chem.* **65**, 1753 (1993).

222. S. J. Pastor, J. A. Castoro and C. L. Wilkins, *Anal. Chem.* **67**, 379 (1995).

223. C. L. Hendrickson, J. J. Drader and D. A. J. Laude, *J. Am. Soc. Mass Spectrom.* **6**, 448 (1995).

224. P. B. O'Connor, J. P. Speir, T. D. Wood, R. A. Chorush, Z. Guan amd F. W. McLafferty, *J. Mass Spectrom.* **31**, 555 (1996).

225. D. L. Rempel and M. L. Gross, *J. Am. Soc. Mass Spectrom.* **3**, 590 (1992).

226. M. V. Gorshkov, S. Guan and A. G. Marshall, *Rapid Commun. Mass Spectrom.* **6**, 166 (1992).

227. S. J. Yu, C. L. Holliman, D. L. Rempel and M. L. Gross, *J. Am. Chem. Soc.* **115**, 9676 (1993).

228. R. C. Dunbar and G. H. Weddle, *J. Phys. Chem.* **92**, 5706 (1988).

229. T. Baer, *Chemtracts: Anal. Phys. Chem.* **1**, 194 (1989).

230. A. R. Woolfitt, R. H. Bateman, M. R. Green, N. H. R. Ordsmith and D. J. Pugh, *44th ASMS Conf Mass Spectrom. Allied Topics,* Portland, OR, poster TPJ 253 (1996).

FOURIER TRANSFORM MASS SPECTROMETRY

MICHAEL L. EASTERLING AND I. JONATHAN AMSTER
Department of Chemistry, University of Georgia
Athens, Georgia 30602-2556

ABSTRACT. Fourier transform mass spectrometry (FTMS) provides high resolution and mass measurement accuracy for the analysis of biomolecules. This chapter provides a tutorial on the basic principles of FTMS. The features and capabilities of FTMS for high molecular weight measurements are introduced. A description of the motion of ions in the FTMS instrument is presented. Mass accuracy and mass calibration issues, and the assignment of molecular weight from high resolution FTMS data are discussed.

1. Introduction

The Fourier transform mass spectrometer is widely recognized as an instrument that provides the highest combination of resolution and mass accuracy of all mass spectrometers. In addition to its high performance capabilities, the Fourier transform mass spectrometer has several other properties that make it powerful tool for chemical analysis. It is an ion trapping instrument that provides simultaneous detection of all ions in a nondestructive manner, and can thus make very sensitive measurements on trace amounts of material. The Fourier transform mass spectrometer is a versatile instrument that can perform a wide variety of experiments with only minor modification to its hardware. These experiments include collisionally activated dissociation,[1-7] photodissociation,[8-12] and gas-phase ion-molecule reactions.[13-18] Collectively, these capabilities make the Fourier transform mass spectrometer a uniquely power instrument for molecular weight and structure determinations.

Fourier transform mass spectrometry (FTMS) is a relatively new type of mass spectrometry, but its roots are well established. FTMS derives from ion cyclotron resonance (ICR), a technology which is more than sixty years old. E. O. Lawrence developed the theory of cyclotron resonance and built the first cyclotron accelerator in the 1930's.[19] Sommer, Thomas, and Hipple built the first ICR mass spectrometer in the 1950's.[20] For twenty years, ICR spectrometry was used by a number of research groups to study gas-phase ion-molecule reactions. In the 1970's Comisarow and Marshall developed image current detection, and adapted Fourier transform methods to ICR spectrometry.[21] The development of Fourier transform ion cyclotron resonance (FTICR) spectrometry, also known as FTMS, as an analytical tool has principally occurred since 1980, when these instruments first became commercially available.

The development and application of Fourier transform mass spectrometry to the analysis of biomolecules has been a particularly active area of research for more than a decade. McIver and Hunt adapted a fast atom bombardment source to the FTMS instrument in 1985 to examine small

R. M. Caprioli et al. (eds.), Selected Topics and Mass Spectrometry in the Biomolecular Sciences, 287–314.

biomolecules with high resolution.[22] Direct laser ionization studies in the 1980's demonstrated FTMS analysis of involatile biomolecules and polymers.[23,24] The most significant advances have followed from the adaptation of electrospray ionization (ESI)[25,26] and matrix-assisted laser desorption/ionization (MALDI)[27,28] to FTMS. These latter methods, which have revolutionized all of mass spectrometry, have been combined with FTMS, and permit the high performance capabilities of this mass spectrometer to be applied to biologically relevant problems.

This chapter provides an overview of the FTMS technique with an emphasis on biological analysis. This material is divided into four sections which cover the basic features of the FTMS instrument, the fundamentals of ion motion in the FTMS analyzer cell, the methods used for mass measurement, and the techniques for analyzing biomolecules. A number of excellent reviews have been published on many of the key features of FTMS, and these are recommended for a more detailed discussion of individual topics.[29-39]

2. Features

2.1 FTMS INSTRUMENTATION

There is a considerable variety in the design of FTMS instruments that are used for examining high molecular weight compounds. However, all Fourier transform mass spectrometers contain four basic components - a magnet, an analyzer cell, an ultrahigh vacuum system, and a data system. In addition, instruments for biological analysis have an ion source that is located either external to the high magnetic field of the analyzer region, or mounted internally, that is inside the homogeneous region of high magnetic field. Both electrospray ionization and matrix-assisted laser desorption/ionization can be used with FTMS. The important considerations in adapting these ionization methods to the FTMS instrument will be presented later in this chapter.

The magnetic field for an FTMS instrument can be produced by a superconducting magnet, an electromagnet, or even a permanent magnet. All performance characteristics of the Fourier transform mass spectrometer improve by increasing the magnetic field strength, including mass range, mass resolution, and sensitivity. For this reason, high field superconducting magnets ranging in strength from 3-9.4 tesla are used for biomolecule analysis. Magnets of even higher field are being designed or constructed at two U.S. national laboratories active in FTMS development, the National High Magnetic Field Laboratory and Battelle Pacific Northwest National Laboratory. The superconducting magnets that are used for FTMS bear some resemblance to nuclear magnetic resonance (NMR) imaging magnets. They typically have a wide bore (150 cm diameter or larger) that is oriented horizontally, in contrast to the narrow, vertical bore magnets used for analytical NMR. The orientation and size of the magnet bore produce a strong fringing field that extends far from the magnet, and is an important consideration in locating equipment and personnel near the instrument. Many high field FTMS magnets are now being constructed with magnetic shielding to significantly reduce the extent of the fringe field.

Although the magnet is the most conspicuous part of the FTMS instrument, the analyzer cell is the central component. It is here that ions are mass analyzed and detected. For some types of ionization, such as electron bombardment or chemical ionization, ions are formed in the analyzer cell as well. For biomolecule analysis, ions are made outside the analyzer cell, and then transported into the cell and trapped. The analyzer cell is located in the homogenous region of the magnetic field.

Once ions are introduced into the analyzer cell, they are trapped by magnetic and electric forces, and can be stored for minutes or even hours. Many different types of analyzer cell designs have been proposed and constructed. The cubic cell, shown in Figure 1, illustrates the basic components found in most FTMS analyzer cells. It is composed of six electrodes called cell plates. Voltages are applied to the cell plates to perform the basic functions required for the FTMS experiment. Two plates are oriented perpendicular to the lines of the magnetic field, and are called trapping plates. A d.c. voltage applied to these plates allows ions to be stored in the cell, as will be described below. Ions that are formed outside the analyzer cell, as is typical for ESI and MALDI, are brought into the cell through an opening in one of the trapping plates. There are four plates that lie parallel to the magnetic field. One opposed pair is called the excitation plates, and the remaining pair of electrodes is called the detection plates. Their functions are described in later sections of this chapter. A second type of analyzer cell is also shown in Figure 1. This cell is composed of cylindrical electrodes, and is open-ended, that is no electrodes are oriented perpendicular to the magnetic field.[40] The cylindrical sections on both ends of the cell function as trapping electrodes. The center cylindrical section consists of two sets of opposed pairs of electrodes that serve as excitation and detection plates. The two cells shown here demonstrate some of the important variations in cell design - open-ended versus close-ended cells, and rectilinear versus cylindrical geometries. The relationship of cell design to performance has been examined by a number of workers.[41-43] Interested readers are referred to recent reviews for further details.[30,44]

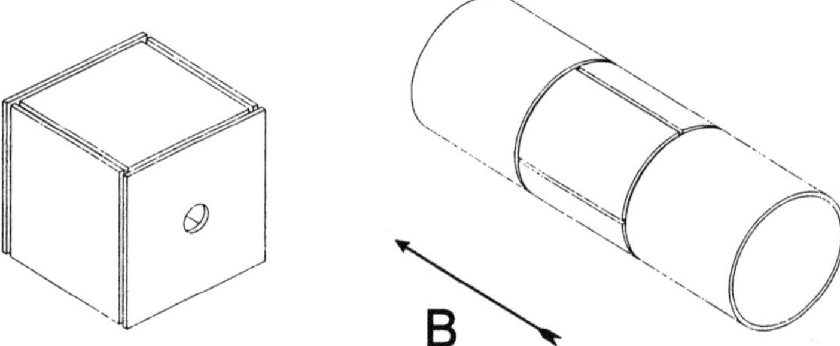

Figure 1. Two examples of analyzer cells used in FTMS. The arrow shows the orientation of the magnetic field with respect to the cells. On the left is a cubic cell, with six plates. The trapping plates are oriented perpendicular to the magnetic field. The four remaining electrodes are the excitation and detection plates. On the right is a open-ended cylindrical cell. The two ends of the cylinder are trapping electrodes. The central cylinder is divided into four sections that form the excitation and detection electrodes.

The third component of the FTMS instrument is an ultrahigh vacuum chamber. All mass spectrometers require vacuum to analyze and detect ions. However, the performance of the FTMS instrument is more sensitive to the pressure in the vacuum chamber than are other types of mass spectrometers. For example, mass resolution decreases in direct proportion to the pressure in the analyzer cell. To achieve ultrahigh resolution, the pressure must be on the order of 10^{-9} to 10^{-10} torr. These low pressures are usually achieved by using cryogenic or turbo molecular pumps rather than diffusion pumps. The low pressure requirement also impacts the manner in which samples are introduced into the mass spectrometer. Atmospheric ionization methods, such as electrospray ionization, require several stages of differential pumping between the ion source and the analyzer cell. Ion formation and ion detection occur at different times in FTMS, and high vacuum is required only during ion detection. The pressure in the analyzer can be raised to allow ion-molecule to reactions to occur or to stimulate ion dissociation by collisions, as long as high vacuum is restored for detection. FTMS vacuum systems are usually equipped with pulsed valves to introduce gas in a short burst, which can then be rapidly pumped away.

The data system is an important component of all FTMS instruments. The requirements for the FTMS data system are generally greater than for other types of mass spectrometers. However, in balance, the complexity of the mechanical hardware and the need for precise alignment and tight mechanical tolerances is generally lower for FTMS than for other types of mass spectrometers. The data system components include a fast transient digitizer with a large buffer memory capacity, a sensitive broad band preamplifier, powerful broad band amplifiers, a frequency synthesizer, and delay pulse generator. A computer coordinates the actions of each piece of hardware, and is used for the acquisition and processing of data. The computers used to process data must be able to rapidly perform the computationally intensive Fourier transform on data sets as large as 10^6 data points (8 MBytes).

2.2 EXPERIMENTAL SEQUENCES

The FTMS instrument is an ion trapping mass spectrometer, which means that ions remain in the instrument on a time scale that is many orders of magnitude longer than for most other types of mass spectrometers. Quadrupole ion traps also share this characteristic. For magnetic sector, time-of-flight, and quadrupole mass spectrometers, ions are formed into a beam that passes from the ion source, through a mass analyzer, and to a detector, a process that lasts on the order of 100 microseconds. In contrast, ions can be trapped for minutes or hours in the FTMS instrument. In contrast to ion beam instruments, where ionization, mass analysis, and detection are distributed through different regions of the instrument, all or most of these processes occur in a small region of the FTMS instrument, but are distributed in time.

The FTMS experiment is best visualized as a sequence of events. Figure 2 shows an experimental sequence that illustrates the events required for simple mass analysis. Experimental sequences usually begin with a quench pulse that removes ions from previous experiments from the cell. The quench is achieved by applying voltages of opposite polarity to the trapping electrodes, which removes ions from the cell on a millisecond or shorter time scale. Ions are formed and trapped during the ionization event. Ions may be formed during a brief ionization pulse, for example, by pulsed laser desorption. Alternatively, ions may be formed continuously, for example, by electrospray ionization, but trapped only during the "ionization" pulse. Once ions are stored in the analyzer cell, they are detected in a two step process. An excite pulse initiates coherent ion motion, and this motion is detected, using methods that are described below. More sophisticated experiments are achieved by adding additional events to the experimental sequence. Other events include voltage pulses applied to various cell plates to manipulate the motion of ions in the analyzer cell, the pulsed introduction of gas for reactions or collisions, or pulsed laser irradiation.

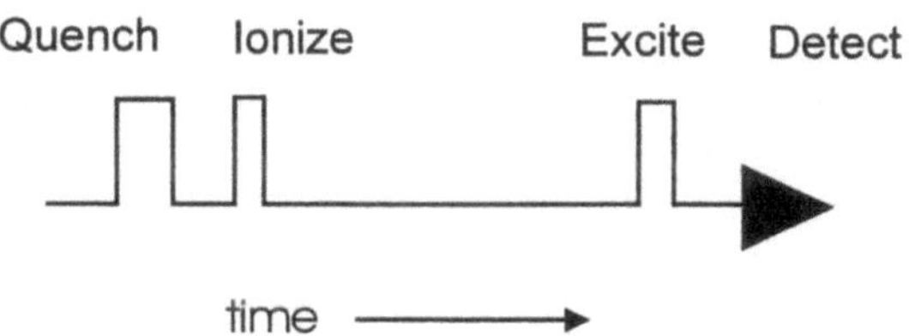

Figure 2. **An experimental sequence that illustrates the four basic events required for mass analysis, quench, ionize, excite, and detect.**

3. Ion Motion

The operation of the Fourier transform mass spectrometer can best be comprehended by considering the motion of ions in the analyzer cell. Ion motion is described by the force expression shown below in Equation 1, where ***F*** is the force experienced by an ion, ***m*** is the ion's mass, ***a*** is the instantaneous acceleration of the ion, ***q*** is the charge, ***E*** is the electric field, ***v*** is the velocity of the ion, and ***B*** is the magnetic field strength. This is a second order differential equation (acceleration is the second derivative of position with respect to time), and it can be numerically integrated to yield an ion trajectory, that is a plot of ion position in the cell as a function of time. The details of such a procedure are beyond the scope of this chapter, however the results of such calculations will be shown to illustrate the types of motion an ion undergoes in the FTMS analyzer cell. An important qualitative point that we can derive from Equation 1 is that there are two types of forces that influence the motion of an ion, those that arise from interaction of the ion with the electric field, ***E***, and those that arise from interaction with the magnetic field, ***B***. Three fundamental ion motions are produced by these forces. *Cyclotron motion* results from the interaction of an ion with the magnetic field. *Trapping motion* results from the force of the electric field on the ion. The combined effect of the magnetic field and the electric field cause *magnetron motion.*

$$\vec{F} = m\vec{a} = q\vec{E} + q(\vec{v} \otimes \vec{B}) \qquad (1)$$

3.1 CYCLOTRON MOTION

The analyzer cell of an FTMS instrument is located within a magnetic field of high homogeneity. This means that the magnetic field is constant in both magnitude and direction within the volume of the analyzer cell. In contrast, the electric field varies in amplitude and direction as a function of position within the cell and additionally, can vary as a function of time. We will begin our analysis of ion motion by considering first the effect of the magnetic field. An ion that moves perpendicular to the magnetic field will experience a force that is perpendicular to both its velocity and the direction of the magnetic field, as indicated by the cross product term in Equation 1. This force causes an ion to follow a circular trajectory in the plane perpendicular to the magnetic field. This is called cyclotron motion, and is illustrated by Figure 3. The frequency of this periodic motion is given by Equation 2. The cyclotron frequency, f_c (with units of Hertz), is proportional to the magnetic field strength, *B*, and is inversely proportion to the mass-to-charge ratio, *m/q*. FTMS measurements are made with instruments in which the magnetic field is a constant. The mass-to-charge of an ion is determined by measuring its cyclotron frequency. The simplicity of this equation is an important feature of the FTMS technique. Mass-to-charge is derived from a measured physical property, cyclotron frequency, which is independent of an ion's kinetic energy. This contrasts sharply with magnetic sector and time-of-flight mass spectrometers, in which the measured mass-to-charge is strongly dependent upon an ion's kinetic energy. The very high resolution and mass accuracy that can be obtained by FTMS arise from the fact that cyclotron frequency is independent of kinetic energy, and from the fact that frequency is a physical property that can be measured with great precision and accuracy. Cyclotron frequencies fall in the range of 5 KHz to 5 MHz. For example, an ion of m/q 1000 at a magnetic field strength of 7 tesla will have a cyclotron frequency of ($1.602*10^{-19}$ C * 7 T/2 * π * 1000 amu * $1.673*10^{-27}$ kg/amu) = 107.4 KHz.

$$f_c = \frac{qB}{2\pi m} \tag{2}$$

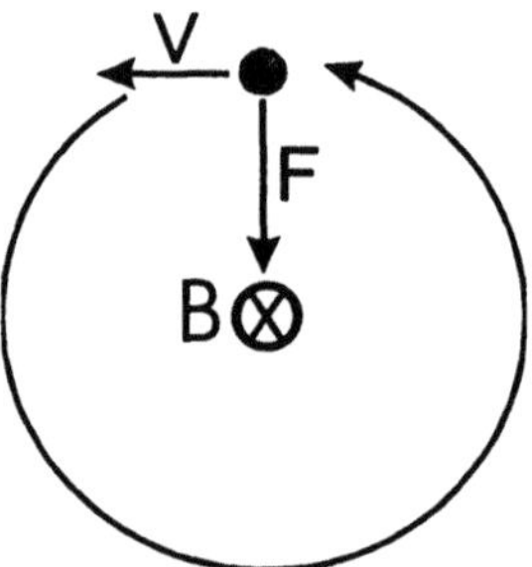

Figure 3. Cyclotron motion of an ion. The magnetic field is directed into the plane of the page. A positive ion moves in a counterclockwise trajectory as a result of the center-directed force from the magnetic field.

The radius of the cyclotron orbit, *r*, depends upon the kinetic energy of an ion, *E*, as shown by Equation 3. When first produced, ions have thermal kinetic energies with small cyclotron orbits. Ions with room temperature kinetic energy distributions have cyclotron orbits on the order of 100 μm. To detect ions, their cyclotron radius is increased to dimensions close to that of the cell. This is accomplished by increasing their kinetic energy. As shown by Equation 3, kinetic energy scales with the square of the magnetic field, the radius of the cyclotron orbit, and the charge present on the ion. The kinetic energy of a singly charged ion of m/q 1000 with a 2 cm orbital radius at 7 tesla is 940 eV. The kinetic energy of a m/q 1000 ion with 10 charges under otherwise identical conditions is 9.4 KeV. The process by which this amount of energy is added to the ions is described below.

$$E = \frac{q^2 B^2 r^2}{2m} \tag{3}$$

3.2 TRAPPING MOTION

The magnetic field exerts no force on an ion that moves parallel to the magnetic field. In the absence of any additional field, ion motion is unconstrained along the magnetic axis, and ions can drift though the analyzer cell within a few milliseconds. In order to keep ions in the analyzer cell for several seconds or longer, a small voltage (ca. 1 V) is applied to the trapping plates of the analyzer cell, which lie perpendicular to the magnetic field, as shown in Figure 1. A positive applied voltage is used to trap positive ions, and a negative voltage is used to trap negative ions. To switch between positive and negative ion analysis, the polarity of the trapping potential is the only cell parameter that is changed. The potential along the magnetic field axis for a cubic analyzer cell is shown in Figure 4. Ions with kinetic energies along the magnetic field axis that are less than the trapping potential undergo a harmonic oscillation between the two trapping electrodes, with a frequency given by Equation 4, where f_t is the trapping frequency, V_T is the applied trapping voltage, a is the separation between the trapping plates, and α is a geometry factor for the analyzer cell (1.387 for the cubic cell). An ion of m/z 1000 in a 5 cm cubic cell with a 1 V trapping potential will undergo a trapping oscillation with a frequency of $(1.387*1.6*10^{-19}\ C*1V/3.1416^2\ *(.05\ m)^2*1000\ amu*1.67*10^{-27} kg/amu)^{1/2} = 2320$ Hz. Trapping motion and cyclotron motion are independent from each other.

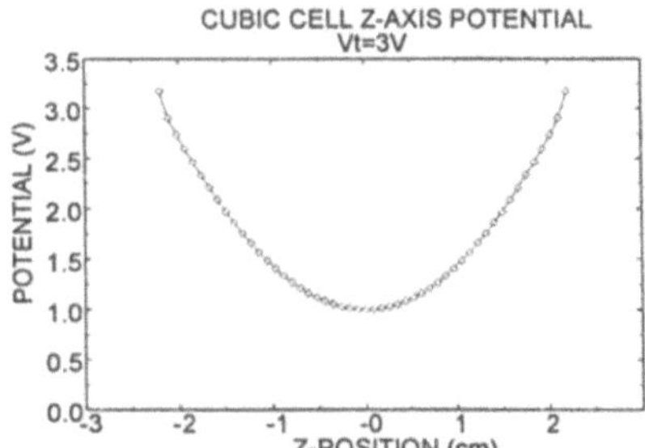

Figure 4. The potential along the magnetic field axis in a 4.4 cm cubic cell with a 3 V trapping potential. The potential at the center of the cell is equal to one-third the applied trapping voltage.

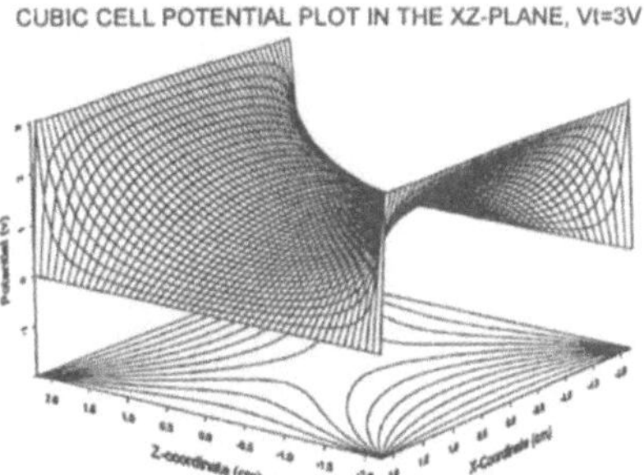

Figure 5. The potential in the plane that lies parallel to the magnetic field axis in the cubic cell of Figure 4. The trapping potential provides a restoring force along the z-coordinate, but repels ions in the plane that lies perpendicular to the magnetic field. The contours of the electric field are projected onto the plane beneath the plot.

$$f_t = \frac{1}{\pi a}\sqrt{\frac{\alpha q V_T}{m}} \tag{4}$$

3.3 MAGNETRON MOTION

Ions undergo a third characteristic motion in trapped analyzer cells. Magnetron motion is a result of the combined force of the electric and magnetic fields. The origin of magnetron motion can be understood by examining the diagram of the trapping potential shown in Figure 4. One observes that the potential at the center of the cell is not zero. In fact for a cubic cell, the potential at the center of the cell is one-third the trapping potential. The potential at the four plates that constitute the excitation and detection electrodes is zero. Figure 5 shows a plot of the electric potential in the plane that passes through the center of the cell and that lies parallel to the magnetic field. This potential provides a restoring force to ions along the magnetic field axis, but repels ions from the center of the cell in a direction that is perpendicular to the magnetic field. In the absence of a magnetic field, this radially repulsive field would eject ions from the cell. However, the magnetic field bends the ion trajectory into a magnetron orbit, a circular path around the center of the cell. The ion trajectory in Figure 6 illustrates magnetron and cyclotron motion of an ion in a cubic analyzer cell. For the purpose of clarity, the radius of the magnetron orbit shown in this figure is greatly exaggerated compared to normal operating conditions. Usually the cyclotron radius is much larger than the magnetron radius. The magnitude of the magnetron radius is controlled by the location of ions as they are formed or injected into the analyzer cell. Ions are usually formed along the central axis of the cell, and so their magnetron radius is usually small.

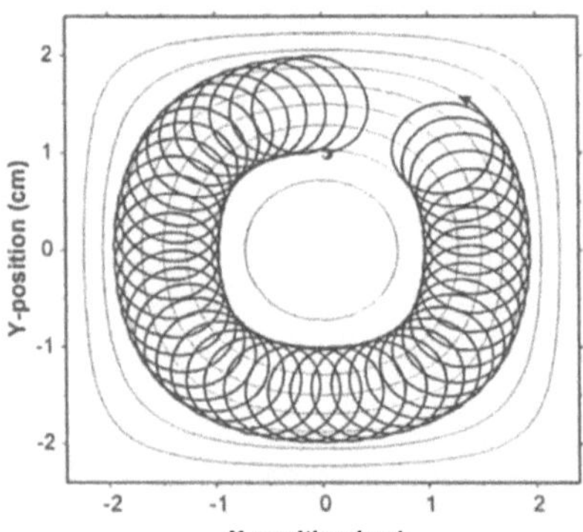

Figure 6. An ion trajectory that shows both cyclotron motion and magnetron motion in the plane perpendicular to the magnetic field. Cyclotron motion appears as the rapidly orbiting component of the trajectory, while magnetron motion can be identified as the slow precession of the orbit around the center of the cell. For purposes of clarity, the magnetron radius shown here is much larger than typical. The isopotential contours that result from a 3 V trapping potential are also shown. The potential due to the trapping voltage is highest at the center of the cell, 1 V, and lowest at the cell's edges, 0 V.

Magnetron motion can be viewed as a slow precession of the guiding center of the cyclotron orbit around the center of the cell along an isopotential contour. Its frequency is given by Equation 5, where f_m is the magnetron frequency in Hertz, α is a geometry factor for the analyzer cell, V is the trapping voltage, a is the separation between the cell plates, and B is the magnetic field strength. It can be seen from this equation that magnetron frequency is independent of an ion's mass-to-charge ratio, and depends principally on the ratio of the trapping voltage to the magnetic field strength. Magnetron motion has a much lower frequency than either cyclotron or trapping motions, on the order of 1-100 Hz. For example, an ion in a cubic analyzer cell with a principal dimension of 5 cm, with a trapping potential of 1 V, and at a magnetic field strength of 7 tesla would exhibit a magnetron frequency of (1.39 * 1 V/π * 0.05² m² * 7 T) = 25.30 Hz. Magnetron motion serves no useful analytical purpose. It is a consequence of the curvature of the trapping electric field due to the finite length of the cell electrodes. This curvature also is responsible for other detrimental effects, such as the collision induced radial diffusion of ions from the analyzer cell,[45] shifts in the observed cyclotron frequency,[45] and sidebands.[46]

$$f_m = \frac{\alpha V}{\pi a^2 B} \tag{5}$$

3.4 COLLISIONAL DAMPING OF ION MOTION

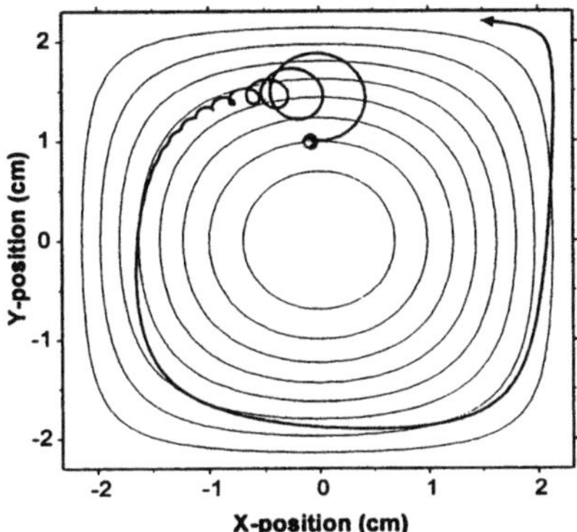

Figure 7. An ion trajectory with the same starting condition as for Figure 6, but with a collision gas present at a pressure of 0.01 Torr. Cyclotron motion is seen to quickly damp with a contraction of the cyclotron radius. Magnetron motion damps more slowly, with an enlargement of its radius as ions move to lower electric potential.

The motions of an ion presented above are characteristic of ion behavior in the absence of collisions, which is typical of the ultrahigh vacuum normally present in the FTMS analyzer region. For some experiments, the pressure inside the analyzer is raised to permit collisions to occur, for example for tandem mass spectrometry, to initiate ion-molecule reactions, or to induce collisional thermalization of excited state ions. It is thus important to consider the effect of collisions on an ion's trajectory. Figure 7 shows an ion trajectory that illustrates the effect of a high collision gas pressure on cyclotron and magnetron motion. Ions lose kinetic energy through the collision process. When an ion undergoes collisional damping, the radius of the cyclotron orbit decreases, as expected from Equation 3 above. However, when magnetron motion is damped, ions move away from the center of the analyzer cell, where the potential due the trapping voltage is the highest, toward the edge of the cell, where the potential is 0 V. Cyclotron motion damps more rapidly than magnetron motion. The damping rate for a particular mode of motion is proportional to an ion's velocity in the selected mode. The cyclotron velocity for an ion with a finite radius of gyration is much higher than the magnetron velocity, as can be inferred from Figure 6. The rate of the radial drift of an ion that results from collisional damping of the magnetron motion is inversely proportional to the magnetic field strength squared. Thus in high field instruments, collision-induced radial ion loss is a slow process. In addition to the damping of cyclotron and magnetron motion, trapping motion is also damped by collisions. This results in the reduction of the amplitude of the oscillation of the ions between the two trapping plates, causing ions to condense to the center of the analyzer cell.

4. Mass Measurement

4.1 ION EXCITATION AND DETECTION

The mass-to-charge ratio of an ion is determined by measuring its cyclotron frequency. This measurement occurs in two steps, ion excitation and ion detection. As presented above, ions move in small cyclotron orbits when they are first created in or introduced into the analyzer cell. The detection process requires that the cyclotron radii of the ions is enlarged to a significant fraction of the dimensions of the analyzer cell. As shown by Equation 4, enlarging the radius of cyclotron motion requires that kinetic energy be imparted to the ion. This is accomplished by exciting an ion with a dipolar radio frequency (RF) electric field that is in resonance with its cyclotron frequency.[47] The excitation signal is applied to a pair of opposed electrodes, called the excitation plates, that lie parallel to the direction of the magnetic field. In response to dipolar excitation at their cyclotron frequency, ions spiral outwards as their cyclotron orbit expands, as shown in Figure 8. If the excitation power is sufficiently high, then ions strike the plates of the analyzer cell, and are lost from the experiment. This can be used to eject unwanted ions, for example to isolate a precursor ion during the first stage of a tandem mass spectrometry experiment. For detection, the dipolar excitation signal is turned off before ions reach the cell plates. Ions then undergo cyclotron motion with a large radius orbit. Ions are excited coherently, which means that all ions of the same mass-to-charge ratio move in phase with each other. The resulting packet of charge induces an image current to flow in the circuit that connects the detection plates, the two remaining electrodes that lie parallel to the magnetic field.[48] The image signal is amplified, digitized, and stored. Ions of one mass-to-charge generate a sinusoidal signal whose frequency is nearly equal to the cyclotron frequency. There is a small difference between the observed frequency and the cyclotron frequency, and the origin of this frequency shift is explained below. The amplitude of the sinusoid is proportional to the number of ions at the mass-to-charge under examination. Ion detection in FTMS is nondestructive, that is the ions still remain in the analyzer cell after detection and are available for further experiments. This feature of FTMS is unique among mass spectrometers. Other mass spectrometers detect ions by allowing them to undergo a destructive collision onto an electron multiplier. Recently, it has been shown that the quadrupole ion trap can detect ions using the image current method, but the practicality of this approach remains to be demonstrated.[49] After detection, ions are spread throughout the analyzer cell, and cannot be redetected unless they are brought back to the center of the cell, a process called axialization. Typically, ions are removed from the cell after detection by a quench event, which is the brief application of an asymmetric trapping voltage pulse, which ejects ions along the magnetic field axis.

Ions of many mass-to-charge ratios can be detected simultaneously in the FTMS instrument. This is achieved by exciting a range of frequencies that corresponds to all the ions of interest. One method to achieve wide band excitation is to sweep the excitation signal rapidly through the desired frequencies. The signal, called a RF chirp, lasts for 1-5 ms, and excites the entire mass range under investigation.[50] The RF chirp produces coherent ion motion over the excitation range, and results in an image signal that is the sum of sinusoids of several frequencies and amplitudes. An example of an image signal produced by excitation of ions of several mass-to-charge values is shown in Figure 9, obtained from the ESI mass spectrum of the protein ubiquitin. On the left side of the figure is a transient, which is the amplified, digitized image signal recorded as amplitude versus time. The transient contains all the frequency and amplitude information from which can be derived the mass-to-charge and relative abundance of the ions in the analyzer cell. The frequency and amplitudes are recovered by performing a Fourier transform on the time-domain transient signal to yield a frequency spectrum, which is then converted to the mass spectrum shown on the right side of Figure 9 by applying a calibration formula.

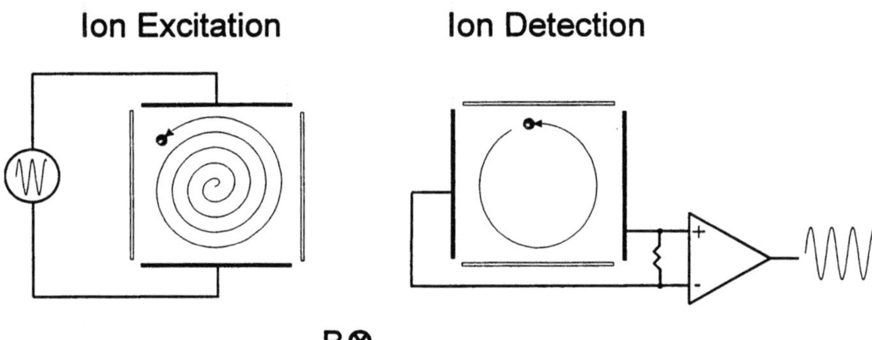

Figure 8. A cross sectional view of an analyzer cell during ion excitation (left) and detection (right). The magnetic field is directed into the plane of the page in this view. A dipolar RF electric field is applied to the excitation electrodes to move ions into a large radius orbit. After excitation, a packet of ions moves coherently in a cyclotron orbit in the analyzer cell. The electric field associated with the packet of charge induces an image current to flow in the external circuit that joins the detection plates. A sinusoidal signal is produced by ions of one mass-to-charge ratio.

There are many types of waveforms that are used to excite ions into larger cyclotron orbits, in addition to the RF chirp described above. A short duration (μs or shorter), high amplitude voltage pulse can be applied to the excitation electrodes to excite ions into higher kinetic energy cyclotron orbits. This method, called impulse excitation, is not mass selective.[51] Ions of all masses down to a low mass limit that is determined by the duration of the impulse waveform are excited to the same cyclotron radius. A RF burst, a short duration excitation at a single frequency, can be used to excite ions over a narrow mass range.[47] The range of masses that are excited is inversely proportional to the duration of the RF burst. The most precise method for the mass selective excitation of ions is the stored waveform inverse Fourier transform (SWIFT) technique.[52] With this method, excitation amplitude is defined as a function of frequency. The user can specify the amount of excitation energy that will be applied at each m/q value. Further details of this method are presented below in the discussion of tandem mass spectrometry. SWIFT simultaneously excites ions for all specified frequencies. The mass selectivity of this method is much higher than for any of the other methods described above.

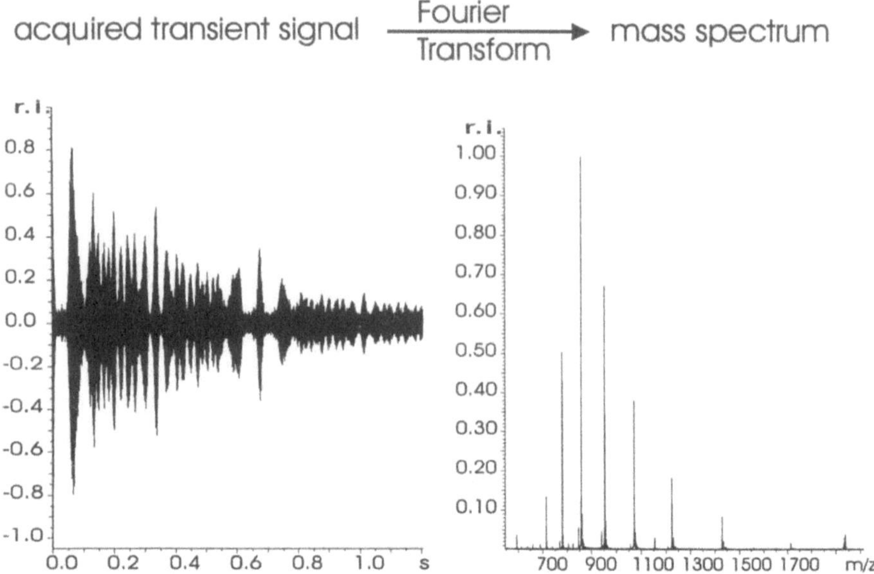

Figure 9. Transient signal (left) obtained from ESI of ubiquitin. A Fourier transform is applied to the transient to produce the mass spectrum shown on the right.

4.2 MASS RESOLUTION

Ions of all mass-to-charge values are detected simultaneously in FTMS. Any time slice that is taken from the transient shown in Figure 9 can be examined by performing a Fourier transform and will be found to contain peaks for all of the ions observed in the mass spectrum shown above. While this may suggest that only a short duration of the transient needs to be recorded to determine the mass of the ions that are present, it should be noted that the appearance of the peaks in the mass spectrum change as the duration of the recorded transient increases. This is demonstrated in Figure 10, which compares mass spectra obtained by applying a Fourier transform to the full transient with that obtained from the first 150 ms of the transient. The mass scale around m/q 857 is expanded to show the appearance of the isotope peaks of the 10+ charge state of ubiquitin in its ESI mass spectrum. The mass resolution is clearly higher for the mass spectrum derived from the 1.2 s transient compared to that obtained from the 150 ms transient. In the absence of other effects that can degrade the signal, such as magnetic field inhomogeneities or space-charge effects, mass resolution is limited by the duration of the transient, T, as shown by Equation 6, where f_c is the cyclotron frequency. Resolution

$$R = \frac{f_c T}{2} \tag{6}$$

increases in direct proportion to the duration of the transient. Ultrahigh resolution mass spectra are obtained by recording transients that last from several seconds up to a minute or longer.

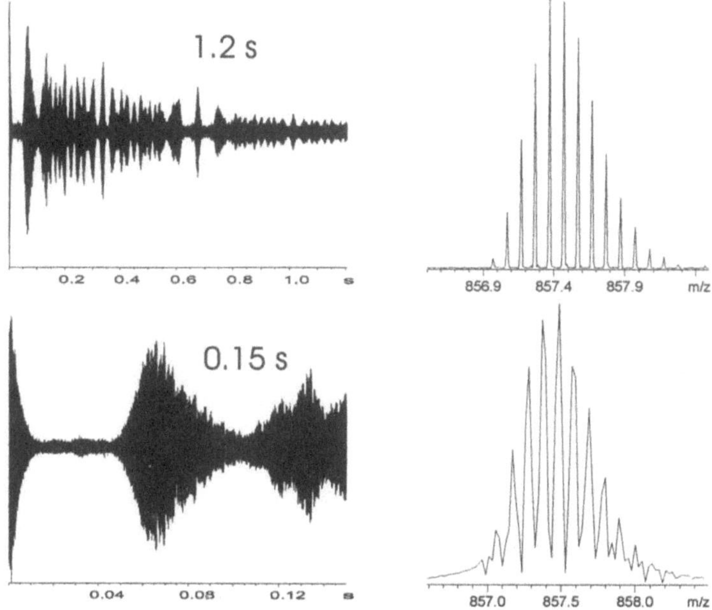

Figure 10. An illustration of the effect of transient duration on mass resolution. Both mass spectra are derived from the transient shown in Figure 9. The bottom mass spectrum is obtained by Fourier transformation of the first 150 ms of the transient. The top mass spectrum is obtained from the full transient. Mass resolution is seen to be proportional to the transient duration.

There are limitations to the length of the transient that can be recorded, and therefore to the mass resolution that can be achieved. The amplitude of the transient shown in Figure 9 decays with time. The image signal is produced by packets of ions of the same mass-to-charge moving coherently. With time, ions experience collisions with background neutral molecules, which destroys the coherence of the ion packet and causes the decay of the image signal. The decay time is inversely proportional to the pressure in the analyzer cell. Extremely high mass resolution can be obtained at pressures of below 10^{-9} torr, where the decay time of the transient is 10 s or more. A second consideration for obtaining high resolution is the number of data points that must be recorded. The transient must be digitized at a rate that is no less than twice the highest frequency that is recorded, so that there are two or more data points per cycle of the sinusoidal signal. This restriction is known as the Nyquist criterion, and is a fundamental rule that applies to the digitization of any waveform. In the case of the m/q 857 ion found in the mass spectrum of ubiquitin in Figure 10, the cyclotron frequency is approximately 125 KHz (for a 7 tesla magnetic field). Digitization of these ions must occur at a rate of 250 KHz or greater to meet the Nyquist criterion. Thus, 250,000 data points must be acquired to accurately digitize a 1 second transient. The transient digitizers used in most FTMS data systems can store up to 1 million data points. (Actually, 1 M sample = 2^{20} data points ≅ 1.05 million points.) For the ubiquitin example, this would restrict the transient length to 4 seconds. Using Equation 6, mass resolution under these conditions is limited to 250,000. This resolution is much higher than can be achieved by any other type of mass spectrometer. However, even higher mass resolution can be achieved by digitizing the signal for a longer period of time using a mode of signal processing called heterodyne detection, a technique which is used in many applications including FTNMR and FM radio.

The principle of heterodyne detection is shown in Figure 11. The sinusoidal signal to be digitized is multiplied times a reference signal that has a well characterized frequency. The resulting signal has two components, a sum and a difference frequency. If the reference signal is close in frequency to the original sinusoid, the difference frequency will be much lower than the sum frequency. For the m/q 857 ion of ubiquitin discussed above, the image signal has a frequency of 125 KHz. If this is mixed with a reference signal of 115 KHz, then the sum and difference frequencies are 240 KHz and 10 KHz, respectively. The difference signal is easily separated from the sum signal by using a low pass filter. The filtered signal can be digitized at twice the rate of the difference frequency, for example, at 20 KHz for the m/q 857 ion of ubiquitin when mixed with a 115 KHz reference signal. At this rate, 1 million data points can be used to digitize the signal for 52 seconds, permitting a mass resolution of 3.3 million. The frequencies of the digitized and reference signals are added to obtain the frequency of the original signal, that is, the cyclotron frequency of the ions.

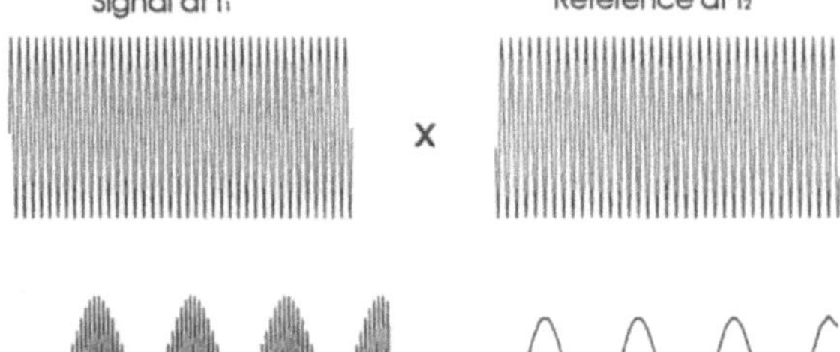

Figure 11. A graphical representation of heterodyne detection. A sinusoidal signal is mulitplied by a reference signal to yield a composite signal that contains sum plus difference frequencies. The sum frequency is filtered away to leave a low frequency difference signal, which can be digitized at a much lower rate than the original signal.

Heterodyne detection reduces the number of data points that are required to digitize a transient of a specified duration by shifting the image frequencies to lower values. For example, the digitization rate (also called the bandwidth) is reduced from 250 KHz to 20 KHz in the case of ubiquitin presented above. For this reason, the heterodyne mode of detection is also called narrowband detection. This mode of detection increases the signal-to-noise ratio compared to wide band detection at the original cyclotron frequency by filtering the random noise that is outside the digitization bandwidth. By narrowing the detection bandwidth, the noise that is present at higher frequencies is rejected. The amplitude of random noise is the same at all frequencies, and therefore the improvement in signal-to-noise is equal to the ratio of digitization rates. For example, narrowband detection at 20 KHz will exhibit one hundred times lower noise than wide band detection at 2 MHz. With these advantages of narrowband detection come the tradeoff of a reduced detected mass range. The exact range of masses that are detected depends upon both the digitization rate (bandwidth) and the reference frequency. For example, at a digitization rate of 20 KHz, a 10 KHz range of frequencies can be detected. If the reference frequency is 1MHz, frequencies from 1-1.01 MHz are sampled, corresponding to a mass-to-charge range of 107.9-108.9, that is only one m/q unit wide. With a reference frequency of 125 KHz, the mass range extends from m/q 857-931, that is 74 m/q units wide. With a reference frequency of 20 KHz, m/q 5400-10700 is sampled, a range of 5300 m/q units.

4.3 CALIBRATION

The frequencies that are measured by image current detection in the trapped analyzer cell are not exactly equal to the cyclotron frequency presented in Equation 2.[53,54] The electric field that is used to trap ions perturbs their motion, causing magnetron motion as explained above. The trapping potential is radially repulsive, that is it repels ions from the center of the cell in the plane that is perpendicular to the direction of the magnetic field. The radially repulsive electric field thus opposes the center-directed force exerted by the magnetic field on an ion, in effect reducing the apparent magnetic field strength. As a result, the observed frequency is reduced from the cyclotron frequency, Equation 2, by an amount equal to the magnetron frequency, Equation 4. In other words, the observed frequency is equal to the difference between the cyclotron frequency and the magnetron frequency. Magnetron frequencies are much smaller than cyclotron frequencies, and so this produces a small but measurable change in the observed frequency compared to the cyclotron frequency. Magnetron frequency has no mass dependence, but is directly proportional to the trapping voltage. The electric field of the ions present in the cell can also produce a shift in the observed frequency of an ion. This space-charge effect is usually small, but grows in proportion to the number of ions in the cell and in proportion to the charge present on the ions. The net effect of trapping voltage and space charge on the observed frequency is shown in Equation 7.[53] The first term on the right side of the equality (qB/m) is the unperturbed cyclotron frequency, the second term is the correction for the trapping voltage, and is equal to the magnetron frequency given in Equation 4, and the third is a correction for space-charge effects. In the space-charge term, ρ is the charge density in the cell, and k is a constant that is related to the geometric shape of the ions distributed in the cell. Under normal circumstances, these terms grow smaller as one goes from left to right. It should be noted that the cyclotron frequency term increases in direct proportion to the magnetic field strength, while the trapping potential and space charge perturbations to the observed frequency have an inverse relationship to magnetic field strength, and therefore decrease at higher fields. Increasing the magnetic field strength leads to higher mass accuracy, and is one of the driving forces behind the development of higher field instruments (9.4 tesla and higher.) In practice, the latter terms of Equation 6 are not calculated explicitly, but are accounted for empirically in the calibration equation. A calibration formula which accounts for these effects is shown in Equation 8.[55] The constants a and b are determined experimentally by fitting data from a standard to the calibration equation. The best mass accuracy is obtained from measurements in which the trapping voltage and the ion density is the same as the conditions used to calibrate the spectrometer, as changes in these parameters alters the value of the constant, b.

$$f_{obs} = \frac{qB}{m} - \frac{\alpha V}{\pi a^2 B} - \frac{kq\rho}{B} \quad (7)$$

$$\frac{m}{q} = \frac{a}{f} + \frac{b}{f^2} \quad (8)$$

5. FTMS Analysis of Biomolecules

5.1 IONIZATION METHODS

Virtually every ionization method known to mass spectrometry has been implemented by FTMS, including electron ionization,[56,57] chemical ionization,[58] laser ionization,[57] glow discharge,[59,60] field desorption, ^{252}Cf plasma desorption,[61] secondary ionization mass spectrometry,[62-64] fast atom bombardment,[22] matrix assisted laser desorption/ionization,[27,28,65] and electrospray ionization.[25,26] The discussion will focus on the two ionization methods that are most successfully applied to biomolecule analysis, namely ESI and MALDI. In coupling an ion source to an FTMS instrument, two chief considerations are the high magnetic field present in the analyzer region and the requirement for ultrahigh vacuum. While some ionization methods are easily implemented in the analyzer region of the mass spectrometer, for example electron ionization, it is advantageous in many cases to form ions outside the bore of the magnet, in an external ion source, where the magnetic field is much lower. By forming ions remotely from the analyzer, differential pumping stages can be placed between the source and analyzer to decouple the pressure requirements of the ion source from the high vacuum of the analyzer. This is particularly useful for implementing electrospray ionization, an atmospheric pressure ionization source. Ions must be focused from the source into the analyzer through a strong magnetic field gradient. Ions that are not actively focused through the region of high magnetic field gradient usually do not reach the analyzer, but are reflected by the magnetic mirror effect. To inject ions into the analyzer region, two general approaches are employed. A RF ion guide (RF-only quadrupole or octupole assemblies) can be used to focus ions of all masses through the fringe field of the magnet. [66,67] The second approach is to use electrostatic fields to guide the ions into the analyzer. [68-70]

Electrospray ionization has been the most effective ionization method for FTMS analysis of biomolecules.[25,26] The multiply-charged ions that are produced by ESI fall in a mass-to-charge range (m/z 500-2000) over which the FTMS instrument exhibits excellent performance. The challenge in combining ESI with FTMS is the large difference in the pressure regimes in which these methods operate. ESI is an atmospheric ionization method, while high resolution FTMS requires analyzer pressures that are lower than 10^{-9} torr. External ion source designs have been adapted to ESI to couple it to the FTMS instrument.[71,72] The electrospray source is separated from the analyzer by three or more stages of differential pumping. Ion guides direct a beam of ions from the source through several conductance limits, past the steep magnetic field gradient, and into the analyzer region. Ions are formed continuously, and can be accumulated in the analyzer cell during the "ionization" pulse of the experimental sequence. Ions must have enough kinetic energy to pass the first trapping electrode, but must lose some of their energy while in the cell so that they do not exit past the second trapping electrode. Collisional damping of an ion's kinetic energy with a buffer gas aids this process.[73] Alternatively, pulsed voltages at the entrance to the cell have been used to assist ion trapping. [74]

The mass spectra that are produced by ESI-FTMS are isotopically resolved, as can be seen in Figure 10 for the protein ubiquitin. The peaks of the isotope distribution result principally from the heavy isotope of carbon, but also have abundance from the isotopes of their other constituent elements, nitrogen, oxygen, sulfur, and hydrogen. The isotope peaks are separated by 1 Da in mass and 1/z units in mass-to-charge. In the ESI-FTMS mass spectrum, the separation between the isotope peaks can be used to assign the number of charges that are present on an ion. For example, in Figure 10, the isotope peaks are observed to be separated by an average of m/z 0.10. The charge present on the ion is calculated to be (1/0.10) = 10. Without isotopic resolution, the assignment of charge requires the identification of two or more peaks in the charge state distribution. For the ESI mass spectra of pure compounds, this task is straightforward. For mixtures of components this task is greatly simplified when isotopic resolution is obtained.

MALDI has also been effectively combined with FTMS.[27,28,65] MALDI ionization can occur in ultrahigh vacuum, and so it couples well with FTMS analysis. Moreover, MALDI produces ions in a short burst, and so it is compatible with the pulsed nature of FTMS detection. Both external[75] and internal[76] ion source designs have been reported. There have been two challenges in coupling MALDI to FTMS. The internal energy of ions formed by MALDI is higher than for those formed by electrospray. The lifetime of ions formed by MALDI can be shorter than the time scale of the FTMS experiment, leading to the observation of fragment ions rather than the molecular species. Wilkins has developed sample preparation methods that increase the lifetime of the ions and permit detection of the intact molecular species.[65] The second challenge arises from difficulties in trapping ions that are formed by MALDI. Ions formed by MALDI have a velocity distribution that appears to be mass independent.[77] This means that ions formed by MALDI have kinetic energies that increase in direct proportion to their mass. These kinetic energies can be quite large, for example a 10000 Da ion will have a kinetic energy of 30 eV. Different schemes have been employed to trap ions with these high axial kinetic energies. Gated trapping has been effective at capturing ions formed by MALDI near the analyzer (internal MALDI).[78-80] With this method, the trapping voltage is lowered for the plate at the entrance to the cell and is raised for the plate at the far end of the cell while ions are formed. The ions enter the cell and are reflected by the high potential of the rear trap plate. Before they can exit the cell, the trapping voltage on the front plate is raised to capture the ions. The ions then undergo collisional damping of their kinetic energy, and the trapping plate voltage is lowered to a value that is compatible with ion detection.

5.2 OBTAINING EXACT MASSES FROM HIGH RESOLUTION DATA

Fourier transform mass spectrometry is capable of providing high resolution mass spectra of biomolecules for which the isotope peaks of the sample are clearly resolved, as can be seen in the mass spectrum of ubiquitin shown in Figure 10. These peaks arise as a result of the natural distribution of isotopes of the elements. For example, 98.9% of all carbon is ^{12}C, while 1.1% of carbon is ^{13}C. For biomolecules composed of hundreds of carbon atoms, several ^{13}C atoms are present on average. The exact isotope distribution for this element and other elements can be described by a binomial distribution. Thus, an isotope distribution can be easily calculated for a given elemental composition. The spacing between isotope peaks is approximately 1 amu. For singly charged ions, such as those produced by MALDI, the isotope peaks are spaced by 1 m/q unit. For multiply charged ions produced by electrospray, the isotope peaks are more closely spaced, 1/z mass-to-charge units, where z is the charge present on the ion. For example, in Figure 10, which shows the isotopically resolved peaks corresponding to ubiquitin with ten charges due to protons, the isotope peaks are spaced by 0.1 mass-to-charge units. For larger proteins which exhibit even higher charge states in their ESI mass spectrum, the isotope peaks will be even more closely spaced. For such samples, high resolution is required to distinguish the isotope peaks in the mass spectrum. A monoisotopic molecular weight can be derived from such isotopically resolved data. The monoisotopic molecular weight corresponds to a molecule composed of the lowest mass, most abundant isotopes (^{1}H, ^{12}C, ^{15}N, ^{16}O, ^{32}S). In the mass spectrum of ubiquitin shown in Figure 10, the monoisotopic mass corresponds to the mass-to-charge of the low intensity peak at the far left of the distribution. If the isotope peaks of the molecular ion are not resolved, as with the data from most other types of mass spectrometers, then the average molecular weight is reported instead.

The average molecular weight corresponds to the centroid of the distribution of all the isotope peaks for the molecular ion, and is the value that would be calculated for a given elemental composition using the chemical masses (averaged over all naturally occurring isotopes) of the elements. In general, measurements of monoisotopic molecular weights are more precise and more accurate than measurements of average molecular weights. Measurements of the average molecular weight require that both mass-to-charge and intensity are determined accurately, since both these values are required to calculate the centroid of the peak. The average molecular weight is sensitive to the isotope distribution that is sampled during the measurement. If the signal intensity is weak, and the number of ions is too small to give an accurate statistical sampling of the isotope distribution, the observed peak shape will be skewed and its centroid will be shifted. The observed isotope distribution also depends upon the isotopic composition of the elements in the sample. While this is normally treated to be a constant, differences in the ratio of ^{13}C to ^{12}C have been reported for biological molecules obtained from different sources, and these can shift the value of the measured average mass.[81] Finally, the procedure of calculating a centroid is susceptible to a determinant statistical error, which causes average mass measurements to underestimate the actual molecular weight.[82] For these reasons, the mass accuracy of an average molecular weight is no less than 0.01% (100 ppm) under the best of circumstances, and is often larger.

In contrast to measurements of average molecular weight, the accuracy of a monoisotopic molecular weight measurement is far less susceptible to fluctuations in the isotope distribution. Monoisotopic molecular weights can be determined to better than 10 ppm accuracy by FTMS. For biomolecules with molecular weights less than 5000, the monoisotopic peak is generally observed in the mass spectrum, and the assignment of the monoisotopic molecular weight is straightforward. For higher molecular weight compounds, the monoisotopic peak is often not observed, and its value must be inferred from the measured values of the ions that contain heavy isotopes. The key to determining the monoisotopic molecular weight from isotopically resolved data is to assign the number of heavy isotopes present in each of the observed peaks. The most abundant heavy isotope in a biomolecule is ^{13}C, and a simplifying assumption can be made that each of the higher mass peaks in the distribution are separated from the monoisotopic peak by a value that is equal to an integer multiplied by the mass difference between ^{12}C and ^{13}C. The monoisotopic molecular weight can be derived by subtracting the mass of the excess number of protons plus the mass difference between ^{13}C and ^{12}C from the mass of the measured peak. For example, in the mass spectrum of ubiquitin, 10+ charge state, shown in Figure 10, the most abundant peak is the fourth peak above the monoisotopic peak. The monoisotopic molecular weight is therefore equal to the mass of the most abundant peak (10 x 857.37) minus the mass of 10 protons (10 x 1.008) minus the mass difference between four ^{12}C and four ^{13}C atoms (4 x 1.003). This calculation yields the value 8559.6, in exact agreement with the calculated monoisotopic molecular weight of bovine ubiquitin.

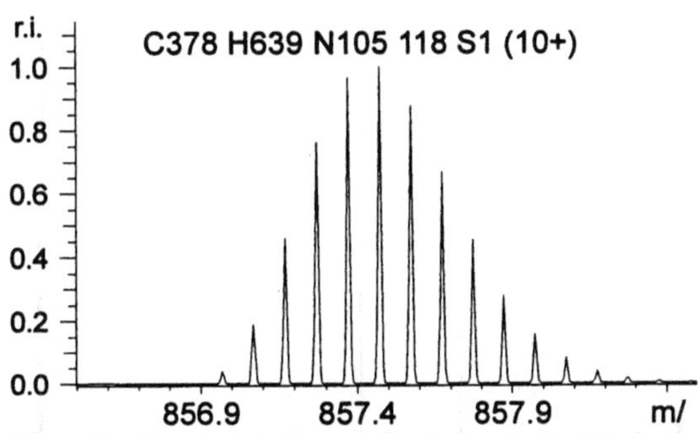

Figure 12. The calculated isotope distribution for the 10+ charge state of ubiquitin. The monoisotopic peak is present on the left side of the distribution. The most abundant peak contains five ^{13}C atoms.

In the case of ubiquitin, the lowest mass isotope peak that is observed is monoisotopic, and therefore the heavy atom content of the most abundant ion is assigned as containing four ^{13}C atoms. Frequently, the abundance of the monoisotopic peak is too small to be observed. In these cases, the isotopic compositions of the molecular ion peaks are assigned by comparing the experimental data to a theoretically derived isotope distribution. Figure 12 shows the isotope distribution that can be calculated for the 10+ charge state of ubiquitin, based on its elemental composition. The most abundant peak in the calculated distribution contains five ^{13}C atoms, rather than the four found in the mass spectrum of Figure 10. This illustrates that the isotope pattern observed for experimental data can exhibit deviations from the theoretical distribution. Such a deviation would cause an error in the determination of the average molecular weight. However, a comparison of all the peaks in the two data sets reveals that the most abundant peak in the experimental data corresponds to the fourth peak in the theoretical isotope distribution. If the correlation between the peaks in the theoretical and experimental data cannot be determined by inspection, the chi-squared statistical test can be used to establish the relationship.[83]

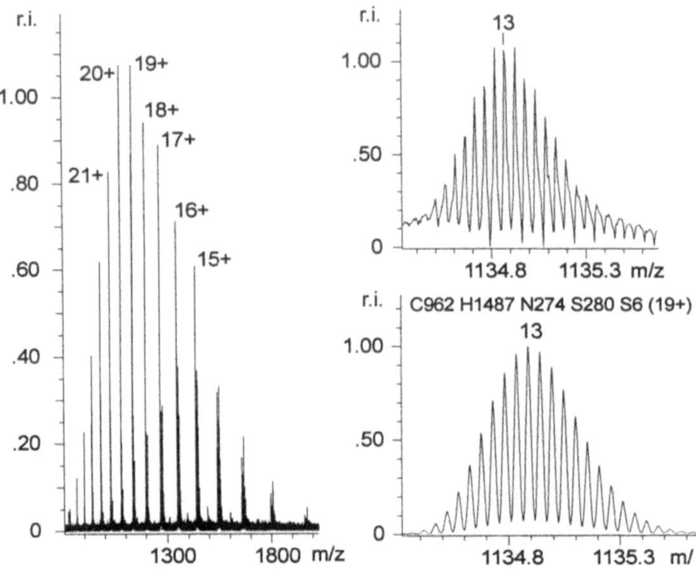

Figure 13. ESI-FTMS mass spectrum of rubrerythrin illustrating the procedure for obtaining a monoisotopic molecular weight. On the left is the complete mass spectrum, exhibiting more than a dozen significant charge states. The top right panel shows a mass scale expansion of the 19+ charge state. The number of heavy isotopes in each peak is assigned by comparison to the theoretically derived isotope distribution, shown in the bottom right panel. The most abundant peak in the theoretical distribution contains 13 ^{13}C-atoms. The peak in the experimental data that contains 13 heavy isotopes is labeled. The correspondence between the theoretically derived data and the experimental data is established by fitting using the chi-squared test.

Assigning the heavy atom content of the isotope peaks becomes more challenging as the molecular weight of the sample increases, when the statistics of the isotope distribution are skewed, and when the elemental composition of the sample is not known *a priori*. Proteins have a fairly constant amino acid composition, and an elemental composition can be estimated by using the average amino acid composition, and scaling it to the average molecular weight of the sample.[83] The procedure for determining a monoisotopic molecular weight for rubrerythrin, a 21.5 KDa protein whose mass spectrum is shown in Figure 13, is presented as an example. An isotopically resolved mass spectrum was obtained by ESI-FTMS. Shown at the left of Figure 13 is a comparison of the 19+ charge state isotope peaks with an isotope distribution based on the known elemental composition of this protein. The most abundant peak in the theoretical data contains thirteen ^{13}C atoms. The monoisotopic peak itself is not observed, as its relative abundance is very small. Therefore, the heavy isotope content of the data can be determined only by comparison to the theoretical isotope distribution. Chi-squared correlation of the theoretical and experimental data allows the assignment of the ^{13}C content of all the isotope peaks. The correlation of the theoretically derived data and the experimental data is based only on a comparison of the relative intensities of the peaks, and not their masses. If the elemental composition of the target molecule was not known in advance, a theoretical distribution could be generated by using the average amino acid composition of all proteins, and scaling this composition to the average molecular weight of this sample. For each isotope peak, we can calculate the monoisotopic molecular weight of rubrerythrin. For example, the mass of the ion that contains 13 ^{13}C atoms (19 x 1134.87 = 21562.53) minus the mass of 19 protons (19 x 1.008 = 19.15) minus the mass difference between thirteen ^{12}C and ^{13}C atoms (13 x 1.003 = 13.04) yields a molecular weight of 21530.31, in error by 20 ppm from the actual value of 21530.75. Each of the isotope peaks of significant abundance for this charge state can be used to extrapolate back to the monoisotopic molecular weight. Furthermore, this procedure can be carried out for several of the different charge states in the mass spectrum. By doing so, several dozen measurements are made of the monoisotopic molecular weight, from which an average monoisotopic value and standard deviation for the measurements can be determined. In the case of the rubrerythrin mass spectrum, the eight most abundant ions in the isotope distribution shown in Figure 13 yield an average value for the monoisotopic molecular weight of 21530.43±0.08, in error by 15 ppm.

5.3 TANDEM MASS SPECTROMETRY

The structural characterization of biomolecules that are ionized by soft methods such as ESI and MALDI is aided by tandem mass spectrometry (MS/MS or MS^n).[84-86] FTMS has unique features which adapt themselves well to tandem mass spectrometry analysis. The capability for simultaneous detection of all the ions that are present in the cell makes FTMS a sensitive method for MS/MS experiments. The nondestructive nature of ion detection provides the capability to form ions by a single ionization event, followed by many stages of MS^n with detection of the products at each step. The latter capability has only been demonstrated, and remains an area of intense research effort.[87] The principles of tandem mass spectrometry and the interpretation of the MS/MS data is presented elsewhere in this volume. The following is a brief presentation of the details relevant to performing tandem mass spectrometry with an FTMS instrument that focuses on collisional dissociation methods.

Collisionally activated dissociation (CAD) is achieved in FTMS by allowing ions to undergo energetic collisions with an unreactive gas. The CAD experiment takes place in the analyzer cell. In order to promote collisions, the pressure must be raised in the analyzer. The subsequent detection of the product ions requires a low pressure. To reconcile these two opposing sets of conditions, the collision gas is usually introduced into the analyzer through a pulsed valve in a short burst that lasts on the order of tens of milliseconds.[88] The pressure in the analyzer remains high for hundreds of milliseconds or seconds, during which time collisional dissociation can occur. The pressure then decreases as the vacuum pumps restores the low pressure necessary to detect ions with high resolution.

Ion activation comes from the conversion of a fraction of the kinetic energy associated with the cyclotron motion of an ion into internal energy. CAD is generally divided into two categories defined by the laboratory energy of the precursor ion. The low energy regime corresponds to kinetic energies less than a few hundreds of electron volts, while the high energy regime is defined by kinetic energies in excess of 1 KeV. The types of fragment ions that are produced varies according to the kinetic energy regime used for CAD. The laboratory frame kinetic energy of an ion in the FTMS instrument is controlled by several factors, as shown in Equation 3. From this equation, we see that high kinetic energies are more easily achieved in instruments with high field magnets. In addition, multiply-charged ions, such as those formed by ESI, achieve much higher kinetic energies than singly-charged ions of the same mass-to-charge. Table I below illustrates the relationship of magnetic field strength and charge on the kinetic energy of an ion of m/q 1000 with a cyclotron radius of 1 cm. The magnetic field strengths listed in the table correspond to values that can be obtained commercially. Surprisingly, multiply charged ions can achieve tens of KeV of kinetic energy. Kinetic energy increases as the square of the cyclotron radius, and so larger cells can be used to promote higher energy collisions. The calculated laboratory energies of the ions in Table I shows that much more energy is available for the dissociation of multiply-charged ions than for singly charged ions. In part, this accounts for the fact that there are many more published accounts of successful CAD of biomolecules by ESI-FTMS than by MALDI-FTMS.

TABLE I. Ion kinetic energy (eV) as a function of charge, z, and magnetic field strength, B (tesla), for an ion of m/q 1000, with a cyclotron radius of 1 cm.

z \ B	1	3	4.7	7	9.4
1	4.8	43	110	240	420
10	48	430	1100	2400	4200
20	480	4300	11,000	24,000	42,000

The entire collisional dissociation experiment takes place in the analyzer cell. For this reason, ion losses from scattering are small in the FTMS instrument compared to ion beam mass spectrometers. Ions can thus be detected even after undergoing "head-on" collisions. Collision gases of high atomic weight (for example, argon or xenon) can be used for CAD experiments in FTMS, and have the advantage of increasing the amount of laboratory energy that is available as center-of-mass energy relative to helium.

The CAD experiment, like other FTMS measurements, is best envisioned as a sequence of events occurring over time. The standard steps of ion isolation, ion activation, and product ion detection occur in one location in the instrument, the analyzer cell, but spread out in time. Conversely, in an ion beam mass spectrometer, such as a magnetic sector or quadrupole instrument, these processes occur simultaneously, but spread out in space. The FTMS arrangement has the advantage that no hardware in addition to that required for standard mass measurements is necessary to perform MS/MS experiments. In contrast, sector and quadrupole mass spectrometers require additional mass filters and collision regions, which adds significantly to the complexity and cost of the instruments.

Ion isolation is easily implemented with the FTMS instrument. Dipolar excitation of an ion at its cyclotron frequency leads to the expansion of its cyclotron orbit. The excitation process can be used to eject ions from the cell by causing them to collide with the cell plates, where they are neutralized and lost from the experiment. In the first step of a CAD experiment, the precursor ion is isolated by exciting all other ions until they leave the cell. This can be accomplished with frequency sweeps to eject ions in the mass range below and above the mass of the precursor. The frequency sweep has the disadvantage that some ion excitation occurs outside the range of interest. Figure 14 shows a frequency sweep in the time domain, as well as in the frequency domain. The excitation power is seen to be uneven throughout, and does not drop to zero at the edges of the excitation range. This means that it is difficult to eject an ion of one mass, while retaining an ion one amu higher or lower in mass with frequency sweep excitation.

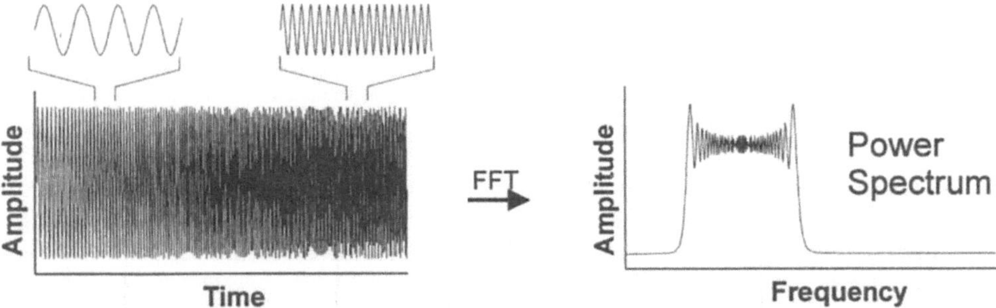

Figure 14. A frequency sweep, or rf chirp, is displayed in the time domain (left). A Fourier transform of the time function shows the power spectrum on the right. Power is not distributed evenly across the excitation range when using a rf chirp. In addition, the power does not drop cleanly to zero at the upper and lower frequency limit of the sweep.

To achieve better selectivity in isolating ions for MS/MS experiments, the excitation waveform can be tailored to provide the optimal characteristics, namely flat response over the excitation range, and sharp cutoffs at the edge of the excitation range. This can be achieved using stored waveform inverse Fourier transform (SWIFT) excitation.[52,89] The basic principles of this technique are illustrated in Figure 15. The desired behavior is defined in the frequency (mass) domain. Excitation of ions to different degrees can be specified. For example, ions in one mass range can receive enough excitation power to be ejected, while in another mass range they are excited enough to undergo CAD or to be detected, while in another mass range, ions are not excited at all. The frequency spectrum is converted to the excitation waveform, a time domain function, by performing an inverse Fourier transform. The excitation waveform is then amplified and applied to the analyzer cell. In practice, the phase shift is also defined as a function of frequency in such a manner as to achieve even excitation amplitude over time. With SWIFT, power is distributed uniformly across the excitation range, and there are sharp transitions between the regions of excitation and no excitation. High resolution precursor ion selection can be achieved with this method. The excitation spectrum shown in Figure 15 can be used for MS/MS experiments. Ions of high and low mass are ejected from the cell by excitation of their cyclotron frequencies, while a selected ion in the middle of the mass range is excited with less power, so that it remains in the cell, but has sufficient kinetic energy to undergo CAD.

Several different methods have been used to energize ions to promote CAD. The most direct method is to excite an ion at its cyclotron frequency for a short duration, to allow the ion to enlarge its cyclotron radius, and thus achieve sufficient kinetic energy to dissociate. A collision gas can then be pulsed into the mass spectrometer to allow collisions to occur. This process is illustrated in Figure 16. Resonant excitation causes the precursor ion to spiral outwards into a larger radius cyclotron orbit. After the excitation is terminated, the ion follows a circular orbit. Once a precursor ion experiences a collision, it dissociates to form a lower mass precursor ion. The velocity of the

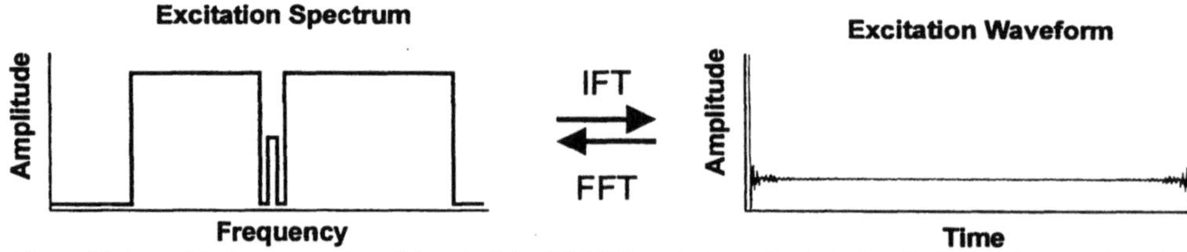

Figure 15. A graphical representation of the principle of SWIFT excitation. The desired excitation performance is defined in the excitation spectrum, a frequency domain function, as shown on the left. An inverse Fourier transform of this function yields the excitation waveform, a time domain function, shown on the right. The excitation waveform is amplified and applied to the analyzer cell excitation plates.

precursor and product are approximately the same for a glancing collision, and so the kinetic energy of the product is related to that of the precursor by the ratio of their masses. This means that the lighter product ion has less kinetic energy than the precursor. Consequently, its cyclotron orbit will be reduced in size. The trajectory of the product lies tangential to that of the precursor. Figure 16 illustrates this process for the dissociation of a singly charge ion of m/q 500 fragmenting to form a product ion of m/q 100. As can be seen, the center of the cyclotron orbit of the product ion is offset from the center of the cell. The magnitude of the offset depends upon the initial radius of the precursor ion's cyclotron orbit after excitation, and the ratio of the masses of the precursor and product ions. Ions that are offset by a considerable amount from the center of the cell cannot be detected efficiently, as excitation may cause them to strike the cell wall and be lost by neutralization. Thus, high kinetic energy collisions which require a large radius of the cyclotron orbit are difficult to achieve by this method. Recent developments of methods to axialize ions, that is to move them to the center of the cell, offers the possibility to perform CAD efficiently for ions with large cyclotron radii.[90-94]

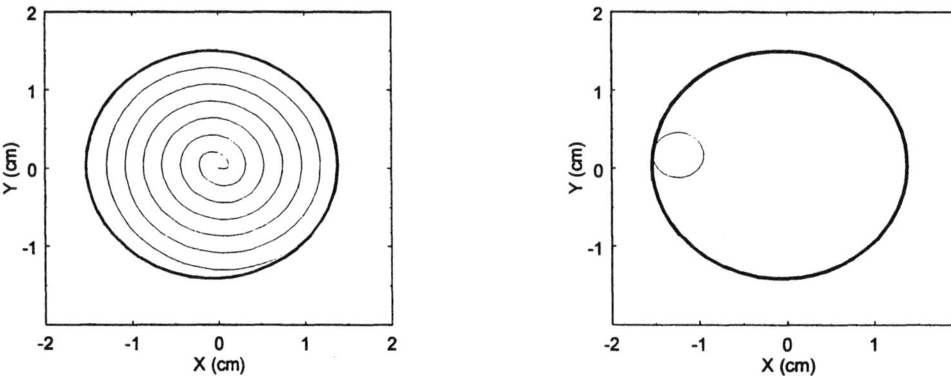

Figure 16. Ion trajectories illustrating ion excitation and dissociation in an FTMS instrument. An ion that undergoes resonant excitation spirals outwards, as shown on the left. After the excitation is terminated, the ion follows a circular cyclotron trajectory. Upon collision, the ion dissociates to form a lower mass product, which undergoes cyclotron motion with a smaller radius orbit, as shown on the right. The small radius product ion orbit is seen to lie tangential to the large radius precursor ion orbit, and is therefore displaced from the center of the cell.

Other methods of ion excitation have been developed to make the detection of product ions more efficient for CAD experiments. Sustained off-resonance irradiation (SORI)[95] has been found to be particularly useful for the dissociation of multiply-charged ions formed by ESI.[96-98] With this method, the precursor ion is excited at a frequency that is slightly different from its cyclotron frequency. The excitation field alternately is in phase and out-of-phase with the motion of the ion. When the excitation signal is in phase, the ion gains energy and expands its cyclotron orbit. When the excitation is out-of-phase, the ion loses energy and its cyclotron orbit contracts. Figure 17 shows the trajectory of an ion that is undergoing SORI excitation. As a result of this cyclical expansion and contraction, the precursor ion is not ejected from the cell by sustained off-resonance excitation, in contrast to sustained resonance excitation. For SORI CAD, a collision gas is introduced through a pulsed valve during the off-resonance excitation. A large number of low energy collisions are sustained while the ions are alternately excited and deexcited. The internal energy of the precursor slowly increases until the ion fragments through its lowest energy channel. SORI dissociation is an efficient process. By sustaining the excitation, all the precursor ion can be dissociated. Most of the product are too far out of resonance with the excitation signal to be excited by this method, and therefore do not undergo further dissociation. This method of CAD has been particularly useful for the dissociation of multiply-charged ions formed by ESI, as the coulombic repulsion resulting from the large number of charges on an ion causes it to fragment to produce complementary pairs of peaks of analytical significance in the mass spectrum.

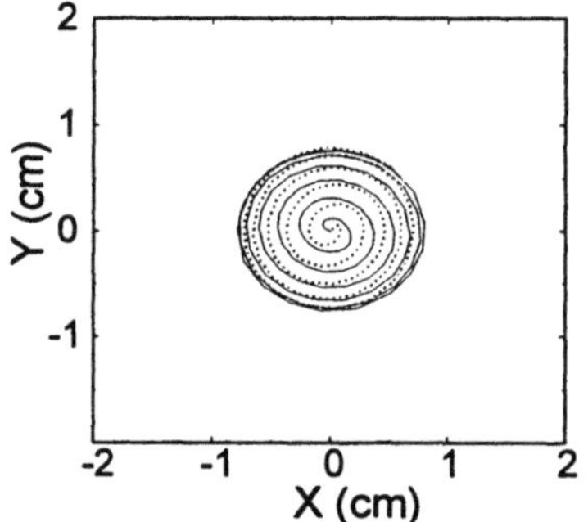

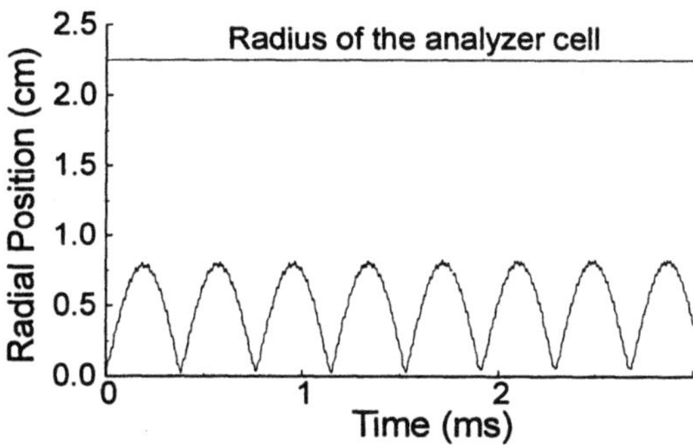

Figure 17. (Left) An ion trajectory that illustrates the effect of off-resonance excitation. An ion that is initially at the center or the cell is excited to a larger cyclotron orbit (solid line). With time, the excitation waveform becomes out-of-phase with the ion motion, and the cyclotron orbit contracts (dashed line). (Right) The periodic expansion and contraction of the cyclotron radius is plotted versus time. The period depends upon the amplitude of the excitation signal, and the frequency difference between the excitation and cyclotron frequencies.

The FTMS instrument has a distinct advantage for CAD experiments of multiply charged ions compared to other mass spectrometers. When multiply-charged ions dissociate, they can produce product ions that have both different mass and different charge than their precursors. Thus the tandem mass spectra of large molecules that produce many different fragments are difficult to interpret. The isotopic resolution that is achieved by FTMS allows the charge of an ion to be assigned directly, without any other information about other members of its charge state distribution. This allows the unequivocal assignment of the mass of each product ion. This is not easily accomplished with other types of mass spectrometers for which isotopic resolution is not achieved.

6. Conclusions

Fourier transform mass spectrometry provides high performance capabilities for the analysis of ions. FTMS measures the molecular weights of proteins and other biomolecules with high resolution and ppm mass accuracy. These high performance capabilities have been demonstrated for ions up to m/z 10000. ESI-FTMS provides the best combination for high performance analysis of biomolecules, since ions are observed at a fairly low mass-to-charge ratio, where the resolution and mass accuracy of the FTMS instrument is high. FTMS is easily adapted to tandem mass spectrometry, and is increasingly being used to provide structural information for biomolecules.

The future of FTMS will be the development of very high magnetic field FTMS instruments, such as are under construction at the National High Magnetic Field Laboratory and Battelle Pacific Northwest National Laboratory. These instruments will yield higher performance and will extend the mass range for analysis to even larger molecules. Methods for manipulating the motion of ions in the analyzer cell and for controlling the ion cloud distribution will continue to develop, and these will lead to sophisticated pulse sequences that permit many sequential stages of tandem mass spectrometry. These high performance capabilities will lead to the widespread application of FTMS for the structural characterization of biomolecules.

7. Acknowledgments

The authors are grateful for support from the National Science Foundation (Grants CHE-9412334 and BIR-9413918).

8. References

1. S. G. Penn, M. T. Cancilla and C. B. Lebrilla, *Anal. Chem.* **68**, 2331 (1996).
2. P. B. M. Andrade and J. M. Riveros, *J. Mass Spectrom.* **31**, 767 (1996).
3. Q. Y. Wu, X. H. Cheng, S. A. Hofstadler and R. D. Smith, *J. Mass Spectrom.* **31**, 669 (1996).
4. K. A. Sannes and J. I. Brauman, *J. Phys. Chem.* **100**, 7471 (1996).
5. C. J. Cassady and S. R. Carr, *J. Mass Spectrom.* **31**, 247 (1996).
6. S. Campbell, M. T. Rodgers, E. M. Marzluff and J. L. Beauchamp, *J. Am. Chem. Soc.* **117**, 12840 (1995).
7. M. W. Senko, J. P. Speir and F. W. McLafferty, *Anal. Chem.* **66**, 2801 (1994).
8. G. S. Gorman and I. J. Amster, *Org. Mass Spectrom.* **28**, 437 (1993).
9. T. Solouki and D. H. Russell, *Appl. Spectr.* **47**, 211 (1993).
10. Y. P. Ho and R. C. Dunbar, *Int. J. Mass. Spectrom. Ion Proc.* **154**, 133 (1996).
11. D. M. Peiris, M. A. Cheeseman, R. Ramanathan and J. R. Eyler, *J. Phys. Chem.* **97**, 7839 (1993).
12. D. P. Little, J. P. Speir, M. W. Senko, P. B. Oconnor and F. W. McLafferty, *Anal. Chem.* **66**, 2809 (1994).
13. T. L. Zhang, L. P. Wang, M. Hashmi, M. W. Anders, C. Thorpe and D. P. Ridge, *Chem. Res. Toxicol.* **8**, 907 (1995).
14. R. M. Li, R. L. Smith and H. I. Kenttamaa, *J. Am. Chem. Soc.* **118**, 5056 (1996).
15. Q. Z. Chen, K. Cannell, J. Nicoll and D. V. Dearden, *J. Am. Chem. Soc.* **118**, 6335 (1996).
16. M. K. Green, E. Gard, J. Bregar and C. B. Lebrilla, *J. Mass Spectrom.* **30**, 1103 (1995).
17. N. G. Alameddin, M. F. Ryan, J. R. Eyler, A. R. Siedle and D. E. Richardson, *Organometallics* **14**, 5005 (1995).
18. J. P. Speir and I. J. Amster, *J. Am. Soc. Mass Spectrom.* **6**, 1069 (1995).
19. E. O. Lawrence and N. E. Edlefsen, *Science* **72**, 376 (1930).
20. H. Sommer, H. A. Thomas and J. A. Hipple, *Phys. Rev.* **76**, 1877 (1949).
21. A. G. Marshall, M. B. Comisarow and G. Parisod, *J. Chem. Phys.* **71**, 4434 (1979).
22. D. F. Hunt, J. Shabanowitz, J. R. T. McIver, R. L. Hunter and J. E. P. Syka, *Anal. Chem.* **57**, 765 (1985).
23. C. L. Wilkins, D. A. Weil, C. L. C. Yang and C. F. Ijames, *Anal. Chem.* **57**, 520 (1985).

24. R. B. Cody, J. A. Kinsinger, S. Ghaderi, I. J. Amster, F. W. McLafferty and C. B. Brown, *Anal. Chim. Acta* **178**, 43 (1985).
25. J. A. Loo, J. P. Quinn, S. I. Ryu, K. D. Henry, M. W. Senko and F. W. McLafferty, *Proc. Natl. Acad. Sci. USA* **89**, 286 (1992).
26. B. E. Winger, S. A. Hofstadler, J. E. Bruce, H. R. Udseth and R. D. Smith, *J. Am. Soc. Mass Spectrom.* **4**, 566 (1993).
27. Y. Z. Li, D. P. Little, H. Koster, R. L. Hunter and R. T. McIver, *Anal. Chem.* **68**, 2090 (1996).
28. C. Koster, J. A. Castoro and C. L. Wilkins, *J. Am. Chem. Soc.* **114**, 7572 (1992).
29. M. B. Comisarow and A. G. Marshall, *J. Mass Spectrom.* **31**, 581 (1996).
30. V. H. Vartanian, J. S. Anderson and D. A. Laude, *Mass Spec. Rev.* **14**, 1 (1995).
31. C. Koster, M. S. Kahr, J. A. Castoro and C. L. Wilkins, *Mass Spec. Rev.* **11**, 495 (1992).
32. M. W. Senko and F. W. McLafferty, *Annu. Rev. Biophys. Biomol. Struct.* **23**, 763 (1994).
33. J. T. Brenna, W. R. Creasy and J. Zimmerman, *Adv. Chem. Ser.* 129 (1993).
34. C. L. Holliman, D. L. Rempel and M. L. Gross, *Mass Spec. Rev.* **13**, 105 (1994).
35. A. G. Marshall and L. Schweikhard, *Int. J. Mass. Spectrom. Ion Proc.* **118**, 37 (1992).
36. A. G. Marshall, *Acc. Chem. Res.* **29**, 307 (1996).
37. S. H. Guan, H. S. Kim, A. G. Marshall, M. C. Wahl, T. D. Wood and X. Z. Xiang, *Chem. Rev.* **94**, 2161 (1994).
38. M. P. Irion, *Int. J. Mass. Spectrom. Ion Proc.* **121**, 1 (1992).
39. L. S. Sheng, S. L. Shew, B. E. Winger and J. E. Campana, *ACS Symposium Series* **581**, 55 (1994).
40. S. C. Beu and D. A. Laude, *Int. J. Mass. Spectrom. Ion Proc.* **112**, 215 (1992).
41. Y. Naito, M. Fujiwara and M. Inoue, *Int. J. Mass. Spectrom. Ion Proc.* **120**, 179 (1992).
42. J. A. Marto, A. G. Marshall and L. Schweikhard, *Int. J. Mass. Spectrom. Ion Proc.* **137**, 9 (1994).
43. V. H. Vartanian and D. A. Laude, *Int. J. Mass. Spectrom. Ion Proc.* **141**, 189 (1995).
44. S. H. Guan and A. G. Marshall, *Int. J. Mass. Spectrom. Ion Proc.* **146**, 261 (1995).
45. R. C. Dunbar, J. H. Chen and J. D. Hays, *Int. J. Mass. Spectrom. Ion Proc.* **57**, 39 (1984).
46. H. P. Allemann, H. P. Kellerhals and K. P. Wanczek, *Chem. Phys. Lett.* **84**, 547 (1981).
47. M. B. Comisarow and A. G. Marshall, *Chem. Phys. Lett.* **25**, 282 (1974).
48. M. B. Comisarow, *J. Phys. Chem.* **69**, 4097 (1978).
49. M. Soni, V. Frankevich, M. Nappi, R. E. Santini, J. W. Amy and R. G. Cooks, *Anal. Chem.* **68**, 3314 (1996).
50. M. B. Comisarow and A. G. Marshall, *Chem. Phys. Lett.* **26**, 489 (1974).
51. R. T. McIver, R. L. Hunter and G. Baykut, *Anal. Chem.* **61**, 489 (1989).
52. A. G. Marshall, T. C. L. Wang and T. L. Ricca, *J. Am. Chem. Soc.* **107**, 7893 (1985).
53. J. B. Jeffries, S. E. Barlow and G. H. Dunn, *Int. J. Mass. Spectrom. Ion Proc.* **54**, 169 (1983).
54. T. J. Francl, M. G. Sherman, R. L. Hunter, M. J. Locke, W. D. Bowers and R. T. J. McIver, *Int. J. Mass. Spectrom. Ion Proc.* **54**, 189 (1983).
55. E. B. J. Ledford, D. L. Rempel and M. L. Gross, *Anal. Chem.* **56**, 2744 (1984).
56. S. H. Guan, A. G. Marshall and S. E. Scheppele, *Anal. Chem.* **68**, 46 (1996).
57. T. L. Spell, S. E. Delong and W. R. Creasy, *Int. J. Mass. Spectrom. Ion Proc.* **124**, 223 (1993).
58. M. Attina, F. Cacace, F. Grandinetti, G. Occhiucci and A. Ricci, *Int. J. Mass. Spectrom. Ion Proc.* **117**, 47 (1992).
59. C. H. Watson, J. Wronka, F. H. Laukien, C. M. Barshick and J. R. Eyler, *Anal. Chem.* **65**, 2801 (1993).
60. C. M. Barshick and J. R. Eyler, *J. Am. Soc. Mass Spectrom.* **3**, 122 (1992).
61. J. A. Loo, E. R. Williams, I. J. Amster, J. J. P. Furlong, B. H. Wang, F. W. McLafferty, B. T. Chait and F. H. Field, *Anal. Chem.* **59**, 1880 (1987).
62. I. J. Amster, J. A. Loo, J. J. P. Furlong and F. W. McLafferty, *Anal. Chem.* **59**, 313 (1987).
63. M. E. Castro and D. H. Russell, *Anal. Chem.* **56**, 578 (1984).
64. M. E. Castro, L. M. Mallis and D. H. Russell, *J. Am. Chem. Soc.* **107**, (1985).
65. J. A. Castoro, C. Koster and C. Wilkins, *Rapid Commun. Mass Spectrom.* **6**, 239 (1992).
66. R. T. McIver, R. L. Hunter and W. D. Bowers, *Int. J. Mass. Spectrom. Ion Proc.* **64**, 67 (1985).
67. Y. L. Huang, S. H. Guan, H. S. Kim and A. G. Marshall, *Int. J. Mass. Spectrom. Ion Proc.* **152**, 121 (1996).
68. P. Kofel, M. Alleman, H. Kellerhals and K. Wancek, *Int. J. Mass. Spectrom. Ion Proc.* **65**, 97 (1985).
69. C. M. Barshick and J. R. Eyler, *J. Am. Soc. Mass Spectrom.* **4**, 387 (1993).
70. P. A. Limbach, A. G. Marshall and M. Wang, *Int. J. Mass. Spectrom. Ion Proc.* **125**, 135 (1993).
71. S. C. Beu, M. W. Senko, J. P. Quinn, F. M. Wampler and F. W. McLafferty, *J. Am. Soc. Mass Spectrom.* **4**, 557 (1993).
72. B. E. Winger, S. A. Hofstadler, J. E. Bruce, H. R. Udseth and R. D. Smith, *J. Am. Soc. Mass Spectrom.* **4**, 566 (1993).
73. S. A. Hofstadler, Q. Y. Wu, J. E. Bruce, R. D. Chen and R. D. Smith, *Int. J. Mass. Spectrom. Ion Proc.* **142**, 143 (1995).

74. C. G. Dekoster, M. C. Duursma, G. J. Vanrooij, R. M. A. Heeren and J. J. Boon, *Rapid Commun. Mass Spectrom.* **9**, 957 (1995).
75. R. T. McIver, Jr., Y. Z. Li and R. L. Hunter, *Int. J. Mass. Spectrom. Ion Proc.* **132**, L1 (1994).
76. M. Dey, J. A. Castoro and C. L. Wilkins, *Anal. Chem.* **67**, 1575 (1995).
77. R. C. Beavis and B. T. Chait, *Chem. Phys. Lett.* **181**, 479 (1992).
78. C. Koster, J. A. Castoro and C. L. Wilkins, *J. Am. Chem. Soc.* **114**, 7572 (1992).
79. J. Yao, M. Dey, S. J. Pastor and C. L. Wilkins, *Anal. Chem.* **67**, 3638 (1995).
80. M. L. Easterling, C. C. Pitsenberger, S. S. Kulkarni, K. L. Taylor and I. J. Amster, *Int. J. Mass. Spectrom. Ion Proc.*, in press.
81. J. T. Brenna, *Acc. Chem. Res.* **27**, 340 (1994).
82. R. A. Zubarev, P. A. Demirev, P. Hakansson and B. U. R. Sundqvist, *Anal. Chem.* **67**, 3793 (1995).
83. M. W. Senko, S. C. Beu and F. W. McLafferty, *J. Am. Soc. Mass Spectrom.* **6**, 229 (1995).
84. K. Biemann and I. A. Papayannopoulos, *Acc. Chem. Res.* **27**, 370 (1994).
85. A. Kuksis and J. J. Myher, *J. Chromatogr., Biomed. Appl.* **671**, 1 (1995).
86. D. Favretto and P. Traldi, *Mass Spec. Rev.* **12**, 313 (1993).
87. T. Solouki, L. Pasa-Tolic, G. S. Jackson, S. Guan and A. G. Marshall, *Anal. Chem.* **68**, 3718 (1996).
88. T. J. Carlin and B. S. Freiser, *Anal. Chem.* **55**, 571 (1983).
89. T. C. L. Wang, T. L. Ricca and A. G. Marshall, *Anal. Chem.* **58**, 2935 (1986).
90. L. Schweikhard, S. H. Guan and A. G. Marshall, *Int. J. Mass. Spectrom. Ion Proc.* **120**, 71 (1992).
91. J. P. Speir, G. S. Gorman, C. C. Pitsenberger, C. A. Turner, P. P. Wang and I. J. Amster, *Anal. Chem.* **65**, 1746 (1993).
92. S. H. Guan, M. C. Wahl, T. D. Wood and A. G. Marshall, *Anal. Chem.* **65**, 1753 (1993).
93. T. D. Wood, C. W. Ross and A. G. Marshall, *J. Am. Soc. Mass Spectrom.* **5**, 900 (1994).
94. S. H. Guan, H. S. Kim, A. G. Marshall, M. C. Wahl, T. D. Wood and X. Z. Xiang, *Chem. Rev.* **94**, 2161 (1994).
95. J. W. Gauthier, T. R. Trautman and D. B. Jacobson, *Anal. Chim. Acta* **246**, 211 (1991).
96. V. C. M. Dale, J. P. Speir, G. H. Kruppa, C. C. Stacey, M. Mann and M. Wilm, *Biochem. Soc. Trans.* **24**, 943 (1996).
97. B. L. Schwartz, J. E. Bruce, G. A. Anderson, S. A. Hofstadler, A. L. Rockwood, R. D. Smith, A. Chilkoti and P. S. Stayton, *J. Am. Soc. Mass Spectrom.* **6**, 459 (1995).
98. J. E. Bruce, S. L. Vanorden, G. A. Anderson, S. A. Hofstadler, M. G. Sherman, A. L. Rockwood and R. D. Smith, *J. Mass Spectrom.* **30**, 124 (1995).

Part III
Application to Biomolecules

PROTEIN CHARACTERIZATION BY MS IN THE PHARMACEUTICAL INDUSTRY

Structural Characterization of Modified Forms of Recombinant Hirudin Sequence Variant 1 and Direct Batch Analysis

D. MÜLLER, P. SCHINDLER, M. COULOT and W. J. RICHTER,
Central Research Services,

W. MAERKI and H. GROSSENBACHER
Core Drug Discovery Technologies

Ciba-Geigy Ltd
CH-4002 Basel, Switzerland

Summary

LC-MS based methodology for the rapid identification of congeners in preparations of the protein *r-hirudin sequence variant 1* (rHV1, structure I) was developed. On-line coupling of reversed phase HPLC with electrospray ionization mass spectrometry (LC-ESMS) was found sufficiently sensitive to allow *direct detection* of such congeners at low concentrations in batch samples despite the presence of the large bulk of r-HV1 and the relatively high concentration of ammonium acetate necessary as a buffer system.

Routine *fingerprinting* of congeners by LC-MS required *full structural characterization* of the modified site, carried out beforehand on sufficient amounts of off-line HPLC-isolated material. For this task, smaller peptides that better exposed their structural detail were produced by combining chemical reactions (reduction of disulfide bonds, pyridylethylation of free SH groups) with enzymatic digestion (Glu-C, Asp-N or trypsin). Peptides carrying structural modifications were either identified in ESMS 'maps' by changes in mass relative to unmodified peptides in reference digests or, when isobaric, by changes in retention behavior in HPLC. In general, the modified peptides were subsequently *sequenced* by tandem mass spectrometry (ES-MSMS) in order to define and locate the sites in question. In a few cases MSMS was ineffective and sequencing by automated Edman degradation was employed.

LC-ESMS on representative batches established the presence of low concentrations of two *isoforms* of rHV1 in addition to the bulk component, of at least two *monodehydration* products, and of several *mono- and dihexosyl*

R. M. Caprioli et al. (eds.), Selected Topics and Mass Spectrometry in the Biomolecular Sciences, 317–345.

derivatives. As illustrated for the off-line isolated Asp33 isoform of rHV1 (*r[iso-Asp33]-HV1*) α->β conversion at the isomerized site can be deduced from the formation of a highly diagnostic β-specific sequence ion in low-energy collision induced dissociation. The two dehydration products were characterized as *r[succinimide33]-* and *r[succinimide53]-HV1* after ring opening at the respective anhydro-positions with $H_2{}^{18}O$ (discussed in detail for the *53*-isomer). For two of the monohexosylated congeners *O*-mannosylation at Thr4 (*r[Man-Thr4]-HV1*) or, alternatively, Thr7 (*r[Man-Thr7]-HV1*) was established by consecutive enzymatic digestions using Glu-C and Asp-N, followed by ESMS, ESMSMS, Edman degradation and monosaccharide analysis. The third isomer was *O*-hexosylated at Thr45 (*r[Man-Thr45]-HV1*). In one of the dihexosylated homologues (*r[Man-Thr4,Man-Thr7]-HV1*) both Thr4 and Thr7 carried separate monosaccharide residues in contrast to its isomer (*r[Man$_2$-Thr4]-HV1*) substituted only at the Thr4 position, but carrying a disaccharide residue.

1. Introduction

Recombinant hirudin sequence variant 1 (rHV1) is a 65 amino acid protein with structural and biological properties very similar to those of *natural hirudin* isolated from the leech *Hirudo medicinalis*. Both the wild type molecule and its genetically engineered analogue are highly potent inhibitors of human α-thrombin. rHV1, expressed in yeast and purified to homogeneity, is currently under development as an agent for treatment of thrombotic disorders. Its primary molecular structure I (Fig. 1) consists of a rather compact *N*-terminal moiety folded by 3 disulfide bonds, and a flexible *C*-terminal part rich in polar acidic amino acids (Asp and Glu residues). In contrast to the natural molecule Tyr63 is not sulfated.

A number of *rHV1-related products* observed during fermentation and purification were isolated as highly enriched fractions from experimental or production batches, or other related matrices, in quantities (usually a few hundred μg) sufficient for structural analysis in the necessary depth. The structures of these congeners formed in low proportions along with rHV1, reflected only minor structural modifications and had retained largely unaltered biological activity. The structural modifications originated from *post-translational* events (mono- and diglycosylation, phosphorylation, *C*-terminal truncation) and intrinsic *chemical* reactions induced by sequence specific structural properties (e.g. succinimide formation at Asp-Gly positions within the amino acid sequence). As the purification process had been optimized in order to remove these congeners from the bulk of unchanged rHV1, only trace amounts of them were expected in samples representative of the final active substance.

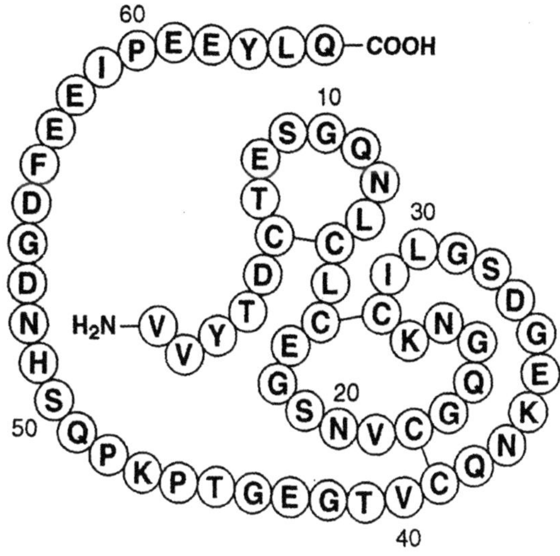

Figure 1. Primary Structure of rHV1 (I).

Mass spectrometry based on *electrospray ionization* as a modern 'soft' ionization technique (ESMS) had proved especially valuable for the *structural characterization* of these modified proteins. The method provided not only molecular mass (M_r) data for the original proteins and corresponding smaller peptides, but allowed, in combination with tandem mass spectrometry (ESMSMS) at the peptide level exact location and characterization of the modifying structural elements. Since in such analyses the backbone structure of the parent protein was *a priori* known, detailed analysis was usually focused only at relatively small segments of the molecule containing the structural deviation.

One of the recent objectives was to develop on-line LC-ESMS methodology for efficient detection and identification of *structurally already known* or only partly characterized low-concentration congeners directly in batches of the active substance. Since an LC-ESMS approach basically relies on component identification by only two criteria (M_r and chromatographic retention time t_R) prior full characterization of the congener structures was indispensable. Carrying out direct LC-MS batch analyses and 'off-line' isolation of HPLC fractions for structural characterization under very similar chromatographic conditions was likewise essential. Only under these conditions can the two t_R scales be directly interrelated and the structural assignments in LC-MS chromatograms be based on the use of 'external' reference structures rather than actual co-injection of authentic materials.

While structural assignments will be reported in this review for *all* congeners so far characterized, the discussion of structure elucidation in detail will limit itself to one or two typical cases illustrating each of the classes of encountered modifications.

2. Methods

LC-ESMS ANALYSIS OF rHV1 BATCHES

Reversed Phase HPLC

A typical rHV1 preparation (lot 800894) was analyzed by rp-HPLC (Applied Biosystems 140B dual syringe pump system equipped with a dynamic mixer of 450 µL, 100 µg injected at a concentration of 1µg/µL in water) using a Hyperchrome Nu C-18 column (Ciba, 5 µm particle size, 4.6 x 250 mm). Solvent A: 0.5% NH_4OAc in 9/1 H_2O/CH_3CN, solvent B: 0.5% NH_4 OAc in 3/1 H_2O/CH_3CN. Gradient: 1 min isocratic at 15% B, in 2 min to 23% B, 1 min isocratic at 23% B, in 13 min to 46% B, 1 min isocratic at 46 % B, in 1 min to 70% B, 1 min isocratic at 70% B, in 2 min to 15% B. Flow rate was 0.75 mL/min and the column temperature was maintained at 50°C. UV detection was performed with an Applied Biosystems 759A detector at 214 nm. The effluent was split 1:124 after UV detection, resulting in a flow of 6 µL/min to the mass spectrometer.

HPLC Coupled ESMS

The HPLC system was interfaced to a PE Sciex API-III triple quadrupole mass spectrometer (Perkin-Elmer Sciex Corp., Thornhill Ontario, Canada) operated under standard conditions (positive electrospray ionization, 70 V orifice potential). The instrument was scanned from m/z 1200 to 2000 with 0.25-Da mass steps and a dwell time of 2 msec (scan time of 6.4 sec). The indispensable use of 0.5% NH_4OAc in the solvent system, while limiting ESMS response, still allowed sensitive detection of products, yet precluded semi-quantitation.

ISOLATION AND STRUCTURAL CHARACTERIZATION OF rHV1-RELATED PRODUCTS

Sample Preparations

rp-HPLC separations for off-line isolation of rHV1 congeners were performed on Merck Hitachi Equipment (Merck cartridge, Supersphere 100 C-18 column; 5 μm particle size; 4.0 x 125 mm). Solvent A: 0.5% NH_4OAc in 9/1 H_2O/CH_3CN, solvent B: 0.5% NH_4OAc in 3/1 H_2O/CH_3CN. Gradient elution from 23% B to 46% B in 15 min. Flow rate 1 mL/min. Sample loads up to 5 mg (enrichment step) and up to 1 mg (isolation for ESMS analysis) were employed. The collected fractions were lyophilized prior to MS analyses.

Chemical and Enzymatic Reactions

Disulfide bond reductions as well as digestions with Glu-C, Asp-N and trypsin were carried out following earlier described procedures [1] and using about 200 μg amounts of rHV1 (reference sample) or the respective structurally unknown congener. Succinimide ring-opening by hydrolyses with $H_2{}^{18}O$ leading to r[iso-Asp33]- and r[iso-Asp53]-HV1 with isotopically labeled modified sites were similarly conducted at the 200 μg level and also reported previously [2].

Automated Edman Degradation.

Amino acid sequencing was performed with either an Applied Biosystems 470A or a Hewlett-Packard G-1000A sequencer. An internal standard, 2-nitroacetophenone, which eluted in between PTH-His and PTH-Tyr, was introduced in order to ensure precise quantitation of PTH derivatives [3].

Off-line Mass Spectrometry

ESMS. All mass spectrometric analyses (unless specified otherwise) were carried out using the PE Sciex API-III triple quadrupole mass spectrometer operated under standard electrospray conditions in the positive ion mode (65 V orifice potential).

ESMSMS. Parent ions of interest were mass selected in MS-1 (quadrupole Q_1, m/z setting fixed), fragmented by CID (argon serving as the target gas in *rf*-only quadrupole Q_2), and mass analyzed for daughter ions in MS-2 (quadrupole Q_3 scanning the required m/z range). Typically, CID experiments of mass selected $[M + 2H]^{2+}$ ions of relevant Glu-C peptides were performed using collision energies of 110 to 150 eV, and target gas densities of several 100 x 10^{16} atoms/cm^2. When required for enhanced sensitivity or precision of mass measurement, narrow scans over pertinent mass regions were performed.

3. Results and Discussion

LC-ESMS ANALYSIS OF rHV1 Batches

Direct analysis of rHV1 batches by on-line LC-ESMS allowed *classification* of congeners according to molecular mass M_r (group assignments) and, after relating retention times with those from their off-line isolation performed for full structural characterization, also individual *identification* (specific structural assignments). For batch 800894 arbitrarily chosen

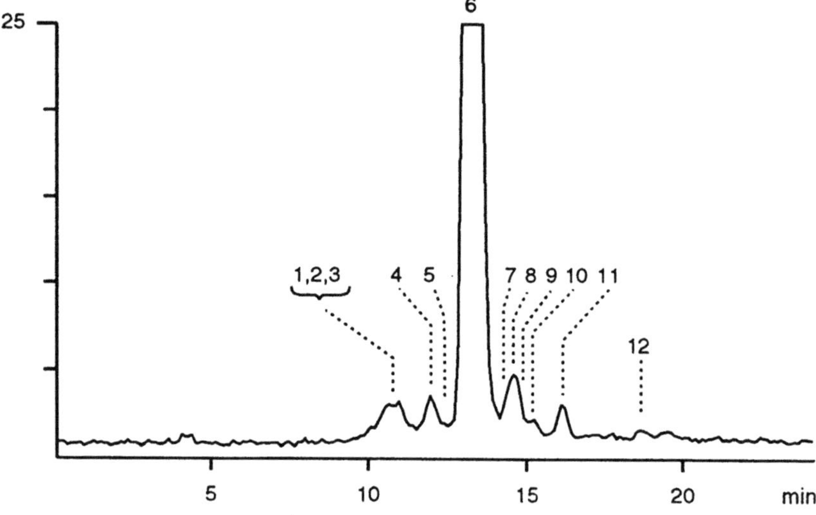

Figure 2. LC-ESMS total ion current chromatogram of rHV1 batch 800894. Structural assignments from off-line isolated congeners to HPLC peaks (n): r[Man-Thr4,Man-Thr7]-HV1 (1), r[iso-Asp53]-HV1 (2, not verified), r[Man-Thr45]-HV1 (3), r[Man-Thr7]-HV1 (4), parent component rHV1 (6), r[Man-Thr4]-HV1 (7), r[succinimide33]-HV1 (8), r[iso-Asp33]-HV1 (9), r[Man_2-Thr4]-HV1 (10), r[succinimide53]-HV1 (11). No assignments were made for peaks 5 and 12. Man = *O*-mannopyranosyl residue.

here for illustration, the total ion current chromatogram (TIC, summed intensities of ions in mass spectra) obtained in such analyses showed, in addition to the intact parent molecule constituting the main component, about six *minor* chromatographic peaks (Fig. 2) under HPLC conditions optimized for most efficient component separation. Despite the necessary high salt concentration (0.5% NH_4OAc in buffers) ESMS spectra and M_r data of good quality were obtained not only for the main component, but also for most of the low-concentration congeners. This is illustrated in Fig. 3 by the ESMS spectrum of the monodehydrated congener eluting at t_R 16.1 min (peak 11 in Fig. 2, mass difference to rHV1 ΔM_r = -18.4 Da).

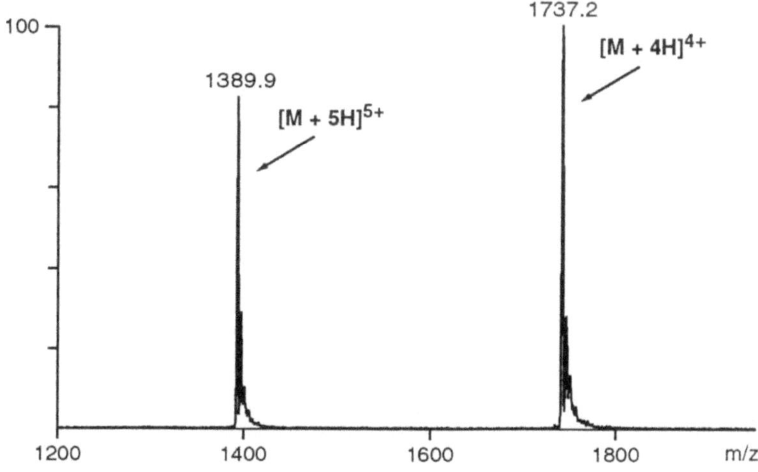

Figure 3. LC-ESMS spectrum of a dehydrated rHV1 product (peak 11, t_R 16.1 min).

First casual inspection of the LC-ESMS spectra of the most prominent chromatographic peaks already permitted group assignments of the eluting species to *4 major classes* of congeners: (i) rHV1 isomers (same M_r), (ii) monodehydrated products (mass differences ΔM_r = -18 Da relative to rHV1), (iii) monohexosylated (ΔM_r = 162 Da) and (iv) dihexosylated rHV1-derivatives (ΔM_r = 324 Da). Systematic search, e.g. by retrieving class-specific selected mass chromatograms for corresponding quadruply protonated species ($[M + 4H]^{4+}$) allowed screening for additional low-concentration congeners of the same type (*cf.* traces A to D in Fig. 4, and the TIC chromatogram in trace E given for reference). By this means the presence of two major singly dehydrated products (Fig. 4, trace A with 'peaks' 8 and 11) and of one minor, presumably also dehydrated species (peak 12) were established. Similarly, two components having the same molecular mass as rHV1 (trace B with peaks 2 and 9), as well as 3 monohexosylated and 3 dihexosylated derivatives of rHV1 (traces C with peaks 3, 4 and 7, and D with peaks 1, 5 and 10, respectively) were detected and identified. Artifact peaks (indicated by *) in the mass chromatogram of the 'rHV1 class' (trace B) resulted from an overlap of $[M + 4H]^{4+}$ ions of rHV1 (m/z 1741.8, $\Delta = 0.5$) with $[M + 3H + NH4]^{4+}$adduct ions of the two major dehyd-

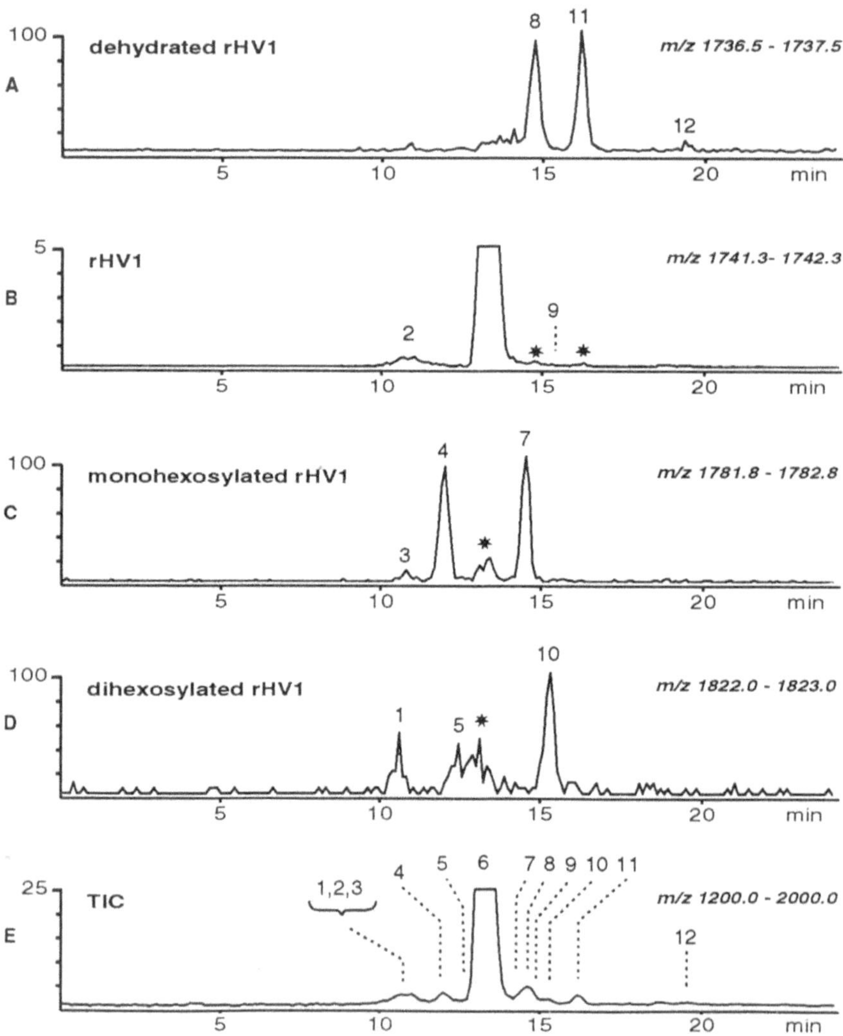

Figure 4. Selected mass chromatograms extracted from LC-ESMS analysis of rHV1 (batch 800894). Signals marked with * indicate chemical background (*cf.* text).

rated products (peaks 8 and 11; m/z 1741.5, Δ = 0.5) and were promoted at the near neutral pH by the necessary high NH_4OAc concentration. Higher order clusters of the overly abundant rHV1 ions (addition of small neutral molecules present as 'chemical noise' at the ultra-trace level) were responsible for the artifact signals in traces C and D. Detailed inspection of series of single mass spectra was necessary in 'dubious' chromatographic regions for avoiding false assignments. Other batches analyzed (data not shown) provided very similar results and thus indicated a high consistency in the congener patterns of batches obtained from different production campaigns.

The ***structural assignments*** resulting from interrelating the on-line LC-MS analysis of this representative batch with analyses of off-line isolated congeners are given in the legend of Fig. 2.

Structural Characterization of Isolated rHV1-Related Products

General Strategy

To enable determination of the nature and site of a greater variety of structural modifications resulting from post-translational events as well as chemical processes, a flexible general strategy based on ESMS and ES-MSMS has been developed (*cf.* Scheme 1).

Following chromatographic isolation, ***M_r determination*** by ESMS provided a first indication of the encountered class of congeners and allowed to locate the isolated fraction within the corresponding retention window of the LC-MS chromatogram (e.g. Fig. 2). In general, for the in-depth struc-tural analysis by MS, peptides (and modified peptides) of manageable size were generated in a first stage by ***Glu-C digestion*** after reduction of the 3 disulfide linkages, pyridylethylation (PE) of the resulting SH groups, and purification of the hexa-PE-derivative by rp-HPLC. In some cases, a second digestion step using another enzyme (Asp-N, trypsin) had to be inserted, yet mostly required no intermediate purification or component separation of the first peptide generation.

Bringing MS into play, the next stage of the procedure engaged direct ESMS analysis of the crude enzymatic digests for *peptide mapping*, i.e. surveying the compositions of the hydrolysates. For non-isobaric congeners, the peptide of interest carrying a modified site will exhibit, in comparison with its analogue in a map of identically processed reference material, a deviating M_r value that allows its immediate detection (comparative mapping). Fig. 5 shows a typical ESMS map for a Glu-C digest of a rHV1 sample (batch 810292) displaying the multiply protonated peptides v_1 to

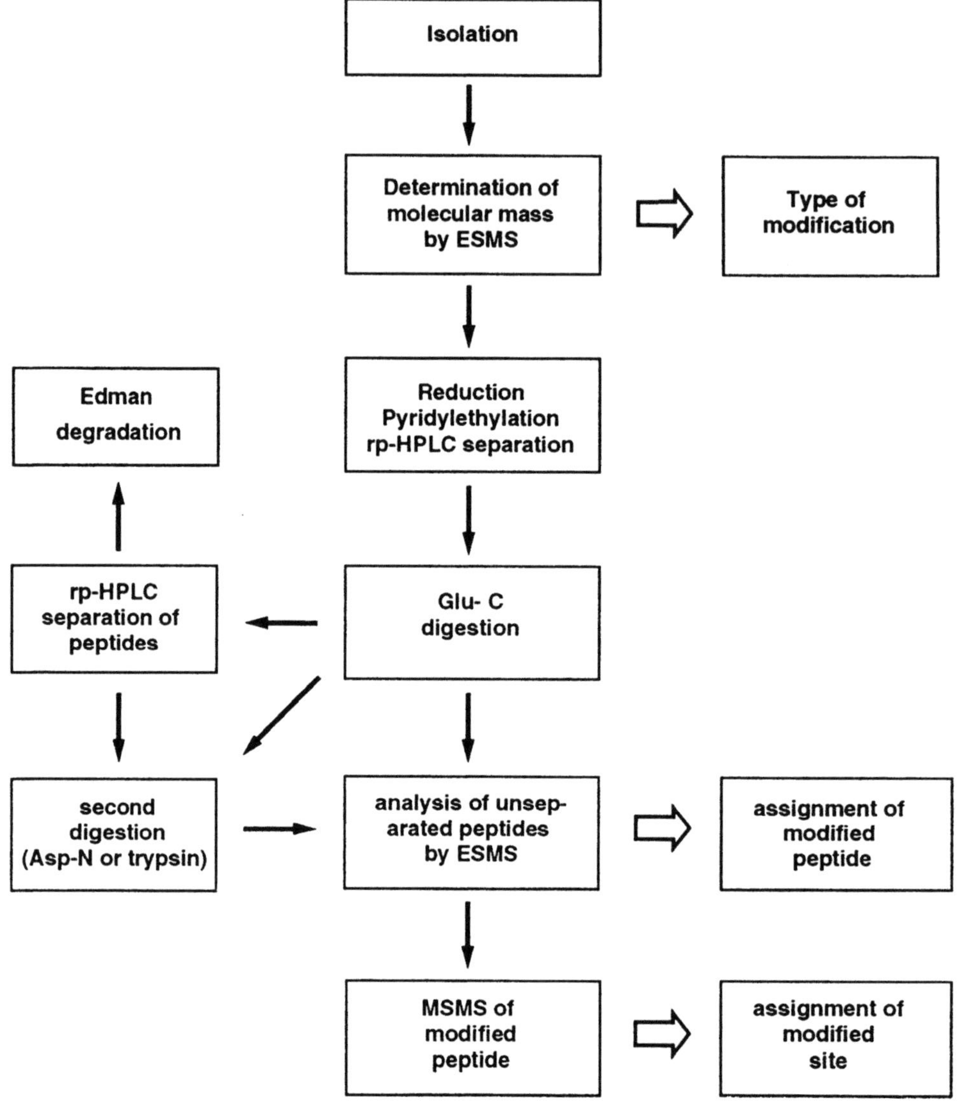

Scheme 1. General strategy for the structural characterization of rHV1-related products

v_6 (together with the incompletely digested peptides v_{3+4} and v_{5+6}) covering the complete amino acid sequence of the parent molecule. Mass measurements on which identification of these peptides was based are compiled in table 1. ESMS maps will be given also in the subsequent sections in such tabular form in which M_r values obtained from the major observed protonated species are assigned and summarized.

At the final stage of the approach, two-dimensional mass spectrometry provided the necessary structural depth in that *sequencing* by MSMS was applied for exactly locating the modified site within the targeted peptide.

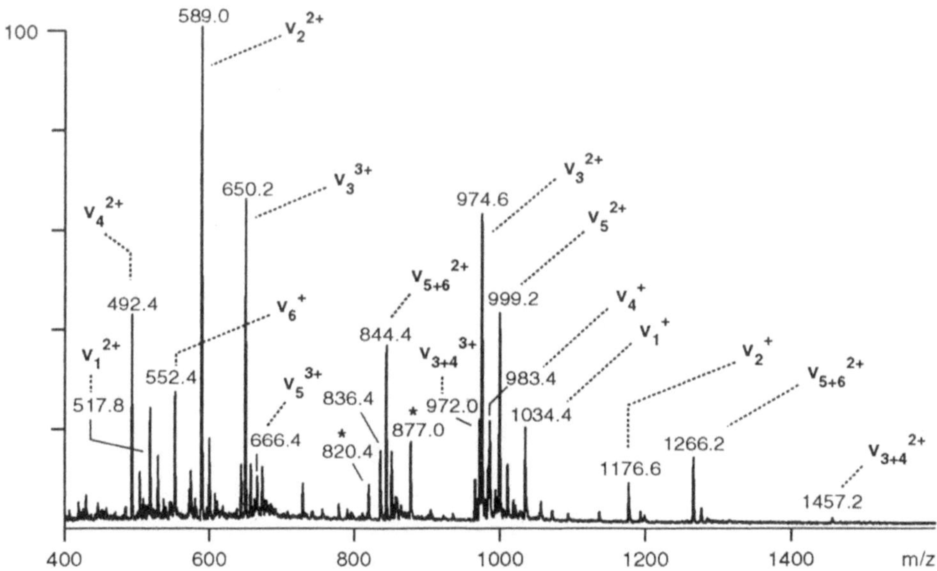

Figure 5 ESMS spectrum of Glu-C digest of rHV1 (PE derivative, batch 810292). Signals marked with * indicate products of MS fragmentation (*cf.* also Fig. 6) induced despite the applied mild desolvation conditions.

In a typical ESMSMS experiment (shown here in Fig. 7 for $[M + 2H]^{2+}$ of v_5, m/z 999.2) parent ions are mass selected and fragmented by low energy CID (*cf.* methods section). In this technique, the most commonly observed daughter ions are formed by heterolytic cleavages of peptidic CO-NH bonds after protonation of the amide nitrogen, either with (*C*-terminal y_n sequence ions) or without concomitant hydrogen transfer (*N*-terminal b_n sequence ions). In most cases the modifying structural elements are readily located as long as such sequence ions shifted in mass relative to a re-

TABLE 1. Molecular masses of Glu-C peptides v_n derived from PE-derivatized rHV1 (batch 810292).

peptide	sequence	M_r calc.	$[M+zH]^{z+}$ (z)	$[M+zH]^{z+}$ (z)	M_r found.
v_1	[1-8]	1033.4[1]	1034.4 (1)	517.8 (2)	1033.5
v_2	[9-17]	1175.5[1]	1176.6 (1)	589.0 (2)	1175.8
v_3	[18-35]	1948.2[2]	974.6 (2)	650.2 (3)	1947.4
v_{3+4}	[18-43]	2913.3[2]	1457.2 (2)	972.0 (3)	2912.7
v_4	[36-43]	982.5[1]	983.4 (1)	492 (2)	982.6
v_5	[44-61]	1997.1[2]	999.2 (2)	666.4 (3)	1996.3
v_{5+6}	[44-65]	2530.6[2]	1266.2 (2)	844.4 (3)	2530.3
v_6	[62-65]	551.3[1]	552.4 (1)		551.4

[1] calculated monoisotopic M_r [2] calculated isotopically averaged M_r.

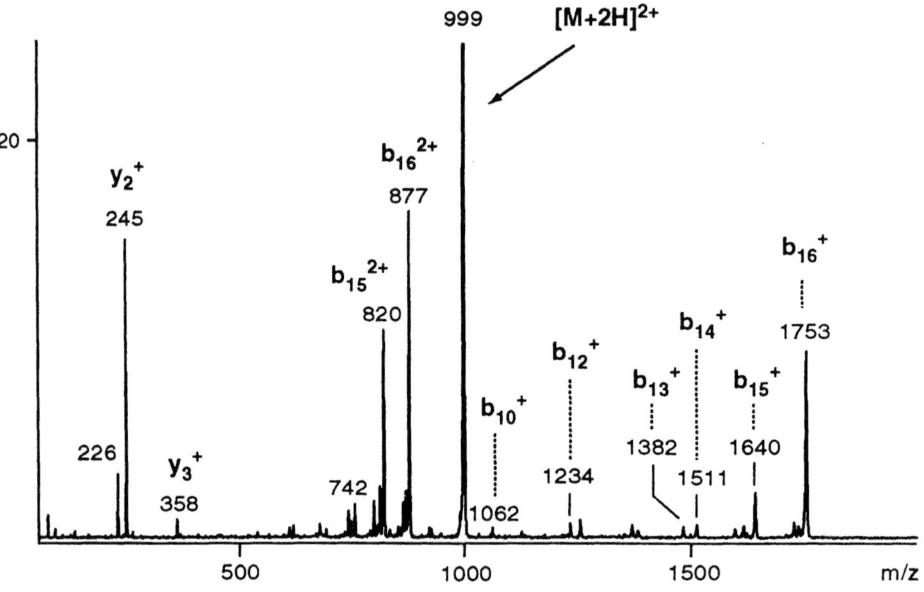

Figure 6. Direct analysis of Glu-C digest of PE-derivatized rHV1 (batch 810292). ESMSMS spectrum of peptide v_5 ($[M + 2H]^{2+}$ ions, m/z 999.2).

ference are detected. Automated Edman degradation after HPLC-isolation of the relevant peptide had to stand in for MSMS in 2 cases where low energy CID failed in this task.

In the remaining sections of this review, applications of this general strategy, with specific variations, will be discussed in some detail. One monodehydration product, one isoform of rHV1, and three mono- or dimannosylated rHV1-derivatives were chosen for illustrating the major types of observed congeners.

Monodehydrated rHV1-Related Products

Two major monodehydration products with retention behaviors closely resembling those of peaks 8 and 11 in the class-specific LC-MS chromatograms of rHV1 (Fig. 4, trace A) had been previously off-line isolated in sufficient quantities for structural characterization [2]. In analogy to the known deamidation at Asn-Gly sequons in variant-2 hirudins [4] dehydration of rHV1 at corresponding Asp-Gly positions was inferred earlier [5] and later shown to proceed by formation of cyclic succinimide structures at Asp33-Gly34 and Asp53Gly54 [2]. For the exact determination of the dehydration sites, ring-opening with $H_2{}^{18}O$ at moderately alkaline pH was utilized in both cases, in order to introduce stable isotope labels marking the positions in question (*cf.* Scheme 2). The mixtures of α- and β-Asp isoforms of rHV1 resulting from this treatment were then directly subjected to the above standard procedure. MSMS sequencing of the modified, i.e.

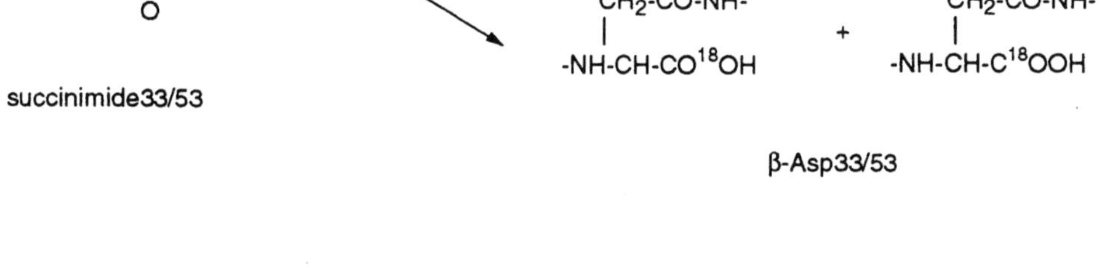

-HN, N- (succinimide ring) + $H_2{}^{18}O$

succinimide33/53

$CH_2CO^{18}OH$ / -NH-CH-CO-NH- + $CH_2C^{18}OOH$ / -NH-CH-CO-NH-

α-Asp33/53

CH_2-CO-NH- / -NH-CH-$CO^{18}OH$ + CH_2-CO-NH- / -NH-CH-$C^{18}OOH$

β-Asp33/53

Scheme 2. Labelling of dehydrated sites in positions 33 or 53 by ring-opening of succinimide group with $H_2{}^{18}O$ (pH 7.8).

^{18}O-labelled peptides indicated the originally dehydrated sites by clear-cut 2-Da mass shifts in sequence ions incorporating the labelled site. In order to illustrate this approach the Asp33 dehydration product (termed Q_5 in the earlier report [2]) is chosen as an example.

r[succinimide33]-HV1. Comparative ESMS mapping for this suspected anhydro-Asp isomer (M_r 6945.0 found by ESMS, M_r 6944.9 peak top mass calculated for [M - H_2O] of rHV1) was carried out after succinimide ring-opening with $H_2{}^{18}O$ and H_2O in two separate, otherwise identical experiments. Results for the unknown and the reference are given in table 2, in which the peptides v_3 ([18-35]) and v_{3+4} ([18-43]) are clearly recognized as the carriers of the suspected modification by their ^{18}O-induced mass shifts (1.7 and 1.8 Da, respectively).

TABLE 2. Comparative mapping of peptides in Glu-C digests of PE-derivatized dehydration product r[succinimide33]-HV1 (peak 8 in LC-MS) after succinimide ring opening with $H_2{}^{18}O$ (unknown, columns B) and H_2O (reference, columns A)

peptide	sequence	M_r calc.	M_r found (A)	M_r found (B)	M_r (B) - M_r (A)
v_1	[1-8]	1033.4[1]	1033.5	1033.4	-0.1
v_2	[9-17]	1175.5[1]	1175.8	1175.8	0.0
v_3	[18-35]	1948.2[2]	1947.6	1949.3	1.7
v_{3+4}	[18-43]	2913.3[2]	2913.0	2914.8	1.8
v_4	[36-43]	982.5[1]	982.4	982.7	0.3
v_5	[44-61]	1997.1[2]	1996.7	1996.3	-0.4
v_{5+6}	[44-65]	2530.6[2]	2530.4	2530.5	0.1
v_6	[62-65]	551.3[1]	551.2	551.2	0.0

[1] calculated monoisotopic M_r [2] calculated isotopically averaged M_r

For sequencing by MSMS, $[M + 2H]^{2+}$ ions of the shorter peptide v_3 were mass selected. In contrast to the results obtained earlier for the other major dehydration product (succinimide53 derivative corresponding to peak 11 in LC-MS), the ESMSMS spectrum of v_3 shown in Fig. 7 displays a relatively uniform distribution of fragment ions due to the absence of proline residues. In this case, the label at Asp33 could not only be identified from the net 1-Da mass shift (2 Da/z shift with z = 2, indicated by

arrows) for b_{16}^{2+}, but also from the 2-Da mass increments observed for all members of the singly charged *C*-terminal y-ion series beyond y_2.

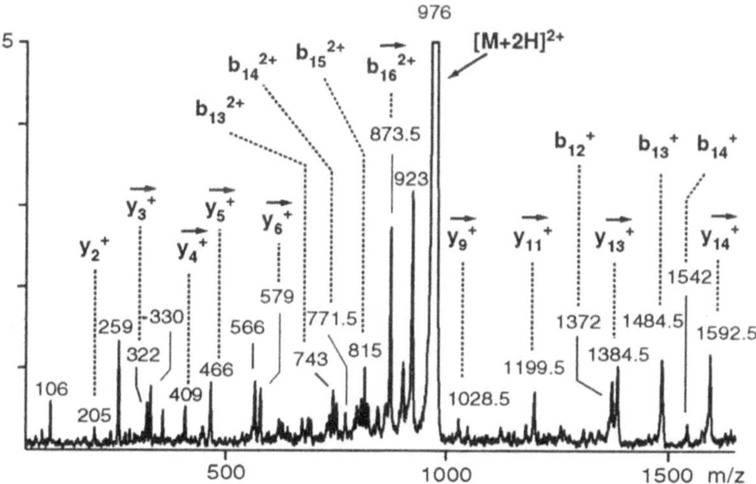

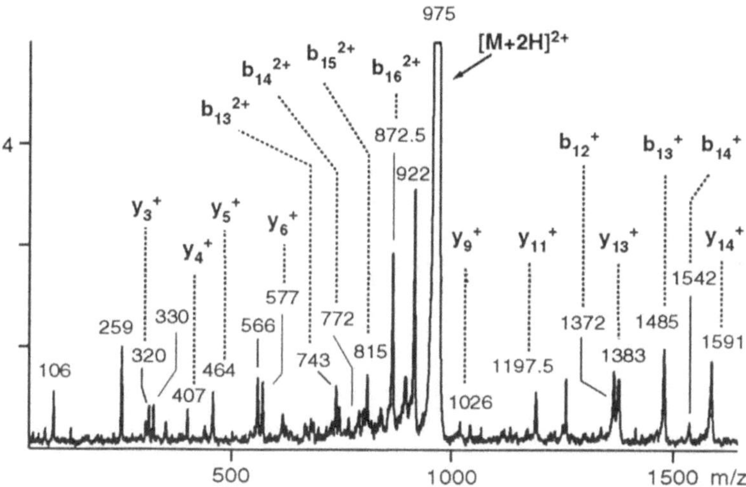

Figure 7. ESMSMS spectra of $[M + 2H]^{2+}$ ions of v_3 ([18-35]) obtained from r[succinimide33]-HV1 after incubation with $H_2{}^{18}O$ (m/z 976, unknown) and H_2O (m/z 975, reference)

β-Asp Isoforms of rHV1

Judged from chromatographic retention the later-eluting of the two rHV1 isomers ('peaks' 2 and 9 in LC-MS analysis, *cf.* Fig. 4) was expected to be a β-Asp congener formed, along with regenerated rHV1, by 'rehydration' of one of the succinimide-type dehydration products in a reaction similar to that used for introducing the ^{18}O labels (*cf.* Scheme 2). The product corresponding in retention time to this isomer had been previously isolated in up-scaled preparative runs conducted under very similar chromatographic conditions. For differentiation between Asp33 and Asp53 as the likely locations of the suspected β-aspartate residue, Edman sequencing stopping at a β-Asp residue represented an obvious alternative, yet would not allow to determine the chemical nature of the structural deviation. According to literature, positive identification together with location of such modifications is, however, offered by two tandem MS approaches employing high-energy CID of $[M + H]^+$ ions of smaller peptides (formerly employing fast atom bombardment for ionization). Carr [6] had, for instance, demonstrated that methyl esters of α-Asp containing peptides can be differentiated from β-Asp isomers by their markedly differing formation of w- and v-type ions. Alternatively, high-energy CID of underivatized β-Asp containing peptides was found to yield highly diagnostic sequence ions, which formally correspond to $[b_n + H_2O]^+$ adducts comprising the complete *N*-terminal sequon adjacent to a β-aspartate located in an *n+1* position [7]. In the course of a recent study on

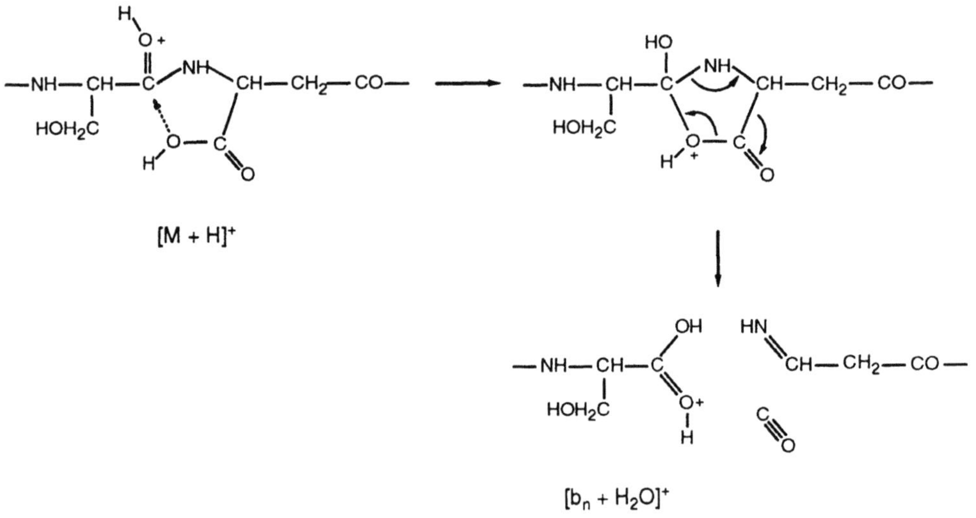

Scheme 3. Formation of $[b_n + H_2O]^+$ ions by charge-promoted OH migration.

the here discussed isoform [8] we were able to show that this diagnostic $[b_n + H_2O]^+$ ion is likewise formed under low-energy CID conditions. That at least in this case a hydroxyl group migrates from the β-aspartyl side-chain (in *n+1* position) to the *N*-terminally neighboring carbonyl group was demonstrated by ^{18}O-labelling (for the proposed mechanism *cf.* Scheme 3). Using this diagnostic tool, characterization of this isoform of rHV1 was achieved without extending the experimental scope of the chosen ESMS and ESMSMS strategy. While the early eluting isoform ('peak' 2 in Fig. 4) has so far not been firmly assigned (an earlier characterized 53-isomer [2] is a candidate), the product corresponding to peak 9 was shown to be α->β iso-merized in the Asp33 position (for structure determination *cf.* [8]).

r[iso-Asp33]-HV1. ESMS analysis of an off-line HPLC isolated fraction corresponding to 'peak' 9 revealed the composite nature of the sample, i.e. a mixture containing the rHV1 isomer under discussion (M_r 6962.0 found, M_r 6962.9 peak top mass calculated for rHV1), and a dihexosylated con-gener of rHV1 (measured mass increment 323.1 Da, *cf.* pertinent section further down) as major and minor components, respectively. The composi-tion of this fraction, which parallels the partly overlapping elution windows of 'peak' 9 (rHV1 mass chromatogram, trace B in Fig. 4) and peak 10

TABLE 3. Identification of v_n peptides HPLC-isolated from Glu-C digest of PE-derivatized rHV1 isoform (corresponding to peak 9 in LC-MS).

peak	t_R [min]	peptide	sequence	M_r calc.	M_r found[3]	M_r found - M_r calc.
1	4.4	v_4	[36-43]	982.5[1]	982.5	0.0
2	5.5	v_6	[62-65]	551.3[1]	551.2	-0.1
3	6.0	v_{1++}	[1-8]	1033.4[1]	1357.6	324.2
4	6.2	v_1	[1-8]	1033.4[1]	1033.5	0.1
5	8.1	v_3^*	[18-35]	1946.9[1]	1946.7[3]	-0.2
6	8.3	v_3	[18-35]	1946.9[1]	1946.8[3]	-0.1
7	8.4	v_{3+4}	[18-43]	2911.3[1]	2911.2[3]	-0.1
8	8.5	v_2	[9-17]	1175.5[1]	1175.4	-0.1
9	9.5	v_5	[44-61]	1997.1[2]	1996.8	-0.3
10	12.9	v_{5+6}	[44-65]	2530.6[2]	2530.5	-0.1

[1] calculated monoisotopic M_r [2] calculated isotopically averaged M_r [3] data obtained at 4000 mass resolution on magnetic sector instrument.

in the LC-MS analysis (mass chromatogram of dihexosylated products, trace D in Fig. 4) required the insertion of an extra HPLC isolation step for the Glu-C digest, in order to separate the expected β-Asp-containing peptides from potential α-isomers originating from the glycosylated minor product. Table 3 summarizes the ESMS-based assignments of the peptides chromatographically isolated after the Glu-C digestion step of the standard procedure.

Most importantly, two separated components (peaks 5 and 6, M_r 1946.7 and 1946.8, respectively), both having molecular masses very close to that expected for v_3 (1946.9 calculated for [18-35]), were detected. Peak 6 was of relatively low intensity and, therefore, assumed to represent the α-isomer derived from the glycosylated minor congener (now apparent as the dihexosylated peptide v_{1++}). The dominant peak 5 represented most probably the β-aspartate isomer. Trypsin digestion of the isolated v_3 peptides in an additional second digestion step provided 'subpeptides' of suitable size ($v_{3.1}$, [18-27]; $v_{3.2}$, [28-35]; *cf.* table 4) for the comparative MSMS analysis.

TABLE 4. Identification of tryptic peptides in digests of HPLC-isolated Glu-C peptides v_3 and v_{3*} (off-line HPLC) derived from PE-derivatized rHV1 isoform (corresponding to peak 9 in LC-MS).

peptide	sequence	M_r calc.	$[M+zH]^{z+}$ (z)	$[M+zH]^{z+}$ (z)	M_r found..	M_r found. - M_r calc.
$v_{3.1}$	[18-27]	1076.5[1]	1068.6 (1)	534.8 (2)	1067.6	0.1
$v_{3.2}$	[28-35]	897.4[1]	898 6 (1)	449.8 (2)	897.6	0.2
$v_{3.1*}$	[18-27]	1067.5[1]	1068.6 (1)	534.8 (2)	1067.6	0.1
$v_{3.2*}$	[28-35]	897.4[1]	898 (1)	449.8 (2)	897.6	0.2

[1] calculated monoisotopic M_r

The tandem mass spectra of the $[M+H]^+$ ions of $v_{3.2}$ shown in Fig. 8 for the suspected α-Asp isomer (peak 6, lower trace) and for $v_{3.2*}$ of the 'unknown' peak 5 (upper trace) appear quite similar. However, in the crucial mass region between m/z 570 and 605, a $[b_n + H_2O]^+$ ion is indeed exclusively observed in the spectrum of $v_{3.2*}$ for $n = 5$ (*cf.* insets in Fig. 8). This clearly established the presence of a β-aspartate residue in the adjacent *C*-terminal position, i.e. position 6 of the peptide v_3 and position 33 in the original protein. Simultaneous isomerization at Asp53 was ruled out by the identity of peptides v_5 (same Edman behavior, identical chromatographic properties).

Glycosylated rHV1-Related Products

All 3 *monohexosylated* congeners of rHV1 detected by LC-MS (peaks 3, 4 and 7 in Fig. 4) were off-line isolated in 3 fractions from batch 810490 (fractions 9, 11 and 7, respectively), again under chromatographic condi-

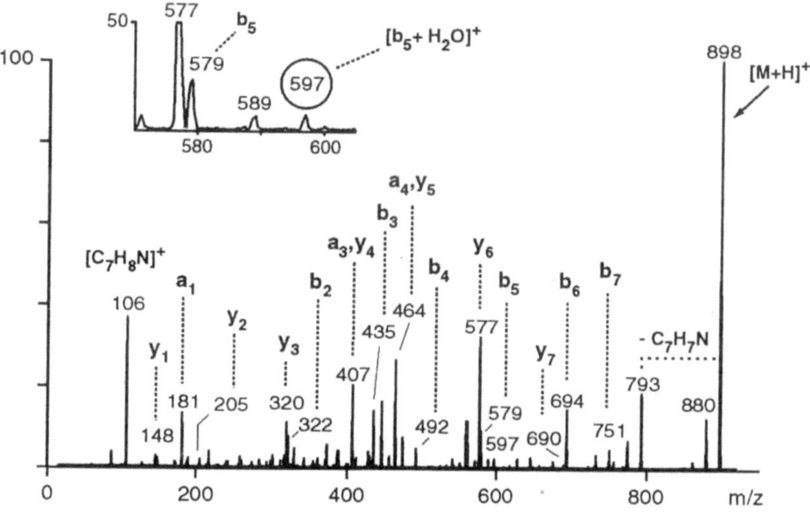

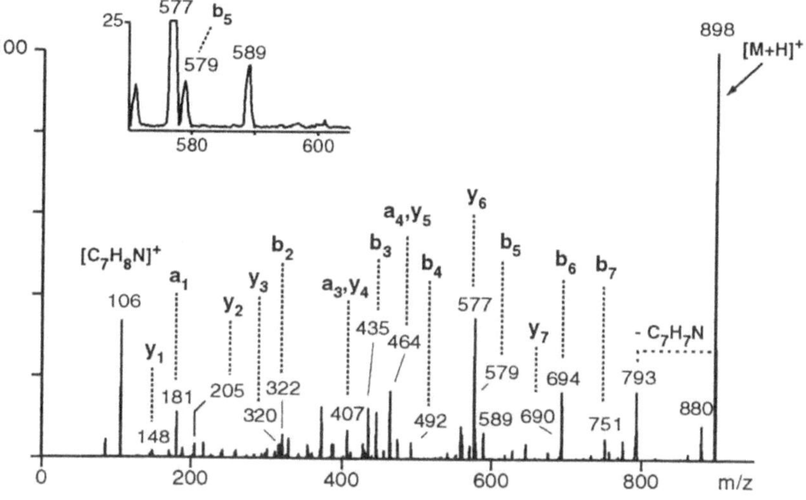

Figure 8. ESMSMS spectra of $[M + H]^+$ ions of tryptic peptides $v_{3.2*}$ and $v_{3.2}$ ([28-35]) obtained from isolated Glu-C peptides v_{3*} and v_3 (upper and lower trace, respectively). The latter peptides were obtained from Glu-C digestion of the PE-derivatized rHV1 isoform corresponding to peak 9 in LC-MS.

tions very similar to those adopted for the LC-MS analysis. The first eluting dihexosylated derivative (peak 1 in LCMS analysis) was contained as a minor product in fraction 9 and the latest-eluting one (peak 10) as a minor component in the β-Asp containing fraction (see above). The third dihexosylated component, i.e. peak 5 in LC-MS eluting close to rHV1 has as yet not been isolated. For establishing the glycosylation sites within the Glu-C derived peptides obtained by the standard procedure, either the isolated glycopeptides or the crude Glu-C hydrolysate had to be 'subdigested' with Asp-N protease in an extra step. ESMSMS and/or automated Edman degradation were mainly used to characterize the products. Routine monosaccharide analysis established mannose ('Man' in the following sections) as the only hexose present in the glycosylated components of the fractions 7 and 11, and in the predominant species in fraction 9 (Oxford Glycosystems, unpublished results). In general, *O*-mannosylation of Ser and Thr residues is observed as the prevailing glycosylation reaction in recombinant proteins expressed in yeast [9].

r[Man-Thr7]-HV1. In a first step the hexose-specific 162-Da mass increment measured in molecular mass determination on fraction 11 (7124.0 Da found by ESMS, M_r 6962.9 peak top mass calculated for rHV1, ΔM_r = 161.1) had to be located within the components of the Glu-C map. A comparison of the observed M_r values of all peptides with those calculated for reference peptides from rHV1 (*cf.* Table 5) unequivocally identified v_{1+}

TABLE 5. ESMS mapping of peptides in Glu-C digest of PE-derivatized fraction 11 (corresponding to peak 4 in LC-MS)

peptide	sequence	M_r calc.	$[M+zH]^{z+}$ (z)	$[M+zH]^{z+}$ (z)	M_r found	M_r found - M_r calc.
v_{1+}	[1-8]	1033.4[1]	1197.0 (1)	599.2 (2)	1196.0	162.6
v_2	[9-17]	1175.5[1]	589.0 (2)	393.0 (3)	1176.0	0.5
v_3	[18-35]	1948.2[2]	874.8 (2)	650.2 (3)	1947.6	-0.6
v_{3+4}	[18-43]	2913.3[2]		972.0 (3)	2931.0	-0.3
v_4	[36-43]	982.5[1]	983.4 (1)	492.4 (2)	982.6	0.1
v_5	[44-61]	1997.1[2]	999.2 (2)	666.6 (3)	1996.6	-0.5
v_{5+6}	[44-65]	2530.6[2]	1266.4 (2)	844.4 (3)	2530.5	-0.1
v_6	[62-65]	551.3[1]	552.4 (1)		551.4	0.1

[1] calculated monoisotopic M_r [2] calculated isotopically averaged M_r

(corresponding to the *N*-terminal peptide [1-8]) as the carrier of the modification (ΔM_r = 162.6 Da). The ESMSMS spectrum of the corresponding $[M + 2H]^{2+}$ ions (m/z 599.0, Fig. 9) confirmed the presence of the hexosyl substituent by the typical pronounced loss of an anhydrohexose unit (m/z 599 - 162/2 -> m/z 518). Since this facile elimination process precedes the formation of actual sequence ions, *no* indication of the position of glycosylation was obtained and alternative methods for the differentiation of the two candidate sites in v_1 (Thr4 or Thr7) had to be considered.

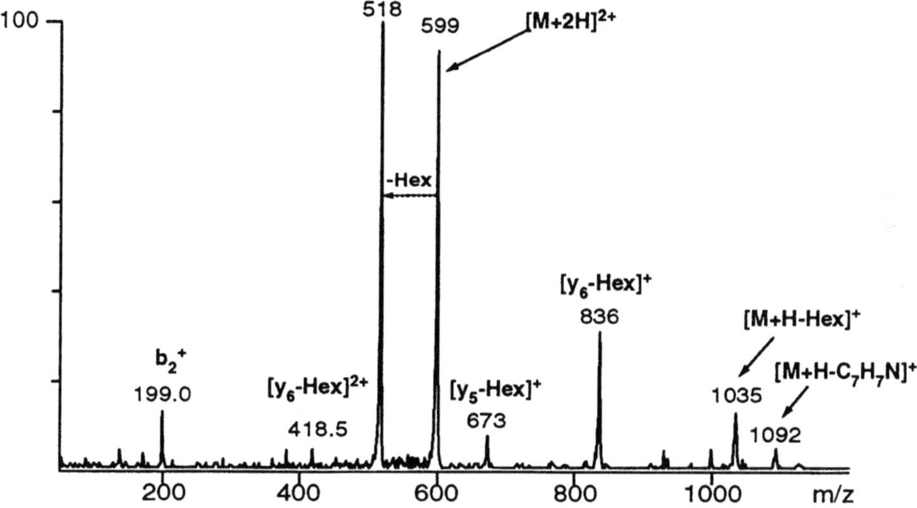

Figure 9. ESMSMS spectrum of v_{1+} ($[M + 2H]^{2+}$, m/z 599.0) derived from fraction 11. (Hex = anhydrohexose $C_6H_{10}O_5$).

The problem was solved by employing Asp-N cleavage of HPLC-isolated v_{1+} dissected the two sites in question and determining which of the resulting tetrapeptides carried the 162-Da increment (cf. Table 6). Unmodified $v_{1.1}$ ([1-4]) together with substituted $v_{1.2+}$ ([5-8], ΔM_r = 162.2 Da) unambiguously identified Thr7 as the site of hexose attachment. The loss of anhydrohexose in the tandem mass spectrum of $v_{1.2}$ confirmed its presence independently (Fig. 10). As an alternative, automated Edman degradation was performed on the HPLC-separated modified peptide v_{1+} ([1-8]). The blank step for position 7 in the sequence VVYTDX?E (X = pyridylethylated cystein) likewise indicated Thr7 to be modified. Monosaccharide analysis of a sample corresponding to fraction 11 isolated from the rHV1 batch 810490 established mannose as the hexosyl substituent in question.

TABLE 6. ESMS mapping of Asp-N subdigestion of Glu-C peptide v_{1+} derived from PE-derivatized fraction 11.

peptide	sequence	M_r calc.	$[M+zH]^{z+}$ (z)	M_r found	M_r .found - M_r calc
$v_{1.1}$	[1-4]	480.3[1]	481.4 (1)	480.4	0.1
$v_{1.2}$	[5-8]	571.2[1]	734.4 (1)	733.4	162.2

[1] calculated monoisotopic M_r

r[Man-Thr4]-HV1. The major component in 'fraction 7' represented (apart from rHV1 present as minor component) another monohexosylated congener according to its M_r (7124.0 Da found by ESMS, M_r 6962.9 peak top mass calculated for rHV1, ΔM_r = 161.1). ESMS mapping of the peptides ob-tained in the Glu-C step of the standard procedure again revealed a mono-glycosylated peptide v_{1+} ([1-8]) in addition to unmodified v_1 due to the presence of rHV1. In this case, the unseparated mixture was subjected directly to Asp-N treatment. As expected, products of subdigestion of the peptides v_5 ($v_{5.1}$, [44-52], $v_{5.2}$, [53-61], v_3. ($v_{3.1}$, [18-32]) and of unmodi-

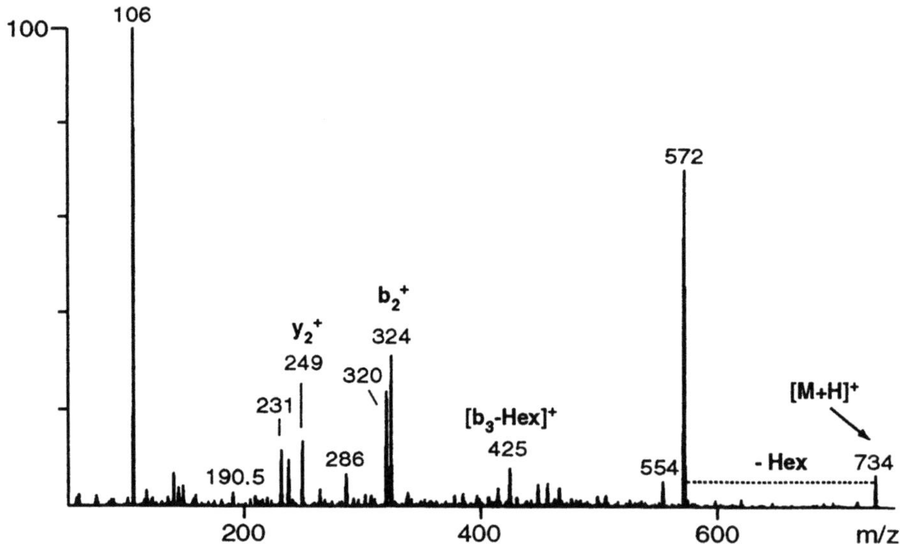

Figure 10. ESMSMS spectrum of $v_{1.2+}$ ($[M + H]^+$, m/z 734.4) derived from fraction 11. (Hex = anhydrohexose $C_6H_{10}O_5$).

fied v_1 ($v_{1.1}$, [1-4]; $v_{1.2}$, [5-8]) derived from rHV1 were detected in the ESMS map in addition to v_2, v_6 and unchanged v_1, v_{1+}, v_3 and v_5 (Table 7). A hexosylated form of $v_{1.2}$ ([5-8]) was definitively absent, but signals for hexosylated $v_{1.1}$ ([1-4]) could be clearly identified ($[M + H]^+$ at m/z 643.4).

TABLE 7. ESMS map of Glu-C/Asp-N digest of PE-derivatized fraction 7 (corresponding to peak 7 in LC-MS).

peptide	sequence	M_r calc.	$[[M+zH]^{z+}$ (z)	$[M+zH]^{z+}$ (z)	M_r found	M_r found. - M_r calc.
$v_{1.1}$	[1-4]	480.3[1]	481.4 (1)		480.4	0.1
$v_{1.1+}$	[1-4]	480.3[1]	643.4 (1)		642.4	162.2
$v_{1.2}$	[5-8]	571.2[1]	572.0 (1)		571.0	-0.2
v_{1+}	[1-8]	1033.4[1]	1196.6 (1)	599.0 (2)	1195.8	162.4
v_2	[9-17]	1175.5[1]	1176.6 (1)	588.8 (2)	1175.6	0.1
$v_{3.1}$	[18-32]	1646.9[2]	824.2 (2)	549.8 (3)	1646.4	-0.5
v_3	[18-35]	1948.2[2]	975.0 (2)	650.4 (3)	1948.1	-0.1
v_4	[36-43]	982.5[1]	983.6 (1)	492.4 (2)	982.7	0.2
v_5	[44-61]	1997.1[2]	999.6 (2)	666.4 (3)	1996.7	-0.4
$v_{5.1}$	[44-52]	964.5[1]	483.4 (2)		964.8	0.3
$v_{5.2}$	[53-61]	1049.4[1]	1050.6 (1)		1049.8	0.2
v_6	[62-65]	551.3[1]	552.4 (1)		551.4	0.1

[1] calculated monoisotopic M_r [2] calculated isotopically averaged M_r

The identity of this latter peptide was further confirmed by the typical loss of anhydrohexose and the formation of b_2, b_3 and deglycosylated y_2 sequence ions in the corresponding MSMS experiment (data analogous to those of Fig. 10). Thus, Thr4 was clearly identified as the substituted site in this glycosylated derivative. Again, Edman degradation of HPLC isolated v_{1+} corroborated the assignment by showing a blank step for position 4 in the sequence VVY-?-DXTE. Also in this case monosaccharide analysis of a corresponding chromatographic fraction (isolated from rHV1 batch 810490) confirmed mannose as being the *O*-glycosylating substituent in question.

r[Man-Thr45]-HV1. The presence of a third monohexosylated isomer as a minor component in fraction 9 (containing a dominating dihexosylated homologue in accordance with LC-MS retention times) was indicated by its M_r (7124.8 Da found by ESMS, M_r6962.9 peak top mass calculated for rHV1 ΔM_r = 161.9). The ESMS map of the Glu-C digest (Table 8) showed, in addition to a dihexosylated peptide v_1 (v_{1++}) clearly derived from the dominating diglycosylated species, also monohexosylated v_5 ([44-61]) obviously originating from the monohexosylated compound under discussion. MSMS of the corresponding triply charged parent ions ($[M + 3H]^{3+}$ at m/z 720.6) indicated the usual loss of anhydrohexose, yet again no hexosyl-containing sequence ions relevant for discerning the two possible substitution sites Thr45 and Ser50 were observed (Fig. 11). Since the sequence Thr45---Ser50 contained no obvious site for further enzymatic cleavage, the modified peptide had to be isolated by HPLC and subjected to automated Edman sequencing. The blank step observed already at position 2 ([Gly44-?-ProLysProAsnSer50His...]) immediately established Thr45 as the carrier of *O*-glycosylation. Monosaccharide analysis of a corresponding chromatographic fraction (isolated from batch 810490) revealed, like the other glycosylated derivatives, mannose as the major constituent (in addition to glucose most likely representing a contaminant).

TABLE 8. ESMS map of Glu-C digest of PE-derivatized fraction 9 (corresponding to peak 3 in LC-MS).

peptide	sequence	M_r calc.	$[[M+zH]^{z+}$ (z)	$[M+zH]^{z+}$ (z)	M_r found	M_r found - M_r calc.
v_1	[1-8]	1033.4[1]	1034.0 (1)	518.0 (2)	1033.5	0.1
v_{1++}	[1-8]	1033.4[1]		680.0 (2)	1358.0	324.6
v_2	[9-17]	1175.5[1]	1176.8 (1)	589.0 (2)	1175.9	0.4
v_3	[18-35]	1948.2[2]	974.6 (2)	650.2 (3)	1947.4	-0.8
v_4	[36-43]	982.5[1]	983.6 (1)	492.4 (2)	982.7	0.2
v_5	[44-61]	1997.1[2]	999.6 (2)	666.4 (3)	1956.7	-0.4
v_{5+}	[44-61]	1997.1[2]	1080.4 (2)	720.6 (3)	2158.8	161.7
v_{5+6}	[44-65]	2530.6[2]	844.4 (3)		2530.2	-0.4
$v_{5.1}$	[44-52]	964.5[1]	483.4 (1)		964.8	0.3
$v_{5.2}$	[53-61]	1049.4[1]	1050.6 (1)		1049.6	0.2
v_6	[62-65]	551.3[1]	552.4 (1)		551.4	0.1

[1] calculated monoisotopic M_r [2] calculated isotopically averaged M_r

r[Man-Thr4,Man-Thr7]-HV1. For this *dihexosylated* rHV1 species dominating in the above sample ('fraction 9', M_r 7287.2 Da found by ESMS, M_r 6962.9 peak top mass calculated for rHV1, ΔM_r 324.3) *both* hexoses were found by the standard Glu-C digestion to be located in the *N*-terminal peptide [1-8].

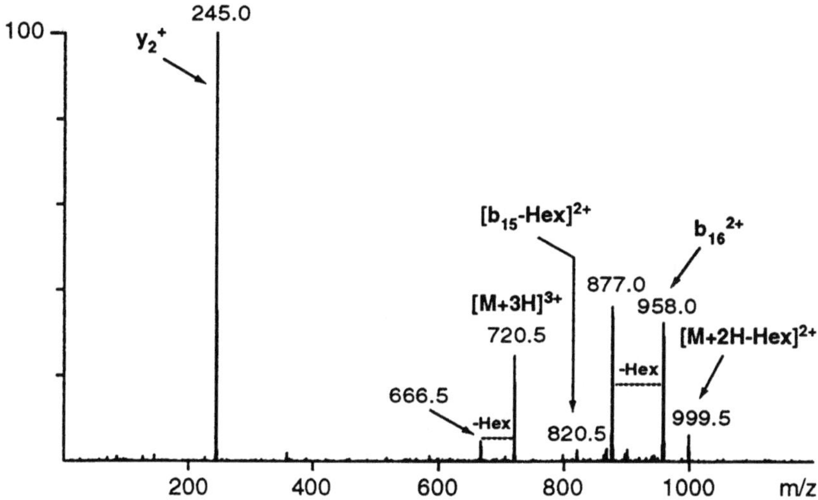

Figure 11. ESMSMS spectrum of v_{5+} ($[M + H]^+$, m/z 720.6, derived from fraction 9. (Hex = anhydrohexose $C_6H_{10}O_5$).

This situation called for subsequent Asp-N digestion as a convenient basis for further detailed analysis. The resulting ESMS map (Table 9) showed significant signals for two singly hexosylated peptides $v_{1.1+}$ (m/z 643.4, [1-4]) and $v_{1.2+}$ (m/z 734.4, [5-8]) without indication of dihexosylation in either tetrapeptide. This immediately excluded substitution of only one site with a dihexosyl unit and established the r[*O*-Hex-Thr4,*O*-Hex-Thr7]-HV1 structure for this component of the mixture. ESMSMS analyses of both monohexosylated tetrapeptides confirmed these assignments despite relatively weak signals by exhibiting prominent losses of anhydrohexose (Fig. 12). Also Edman sequencing conducted after HPLC isolation of peptide v_1 was in agreement with this structure in that blank steps for both positions in a VVY-?-DX-?-E sequence were obtained. As mentioned above, monosaccharide analysis had identified mannose as the main sugar constituent released by methanolysis of this fraction.

r[Man₂-Thr4]-HV1. This late-eluting *dihexosylated* isomer (M_r 7286.0 Da found by ESMS, M_r 6962.9 peak top mass calculated for rHV1, ΔM_r 323.1) corresponding to peak 10 in Fig. 4 was contained in the sample off-line isolated for structure analysis of r[iso-Asp33]-HV1 (see above).

TABLE 9. ESMS map of Glu-C digest of PE-derivatized fraction 9 (corresponding to peak 1 in LC-MS).

peptide	sequence	M_r calc.	$[[M+zH]^{z+}$ (z)	$[M+zH]^{z+}$ (z)	M_r found.	M_r found - M_r calc.
$v_{1.1+}$	[1-4]	480.3[1]	643.4 (1)		642.4	162.1
$v_{1.2+}$	[5-8	571.2[1]	734.4 (1)		733.4	162.2
v_{1++}	[1-8]	1033.4[1]	680.2 (2)		1358.4	325.0
v_2	[9-17]	1175.5[1]	1176.6 (1)	589.0 (2)	1175.8	0.3
$v_{3.1}$	[18-32]	1646.9[2]	824.2 (2)	549.8 (3)	1646.4	-0.5
v_3	[18-35]	1948.2[2]	975.0 (2)	650.2 (3)	1847.8	-0.4
v_4	[36-43]	982.5[1]	983.8 (1)	492.4 (2)	982.8	0.3
$v_{5.1}$	[44-52]	964.5[1]	483.4 (2)		964.8	0.3
$v_{5.2}$	[53-61]	1049.4[1]	1050.8 (1)		1049.8	0.4
v_6	[62-65]	551.3[1]	552.4 (1)		551.4	0.1

[1] calculated monoisotopic M_r [2] calculated isotopically averaged M_r

Dihexosylation was again found to be entirely located within the *N*-terminal peptide [1-8] (*cf.* peptide v_{1++} in map of Glu-C digest of this sample in table 3). In this case enzymatic cleavage with Asp-N failed to produce the 2 expected tetrapeptides that are essential for the assignment of the substituted site(s). Direct Edman sequencing of the HPLC isolated peptide v_{1++} provided, however, a successful alternative and revealed Thr4 as being the only modified amino acid within the sequence VVY-?-DXTE. Hence, the modifying group responsible for the 324 Da mass increment in this isomer is most likely a disaccharidic, i.e. dimannosidic substituent *O*-linked to Thr4.

rHV1-related products not detected in LC-MS analyses

Some rHV1-related products had been earlier identified in matrices such as fermentation broths or enriched chromatographic fractions isolated from diverse rHV1 batches, but were not detected in the LC-MS analysis of

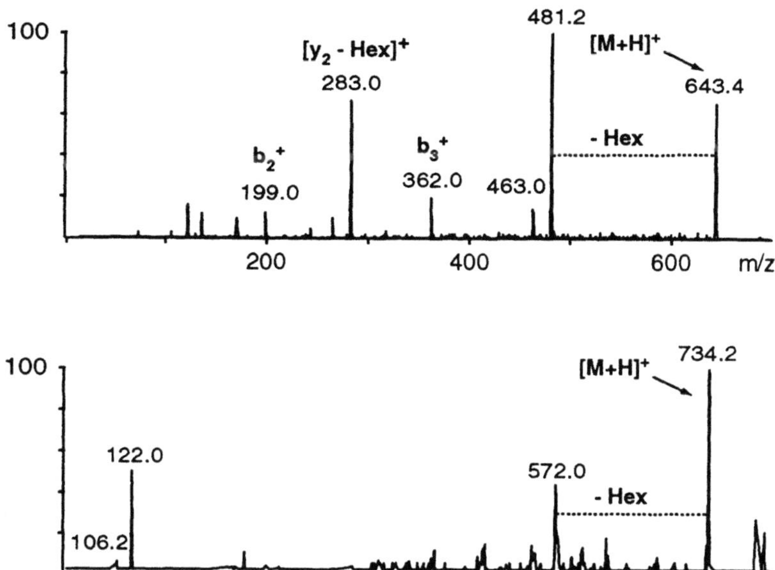

Figure 12. ESMSMS spectra of $[M + H]^+$ of $v_{1.1+}$ (m/z 643.4) and $v_{1.2+}$ (m/z 734.4) derived from fraction 9. (Hex = anhydrohexose $C_6H_{10}O_5$).

the representative batch (batch 800894). The reason for non-detection may be too low concentrations and/or changes in expression and fermentation conditions (different strains of yeast) or optimized purification.

r[phospho-Thr45]-rHV1 showed a highly suggestive 80-Da mass increment over M_r of rHV1 and represented a class of products no longer detectable by LC-MS. Its structural characterization involved extension of the general strategy by inserting 'comparative LC-ESMS mapping' at the stage of Glu-C digestion, performed in addition to ESMS mapping of the peptides obtained from reference and unknown. Moreover, mapping of the unseparated peptide mixtures was not only performed in the positive, but also in negative ion mode. *Positive* ESMSMS of peptide v_5 ([44-61]) proved successful in locating the unknown substituent at the Thr45 position, whereas *negative* ESMSMS provided differentiation of *O*-sulfate and *O*-phosphate as the modifying substituent in favor of the latter. Clear distinction between these two possibilities (ΔM_r = 80 Da for either case) was achieved by CID inducing phosphate-specific fragmentation $[M - 2H]^{2-}$ -> m/z 97 ($H_2PO_4^-$) -> m/z 79 (PO_3^-). A detailed report on this work was given in ref. [1].

Also belonging to this category is a series of *C*-terminally truncated rHV1 analogues such as r[*des*-Gln65]-rHV1 (M_r 6832.9), r[*des*-Leu64,Gln65]-rHV1 (M_r 6721.4) and r[*des*-Tyr63,Leu64,Gln65]-rHV1 (M_r 6558.0). Intact *N*- and truncated *C*-termini were confirmed by Edman degradation and carboxypeptidase Y degradation, respectively [10].

4. Conclusion

The LC-ESMS method developed for direct analyses of rHV1-related products in preparations of rHV1 without prior enrichment was successfully applied to the characterization of various batches for quality control. The analysis of representative rHV1 preparations (e.g. the batches 800894 and 810490 discussed in this review) established the presence of 3 monodehydrated forms of rHV1, of 2 isomers of rHV1, as well as of 3 monohexosylated and 3 dihexosylated congeners, all invariably present at low concentration levels. Interrelationships between chromatographic peaks observed in on-line LC-ESMS and corresponding fractions independently isolated for structure elucidation were established by performing both types of analyses under very similar chromatographic conditions. Within this review the structural characterization of the individual congeners is summarized and illustrated in greater detail for selected representatives of each class of congeners. The chosen combination of ESMS with chemical and enzymatic methods (supplemented, when required, by automated Edman degradation and monosaccharide analyses) demonstrated the broad potential of the chosen approach.

Acknowledgement

The authors are grateful to Mr. R. Knecht for performing sequence analyses by automated Edman degradation.

5. References

1. Coulot, M., Domon, B., Grossenbacher, H., Guenat. C., Märki..W., Müller D. and Richter, W.J. (1993) LCMS and MS/MS in the analysis of recombinant proteins. J. Mol. Struct. 292, 89-104.

2. Grossenbacher, H., Märki, W., Coulot, M., Müller, D.R. and Richter, W.J (1993) Characterization of succinimide-type dehydration products of recombinant hirudin variant 1 by electrospray tandem mass spectrometry, *Rapid Commun. Mass Spectrom.* 7, 1082-1085.

3. Ramseier, U., and Chang, J.-Y. (1994) Modification of cysteine residues with *N*-methyl iodoacetamide, *Anal. Biochem.* 221, 231-233.

4. Bischoff, R., Lepage, P., Jaquinod, M., Cauet, G., Acker-Klein, M., Clesse, D., Laporte, M., Bayol, A., Van Dorsselaer, A., and Roitsch, C. (1993) Sequence-specific deamidation, isolation and biochemical characterization of succinimide intermediates of recombinant hirudin. *Biochem.* **32**, 725-734.

5. Märki, W., Grossenbacher, H., Raschdorf, F. and Liersch, M., (1990) Proceedings of the 5th European Congress on Biotechnology, Copenhagen, July 8-13, pp 945-948.

6. Carr, S.A., Hemling, M.E., Bean, M. and Roberts, G.D., (1991) Integration of mass spectrometry in analytical biotechnology. *Anal. Chem.* 63, 2802 -2824.

7. Papayannopoulos, I.O. and Biemann, K. (1992), Fast atom bombardment and tandem mass spectrometry of synthetic peptides and by-products. *Peptide Research* 5, 83-89.

8. Schindler, P., Müller, D., Grossenbacher, H., Märki, W. and Richter, W.J. (1996) Characterization of a β-Asp33 isoform of recombinant hirudin sequence variant 1 by low energy collision induced dissociation. *J. Mass Spectrom* **31**, in press.

9. Lehle, L. and Tanner, W. Protein glycosylation in yeast in "Glycoproteins 1995", pp 475-506, Montreuil J, Schachter H, Vliegenthart JFG, ed. Elsevier Science.

10. Heim, J., Takabayashi, K., Meyhack, B., Märki, W. and Pohlig, G. (1994), *C*-terminal proteolytic degradation of recombinant desulfato-hirudin and its mutants in the yeast *Saccharomyces cerevisiae. Eur. J. Biochem.* **226**, 341-353.

FOLDING PATHWAYS OF DISULPHIDE CONTAINING PROTEINS

Margherita Ruoppolo#,^, Fabio Talamo#, Piero Pucci#,° and Gennaro Marino#,°,*
#Servizio Spettrometria di Massa, CNR-Università di Napoli, Via Pansini 5, 80131, Naples, Italy
^Dipartimento di Chimica, Università degli Studi di Salerno, Salerno, Italy
°Dipartimento di Chimica Organica e Biologica, Università Federico II, Naples, Italy

ABSTRACT

Protein folding, associated with oxidation and isomerization of disulphide bonds was studied using reduced and denatured RNase T_1 (rd-RNase T_1) or RNase A (rd-RNase A) and mixed disulphide between glutathione and reduced RNase T_1 (GS-RNase T_1) as starting protein materials. Folding was initiated by addition of free glutathione (reduced glutathione (GSH) + oxidised glutathione (GSSG)) and was monitored by electrospray mass spectrometry (ES-MS) time-course analysis. This permitted both the identification and quantitation of the population of intermediates present during the refolding process. All the analyses indicate a pathway of sequential reactions in the formation of native RNase T_1 or RNase A which occurs via the reiteration of two steps: i) formation of a species containing both mixed disulphides with glutathione and free protein thiols, ii) formation of an intramolecular disulphide via thiol-disulphide interchange reaction betwen them. Refolding of rd-RNase T_1 and GS-RNase T_1 was also performed in the presence of Protein Disulphide Isomerase (PDI). Addition of PDI led to a catalysis of each individual reaction of the entire process without altering the refolding pathway. On the basis of these experiments an hypothesis of a general pathway for folding of S-S containing proteins is proposed.

Introduction

Folding of disulphide-containing proteins is thermodynamically associated with formation of the unique set of native disulphide bonds (1). As a result, many theoretical and experimental studies have addressed to the characterization of disulphide bonded intermediates in order to elucidate protein folding pathways by determining the appearance of these intermediates and their interconversion rates (2, 3). Some of these studies have focused on the process of initial biosynthetic formation of native disulphides which *in vivo* takes place in the endoplasmic reticulum (ER) (4). According to present evidence, the major potential source of oxidising equivalents for protein disulphide formation within the endoplasmic reticulum (ER) is oxidised glutathione (5).

The simplest realistic model for studying the folding of disulphide containing proteins is RNase T_1 since it contains only two disulphide bonds (Cys2-Cys10 and Cys6-Cys103). When the disulphide bonds are reduced, RNase T_1 is essentially unfolded but can adopt compact native-like structure in the presence of high concentrations of NaCl (6, 7). Kinetic analysis of rd-RNase T_1 refolding in the presence of a glutathione redox system was based mainly on fluorescence analysis reporting an initial fast-phase in the process (8). Recently, the folding of a novel substrate GS-RNase T_1, the product of converting each free Cys residue of reduced and denatured RNase T_1 to a mixed disulphide with glutathione, has been described (9, 10). GS-RNase T_1 is fully

R. M. Caprioli et al. (eds.), Selected Topics and Mass Spectrometry in the Biomolecular Sciences, 347–359.

unfolded in "native" conditions, but could be refolded in high yield to native RNase T_1 in conditions which permit thiol-disulphide interchange, i. e., in the presence of a thiol-disulphide redox buffer comprising GSH $\pm$ GSSG. GS-RNase T_1 refolding was monitored by a number of independent techniques, such as recovery of enzyme activity and intrinsic fluorescence. However these methods mainly report the formation of the final product, although fluorescence analysis detected partially folded intermediates whose presence was confirmed by HPLC analysis.

The study of the protein folding pathway becomes much more difficult with the increase in the number of cysteine residues in the protein due to the rapid increase in the number of possible disulphide containing intermediates. The folding studies in fact led to unsatisfactory results when applied to bovine pancreatic RNase A which has four disulphide bonds (Cys26-Cys84, Cys40-Cys95, Cys58-Cys110 and Cys 65-Cys72). Although RNase A has been one of the most thoroughly studied proteins with regard to refolding (11-15), several aspects of this pathway have not yet been clarified.

In the endoplasmic reticulum of eucaryotic cells, disulphide bond formation is dependent on protein disulphide isomerase (PDI) which is present at high concentrations and functions in the presence of millimolar concentrations of GSH and GSSG (16, 17). Previous studies showed that the addition of PDI catalyses the refolding both of reduced and denatured RNase T_1 (8) and of GS-RNase T_1 in which refolding involves a series of disulphide isomerizations without net oxidation of protein thiols (10).

Here we describe the use of electrospray-mass spectrometry (ES-MS) time-course analysis to obtain significant data for kinetic and structural studies on the refolding pathway of a disulphide containing protein. The experimental approach is based upon the determination of the molecular weight of the alkylated disulphide-bonded intermediates through a time-course analysis of the refolding process. The alkyl group introduced with the quenching reaction, increased the molecular weight of the intermediates by a fixed mass, thus allowing the separation by mass of intermediates containing different numbers of disulphide bonds and the determination of their relative concentration during the refolding. A further step would consist, once the pathway(s) have been described, in the structural assignment of each single component of intermediate populations. Figure 1 shows the outline of the general strategy making extensive use of mass spectrometry. In this paper we describe the application of the ES-MS time-course analysis as a powerful tool in the description of the refolding pathways of typical disulphide containing proteins.

Figure 2 and 3 represent all possible intermediates and equilibria in the refolding of a two-disulphide and a four-disulphide containing protein, respectively, in the presence of a glutathione redox system. Figures in parentesis show the average molecular weight of the carboxyamidomethylated intermediates. Each population of intermediates is described in terms of the number of intramolecular disulphides (indicated as nS), mixed disulphides with glutathione (nG) and free cysteine residues (nH). The schemes shown in Figure 2 and 3 are derived from an analogous scheme previously reported (18).

We used two derivatives of RNase T_1, rd-RNase T_1 and GS-RNase T_1 (named 4H and 4G, respectively, in figure 2) in order to evaluate the effect of different starting materials on the refolding pathway and on the distribution of the intermediates along the whole process. The refolding process was also studied in the presence of PDI in order to define the effect of the enzyme as a catalyst on the refolding pathway. The addition of PDI resulted in a catalysis of each individual step of the refolding process.

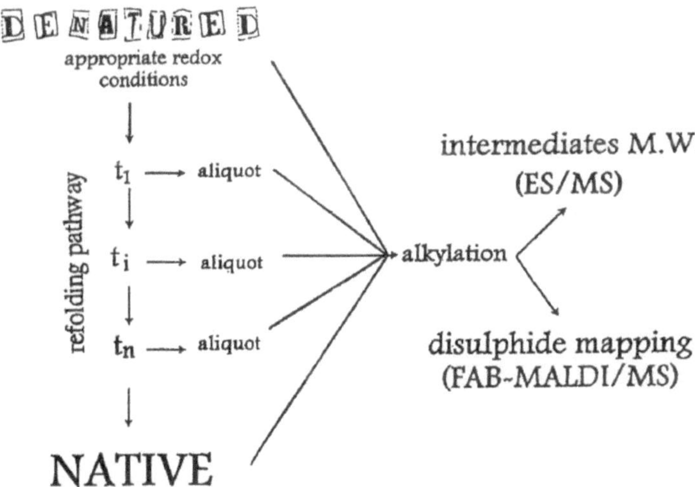

Figure 1: Experimental approach to use disulphide bonds to trap intermediates in protein folding making extensive use of mass spectrometry

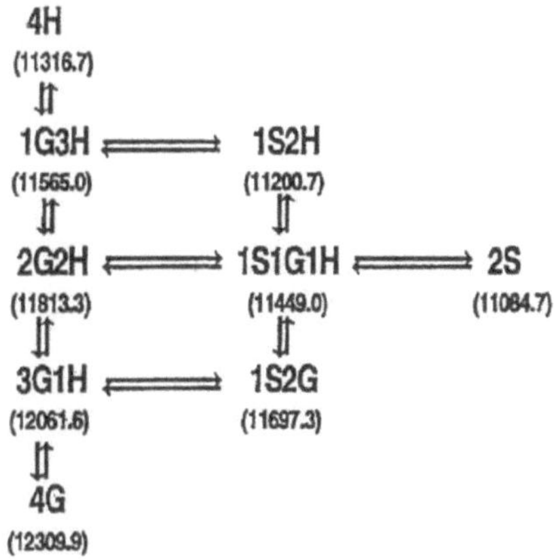

Figure 2: Representation of all possible protein intermediates and equilibria in the refolding of a two-disulphide bond containing protein in the presence of a glutathione redox system. Intramolecular disulphide bonds are indicated as nS, mixed disulphide with glutathione as nG and free cysteine residues as nH. Free glutathione (GSH/GSSG) is not represented. Downward vertical arrows represent reactions with GSSG to generate protein-SSG + GSH, upward vertical arrows represent reactions with GSH to reverse this process. Horizontal arrows to the right represent unimolecular reactions to convert protein-SSG mixed disulphide into protein disulphide + GSH; horizontal arrows to the left represent reduction of protein disulphide by GSH to generate protein -SSG. Figures in parenthesis show the average molecular weight of the carboxymethylated intermediates.

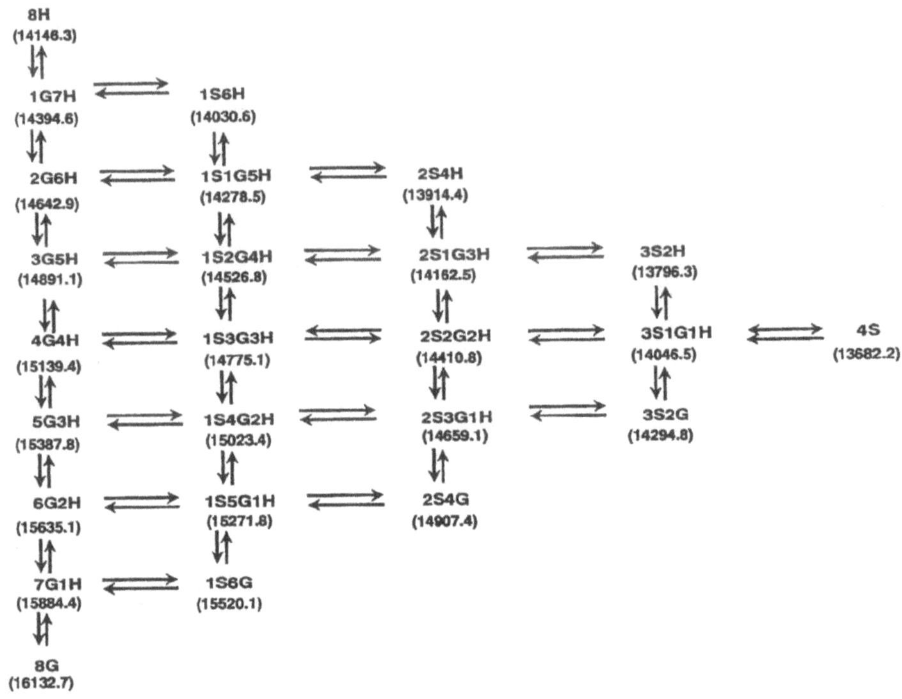

Figure 3: Representation of all possible protein intermediates and equilibria in the refolding of a four-disulphide bond containing protein in the presence of a glutathione redox system.

Results

Refolding by disulphide oxidation.

RNase T_1 was reduced in denaturing conditions to generate rd-RNase T_1 which was shown to have 3.95 SH/molecule by DTNB titration. The reduced protein was immediately lyophilized and always used within 1-2 days to avoid any spontaneous reoxidation.

rd-RNase T_1 was incubated in the presence of 2 mM GSH/1.5 mM GSSG. Aliquots of the refolding processes were withdrawn at different intervals, trapped by alkylation of the free thiol groups as described, and analysed by ES-MS to identify the disulphide bonded intermediates formed. The carboxyamidomethylation reaction used to trap the free SH groups increased the molecular weight of the intermediates by a fixed amount, 57 Da for each free SH group, thus allowing the separation by mass of intermediates containing different numbers of disulphide bonds and the determination of their relative concentration during the refolding.

Figure 4A shows the deconvoluted ES spectra of the mixtures of species sampled at time 0, 2 min, 2 h and 5 h from the refolding of rd-RNase T_1 in the presence of 2 mM GSH/1.5 mM GSSG. At different times, different populations of disulphide intermediates are present which were identified on the basis of their accurate molecular weight. Each population of trapped intermediates is characterized by a different number of intramolecular disulphide bonds (indicated as nS in Figure 2), mixed disulphides with the exogenous glutathione (nG) and carboxyamidomethyl groups (nCAM). The number of CAM groups corresponds to the number of free thiols present in the refolding intermediates and are therefore indicated as nH. It is important to emphasise that each observed molecular weight corresponds to a population of isomers, in which the number of protein disulphides, protein free thiols and protein-SSG mixed disulphides is defined, but not the positions of the cysteine residues involved.

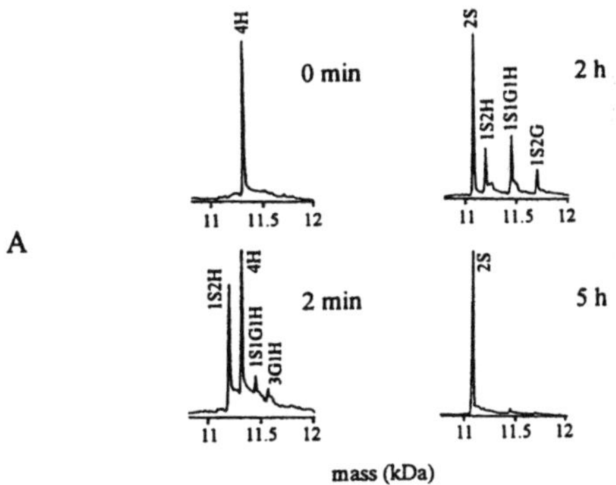

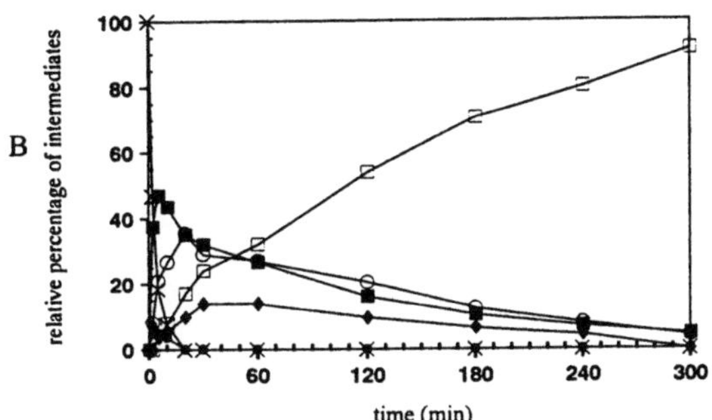

Figure 4: (A) ES-MS analysis of rd-RNase T1 refolding carried out at 1 mg/mL in 0.1 M Tris-HCl, 1 mM EDTA (pH 7.5) at 25°C in the presence of 2 mM GSH/1.5 mM GSSG. Aliquots were withdrawn from the refolding mixture at time 0, 2 min, 2h and 5h. (B) Time-course analysis of the refolding of rd-RNase T1 in the presence of 2 mM GSH/1.5 mM GSSG . (×) 4H; (●) 1G3H; (■) 1S2H; (O) 1S1G1H; (◆) 1S2G; (□) 2S.

The ES spectrum of the aliquot withdrawn at time 0 (Figure 4A) shows the presence of a single component (A), exhibiting a molecular weight of 11316.75 ± 0.40 Da corresponding to the protein carrying 4 carboxyamidomethyl groups (expected value 11316.66 Da). After two min of the refolding process, besides the reduced protein (4H), the spectra revealed the simultaneous presence of three species, i. e., B, C and D. On the basis of their molecular weights, B corresponds to a population of isomers containing one mixed disulphide with the exogenous glutathione and three free SH groups (1G3H), C to intermediates containing an intramolecular disulphide bond and two free SH groups (1S2H) and D to species having an intramolecular disulphide bond, a mixed disulphide with the exogenous glutathione and a free thiol group (1S1G1H).

The presence of intermediates containing mixed disulphides with the exogenous glutathione confirmed that formation of an intramolecular disulphide bond proceeds via the formation of a protein-glutathione mixed disulphide as shown in the case of BPTI (19) and of a simple model peptide (20). The analysis of the sample withdrawn at 2 h reveals, in addition to species C and D already observed, the presence of species F containing two intramolecular disulphides (2S) and E having one intramolecular disulphide and two mixed disulphides with the exogenous glutathione (1S2G); components A and B corresponding to species 4H and 1G3H respectively, have disappeared at this stage of the process. Finally, after 5 h of refolding only component F was detected in the mixture. Fluorescence and activity analyses (data not shown) indicate an almost complete recovery of the native intrinsic fluorescence and catalytic activity at this stage of the process, thus supporting the conclusion that the refolding generates native RNase T_1 (i. e. component F is predominantly the native species, rather than containing significant amounts of the two alternative 2S isomers).

A further bonus in the use of ES-MS is that each population of intermediates can be quantitated by measuring the total current of ions produced by each species (21). The time course plotted in Figure 4B shows very clearly that intermediates 1S2H and 1S1G1H are predominant from the beginning of the reaction up to about 1 h when the relative concentration of the species 2S increases. A steady state is established between intermediates 1S2H and 1S1G1H; from 30 mins onwards, the concentration of these species is approximately equal. The component 1G3H is present only in the first 20 min of the reaction and it rapidly disappears. The species 1S2G is present at very low concentration at the early stages of the reaction, reaching a constant concentration (less than 20%) and disappearing at the late stages of the refolding.

RNase A was reduced in the presence of 6 M guanidinium chloride to produce rd-RNase A. Before starting each refolding experiment, the reduced and denatured RNase A sample was tested for the content of free SH groups and enzymatic activity, exhibiting 8.1 SH/molecules and no detectable enzymatic activity, thus showing that the reduction and denaturation steps led to the complete loss of enzymatic activity. The fully reduced and denatured RNase A was then allowed to refold at 1 mg/mL in the presence of 1.5 mM GSH/0.3 mM GSSG in 0.1 M Tris-HCl, pH 7.5, containing 1 mM EDTA, at room temperature under a nitrogen atmosphere. Aliquots of the refolding process were withdrawn at different intervals, trapped as described and then analysed by ES/MS to identify the intermediates formed. Figure 5A shows the spectra of the sample withdrawn at time 0, 1', 120' and 24 h of the refolding process. The spectrum of the aliquot withdrawn at time 0 shows the presence of a single component (A), corresponding to the protein carrying 8 carboxyamidomethyl groups. After 1 min of the refolding process, besides the reduced protein (8H), the spectra revealed the simultaneous presence of two species, i. e., B and C. On the basis of their molecular weights, B corresponds to a population of isomers containing an intramolecular disulphide bond and six free SH (1S6H) and C to intermediates having one mixed disulphide with the exogenous

glutathione and seven free SH groups (1G7H). After 120 min, the ES/MS spectra revealed the enormous complexity of the refolding mixture, showing the simultaneous presence of several population of disulphide intermediates.
Finally, the ES/MS spectrum recorded at 24 h shows the presence of a major component, I, containing four S-S bonds; at this stage of the process the sample is showing a recovery of more than 80% of enzymatic activity, thus supporting the conclusion that the refolding generates native RNase A (i. e. component I is predominantly the native species rather than containing significant amounts of the alternative isomers). As underlined for the rd-RNase T_1 refolding, it has to be noted that each observed molecular weight intermediate could correspond to a population of disulphide isomers.

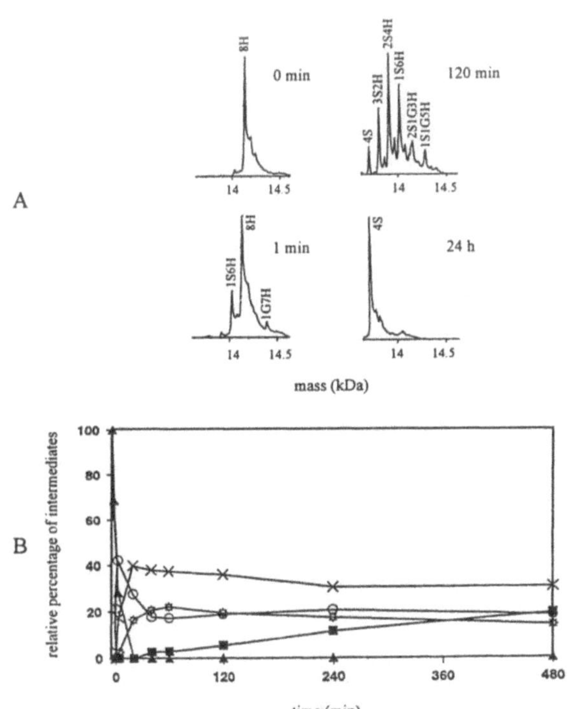

Figure 5: (A) ES-MS analysis of rd-RNase A refolding carried out at 1 mg/mL in 0.1 M Tris-HCl, 1 mM EDTA (pH 7.5) at 25° C in the presence of 1.5 mM GSH/0.3 mM GSSG. Aliquots were withdrawn from the refolding mixture at time 0, 1 min, 2h and 24h. (B) Time-course analysis of the refolding of rd-RNase A in the presence of 1.5 mM GSH/0.3 mM GSSG. (▲) 8H; (O) 1S6H; (×) 2S4H; (○) 3S2H; (■) 4S.

Figure 5B shows the relative intensities of the refolding intermediates detected by ES/MS analysis plotted vs. the reaction time. The time-course shows very clearly that the intermediates containing two intramolecular disulphide bonds are predominant from the beginning of the reaction.

Refolding by disulphide isomerization.

The unfolded mixed disulphide derivative GS-RNase T_1 can be refolded to native RNase T_1 in the presence of different ratios of GSH/GSSG and the refolding is strictly linked to disulphide isomerization (10). GS-RNase T_1 was incubated in the presence of the same redox buffer used for the refolding of the reduced and denatured RNase T_1. Aliquots were sampled from the refolding mixture at different intervals, alkylated with IAM and analysed by ES-MS, as described.

Figure 6 shows the time course of the GS-RNase T_1 refolding in the presence of 2 mM GSH/1.5 mM GSSG. The spectrum of the aliquot withdrawn at time 0 shows the presence of a single component K, exhibiting a molecular weight of 12309.45 $\pm$ 0.37 Da corresponding to the protein carrying 4 exogenous glutathione groups (4G).

During refolding with this substrate, the same species were detected as in the refolding of rd-RNase T_1, with the addition of a new component J corresponding to the protein with three mixed disulphides with glutathione and a single protein thiol (3G1H). The intermediate 3G1H is present at very low concentration only at the early stages of the reaction and it then disappears. After 5 min, the concentrations of species 4G and 1S2G are of similar magnitude while species 1S1G1H and 3G1H are present at much lower concentration. The time-course analysis shows clearly that the population of intermediates 1S2G is predominant from the beginning of the reaction up to 2 h when the concentration of the species 2S increases. The concentration of the other major population of intermediates, 1S1G1H, remains lower than that of the species 1S2G throughout. Finally, the species 1S2H is present at very low concentration, never greater than 10% and it disappears at 5 h. At 5 h, species 2S is predominant (75%) with a minor presence of species 1S2G and 1S1G1H. Fluorescence and activity data (10) indicate after 5 h a recovery of about 80% of the native intrinsic fluorescence and catalytic activity thus supporting the conclusion that species 2S is only native RNase T_1. As underlined for the rd-RNase T_1 or rd-RNase A refolding, it has to be noted that each observed molecular weight intermediate could correspond to a population of disulphide isomers.

Refolding assisted by PDI.

When refolding was performed in the presence of protein disulphide-isomerase (PDI), the enzyme was pre-incubated in the redox buffer and then added to the rd-RNase T_1 or GS-RNase T_1 solution at a molar ratio of 0.25=[PDI]/[substrate]. The PDI-assisted refolding reactions were followed by ES MS analysis as described.

Figure 7A shows the time-course analyses of PDI-assisted rd-RNase T_1 refolding in the presence of 2 mM GSH/1.5 mM GSSG. Addition of PDI led to a net catalysis of the refolding reaction. The time-course analysis shows the absence of the intermediate 1G3H detected in the uncatalysed reaction; in the presence of such a high concentration of PDI it is possible that the reaction is too fast at the early stages to allow detection of species 1G3H even when the reaction mixture was sampled after only 30 sec. The reduced protein disappears after only 30 sec when species 1S2H, 1S1G1H and 2S are present to a similar extent. At the one disulphide level, the species 1S2G is present at a concentration never greater than 20% and the steady-state approximately equal concentration of species 1S2H and 1S1G1H is established more rapidly in the presence of PDI than in its absence. The catalysed refolding was completed after only 45 min.

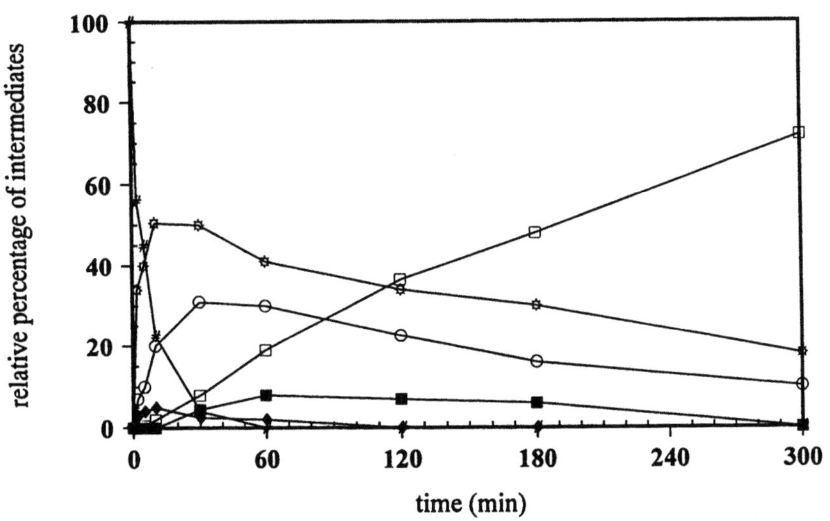

Figure 6: Time-course analysis of the refolding of GS-RNase T1 in the presence of 2 mM GSH/1.5 mM GSSG . (#) 4G; (◆) 3G1H; (✡) 1S2G; (O) 1S1G1H; (■) 1S2H; (□) 2S.

The effect of PDI on the refolding of GS-RNase T_1 was also investigated. PDI was incubated in the presence of the redox buffer and then added to the GS-RNase T_1 at a molar ratio of 0.25=[PDI]/[GS-RNase T_1]. Figure 7B shows the time-course analysis of PDI-assisted GS-RNase T_1 refolding in the presence of 2 mM GSH/1.5 mM GSSG. Addition of PDI led to a net catalysis of the refolding reaction, without altering substantially the distribution of the intermediates along the refolding pathway (note that the distribution after 2 min in the presence of PDI is closely comparable to that after 60 min in the absence of PDI). GS-RNase T_1 disappears after only 1 min when species 1S2G, 1S1G1H constitute the two major populations of intermediates. Species 2S is predominant after only 5 min and its appearance is much faster than in the absence of PDI, as expected. Finally, species 3G1H and 1S2H are present at very low concentration only at the first stages of the refolding.

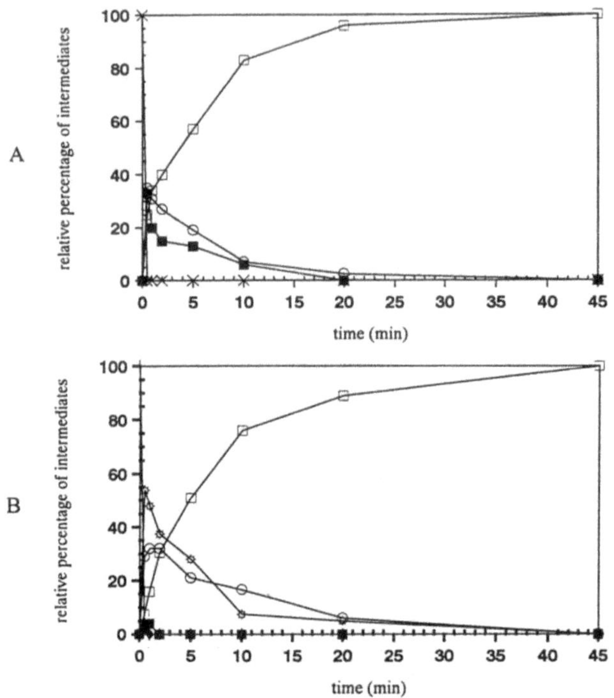

Figure 7:
(A) Time-course analysis of the refolding of rd-RNase T1 in the presence of PDI. PDI was incubated with 2 mM GSH/1.5 mM GSSG for 10 min at 25 °C and then added to rd-RNase T1 (1 mg/mL) at a molar ratio of 0.25= [PDI]/[rd-RNase T1]. (×) 4H; (■) 1S2H; (O) 1S1G1H; (□) 2S.
(B) Time-course analysis of the refolding of GS-RNase T1 in the presence of PDI. PDI was incubated with 2 mM GSH/1.5 mM GSSG for 10 min at 25 °C and then added to GS-RNase T1 (1 mg/mL) at a molar ratio of 0.25=[PDI]/[GS-RNase T1]. (+) 4G; (◆) 3G1H; (☼) 1S2G; (O) 1S1G1H; (■) 1S2H; (□) 2S

Discussion

Formation of native disulphide bonds is a rate limiting process in the folding of disulphide containing proteins during their residence in the endoplasmic reticulum (22, 23) where the major thiol redox system is provided by glutathione (5). Folding studies *in vitro* have been performed using reduced and oxidised glutathione with different ratios and different concentrations and several folding pathways have been deduced depending on the physicochemical conditions and the protein under study (24). Most studies focused on the refolding of fully reduced, denatured proteins taking these derivatives as a model of the denatured state. In this work we have studied the refolding of RNase T_1 and RNase A. In the case of RNase T_1 the refolding was started from both the reduced protein and a mixed disulphide between the protein and glutathione, with the aim of investigating the effect of different denatured starting materials on the refolding pathways and on the relative abundance of refolding intermediates. It seems conceivable that mixed disulphides between proteins and glutathione are important folding intermediates *in vivo*. Since the nascent chain emerges into the ER lumen into conditions of high oxidised glutathione concentration, it may well be that unpaired cysteine residues are converted to mixed disulphides before the complete translocation.

We employed a novel approach based on ES-MS time-course analysis which provides crucial data for kinetic and structural studies of the refolding processes. The definition of the precise molecular weight of the refolding intermediates allowed us to establish the isomeric species formed during the process and to determine a pathway of sequential reactions for all the refolding processes analysed. All the analyses demonstrate that only a limited number of intermediates is formed in the process as compared to all those theoretically possible. In particular one population of intermediates containing more than one mixed disulphide with glutathione was detected in folding from the reduced species and one population of intermediates with more than one free -SH was detected in folding from GS-RNase T_1. These observations allowed us to exclude some possible refolding pathways for rd-RNase T_1, rd-RNase A and GS-RNase T_1, thus significantly decreasing their number.

The first step in the rd-RNase T_1 or rd-RNase A refolding is the formation of a mixed disulphide, producing species 1G3H or 1G7H while the first step in the GS-RNase T_1 refolding is a reduction reaction producing species 3G1H. Then, all the species, 1G3H, 1G7H and 3G1H, undergo an internal attack by a free SH group to form an intramolecular disulphide. Formation of an intramolecular disulphide is in all cases favoured over a further reaction with exogenous GSH or GSSG and this significantly restricts the pathway: a species 2G2H or 2G6H was in fact never found.

The time course analysis of RNase T_1 refolding clearly indicates that species 1S1G1H is the immediate precursor of species 2S from both starting materials, but interconversion between species 1S2H, 1S1G1H and 1S2G is observed whether the refolding starts from the 4H state or from the 4G state. Species 1S2G and 1S2H are dead-end intermediates in the refolding of rd-RNase T_1 and GS-RNase T_1, respectively; they in fact, have to be transformed back to 1S1G1H to form the final product 2S. Formation of species 2S from 1S1G1H is the rate limiting step for all the processes studied.

The time-course analysis of RNase A refolding clearly reveals that the pairing of the cysteine residues proceeeded via three groups of balanced intermediates containing intramolecular disulphide bonds, namely one-disulphide, two-disulphides and three disulphides. The species containing two disulphide bonds predominate until the late stages of the process when the component containing four disulphide bonds increases. The FAB/MS analysis (25) of the sample withdrawn at 24 h showed the existence of all four native disulphide bonds and very few weak signals corresponding to non-native disulphides, confirming that some molecular species exist which have not completed the refolding process. The predominance of the species containing two disulphide bonds along the pathway appears significant. On a purely statistical basis, in fact, a larger number of intermediates containing three S-S bonds was expected, whereas their concentration remains lower than that of the components containing two disulphides along the entire process. Therefore, the accumulation of the two disulphides-containing intermediates along the refolding pathway could be ascribed either to their intrinsic thermodynamic stability compared to other species or to a kinetic control of the whole process.

It should be noted that the intermediates identified by the ES-MS analysis are populations of molecular species characterised by the same number of disulphide bonds and GS-mixed disulphides. These intermediates may include nearly all possible disulphide bond isomeric species. However, we cannot exclude the possibility that only a limited number of isomers is formed within an intermediate population due to kinetic or thermodynamic constraints, thus further restricting the possible folding pathways. The identification of disulphide bonds within each intermediate population was out of the scope of the present study but it is clear that the definition of such isomers within a population will allow the definitive and complete description of the folding pathway in terms of individual molecular species.

The refolding experiments were performed using the same approach in the presence of PDI to understand the role of PDI catalysis in the formation of disulphide bonds. The addition of PDI results in a net catalysis of the rd-RNase T_1 and GS-RNase T_1 refolding without altering the pathway but causing small changes in the relative distribution of the various population of intermediates. All the experiments suggest that the addition of PDI results in a catalysis of each individual reaction of the folding processes, thus PDI seems to catalyse: i) formation of mixed disulphide with glutathione, ii) reduction of mixed disulphide, iii) formation of intramolecular disulphide bonds. These results are not surprising considering the broad range of activities shown by PDI (16, 26). The small changes in relative concentrations of intermediate found in the presence of PDI indicate that direct transfer of disulphide from PDI-active site to protein dithiol (followed by reoxidation of PDI by GSSG) can only be a very small component of the overall PDI-catalysed reaction (cf. 20, 27). It is important to underline that by this approach we cannot detect any effect of PDI on the isomerization of disulphide bonds within a single population of intermediates. Actually, PDI could possibly catalyse the formation of only native disulphide bonds in both processes, bypassing disulphide rearrangements thus further restricting the folding pathways (28).

The results presented in this paper may lead to more general conclusion on refolding mechanisms. In quasi-physiological conditions, it appears that the refolding of single-domain disulphide-containing proteins occurs via reiteration of two sequential steps: i) formation of a mixed disulphide with glutathione, ii) internal attack of a free SH group to form an intramolecular disulphide bond. This hypothesis is supported by recent results obtained by Weissman and Kim (1995) on the refolding of bovine pancreatic trypsin inhibitor (BPTI) where they demonstrate by using a different analytical approach that the formation of an intramolecular disulphide bond occurs via a protein-glutathione mixed disulphide. The proposed mechanism predicts that only a limited number of intermediates actually accumulate and that isomerization between species with the same number of disulphides is only extensive at late stages of the process where slow conformational transitions become significant.

Acknowledgments

We acknowledge funding from CNR-Progetto Strategico "Biologia Strutturale" (Rome). We thank Dr. S. H. McLaughlin (University of Kent, Canterbury) for providing pure PDI and Prof. R. B. Freedman (University of Kent, Canterbury) for many helpfull discussion. We are indebted to Prof. N. Pace (Texas A & M University, Health Science Center) for generously providing RNase T_1.

REFERENCES

(1) Creighton, T. E. (1984) *Methods Enzymol.* **107,** 305-329
(2) Creighton, T. E. (1992) in *Protein Folding* (Creighton, T. E., eds) pp. 301-351, W. H. Freeman and Company
(3) Gilbert, H. F. (1994) in *Mechanism of Protein Folding* (Pain, R. H., eds.) pp 104-136, Oxford University Press
(4) Freedman, R. B., Hawkins, H. C., & Mc Laughlin, S. H. (1995) *Methods Enzymol.* **251**, 387-406
(5) Hwang, C., Sinskey, A. J., & Lodish, H. F., (1992) *Science* **257,** 1496-1502
(6) Oobatake. M., Takahashi, S., & Ooi, T. (1979) *J. Biochem. (Tokyo)* **86,** 65-70
(7) Pace, N.C., Grimsley, G. R., Thomson, J. A., & Arnett, B. J. (1988a) *J. Biol. Chem.* **263,** 11820-11825
(8) Schönbrunner, E. R., & Schmid, F. X., (1992) *Proc. Natl. Acad. Sci. USA,* **89,** 4510-4513
(9) Ruoppolo, M., & Freedman, R. B. (1994) *Biochemistry* **33,** 7654-7662
(10) Ruoppolo, M., & Freedman, R. B. (1995) *Biochemistry* **34,** 9380-9388
(11) Anfinsen, C. B., Haber, E., Sela, M. & White Jr., F. H. (1961) *Proc. Natl. Acad. Sci USA* **47**, 1309-1314
(12) Rothwarf, D. M., & Scheraga, H. A. (1993a) *Biochemistry* **32,** 2671-2679
(13) Rothwarf, D. M., & Scheraga, H. A. (1993b) *Biochemistry* **32,** 2680-2689
(14) Rothwarf, D. M., & Scheraga, H. A. (1993c) *Biochemistry* **32,** 2690-2697
(15) Wearne, S. J. & Creighton, T. E. (1988) *Proteins: Struct. Funct. Genet.* **4**, 251-261
(16) Freedman, R. B., Hirst, T. R., & Tuite, M. F. (1994) *Trends Biochem. Sci.,* **19,** 331-336
(17) Freedman, R. B. (1995) *Curr. Opin. Struc. Biol.* **5,** 85-91
(18) Konishi, Y., Ooi, T., & Scheraga, H. A. (1981) *Biochemistry* **20**, 3945-3955
(19) Weissman, J. S., & Kim, P. S. (1995) *Nature Struct. Biol.* **2,** 1123-1130
(20) Darby, N. J., Freedman, R. B., & Creighton, T. E. (1994) *Biochemistry 33*, 7937-7947
(21) Miranker, A., Robinson, C. V., Radford, S. E., Aplin, R. T. & Dobson, C. M. (1993) *Science* **262,** 896-899
(22) Creighton, T. E. (1986) *Methods Enzymol.***131,** 83-106
(23) Gething, M. J., & Sambrook, J. (1992) *Nature* **355,** 33-45
(24) Creighton, T. E., Zapun, A., & Darby, N. J. (1995) *TIBTech 13*, 18-23
(25) Ruoppolo, M., Torella, C., Kanda, F., Panico, M., Pucci, P., Marino, G. & Morris, H. R. (1996) *Folding & Design*, 1, 381-390
(26) Freedman, R. B., Brockway, B. E., & Lambert, N. (1984) *Biochem. Soc. Trans.* **12,** 929-932
(27) Ruddock, L. W., Hirst, T. R., & Freedman, R. B. (1996) *Biochem. J.* , in the press
(28) Creighton, T. E., Bagley, C. J., Cooper, L., Darby, N. J., Freedman, R. B., Kemmink, J., & Sheikh, A. (1993) *J. Mol. Biol.* **232,** 1176-1196

INTEGRATION OF CHEMICAL MODIFICATION REACTIONS AND MASS SPECTROMETRIC METHODS IN PROTEIN CHARACTERISATION

S. FOTI
Dipartimento di Scienze Chimiche, Università di Catania
Viale A. Doria, 6 95125 Catania (Italy)

1. Introduction

The recent development of new "soft" desorption-ionization methods has substantially increased the potential of mass spectrometry (MS) for the structural analysis of biomacromolecules, particularly proteins. Fast atom bombardment (FAB) and ^{254}Cf-plasma desorption (PD) are currently being used for accurate molecular mass determination of peptides. Electrospray (ES) and matrix-assisted laser desorption ionization (MALDI) have provided the possibility of determining the molecular mass of large proteins (> 100 kD) with an accuracy by far greater than classical methods.

Although these ionization-desorption methods are designed, and are particularly useful, for their capability of dealing with intact peptides and proteins, it has been clear since their initial introduction that the amount of structural information that can be derived from their use can be considerably increased when their application is combined with appropriate chemical (or enzymatic) structural modification and that, besides being utilized for the detection of natural occurring modifications in proteins (protein variants, post-transdutional modifications), they can also be used for characterising chemically introduced covalent modification of peptides and proteins, thus providing a powerful tool for the investigation of structure-properties relationships.

"Chemical modification" here indicates any chemically induced structural modification. More specifically chemical modification reactions include:

- Chemical reactions for cleavage of disulphide bridges and peptide bonds.
- Site-specific chemical modifications.

R. M. Caprioli et al. (eds.), Selected Topics and Mass Spectrometry in the Biomolecular Sciences, 361–376.

Reactions in the first group introduce chemical modification in the primary and secondary structure of proteins. Reactions in the second group are directed to modify amino acid residues and can be used to investigate the tertiary and quaternary structure of proteins.

Chemical modification reactions have long been developed and used, in combination with various analytical techniques. Various aspects of this field of protein chemistry and its applications have been discussed in several reviews [1- 11].
It is well known that enzymatic methods of protein modification combined with mass spectrometry also play a key role in protein characterisation. The following discussion, however, will be restricted only to the integration of mass spectrometry with chemical methods of protein modification. This approach will be illustrated with a few examples, which, of course, do not pretend to cover all the application in this area.

2. Chemical cleavage of disulphide bridges and peptide bonds.

2.1. REDUCTIVE CLEAVAGE OF DISULPHIDE BONDS AND BLOCKING OF THIOL GROUPS

There are several methods for disulphide bond cleavage [5]. Reduction with dithiothreitol (DTT, Cleland reagent) and subsequent blocking of the free thiol groups with iodoacetic acid, iodoacetamide, methyl iodide or 4-vinylpyridine has been extensively used in combination with mass spectrometry for protein analysis. For instance, it has been used in the mass spectrometric structural investigation of human serum albumin (HSA), a protein which contains seventeen disulphide bridges [12,13]. Mass spectrometric analysis has always confirmed that this reductive procedure is by far the most useful for splitting the disulphide bond of proteins because of its high selectivity and absence of secondary reactions.

2.2. PARTIAL ACID HYDROLYSIS AND METHANOLYSIS

Partial acid hydrolysis has been extensively used, especially in the early period of protein sequence analysis, to cleave proteins into smaller peptides [2-4]. When the reaction is carried out in 70% formic acid, cleavage occurs mainly at the aspartyl-prolyl bond. With dilute acetic or hydrochloric acid cleavage occurs before and after aspartic residues and, to a minor extend, after asparagin (Table 1).

TABLE 1. Partial acid hydrolysis.

Reaction conditions	Site of cleavage
70% formic acid 40 °C, 24 h	Asp-Pro
2% acetic acid or 0.03 N HCl 110 °C, 15 h	X-Asp-X; Asn-X

This degradation methods is scarcely selective, so that its application results in the formation of complex mixtures of overlapping peptides. Hence, it has been rarely used in combination with mass spectrometric analysis.

In contrast, taking just advantage of the poor selectivity, partial methanolysis coupled with FAB-MS analysis offers the possibility of obtaining sequence data for relatively low molecular weight peptides [14-16]. Treatment of small peptides with 5N HCl in dry methanol, in addition to the esterification of the free carboxylic groups, also produces the methanolysis of peptide bonds and a mixture containing almost all the methyl ester fragments that can be produced by cleavage of a single peptide bond in the original peptide, is obtained.

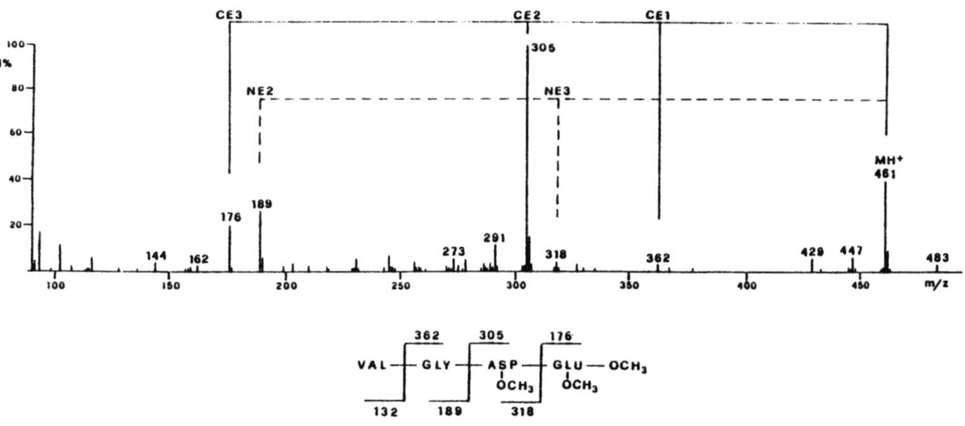

Figure1. FAB mass spectrum of the mixture of partial methanolysis of the peptide Val-Gly- Asp-Glu.

The FAB mass spectrum of the resulting mixture contains the MH^+ ions of all these fragments, thus considerably increasing the amount of sequence information in comparison with the spectrum of the original peptide. Analysis of the FAB spectrum of the partial methanolysis mixture usually allows the unequivocal reconstruction of the peptide sequence. A typical result is shown in Fig. 1.

2.3. CYANOGEN BROMIDE CLEAVAGE

Cyanogen bromide is considered the most useful reagent for chemical cleavage of polypeptides for its high specificity and quantitative cleavage yield. This reagent cleaves polypeptide chains at methionyl bonds and converts methionine residues into an equilibrium mixture of C-terminal homoserine and homoserine lactone residues. Since its introduction in 1961 this reaction has been extensively employed and studied [1-4]. Usual reaction conditions are 70% aqueous formic acid or trifluoroacetic acid, but 0.1% trifluoroacetic acid in water/acetonitrile (6:4 v/v) has also been used.

It has been observed that Met-Ser and Met-Thr linkages are particularly resistant to cleavage and this has been explained with the partecipation of the hydroxyl groups of Ser and Thr to the cyanogen bromide reaction with conversion of Met residues into uncleaved homoserine residues [see 3 for references].

A recent ES investigation of this reaction has lead to the proposal of a different mechanism for this process [17]. Moreover this work has shown that the intermediates containing homoserine residues along the chain are unstable and, providing that enough time is allowed for the hydrolysis reaction, they can be cleaved through an intermediate hemiacetal-like structure to give the usual homoserine end-group. Further information gained from this investigation is that formilation may occur when the reaction is carried out in 70% aqueous formic acid and that oxidation of methionine to methionine S-oxide is produced when 1% TFA water/acetonitrile is used.

It was recognised earlier that, for peptides of relatively low molecular mass, the homoserine and the corresponding homoserine lactone ending peptides give rise to separate peaks under the condition normally used for the high performance liquid chromatography (HPLC) fractionation of peptide mixtures, the higher retention time peak corresponding to the lesser polar homoserine lactone ending peptide [18].

In contrast, for higher molecular mass cyanogen bromide fragments the pair homoserine-homoserine lactone containing peptides cannot be separated as they co-elute in a single HPLC peak as illustrated by the HPLC separation of the fragments originated by cyanogen bromide fission of reduced and carboxymethylated HSA [19]. This protein contains six methionines in its sequence and therefore cyanogen bromide cleavage produces seven independent fragments (Table 2).

TABLE 2. Sequence and average molecular mass of CNBr fragments of reduced and carboxymethylated HSA and molecular mass determined by MALDI spectra.

Fragment	Sequence	Average Mr (homoserine lactone)	Mr from MALDI spectra
I	1-87	10023.1	10025
II	88-123	4336.7	4337
III	124-298	20681.3	20687
IV	299-329	3354.7	3358
V	330-446	13887.9	13889
VI	447-548	11941.6	11942
VII	549-585	4098	4098

These fragments can be completely separated by HPLC, except for fragments I and V which co-elute in the last HPLC peak (Fig. 2).

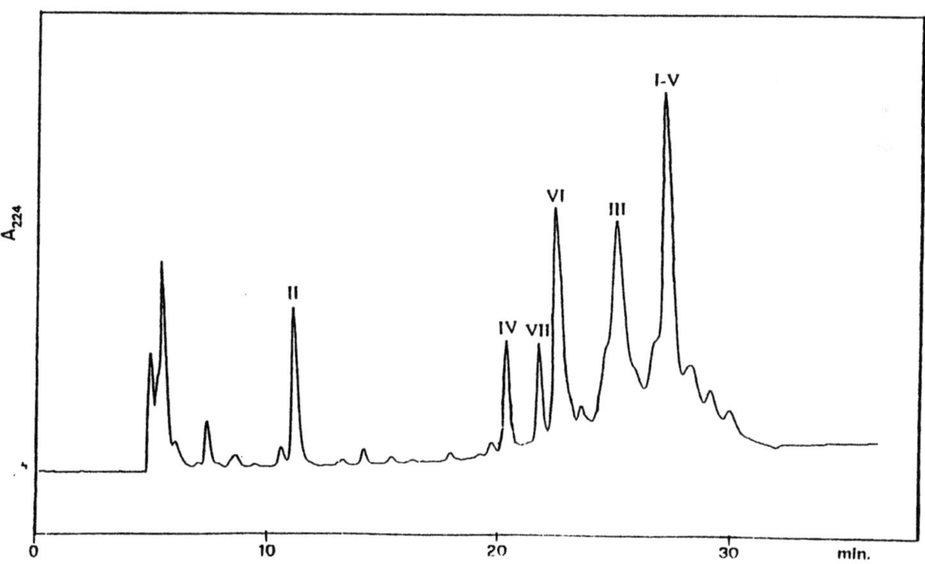

Figure 2. HPLC separation of CNBr fragments from reduced and carboxymethylated HSA.

Although homoserine and homoserine lactone ending species should be present for fragments I-VI, only a single HPLC peak is observed. MALDI-MS analysis of the six fractions showed sharp molecular ion signals corresponding closely to the homoserine lactone form (Table 2). This finding may derive from the preparation of the samples or from the conditions of MALDI analysis, in a way that deserves further investigation.

When dealing with cyanogen bromide produced fragments, it should be remembered that the homoserine lactone ring reacts readily with nucleophilic reagents. Hence, if peptide fragments obtained by cyanogen bromide fission are exposed to nucleophilic agents in subsequent chemical or enzymatic reaction, ring opening with formation of homoserine derivatives may occur.

This kind of reaction was observed when, in an alternative degradation procedure of HSA, the cyanogen bromide cleavage was performed directly on the "native" (i.e. ureduced) HSA. Direct CNBr cleavage of HSA produces only three independent fragments which, according to a generally accepted naming convention, are indicated as fragments A, B, and C [20]. Fragment C consists of a single polypeptide chain. Fragment A and B are composed of four and two subfragments, respectively, held together by disulphide bonds (Table 3).

TABLE 3. CNBr fragments of unreduced HSA.

Fragment	Sequence	Subfragments	Sequence
A	299-585	A1	299-329
		A2	330-446
		A3	447-548
		A4	549-585
B	1-123	B1	1-87
		B2	88-123
C	124-298	-	-

Thus, subsequent reduction and carboxymethylation of fragment A originates four independent fragments. When fragment A was reduced by DTT at pH 8.5 in Tris(hydroxymethyl)aminomethane hydrochloride (TRIS hydrochloride) buffer and carboxymethylated, the resulting reaction mixture gave the HPLC separation shown in Fig. 3 [19]. Peaks X1 and X2 were identified by MALDI analysis as partially cleaved fragments derived from incomplete fission of methionyl bonds.

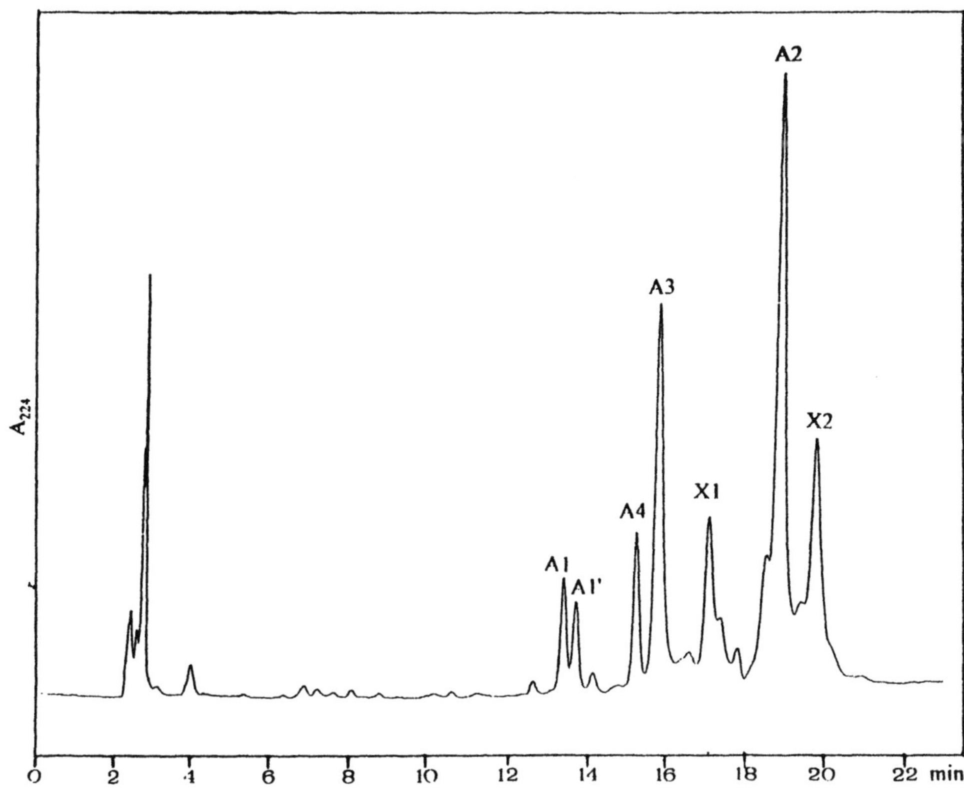

Figure 3. HPLC separation of the subfragments obtained by reduction and carboxymethylation of CNBr fragment A of HSA.

The MALDI spectrum of the first two peaks A1 and A1', collected together, showed two signals at m/z 3374 and 3478 (Fig. 4), corresponding to the reaction products of the homoserine lactone end-group with the components of the buffer, ammonia and TRIS (Scheme 1). In the higher molecular mass fragments, the two reaction products may not be separated by HPLC. For instance, the chromatogram in Fig. 3 shows a single HPLC peak for fragment A3, which has a molecular mass of about 12000 daltons. The MALDI spectrum of this peak gave a large molecular ion signal, which appears as a superimposition of the molecular ion signals corresponding to the two components (Fig. 5).

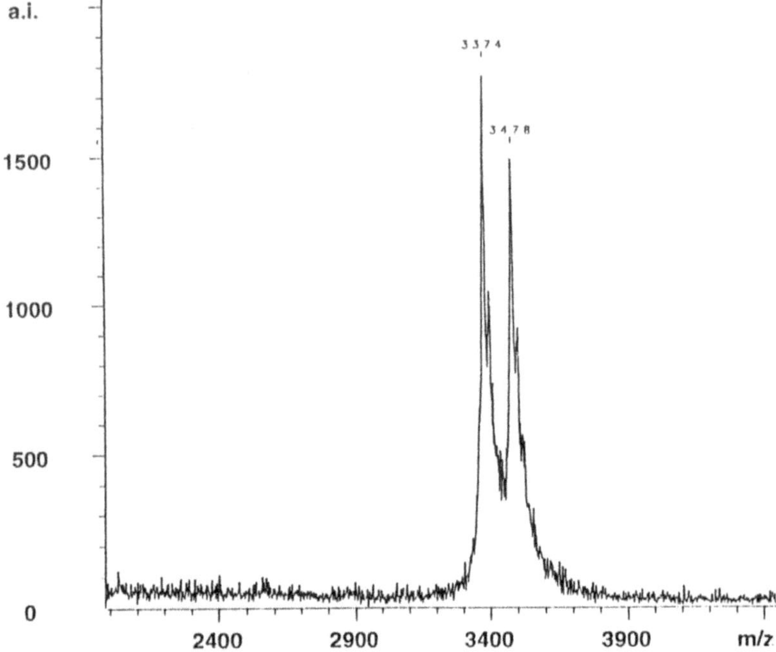

Figure 4. MALDI-TOF spectrum of fragments A1 and A1'

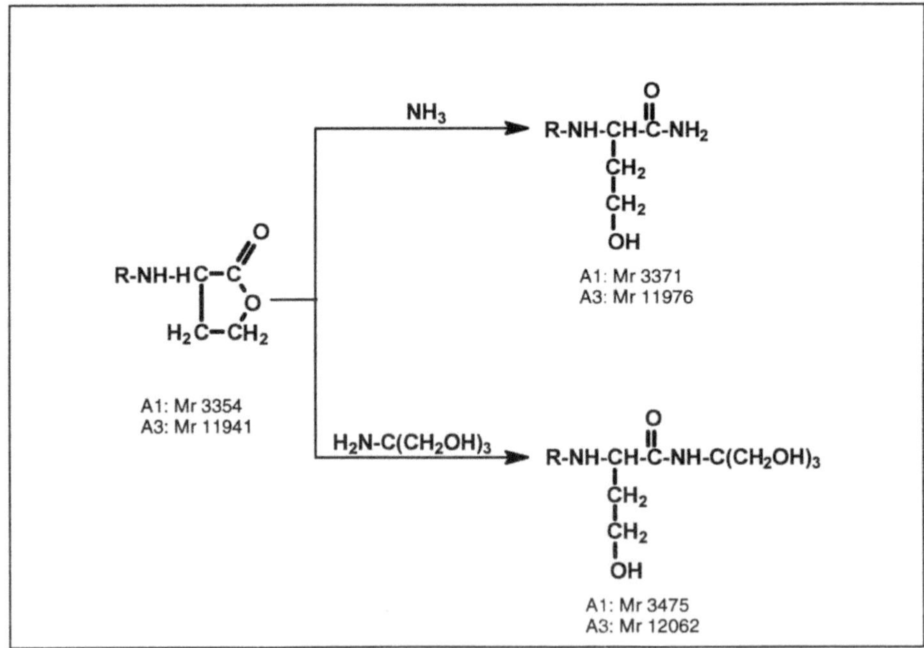

Scheme 1. Reaction between homoserine lactone end-group and Tris.

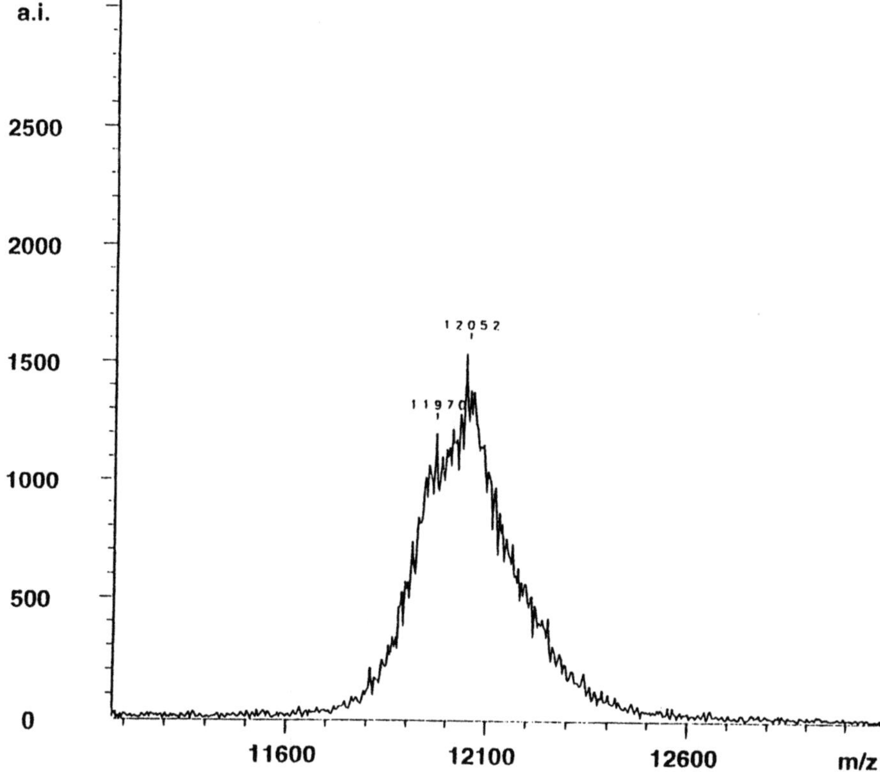

Figure 5. MALDI-TOF spectrum of fragment A3

2.4. OTHER CLEAVAGES

Several other useful methods for selective chemical cleavage of peptide bonds have been reported [1-4]. For instance, cleavage of tryptophyl linkages with 2-(nitro phenylsulfenyl)-3-methylindolenine (BNPS-skatole) or dimetyl sulfoxide and alogen acids and fission of tyrosil peptide bond by N-bromosuccinimide (NBS) have been successfully used for protein degadation, but their usefulness in combination with MS methods has not been investigated.

3. Site-specific chemical modification.

Chemical covalent modification of side-chain residues of proteins by site-specific reagent is a method which is currently used in structure-function studies for a variety of applications which include identification of "essential groups" for enzyme activity, immunological reactivity and protein interaction with low-molecular mass compounds. Besides being used for revealing amino acids residues primarily involved in these activities, these chemical reactions can be used for introducing a convenient alteration of the residue(s) itself in order to modify the activity of the protein. Chemical covalent modifications have also been used to modify proteolytic digestion pattern. Furthermore, site specific chemical modification can be employed for investigation of protein tertiary structure and surface accessibility [6,7].

A large number of reagents can be employed for modification of side chain in proteins and in principle all functional groups are amenable to chemical alteration [6-11]. Pratical limitation of the use of site-specific reagent is that some of them are not absolutely specific because they may react with more than a determined residue or even a single class of residues. This may complicate the interpretation of chemical modification studies. In site-specific covalent modification studies, it is of key importance to localize exactly the site of modification, to measure the extend of the modification, and to ascertain that unwanted side-reactions have not taken place. The characterization of the modified protein using classical analytical methods generally requires laborious procedure and, in some instances, radioactive labelling. Very often unequivocally results are not achieved. Instead, application of MS to the characterization of chemically modified protein, has proved that this technique is able to produce fast and reliable results.

A selection of the most commonly used site-specific reagents are reported in Table 4.

The reaction of diethylpyrocarbonate (DEP), a site-specific reagent for histidine residues and amino groups, has been investigated by HPLC and FAB-MS using model peptides. All the reaction products of the peptides with DEP were isolated and characterised, including the products of the Bamberger cleavage of the imidazole ring of histidine [21].

An analogous investigation of the reactivity of phenylglioxal (PGO) has revealed that this reagent gives at least two HPLC-isolable derivatives corresponding to the addition of two PGO units to the guanidino group of HSA with elimination of water [22].

TABLE 4. Site-specific reagents commonly employed for covalent chemical modification of proteins. See 6-11 for references.

Target a.a. residue	Reagent	Interfering groups	Product
Lys α-NH_2	Acetic anhydride	-SH; ROH; ArOH; imidazole	CH_3-CO-NH-
	Succinic anhydride	As above	HOOC-$(CH_2)_2$-CONH-
	Diethylpirocarbonate	imidazole	see ref. 21
Glu, Asp	CH_3OH/HCl		-$COOCH_3$
	Carbodiimide/GlyOCH_3		-CO-NH-CH_2-$COOCH_3$
Tyr	I_2		mono- and di-iodo-Tyr
	Tetranitromethane		nitro-Tyr
Arg	1,2-Cyclohexanedione		see ref. 7, 10, 11
	Phenylglyoxal		see ref. 7, 10, 11

3.1. APPLICATION TO PROTEIN STRUCTURE-PROPERTIES INVESTIGATION

PGO derivatization has been used to investigate the interaction of HSA with Cibacron Blue F3GA, a syntetic dye used in dye-protein affinity chromatography [23]. Reaction under mild conditions of PGO with HSA resulted in the complete inactivation of HSA with respect to the dye. MALDI-MS analysis of the cyanogen bromide fragment of modified and inactive HSA allowed the localization of modified residues within the sequences 124-298 and 330-446 of HSA.

Succinylation has been employed to investigate the structure-property relationships of porin from *Rhodobacter capsulatum 37b4* [24]. This reagent introduces a negative charge in place of the original amino group. Measurement of the molecular mass of the modified protein by MALDI-MS showed an increment of 400 daltons, thus indicating four sites of modification. A precise localization of the four succinylated amino groups was provided by MALDI-MS peptide mapping of cyanogen bromide and tryptic fragments. Mass spectrometric data were complemented by X-ray crystallography and this information constituted the basis for the correlation of structural modification with activity alteration.

3.2. APPLICATION TO TERTIARY PROTEIN STRUCTURE INVESTIGATION

The possibility to detect differential reagent accessibility to specific residues in protein upon chemical modification and MS analysis was shown in a study in which the iodination reaction of tyrosine residues in Citochrome *c* was used as model system [25]. The analytical procedure involved the reaction of the native protein followed by characterization of the modified structure through enzymatic digestion and HPLC/FAB-MS identification of the proteolytic peptides. This established a selective iodination of Tyr 74 among the four tyrosine residues present in the Citochrome *c* sequence, a result which was consistent with the known X-ray structural data of this protein (Fig. 6).

Figure 6. X-ray structure of Citochrome *c* showing the different accessibility of the four Tyr residues.

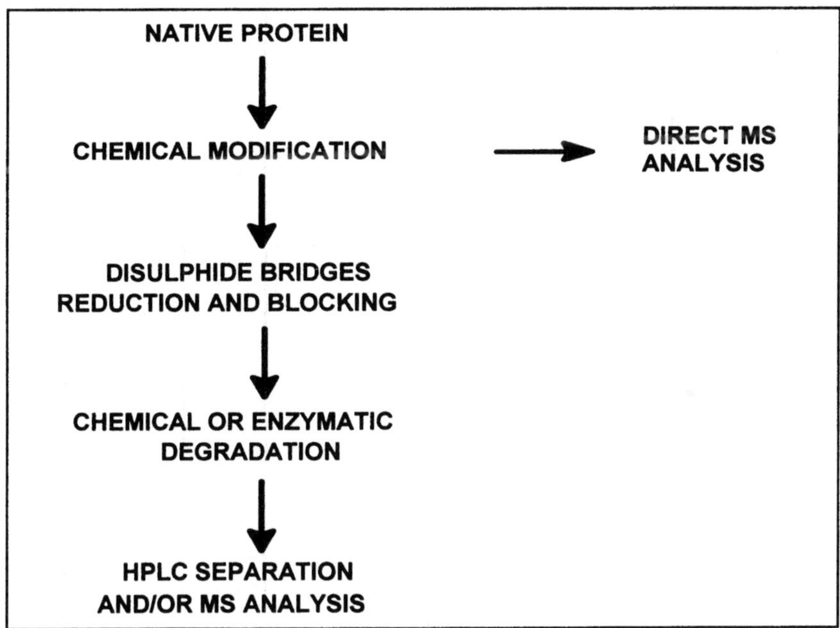

Scheme 2

With the capability of ES- and MALDI-MS to measure directly the molecular mass of proteins of considerable size, direct analysis of the chemically modified protein is now feasible. This allows not only the monitoring of the extent of modification, but also gives the possibility to follow the time-course of the modification reaction. thus providing a tool for the characterization of the tertiary structure [26-30]. An analytical procedure for the investigation of tertiary structure of proteins by selective covalent chemical modifications and MS analysis is summarized in Scheme 2.

In a recent paper [30], a clear correlation of the relative reactivities of amino groups upon acetilation and their accessibility was demonstrated. A different distribution of multi-charged ions was observed in the ES mass spectra of extensively modified proteins, whereas specific modification did not alter the charge distribution. In another paper lysine acetilation and arginine modification with 1,2-cyclohexanedione as function of time and/or molar reagent concentration was followed by PD-MS and ES-MS determination of the molecular mass of the protein derivatives [28]. Therefore these model studies suggest the possibility of using ES to monitor conformational changes following chemical modification, that are difficult to assess by classical technique.

4. Conclusion and perspectives

With the introduction of new, accurate and sensitive desorption-ionization methods, mass spectrometric characterisation of chemically modified large polypeptides and proteins is nowadays feasible just as the mass spectrometric characterisation of structural modification introduced in low molecular mass compounds was in the past. Whereas protein chemistry will benefit from this possibility, applications of mass spectrometry to protein characterisation will gain advantage from the integration with chemical modifications.

In particular, chemical cleavage of polypeptide chains appears valuable for MS coupling since, in contrast with proteolytic methods, it is directed to produce large fragments suitable for ES or MALDI analysis.

Classical chemical modification reactions were primarily developed to be used in conjunction with classical analytical techniques. In some instances, coupling of known protein chemistry reactions with MS requires a re-evaluation of the chemical reaction adopted. Chemical modification reactions that have resulted of little value within the classical analytical schemes may prove to be valuable when integrated with MS, whereas other reactions, largely used in protein modification, need to be critically considered when coupled with MS. Moreover, it appears that development of new chemical modification reactions especially designed for MS coupling is of interest.

The combined use of selective chemical modification and MS analysis is now being regarded as a new analytical tool that can produce information on the tertiary structure of protein in solution . In fact, despite the great capability of X-ray crystallography to determine proteins structures in the solid state, it should be considered that techniques that can give information to complement these data in solution are highly desirable.

Examining the most recent trend in this field, is appears that, although the integration of chemical modification reactions with the more recent ionisation methods is at the initial stage, it is gaining growing attention in view of its broad perspectives of potential applications in protein structure and function investigation.

5. References

1. Hermodson, M.A. (1982) in M. Elzinga (ed.), *Methods in Protein Sequence Analysis*, Humana Press, Clifton , pp.313-322.

2. Allen, G. (1981) *Sequencing of Proteins and Peptides*, North-Holland Publishing Company, Amsterdam.
3. Fontana, A. and Gross, E. (1986) in A. Darbre (ed.), *Pratical Protein Chemistry - A Handbook,* John Wiley & Sons, Chichester, pp. 83-88.
4. Kamp, R.M. (1986) in B. Wittmann-Liebold, J. Salnikow and V.A. Erdmann (eds.), *Advanced Methods in Protein Microsequence Analysis*, Springer-Verlag, Berlin, pp.8-20.
5. Henschen, A. (1986), *ibidem,* pp. 244-255.
6. Glazer, A.N. (1976) in N. Neurath and R.H. Hill (eds.), *The Proteins*, Academic Press, London, pp. 1-103.
7. Means, G.E., and Feeney, R.E. (1971) *Chemical Modification of Proteins*, Holden-Day, San Francisco.
8. Riordan, J.F. and Sokolowsky, M. (1971) *Accounts Chem. Res.* **4**, 353-360.
9. Cohen, L.A. (1968) *Annu. Rev. Biochem.* **37**, 695-726.
10. Lundblad, R.L. and Noyes, C.M. (1984) *Chemical Reagents for Protein Modification*, CRC Press, Boca Raton.
11. Lundblad, R.L. (1995) *Techniques in Protein Modification*, CRC Press, Boca Raton.
12. Compagnini, A., Fisichella, S., Foti, S., Saletti, R. and Sardo, L. (1994) *Rapid Commun. Mass Spectrom.* **8**, 459-464.
13. Fisichella, S., Foti, S., Maccarrone, G. and Saletti, R. (1995) *J. Chromatogr.* **693**, 33-44.
14. Foti, S. and Saletti, R. (1989) *Biomed. Environ. Mass Spectrom.* **18**, 168-173.
15. Foti, S. and Saletti, R. (1991) *Biol. Mass Spectrom.* **20**, 345-350.
16. Foti, S., Saletti, R. and Marletta, D. (1991) *Org. Mass Spectrom.* **26**, 903-907.
17. Zhang, X., Dillen, L., Vanhoutte, K., Van Dongen, W., Esmans, E. and Claeys, M. (1996) *Anal. Chem.*, in press.
18. Pucci, P., Borgia, R. and Marino, G. (1985) *J. Chromatogr.* **331**, 425-431.
19. Compagnini, A., Fisichella, S., Foti, S., Jetschke, M., Maccarrone, G., Przybylski, M. and Saletti, R. (1996) *3rd EU Network Meeting: Peptide and Protein Structure Elucidation by Mass Spectrometry*, Vienna.
20. Compagnini, A., Fichera, M., Fisichella, S., Foti, S. and Saletti, R. (1993) *J. Chromatogr.* **639**, 341-345.
21. Foti, S., Marletta, D., Saletti, R. and Petrone, G. (1991) *Rapid Commun. Mass Spectrom.* **5**, 336-339.

22. Fisichella, S., Foti, S., Marletta, D. and Saletti, R. (1992) *Massa 92*, Milano, p.65.
23. G. Maccarrone (1995) Ph. D. Thesis, Dipartimento di Scienze Chimiche, University of Catania
24. Przybylski, M., Glocker, M.O., Nestel, U., Schnaible, V., Bluggel, M., Diederichs, K., Weckesser, J., Schad, M., Schmid, A., Welte, W. and Benz, R. (1996) *Protein Science* **5**, 1477-1489.
25. Amico, V., Foti, S., Saletti, R., Cambria, A. and Petrone, G. (1988) *Biomed. Environ. Mass Spectrom.* **16**, 431-437.
26. Przybylski, M. (1995) in I. Cornides, Gy. Horváth and K. Vekey (eds.), *Advances in Mass Spectrometry* **13**, Elsevier, pp. 257-285.
27. Nielsen, P.F., Schneider, K., Suckau, D., Landis, B. and Przybylski, M. (1990) in E. Hilf (ed.), *Mass Spectrometry of Large Non-Volatile Molecules,* Vol. 3, World Scientific, London, pp. 194-207.
28. Suckau, D., Mák, M. and Przybylski, M. (1992) *Proc. Natl. Acad. Sci. USA* **89** 5630-5634.
29. Przybylski, M., Borchers, C., Suckau, D., Mák, M. and Jetschke, M. (1993) in C.H. Schneider and A.N. Eberle (eds.), *Peptides 1992*, Escom Science Publishers, Amsterdam, pp. 915-916.
30. Glocker, M.O., Borchers, C., Fiedler, W., Suckau, D. and Przybylski, M. (1994) *Bioconj. Chem.* **5**, 583-591.

PSD-MALDI ANALYSIS OF PEPTIDES

B. SPENGLER
Institute of Laser Medicine
Heinrich-Heine-University of Duesseldorf
P.O. Box 101007
D-40001 Duesseldorf
Germany

1. Introduction

Postsource decay - matrix assisted laser desorption ionization ("PSD-MALDI") [1-7] has become one of the most powerful methods for mass spectrometrical structure analysis of biopolymers. Due to its high sensitivity in the range of 100 fmol and its high tolerance of sample impurities and sample inhomogeneity, the method found broad acceptance and application rather rapidly [8-16].

Automated Edman degradation, which has routinely been employed over the years for sequence analysis of peptides, is now more and more shored up or even replaced by mass spectrometrical methods. These methods are almost exclusively electrospray ionization [17] and PSD-MALDI. The main advantage of PSD-MALDI over classical Edman sequencing is, besides its much higher sensitivity, the ability to analyze peptides which are chemically modified or blocked, which are part of inhomogeneous samples or which are just barely purified from salts or detergents. The limited instrumental requirements of these modern mass spectrometrical techniques, in comparison to e.g. established four-sector instruments, have led to a real conquest in the field of molecular biology.

The topic of this chapter is to summarize the basic mechanisms and operational methods, to give an overview of the possibilities and limitations of peptide sequence analysis by PSD-MALDI mass spectrometry and to describe strategies for chemical and computational assistance in this field.

2. Principles of PSD-MALDI mass spectrometry

Fast ion fragmentations in the ion source, taking place within the first few percent of ion acceleration time, can be observed in MALDI under certain instrumental

R. M. Caprioli et al. (eds.), Selected Topics and Mass Spectrometry in the Biomolecular Sciences, 377–398.

conditions [18], but in general are rather rare. MALDI spectra thus typically show the molecular ion signal only, often accompanied by signals of adducts, multiply charged molecular ions or multimeric molecular ions. Delayed ion fragmentation taking place in the field-free drift region of the instrument, on the other hand, has been found to be a rather common event and is used in PSD-MALDI for structure analysis of biomolecules.

The degree of fragmentation can be controlled widely by variation of instrumental parameters. As described in the following, fragmentation is taking place as a result of three basic activation mechanisms.

2.1. CONTROL OF FRAGMENTATION

The dependence of ion fragmentation on various instrumental conditions can be investigated using a rather simple setup [1]. The instrument used is a linear Time-of-Flight mass spectrometer, additionally equipped with an extended deceleration stage (Fig. 1).

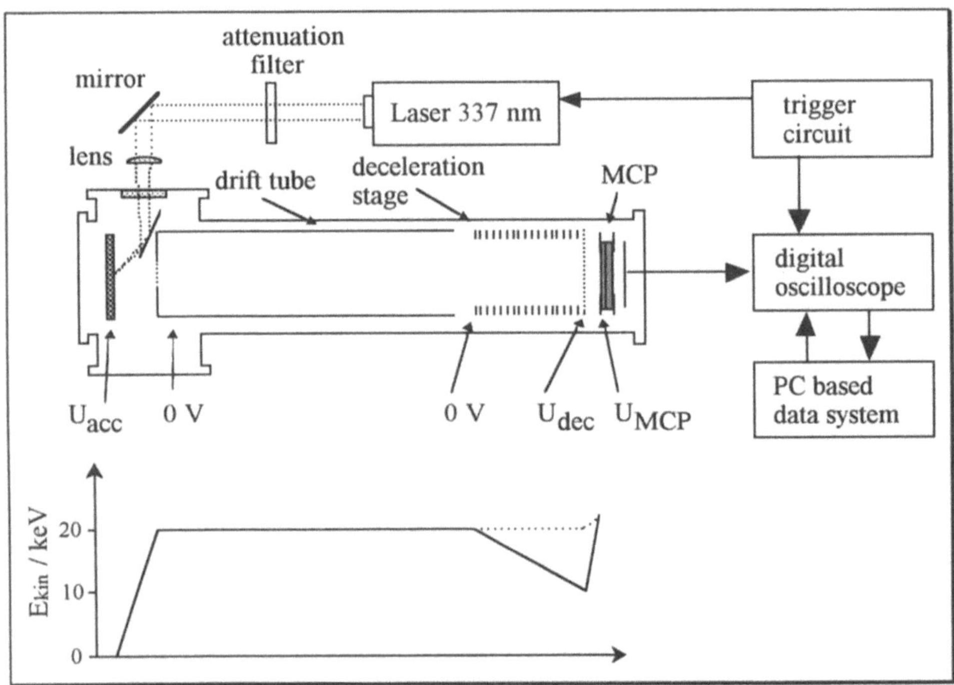

Figure 1: Linear Time-of-Flight mass spectrometer for investigation of postsource decay. The deceleration stage is used to separate in time stable ions and neutral fragmentation products. The variation of kinetic energies through the instrument is drawn in the lower part.

The effect of using the deceleration stage is shown in Fig. 2. Stable ions are decelerated from e.g. 16 keV to 6 keV and reaccelerated again to 16 keV (retaining the original detection probability). By this procedure their total flight time through the instrument is increased considerably. Neutral products from delayed ion fragmentation in the field-free drift region, on the other hand, are not affected by the deceleration stage and thus keep the original total flight time.

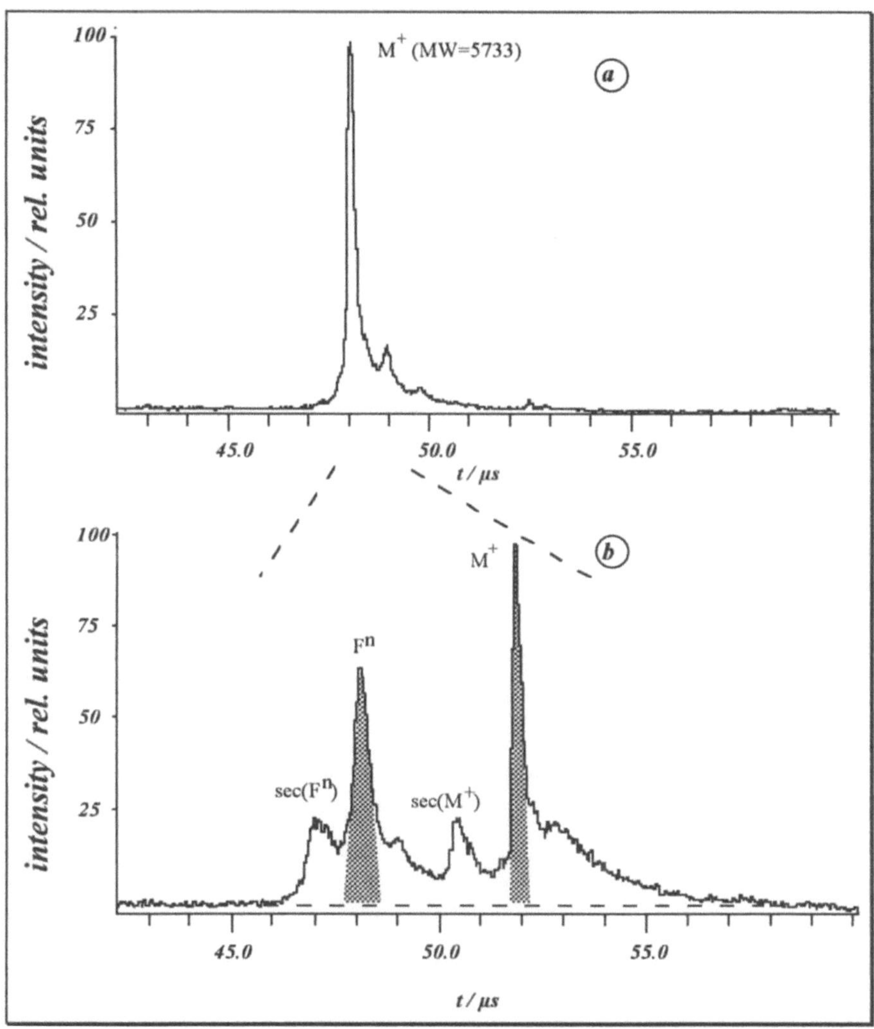

Figure 2 : The peak of bovine insulin (a) is split into two peaks (b) by applying the deceleration field. While stable precursor ions shift to longer total flight times, neutral products from ion decay keep their original velocity and are detected at the original total flight time. Integration of the two peaks (shaded areas) can be used to calculate decay rates.

Both signals are detected as slightly broadened peaks in the deceleration spectrum and can be used to evaluate the fragmentation behavior. While the shifted peak corresponds to stable, unfragmented molecular ions, the non-shifted peak can be taken as a measure for the number of fragmentation events. The integrated signals are used to derive values for relative ion stabilities and are plotted versus the residual gas pressure of the instrument (Fig. 3).

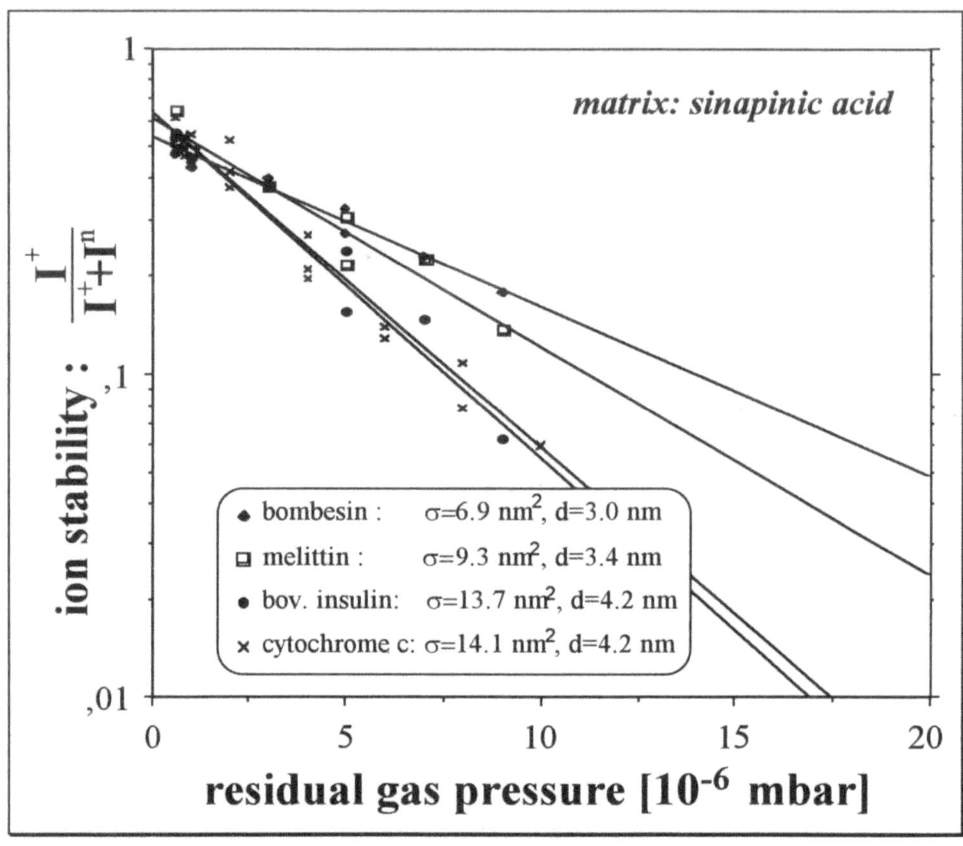

Figure 3 : Observed relative stabilities of precursor ions after a 0.72 m field-free drift path. The residual gas was dominated by nitrogen. The acceleration voltage was 16 kV (20kV for cytochrome C). The deceleration voltage was 10 kV.

As shown for peptides in sinapinic acid matrix, the residual gas pressure has a considerable influence on postsource decay, meaning that high-energy collisions of accelerated molecular ions with residual gas molecules in the instrument can be an important pathway of ion activation. From the slope of the curves, reaction cross sections for ion fragmentation can be derived as described by:

$$\log\frac{I^+}{I^+ + I^n} = -\frac{L\cdot\sigma}{2.3\cdot k\cdot T}\cdot p_g \qquad (1)$$

where

$$\frac{N_{ion}}{N^0_{ion}} = \frac{I^+}{I^+ + I^n}, \tag{2}$$

N_{ion} is the number density of stable parent molecular ions at the exit of the decay region, N^0_{ion} is the initial number density of stable parent molecular ions, I^+ is the ion abundance of stable parent molecular ions, obtained from the spectra by peak integration of M^+, I^0 is the abundance of neutral PSD products, obtained from the spectra by peak integration of F^n, L is the length of the field-free drift region (the decay region), σ is the reaction cross section for fragmentation by collision of parent molecular ions with residual gas molecules, k is Boltzmann's constant, T is the absolute Temperature and p_g is the residual gas pressure.

It has been found that reaction cross sections are rather high for peptides and are still increasing with mass even in the range above 10000 u. This observation is in sharp contrast to the results of high-energy CAD in sector-field instruments where the fragmentation yield dramatically decreases above 5000 u.

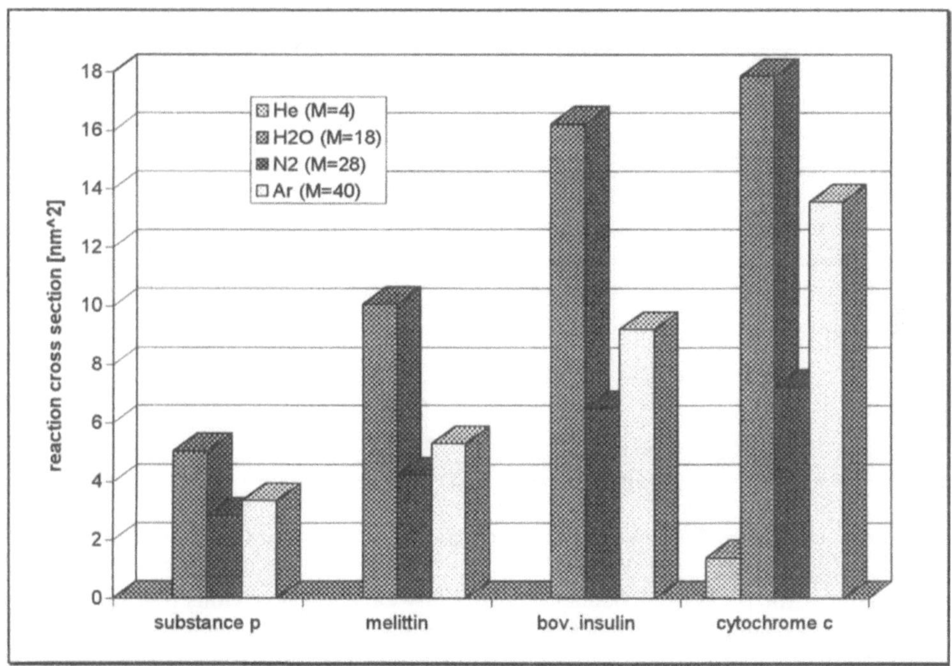

Figure 4 : Reaction cross sections for fragmentation of peptide ions by high-energy collision (20 keV) with various target gases.

Using different target gases for ion activation, the fragmentation yield can even be increased. Fig. 4 shows a comparison of reaction cross sections for ion fragmentation using argon, nitrogen, water and (in the case of cytochrome c) helium as the dominating contributor to the residual gas pressure of the instrument. In all cases the reaction cross sections increase with mass over the mass range tested. The behavior of the different target gases is in general as one would expect from the increase of collision energy due to increase of the mass of the target molecules/atoms. In the case of water as a target gas, however, a fragmentation yield much higher than expected has been observed. The reason for this exceptional behavior has not been discovered yet. It can be suspected that peptide ion discharging by fast proton exchange reactions between the water molecule and the protonated peptide ion contributes considerably to the overall ion decay measured. In any case, the strong effect of interactions of water molecules with peptide ions is an important instrumental factor, keeping in mind that the residual gas of poorly baked vacuum systems is usually dominated by water.

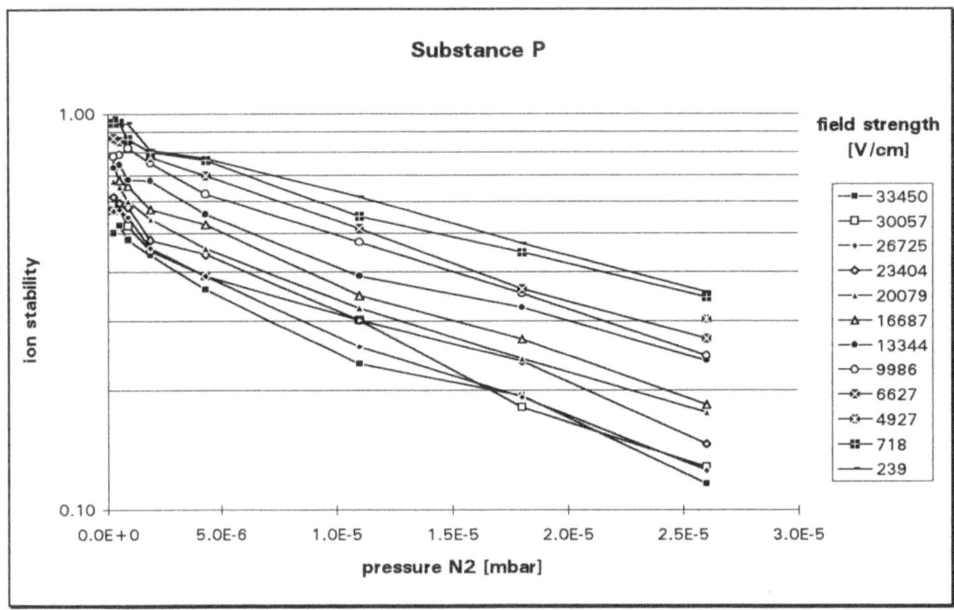

Figure 5 : Ion stability of substance P as a function of residual gas pressure and initial acceleration field.

Looking at the pressure dependent ion stability curves it is obvious that high-energy collisions with residual gas can not be the only contributors to ion activation and might even be of rather low significance in typical instrumental setups. (The use of the high-energy mode of postsource decay in collision cells will be described later.) Extrapolation of the ion stability curves to zero residual gas pressure shows that, under the experimental conditions used for this investigation, almost 40% of the molecular ions decay without undergoing collisions with residual gas molecules. The activation mechanism of this mode of operation has been investigated using a two-stage acceleration system in the ion source. Peptide ions enter the field-free drift region with a constant kinetic energy supplied by the second stage, but are initially accelerated (immediately after ion formation) by the variable acceleration field of the first stage. Fig. 5 describes the behavior of the ion stability curves as a function of the initial acceleration field strength. While the slope of the curves is almost unaffected, the extrapolated ion stabilities at zero residual gas pressure vary considerably.

This characteristic behavior can only be explained by assuming multiple collisions of low to medium energy between peptide ions and neutral species within the desorption plume. An approximation of particle densities above the surface after laser desorption, assuming an ejection of 1 μm^3 of material, can be used to derive curves of pressure and mean free path as functions of the distance from the surface (Fig. 6 and 7). It shows that at a distance of 100 μm the peak pressure is still in the range of a mbar and the corresponding mean free path is in the range of 10 μm. It is thus obvious that within the first 100 to 200 ns after the laser pulse, multiple collisions between accelerated ions and neutral particles will take place, leading to considerable ion activation. Analyte ions are drawn through the cloud of neutral particles by the accelerating electrical field. The collision energy (and thus the degree of ion activation) is mainly controlled by the effective field strength in this region.

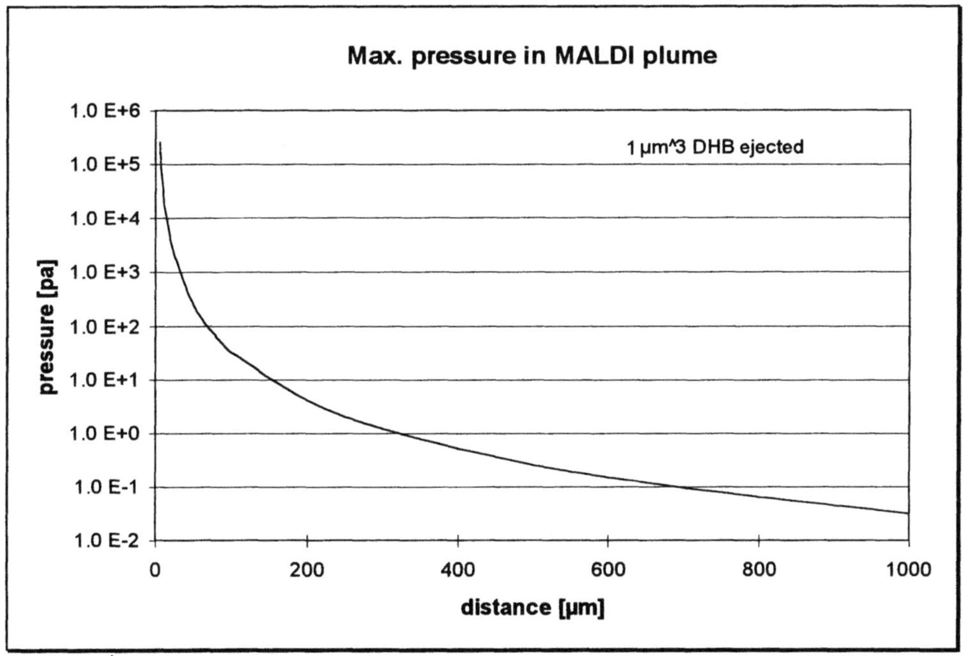

Figure 6 : Calculated peak pressures of a MALDI plume above the surface after laser irradiation by a single laser pulse, assuming 1 μm^3 of material to be ablated. The temporal and spatial distribution used for calculation was taken from experimental data as reported in [26].

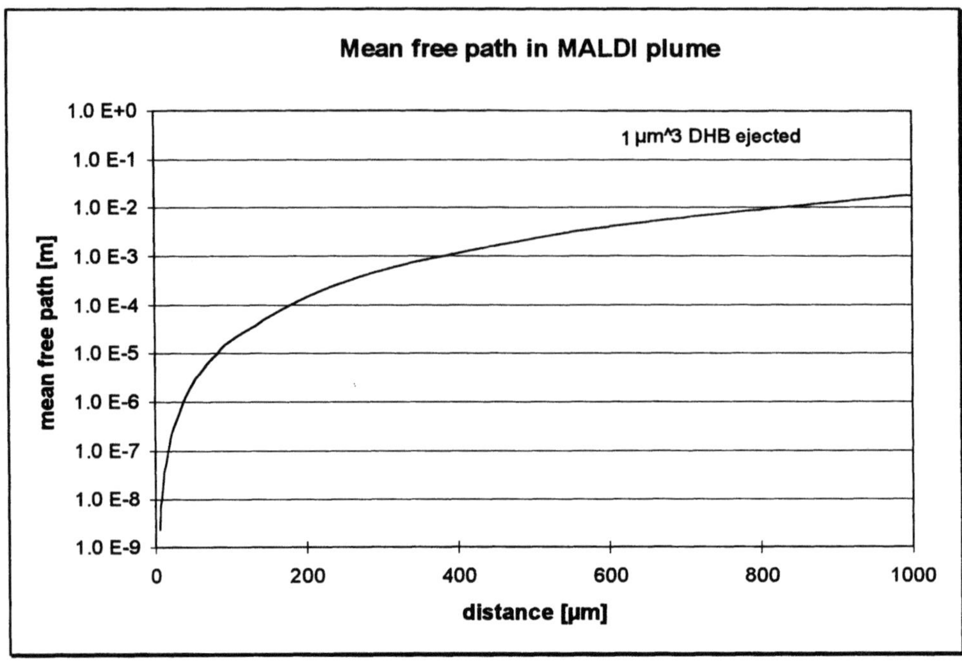

Figure 7 : Calculated mean free path (peak minimum) of peptide ions within a MALDI plume according to Fig. 6.

The values of ion decay at zero residual gas pressure can easily be transformed into unimolecular rate constants, which are then independent of the instrumental conditions chosen for the experiments described here. These unimolecular rate constants k can be derived from the measured decay curves according to

$$k = -\log\left(\frac{I^+}{I^+ + I^n}\right) \cdot \frac{2.3}{L} \cdot \sqrt{2 \cdot U_{acc} \cdot \frac{e}{m}} \tag{3}$$

where U_{acc} is the full acceleration voltage of the two stages and $I^+ / (I^+ + I^n)$ is the ion stability extrapolated to zero residual gas pressure. A plot of the resulting unimolecular rate constants k versus the initial field strength (at constant total acceleration energy) emphasizes the strong dependence of ion stability on in-source collisions (Fig. 8).

It shows a linear relationship, meaning that a doubling of the initial field strength roughly leads to a doubling of the unimolecular rate constant.

Even at zero initial field strength a considerable amount of unimolecular decay is observed in Fig. 8. It is thus obvious that acceleration-induced activation and residual gas collision-induced activation are not the only two mechanisms of importance in PSD-MALDI. Decay of molecular ions at zero residual gas pressure and at zero initial field strength can be attributed to the internal energy of molecular ions right after ion formation (in the gas phase) or ion desorption (in the case of desorption of preformed ions). Several activation pathways can be assumed for this process, such as direct photon-molecule interaction, solid state activation, temperature effects in the solid state or excess energy from the ionization process.

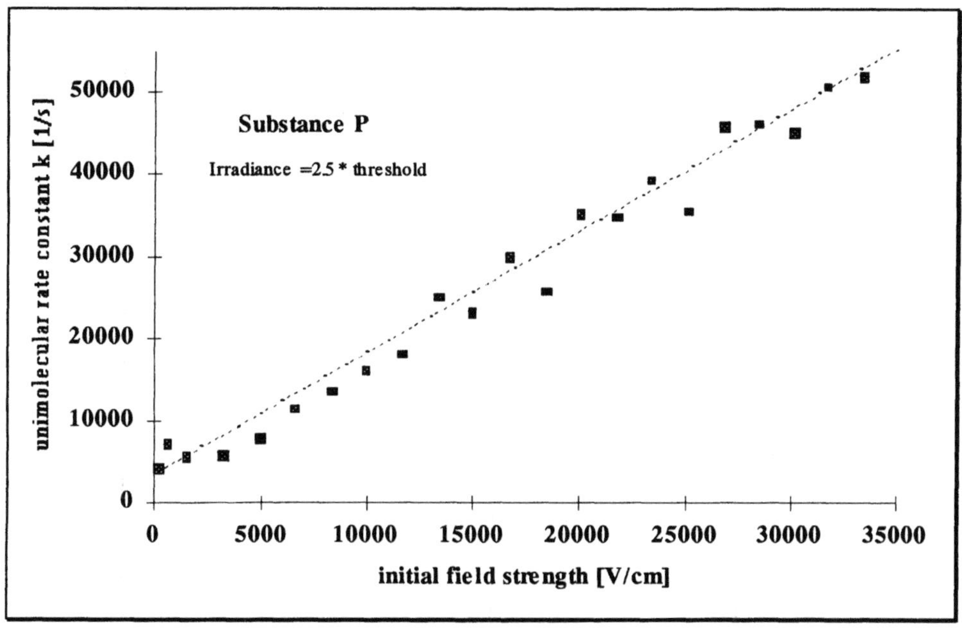

Figure 8 : Field dependence of the unimolecular rate constant k of substance P at zero residual gas pressure (extrapolated).

This third mode of ion activation in PSD-MALDI is of high importance in instruments using delayed ion extraction as a tool for increasing the mass resolution. In these instruments the initial field strength is almost zero for several hundred nanoseconds after the laser pulse. In-source activation by low-energy collisions can thus not be used to activate molecular ions. Fig. 8 shows that under typical instrumental conditions of prompt-extraction and delayed-extraction instruments (initial field strength 10000 V/cm compared to zero initial field strength), a reduction of the fragmentation rate by a factor of about 5 is expected. This is consistent with experimental observations. Fig. 9 illustrates the decrease of ion fragmentation under delayed extraction conditions (B) compared to prompt ion extraction (A). Loss of fragmentation would be expected to result in a decrease of analytical sensitivity. It has been shown, however, that reduction of the fragmentation rate is in general not a disadvantage for practical PSD-MALDI analysis under delayed extraction conditions but is compensated by an increase in mass resolving power and by a reduction of noise [19].

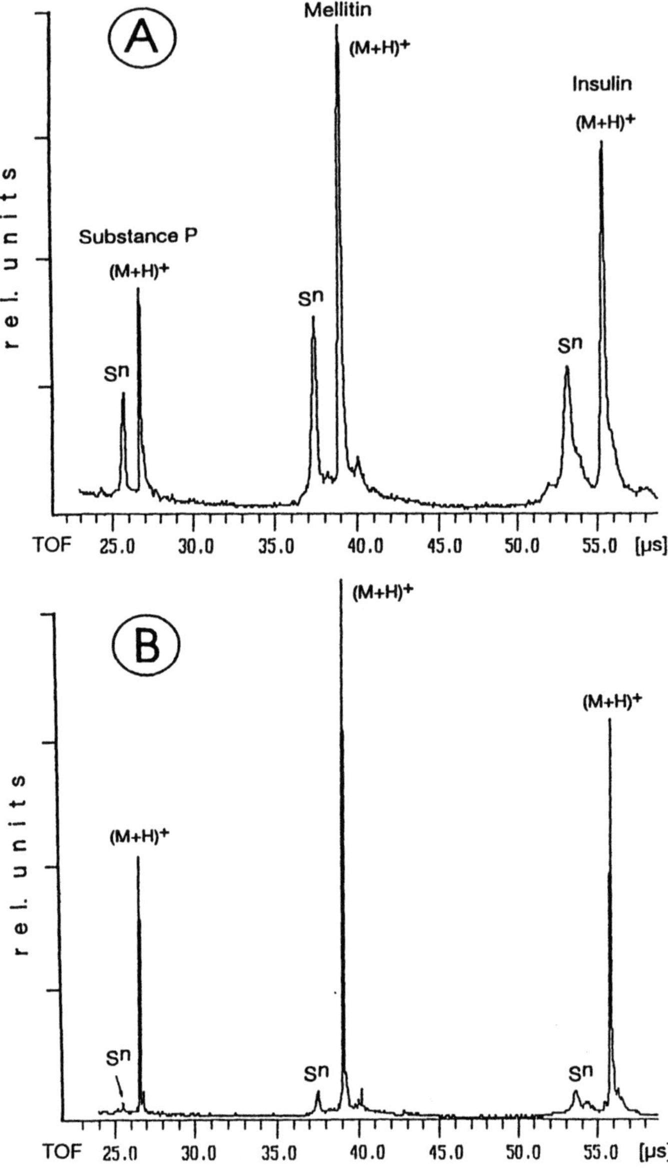

Figure 9 : Reduction of postsource decay under delayed extraction (B) versus prompt extraction (A) conditions. S^n are the signals from neutral products of postsource decay.

Table 1 summarizes the different activation mechanisms available for PSD-MALDI analysis. The activation mode mostly used in commercial PSD-MALDI instruments is mode 2, employing in-source activation by multiple low-to-medium-energy collisions. The interaction region of this mode can be approximated to be within 200 μm above the surface and interaction will take place in the ns to μs time frame. Under delayed extraction conditions mode 2 is almost neglectable and mode 1 becomes the more effective way of activation. Interaction can be assumed to take place in the solid phase or on the surface in a time frame of ps to ns. Mode 3 can be employed intentionally or (in long instruments with non-optimized vacuum conditions) unintentionally for ion activation through high-energy collisions of molecular ions with residual gas molecules in the field-free drift region or with target gas molecules in a collision cell.

It has to be kept in mind that for all three activation modes the location of ion decay is always in the field-free drift region, regardless the location of ion activation.

Mode	1	2	3
activation mechanism	sample activation	in-source activation	post-source activation
activation region	surface / sample	selvedge	vacuum
activation processes	direct photon-molecule interactions, solid state activations, temperature effects, excess energy from ionization	multiple low-to-medium-energy collisions	high-energy collisions
time scale	ps .. ns	ns .. μs	μs .. ms
distance from surface	0	< 200 μm	> acceleration zone

Table 1: Activation mechanisms in PSD-MALDI

2.2. PSD-MALDI INSTRUMENTS

The principle setup of a PSD-MALDI instrument is shown in Fig. 10 [2]. The sample is irradiated by a pulsed UV-laser beam of typically 337 nm wavelength (N_2 laser). Ions are formed and accelerated in the two-stage acceleration system to a mass-independent uniform kinetic energy. In the field-free drift region, part of these ions decay into smaller product ions that have a lower kinetic energy, according to their mass-fraction of the precursor ions. The reflector that, in usual time-of-flight instruments, is used for improving mass resolution, is used here as an energy analyzer. Due to their mass dependent kinetic energy, PSD ions are reflected at different positions within the reflector and thus have mass dependent total flight times through the instrument. Acquisition of a complete PSD spectrum has to be performed in several steps with varied reflector potentials in order to image the complete PSD ion mass range with a sufficient mass resolution. In a two-stage gridded reflector, as shown here, PSD ions of the chosen mass range appear as well-resolved signals in the spectrum, accompanied by a broad signal resulting from all PSD ions that, under these conditions, do not have enough kinetic energy to enter the second stage of the reflector but are reflected in the short first stage already.

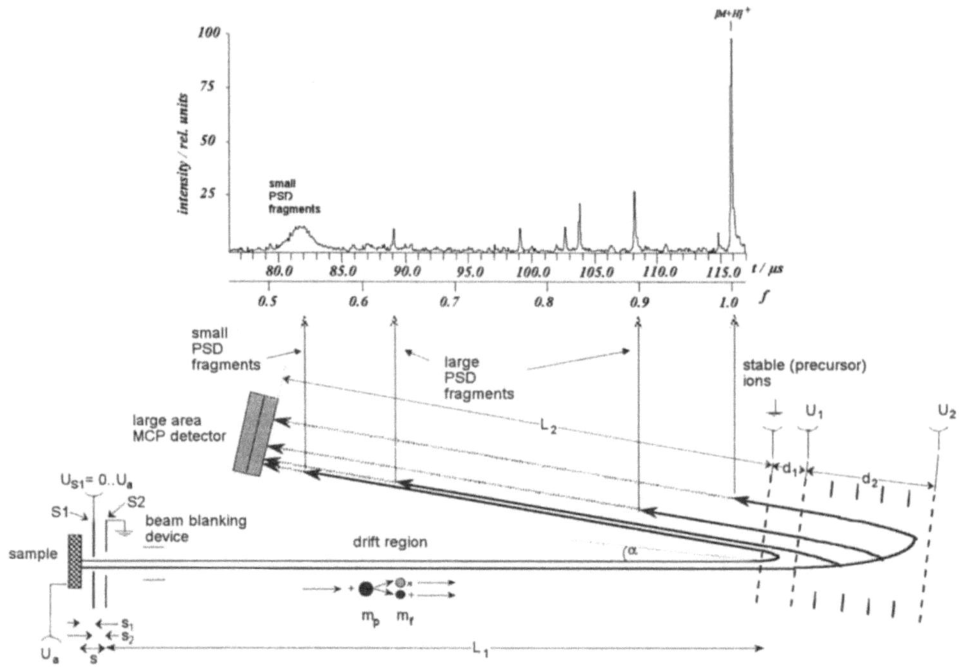

Figure 10 : Scheme of the instrument for PSD-MALDI mass analysis, equipped with a two-stage gridded ion reflector. See text for further explanation.

After acquisition of 10 to 15 spectra for varied reflector potentials covering the entire PSD ion mass range, a complete PSD spectrum is generated in the computer by concatenation of all sections.

After the ion source a beam blanking device is located acting as a precursor selector. It allows the operator to deflect all ions except a chosen mass window in order to transmit only one type of molecular ions. This MS-MS mode can be used to perform mixture analysis and to suppress adduct peaks.

2.3. ACQUISITION OF PSD SPECTRA

A typical mixture analysis by PSD-MALDI is illustrated in Fig. 11. A tryptic digestion of a protein is analysed by MALDI in spectrum A. By means of the precursor selector any of the peptides can be chosen for PSD analysis (spectra B to E). PSD analysis of the selected peptide gives a series of section spectra (F1 to F8) that can then be concatenated to a complete PSD product ion spectrum (spectrum G).

The sensitivity of the method is in the range of 100 fmol prepared sample and is mainly limited by the preparation techniques available. Mass resolving power in delayed-extraction mode is typically 6000 to 10000 for precursor ions and is in the range of 1500 to 2000 for PSD ions.

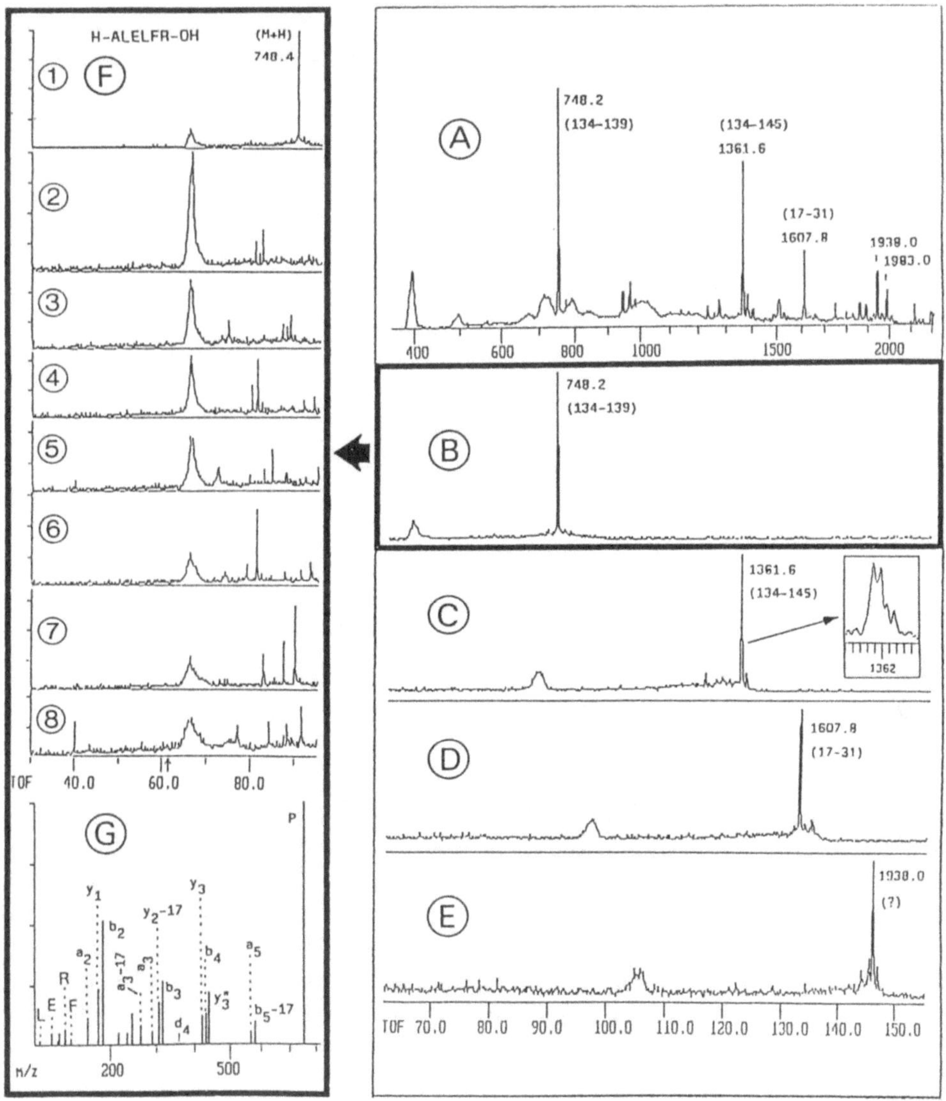

Figure 11 : Procedure of a PSD analysis of peptides from a tryptic digestion of myoglobin. See text for details.

3. Strategies for PSD-MALDI analysis of peptides

3.1. CHARACTERISTICS OF PSD SPECTRA OF PEPTIDES

Fragmentation of peptides leads to ions of characteristic chemical structure (Fig. 12). Two classes of fragment ion types are distinguished, called N-terminal („a“, „b“, „c“, „d“) and C-terminal ions („v“, „w“, „x“, „y“, „z“). N-terminal ions contain the amine end of the peptide, while C-terminal ions contain the carboxy end of the peptide.

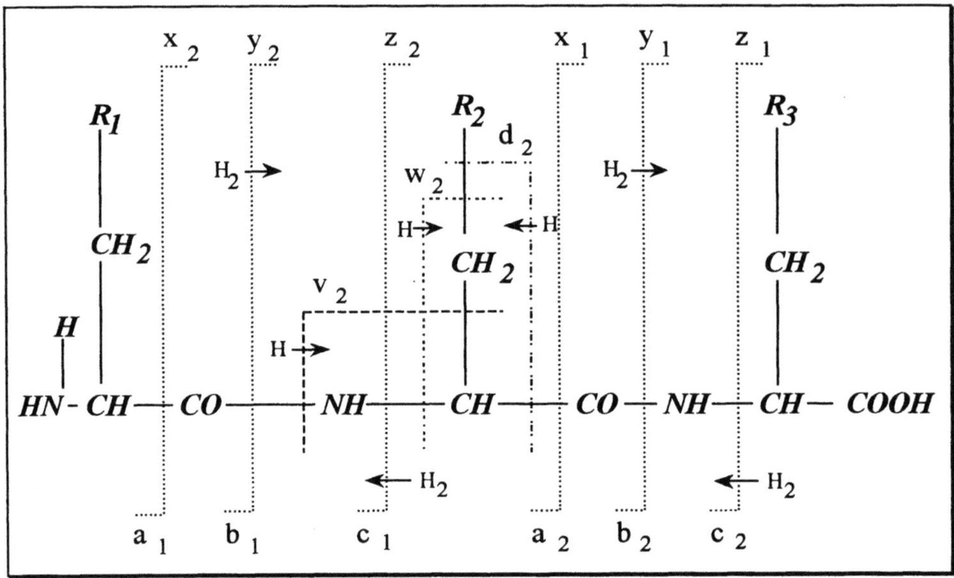

Figure 12 : Nomenclature of peptide fragment ions.

Ions of type a, b, c, x, y, z are due to backbone cleavages, while d-, v- and w-ions are the result of additional side chain cleavages. A typical example of a scheme of possible fragment ion masses is shown in Fig. 13 for [Glu[1]]-fibrinopeptide B. The number of fragment ion masses that can be observed in principle is rather high. Fortunately not all of these ions actually appear in PSD spectra, but rather a limited number of characteristic ions, following certain rules of favored cleavages and charge retention. Ions of highest abundance are usually of type a, b, and y in PSD analysis, while side chain cleavages are rather rare. In addition to N- and C-terminal ions, internal ions (containing neither end of the peptide chain) and amino acid immonium ions are observed in high abundance.

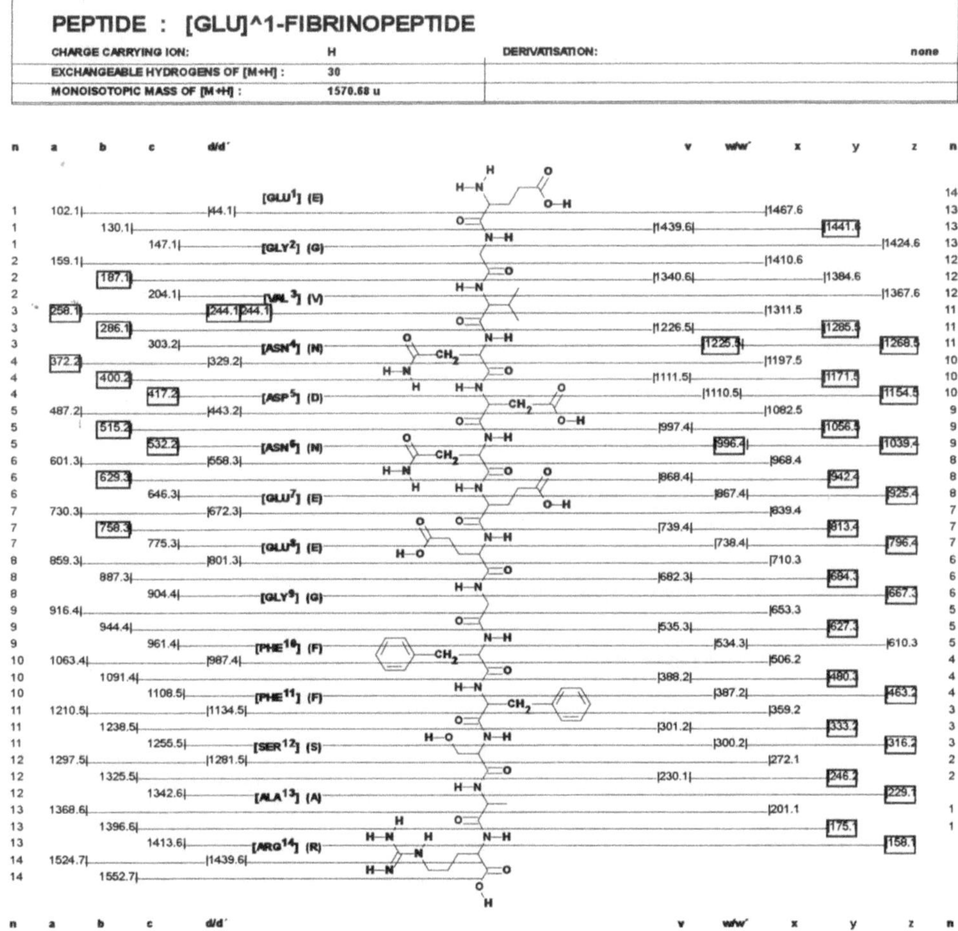

Figure 13 : Possible fragment ion masses from [Glu^1]-fibrinopeptide B. Fragment ion signals observed by PSD-MALDI are marked.

3.2. COMPUTER ASSISTED PEPTIDE SEQUENCE ANALYSIS

Sequence analysis of unknown peptides by mass spectrometry is possible, but is not obvious in all cases. In the example of the [Glu^1]-fibrinopeptide B the PSD spectrum (Fig. 14) is rather easy to interpret and the sequence would have been identified if the peptide were unknown.

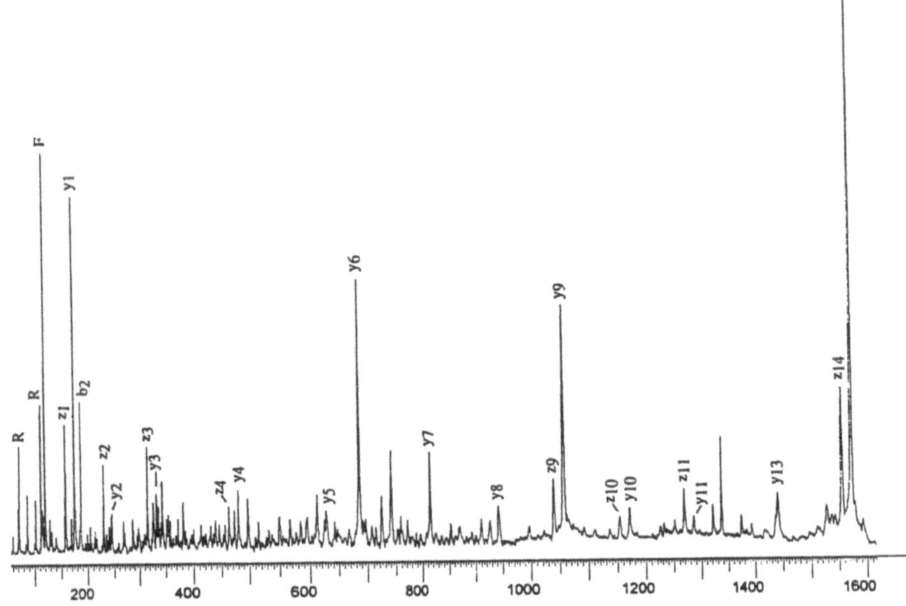

Figure 14: PSD spectrum of [Glu1]-fibrinopeptide B (EGVNDNEEGFFSAR-OH), M_{mono}=1569.7 u. 2,5-dihydroxybenzoic acid was used as a matrix.

In the general case, the mixture of N-terminal, C-terminal and internal ions obtained in product ion analysis prevents to read the amino acid sequence directly from the spectrum. Two approaches are used to identify peptides from mass spectral data. One is to assume that the peptide under investigation is part of a protein which is listed in a protein database and to compare experimental data by computer assistance with data expected from database structures. The other is to develop skilful computer programs which are able to decipher the sequence on the basis of the product ion spectrum, using pattern based, rule based or combinatorial algorithms.

The first approach is generally working with much success [20,21], but fails if the proteins are heavily modified or not listed in the data base.

Computer based sequence analysis of unknowns without the aid of databases is still an underdeveloped area. A possible strategy is described in Fig. 15. The goal of this iterative process is to set up a list of possible amino acid sequences which should ideally contain just one (the correct) entry. To reach that goal, a series of filters is used that interactively and iteratively reduce the number of possibilities. One of the most effective and most obvious filters is to interpret the PSD mass range of small fragment ions. This mass range below 200 u contains valuable information on the amino acid composition (not the sequence) due to characteristic amino acid immonium ions and several other amino acid-specific fragment ions.

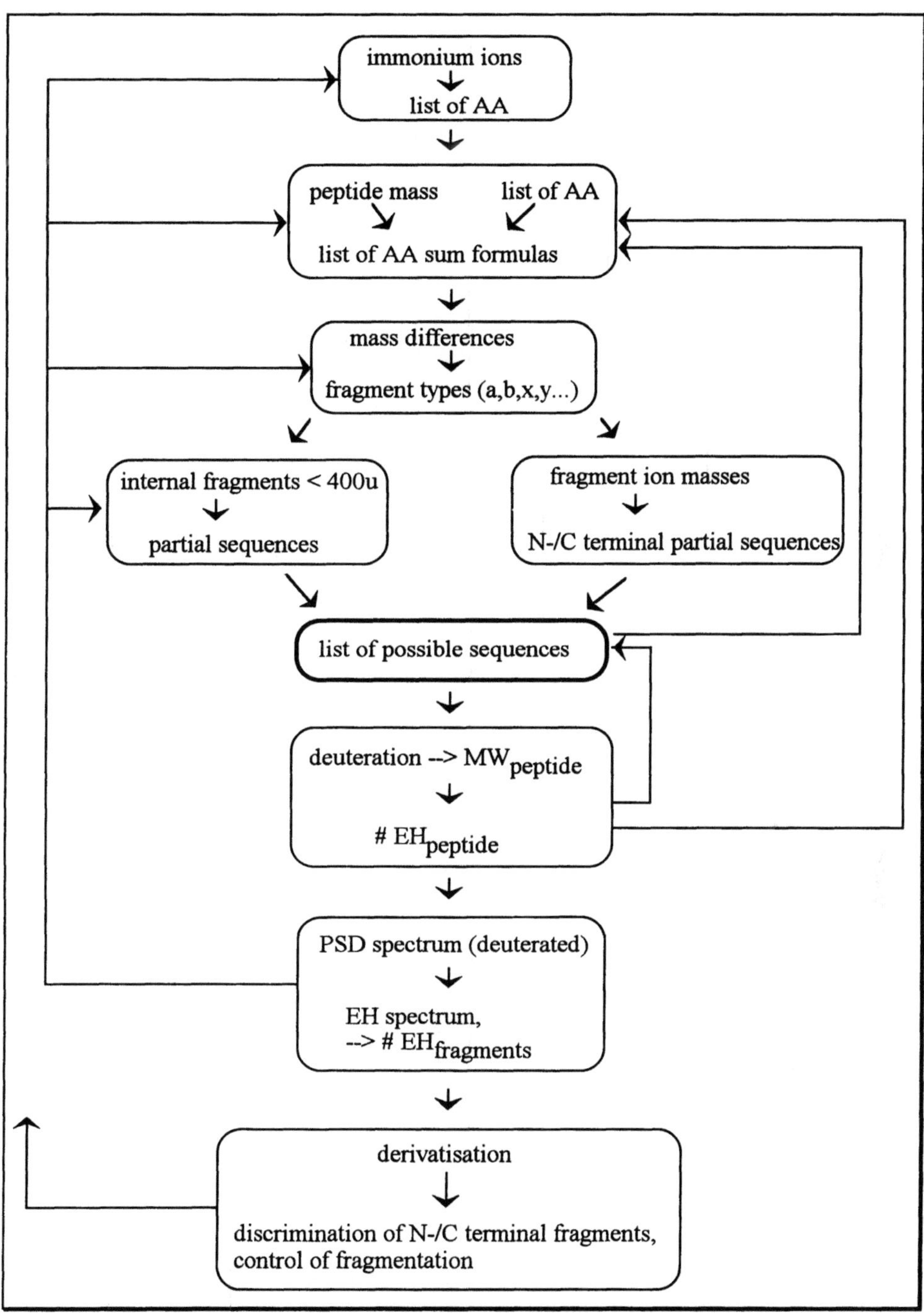

Figure 15 : Schematic diagram for the sequence analysis of unknown peptides.

The discriminative information in combination with the measured peptide mass is used to set up a list of possible amino acid compositions (sum formulae). In a third filter, PSD ion signals can be interpreted with respect to their fragment type (a, b, y, ...) by evaluation of mass differences to neighboring signals. Partial sequences can be identified from characteristic peaks in the mass range below 400 u, belonging to internal fragments. Finally, starting at the N- and the C-terminus, possible sequences are built up from the list of observed fragment ion masses.

The list of possible sequences resulting from this interactive filtering procedure is usually still rather long and does not necessarily give the highest score to the correct sequence. Simple derivatisation methods are therefore used routinely which provide a further discriminative effect on the described filters. The most effective and easiest derivatisation procedure is hydrogen-deuterium exchange, as described in the following.

3.3. HYDROGEN-DEUTERIUM EXCHANGE AS A ROUTINE TOOL FOR SEQUENCE DETERMINATION

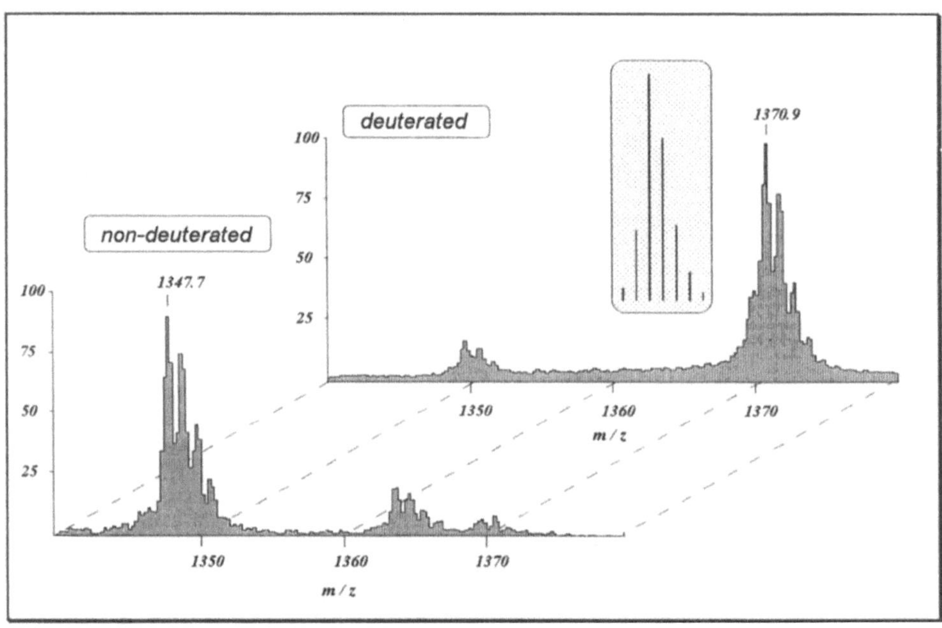

Figure 16 : Molecular ion region of substance P (RPKPQQFFGLM), $[M+H]^+$=1347.7 (monoisotopic mass of protonated molecule). By hydrogen-deuterium exchange, 23 protons are replaced by deuterium, resulting in a mass shift to 1370.9 u. The insert shows the theoretical isotope distribution of 98.5 % H-D exchange.

Hydrogen-deuterium exchange of peptides is a procedure that can be performed in a 3-minute step without further sample consumption, using the prepared sample which has already been analyzed by PSD-MALDI [22]. In solution all exchangeable hydrogens (EH), i.e. hydrogens bound to heteroatoms, can be exchanged by deuterium.

Exchange efficiencies of 98.5 % are usually attainable. Fig. 16 shows as an example the parent ion region of PSD-MALDI spectra of deuterated and non-deuterated substance P. A shift of the parent ion mass of 23 mass units is observed meaning that the protonated peptide contains 23 exchangeable hydrogens. It is obvious that this information alone is already a strong discriminator for the list of possible amino acid sum formulae, and by that for the list of possible sequences.

A much stronger discrimination is obtained by evaluation of the PSD ion mass shifts observed from the deuterated sample. As demonstrated in Fig. 17, any of the PSD ion signals can be correlated with a distinctive number of exchangeable hydrogens of the corresponding fragment ion. This additional „exchangeable-hydrogen spectrometry“ has been found to be extremely helpful in structure analysis of peptides and other biopolymers. For computer assisted sequence analysis it usually results in a reduction of possibilities to only very few or even just one sequence.

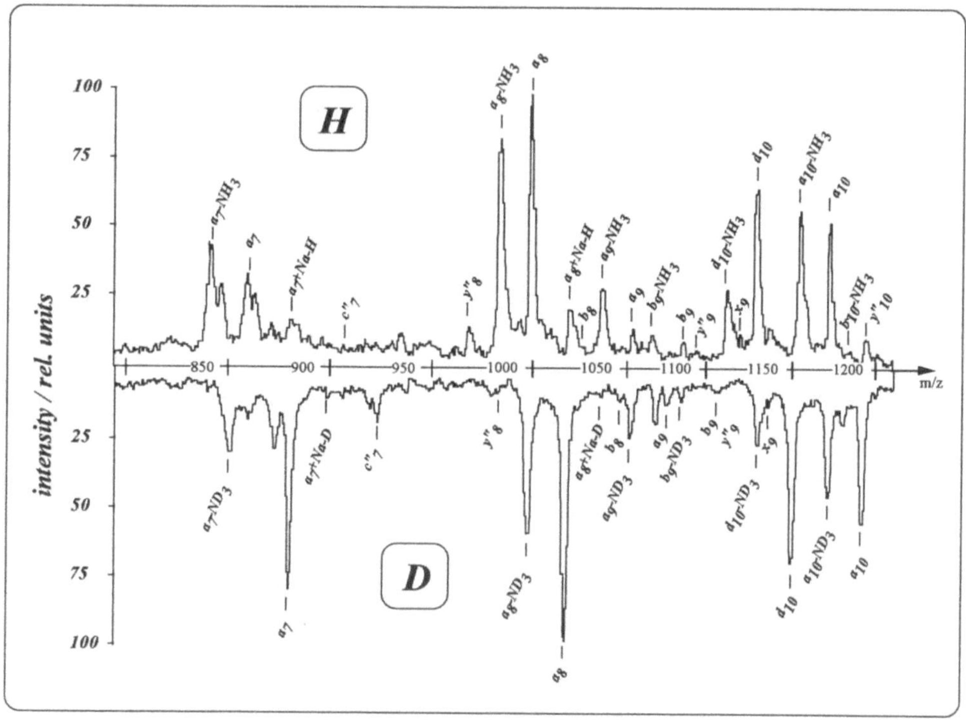

Figure 17: Section of the PSD spectra of deuterated (lower inverted part) and protonated (upper non-inverted part) substance P.

3.4. DERIVATISATION FOR FRAGMENTATION CONTROL

In certain cases the above method of sequence determination of unknown peptides does not lead to an unambiguous proposition. This might be due to a too large number of different fragment ion signals, leading to frequent overlapping, poor signal-to-noise ratios or ambiguities in the interpretation of fragment ion types. Chemical derivatisation of one end of the peptide chain often improves the spectrum quality or allows to identify fragment ion types.

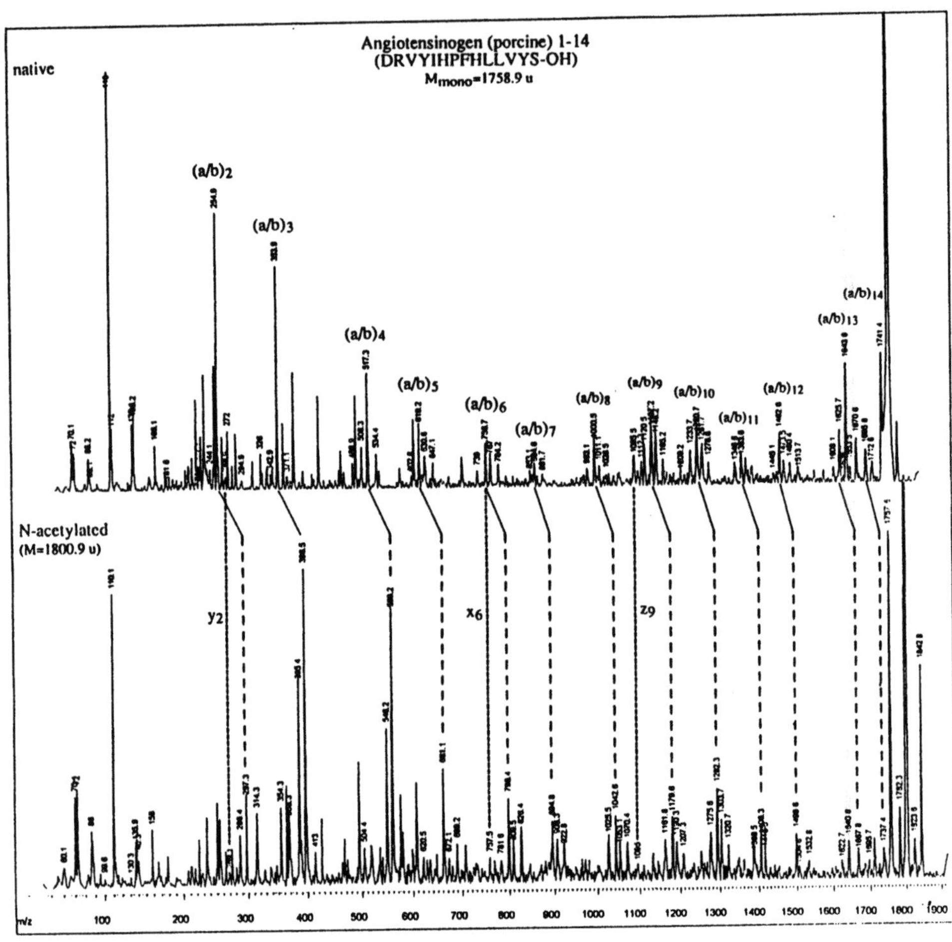

Figure 18: PSD spectrum of native (upper part) and acetylated (lower part) porcine angiotensinogen 1-14.

One of these methods is to use a mass shift effect by acetylation of the N-terminus (Fig. 18). All ions containing the N-terminus are shifted by 42 mass units, while C-terminal ions keep their original mass. A distinction is thus very easy using this derivatisation procedure.

Another approach is to control the formation of certain types of fragment ions. It has been found that in PSD-MALDI, fragmentation is not necessarily controlled by the location of a charge as in low-energy CID of electrospray ions [23], but primarily follows a charge-remote mechanism [24]. This means that fragmentation in PSD-MALDI takes place independent of the presence or location of a charge, and that the fragment ion type (N- or C-terminal) formation is simply controlled by the initial location of the charge. In accordance with this assumption, peptides derivatized at the N-terminus with a fixed charge (a quarternary ammonium ion) gave PSD spectra containing only N-terminal fragment ion signals with a superior signal-to-noise ratio [25].

4. Conclusion

Postsource decay MALDI has become a method of high reliability, high sensitivity and high versatility for chemical structure analysis of biopolymers. With the knowledge of the mechanisms of ion formation, ion activation and ion decay, a wide controllability of the method has been attained that allows to investigate almost any kind of bioorganic substances under various preparational conditions. With the availability of delayed ion extraction technology the performance of PSD-MALDI instruments has considerably improved in terms of mass resolving power, mass accuracy and parent ion sensitivity.

Complete interpretation of PSD spectra of peptides is usually circumvented by employing protein database searching. Fully unknown peptides or heavily modified peptides have to be analyzed manually in general, asking for an intelligent, future computer program to do this job.

Acknowledgement

Financial support by the Bundesministerium für Bildung, Wissenschaft, Forschung und Technologie (BMBF, contract No. 0310709), by the European Community (contract No. MAT1-CT94-0034) and by Thermo Bioanalysis Ltd., Hemel Hempstead, UK is gratefully acknowledged.

5. References

1. Spengler, B., Kirsch, D., Kaufmann, R. (1992) Fundamental Aspects of Postsource Decay in Matrix-Assisted Laser Desorption Mass Spectrometry. *J. Phys. Chem.* **96**, 9678-9684.
2. Kaufmann, R., Spengler, B., Lützenkirchen, F. (1993) Mass Spectrometric Sequencing of Linear Peptides by Product-ion Analysis in a Reflectron Time-of-flight Mass Spectrometer Using Matrix-assisted Laser Desorption Ionization. *Rapid Commun. Mass Spectrom.* **7**, 902-910.
3. Kaufmann, R., Kirsch, D., Spengler, B. (1994) Sequencing of peptides in a time-of-flight mass spectrometer: evaluation of postsource decay following matrix-assisted laser desorption ionization (MALDI). *Int. J. Mass Spectrom. Ion Proc.* **131**, 355-385.
4. Spengler, B., Kirsch, D., Kaufmann, R. (1991) Metastable Decay of Peptides and Proteins in Matrix Assisted Laser Desorption Mass Spectrometry. *Rapid Comm. in Mass Spectrometry*, **5**, 198-202.

5. Spengler, B., Kaufmann, R. (1992) Gentle Probe for Tough Molecules: Matrix-Assisted Laser Desorption Mass Spectrometry. *ANALUSIS*, **20**, 91-101.
6. Spengler, B., Kirsch, D., Kaufmann, R., Jaeger, E. (1992) Peptide Sequencing by Matrix Assisted Laser Desorption Mass Spectrometry. *Rapid Comm. in Mass Spectrometry*, **6**, 105-108.
7. Spengler, B., Kirsch, D., Kaufmann, R., Lemoine, J. (1995) Structure analysis of branched oligosaccharides using post-source decay in matrix-assisted laser desorption ionization mass spectrometry. *J. Mass Spectrom.* **30**, 782-787.
8. Yu, W., Vath, J.E., Huberty, M.C., Martin, S.A. (1993) Identification of the facile gas-phase cleavage of the Asp-Pro and Asp-Xxx peptide bonds in matrix assisted laser desorption time-of-flight mass spectrometry. *Anal. Chem.* **65**, 3015-3023.
9. Huberty, M.C., Vath, J.E., Yu, W., Martin, S.A. (1993) Site-specific carbohydrate identification in recombinat proteins using MALD-TOF MS. *Anal. Chem.* **65**, 2791-2800.
10. Critchley, W.G., Hoyes, J.B., Malsey, A.E., Wolter, M.A., Engels, J.W., Muth, J. (1994) Structural analysis of mixed base oligonucleotides using negative ion matrix assisted laser desorption ionisation time of flight (MALDI-TOF) mass spectrometry metastable analysis. *Proceedings of the 42nd Annual Conference on Mass Spectrometry and Allied Topics*, Chicago, Illinois, May 29 - June 3.
11. Hoyes, J.B., Curbishley, S.G., Doorbar, P., Tatterton, P., Bateman, R.H. (1994) A new high performance MALDI-TOF instrument for PSD and CID metastable ion analysis. *Proceedings of the 42nd Annual Conference on Mass Spectrometry and Allied Topics*, Chicago, Illinois, May 29 - June 3.
12. Talbo, G., Mann, M. (1994) Localization of phosphorylation sites in peptides by MALDI metastable decay at the subpicomole level. *13th International Mass Spectrometry Conference*, Budapest, Hungary, August 29 - September 2.
13. Stahl-Zeng, J, Hillenkamp, F., Karas, M. (1996) Metastable fragment-ion analysis in a reflectron instrument with a gridless ion mirror, *Eur. Mass Spectrom.* **2**, 23-32.
14. Medzihradszky, K.F., Adams, G.W., Burlingame, A.L., Bateman, R.H., Green, M.R. (1996) Peptide Sequence Determination by Matrix-Assisted Laser Desorption Ionization Employing a Tandem Double Focusing Magnetic-Orthogonal Acceleration Time-of-Flight Mass Spectrometer. *J. Am. Soc. Mass Spectrom.* 7, 1-10.
15. Cornish, T.J., Cotter, R.J. (1994) A Curved Field Reflectron Time-of-Flight Mass Spectrometer for the Simultaneous Focusing of Metastable Product Ions. *Rapid Commun. Mass Spectrom.* **8**, 781-785.
16. Kaufmann, R., Kirsch, D., Tourmann, J.L., Machold, J., Hucho, F. (1995) Matrix-assisted laser desorption ionization (MALDI) and post-source decay (PSD) product ion mass analysis localize a photolabel crosslinked to the delta-subunit of nAChR protein by neurotoxin II. *Eur. Mass Spectrom.* **1**, 313-325.
17. Wilm, M., Shevchenko, A., Houthaeve, T., Breit, S., Schweigerer, L., Fotsis, T., Mann, M. (1996) Femtomole sequencing of proteins from polacrylamide gels by nano-electrospray mass spectrometry. *Nature* **379**, 466-469.
18. Brown, R.S., Carr, B.L., Lennon, J.J. (1995) Factors That Influence the Observed Fast Fragmentation of Peptides in Matrix-Assisted Laser Desorption. *J. Am. Soc. Mass Spectrom.* **7**, 225-232.
19. Kaufmann, R., Chaurand, P., Kirsch, D., Spengler, B. (1996) Post-source Decay and Delayed Extraction in Matrix-assisted Laser Desorption/Ionization-Reflectron Time-of-Flight Mass Spectrometry. Are There Trade-offs? *Rapid Commun. Mass Spectrom.* **10**, 1199-1208.
20. Eng, J.K., McCormack, A.L., Yates III, J.R. (1994) An Approach to Correlate tandem Mass Spectral data of Peptides with Amino Acid Sequences in a Protein Database. *J. Am. Soc. Mass Spectrom.* **5**, 976-989.
21. Mann, M., Wilm, M. (1994) Error-Tolerant Identification of Peptides in Sequence Databases by Peptide Sequence Tags. *Anal. Chem.* **66**, 4390-4399.
22. Spengler, B., Lützenkirchen, F., Kaufmann, R. (1993) On-target Deuteration for Peptide Sequencing by Laser Mass Spectrometry. *Org. Mass Spectrom.* **28**, 1482-1490.
23. Cox, K.A., Gaskell, S.J., Morris, M., Whiting, A. (1996) Role of the Site of Protonation in the Low-Energy Decompositions of Gas-Phase Peptide Ions. *J. Am. Soc. Mass Spectrom.* 7, 522-531.
24. Spengler, B., Luetzenkirchen, F., Metzger, S., Kaufmann, R., Jeffery, W., Pappin, D.J.C. (1996) Directing Fragmentation for Primary Structure Elucidation of Peptides by Postsource decay MALDI. *Proceedings of the 44th Annual Conference on Mass Spectrometry and Allied Topics*, Portland, Oregon, May 13 - 17.
25. Spengler, B., Luetzenkirchen, F., Metzger, S., Kaufmann, R., Jeffery, W., Pappin, D.J.C., paper in preparation.
26. Bökelmann, V., Spengler, B., Kaufmann, R. (1995) Dynamical parameters of ion ejection and ion formation in matrix-assisted laser desorption/ionization. *Eur. Mass Spectrom.* **1**, 81-93.

DEVELOPMENT OF A MASS SPECTROMETRIC APPROACH FOR THE CHARACTERISATION OF HEMOGLOBIN ADDUCTS

Pasquale Ferranti
Dipartimento di Scienza degli Alimenti, Università Federico II, Italy. Servizio di Spettrometria di Massa del CNR, Napoli, Italy

Nicola Sannolo
Sezione di Medicina Occupazionale e Igiene Industriale, Dipartimento di Biochimica e Biofisica, II Università, Napoli, Italy

Gianfranco Mamone and Giuseppe Forte
Servizio di Spettrometria di Massa del CNR, Napoli, Italy

Antonio Malorni
Servizio di Spettrometria di Massa del CNR, Napoli, Italy. Istituto Chimica Molecole Interesse Biologico, CNR, Arco Felice, Italy

It has been observed since the 1950's that a wide variety of structurally diverse chemicals can produce cancer in humans and animals. For example, the following associations have emerged between exposure and the development of specific cancers: vinyl chloride and hepatic cancer, amine dyes and bladder cancer, benzene and leukaemia. Naturally occurring chemicals or agents such as asbestos, aflatoxin B_1, betel nut, nickel and arsenite are also associated with an increased incidence of human cancer. Many scientists propose that naturally occurring and endogenous substances contribute more to human cancer incidence and mortality than synthetic chemical carcinogens. Many of these are electrophiles or require metabolic activation to electrophilic reactive intermediates that bind covalently to nucleophilic centres on proteins, RNA or DNA.

The request for the development of biological exposure tests for carcinogens has stimulated studies on their reactivity with biomolecules. The determination of hemoglobin (Hb) adducts was found a suitable method to evaluate the exposure to carcinogens and for genotoxic risk assessment. As there is a precise relationship between the reaction of several chemical carcinogens with DNA and hemoglobin (1,2), the quantification of amino acid adducts produces useful information on the genotoxic risk. Ideally, methods are desirable which detect any modification in the intact protein easily and rapidly and identify the specific nature and sites of modification. The recent developments have made mass spectrometry the method of choice to address structural problems involving covalently modified proteins. Furthermore, mass spectrometric methods are virtually independent of the type of modification to be identified and therefore ideally suited to the study of covalent xenobiotic modifications of proteins.

The early studies on the quantification of the modified aminoacids in Hb were performed by analysing the adducts to cysteine and histidine residues. This approach was

R. M. Caprioli et al. (eds.), Selected Topics and Mass Spectrometry in the Biomolecular Sciences, 399–411.

rapidly dropped up because of the tedious work-up of samples, the time consuming and the low sensitivity.

Afterwards a new analytical method was set up based on the specific determination of alkylated N-terminal valine in Hb by means of the modified Edman degradation reaction (4,5). Because it is easy, rapid to perform and very sensitive, it is by far, the most useful method for *in vivo* dosimetry of genotoxic compounds. Unfortunately this method gives information only on the modification of N-terminal residues whereas in the study of interactions between genotoxic compounds and cellular macromolecules, the best and most desirable method should be based on procedures able to detect all the modifications in the intact protein, identifying also the nature and sites of such modifications.

The specific objective of our research has been the setting up of an analytical strategy for the structural characterisation of the adducts formed by interaction of hemoglobin with agent present in the environment as well as those produced by endogenous compounds. To this purpose we developed a general strategy for the study of chemical interaction of xenobiotics with proteins combining the use of Electrospray/Mass Spectrometry (ES/MS) and Tandem Mass Spectrometry (MS/MS). Furthermore, considering that the protein adducts can be stable or unstable under the standard acidic hydrolysis, we used specific enzymes to obtain soft hydrolysis of the globin under study to analyse all the possible Hb adducts, even labile ones.

Here we report the structural characterisation of the adducts formed by hemoglobin with the pesticide methyl bromide, chosen as a model alkylating compound, and of the species formed on hemoglobin by the action of nitric oxide, an endogenous messenger molecule.

EXPERIMENTAL PROCEDURES

In vitro alkylation with MeBr The solution of MeBr for *in vitro* incubation was prepared by bubbling gaseous MeBr in cold pure ethanol. The concentration of the MeBr solution was determined from the increase in the weight. Red cells from a normal volunteer were washed three times with isotonic 0.9% NaCl solution and 2 ml of erythrocytes were suspended in 8 ml of 10 mM sodium phosphate, pH 7.0, to reach a hemoglobin concentration of about 1mM. Aliquots of ethanolic MeBr solution were added to 10 ml of red cell sample prepared as above. The reaction mixture was left at 37°C for 16 h, added to a solution of 2% HCl in isopropanol and centrifuged for 10 min at 5000 rpm. Precipitated globins were collected by centrifugation at 10.000 rpm for 30 min, washed with ethyl acetate and dried under a nitrogen stream.

Preparation of S-nitrosothiols. Solutions (50 mM) of Cys-NO were prepared by mixing equal volumes of 100 mM L-cysteine in 250 mM HCl, 0.1 mM Na_2EDTA with 100 mM $NaNO_2$ in water at room temperature. Previous studies have established that essentially no nitrite remains in the nitrosothiol solutions using this synthetic method (6). Human erythrocytes from a normal volunteer were washed in isotonic saline as described previously (7) and then lysed by addition of a equal volume of water. This oxy-Hb (about 1 mM) was estimated by SDS-gel electrophoresis and by isoelectric focusing (IEF) to be about 95% pure and prepared immediately prior to its use. Incubations of Cys-NO with these fresh hemolysates (200 µl) were performed at 25°C for 30 min, using 1:10 and 1:100 haemoglobin/Cys-NO respectively, in 1ml total volume by addition of aerate phosphate buffer pH 7.4, 0.5 mM EDTA and subsequently of the

RSNO solution. The reactions were stopped by dilution of the samples with 400 µl 0.1% TFA. 200 µl of this solution were injected on the HPLC column.

HPLC separation of globin chains. Globin chains were purified by RP-HPLC using the procedure developed by Shelton et al. (8) with the following modifications: the samples (500g of globin) were loaded onto a Vydac C4 large pore column equilibrated in the following buffer: [A] 80:20 water-acetonitrile, 0.1% TFA; [B] 40:60 water-acetonitrile, 0.1% TFA. The column was equilibrated at 52% of Buffer B. After two min hold, buffer B was raised to 69% in 24 min. Before analysis globin sample were filtered through a 45m filter [Millipore]. For analytical runs 100 µg samples were used. The column effluent was monitored at 280 nm. Protein fractions were manually collected and directly injected into the ES/MS source.

ES/MS and LC/ES/MS peptide mapping. Liquid chromatography was performed using a 2.1 mm i.d. X 250 mm, C18, 5 µm (Vydac) reverse-phase column with a flow rate of 0.2 ml/min on a Kontron modular system. The column effluent was split 1:25 with a Valco tee, to give a flow rate of about 8 µl/min into the electrospray nebulizer. The bulk of the flow was run through the detector for peaks collection as measured by following the absorbance at 220 nm. Solvent A was 0.03% TFA (v/v) in water; solvent B was 0.02% TFA in acetonitrile. Separation of the peptides was effected with a gradient 5-45% B in 60 min. ES/MS analyses with a PLATFORM mass spectrometer (Fisons, Manchester, UK). Samples from HPLC separation (10 µl, 50 pmol) were injected into the ion source at a flow rate of 10 µl/min; the ES-mass spectra were scanned from 1600 to 600 u at scan cycle of 10 s/scan. ES-mass spectra were scanned from 1800 to 400 u at a scan cycle of 5 s/scan. The source temperature was 120°C. Mass scale calibration was carried out using myoglobin as reference compound.

FAB mass spectrometry. FAB mass spectra were performed on a ZAB 2SE (Fisons, Manchester, U.K.) double-focusing, reversed geometry mass spectrometer equipped with a caesium gun operating at 25 kV (2mA). Sample was dissolved in 0.1 M HCl and loaded on to a glycerol-thioglycerol coated probe tip. High-energy collision induced-dissociation (CID) mass spectra were obtained on a ZAB-T (Fisons, Manchester, U.K.) four-sector (B1E1B2E2) mass spectrometer under control of an OPUS V3.1X data system and equipped with a 2048 channel-linear photo diode focal plane array detector. The sample was bombarded with a beam of Cs^+ ions having an energy of 30 KeV. Analyses were performed at 8 KV accelerating potential. For collision induced decomposition experiments, argon collision gas was used until 50% attenuation of the parent ion beam. The collision cell was held on 50% of the accelerating potential. For each spectrum, 100-200 pmol of sample were dissolved in 1 µl of 5% acetic acid, and this solution was placed on the glycerol/thioglycerol (1:1) matrix on the probe tip.

Amino acid analysis - Amino acid analyses were performed by a TRIO 2000 (Fisons Manchester, UK) GC/EI/MS apparatus equipped with a gaschromatograph 5890 Series II (Hewlett-Packard, USA). 10 nmol of globins were hydrolysed with 300 µl HCl 6 N at 120°C for 18 h; the dry samples were derivatized to tert-butyldimethylsylil (TBDMS)-derivatives with 40 µl of a mixture of acetonitrile /MTBSTFA (50/50 v/v) at 100°C for 2 h (9). The solution was diluted to 150 µl with acetonitrile and 0.5 µl were used for the analysis. The following conditions were used: Injector 260°C; Ion source 180°C; GC-MS Interface 260°C; GC column raising from 60°C to 280°C at 3°C/min. Mass spectrometric analysis was performed at 70eV ionization energy in both scan and selected ion monitoring mode. In scan mode the range 50-600 u was analysed using a scan time of 0.6 s and an interscan-time of 0.1 s. Quantitative data were obtained performing calibration curves using normal and methylated amino acid standards.

RESULTS

1 Structural analysis of methyl bromide adducts with hemoglobin

Methyl bromide (MeBr) is a highly toxic gas widely used as a fumigant of field soil for control of a wide spectrum of pests and diseases. Many accidental exposures to MeBr with lethal effects in humans have been reported (12-15). MeBr reacts *in vitro* and *in vivo* with proteins; the cysteine-thiol group was indicated in several studies as the main site of electrophilic reaction (16,17). We studied the *in vitro* interaction between MeBr and hemoglobin in order to either identify the modification induced on the protein or assess the possibility to use the hemoglobin adducts analysis as a biological index of exposure to MeBr (18-19). The reaction products after *in vitro* incubation of erythrocytes with different concentrations of MeBr have been characterised; the modification degree was evaluated by ES/MS and GC/MS analysis, whereas the reactive amino acid within the polypeptide globin chains were identified by tandem FAB/MS analysis.

Hemolysates exposed *in vitro* to 10 and 1mM MeBr were directly analysed by ES-MS to figure out the average methyl group number for each globin. A molecular mass of 16010.5±2.6 was measured for the modified β-globin corresponding to the introduction of 10 methyl groups in the native protein chain (molecular mass 15867.2). As for the α-chain, a molecular mass of 15238.9±4.1 was recorded with an increase of 112.5 mass units over the unmodified globin chain (15126.4), corresponding to the addition of 8 methyl groups. Only one main species was observed for each globin; less alkylated minor components failed to be detected possibly owing their lower concentration. A very different pattern was observed at lower MeBr concentrations (1mM). The main components detected by ES/MS analysis of the HPLC peaks were the native globin chains. Minor peaks were also present 14, 28 and 42 mass units above in the β-chain and 14 and 28 mass units above in the α-chain, indicating the introduction of up to 3 and 2 methyl groups respectively.

The same samples were submitted to acid hydrolysis and then amino acids were analysed by GC-MS (Fig. 1) as TBDMS-derivatives, in order to identify those residues within the polypeptide chains involved in the formation of the adducts. The results showed the presence of modifications on the imidazolic ring of histidine (both 1- and 3-methyl isomers), the thiol group of cysteine and the α-amino group of N-terminal valine in both globins. The identity of the amino acid adducts was established by comparison of the corresponding mass spectra with those of synthetic standards. It must be observed that N-methyl valine and methyl cysteine can occur *in vivo* at three order magnitude lower as background. The sensitivity level required can be achieved by using the same GC-MS approach working in single ion monitoring (SIM) mode.

Identification Of Methylated Amino Acid Residues - Methylated globin chains were digested with trypsin and the resulting peptides analysed by tandem FAB/MS of daughter ions of individual peptide peaks. This allowed to define the modified residues.

The N-terminal peptide 1-8 showed a high heterogeneity of methylation ranging from the unmodified to the trimethylated species, the most abundant being the monomethylated. In the monomethyl-peptide the modified group is the N-terminal α-amino group; the dimethyl-peptide resulted to be a mixture of isomers where the second methyl group was alternatively on histidine 2 and on the already methylated α-amino group, still reactive towards methyl-bromide. In the trimethyl-peptide two methyl groups were located on the α-amino group whereas the third one on histidine 2.

Methylated residues were the protein α-amino group, cysteines (fully alkylated), histidine residues (at a very variable level depending on their position within the sequence), lysine 17, aspartic acid 94 and tyrosine 35 and 145. Taking into account all the modified sites on the single peptides a total number of 19 was obtained (including unidentified residues in fractions 8 and 14 as well as double methylation of the α-amino group), which was not in contrast with the average number measured by ES/MS because

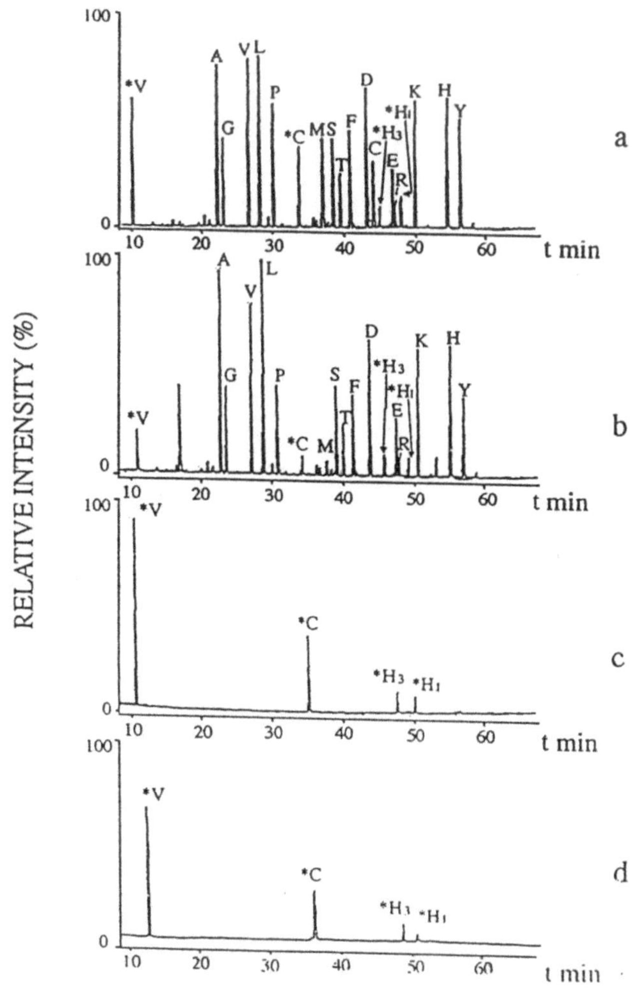

Fig. 1. Quantitative analysis of methylated residues in hemoglobin. GC/MS analysis of amino acids obtained from MeBr-modified globin chains (for conditions see text). (a) standard amino acid mixture; (b) amino acids from erythrocytes exposed to 10 mM MeBr (total ion current); (c) amino acids from erythrocytes exposed to 10 mM MeBr (selected ion monitoring of the peaks at m/z 245, 230, 202, 188, 160, 86 (*V=N-Me-Val); 363, 348, 320, 306, 278, 204 (*C=Me-Cys); 397, 382, 354, 340, 312, 238 (*H_3=N_3-Me-His and *H_1=N1-Me-His); (d) SIM of amino acids from erythrocytes exposed to 1mM MeBr.

methylation at a given site was never quantitative but for cysteine residues.

When the above analysis was extended to the β globin purified from red cells incubated with 1mM MeBr, the decrease in the modification pattern predicted by GC/MS and ES/MS data was confirmed. The only modified peptides observed in the HPLC map where the N-terminal 1-8 (m/z 966, methylation of the α-amino group), peptide 83-104 (m/z 2542, methylation of Cys 93; m/z 2566, methylation of Cys 93 and His 97), peptide 83-95 (m/z 1435, methylation of Cys 93), peptide 96-104 (m/z 1140, methylation of His 97). Peptides 83-104 and 83-95 were both modified at level of Cys 93. At difference with these, the other peptide containing a cysteine residue, 105-120, showed no detectable modification (m/z 1719). This finding demonstrates the occurrence of selective modification at Cys 93 compared to Cys 112.

Analogous results were obtained for the α globin. Of particular interest are the following results:

- the signal at m/z 2966, related to peptide 100-127, was absent due to its

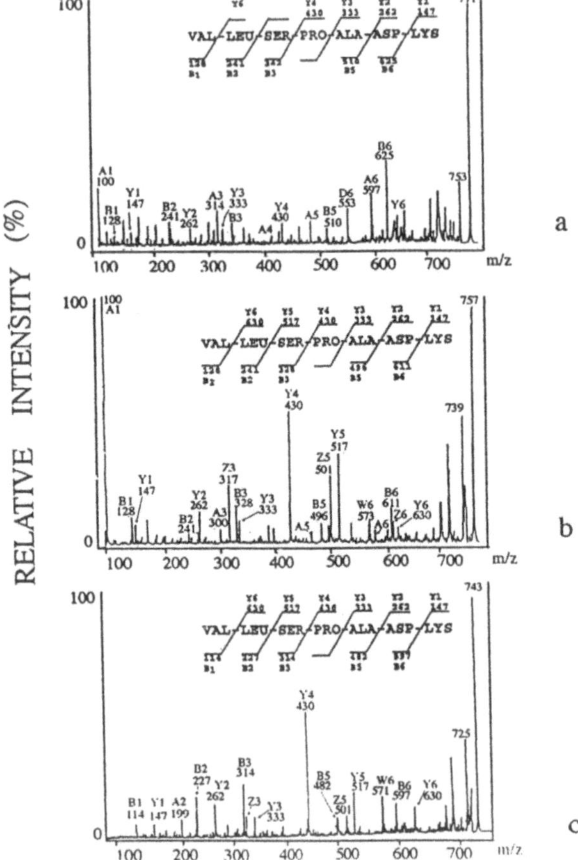

Fig. 2. Tandem FAB/MS spectra of the N-terminal peptide of α-chain. Analysis of HPLC fraction revealed adducts at (a) m/z 743, (b) m/z 757 and (c) m/z 771 corresponding to the mono-, di- and tri-methylated species. Methyl groups were located by the fragmentation pattern on the α-amino group (double methylated) and on Ser 2 respectively.

complete modification with relative mass shift. It corresponds to the only cysteine-containing peptide, which confirms amino acid analysis indicating the thiol groups of cysteines as the most reactive sites.

- the dimethylated form of the N-terminal peptide 1-7 contained N-, N-dimethyl-valine, similarly to the β-globin (Fig. 2). Only in the tri-methylated species, together with N-,N-dimethyl-valine, methylation of serine 3 was observed, indicating that the latter species was formed after double methylation of the N-terminus occurred.

When MeBr concentration was lowered to 1mM, the FAB/HPLC map showed the expected decrease of methylated peptide peaks. The only modified peptides were the monomethylated peptide containing cysteine (100-127, m/z 2980, methylation of Cys 104), the peptides containing the α-amino group (1-7, m/z 743, 1-11, m/z 1185), peptide 17-31 (His 20) and peptide 62-90 (m/z 3009), as shown in Figure 4. This last peptide was indeed a mixture of methyl isomers at level of His 72 and 89 respectively (spectrum not shown) whereas His 87 was not reactive towards MeBr.

Methyl Bromide induces modification of specific residues in hemoglobin - The results described demonstrate that protein side chains can be modified by MeBr or by similar compounds. However, the susceptibility of a particular functional group to modification by MeBr probably depends on its pKa value and its particular environment and accessibility. We aimed to reproduce *in vitro* conditions to mimic the physiological environment: whole blood red cells were used to preserve the selective shielding and transport effects of the cell membrane; erythrocytes were washed, then diluted to obtain a concentration of hemoglobin (about 1mM) comparable to that *in vivo*, and incubation were carried out in such conditions and at the body temperature (37°C).

The experimental conditions used to achieve hemoglobin modification led in one case (high ratio MeBr: hemoglobin) to extensive chemical modification and in the second (low ratio MeBr: hemoglobin) resulted in a more selective modification. We did not attempt to modify a given site in every protein molecule, but our objective was rather to identify functional amino acid side chains in a qualitative manner by mass fingerprinting of the modified protein chains. A quantitative evaluation was instead obtained by coupling of the fingerprint approach with GC/MS analysis of modified amino acids. This combined strategy does not require extensive purification procedures, as the only separation step performed was the isolation of the α and β subunits. Hence, in the present study, all the analytical determinations were performed on a mixture of modified products so that the primary as well as the minor sites of protein modification could be ascertained.

Both α and β chain of hemoglobin showed a comparable reactivity either for the amino acids residues involved in the adduct formation and for the extent of modification. In the case of 10 mM MeBr, in front of α-global addition of ten methyl groups/globin observed by ES/MS measurement, an average number of 17 (α chain)-19 (β chain) methylated amino acid residues was determined. The modified residues within the globin sequence are reported in Fig. 3

A decise reactivity of the N-terminal portion of the globin chains was also shown for both α and β globin chains. N-terminal peptides were actually modified by incorporation of multiple methyl groups including double reaction of the α-amino group, methylation of lysine 7, serine 3 (α-globin) and of lysine 17 and histidine 2 (β-globin).

Extensive modification was also present in internal portions of the molecule, particularly in the β-globin "inner core" (sequence 83-120); it is interesting to note that such reactivity is not related to the amount of MeBr, because our results suggest that modification of these peptides takes place also at 1mM MeBr. In this last case, however, a very different selectivity was observed in the modification pattern. ES/MS data demonstrating the occurrence of multiple modified species containing 1, 2 and 3 methyl groups for globin were confirmed by the structural characterization of modified residues. Only specific residues were methylated and only at low extent, namely the α-amino group of both chains, Cys 104, His 20, 72 and 89 in the α chain and Cys 93 and His 97 in the β chain. In this respect, a different behaviour was exhibited by the two cysteine residues in the β chain, which is in perfect agreement with the already reported high reactivity of cysteine 93 in the β chain which *in vivo* reacts in oxy-hemoglobin (20) and readily undergoes oxidation to form mixed disulphides and other thioethers due to its position within the protein quaternary structure and has been related to hemoglobin catabolism. On the contrary, cysteine 112, appears to be unreactive at 1mM MeBr, possibly due to its position in a pocket at the $\alpha_1\beta_2$ subunit contact, environment which prevents modification at the lower MeBr concentration. Quantitative methylation of this residue is instead achieved at 10mM MeBr.

α-Chain

VLSPADKTNV^{10} KAAWGKVGAH^{20} AGEYGAEALE^{30}

RMFLSFPTTK^{40} TYFPHFDLSH^{50} GSAQVKGHGK^{60}

KVADALTNAV^{70} AHVDDMPNAL^{80} SALSDLHAHK^{90}

LRVDPVNFKL^{100} LSHCLLVTLA^{110} AHLPAEFTPA^{120}

VHASLDKFLA^{130} SVSTVLTSKY^{140} R^{141}

β-Chain

VHLTPEEKSA^{10} VTALWGKVNV^{20} DEVGGEALGR^{30}

LLVVYPWTQR^{40} FFESFGDLST^{50} PDAVMGNPKV^{60}

KAHGKKVLGA^{70} FSDGLAHLDN^{80} LKGTFATLSE^{90}

LHCDKLHVDP^{100} ENFRLLGNVL^{110} VCVLAHHFGK^{120}

EFTPPVQAAY^{130} QKVVAGVANA^{140} LAHKYH^{146}

Fig. 3. Reactive sites towards MeBr in the amino acid sequence of α and β globin chains. Reactive amino acids at 10mM are in bold and those reactive at 1mM are underlined.

At difference with cysteines, histidine residues did not show an identical reactivity even at 10mM MeBr, and modification appears to be largely directed by their environment. The most striking example is peptide 62-90 of the α globin, where one out of three histidine residues, the heme-proximal histidine 87, was never observed as a reactive site, whereas the other two residues on the peptide, histidine 72 and histidine 89 were among the most reactive, being modified even at 1mM MeBr. An identical behaviour was observed for the proximal histidine 92 in the β globin. On the contrary, distal histidines of both globin chains displayed partial modification in the presence of 10mM MeBr. A possible explanation of the lack of reactivity of proximal histidines compared to the other histidines might therefore be found in their peculiar binding to the heme group (20).

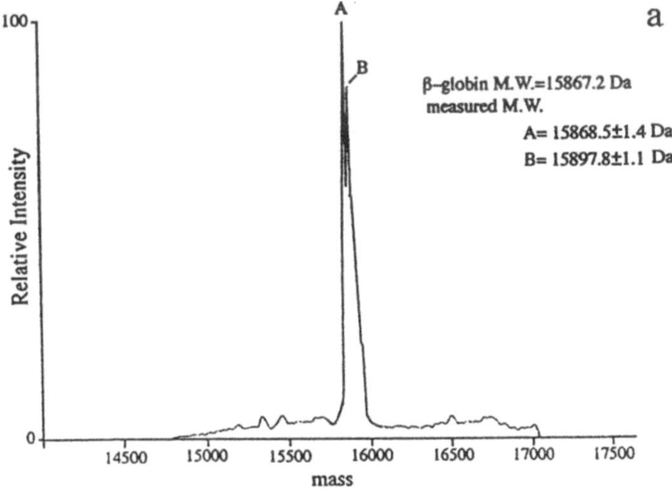

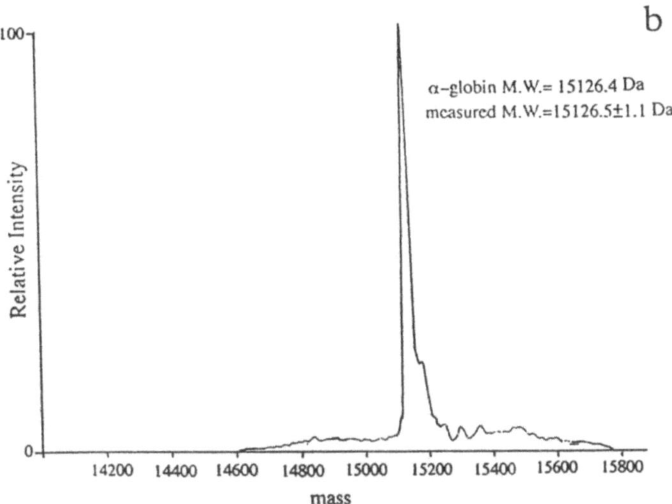

Fig. 4. ES/MS spectra in the transformed mass scale of globin chains from hemoglobin incubated with Hb/Cys-NO ratio (1:10). a) β-globin chain; b) α-globin chain.

2 Characterisation of S-Nitrosohaemoglobin

The interaction of nitric oxide (NO) with proteins is known to play a critical role in several different physiological systems extending from blood pressure regulation to neurotransmission (21-23). In addition, NO produced in infected and inflamed tissue could contribute to the process of carcinogenesis (24, 25). Although NO was known since the late 1970s to be among the ligands that activate soluble guanylyl cyclase and cause vascular smooth muscle relaxation, it was not appreciated until more recently that NO is also an endogenous vasorelaxant. NO and endothelium-derived relaxant factor (EDRF) comparably relax vascular smooth muscle and are similarly quenched by haemoglobin and superoxide. Recent studies have demonstrated the biological importance of the interaction of S-nitrosothiols, which can be considered as NO donors, with proteins, especially haemoglobin, at the level of Cys residues. It was recently proposed that S-nitrosohemoglobin forms within red blood cells and serves as a regulatory function. In human haemoglobin the a subunit contains one Cys residue and the β subunit contains two Cys residues, one of which (β-Cys93) highly reactive and conserved among species, although their function has remained unknown. It has been recently proposed that it can interact with endogenous RSNOs to form NO-Hb, suggesting that this may serve for regulation of blood pressure and facilitates efficient delivery of oxygen to tissues (26). However, no structural data have been produced yet to support the preferential formation of NO-Hb at β-Cys 93.

We used ES/MS (Fig. 4) to monitor the exposure of hemoglobin to Cys-NO under different conditions and to determine the cysteine residue involved in the formation of NO-Hb. When an equimolar ratio of S-nitrosothiol was added to haemoglobin only a single NO molecule was added. The mass measured for the α-chain (15126.5±1.1) perfectly matched with that of the native globin (15126.4). Of the two species present at the β position, the first one was the native β-globin (15868.5±1.4, expected mass 15867.2) whereas the second had a mass of 15897.8±1.1 with a shift of +29 units compared with that measured for the β-globin, indicating the addition to the protein of a single NO moiety.

Peptide mapping by liquid chromatography-mass spectrometry located the nitrosyl group at the level of β-Cys93 demonstrating that this was the preferred site of formation of S-nitrosohemoglobin. The chromatogram obtained for the peptic digest of modified β-globin is shown in fig. 5a. The peak at 30.9 min in the chromatogram contained a peptide with a molecular weight of 2052.6+0.2 was assigned to peptide 89-105 (molecular weight 2052.3) containing β-Cys93. This peak was also present in the spectrum of unmodified β-globin (Fig. 5b). A novel peak was instead found at an higher retention time (31.9 min) containing two mass signals: the first one at 2081.3+0.7 corresponded to the mass of peptide 89-105 increased of 29 mass units, possibly due to nitrosylation at the level of β-Cys93. A second species was observed at 2052.3±0.7.

The only modified peptides were those including β-Cys93 (peptides 92-102, 89-102 and 89-105). No trace of modified peptides containing β-Cys112 was detected, which indicated that nitrosylation selectively occurred at the level of β-Cys93.

Nitric oxide induces modification of a specific cysteine residue in Hemoglobin - A recent study (26) reported the *in vitro* and *in vivo* formation of NO-Hb in the presence of RSNO and NO-Hb has been recently involved in blood pressure regulation. It was earlier reported that no modified residue occurred different from β-Cys93, but no structural data were available on the nature of the NO-Hb. These conflictory observations prompted us to undertake this study. To selectively nitrosylate the thiol group of cysteine residues a procedure involving RSNO intermediates was chosen. This allowed us to monitor the reactivity of cysteine residues within the hemoglobin sequence. We varied the ratio of reagent to protein to determine some degree of selectivity. In fact, the use of NO at two different concentrations allowed determination of the level at which side chain reactivity was modulated by the ratio NO: hemoglobin. We actually demonstrated that under the conditions used by Jia et al only a single cysteine residue, namely β-Cys93 is almost completely modified *in vitro* after 30 min. At longer times [2h] the reaction is complete and no further modification was observed. Therefore one can reasonably conclude that only β-Cys93 is actually reversibly modified in red blood cells. The different behaviour exhibited by β-Cys93 compared to the other cysteines is in agreement with the already reported *in vivo* reactivity of this residue in oxy-hemoglobin which readily undergoes oxidation to form mixed disulphides and other thioethers due to its position within the protein quaternary structure.

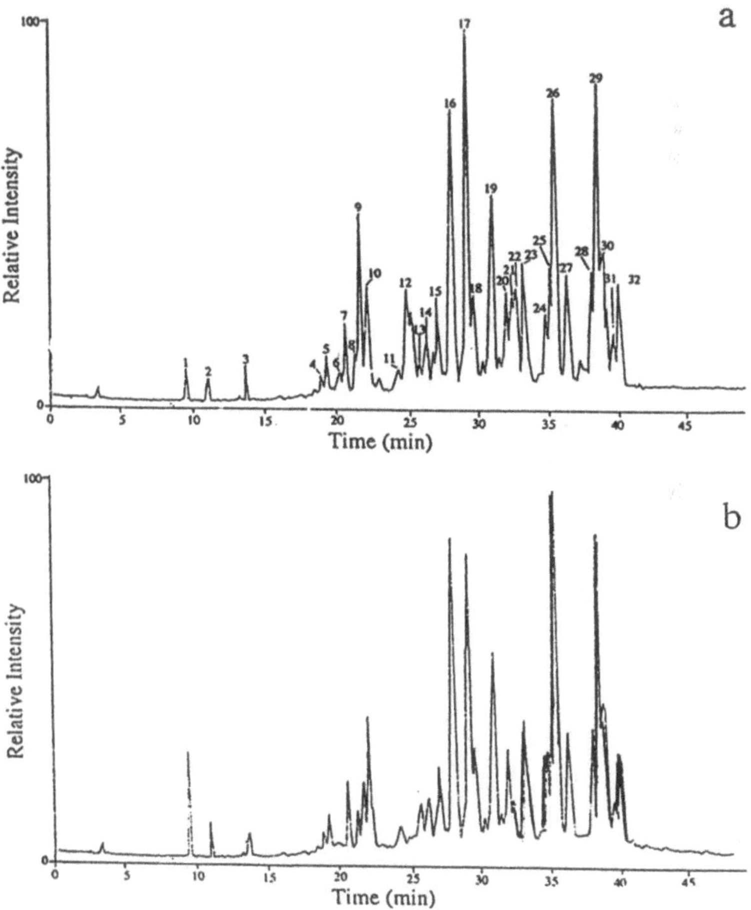

Fig. 5. LC/ES/MS of the peptic digest of the mononitrosyl β-globin (a), compared with that of native β-globin (b).

Once formed, Hb-NO adducts are readily degraded at neutral or alkaline pH. For this reason nitrosylated peptides were not observed after tryptic hydrolysis [data not shown]. A much greater stability of the nitrosylated globins was instead observed at low pH. Samples stored at pH<3 showed minimal degradation over a period of several weeks. For this reason a procedure based on the use of pepsin as the proteolytic agent was chosen, which allowed to preserve nitrosylated peptides during hydrolysis at low pH values.

The peptic mixture was directly analysed in LC/ES/MS. Enzymatic digestion of modified hemoglobin resulted in a mixture of native and nitrosylated peptides, the relative abundance depending on the extent of overall nitrosylation. The presence of overlapping peptides due to incomplete digestion was useful in obtaining a complete peptide map and in the accurate location of the modified cysteines. Nitrosylated peptides were readily identified from the molecular weight compared to the native ones and their identity was confirmed by selective fragmentation at the S-NO bond by increasing the cone voltage.

CONCLUSION

The present study based on the analysis of the hemoglobin adducts was aimed to (i) verify the overall reactivity of hemoglobin and evaluate the level of modification; (ii) identify the effectively modified sites along the nucleophilic sites potentially reactive.

Molecular weight measurement by ES/MS was instrumental for an immediate evaluation of the alkylation degree of the globin chains. This technique can be applied to the analysis of purified globins as well as to direct examination of hemolysates without the need for adduct purification. Modified species can be identified even in the case of MeBr where the mass difference due to the reaction is only 14 mass units. Better resolution is expected in the case of modifying agents causing an higher mass shift.

The proposed approach is not intended to be applied in the analysis of real exposures, which are in most cases at very low doses. However, it can identify the reactive groups and the corresponding molecular adduct which are the most suitable to monitor for real exposures. Considering the extremely variable and not sufficiently known *in vivo* and *in vitro* reactivity of electrophilic agents with proteins, the proposed methodology may represent a practical approach either for the complete analysis of protein adducts formed following exposure to alkylating compounds and for the unambiguous structural determination of modified sites in polypeptide chains.

REFERENCES

1. Calleman, C.J., Ehrenberg, L., Jansson, B., Osterman-Golkar, S., Stevenson, K. and Watchmainster, C.A. (1978) *J. Environ. Pathol. Toxicol.*, **2**, 427-442.

2. Neumann, H.G. (1984) *Arch. Toxicol.*, **56**, 1-6 and references therein.

3. Tornqvist, M, (1993) In *Use of biomarkers in Assesing Health and Enviromental Impacts of Chemical Pollutants,* Travis C.C. (ed.), Plenum Press, New York.

4. Tornqvist, M. , Mowrer, J., Jensen, S. and Ehrenberg, L. (1986) *Anal. Biochem.*,**154** , 255-266.

5. Tornqvist, M., Osterman-Golkar, S., Kautiainen, A., Naslund, M Calleman, C.J. and Ehrenberg, L. (1988) *Mutation Res.*, **204**, 521-529.

6. Stamler, J. S. (1994) Cell 78, 931-936.

7. Moriguchi M., Manning L. R. and Manning M. (1992) *Biochem.Biophys. Res. Commun.* 183, 598-604.

8. Shelton, J. B., Shelton, J. R. and Schroader, W. A. (1984) *J. Liquid Chromatogr., 7* , 1969-1977.

9. Ferranti, P., Milone, A., Ruoppolo, M., Gallo, M., Malorni, A., Sannolo, N., Carbone, V. and Fiume, A. (1992) *Carcinogenesis,* **13**, 1397-1401.

10. Chavas Das Neves, H.J. and Vasconcellos, A.M.P. (1987) *J. Chromatogr.*, **392**, 249-258.

11. Sannolo, N., Carbone, V., Ferranti, P., Fiume, I., Mamone, G. and Malorni, A. (1995) *J. Chromatogr. B* , **670,** 349-353.

12. Alexeeff, G.V. and Kilgore, W.W. (1983) *Residue Rev.*, **88**, 101-153.

13. Smart, N.A. (1990) *Adv. Environ. Sci. Technol.*, **23**, 227-255.

14. Commission of the European Communities Scientific Committee on Pesticides (1985) In *Reports of the Scientific Committee for Pesticides* (Second Series) Louxembourg (EUR 10211 EN), pp 15-32.

15. Van den Oever, R. Roosels, D. and Lehaye, D. (1982) *British J. Ind. Medicine*, **39**, 140-144.

16. Djalali-Behzad, G., Hussain, S., Osterman-Golkar, S. and Segerback, S. (1981) *Mutation Research,* **84**, 1-9.

17. Goergens, H.W., Hallier, E., Muller, A. and Bolt, H.M. (1994) *Toxicology Letters,* **72**, 199-203.

18. Ehrenberg, L. and Osterman-Golkar, S. (1980)*Teratogenesis, Carcinogen. Mutagen.*, **1**, 105-127.

19. Ferranti, P., Sannolo, N., Mamone, G., Fiume, I., Carbone, V., Tornqvist, M., Bergman, A. and Malorni, A. (1996) *Carcinogenesis* in press.

20. Bunn, H.F. and Forget, B.G. (1986) In *Hemoglobin: Molecular, Genetic and Clinical Aspects* Dyson, J. (ed.), W.B. Saunders Company, Philadelphia.

21. Umans, G. and Levi R. (1995) *Annu. Rev. Physiol.* **57**, 771-790.

22. Schmidt, H. H. H. W., and Water, U. (1994) *Cell* **78**, 919-925.

23. Rand, M. J. and LI, C. J. (1995) *Annu. Rev. Physiol.* **57**, 659-682.

24. 0hshima, H. and Bartsch, H. (1994) *Mutat. Res.* **305**, 253-264.

25. Ohshima, H., Bandaletova, T.Y., Brouet, I., Bartsch, H., Kirby, G., Ogunbiyi, F., Vatanasapt, V. and Pipitgool, V. (1994) *Carcinogenesis* **15**, 271-275.

26. Jia, L., Bonaventura, C., Bonaventura, J. and Stamler, J, S., (1996) *Nature* **380**, 221-226.

REACTIVITY OF ANTINEOPLASTIC DRUGS WITH MODEL PEPTIDES STUDIED BY ADVANCED MASS SPECTROMETRY METHODOLOGIES

VIRGINIA CARBONE, GABRIELLA POCSFALVI, NICOLA SANNOLO AND ANTONIO MALORNI
International Mass Spectrometry Facilities Centre-National Research Council
Via S. Pansini 5,
80131 Naples (Italy)

ABSTRACT. The *in vivo* interaction of the antineoplastic drug 1-3-bis-(2-chloroethyl)-1-nitrosourea (BCNU) and acrolein with model peptides has been investigated in order to provide a detailed description of their electrophilic reactivity towards biological macremolecules. Following incubation with these substances, the modified species were separated by HPLC and identified by fast atom bombardment mass spectrometry, whereas the reactive amino acids within the peptides structure were assigned by tandem mass spectrometry. Incubation with BCNU led essentially to the formation of an N-terminal carbamoil-derivative that slowly decomposed to form three isomeric structures and a very minor ethylated adduct. Alkylation with acrolein gives rise to a mixture of different adducts due to the reaction of both the double bond and the carbonyl group. Two species containing intramolecular cross-links were also observed. These results constitute the pre-requisite for *in vitro* and *in vivo* studies on the modification of hemoglobin in patients following treatment with antineoplastic drugs.

Introduction

Most genotoxic xenobiotics show an electrophilic reactivity or can acquire it after metabolic activation (1). They are able to bind covalently to DNA, the target for the effects, and to nucleophilic sites present in proteins, especially haemoglobin (2, 3). Modifications product by DNA adduct formation are considered responsible for alteration of DNA replication and, therefore, for development of mutations and the degenerative process leading to cancer formation (4). It has been demonstrated that a constant ratio exists between DNA and haemoglobin adduct formation over a wide range of doses for some compounds (3, 5, 6, 7, 8). The amount of haemoglobin adducts is directely related to the tissue dose, that is the actual dose responsible for the macromolecular damage; so the tissue doses absorbed in working place and in the environment can be determinate through biomonitoring of haemoglobin adducts. The advantages of haemoglobin dosimetry compared to the determination of the DNA adducts rely on the large amount of protein available in the blood, the easy sampling and the integration of the tissue dose over the erythrocytes life-span allowing the evaluation of an exposure protracted in time.

Alkylating agents, a class of antineoplastic drugs, perform their therapeutic action react with DNA causing different modification, often enzyme mediated. Of course alteration of genetic information is direct not only to neoplastic cells but also to normal ones; this mechanism is responsible for side effects during a chemotherapy.

R. M. Caprioli et al. (eds.), Selected Topics and Mass Spectrometry in the Biomolecular Sciences, 413–425.

In order to verify the electrophilic reactivity to macromolecules of some antineoplastic drugs the authors carried out a study on the *in vitro* interaction of 1-3-bis-(2-chloroethyl)-1-nitrosourea (BCNU) and Cyclophosphamide with angiotensin I, chosen as a model substrate.

BCNU is an antineoplastic agent which possesses a broad spectrum of activity against human malignancies. It is used in the treatment of malignant brain tumors because it crosses the blood-brain barrier for its high lipophilicity (9). Under physiological conditions BCNU decomposes to give two electrophilic intermediates: 2-chloroethyl diazohydroxide and 2-chloroethyl isocyanate (CEIC) (10) (Fig. 1). 2-Chloroethyl diazohydroxide can react directly with nucleophiles to form 2-chloroethylated products or can rearrange to give 2-chloroethyl corbonium ion after loss of nitrogen (11). 2-Chloroethyl isocyanate is involved in carbamoylation reactions of amino groups in macromolecules (12, 13, 14, 15). That can explain the cytotoxic action of BCNU.

Cyclophosphamide is a widely used drug against many animal and human tumors, particularly in combination with other agents (16). The methabolism of this drug has been largely investigated (17; 18): it is activated *in vivo* bv the hepatic cytochrome P-

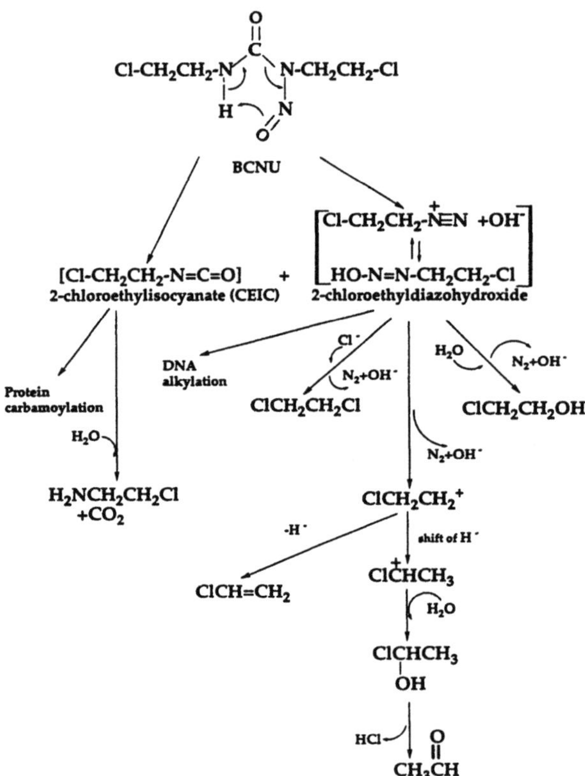

Figure 1: Metabolic scheme of BCNU.

450 to 4-hydroxy cyclophosphamide, which breaks down into the reactive metabolites phosphoramide mustard and acrolein (Fig. 2). Phosphoramide mustard is a bifunctional alkylating agent; it has been demonstrated that it binds to the N^7 position of guanine (19, 20, 21, 22), to the phosphate backbone of DNA (23, 24, 25) and interacts with the O^6 position of guanine (26). It also can induce DNA cross-linking (27). Acrolein binds to proteins (28), forms adducts with DNA (29, 30) and induces DNA single-strand breaks (31, 32). It has been suggested that acrolein-induced damage both on DNA and on proteins can contribute to the toxic effects of Cyclophosphamide (28, 29, 30, 33).

This paper shows that both agents can interact with the model peptides although in a very different manner; BCNU essentially forms a single adduct which slowly decomposes to produce other minor species whereas acrolein gives rise to a wide number of species present in almost equivalent amount.

Results

Incubation of model peptides with BCNU.

Angiotensin I (DRVYIHPFHL) is a decapeptide containing three main nucleophilic sites, the terminal amino group and the two histidine residues at positions 6 and 9. Model peptide (80-100nmol) was incubated in a 1:10 molar ratio with BCNU in 50 mM phosphate buffer, pH 7,8, at 37°C for 24 h. Stock solution of alkylating agent

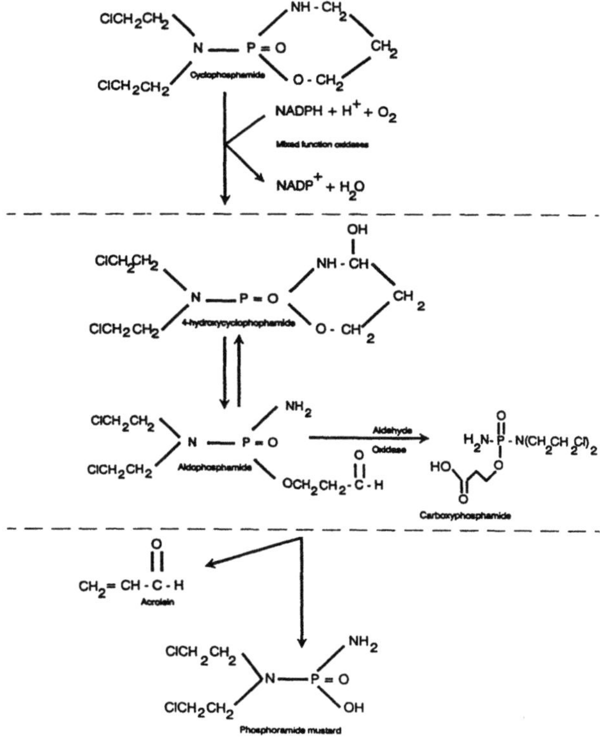

Figure 2: Metabolic scheme of Cyclophosphamide.

was prepared dissolving BCNU in 13% ethanol in 50 mM phosphate buffer; the desired amount of the stock solution was then added to the peptide solution. Time-course experiments were performed by withdrawing aliquots of the reaction mixture at different interval times which were then analysed by HPLC (Fig.3). Modification of angiotensin I resulted in the rapid formation of a main adduct, designed as component 2, together with other minor species (components 3 and 4). As the incubation time increases, the concentration of the main adduct decreases whereas the minor components tend to accumulate, as shown in Figure 3B, 3C and 3D. After 24h of incubation, a consistent amount of unmodified angiotensin I (designed as component 1 in Figure 3), about 50%, is still present in the reaction mixture.

Individual components were collected and submitted to mass spectral analyses. Fast atom bombardment (FAB) mass spectrometric analyses were carried out on a ZAB 2SE double focusing mass spectrometer (Micromass, Manchester, UK) equipped with caesium guns operating at 25 kV (2 mA).Figure 4A shows the molecular ion region of the FAB spectrum of the main adduct (compound 2). A strong signal was recorded at m/z 1401, with an increment of 105 Da as compared to native angiotensin; moreover, an abundant +2 isotopic peak, clearly indicating the presence of a chlorine atom, was detected (Fig. 4B). On the basis of these observations, the main adduct was identified as the 2-chloroethylcarbamoyl derivative of angiotensin. This result confirms previous findings that BCNU undergoes a rapid and spontaneous decomposition in aqueous media producing different species, among which 2-chloroethylisocyanate was indicated as responsible for the carbamoylation of proteins (12, 13, 14, 15).

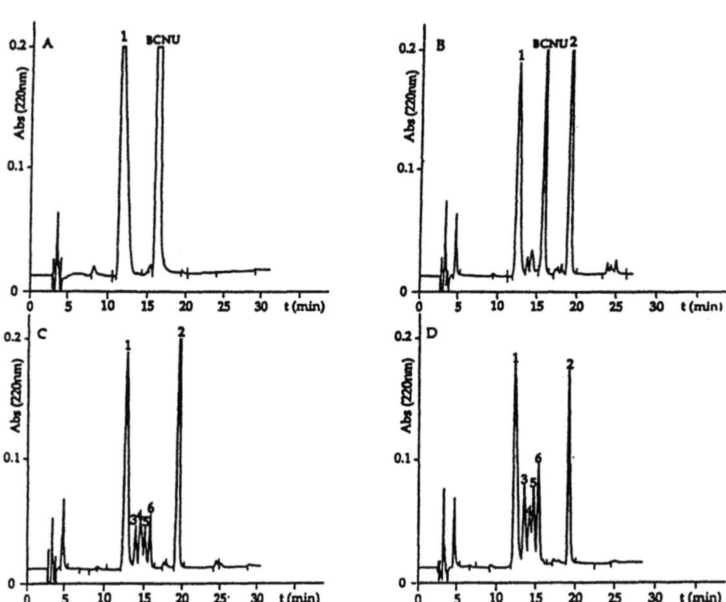

Figure 3: HPLC time-course analysis of angiotensin I incubated with BCNU. Aliquots of the reaction mixture were withdrawn at time 0 (A), 2h (B), 8h (C) and 24h (D).

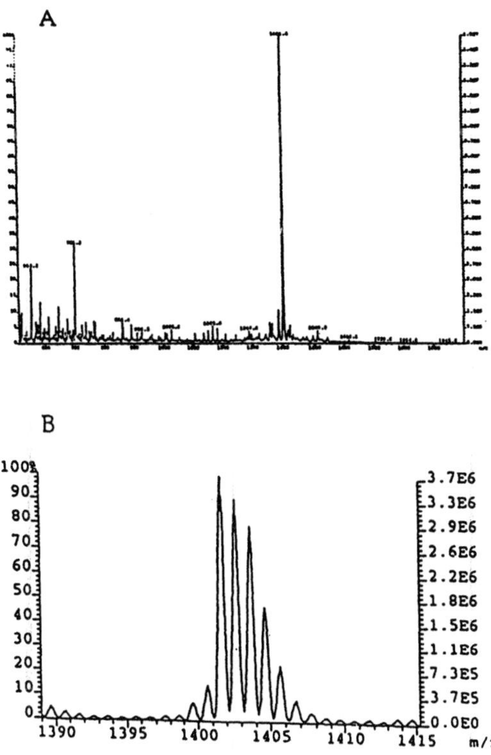

Figure 4: FAB mass spectrometric analyses of the angiotensin main adduct (compound 2). A: partial FAB spectrum B: molecular ion region.

The identification of the modification site within the angiotensin molecule was achieved by tandem mass spectral analysis. Analyses were carried out on a ZAB T tandem four sector instrument (Micromass, Manchester, UK) equipped with caesium guns operating at 25 kV (2 mA). FAB high energy collision induced dissociation (CID) experiments were performed on the tandem four sector instrument by selecting the ion of interest in the first mass spectrometer. Spectra were recorded in positive ionization mode at 8kV of accelerating potential; the variable potential collision cell was floated at 50% of accelerating voltage. The precursor ion was activated in the collision chamber and fragmented using argon as collision gas. The daughter ions were separated in the second mass spectrometer and detected by a focal plane array detector (34).

Figure 5 shows the daughter spectrum of the modified peptide ion at m/z 1401 which is dominated by fragment ions belonging to the "a" series (35). This fragmentation allowed us to locate the carbamoyl moiety on the terminal amino group of angiotensin.

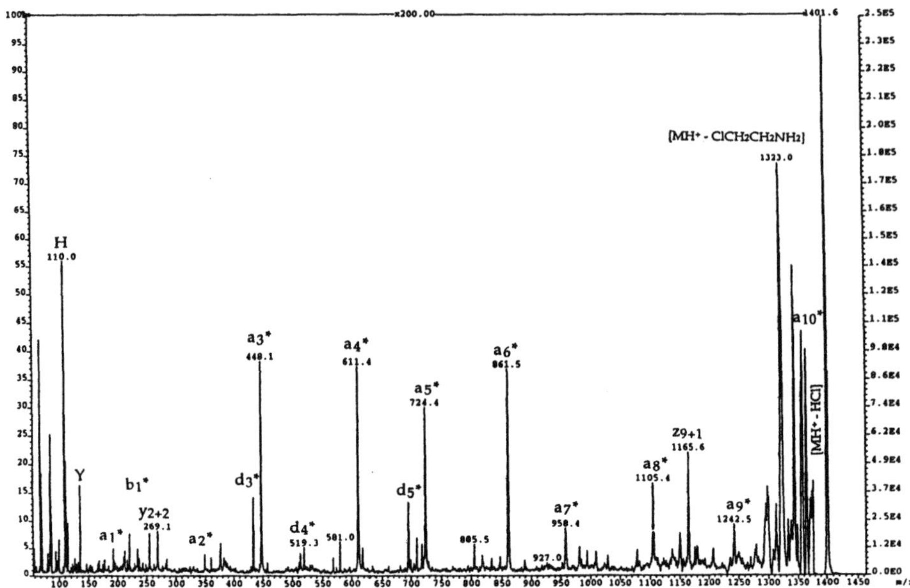

Figure 5: FAB MS/MS spectrum of the angiotensin main adduct (compound 2) obtained using argon as collision gas.

This assignment was confirmed by the presence of the z_9+1 ion (loss of Val + 2-chloroethylcarbamoyl group) and the two intense fragments occurring in the CID spectrum at m/z 1365 and 1323 originated from the loss of HCl and 2-chloroethylamino moieties respectively, thus confirming the nature of the modifying group.

Similar analyses were carried out on the minor components detected in the time-course experiment and the results are summarised in Table 1. Components 3, 5 and 6 exhibited the same molecular mass, MH^+ = 1365, corresponding to a 69 Da increase over the mass value of native angiotensin. The absence of any significant +2 isotopic peak indicated the absence of chlorine in these species. These results together with the HPLC time-course experiment suggested that under the experimental conditions used, the 2-chloroethylcarbamoyl derivative of angiotensin I undergoes a slow decomposition leading to the formation of three isomeric species having identical molecular mass but different chromatographic behaviour.

This hypothesis was confirmed by incubating the (Ang+105) derivative in 100mM borate buffer, pH 10.5, at 37°C for 24h. The HPLC time-course analysis indicated a slow decrease in the concentration of the derivative with the corresponding formation of three minor species exhibiting retention times coincident with those of components 3, 5 and 6. The FAB/MS analysis of these species confirmed their identity (data not shown).

The structure of the three isomeric adducts was investigated by MS/MS analysis. The CID spectrum of compound 3 showed the occurrence of abundant fragment ions of the "a" series which identified the terminal NH_2 group as the modification site. The daughter ion spectra of compound 5 and 6 were very similar to that of component 3, the only difference being the absence of the a_1 and a_2 ions. This finding might indicate that in

Fraction	MH^+	Δm	Peptide modification site
1	1296	-	-
2	1401	105 Da	N-term
3	1365	69 Da	N-term
4	1322	26 Da	N-term-Arg-2
5	1365	69 Da	N-term-Arg-2
6	1365	69 Da	N-term-Arg-2

Table 1: Mass spectrometric analysis of angiotensin adducts following incubation with BCNU and separation by HPLC (Fig.3).

these species both Asp-1 and Arg-2 had been involved in the modification. However, on these bases no definitive conclusions can be drawn on the structural differences existing among the three isomers. It has been recently reported that, during the reaction of BCNU with glutathione, the S-2-chloroethylcarbamoyl derivative undergoes an intramolecular cyclization with the concomitant release of HCl (36). Therefore, at least one of the three isomers could be tentatively identified as the N-(2-oxazolin-2-il-)-angiotensin; the structure of the remaining two adducts is currently under investigation.

The FAB spectrum of compound 4 showed a molecular ion at m/z 1322 corresponding to a 26 Da increase over the mass value of native angiotensin. This mass difference might be related either to a CH_2=CH- moiety attached to the angiotensin molecule or to an intramolecular ethylenic bridge (-CH_2-CH_2-) connecting two nucleophilic sites on the same peptide molecule. The MS/MS spectrum of this adduct showed the occurrence of the entire series of "a-type" fragments with the exception of the a_1 ion, leaving the ambiguity still unsolved. This fragmentation pattern, in fact, can correspond either to a modification of the N-terminus or to an ethylenic bridge involving the N-terminus and the adjacent arginine residue.

However, these results demonstrated that the (Ang+26) adduct is originated by reaction of the peptide with the 2-chloroethyldiazohydroxide moiety of BCNU and it is not produced by an intramolecular rearrangement of the (Ang+105) derivative. This observation is further supported by the absence of compound 4 in the alkaline decomposition of the (Ang+105) adduct. This finding was definitively confirmed by direct incubation of angiotensin I with 2-chloroethylisocyanate under the same conditions used for the reaction with BCNU. The time-course analysis (Figure 6) showed the rapid formation of the (Ang+105) adduct and its slow rearrangement to yield the three (Ang+69) isomers. No trace of the (Ang+26) derivative was detected. This latter adduct is then produced by alkylation of the peptide possibly via a mechanism involving a chloroethyl carbonation as reactive intermediate followed by intramolecular rearrangements (37). The occurrence of intramolecular ethylene bridges in peptides or proteins has never been observed, although an intermolecular bridge connecting two glutathione molecules was recently detected (36).

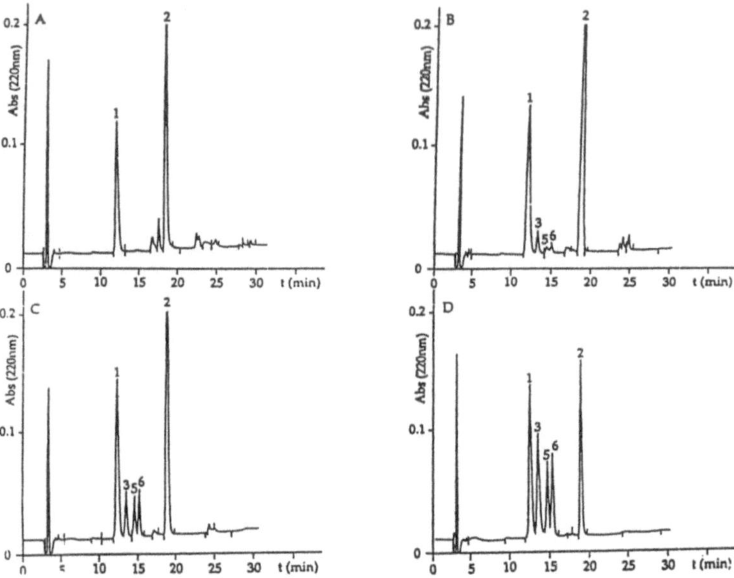

Figure 6: HPLC time-course analysis of angiotensin I incubated with 2-chloroethylisocyanate. Aliquots of the reaction mixture were withdrawn at time 0 (A), 2h (B), 7h (C) and 24h (D).

However, it should be underlined that the (Ang+26) adduct represents a minor component in the incubation mixture. The relative amounts of the carbamoyl- versus the alkyl-derivative clearly indicate that in aqueous media BCNU preferentially acts as a carbamoylating agent. Carbamoylation of the N-terminal group of angiotensin was substoichiometric as a consistent amount of the peptide remained unmodified even at late stages of the process.

The reactivity of BCNU was further investigated by using a model peptide having different nucleophilic groups. Synthetic peptide 128 consists of an N-acetylated decapeptide containing one serine residue and one cysteine in the free thiol form (Ac-DEMEECASAL-NH_2). Incubation of this peptide with BCNU under the usual experimental conditions resulted in the almost instantaneous formation of a main derivative which was characterised by mass spectral analyses. The FAB spectrum revealed a molecular ion at m/z 1243 corresponding to a 105 Da increase in the mass value of the unmodified peptide. The isotopic pattern confirmed the presence of a chlorine atom and the CID spectrum identified the modification of the cysteine residue to an S-carbamoyl derivative. Again, as the incubation time increases, a slow decomposition of the main adduct was observed although to a much lesser extent than in the angiotensin adduct. The unmodified peptide rapidly disappeared indicating a complete reaction of the

SH group. These findings indicate that carbamoylation of cysteine residues occurs much faster than the reaction with amino groups and results in a complete modification of the thiol group.

Incubation of acrolein with model peptides

Reaction of angiotensin with acrolein was carried out at pH 7.8 using a 10-fold molar excess of the alkylating agent. Aliquots of the incubation mixture were withdrawn at different interval times and directly analysed by HPLC. Figure 7 shows the corresponding HPLC time-course analysis; reaction of the peptide with acrolein resulted in the formation of a complex mixture of different species whose relative abundances remained almost unchanged with the time. Acrolein is an α-β unsaturated aldehyde which can display different reactivities towards nucleophilic groups present in peptide sequence by either reacting with the electrophilic double bound or giving rise to nucleophilic addition on the carbonyl group to form a Schiff's base. The former reaction leads to an increment of 56 Da in the mass value of the peptide whereas formation of an aldimine bond increases the peptide molecular mass by 38 Da.

The individual adducts were then analysed by mass spectrometric techniques and the results are summarised in Table 2. Compounds 3 and 7 exhibited the same molecular mass, MH^+ = 1390, corresponding to a 94 Da increase in the mass value of unmodified angiotensin, although they displayed a very different chromatographic behaviour. On the basis of their mass value, these isomeric species could represent doubly modified angiotensin molecules with one acrolein forming a Schiff's base and the other reacting via a nucleophilic addition on the double bound (38+56=94 Da). The MS/MS analysis of these two compounds eventually led to the identification of the two isomers. The CID spectrum of component 7 indicated that in this species both the N-terminal amino group and the His residue at position 9 had been modified by acrolein. The daughter ion spectrum of compound 3, on the contrary, demonstrated the occurrence of a single modifying group located at the N-terminus of the peptide and accounting for the entire mass difference. This finding suggested that some acrolein dimerization had occurred within the original sample or under the incubation conditions and that adduct 3 was very likely originated from this dimeric acrolein species which formed an aldimine bond with the angiotensin N-terminal group. This interpretation was confirmed by reduction experiments carried out with sodium cyanoborohydride ($NaBH_3CN$) which shifted the mass signal of the peptide by 2 Da to m/z 1392 as the aldimine bond was selectively converted to an amine bond.

The most interesting and intriguing components among the adducts mixture were those corresponding to peaks 6, 10 and 11 in the HPLC profile. The FAB spectra of these species showed the presence of the same mass signal at m/z 1334, indicating a mass difference of 38 Da as compared to native angiotensin. As stated above, this mass difference can be related to the simple reaction between acrolein and the peptide to form a Schiff's base. However, none of the three isomeric adducts eventually resulted to contain this modification. The MS/MS spectrum of component 6 contains strong "a-type" fragments at m/z 1175.7 (a_9), 1038.1 (a_8), 891.5 (a_7) and 794.4 (a_6) corresponding to the unmodified angiotensin C-terminal sequence -Pro-Phe-His-LeuCOOH. No further fragmentation was detected along the N-terminal segment 1-6 which, according to the mass value of the a_6 ion (794.4 Da), is still carrying the modification. These resultsstrongly suggest that compound 6 contains an intramolecular cross-link involving the terminal amino group and the side chain of His-6 bridged by an acrolein molecule. The formation of such an adducts was very likely originated via addition of the imidazole ring of His-6 on the double bond of acrolein followed by an intramolecular reaction

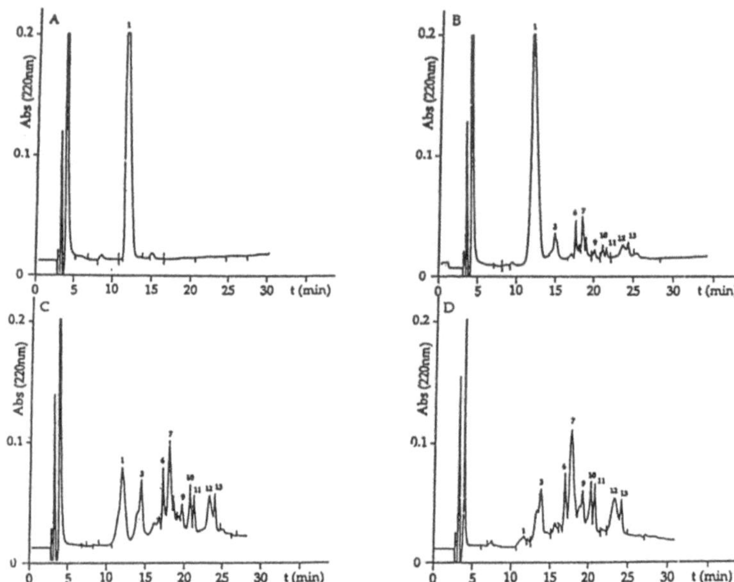

Figure 7: HPLC time-course analysis of angiotensin I incubated with acrolein. Aliquots of the reaction mixture were withdrawn at time 0 (A), 2h (B), 7h (C) and 24h (D).

Fraction	MH^+	Δm	Hypothesized structure
1	1296	-	Angiotensin
3	1390	94 Da	NH_2-term+94
6	1334	38 Da	NH_2-term-His-6 cross link
7	1390	94 Da	NH_2-term+38 His-9+56
10	1334	38 Da	NH_2-term-His-9 cross link
11	1334	38 Da	NH_2-term-His-9 cross link
12	1370	74 Da	?
13	1370	74 Da	?

Table 2: Mass spectrometric analysis of angiotensin adducts following incubation with acrolein and separation by HPLC (Fig.7).

between the NH_2 terminal group of the peptide and the free carbonyl group of the aldeyde.

This mechanism is supported by the knowledge that nucleophilic addition on the double bond of α-β unsaturated carbonyl compounds occurs much faster than reaction with the aldehydic or ketonic group (38). Furthermore, in this particular case, the formation of the intramolecular aldimine bond should be favoured by entropic factors. This reactivity of acrolein is not surprising as this compound has been widely used as a cross-linking reagent in protein chemistry to covalently link different subunits in oligomeric proteins (39).

The formation of a macrocyclic ring within the 1-6 segment of angiotensin might explain the absence of any fragmentation in this region of the peptide (40). The decrease in flexibility of the modified peptide was also confirmed by tryptic hydrolysis experiments carried out in comparison with native angiotensin. The unmodified peptide is easily cleaved by the enzyme at Arg-2 and Phe-8, being this latter an aspecific cleavage site. The structure of the macrocyclic ring at the N-terminal region of compound 6 prevented the protease to cleave the peptide chain at its preferential site Arg-2 and the tryptic digestion produced only the modified fragment 1-8 at m/z 1084.

The MS/MS analysis of components 10 and 11 showed nearly identical daughter ion spectra characterised by a single intense a_9 fragment at m/z 1175.7. On the basis of the above considerations, these species were identified as modified Angiotensin molecules containing an intramolecular acrolein cross-link bridging the N-terminal group and the site chain of the His residue at position 9. This hypothesis was confirmed by tryptic hydrolysis of the two compounds; the enzymatic digest did not produce any cleavage of the peptide chain since both Arg-2 and Phe-8 are located within the structure of the macrocyclic ring.

The presence of two isomers was related to the proline residue at position 7 which can occur either in the *cis* or *trans* configuration. The decrease in flexibility of the peptide chain due to the presence of the macrocyclic ring prevented the two configurational isomers to interconvert giving rise to two different chromatographic peaks. The other compound containing a macrocyclic ring, component 6, did not show similar isomer configuration structures since Pro-7 is located outside the 1-6 ring where the flexibility of the peptide chain is still allowing a rapid *cis-trans* equilibrium.

It should be noted that an angiotensin adduct having both the N-terminus and the His-6 residue modified by acrolein molecules was never observed, although the isomeric peptide modified at His-9 was indeed detected. This is probably due to a more favourable interaction between modified His-6 and the N-terminal group. Once His-6 has reacted with an acrolein molecule, the modified peptide undergoes an intramolecular rearrangement with the rapid formation of the 1-6 macrocyclic ring which is entropically favoured over the reaction of the terminal amino group with a second acrolein molecule to yield the doubly modified adduct. Cyclization of modified His-9 with the terminal amino group seems to occur at a slower rate thus making the reaction of the terminal NH_2 with a second acrolein molecule competitive.

The reactivity of acrolein with different nucleophilic groups was investigated by incubation of the alkylating agent with the synthetic peptide 128 under the same experimental conditions. Again a much faster reaction was observed which led to the formation of a single adduct. The FAB/MS analysis of this species showed a 56 Da increase in the molecular mass of the native peptide indicating that the adduct was generated by a nucleophilic addition on the acrolein double bond presumably carried out by the thiol group. The MS/MS analysis, in fact, confirmed the modification of the cysteine residue: the presence of essentially a single nucleophilic group in this peptide prevented the occurrence of intramolecular rearrangements leading to the formation of different adducts. It should be again noted that the reaction involving the SH group resulted in a complete modification of the peptide.

Discussions

Both BCNU and acrolein react *in vitro* with nucleophilic groups present in peptides and proteins under physiological conditions although they exhibit different reactivities. BCNU preferentially acts as a carbamoylating agent leading to the formation of adducts which slowly decompose into a mixture of isomers. This reaction occurs much faster with thiol groups than with amino groups and results in a complete modification of cysteine residues. Acrolein gives rise to a more complicate mixture of adducts including species containing intra molecular cross-links. As expected, addition of nucleophilic groups to the activate double bond is kinetically favoured over the reaction with the carbonyl group. As in the case of BCNU, modification of the cysteine residues is faster than reaction with other nucleophiles; however, acrolein reacts also with the imidazole ring of histidine residues. In the light of these observations, it is conceivable that whenever a free SH group is accessible within the protein structure, the modification induced by BCNU and/or acrolein will preferentially take place at the cysteine residue.

These results constitute the pre-requisite for *in vitro* and *in vivo* studies on the modification of haemoglobin in patients following treatment with, and in subjects professionally exposed to, antineoplastic drugs. Biomonitoring of BCNU-haemoglobin adducts, in fact, can be addressed towards the detection of carbamoylated species showing single or multiple increments of 105/69 Da in their molecular mass. Modification induced by acrolein can be revealed by looking for protein components with mass values increased by 38/56 Da or multiples, although in this particular case attention should be payed to the occurrence of intramolecular, or possibly intermolecular, cross-links.

References

1. J. A. Miller and E. C. Miller in *Origins of human cancer*, Book B, "Mechanisms of carcinogenesis", H. H. Hiatt, J. D. Watson and J. A. Winsten Eds, Cold Spring Harbor Laboratory, p. 605, 1977.
2. P. Brookes, *Cancer Res*, 26, p. 1994, 1966.
3. L. Ehrenberg and S. Osterman-Golkar, *Teratog. Carcinog. Mutagen.*, 1, p. 105, 1980.
4. E. C. Miller and J. A. Miller, *Cancer*, 47, p. 1055, 1981.
5. H. G. Neumann, *Arch. Toxicol.*, 56, p. 1, 1984.
6. J. Lewalter and U. Korallus, *Int. Arch. Occup. Environ. Health*, 56, p. 179, 1985.
7. M. Törqvist in *Mass Spectrometry in Biomolecular Sciences*, R. M. Caprioli, A. Malorni and G. Sindona Eds, Kluwer Academic Publishers, Netherlands, Vol. 475, p 381, 1996.
8. N. Sannolo, V. Carbone, P. Ferranti, I. Fiume, D. Santoro and A. Malorni in *Mass Spectrometry in Biomolecular Sciences*, R. M. Caprioli, A. Malorni and G. Sindona Eds, Kluwer Academic Publishers, Netherlands, Vol. 475, p 397, 1996.
9. E. P. Mitchell and P. S. Schein, *Cancer Treat. Rep.*, 70, p. 31, 1986.
10. J. A. Montgomery, R. James, G. S. McCaleb and T. P. Johnston, *J Med. Chem.*, 10, p. 668, 1967.

11. R. B. Brundrett, J. W. Cowens, M. Colvin and I. Jardine, *J. Med. Chem.*, 19, p. 958, 1976.
12. J. R. Babson, D. J. Reed and M. A. Sinkey, *Biochemistry*, 16, p. 1584, 1977.
13. R. B. Brundrett, M. Colvin, E. H. White, J. McKee, P. D. Hartman and D. L. Brown, *Cancer Res*, 39, p. 1328, 1979.
14. A. J. Fornance, K. W. Kohn and H. E. Kann, *Cancer Res*, 38, p. 1064, 1978.
15. H. E. Kann, M. A. Schott and A. Petkas, *Cancer Res*, 40, p. 50, 1980.
16. M. Colvin in *Cancer and Chemotheraphy* S. T. Crooke, A. W. Prestayko Eds., Academic Press, New York, Vol.III, p. 25, 1981.
17. T. A. Connors, P. J. Cox, P. B. Farmer, A. B. Fosterand M. Jarman, *Biochem. Pharmacol.*, 23, p. 115, 1974.
18. D. L. Hill, W. R. Jr. Laster, R. F. Struck, *Cancer Res*, 32, p. 658, 1972.
19. A. J. Benson, C. N. Martin, R. C. Garner, *Biochem Pharmacol*, 37, p. 2979, 1988.
20. K. Hemminki, *Cancer Res*, 45, p. 4237, 1985.
21. K. Hemminki, *Chem Biol Interact*, 61, p. 75, 1987.
22. J. R. Mehta, M. Przybylski, D. B. Ludlum, *Cancer Res*, 40, p. 4183, 1980.
23. H. Lindemann, *Anticancer Res*, 4, p. 53, 1984.
24. A. E. Maccubbin, L. Caballes, G. B. Cheda, R. F. Struck, H. L. Gurtoo, *Biochem Biophys Res Commun*, 163, p. 843, 1989.
25. A. E. Maccubbin, L. Caballes, J. M. Riordan, D. H. Huang, H. L. Gurtoo, *Cancer Res*, 51, p. 886, 1991.
26. L. Meer, S. C. Schold, P. Kleihues, *Biochem Pharmacol*, 38, p. 929, 1989.
27. M. Colvin and B. A. Chabner, in *Cancer chemotheraphy: principles and practice*, B. A. Chabner, J. M. Collins Eds , J. B. Lippincott, Philadelphia, p. 276, 1990.
28. A. J. Marinello, S. K. Bansal, B. Paul, P. L. Koser, L. Love, R. F. Struck, H. L. Gurtoo, *Cancer Res*, 44, p. 4615, 1984.
29. A. E. Maccubbin, L. Caballes, F. Scappaticci, R. F. Struck, H. L. Gurtoo, *Cancer Commun*, 2, p. 207, 1990.
30. R. A. Smith, D. S. Williamson, R. L. Cerny, S. M. Choen, *Cancer Res*, 50, p. 3005, 1990.
31. T. R. Crook, R. T. Souham, A. E. M. Mclean, *Cancer Res*, 46, p. 5029, 1986.
32. L. C. Erickson, L. M. Ramonas, D. S. Zaharko, K. W. Kohn, *Cancer Res*, 40, p. 4216, 1980.
33. R. A. Smith, I. A. Sysel, T. S. Tibbels, S. M. Choen, *Cancer Lett*, 40, p. 103, 1988.
34. M. L. Gross in *Methods in Enzymology*, J. A. Mc Closkey Ed, Academic Press, Inc., Vol. 193, p. 131, 1990.
35. P. Roepstorff and J. Fohlman, *Biomed. Mass Spectrom.* 11, 601, 1984.
36. M. R. Davis and T. A. Baillie *J. Mass Spectrom* 30, p. 57, 1995.
37. K. W. Kohn *Cancer Res* 37, 1450, 1977.
38. G. Witz *Free Radic Biol. Med.* 7, 333, 1989.
39. H. D. Hoberman, R. C. San George *J. Biochem. Toxicol.* 3, 105, 1988.
40. A. A. Tuinman and G. R. Pettit *Int. J. Peptide Protein Res.* 36, 331, 1990.

APPLICATIONS OF MASS SPECTROMETRY IN BIOCHEMICAL STUDIES OF NUCLEOSIDES, NUCLEOTIDES AND NUCLEIC ACIDS.

Russell P. Newton.
Biochemistry Group, School of Biological Sciences, University of Wales Swansea, Singleton Park, Swansea SA2 8PP, United Kingdom.

ABSTRACT. The advent of soft ionization procedures such as fast-atom bombardment, electrospray and thermospray, and, more recently, matrix-assisted laser desorption - time of flight, has greatly extended the utilization of mass spectrometry in analyses of nucleobases, nucleosides, nucleotides and nucleic acids. Here specific examples of the application of MS techniques to studies of nucleic acid base composition, sequence and modifications, of nucleoside and nucleotide antibiotic activity, and of nucleotide cell signalling functions are reviewed and potential future advances discussed.

1. Introduction.

The application of MS analysis to nucleic acids and related compounds is of great potential given the explosion of interest in the field of molecular biology over the last decade. Productive study of the DNA and RNA macromolecules without degradation prior to MS has until recently proved impracticable and most such research programmes have consequently been confined to examination of the released, constituent nucleotides, nucleosides or bases. The major impetus of these investigations has lain in the structural elucidation of nucleic acids in respect of sequencing for the purposes of elaborating the genome, or for identifying adducts to determine modifications caused by carcinogens. However, while many analyses of free bases, nucleosides and nucleotides [fig. 1] reflect their position as precursors and/or breakdown products of the nucleic acids, these compounds are also of great significance *per se*; endogenous nucleotides for example play a vital role in cellular signal transduction and metabolic regulation, while synthetic modified nucleosides and bases and their analogues have application in cancer therapies and as anti-bacterial and anti-viral agents.

The first mass spectra of nucleosides and nucleotides were published in the late 1960s and early 1970s [1-3], and elegant analyses of nucleotides, nucleosides and the related heterocyclic bases have been carried out subsequently using the more 'traditional and established' forms of MS. The advent of 'soft ionization' procedures in the form of fast-atom bombardment (FAB), electrospray (ES) and thermospray (TSP), enabling MS analyses of polar, non-volatile compounds, and their application to nucleosides and nucleotides has greatly enhanced progress in this area; still more recently the use of

R. M. Caprioli et al. (eds.), Selected Topics and Mass Spectrometry in the Biomolecular Sciences, 427–453.

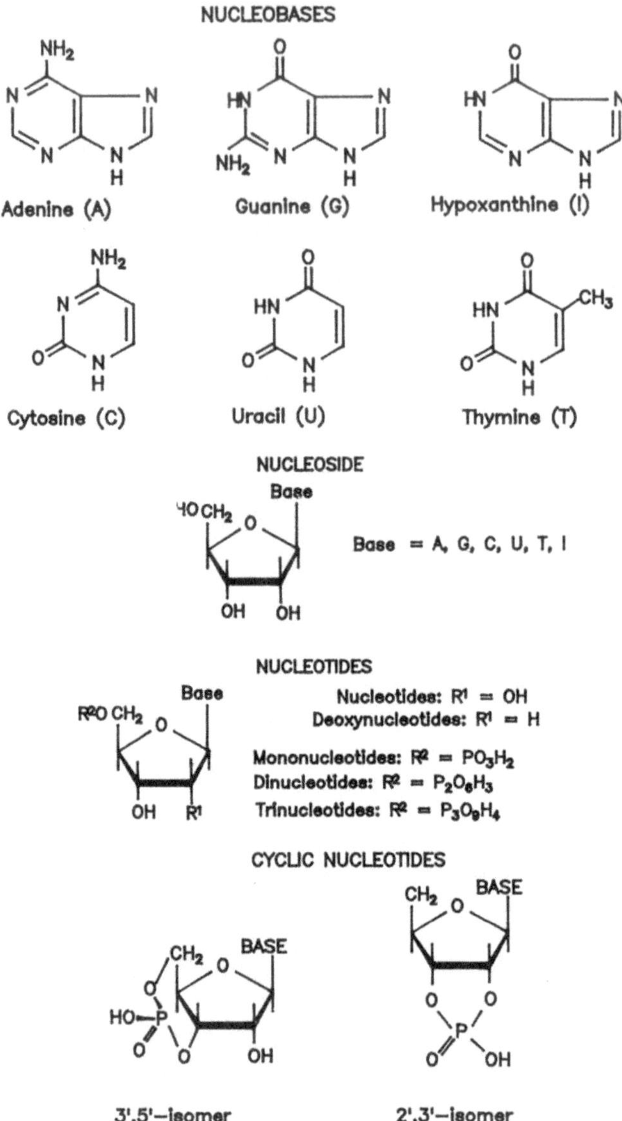

Figure 1a. Chemical structures of nucleobases, nucleosides and nucleotides.

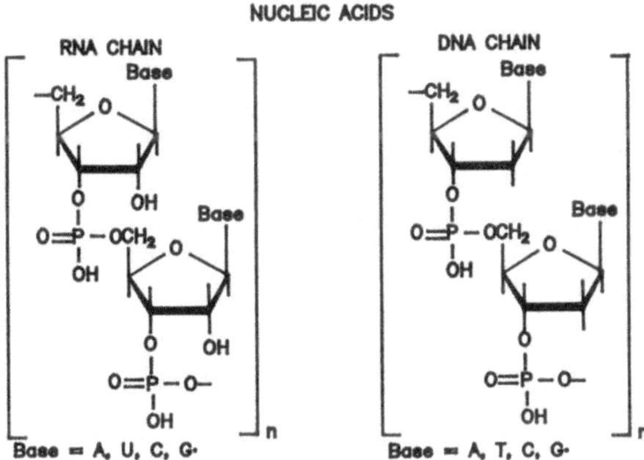

Figure 1b. Chemical structures of nucleic acids.

matrix-assisted laser desorption - time of flight (MALDI-TOF) instruments has been successfully applied to studies of larger nucleic acid components. In an article of this length it is impossible to produce a comprehensive review of so broad and rapidly developing a research area and the reader will be referred to other specific reviews where appropriate. This chapter will thus confine itself to a general overview containing a few detailed, specific, examples of the contributions of the various MS technologies at the levels of base, nucleoside, nucleotide and nucleic acid, with the objective of illustrating the experimental rationale, the success of the various approaches, and their future potential.

2. MS of bases.

While the free base can be regarded as the first component of nucleosides, nucleotides and nucleic acids, and as such the simplest entity, the fragmentation behaviour of such heterocyclic compounds is arguably one of the most complex processes in mass spectrometry [4]. Before the use of soft ionization MS techniques, fragmentation profiles of a number of pyrimidines and purines had been described [5]. While the fragmentation of pyrimidines proceeded primarily via losses of HCN and CH_3CN, purine fragmentation involved molecular rearrangement and was consequently much more difficult to interpret and apply. The use of FABMS, together with collisionally induced dissociation (CID) in conjunction with tandem MS (MSMS) has provided a very powerful means of promoting extensive fragmentation of free nucleobases. Mass selection of the free base fragment ion from a larger molecule allows the collision energy to be directed into the fragmentation of the base, rather than the cleavage of the glycosidic bond, producing extensive dissociation of the heterocyclic ring [6,7]. This provides a much more useful basis for structural elucidation and hence identification of, for example, the nucleobases and modified nucleobases in nucleic acid hydrolysates. Examples of such fragmentations are depicted for a pyrimidine, uracil, [fig. 2] and a purine, adenine, [fig. 3], as proposed by Nelson and McCloskey [8,9]. By the use of analytes labelled with the stable isotopes, ^{13}C, ^{2}H and ^{15}N, and generation of CID mass spectra, several pathways of minor or non-occurrence in the corresponding electron impact (EI) mass spectra have been elaborated.

With protonated uracil the FAB/CID mass spectrum, with major peaks at *m/z* 96(59%), 95(7), 70(100), 68(8) and 67(2), indicates two principal pathways [fig. 2] not present in the EI mass spectra, the elimination of NH_3 followed by the loss of CO, and loss of H_2O. Loss of HNCO, the principal process in EI, also occurs, together with secondary dissociations yielding C_3HO^+, H_2CN^+ and $C_2NH_2^+$ [8]. With protonated

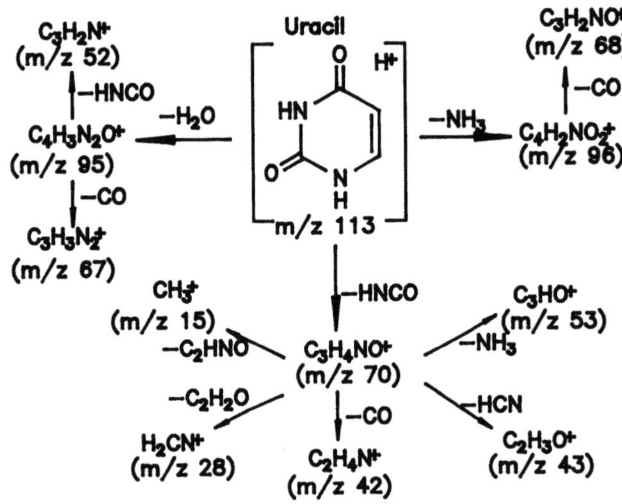

Figure 2. Fragmentation of uracil. (Adapted from [9]).

adenine the major peaks of interest in the FAB/CID mass spectrum are at *m/z* 119(100), 109(31), 94(34), 92(43), 82(27), 77(3), 67(43), 65(29), 55(48), 40(3), 28(6) and 18(18): these fragments reflect three major pathways [fig. 3] which are of minor significance in the EI mass spectra, loss of NH_3, loss of NH_2CN and formation of NH_4^+, while the fourth pathway, the sequential expulsion of three molecules of HCN, is also evident in the EI spectrum [9]. Understanding of these fragmentation pathways has greatly enhanced the ability to deduce the structure of modified bases in larger molecules, as will become evident later in this article.

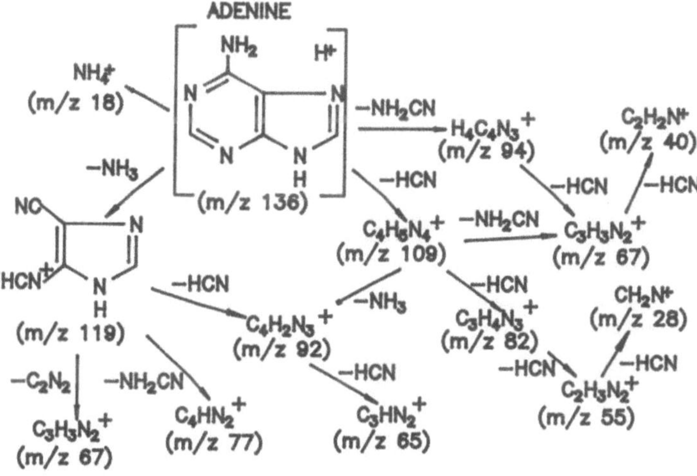

Figure 3. Fragmentation of adenine. (Adapted from [8]).

3. MS of nucleosides.

Nucleosides are examined by MS analysis as products of enzymic digests of DNA, RNA and nucleotides, as antibiotics or other pharmacological agents of novel structure, and as diagnostic components of tissues, particularly urine. EI, FAB, ES, TSP and MSMS have each made valuable contributions to such nucleoside analysis. Identification of nucleosides and modified nucleosides of previously established structure can now often be made on the basis of chromatographic behaviour, UV absorbance and MS. This is of use in clinical diagnosis, since elevated levels of modified nucleosides are excreted

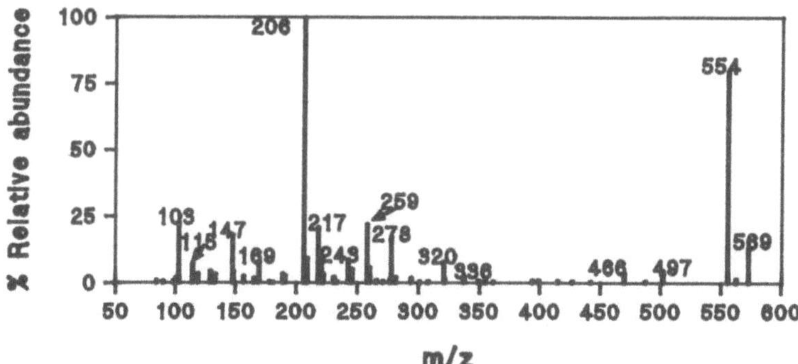

Figure 4. EI mass spectrum of 1-methyladenosine extracted from human urine. (Adapted from [12,13]).

in the urine of cancer patients; these levels of modified nucleosides, originating from tRNA, result from the abnormally high tRNA turnover rate [10]. In this context GCMS of the trimethylsilyl derivatives of nucleosides has provided a rapid and reliable means of identifying and quantifying urinary nucleosides from EI mass spectra [11]. As an example the EI mass spectrum of trimethylsilylated 1-methyladenosine extracted from human urine is shown in fig. 4, with the characteristic molecular-ion-derived species at *m/z* 569, 554 and 497 corresponding to M^+, $[M-15]^+$ and $[M-17]^+$, base-derived ions at *m/z* 206, 278, 320 and 336 corresponding to $[B-14]^+$, $[B+58]^+$, $[B+100]^+$ and $[B+116]^+$, and ribose-derived ions at *m/z* 103, 169, 217 and 259 corresponding to CH_2OTMS, ribose-180, $C_3H_3O_2TMS_2$ and S-90 respectively [12,13]. Comparison of the peak areas of the expanded, nucleoside, regions of the reconstructed ion chromatographs of pooled normal urine and the urine of cancer patients [fig. 5] with standards enabled the identification of 14 urinary nucleosides including several modified nucleosides.

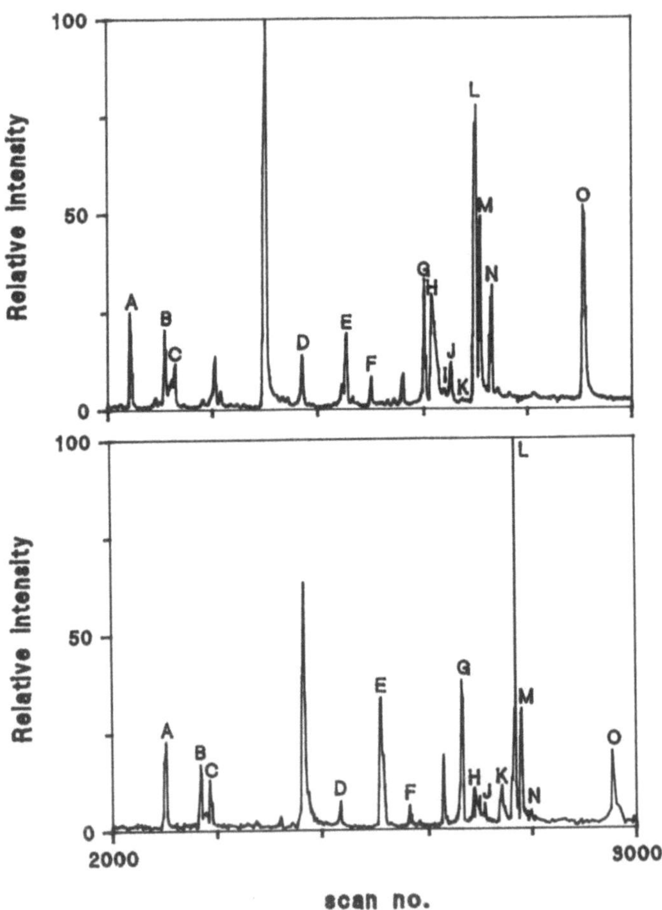

Figure 5. Expanded nucleoside region of the reconstructed ion chromatogram from extract of (a) normal urine and (b) urine from a cancer patient. (Adapted from [12,13]).

Levels of the modified nucleosides 1-methyladenosine, 2-methylguanosine, N^2,N^2-dimethylguanosine and 1-methylinosine as well as the total nucleoside level were elevated in the urine when a malignant tumour was present, and levels of N^2,N^2-dimethylguanosine were found to correlate with the stage of cancer, illustrating the potential of MS analysis of urinary nucleosides as a diagnostic tool [12,13].

To identify unknown nucleosides, a successful interpretation protocol has been designed by McCloskey [14]. The characteristic ions from trimethylsilylated nucleosides are recognised as M-CH_3, M-$C_3H_{10}OSi$, M-$C_4H_{11}OSi$, M-$C_4H_{13}OSi$, M-$C_4H_{10}O_2Si$, M-$C_5H_{11}O_2Si$, M-$C_6H_{20}O_2Si_2$ and M-$C_7H_{23}O_2Si_2$ [15], and knowledge of these fragments is utilized in the following procedure [14]:

(i) Assign molecular ion and molecular ion series, and estimate the extent of contributions from other components, including derivatives which have incorporated more or fewer silyl groups than the principal derivative.

(ii) Determine the value of the base fragment, by subtracting the value of the sugar component.

(iii) Assign ions of the sugar series.

(iv) Seek evidence for side-chain constituents.

(v) Consider information from additional experiments using stable isotope-labelled compounds.

Utilization of this approach has led to the identification of over 90 modified nucleosides [16] and there is now a database recording such modificatios [17]. An excellent example of the application of MS analysis, in combination with magnetic resonance studies, is the elucidation of the structure of archaeosine [fig. 6] shown to

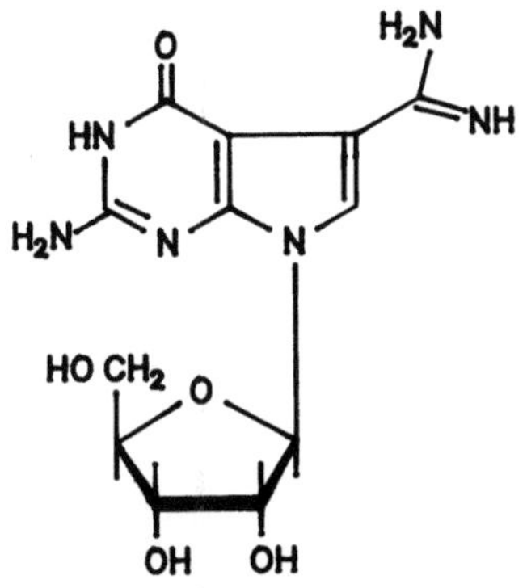

Figure 6. Chemical structure of archaeosine [18].

be 2-amino-4,7-dihydro-4-oxo-7-6-D-ribofuranosyl-1Hpyrrolo[2,3-d]pyrimidine-5-carboximidamide by EIMS, GCMS, FAB/CID MSMS and use of deuterium exchange labelling [18]. This nucleoside was first reported in 1982 [19] as nucleoside G*-15 and its recent structural elucidation provides insight into the evolution of one of the most conserved nucleoside modifications in RNA, with the suggestion that the unusual, charged, side chain reflects a role in stabilization of tertiary interactions [18].

In addition to nucleosides yielding information relating to nucleic acid structures, MS analysis has made important recent contributions to our knowledge of other categories of nucleoside. Modified nucleosides in some cases are potent antibiotics, acting by exploiting the difference in specificity of host and pathogen enzymes toward the natural substrate nucleoside. FAB/CID mass spectra of the protonated trimethylsilyl derivatives, in concert with NMR and chemical degradation analyses, have enabled the recent elucidation of the structure of liposidomycins, novel lipid nucleoside antibiotics produced by *Streptomyces griseoporeus*, which inhibit cell wall synthesis in some species of pathogenic bacteria.

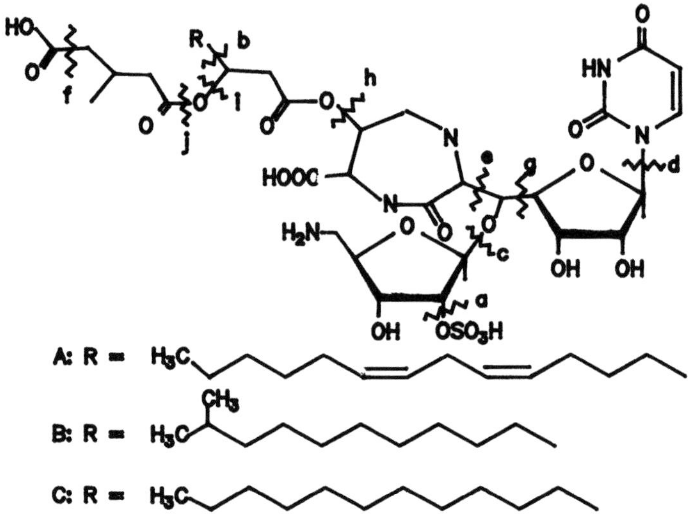

Figure 7. Structures of liposidomycins A, B and C, with major MS cleavages a-j. (Adapted from [20]).

The structure of three liposidomycins, A, B and C [fig. 7] shows that they each contain 5-amino-5-deoxyribose-2-sulphate, perhydro-1,4-diazepine and 5'-substituted uridine. They differ in the lipid side chain: liposidomycin A contains 3-(3'-methylglutaryl)-7,10-hexadecadienoic acid, liposidomycin B contains 3-(3'-methylglutaryl)-12-methyltridecanoic acid, and liposidomycin C contains 3-(3'-methylglutaryl)-tetradecanoic acid. These structures were identified following FAB/CID MS [20]: the cleavages in fig. 7 were deduced from a series of MSMS experiments in which spontaneous dissociation fragments in MS-1 were mass selected and subjected to collisional activation and the fragments measured in MS-2, with the complete intercalating network of subunit ions observed by sequential analysis of increasing smaller precursor ions. This enabled partial characterization of the structural subunits and of the interconnection sites between the subunits in the intact molecule [20].

Electrospray MS has provided an immense improvement in the application of MS to studies of another group of nucleoside derivatives, the cytokinins, which are plant hormones. Previous analyses using GCMS required derivatization, while LCMS with FAB or TSP lacked sensitivity. ES LC-MSMS has recently been used to analyse cytokinins [fig. 8], this technique demonstrating high resolution and the high level of

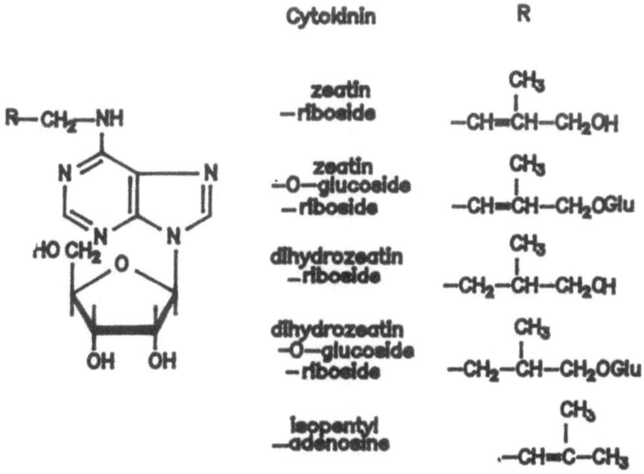

Figure 8. Structures of some cytokinins containing a ribonucleoside component [21].

sensitivity required for analyses of extracts from small quantities of plant material [21]. As would be predicted the major cleavage is of the glycosidic bond: other fragmentations allow the identification of the side chains and substituents of the various cytokinin forms. By optimization of chromatographic and MS conditions it is now possible to quantitate all the constituent cytokinins, free bases, ribosides and glucosides in a single extract within a range of 1-100 pmol, and at a rate of 4 minutes per compound [21]. Application of this technique will facilitate future kinetic analyses of cytokinin metabolism on a scale previously impossible.

Quantitation of some modified nucleoside antibiotics, including several anti-viral agents, is also carried out by MS-based techniques . A topical example is that of 2',3'-dideoxycytidine (ddC: zalcitabine), used in anti-AIDS therapies: ddC can be determined in plasma samples by LC-TSPMS, using a stable isotope-labelled internal standard, $[^{15}N_2,^{2}H_2]$-ddC. Selected ion monitoring of the protonated molecules of ddC and of the internal standard is used to record mass chromatograms and a linear response over the desired physiological range of 0.25-20 ng/ml is obtained [22].

4. MS of nucleotides.

FAB/CID MSMS had an immediate, dramatic, effect upon nucleotide analysis after its first applications in the 1980s [23]. The great advantages that soft ionization MS has conferred upon studies of the involatile, polar, nucleotides can be exemplified by its use in cyclic nucleotide analysis.

Cyclic nucleotides are components of biological inter- and intra-cell signal transduction systems, and as such are important pharmacological targets [24]. The biochemical second messengers are 3',5'-cyclic nucleotides [fig. 1], with the 2',3'-isomers, intermediates of nucleic catabolism, also occurring but not fulfilling this same regulatory

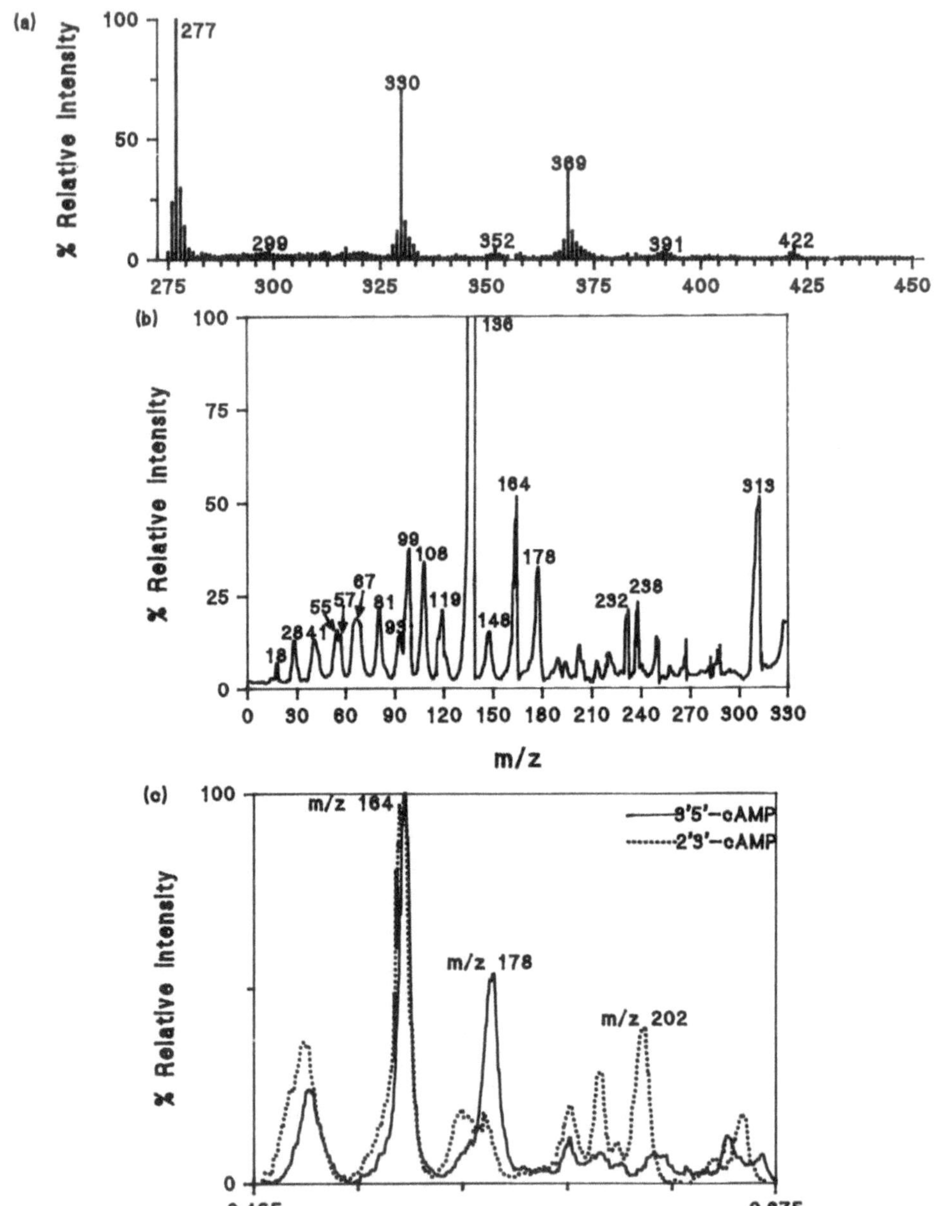

Figure 9. (a) FAB mass spectrum of 3',5'-cyclic AMP: (b) CID/MIKE spectrum of $[MH]^+$ of 3',5'-cyclic AMP: (c) Expanded part CID/MIKE spectra of $[MH]^+$ of 3',5'- and 2',3'-cyclic AMP. (Adapted from [25,26]).

function. While both sets of isomers yield a FAB mass spectrum with the major peaks corresponding to the protonated molecule, its sodium and glycerol adducts, and to glycerol matrix peaks [fig. 9] [25,26], CID of the protonated molecule yields a mass-analysed ion kinetic energy (MIKE) spectrum which readily differentiates the two isomers [fig. 9b,c]. In both cyclic nucleotide isomers the major peak is at *m/z* 136, corresponding to protonated base ($[BH_2]^+$), but while the 3',5'-cyclic nucleotide undergoes both S_1 and S_2 cleavages [fig. 10] the 2',3'-cyclic nucleotide does not

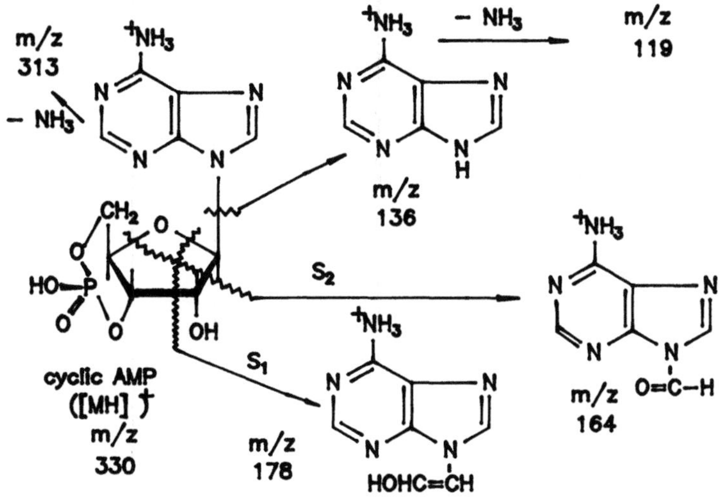

Figure 10. Major fragmentation pathways of 3',5'-cyclic AMP. (Adapted from [25,26].

undergo significant S_1 cleavage due to the blocking of the 2'-O-position [fig. 11]. As will become apparent, this characteristic fragmentation of nucleotides (and nucleosides) unsubstituted at the 2'-OH, with major peaks at $[BH_2]^+$, $[BH_2+28]^+$ and $[BH_2+42]^+$, together with peaks at each of these *m/z* values less NH_3, is of crucial diagnostic value in nucleotide analysis. By application of FAB/MIKES analysis to nucleotide fractions in tissue extracts the debate over the natural occurence in plants of adenosine- and guanosine-3',5'-cyclic monophosphates [fig. 1] has been resolved [27,28], and cytidine-, uridine-, inosine- and thymidine-3',5'-cyclic monophosphates have also been shown to be endogenous components of both plant and animal cells [29-31]. Synthetic cyclic nucleotide derivatives, used in cell permeation studies [32] and in radioimmunoassay [33] have also been unabiguously identified by this approach.

The majority of FABMS studies of nucleotides have been qualitative. There are a number of reasons for this: for cyclic nucleotides and nucleotides there is only a small concentration range over which peak intensity is directly proportional to concentration, and this can be compounded by the presence, for example in enzyme incubation mixtures, of a complex mix of compounds where the movement of an analyte molecule to the surface of the matrix is affected by the presence of other molecules. However one enzyme quantitation approach recently used successfully in our laboratories has been to

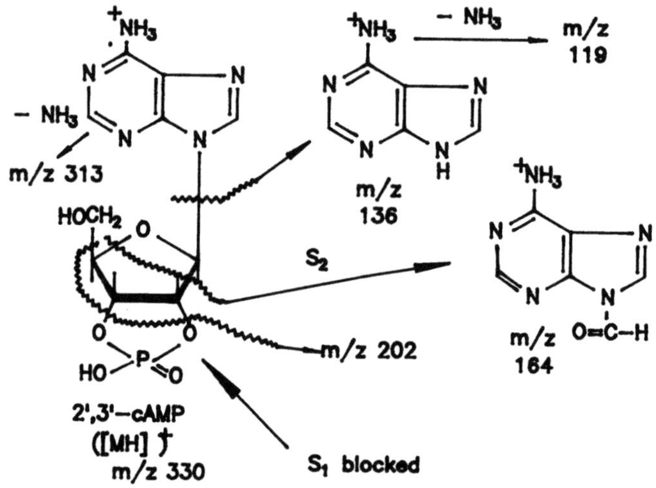

Figure 11. Major fragmentation pathways of 2',3'-cyclic AMP. (Adapted from [25,26]).

obtain the CID/MIKE spectra from the protonated molecules of both substrate and product in the incubation, then to obtain a second pair of spectra after the sample has been spiked with a known quantity of substrate and product. The calculation of activity can then be made from:

$$\frac{[P]_t}{[S]_t} = \frac{\text{Sum of relative intensities of product diagnostic ions}}{\text{Sum of relative intensities of substrate diagnostic ions}}$$

$$\text{and} \quad \text{Rate} = \{([S]_0 /t) * ([P]_t / [S]_t)\}$$

where $[S]_0$ = initial substrate concentration, $[P]_t$ and $[S]_t$ are the product and substrate concentrations at time t, and relative intensity is the peak intensity at *m/z n* in the MIKE spectrum divided by the change in peak intensity after spiking, with spikes of equal magnitude, of standard substrate and product. This has been applied to good effect in phosphodiesterase kinetic analyses, and provides data showing good correlation [34,35] with data from conventional radiometric assays.

Such quantitative CID/MIKES analyses are demanding in respect of quantities of substrate and product required, thus whenever feasible quantitation via the FAB mass spectrum is utilized instead. In this method the peak heights of the protonated molecules, their adducts, and characteristic fragments of the compounds of interest are determined, together with major peaks emanating from the glycerol matrix: quantitation is carried out by a proportionation of the sums of the relative peak intensities. Since the mass spectra from enzyme incubates are very complex due to the presence of several different molecular species, it is neccessary to carry out parallel series of incubations containing active and inactive enzyme preparations to obtain viable kinetic data. Change in the concentration of analyte A in the enzyme-catalyzed reaction is calculated from the intensities of diagnostic peaks by

$$\Delta[A] = \frac{(\Sigma InA/\Sigma Inmatrix)_{expt} - (\Sigma InA/\Sigma Inmatrix)_{control}}{(\Sigma InA/\Sigma Inmatrix)_{control}}$$

where Σ*In*A and Σ*In*matrix are the sum of the relative peak intensities of the characteristic ions of analyte A and the glycerol matrix respectively. This procedure has been used successfully with several cyclic nucleotide-related enzymes, for example cyclic nucleotide-responsive protein kinases [36,37]. It can be seen directly from the FAB mass spectra of active and inactivated protein kinase preparations [fig. 12] that the peak intensity of *m/z* 330 decreases in the active preparation reflecting cyclic AMP binding to the active enzyme, and that the ATP peak series at *m/z* 508, 530, and 552 decreases with a concommitant increase in ADP at *m/z* 428, 450 and 472, reflecting the consumption of ATP to phosphorylate the substrate protein. Computation of the

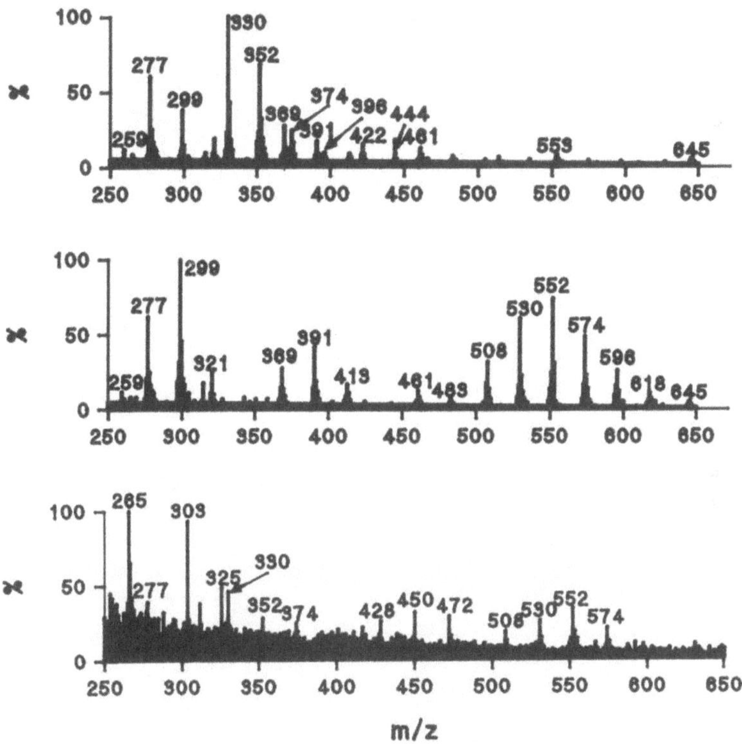

Figure 12. FAB mass spectra of
(a) cyclic AMP added to kinase incubation buffer
(b) ATP added to kinase incubation buffer
(c) Active protein kinase incubation mixture
(Adapted from [37]).

characteristic peak heights into the equation above yields kinase data with good correlation to those obtained by the conventional radiometric assay [36,37], with the inherent advantage that the MS-based assay has the additional capacity to monitor simultaneously the cyclic nucleotide binding capacity of the kinase, its specificity, any integral nucleotidase and phosphodiesterase activities, and the turnover of any effectors added.

This facility to simultaneously monitor multiple components of an enzyme incubation by FABMS has been used in combination with CID/MIKES analysis to differentiate isomers. With phosphodiesterase activity it is possible for the 3',5'-cyclic nucleotide substrate to be cleaved at either of two positions to yield two isomeric products:

Cyclic Nucleotide Phosphodiesterase

Identification of these products yields valuable information about the mechanism of action of the enzyme and facilitates development of effectors with which to manipulate its activity. The FAB mass spectra of active and inactive cyclic CMP phosphodiesterase incubates [fig. 13] show the decrease in $[MH]^+$ of the substrate cCMP at *m/z* 306 and

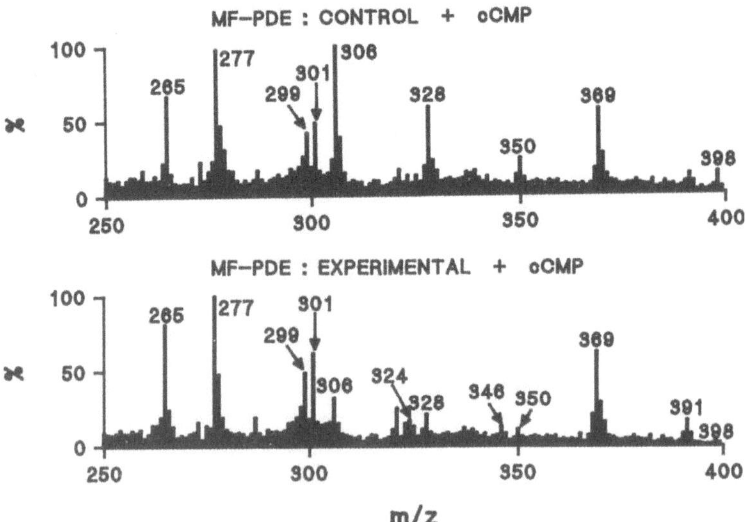

Figure 13. FAB mass spectra of (a) Inactive phosphodiesterase incubation mixture and (b) Active phosphodiesterase incubation mixture.

increase in $[MH]^+$ of the product CMP at *m/z* 324. The CID/MIKE spectra of the isomers of CMP, 5'-CMP and 3'-CMP, can be differentiated by the relative intensities of *m/z* 213, 226, 232, 263 and 268 [fig. 14]; simple visual comparison of the product

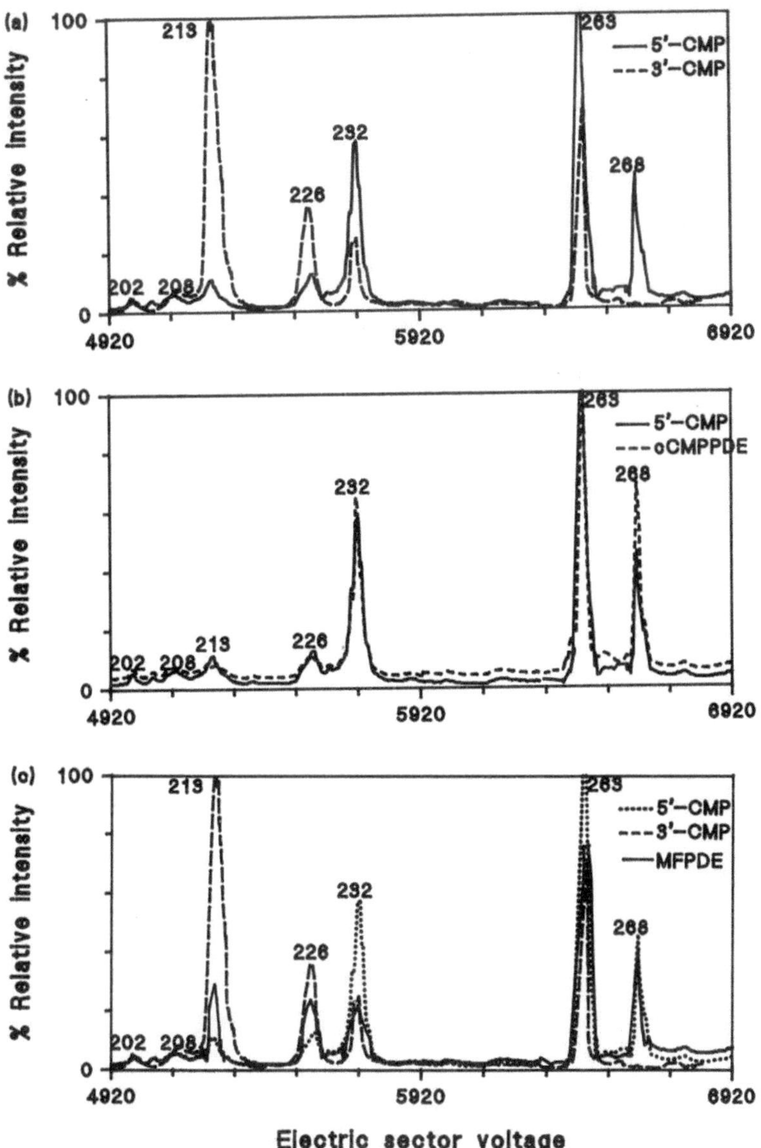

Figure 14. CID/MIKE spectra of the protonated molecules of (a) 5'- and 3'-CMP, (b) 5'-CMP and the product of cCMP-phosphodiesterase activity, (c) 5'-CMP, 3'-CMP and the product of multifunctional phosphodiesterase activity.

of one phosphodiesterase, cCMP-specific phosphodiesterase, shows that the CID/MIKE spectrum is essentially identical with that from 5'-CMP, indicating that 5'-CMP is the sole or major product of this enzymes activity. Comparison of the CID/MIKE spectrum from $[MH]^+$ of the product of a second enzyme, multifunctional phosphodiesterase, shows it is intermediate between those of 5'-CMP and 3'-CMP, indicating that a mixture of products are present. Differentiation of the 2'-, 3'- and 5'-mononucleotide isomers has been carried out [38,39] and is being applied to enzyme product analysis: differentiation of *R* and *S* isomers of cyclic nucleotide thioate analogues used as site-selective protein kinase effectors is also being utilized [40].

The major contribution of FABMS analysis to this field has been in the identification

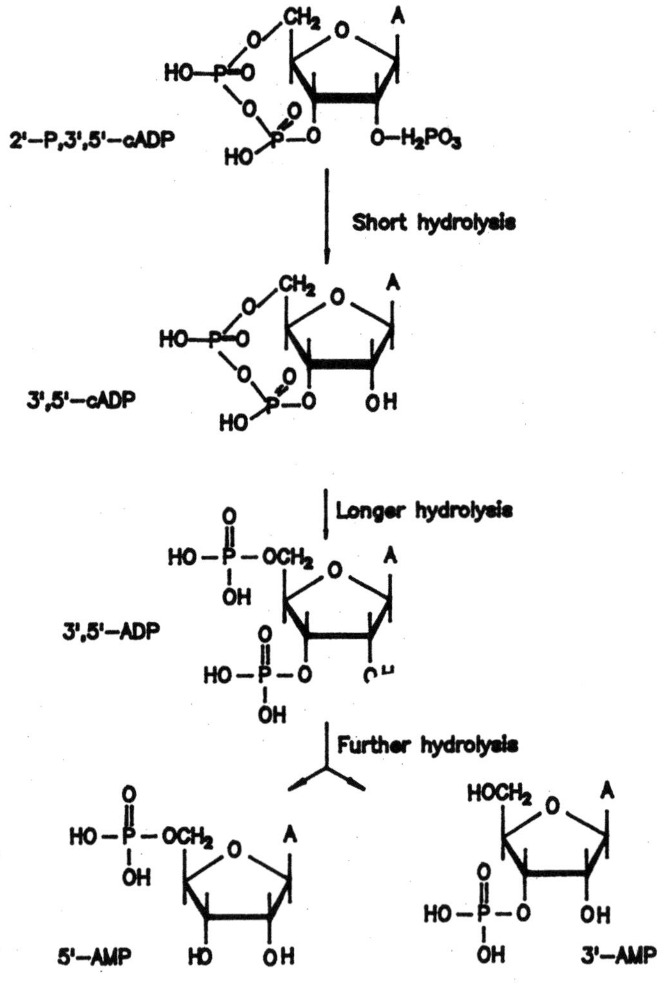

2'-P,3',5'-cADP

Short hydrolysis

3',5'-cADP

Longer hydrolysis

3',5'-ADP

Further hydrolysis

5'-AMP

3'-AMP

Figure 15. Identification of 2'-phosphoadenosine-3',5'-cyclic pyrophosphate from extracts of *Porphyra*: hydrolysis yields products readily identifiable by MS. (Adapted from [41]).

of unknown nucleotides. Often simple chemical or enzymic modification can yield products from which MS data, together with those from the original compound, provide conclusive structural information. The novel compound 2'-phosphoadenosine-3',5-cyclic pyrophosphate [fig. 15] has been identified by such rationale in extracts from the seaweed *Porphyra*. It possessed a phosphate/adenine ration of 3:1 as determined spectrophotometrically, but the mass spectrum contained $[MH]^+$ at m/z 490, not consistent with the unknown being the previously suspected adenosine triphosphate (M_r 507). The CID/MIKE spectrum contained $[BH_2]^+$ and $[BH_2+28]^+$ at m/z 136 and 164, but $[BH_2+42]^+$ was absent, indicating that the 2'-O position was blocked. A short acid hydrolysis liberated a compound with $[MH]^+$ at m/z 410, consistent with the loss of a phosphate, which was confirmed spectrophotometrically, and a CID/MIKE spectrum which now contained $[BH_2+42]^+$ at m/z 178, indicating that the phosphate moiety had been removed from the 2'-O-position. Acidic hydrolysis for an extended period liberated three further products, one with $[MH]^+$ at m/z 428, suggesting addition of H_2O to the previous 'dephosphorylation' product, and two with $[MH]^+$ at m/z 348, indicating loss of a further phosphate. The CID/MIKE spectra of the latter two compounds were identical to those of 5'-AMP and 3'-AMP, the data collectively reflecting successive hydrolyses of the parent compound [41]. A similar rationale has been used to identify several other novel nucleotides [fig. 16] in seaweed [41], and four novel side-products

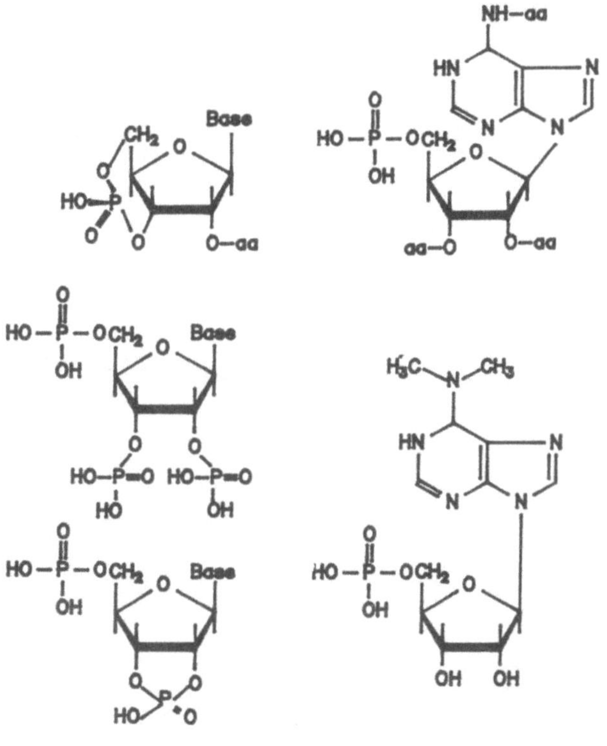

Figure 16. Novel nucleotides from *Porphyra* identified by MS [41].

of the activity of the enzyme cytidylyl cyclase [42]. In the latter case hydrolysis was carried out with a phosphodiesterase specific for the cyclic phosphodiester bond, a nucleotidase specific for the monoester bond, and nonspecific acid hydrolysis [fig. 17]. Although the MS data of the parent compound was not immediately comprehensible, the CID/MIKE spectra of the hydrolytic products were easily recognisable and led to identificationof the parent compound as 2'-phosphocytidine-3',5'-cyclicmonophosphate. Similar rationale led to the identificationof three other side products of the reaction [fig. 18] [42].

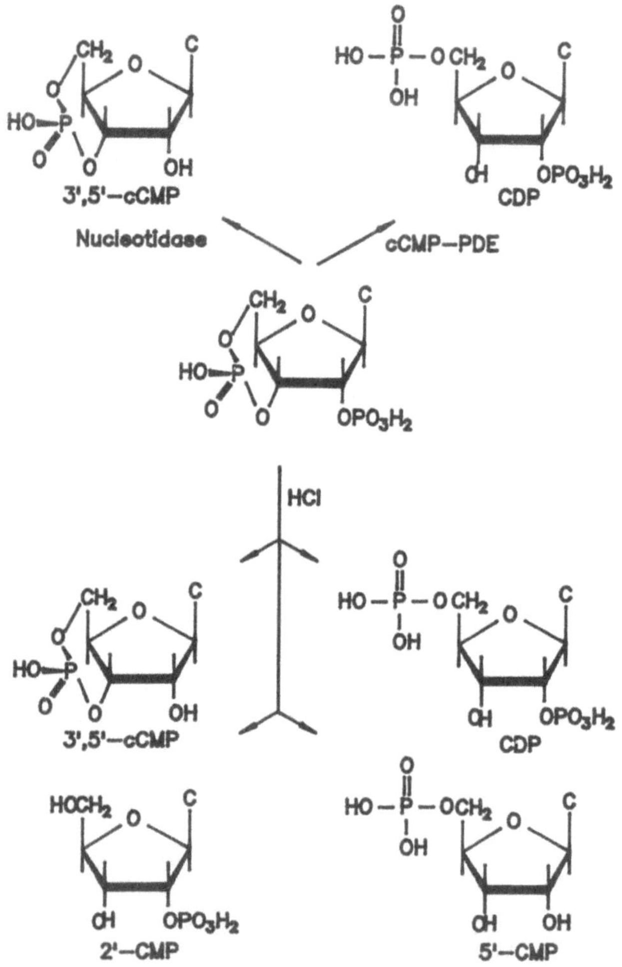

Figure 17. Identification of 2'-phosphocytidine-3',5'-cyclic monophosphate as a side product of cytidylyl cyclase activity [42]. Selective hydrolysis yields products readily identifiable by MS.

The examples of nucleotide MS analysis detailed above are primarily confined to the application of FABMS and FAB/MIKES. A number of laboratories successfully employ

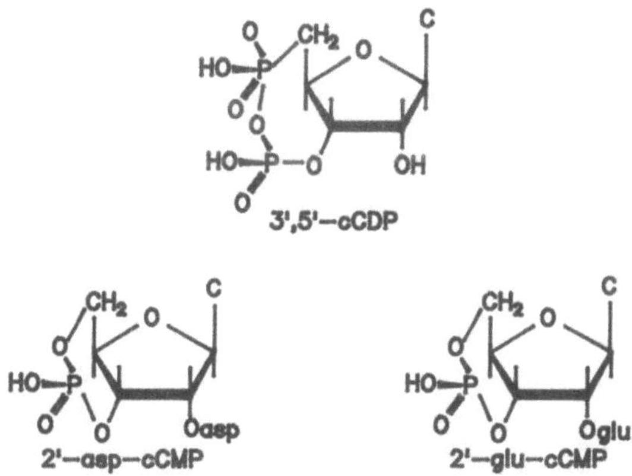

Figure 18. Additional side products of cytidylyl cyclase reaction, identified by MS [42].

MS analysis not of the original analyte but of the trimethylsilyl derivatives of nucleotides [43] which offers greater sensitivity, as also does continuous flow FAB [44,45]. Electrospray MS has also recently been applied to cyclic nucleotide identification and quantitation to great effect; as a result of combination with LC, it offers not only greatly enhanced sensitivity but has the additional advantage of a non-requirement for extensive purification of an extract prior to MS analysis [46]. MS analysis of novel nucleotides often benefits from a combination of approaches and, as above, from modification and re-examination. This 'combined attack' approach is exemplified by the recent elucidation of the structure of phosmidosine [fig. 19] an antifungal nucleotide antibiotic identified by LC-MS of a crude isolate, MSMS of the isopropylidine derivatives, and FAB-deuterium exchange MS [47].

Figure 19. Chemical structure of phosmidosine [47].

5. MS of nucleic acids.

The theoretical ideal for MS analysis of nucleic acids is the fragmentation depicted in fig. 20: a series of cleavages a, b, c, etc., generating a mass spectrum with the molecular ion then a series of fragments whose decreasing *m/z* ratio reflects the sequence of bases present. Selection of these fragments in MSMS followed by CID would enable

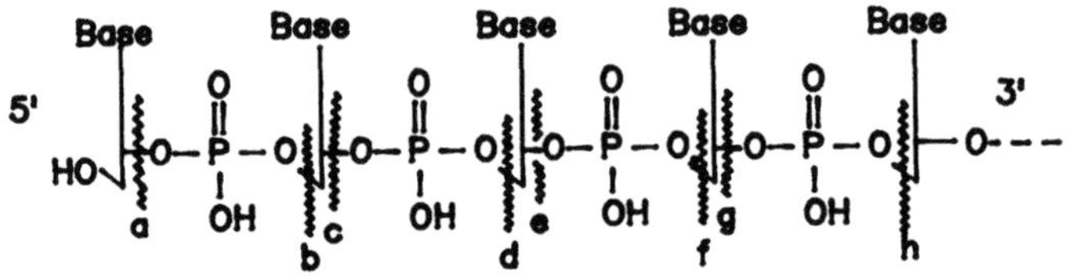

Figure 20. Idealized fragmentation of nucleic acids.

identification of any base modifications. While this is not (yet) attainable for nucleic acids of macromolecular size, recent developments in electrospray and MALDI-TOF procedures have made dramatic advances possible; this has contributed to knowledge of base composition and base modifications, enhanced sequencing facilities for both RNA and DNA, and facilitated identification of adducts formed between DNA and carcinogens, mutagens and other environmental pollutants.

Determination of oligonucleotide composition can be made directly from accurate mass determination by ESMS of the intact oligonucleotide [48]. In the absence of base modifications there are only four principal constituent residues that are required to be montiored in DNA and RNA: for DNA they are pdC (M_r=289.185), pdT (304.196), pdA (313.210) and pdG (329.209), and for RNA pC (305.184), pU (306.169), pA (329.209) and pG (345.209) [48]. Although mass determination of within 0.01% error is now attainable by ESMS, the values of M_r for each of the nucleotide subunits are sufficiently close that even in small oligonucleotides more than one base composition will be compatible with a specific molecular mass. For example d(A_1G_7) will have a mass of 2617.707 and d(C_8T_1) a mass of 2617.711, and the higher the value of M_r the more ambiguous the data becomes. It is calculated [48] that above 4900 Da there is at least one possible combination of nucleotides associated with every mass value and above 6000 Da there are at least two compositions with every mass unit interval. McCloskey [48,49] has reduced this problem by determining the composition in respect of one base, either by selective chemical modification or cleavage reactions, or by use of site-specific ribonucleases. When G or A is fixed in this manner, all possible oligoribonucleotide compositions can be determined specifically up to the 14-mer level, and there is a substantial reduction in ambiguity of composition of larger oligonucleotides. Thus in the example above, after the hydrolysis of the oligonucleotides with M_r of 2617.7 with ribonuclease U_2 which cleaves selectively at A, ESMS of dCdCdCdTdCdCdCdC would indicate M_r still at 2617.707 while dGdGdGdAdGdGdGdG would now be present as two smaller fragments. Any remaining ambiguities at the higher mass level can potentially be solved by use of LC-ESMS.

From the initial use of ESMS for oligonucleotides [50] resolving power has been increased by the use of trimethylammonium salts [51] and precipitation [52] to reduce cation adductation, and combination of precipitation and metal chelators has been shown to reduce peak widths of the individual charge states from clusters of 10 mass units to 2 mass units. Application of this approach has enabled *E. coli* tRNAs of 23-28 x 10^3 Da and 5S rRNA of 38 x 10^3 Da to be mass-determined within 0.01 % error [49]. This high level of accuracy of mass-determination and comparison of experimentally-determined M_r with the predicted mass for an oligonucleotide can be used for the identification of post-transcriptional modifications [53], for example enabling identification of modifications occurring in domains V and IV of the *E. coli* 23S rRNA [54,55]. These two domains play an integral role in peptidyl transferase activity [56] in the translation process; from oligonucleotides generated by enzymic cleavage, two segments (51-mer and 39-mer) of domain V were examined by ESMS and found to have molecular masses differing from those predicted from the rDNA. The 51-mer was found to have M_r 2 Da greater and LC-TSPMS of an enzymic digest confirmed the modification to be 5,6-dihydrouridine at position 2449 [54]. Similar strategy enabled identification of 2-methyladenosine at position 2503 in the 39-mer oligonucleotide from domain V [54], and 3-methyluridine at position 1915 in domain IV [55].

Just as the resolving power and consequently the mass range of nucleic acid components examinable has been greatly increased by improvement in methods of cation adductation reduction, the value of MALDI-TOF to nucleic acid biochemistry has been progressively enhanced by optimization of sample preparation methodology and of matrix-laser combinations. Replacement of metal cations with ammonium ions [52,57] again reduces peak broadening, and use of specific matrices with IR MALDI and UV MALDI has improved signal-to-noise ratios [58]. Better spectra in respect of signal strength and mass resolution are obtained in negative-ion mode relative to positive-ion mode with most wavelength-laser combinations, although with UV MALDI using a 3-hydroxypicolinic acid matrix comparable spectra are obtained in both modes [59]. The other major problem to be overcome in nucleic acid MALDI-TOF MS is the predilection of the *N*-glycosidic bond between the deoxyribose and nucleobase to fragment, this tendency increasing with increased M_r [60,61]. Since in linear time-of-flight mass spectrometers fragment ions formed outside the ion source are not separated from their intact precursors, linear mass spectrometers are potentially vastly superior to reflectron mass spectrometers for the analysis of larger oligodeoxyribonucleotides [59,62]. With IR MALDI of oligodeoxyribonucleotides, well resolved peaks then occur resulting from cleavages of the deoxyribose phosphate backbone, forming the major fragment ion series [fig. 21] designated X, X^* and Y [58,63].

Such fragmentations lend credence to the potential of direct MS sequencing of DNA and RNA by ion fragmentation as already demonstrated by the MALDI sequencing of the first 24 DNA fragments generated in typical sequencing reactions using standard M13mp19 template and primer [64], although an order of magnitude increase in both sensitivity and mass range is necessary before parity will be reached with competing methodologies [65]. These limitations can be at least partially overcome by the use of sequence ladders released by time-limited degradation with 3'- and 5'-specific

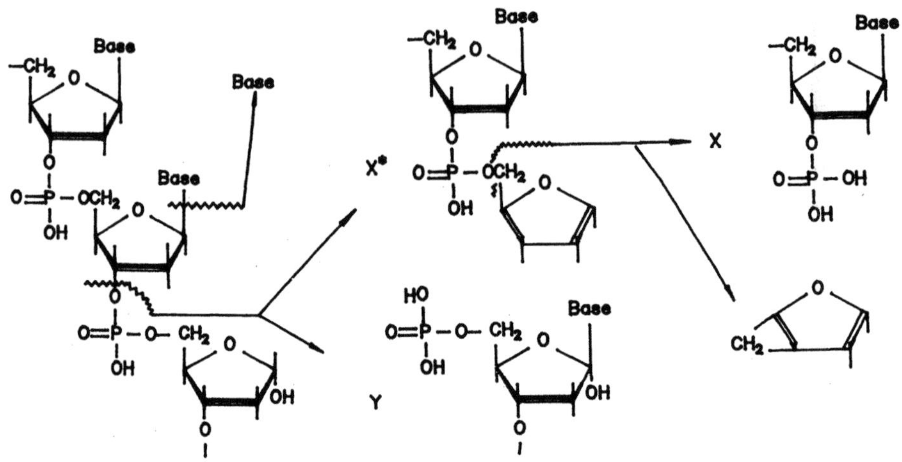

Figure 21. Fragmentation of deoxyribose phosphate backbone during IR MALDI of oligodeoxyribonucleotide. (Adapted from [58,63]).

exonucleases [66] [fig. 22], and the stability increase gained when adenine and guanine bases in the chain are converted to their 7-deazapurines [67,68]. This approach of selective enzymic or chemical fragmentation followed by MS as a sequencing protocol is equally applicable to combination with ES or other forms of MS, and to the MSMS structural elucidation procedures used in the analysis of nucleobases, nucleosides and nucleotides discussed earlier. For example DNA oligomers have recently been examined by ES MSMS with a quadrupole ion trap [69,70] and the resultant fragmentation allowed sequence information to be obtained [71]: two series of synthetic self-complimentary isomeric DNA hexamers, d(CAGCTG), d(CGATCG) and d(CGTAGC), and d(CATACG), d(TGATCA) and d(TGTACA) exhibited identical series of multiply-charged deprotonated molecular ions, as would be predicted, but had diagnostic low energy CID spectra from $[M-3H]^{3-}$ and $[M-4H]^{4-}$ which allowed discrimination between the individual isomeric DNA hexamers and permitted direct sequence determination [72].

Analysis of xenobiotically modified DNA is a further area in which recent developments in MS technologies have provided new impetus. In contrast to sequencing of oligonucleotides, CID is often desirable for analysis of DNA adducts, which undergo insufficient dissociation other than loss of the sugar moiety without collisional activation [73]. The intrinsic problem of studying such adducts is the low frequency with which they occur: while several conventional technologies including ^{32}P-postlabelling and immunoassay have the sensitivity to monitor covalent modification of DNA in the requisite range, they are unable to provide structural information. FABMS, EIMS, GCMS, laser desorption Fourier Transform MS, LC-TSPMS, LC-MSMS and accelerator MS have all made useful contributions to DNA-adduct structural elucidation as reviewed in [73], but suffer from the reciprocal problem, inadequate sensitivity. In some recent

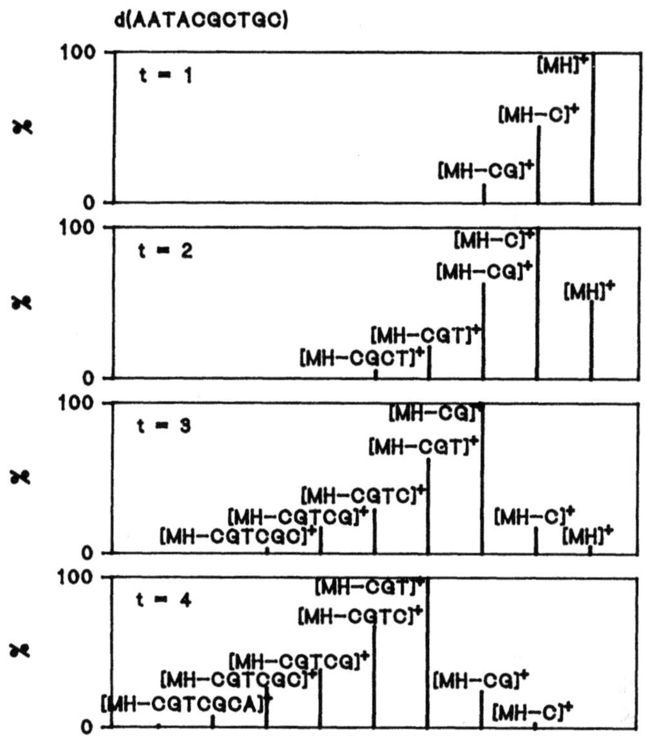

Figure 22. Schematic of MS from time-limited degradation of oligonucleotide by exonuclease.

applications however, in addition to providing this structural information, continued improvements in analytical processes have engendered significant increases in sensitivity. This can be illustrated by consideration of the adducts formed between DNA and bisphenol A diglycidyl ether (BPADGE) [fig. 23], an epoxide capable of entering the cardiovascular system from the environment and causing DNA misreplication and eventual neoplasm [74]. FABMS and MSMS studies [75-77] yieded information on the structure of the adducts formed, but LC-TSPMS has enabled the development of a sample enrichment procedure and identified adducts formed between BPADGE and the deoxyribonucleosides dAdo, dGuo, dCyd and dUrd [78,79]. The sensitivity was greatly enhanced by replacing a conventional LC with a capillary LC-ESMS, which enabled 600 fg of dAdo and 1 pg of BPADGE-dAdo adduct to be detected [80], while the CID spectrum of the adduct protonated molecule [fig. 23] had major peaks at m/z 476(24%), 192(62) and 136(100), compatible with the cleavages a, b and c, confirming adductation on the purine ring, although the position of alkylation is not yet completely established [80].

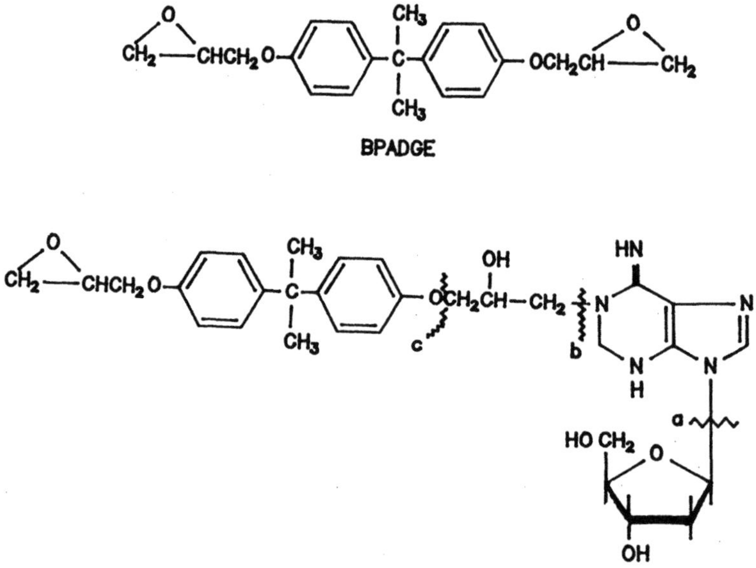

Figure 23. Structure of bisphenol A diglycidyl ether (BPADGE) and of its adduct formed with dAdo, with probable cleavages indicated at a, b and c. (Adapted from [80]).

6. Concluding comment.

Traditional MS methods made a valuable contribution to studies of nucleobases, nucleosides and to a lesser extent nucleotides: soft ionization methods instigated an exponential growth in the applications of MS in studies of the more polar nucleotides, oligonucleotides and nucleic acids. While currently MS technologies have not replaced conventional biochemical techniques used for example in nucleic acid sequencing, they do provide an increasingly valuable complement. Progressive increases in sensitivity and mass range of ES, TSP and MALDI-TOF MS indicate a further rapid expansion in their application to biomolecular studies, while other MS technologies continue to find novel niches of application, for example low temperature FABMS of nucleobases [81], GCMS of nucleic acid components in ancient seeds [82] and mass isotopomer distribution analysis to monitor nucleic acid turnover [83]. The success claimed by MS in this field of biomolecular analysis in the last decade thus appears set for further dramatic expansion in the next.

References.

1. S. Hanessian, D.C. DeJongh & J.A. McCloskey, *Biochim. Biophys. Acta* **117**, 480 (1966).
2. K. Bieman, S. Tsunakawa, J. Sannen, H. Feldmann, D. Dütting & H.G. Zachau, *Agnew. Chem. Int. Ed. Engel.* **5**, 560 (1966).
3. A.M. Lawson, R.N. Stillwell, M.M. Tacker, K. Tsuboyama & J.A. McCloskey, *J. Amer. Chem. Soc.* **93**, 1014 (1971).
4. Q.M. Porter, *Mass Spectrometry of Hererocyclic Compounds*, Wiley, New York (1985).
5. C. Hignite, in G.R. Waller (ed) *Biochemical Applications of Mass Spectrometry*, Wiley, New York pp 527-566 (1979).
6. P.B. Farmer, *Biomed. Environ. Mass Spectrom.* **17**, 143 (1988).
7. C.H. Hocart & J.P. Schlunegger, *Rapid Commun. Mass Spectrom.* **3**, 143 (1989)
8. C.C. Nelson & J.A. McCloskey, *J. Amer. Soc. Mass Spectrom.* **5**, 339 (1994).
9. C.C. Nelson Sr J.A. McCloskey, *J. Amer. Chem. Soc.* **114**, 3661 (1992).
10. E. Borck, B.S. Baliga & C.W. Gehrke, *Cancer Res.* **37**. 3362 (1977).
11. T.D. McClure, K.H. Schram, K. Nakano & T. Yasaka, *Nucleosides & Nucleotides* **8**, 1399 (1989).
12. J.I. Langridge, T.D. McClure, S. El-Shakawi, A. Fielding, K.H. Schram & R.P. Newton, *Rapid Commun. Mass Spectom.* **7**, 427 (1993).
13. J.I. Langridge, T.D. McClure, K.H. Schram. S. El-Shakawi, A. Fielding & R.P. Newton, in E. Constantin (ed) *Therapeutic Aspects and Analytical Methods in Cancer Research*, Amudes, Strasbourg, pp 96-117 (1994).
14. A. McCloskey, *Methods in Enzymol.* **193**, 825 (1990).
15. H. Pang, K.H. Schram, D.L. Smith, S.P. Gupta, L.B. Townsend & J.A. McCloskey, *J. Org. Chem.* **47**, 3923 (1982).
16. P.A. Limbach, P.F. Crain & J.A. McCloskey, *Nucleic Acids Res.* **22**, 2183 (1994).
17. P.F. Crain & J.A. McCloskey, *Nucleic Acids Res.* **24**, 98 (1996).
18. J.M. Grgson, P.F. Crain, E.G. Edmonds, R. Gupta, T. Hashizume, D.W. Phillipson & J.A.McCloskey, *J. Biol. Chem.* **268**, 10076 (1993).
19. M.W. Kilpatrick & R.T. Waller, *Zentrabl. Bakteriol. Mikrobiol. Hyg. 1 Abt Orig. C.* **3**, 79 (1982).
20. M. Ubukutata, K. Kimura. K. Isono, C. Nelson, J.M. Gregson & J.A. McCloskey, *J. Org. Chem.* **57**. 6392 (1992).
21. E. Prinsen, P. Redig, W. Van Dongen, E.L. Esmans & H. Van Onckelen, *Rapid Commun. Mass Spectrom.* **9**, 948 (1995).
22. H.K. Jajoo, S.M. Bennett & D.M. Kornhauser, *J. Chromatog.* **577**, 299 (1992).
23. D.L. Slowikowski & K.H. Schram. *Nucleosides & Nucleotides* **4**, 309 (1985).
24. R.P. Newton, in G. Dealty & D. Rickwood (eds) *Cell Biology Labfax*, Bios Scientific, Oxford. pp 153-166 (1992).
25. E.E. Kingston, J.H. Beynon & R.P. Newton, *Biomed. Mass Spectrom.* **11**, 367 (1983).
26. E.E. Kingston. J.H. Beynon, R.P. Newton & J.G. Liehr, *Biomed. Mars Spectrom.* **12**, 525 (1985).
27. R.P. Newton, N. Gibbs, C.D. Moyse, J.L. Wiebers & E.G. Brown, *Phytochem.* **19**. 1909 (1980).
28. R.P. Newton, E.E. Kingston, D.E. Evans, L.M. Younis & E.G. Brown, *Phytochem.* **23**, 1367 (1986.
29. R.P. Newton, S.G. Salih, B.J. Salvage & E.E. Kingston, *Biochem. J.* **221**, 665 (1984).
30. R.P. Newton, E.E. Kingston, N.A. Hakeem, S.G. Salih, J.H. Beynon & C.D. Moyse, *Biochem. J.* **236** 431 (1986).
31. R.P. Newton, D. Chiatante, D. Ghosh, A.G. Brenton, T.J. Walton, F.M. Harris & E.G. Brown, *Phytochem.* **28**. 2243 (1989).
32. R.P. Newton, T.J. Walton, S.A. Basaif, A.M. Jenkins, A.G. Brenton, D. Ghosh & F.M. Harris, *Org Mass Spectrom.* **24**. 679 (1989).
33. R.P. Newton, A.M. Evans, H.G. Hassan, F.M. Harris, A .G. Brenton & T.J. Walton, *Org. Mass Sprctrom.* **28**, 899 (1993).
34. R.P. Newton, T.J. Walton, A.G. Brenton, E.E. Kingston & F.M. Harris, *Rapid Commun. Mass Spectrom.* **3**, 178 (1989).

35. R.P. Newton, J.A. Khan, D. Ghosh, J.I. Langridge, A.G. Brenton, F.M. Harris & T.J. Walton, *Org. Mass Spectrom.* **26**, 447 (1991).
36. R.P. Newton, J.A. Khan, A.G. Brenton, J.I. Langridge, F.M. Harris & T.J. Walton, *Rapid Commun. Mass Spectrom.* **6**, 601 (1992).
37. R.P. Newton, A.M. Evans, J.I. Langridge, T.J. Walton, F.M. Harris & A.G. Brenton, *Analyt. Biochem.* **224**, 32 (1995).
38. T.J. Walton, D. Ghosh, R.P. Newton, A.G. Brenton & F.M. Harris, *Nucleosides & Nucleotides* **9**, 365 (1990).
39. D. Ghosh, R.P. Newton, A.G. Brenton, F.M. Harris, M.P. Donovan, E.G. Brown & T.J. Walton, *Analyt. Chimica Acta* **247**, 187 (1991).
40. R.P. Newton, M.A. Bayliss, A.C.R. Wilkins, D.E. Games, A.G. Brenton, J.I. Langridge, T.J. Walton & F.M. Harris, *J. Mass Spectrom. & Rapid Commun. Mass Spectrom.* S107 (1995).
41. R.P. Newton, E.E. Kingston & A. Overton, *Rapid Commun. Mass Spectrom.* **9**, 305 (1995).
42. R.P. Newton, N.A. Hakeem, B.J. Salvage, G. Wassenaar & E.E. Kingston, *Rapid Commun. Mass Spectrom.* **2**, 118 (1988).
43. K.H. Schram, *Methods in Enzymol.* **193**, 791 (1990).
44. J.I. Langridge, A.M. Evans, D. Ghosh, T.J. Walton, A.G. Brenton, F.M. Harris & R.P. Newton, *Analyt. Chimica Acta* **247**, 177 (1991).
45. J.I. Langridge, A.G. Brenton, T.J. Walton, F.M. Harris & R.P. Newton, *Rapid Commun. Mass Spectrom.* **7**, 293 (1993).
46. E. Witters, L. Roef, R.P. Newton, W. Van Dongen, E.L. Esmans & H. Van Onckelen, *Rapid Commun. Mass Spectrom.* **10**, 225 (1996).
47. D.R. Phillips, M. Uramoto, K. Isono & J.A.McCloskey, *J. Org. Chem.* **58**, 854 (1993).
48. S.C. Pomerantz, J.A. Kowalak & J.A. McCloskey, *J. Amer. Chem. Soc.* **4**, 204 (1993).
49. P.A. Limbach, P.F. Crain & J.A. McCloskey, *J. Amer. Soc. Mass Spectrom.* **6**, 19 (1995).
50. T.R. Covey, R.F. Bonner, B.I. Shushan & J. Henion, *Rapid Commun. Mass Spectrom.* **2**, 249 (1988).
51. L. Grothajn, H. Blocher & R. Frank, *Biomed. Mass Spectrom.* **12**, 514 (1985).
52. J.T. Stults & J.C. Masters, *Rapid Commun. Mass Spectrom.* **5**, 359 (1991).
53. J.A. Kowalak, S.C. Pomerantz, P.F. Crain & J.A. McCloskey, *Nucleic Acids Res.* **21**, 4577 (1993).
54. J.A. Kowalak, E. Bruenger, & J.A. McCloskey, *J. Biol. Chem.* **270**, 17758 (1995).
55. J.A. Kowalak, E. Bruenger, T. Hashizume, J.M. Peltier, J. Ofengand & J.A. McCloskey, *Nucleic Acids Res.* **24**, 688 (1996).
56. H.F. Noller, *J. Bacteriol.* **175**, 5297 (1993).
57. E. Nordhoff, A. Igendoh, R. Cramer, A. Overburg, B. Stahl, M. Karas, F. Hillencamp & P.F. Crain, *Rapid Commun. Mass Spectrom.* **6**, 771 (1992).
58. E. Nordhoff, F. Kirpekar, S. Hahner, F. Hillencamp & P. Roepstorff, in R.P. Newton & T.J. Walton (eds) *Applications of Mass Spectrometry in Plant Science Research*, Oxford University Press, Oxford, pp 86-101 (1996).
59. E. Nordhoff, F. Kirpekar, M. Karas, R. Cramer, S. Hahner, F. Hillencamp, K. Kristiansen, P. Roepstorff & A. Lezius, *Nucleic Acids Res.* **22**, 2460 (1994).
60. E. Nordhoff, R. Cramer, M. Karas, F. Hillencamp, F. Kirpekar, K. Kristiansen & P. Roepstorff, *Nucleic Acids Res.* **21**, 3347 (1993).
61. K. Schneider & B.T. Chait, *Proc. XXXXth ASMS Conf. Mass Spectrom. Washington*, 144 (1994).
62. K.J. Wu, T.A. Shalter & C.H. Becker, *Analyt. Chem.* **66**, 1637 (1994).
63. E. Nordhoff, M. Karas, R. Cramer, S. Hahner, F. Hillencamp, F. Kirpekar, A. Lezius, J. Muth, C. Meier & J.W. Engels, *J. Mass Spectrom.* **30**, 99 (1995).
64. M.C. Fitzgerald, L. Zhu & L.M. Smith, *Rapid Commun. Mass Spectrom.* **7**, 895 (1993).
65. M.C. Fitzgerald & L.M. Smith, *Annu. Rev. Biophys. Biomol. Struct.* **24**, 117 (1995).
66. U. Pieles, W. Zurcher, M. Schär & H.E. Moser, *Nucleic Acids Res.* **21**, 3191 (1993).
67. K. Schneider & B.T. Chait, *Nucleic Acids Res.* **23**, 1570 (1995).
68. F. Kirpekar, E. Nordhoff, K. Kristiansen, P. Roepstorff, S. Hahner & F. Hillencamp, *Rapid Commun. Mass Spectrom.* **9**, 525 (1995).
69. S.A. McLuckey, G.J. van Berkel & G.L. Glish, *J. Amer. Mass Spectrom.* **3**, 60 (1992).

70. M.J. Doktycz, S. Habibi-Gondarzi & S.A. McLuckey, *Analyt. Chem.* **66**, 3416 (1994).
71. S.A. McLuckey & S. Habibi-Gondarzi, *J. Amer. Chem. Soc.* **5**, 740 (1993).
72. E. Gentil & J. Banoub, *J. Mass Spectrom.* **31**, 83 (1996).
73. J.A. McCloskey & P.F. Crain, *Int. J. Mass Spectrom. Ion Processes* **118**, 593 (1992).
74. M.M Manson, *Brit. J. Ind. Med.* **37**, 317 (1980).
75. E.G. Van den Eeckhout, A. De Bruyn, H. Pepermans, E.L. Esmans, I. Vryens, J. Claereboudt, M. Claeys & J.E. Sinsheimer, *J. Chromatog.* **504**, 113 (1990).
76. E.G. Van den Eeckhout, J. Coene, J. Claereboudt, F. Borremans, M. Claeys, E.L. Esmans & J.E. Sinsheimer, *J. Chromatog.* **514**, 317 (1991).
77. J. Claereboudt, E.L. Esmans, E.G. Van den Eeckhout & M. Claeys, *Nucleosides & Nucleotides* **9**, 333 (1990).
78. F. Lemière, E.L. Esmans, W. Van Dongen, E. Van den Eeckhout & H. Van Onckelen, *J. Chromatog.* **647**, 211 (1993).
79. K. Vanhoutte, P. Joos, F. Lemière, W. Van Dongen, E.L. Esmans, M. Claeys & E. Van den Eeckhout, *J. Mass Spectrom.* **30**, 1453 (1995).
80. K. Vanhoutte, P. Joos, F. Lemière, W. Van Dongen, E.L. Esmans & E. Van den Eeckhout, *J. Mass Spectrom. & Rapid Commun. Mass Spectrom.* S143 (1995).
81. O.A. Boryak, M.V. Kosevich, V.S. Shelkovsky and Y.P. Blagoy, *Rapid Commun. Mass Spectrom.* **10**, 197 (1996).
82. K. O'Donaghue, T.A. Brown, J.F. Carter & R.P. Evershed, *Rapid Commun. Mass Spectrom.* **8**, 503 (1994).
83. M.K. Hellerstein & R.A. Neese, *Amer. J. Physiol.* **263**, E988 (1992).

MALDI MASS ANALYSIS OF OLIGONUCLEOTIDES

U. BAHR and M.KARAS
Institute of Physical and Theoretical Physics
Dep. of Instrumental Analytical Chemistry
University of Frankfurt
Theodor-Stern-Kai 7, 60590 Frankfurt
Germany

1. Introduction

Whereas MALDI-MS is almost universally applicable for the analysis of proteins and glycoproteins with detection limits of some fmol and upper mass limits of several hundred kDa the MALDI analysis of RNA or DNA has found to be more difficult in terms of sensitivity and accessible mass range. Progress in MALDI analysis of this class of compounds in the last years is mainly due to the development of new matrices and preparation techniques.
In the following the sample preparation for MALDI of nucleic acids is described and a short overview over the potential of the method for rapid screening and sequencing is given.

2. Sample preparation procedures

One problem in the analysis of oligonucleotides and nucleic acids arises from the polyanionic structure of the samples. Due to this nature, oligonucleotides are chemically stable only in salt form and are particular susceptible to collect impurity cations which are inevitably introduced during sample preparation steps such as extraction, digestion or purification. It is therefore not a surprise that oligonucleotide ions generated by MALDI show a distribution of molecular ions arising from varying number and kind of counter ions which are only partially exchanged against protons by the MALD ionization process. This simple

R. M. Caprioli et al. (eds.), Selected Topics and Mass Spectrometry in the Biomolecular Sciences, 455–464.

chemical approach also rationalizes that, even though oligonucleotides exist as polyanions in solution, they are equally well detected as singly charged negative or positive ions. Fig. 1a shows as example the spectrum of a 10-mer DNA in the

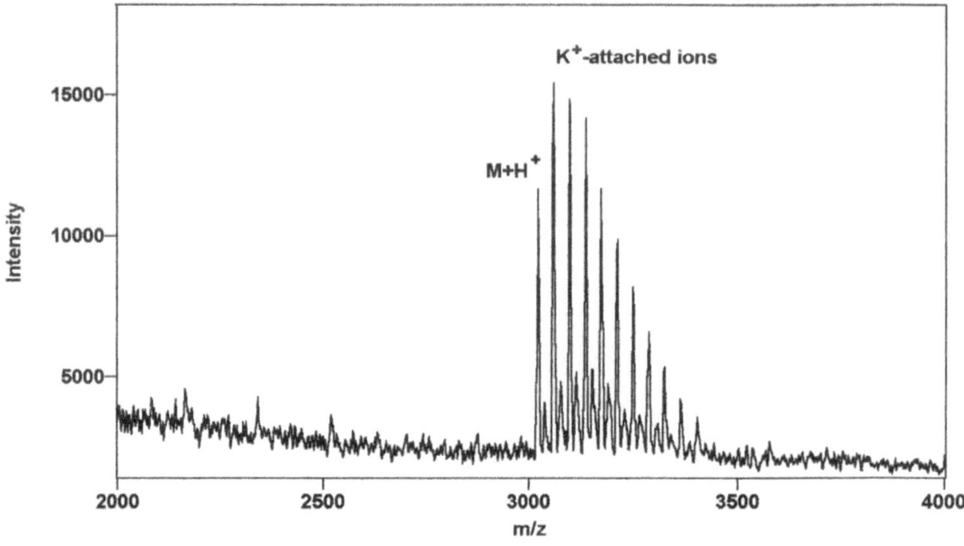

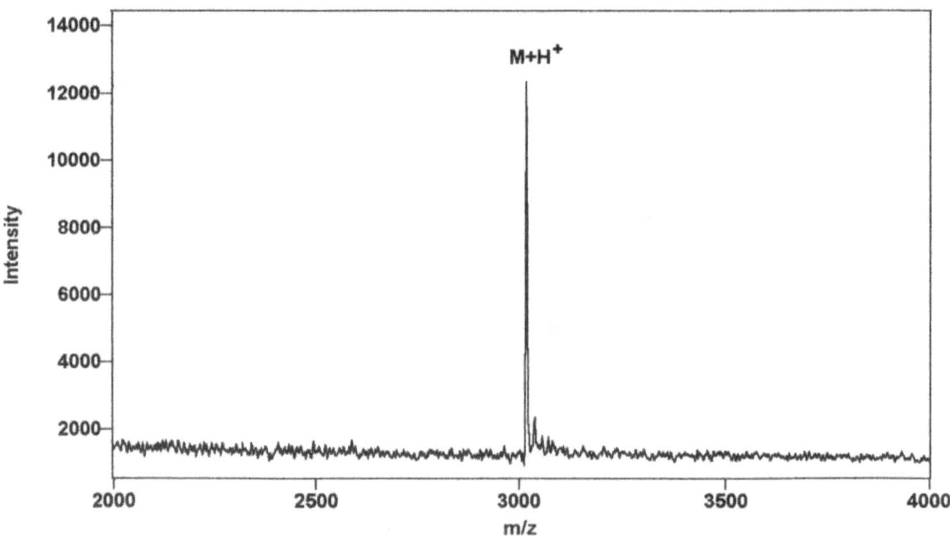

Fig. 1: MALDI spectrum of a DNA 10mer in the presence of 20 mM potassium chloride (top) and after desalting with cation exchange beads (bottom).

presence of 20 mM K^+-ions. The different ion peaks represent $[M-(n+1)H+nK]^+$ species, the protonated molecular ion peak is only of low intensity. Often additional substitution of H by sodium occurs. These effects complicate or prevent an accurate molecular mass determination especially in the higher mass range where the different species are no longer resolved.

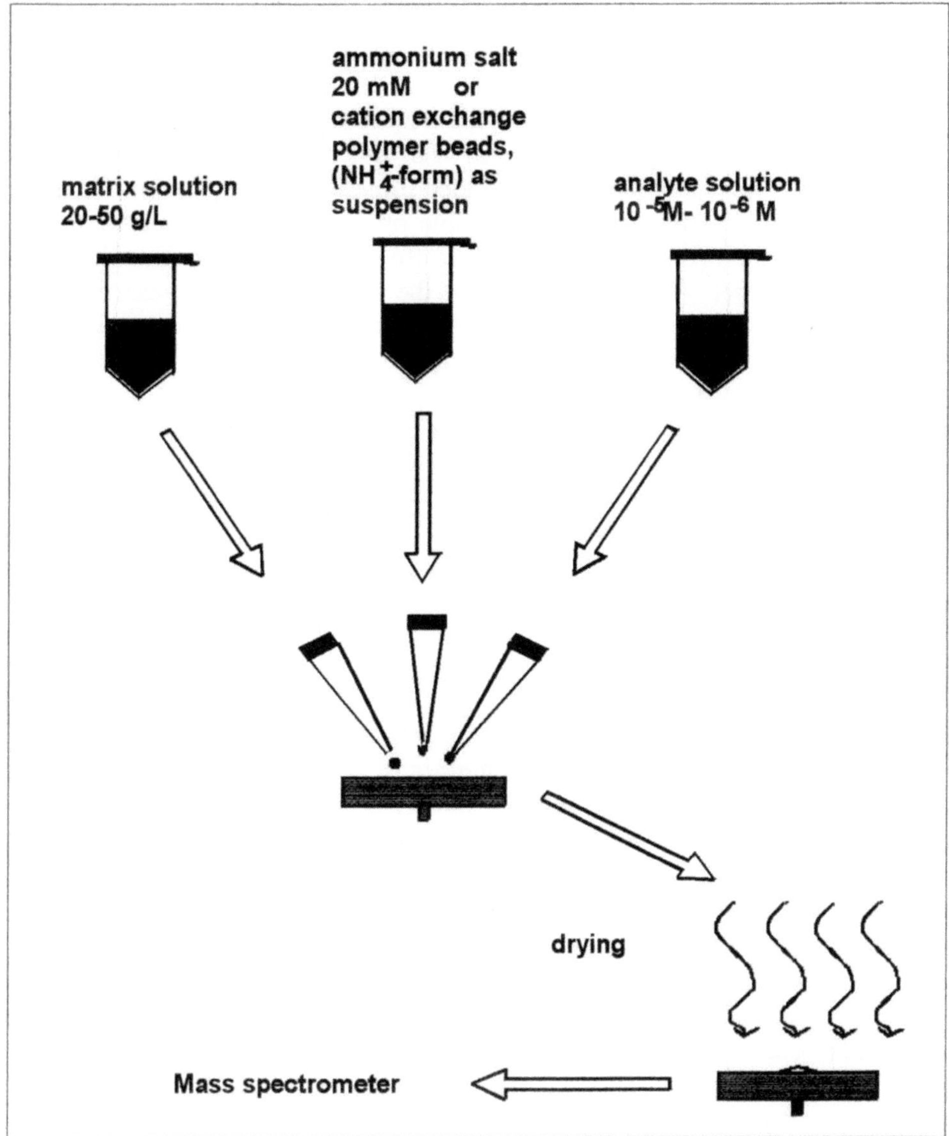

Fig. 2: Sample preparation procedure for MALDI mass analysis of oligonucleotides

There are different approaches to solve this problem. One is the addition of ammonium salts such as ammonium acetate, diammonium hydrogen citrate or diammonium-L-tartrate [1,2]. By addition of 20-100 mM ammonium salts no or only minor counter ion distribution is observed. This effect can be explained by formation of ammonium phosphate ion pairs by cation exchange in solution and dissociation into the free acid form of the phosphodiester group and the ammonium via proton transfer during the desorption process. Such a neutralization reaction by dissociation is not possible for alkali phosphate ion pairs.

The other possibility is the desalting of the matrix and analyte solutions with an appropriate cation exchange material [3]. For this NH_4^+-loaded cation exhange polymer beads are used which can be added directly to the solutions. To avoid salt contamination in the last step of sample preparation, e.g. from the sample target or the pipette tips some of these beads are added to the final sample/matrix droplet on the target. After evaporation of the solvent the beads can easily be removed prior to mass spectrometric analysis. Fig. 1b shows the spectrum of the 10mer after this procedure. Almost all cations are removed and only the protonated molecular ion is to be seen.

Another approach is the cation exchange using a „Nafion"-coated sample support [1,4,5] which is treated with concentrated ammonia to convert the surface-linked acidic functions ($-SO_3H$) into their ammonium salts or the use of nitocellulose instead of „Nafion" [6].

Fig. 2 shows schematically the preparation procedure used for MALDI mass analysis of RNA/DNA.

2.1 MATRICES

The matrices used for oligonucleotide analysis differ from the ones used for peptide and protein analysis. A large number of different substances has been tested and good results have been obtained at 337 nm wavelength with 2-aminobenzoic acid [3,7], 3- aminopicolinic acid [8], 3- hydroxypicolinic acid (3-HPA) [9], a 4:1 mixture of 3-HPA and picolinic acid [10]. Moreover 2,4,6 trihydroxyacetophenone [1], a 2:1 mixture of 2,4,6- and 2,3,4 trihydroxy-acetophenone [11] and 6-aza-thiothymine [8] all in combination with organic ammonium salts have been successfully applied for oligonucleotide analysis. The reason for the need of other matrices is higher lability of the oligonucleotides compared to other biomolecules leading to fragmentation, the amount of which is dependant on the matrix. The lability increases with increasing mass and is higher for DNA than for RNA [2] and strongly dependent on the base composition [2,12-14]. Fig. 3 shows a comparison of spectra of a 20mer with different matrices in a reflector TOF mass spectrometer. Whereas 3-HPA shows the protonated molecular ion peak and only small amounts of base cleavage, with 2,5-dihydroxybenzoic acid, the matrix most widely used for protein analysis,

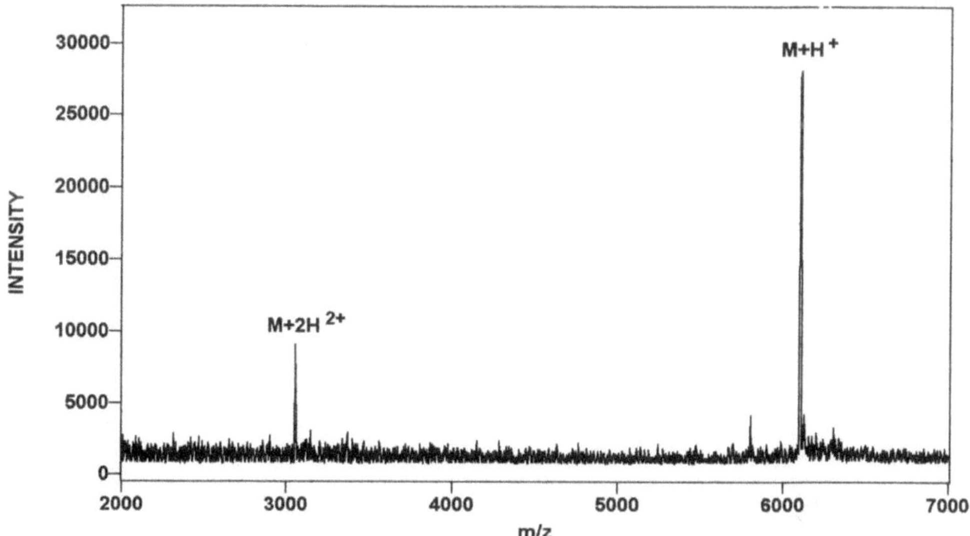

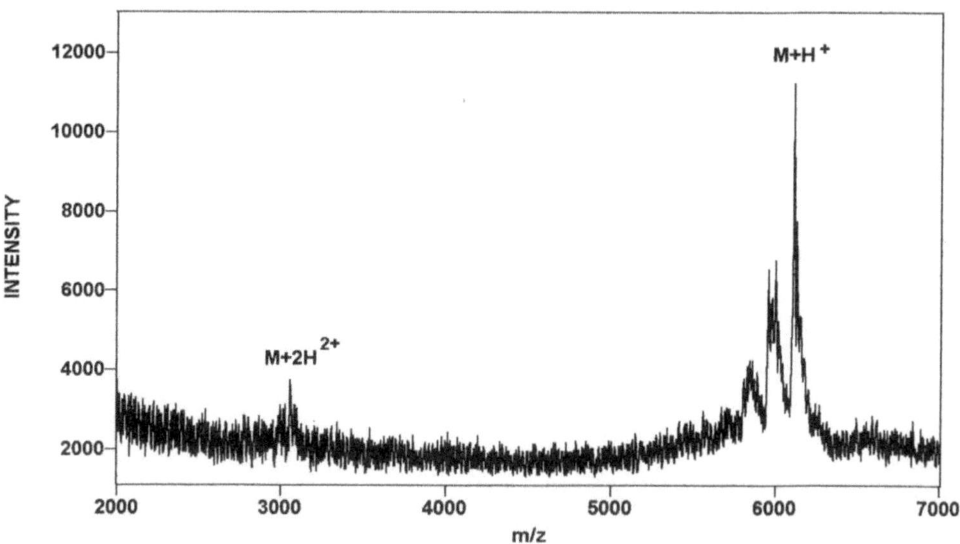

Fig. 3: MALDI mass spectra of a DNA 20mer obtained with a reflector instrument with 3-HPA (top) and 2,5-dihydroxy benzoic acid (bottom) as matrix

the fragmentation is drastically enhanced which is indicated by a broadening to the lower mass side of the peak and a decrease in peak intensity. Using a linear mass analyzer, the fragmentation cannot be observed because the fragments from metastable decay have the same velocity than the intact molecular ion and thus arrive at the same time at the detector [15]. Only when the fragmentation occurs at a shorter time scale which is the case for higher mass DNA fragmentation already occurs already in the ion source (prompt fragmentation) and can be observed in the linear mass analyzer. The significant reduction in ion fragmentation with 3-HPA made this matrix to the most widely used one for nucleotide analysis.
The largest DNA measured by MALDI-MS with 3-HPA so far has reached a 500mer [16].

2.2. POSITIVE/NEGATIVE ION MODE

As pointed out above oligonucleotides yield strong positive as well as negative ions. Fig. 4 shows the positive (top) and negative (bottom) ion spectrum of a 30mer DNA.

2.3. WAVELENGTHS

Although most applications for oligonucleotides have been performed using a N_2-Laser at a wavelength of 337 nm there are some applications with 266 nm, the quadrupled wavelength of a Nd-YAG-laser with picolinic acid [17], a 8:1 mixture of picolinic acid and 3-HPA [16] or 6-azathiothymine [8] as matrices. Good results have been obtained with an infrared laser emitting a wavelength of 2,94 µm. Using succinic acid or urea as matrix, some improvements in mass resolution and mass accuracy were received [3]. Another approach is the use of ice as matrix [18]. Also laser light in the visible range has been used [19,20].

3. Applications

3.1. MOLECULAR MASS DETERMINATION

MALDI mass spectrometry can be used as a quick and acurrate method for molecular mass determination. This has been demonstrated for a large number of synthetic oligonucleotides and nucleic acids. Among the modifications analyzed are different nucleobase modifications [21-26], conjugates with polyethylene glycol [27], oligonucleotide-peptide hybrids [28], phosphorothioate DNA [29,30,] and oligodeoxyribonucleotides with deoxyribose-methylphosphonate backbones [31]. With delayed extraction it is much easier now to

verify small differences in the sequence of oligonucleotides in the mass range between 1000-10000 Da [32]. Single-stranded as well as intact double-stranded DNA [33,34] have been analyzed.
A mass accuracy of 0.01 - 0.1% for samples up to 50mers can be obtained. The

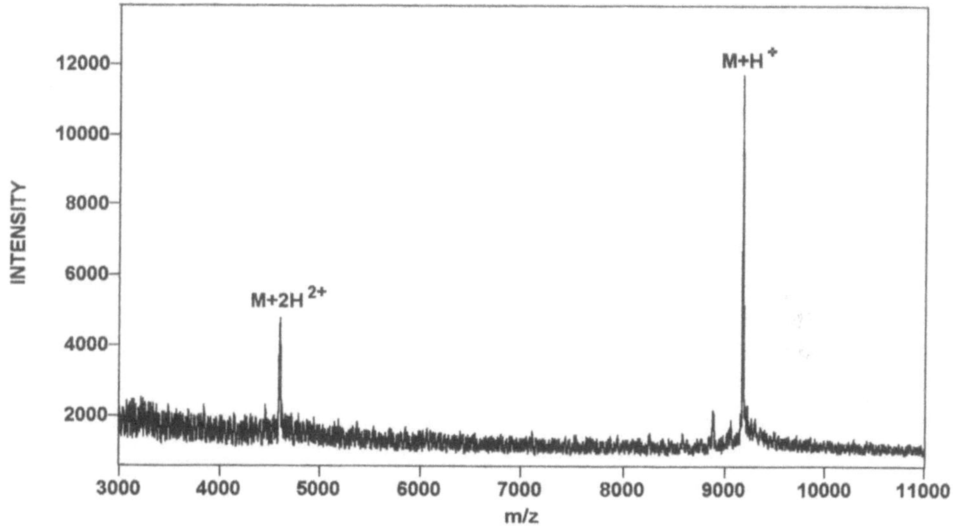

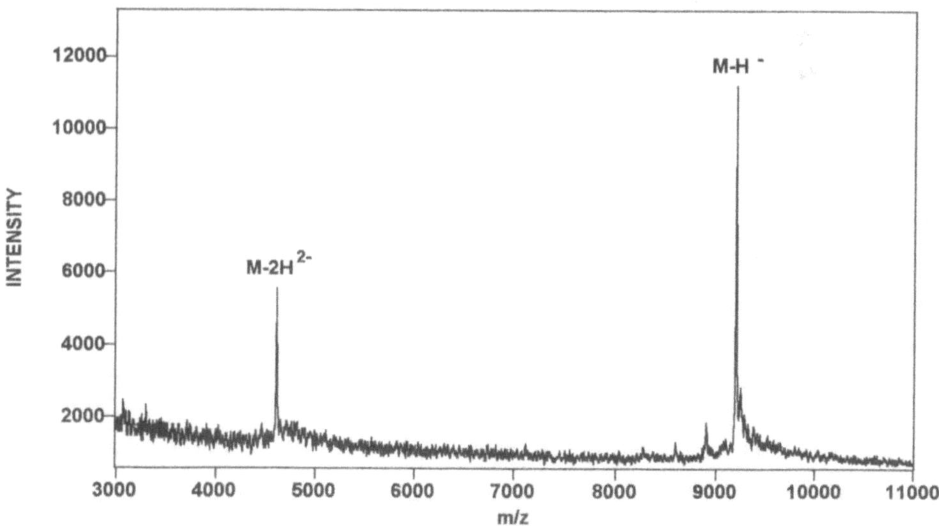

Fig. 4: MALDI mass spectra of a DNA 30mer in the positive ion mode (top) and the negative ion mode (bottom)

sensitivity of the method depends on the size of the molecule, only some fmol are needed for small oligonucleotides [35] whereas some pmol are needed for larger molecules. Here typically 1-30 pmol are loaded [2,17, 36,37].

3.2. OLIGONUCLEOTIDE SEQUENCING

As MALDI mass spectra show predominantly singly and doubly charged molecular ion species the method is ideally suited for analysis of mixtures and can be used for sequence analysis of oligonucleotides. There are different strategies for sequencing of oligonucleotides. One is the enzymatic digestion of the sample by exonucleases [38,39]. When the digestion is stopped after several minutes a sequence ladder of oligonucleotides differing in one nucleotide is obtained and can be analyzed by MALDI-MS. From the mass differences of adjecent peaks one can directly read the sequence. As the digestion usually is incomplete, always the last 3-5 nucleotides are not generated, two digestions have to be performed one from the 3' and one from the 5' end. It has been shown that using delayed extraction all fragments can be analyzed with a mass accuracy of ± 0.8 Da and the complete sequence can be red [39].
The sequence ladder generated by Sanger reaction can be analyzed by MALDI-MS as has been shown for synthetic oligonucleotide mixtures [40,41]. But since conventional Sanger reactions produce only low femtomoles per oligonucleotide, the detection limit of MALDI-MS has to be improved for routine analysis.
Another approach to obtain sequence information is the analysis of the metastable fragments (PSD-, post source decay analysis). This method which is often used for peptide sequencing has been used for analysis of a 5mer [42]. However, the dissociation is blocked in the presence of T and than the sequence information is incomplete.

4. References

1. Pieles, U., Zürcher, W. , Schär, M., and Moser, H.E. (1993) *Nuc. Acid Res.* **21**, 3191
2. Nordhoff, E., Cramer, R., Karas, M., Hillenkamp, F., Kirpekar, F., Kristiansen, K., and Roepsdorff, P. (1993) *Nuc. Acid Res.* **21**, 3347
3. Nordhoff, E., Ingendoh, A., Cramer, R., Overberg, A., Stahl, B., Karas, M., Hillenkamp, F., Crain, P.F. (1992) *Rapid Commun. Mass Spectrom.* **6**, 771
4. Bai, J., Liu, J.H., Lubman, D.M., Siemieniak, D. (1994) *Rapid Commun. Mass Spectrom.* **8**, 687
5. Bai, J., Liu, J.H., Liang, X., Zhu, Y., Lubman, D.M. (1995) *Rapid Commun. Mass Spectrom.* **9**, 1172
6. Liu, J.H., Bai, J., Zhu, Y., Liang, X., Siemieniak, D., Venta, P.J., Lubman, D.M. (1995) *Rapid Commun. Mass Spectrom.* **9**, 735

7. Wang, B.H., Biemann, K. (1994) *Anal. Chem.* **66,** 1918
8. Taranenko, N.I. Tang, K., Allman, S.L. Ch'ang, L.Y., Chen, C.H. (1994) *Rapid Commun. Mass Spectrom.* **8** , 1001
9. Wu, K.J., Steding, A., Becker, C.H. (1993) *Rapid Commun. Mass Spectrom.* **7**, 142
10.Liu, Y.H., Bai, J., Lubman, D.M., Venta, P.J. (1995) *Anal. Chem.* **67**, 3482
11.Zhu, Y.F., Chung, C.N., Taranenko, N.I., Allman, S.L., Martin, S.A., Haff, L., Chen, C.H. (1996) *Rapid Commun. Mass Spectrom.* **10**, 383
12. Schneider, K., Chait, C.T. (1993) *Org. Mass Spectrom.* **28**, 1353
13. Zhu, L., Parr, G.R., Fitzgerals, M.C., Nelson, C.M., Schmith, L.M. (1995) *J. Am. Chem. Soc.* **117**, 6048
14.Nordhoff, E., Karas, M., Cramer, R., Hahner, S., Hillenkamp, F. Kirpekar, F., Lezius, A., Muth, J., Meier, C., Engels, J.W. (1995) *J. Mass Spectrom.* **30**, 99
15.Karas, M., Bahr, U., Stahl-Zeng, J. (1996) in *Large Ions: The Vaporization, Detection and Structural Analysis*, Ed.: Bear, T., Ng, C.Y., Powis, I. p. 27
16.Tang, K., Taranenko, N.I., Allman, S.J., Chang, L.Y., Chen, C.H. (1994) *Rapid Commun. Mass Spectrom.* **8**, 727
17.Tang, K., Taranenko, N.I., Allman, S.J., Chen, C.H., Chang, L.Y., Jacobson, K.B., (1994) *Rapid Commun. Mass Spectrom.* **8**, 673
18.Williams, P. (1994) *Int. J. mass Spectrom. Ion Proc.* **131**, 335
19.Tang, K., Allman, S. L., Chen, C.H. (1992) *Org. Mass Spectrom.* **27**, 1389
20.Tang, K., Allman, S.L., Chen, C.H. (1992) *Rapid Commun. Mass Spectrom.* **6**, 365
21.Wang, B.H., Biemann, K., (1994) *Anal. Chem.* **66**, 1918
22.Christian, N.P., Colby, S.M., Giver, L. Houston, C.T., Arnod, R.J., Ellington, A.D., Reilly, J.P. (1995) *Rapid Commun. Mass Spectrom.* **9**, 1061
23.Hettich, R.L., Buchanan, M.V. (1991) *Int. J. Mass Spectrom. Ion Proc.* **111**, 365
24.Hettich, R.L., Stemmler, E.A., (1996) *Rapid Commun. Mass Spectrom.* **10**, 321
25.Stemmler, E.A., Buchanan, M.V., Hurst, G.B., Hettich, R.L. (1994) *Anal. Chem.* **66,** 1274
26.Stemmler, E.A., Buchanan, M.V., Hurst, G.B., Hettich, R.L. (1995) *Anal. Chem.* **67,** 2924
27.JäschkeA., Furste, J.P., Nordhoff, E., Hillenkamp. ., Cech, D., Erdmann, V.A. (1994) *Nucleic Acid Res.* **22**, 4810
28.Sochacki, M., Bergmann, F., Tam, S., Michel, H. (1995) *Proc.of the 43th ASMS Conf.on Mass Spectron and Allie Topics,* Atlanta, p.598
29.Wang, B.H.,Bourque, A.J.,Belenky A.B.Cohen, A.S., (1994) *Proc.of the 42th ASMS Conf.on Mass Spectron and Allie Topics,* Chicago, p. 145
30.Wang, B.H.,Bourque, A.J.,Belenky A.B.Cohen, A.S., (1995) *Proc.of the 43th ASMS Conf.on Mass Spectron and Allie Topics,* Atlanta, p.580

31. Keough, T., Baker, T.R., Dobsen, R.L. Lacey, M.P. Riley, T.A. Hasselfield, J.A., Hesselberth, P.E. (1993) *Rapid Commun. Mass Spectrom.* **7**, 195
32.Juhasz, P., Roskey, M.T., Smirnov, I.P., Lawrence, A.H., Vestal, M.L., Martin, S.A. (1996) *Anal. Chem.* **68**, 941
33.Lecchi, P., Le, H.M.T., Panell, L.K. (1995) *J.Am. Soc. Mass Spectrom.* **6**, 972
34.Benner, W.H. Horn, D., Katz, J., Jaklevic J. (1995) *Rapid Comm. Mass Spectrom.* **9**, 537
35. Shaler, T.A., Tan, Y., Wickham, J.N., Wu, K.J., Becker, C.H. (1995) *Rapid Commun. Mass Spectrom.* **9**, 942
36.Wu, K.J., Shaler, T.A., Becker, C.H. (1994) *Anal. Chem.* **66**, 1637
37.Kirpekar, F., Nordhoff, E., Kristiansen K., Roepsdorff, P., Lezius, A., Hahner, S., Karas, M., Hillenkamp F. (1994) *Nuc. Acid Res.* **22**, 3866
38.Pieles, U., Zurcher, W., Schar, M., Moser, H. (1993) *Nuc. Acid Res.* **21**, 3191
39.Smirnov, I.P., Roskey, M.T., Juhasz, P., Takach, E.J., Martin, S.A., Haff, L.A. (1996) *Anal. Biochem.* **238**, 19
40. Fitzgerald, M.C., Zhu, L., Smith L.M. (1993) *Rapid Commun. Mass Spectrom.* **7**, 895
41.Cohen, A.S., Smisek, D.L., Wang, B.H. (1996) *Advances in Chromatography, Vol. 36,* (P.R. Brown, E. Grushka, Eds.) Marcel Dekker, New York, p 127
42-Kirsch, D., Spengler, B., Kaufmann, R. (1994) *Proc. 42nd ASMS Conf. on Mass Spectrom. and Allied Topics,* Chicago, p. 821

DNA ADDUCTS IN HUMAN CARCINOGENESIS

Clues to Etiology and Mechanisms

H. BARTSCH

German Cancer Research Center
Im Neuenheimer Feld 280
D-69120 Heidelberg, Germany

Abstract

The rationale and biological basis for measuring DNA adducts as relevant biomarkers of the adverse health effects of carcinogens in humans are summarized. By integrating highly sensitive and specific analytical detection methods into well-designed epidemiological studies, the following contributions to etiology and mechanisms in human carcinogenesis can be gained: (i) Adduct measurement allows new insights into disease etiology as the adduct analysis permits the identification of hitherto unknown DNA-reactive chemicals and of carcinogenic components in complex exposures, thus increasing the power to establish causal relationships. (ii) Highly exposed individuals can be more timely identified, leading to avoidance or lowering of exposure to carcinogenic risk factors. (iii) Subgroups in the population (pharmacogenetic variants) that are more susceptible to the action of environmental carcinogens, due to genetic polymorphism of xenobiotic-metabolizing enzymes are identifiable by a combination of genotyping and DNA-adduct measurements. (iv) Repeated applications of dosimetry methods for macromolecular adducts can evaluate the effectiveness of intervention studies, either by reduction of carcinogen exposure or through chemopreventive strategies. (v) Incorporation of DNA-adduct measurements (and other data on critical parameters involved in carcinogenesis) will reduce (a) the enormous uncertainties associated today with high-to-low dose and species-to-species extrapolation and (b) yield information on interindividual and interethnic differences among human populations that can be integrated into risk assessment procedures. (vi) the role of specific carcinogens may be implicated in the etiology of cancers retrospectively by analyzing mutational spectra in tumors that arise from exogenous and endogenous mutagens by reaction with DNA; this is based on the assumption that carcinogens and their adducts leave specific mutational fingerprints in critical genes involved in human carcinogenesis, and that fingerprints are still found in tumors or precursor lesions so that they could be used to trace etiological agents that caused the tumors several decades ago. (vii) Adducts may be useful to study pathogenesis and preventive

R. M. Caprioli et al. (eds.), Selected Topics and Mass Spectrometry in the Biomolecular Sciences, 465–476.

approaches of degenerative diseases other than cancer. Recent advances in these areas are illustrated by specific examples from the author's own research activities, covering benzo[a]pyrene diolepoxide-, etheno- and 3-alkyladenine-DNA adducts.

1. Rationale for Adduct Measurements

DNA- and protein-bound carcinogen adducts reflect the amount of carcinogen that interacted with cellular macromolecules at the target site or within established surrogate (biologically effective dose). Protein and in particular DNA adducts are mechanistically more relevant to carcinogenesis than internal dose, since it takes into account differences in metabolism of the carcinogen in question as well as the extent of repair of DNA adducts. The biological basis for measuring DNA adducts as relevant dosimeters of biological effects is derived from extensive experimental data supporting their role in the initiation and possibly the progression of cancer. Supportive evidence includes: (i) 80-90% of identified (or suspected) human chemical carcinogens (after metabolic conversion) react with nucleic acids and proteins to form macromolecular products (adducts); (ii) carcinogen-DNA adducts represent the initiating events leading to mutations and malignant transformation which is the central theorem of carcinogenesis; (iii) the carcinogenic potency of a large number of chemicals was shown to be proportional to the ability to bind to liver DNA (termed covalent binding index).

2. DNA Adducts as Biomarkers for Exposure, Effects or Both

Biological effect markers are defined as indicators of irreversible damage resulting from toxic interactions at the target (or at a surrogate) organ site; such adverse effects should be known or believed to be pathologically linked to cancer development. As DNA adducts do not represent irreversible lesions (because they undergo repair which however may not be complete), they are not in the strict sense biological effect markers. However, as carcinogen exposure is linked to cancer risk, DNA adducts as markers of exposure (of which a part is converted into permanent mutations) must per definition also be associated with cancer risk. This has been shown for certain carcinogens and their adducts, when certain critical toxico-kinetic parameters are taken into account (steady state of concentration of the adduct, measurement of the critical adduct among other biologicallly less relevant adducts, organ- and cell-specificity and persistence of adducts after cessation of exposure).

3. Advantages and Disadvantages of DNA- Compared to Protein-Adduct Measurement

For the purpose of exposure monitoring both endpoints can be used as long as the proportionality of response in target organ(s) *vs.* surrogate tissue(s) is shown (which has to be determined for each carcinogen separately). The advantage of some protein adduct measurements is that they often reflect distant or past exposure (of several months), while DNA adducts (at least the majority) are rapidly removed or lost after cessation of exposure within days. However, a small fraction of DNA adducts may persist for several months or years; thus, subpopulations of non-dividing cells or those with slow repair seem to exist but need to be characterized.

4. Value of Macromolecular Adducts Compared to Other Biomarkers

None of the somatic genetic or cytogenetic effect markers is chemical- or exposure-specific. Thus, only macromolecular adducts allow the identification of the structure and elucidation of the genotoxic exposure sources in question. Also, other cytogenetic markers are more easily affected by lifestyle and environmental confounders that often act as uncontrolled (or uncontrollable) variables in molecular epidemiologic studies. In addition, at equal exposure DNA and protein adducts represent measures for the hosts's metabolic activating capability of carcinogens and can now be used to determine the effect of polymorphic enzymes involved in carcinogen metabolism on macromolecular interactions by a given class of carcinogens.

5. Incorporation of Adducts into Disease Epidemiological Studies

Cross-sectional and longitudinal studies in disease epidemiology allow to assess the relationship between carcinogen exposures and biomarker (adduct) levels. These types of study design preclude to establish a causal relationship betweeen exposure or biomarker and disease. However, the studies allow to demonstrate the ability of adduct measurements to detect, quantify and structurally elucidate specific DNA damage in humans. The findings in these studies include the importance of background exposures manifested in so-called 'unexposed' control populations and a significant interindividual variation in adduct levels in persons with comparable exposure. In occupational settings dramatic effects of exposure on adduct levels have been observed, and significant correlations have been seen between DNA adducts and effect markers like mutations in various critical cell growth-controlling genes. Studies on geographical variations of a disease, for example, hepatocellular carcinoma and aflatoxin B_1 dosimetry (albumin adduct, urinary aflatoxin B_1-N^7-guanine adduct) have allowed to assess the contribution of chemical carcinogens *vs.* other cancer risk factors, i.e. hepatitis B virus infection.

Case-control studies in disease epidemiology provide a good opportunity to explore the role of markers as risk factors for cancer as well as in evaluating their value in assessing mechanisms of carcinogenesis. Again, such studies cannot establish causal sequence between marker formation and cancer. Especially as the latency periods from exposure to cancer induction extend over several decades, adduct measurements are not relevant to the present risk of cancer, unless exposure has been continuous.

An optimal study design to be used to establish causation is a nested case-control study, using questionnaire data and biological sample collection prior to disease manifestation. Once diagnosis of cancer is made, cases are matched to appropriate controls and their stored samples analyzed. The predictive value of the specific markers can thus be determined in biological samples collected prior to clinical disease.

6. Contributions of Adduct Measurements to Disease Etiology and Pathogenesis

(i) Adduct measurement allows new insights into disease etiology as the adduct analysis permits the identification of hitherto unknown DNA-reactive chemicals and of carcinogenic components in complex exposures, thus increasing the power to establish causal relationships in molecular epidemiology. (ii) Highly exposed individuals can be more timely identified, leading to avoidance or lowering of exposure to carcinogenic risk factors. (iii) Subgroups in the population (pharmacogenetic variants) that are more susceptible to the action of environmental carcinogens, due to genetic polymorphism of xenobiotic-metabolizing enzymes are identifiable by a combination of genotyping and DNA-adduct measurements. (iv) Repeated applications of dosimetry methods for macromolecular adducts can evaluate the effectiveness of intervention studies, either by reduction of carcinogen exposure or through chemopreventive strategies. (v) Incorporation of DNA-adduct measurements (and other data on critical parameters involved in carcinogenesis) will reduce (a) the enormous uncertainties associated today with high-to-low dose and species-to-species extrapolation and (b) yield information on interindividual and interethnic differences among human populations that can be integrated into risk assessment procedures. (vi) the role of specific carcinogens may be implicated in the etiology of cancers retrospectively by analyzing mutational spectra in tumors that arise from exogenous and endogenous mutagens by reaction with DNA; this is based on the assumption that carcinogens and their adducts leave specific mutational fingerprints in critical genes involved in human carcinogenesis, and that fingerprints are still found in tumors or precursor lesions so that they could be used to trace etiological agents that caused the tumors several decades ago. (vii) Adducts may be useful to study pathogenesis and preventive approaches of degenerative diseases other than cancer.

7. Effect of Genetic Polymorphisms on Carcinogen Metabolism, Adducts and DNA Repair

Increased susceptibility to cancer may result from several host factors, including differences in metabolism and DNA repair. Since many human carcinogens require metabolic activation before binding to DNA, individual features of carcinogen metabolism could play an essential role in the development of environmentally and occupationally induced cancer. Variations in an individual's metabolic phenotype have been detected in a variety of enzymes involved in activation and detoxification of chemical carcinogens. This phenotypic metabolic variation has now been related to genetic polymorphism. A growing number of genes encoding carcinogen-metabolizing enzymes have been identified and cloned. Consequently, there is increasing knowledge of the allelic variants of genetic defects which give rise to the observed variation. Development of rather simple new techniques has enabled identification of an individual's genotype for a variety of metabolic polymorphisms with precision. Also, new polymorphisms have been recognized. Thus, recent knowledge of the genetic basis for individual metabolic variation has opened new possibilities for studies focusing on increased susceptibility to environmental cancer. Remarkable variation in metabolic phenotypes and genotypes has also been reported for different ethnic and/or geographic populations. The strong interethnic variation has been underlined as a major obstacle for extrapolation of results between different ethnic groups.

Given the number and variability in expression of carcinogen-metbaolizing enzymes now identified and the complexity of chemical exposures, assessment of a single polymorphic enzyme or genotype may not be sufficient. A number of studies have tried to relate metabolic or, more recently, genotype to cancer risk. These efforts are presently extended to studies on various other biomarkers or carcinogen exposure and effects to include DNA adducts, urinary mutagens or mutations in p53. Only a few studies have so far been focused on the analysis of combined genotypes. These new developments and early findings suggest that determination of metabolic 'at risk' genotypes in combination with levels of DNA adducts in target (surrogate) tissues should allow to identify more susceptible individuals/subgroups in carcinogen-exposed populations.

DNA repair systems such as nucleotide excission repair, O^6-methylguanine-DNA methyltransferase and mismatch repair systems are operating in human cells. Recent molecular analysis has shown that mutated genes involving nucleotide excission repair cause Xeroderma pigmentosum and its high-rate occurrence of skin cancers. Also, XPA repair gene deficient transgenic animals are highly susceptible to ultraviolet B and 7,12-dimethylbenz[a]anthracene-induced carcinogenesis. A defective mismatch repair system is thought to be related to hereditary non-polyposis colorectal cancer, where patients inherit (a) mutated mismatch repair gene(s) through the germ line.

Genetic polymorphism in these DNA repair genes leading to impaired functions may lead to dramatic consequences when DNA alkylation damage, DNA mismatches and DNA loops are not repaired prior to cell replication. Thus, a search for germ line mutation or for somatic mutations caused by exposure to carcinogens in these repair genes would yield another strategy to identify high-risk-subjects that especially in the case of bi-allelic mutation may lead to functional impairment of the repair protein.

8. Association of DNA Adducts with Cancer Risk

Clearly not all DNA adducts are associated with the same cancer risk. A comparison of DNA adducts derived from alkylating agents, aflatoxins and aromatic amines in target tissues of experimental animals (that induced 50% tumor incidence, TD_{50}) revealed a hundredfold range in the ability of DNA adducts to induce the same tumor incidence. This makes a prediction of tumor induction potential of unknown DNA adducts difficult. Unfortunately, most available assays for DNA adduct determination provide information on the total amount of adducts and rarely are capable of pinpoiting the critical adducts from DNA.

Because of the multistage and complex nature of human carcinogenesis carcinogen-macromolecular adducts *per se* will not be a precise and quantitative predictor of individual cancer risk. Therefore, risk estimation is confined to a group level.

9. Background Adduct Levels: Sources, Variations and Implications with respect to Exposure and Cancer Risk

As more sensitive analytical methods are becoming available that allow to examine DNA and protein adducts (previously thought to be chemical-specific), these are now being detected in unexposed humans and untreated animals. Lipid peroxidation products often occur in response to oxidative stress and have been associated with a variety of pathological conditions including cancer. Recently developed sensitive methods have shown that lipid peroxidation products are formed in human tissues that can react with DNA to yield adduct levels that are close to those formed by environmental carcinogens like polycyclic aromatic hydrocarbons (PAH) or aromatic amines. These background adducts have been shown in general to increase with age (both in humans and in experimental animals). There seems to be a great variation between human subjects in particular, if they were affected by disease states producing oxidative stress, including chronic inflammatory processes, nutritional deficiencies, alcohol abuse and metal storage disorders. In addition to DNA-reactive lipid peroxidation products, oxidized DNA bases have been recognized prior as common lesions that occur more frequently in cells where antioxidant defence mechanisms are impaired. The biological relevance of these DNA lesions like exocyclic DNA adducts and oxidatively damaged bases (8-oxo-guanine) is underlined by the fact that they are highly miscoding lesions and specific DNA repair enzymes are recognizing these promutagenic DNA modifications. Moreover, there is increasing evidence that carcinogens themselves induce oxidative stress and as a consequence, in addition to agent-specific DNA adducts, an array of oxidative damage DNA base are produced. Although it must be assumed that both types of lesions play a role in carcinogenesis, at present their individual contribution (e.g., on the shape of the dose-response curve for a given carcinogen in a given organ) cannot be discerned.

Background adducts may not only arise from endogenous sources but also fiom cryptic genotoxic environmental exposures, and thus DNA and protein adducts offer the possibility of identifying hitherto unknown sources of potential carcinogen exposures.

Some questions that have not been answered are: (i) the significance of endogenously formed adducts in human cancer in particular in relation to spontaneous tumors; (ii) whether the number of environmentally induced cancers has been overestimated and those arising from endogenous processes have been ignored; (iii) if chemical exposure does not increase background adducts, does the presence of endogenous DNA adducts affect the mathematical low-dose extrapolation (assuming low-dose linearity) that is currently being used in mathematical risk extrapolation procedures?

10. Research Results on Specific DNA Adducts

Sensitive methods for quantifying DNA adducts from (i) benzo[*a*]pyrene (BP), (ii) alkylation exposure, and (iii) etheno(ε)-DNA adduct-forming chemicals were developed and applied to human subjects. The aims were to identify hitherto unknown sources and mechanisms of exogenous and endogenous DNA damage and to examine the effect of drug polymorphism on carcinogen-DNA adduct levels. A full acount of these studies has been published elsewhere [1].

10.1. A FLUORIMETRIC ASSAY FOR BENZO[A]PYRENE DIOLEPOXIDE (BPDE) ADDUCTS IN LUNG TISSUE AND HUMAN WHITE BLOOD CELL DNA: EFFECT OF ENZYME POLYMORPHISM

The formation of smoking-related DNA adducts in human lung tissue appears to be a good dosimetric exposure marker. Smokers have significantly elevated levels of aromatic and/or hydrophobic adducts compared with nonsmokers and from some studies it is evident that adduct levels are linearly related to total smoking exposure. A major representative of PAH is BP, present in tobacco smoke and ambient air in industrialised areas. BP is converted into phenolic metabolites, such as 3-OH-BP and BP-7,8-diol, in human lung mainly by a cytochrome P450(CYP)1A1-mediated process. Secondary metabolism, involving epoxide hydrolase and other CYP isoforms, leads to the formation of the highly reactive (+)-*anti*-BPDE. This metabolite has been shown to bind to genomic DNA, activate oncogenes or other critical genes and is likely to be a causative factor in several types of smoking-related cancers. Using an improved HPLC-fluorescence assay [2], the levels of the biologically most active (+)-*anti*-BPDE bound to DNA can now be quantified through the release of the respective BP-tetrol from lung tissue DNA and from lymphocyte DNA [3].

This assay with a detection limit of 1 BPDE adduct/10^8 unmodified nucleotides in WBC DNA was validated for quantitation of BPDE adducts in DNA of human white blood cells (WBC). The quantity of WBC DNA required depends on the level of modification and can vary from 5 to 500 μg of DNA.

With these improved methods to quantify very low levels of BP-DNA adducts, their dependence on a metabolic phenotype/genotype related to cytochrome P450 (1A1) and glutathione *S*-transferase (μ1) polymorphism was examined. BPDE-DNA adduct levels in bronchial tissue of subjects with high CYP1A1 inducibility (phenotype determined by immunohistochemistry for cytochrome P4501A1 [4] and GSTM1 genotype status (null or wild type determined by PCR according to [5]) were measured in lung cancer patients who were recent smokers. The combination of 'at risk' geno/phenotypes had ~ 100-fold higher BPDE-DNA adduct levels in bronchial tissue than the GSTM1 wild type at a similar smoking dose [1]. Such a difference was not seen in lung parenchyma tissue. Although these preliminary results need to be verified in a larger cohort, they suggest that genetic polymorphism of carcinogen-metabolizing enzymes effects PAH-DNA binding in the target tissue of smokers and likely cancer risk.

BPDE-DNA adduct levels in peripheral mononuclear cells form coke oven workers were determined and an enhancing effect of smoking could be demonstrated [6]. BPDE-DNA adducts were measured in coke oven workers (exposed to PAH) and non-exposed controls (each group consisting of smokers and nonsmokers). The mean level in workers was ~ 15 times higher than in non-exposed controls. The interindividual variation was ~ 200-fold in workers and only sixfold in controls. Smoking workers had ~ 3.5 times more adducts than nonsmokers. The increased formation and high variability of BPDE-DNA adducts in smoking workers suggest variations in genetically controlled PAH metabolism, leading to synergistic effects in PAH-DNA binding and possibly also in cancer risk.

This type of PAH-DNA modification may not only initiate carcinogenesis, but could also be related to the pathogenesis of smoking-related degenerative cardiovascular diseases. BPDE- and other PAH-DNA adducts have for the first time been detected in human atherosclerotic lesions [7].

Thus, by establishing a 'risk profile', there is hope that more susceptible subjects in carcinogen-exposed populations may be identifiable through genotyping assays for polymorphic drug-metabolizing enzymes, in combination with molecular dosimetry of critical carcinogen-DNA adducts in target (surrogate) cells and analyses of the resulting mutational spectra in critical cancer target genes.

10.2 HUMAN EXPOSURE TO ALKYLATING AGENTS USING EXCRETED DNA-BASE ADDUCTS

Alkylation at N-3 of adenine is a major route of DNA adduct formation for many alkylating carcinogens. The resulting 3-alkyldeoxyadenosines are unstable and rapidly depurinate either spontaneously, or *via* the action of specific DNA glycosylases to give the corresponding 3-alkyladenines (3-alkAde) that are excreted in urine. This phenomenon formed the basis of an approach which involves the determination of urinary 3-alkAde using a combination of immunoaffinity purification and gas chromatography-mass spectrometric (GC-MS) analysis [8-10], and the subsequent application of this noninvasive methodology for the measurement of human exposure to alkylating carcinogens for use in epidemiological studies.

A method for the simultaneous analysis of several different 3-alkyladenines has been developed. Immunoaffinity gel was prepared by coupling a monoclonal antibody, which cross-reacts with a range of 3-alkyladenines [11], to protein A-Sepharose CL-4B [12]. Columns were prepared in a similar way to that described for 3-MeAde [13] and analytical procedures for 3-MeAde, 3-ethyladenine (3-EtAde), 3-(2-hydroxyethyl)adenine (3-HOEtAde) and 3-benzyladenine (3-BzAde) have been developed. Deuterated internal standards were synthesized and were routinely added to urine samples prior to immunoaffinity purification. Selected ion monitoring using a low resolution quadruple GC-MS system was used to quantitate the 3-alkAde. Using this approach, background levels of 3-alkAde in urine of subjects on free-choice and standardized diets were determined.

To examine whether tobacco smoking increased 3-EtAde excretion, 24-hr urine samples were collected over a 15-day period from two smokers, and the number of cigarettes smoked during each 24-hr period was recorded. A dose-dependent increase of 3-EtAde excretion with cigarette consumption (nicotine intake) was observed. Analysis of tobacco smoke only revealed the presence of trace levels of preformed 3-EtAde (<1% of observed urinary levels). When cigarette smoke was passed through an aqueous solution of calf thymus DNA, a dose-dependent increase of 3-EtAde formation in DNA was observed which indicated the presence of (a) hitherto unknown direct-acting ethylating agent(s). The identity of the(se) agent(s) is currently under investigation [14]. Ethylated DNA bases may be important lesions in tobacco-associated lung cancer as smoking-related O^4-ethylthymine was detected in human lung tissue for the first time [15]. Mean levels were twofold higher in current smokers than in ex-smokers. Thus, the use of excreted DNA adducts such as the 3-alkAde is now both technologically feasible and informative in relation to human exposure to alkylating agents. In situations where exposure to alkylating agents may be from both exogenous and endogenous sources the determination of urinary DNA adducts should allow a reliable determination of the whole body dose of active agent(s).

10.3 ETHENO-DNA ADDUCTS FROM ENDOGENOUS AND EXOGENOUS SOURCES

The etheno adducts of deoxyadenosine (εdA), deoxycytidine (εdC) and deoxyguanosine (εdG) in DNA are formed from the carcinogens vinyl chloride, urethane (via their epoxy intermediates) and other chemicals such as α-haloacetaldehydes [16]. The etheno ring system together with other exocyclic DNA base adducts is also generated by the attack of reactive bifunctional epoxides or aldehydes derived from lipid peroxidation (LPO) such as *trans*-4-hydroxy-2-nonenal. We recently presented evidence [17] that LPO products react with nucleic acid bases to form εdA and εdC, providing an explanation for the existence of a low background level of etheno residues in tissue DNA of unexposed animals and humans [18]. An ultrasensitive detection method based on immunoaffinity purification, ^{32}P-postlabelling and subsequent separation of 5'-monophosphates of εdA and εdC has been developd. The detection limit is four etheno adducts in 10^{10} normal bases requiring 50-μg DNA samples [18]. Using this method, the formation of etheno-DNA adducts in the liver was investigated in Long Evans Cinnamon (LEC) rats, a strain with hereditary abnormal copper metabolism, that develop spontaneous hepatitis and later hepatocellular carcinoma. Our results demonstrated for the first time a copper- and age-dependent accumulation of highly miscoding etheno-DNA adducts in the liver of LEC rats [19], which is an animal model for human Wilson's disease. These adducts could arise from oxygen radicals generated by copper-catalyzed Fenton-type reactions. Wilson's disease and primary hemochromatosis are human genetic disorders that cause copper or iron accumulation resulting in a high risk for liver cancer. We could now show [20] that the level of etheno adducts is also significantly elevated in the liver of these patients. Etheno-DNA adducts along with other oxidative DNA base damages may thus be involved in the initiation of experimental and human liver cancers. In a group of volunteers kept on controlled diets, the effect of dietary fatty acid composition on the endogenous formation of lipid peroxidation-derived DNA adducts was investigated in DNA from white blood cells [21]. Our results showed that dietary polyunsaturated fatty acids increase lipid peroxidation-derived etheno-DNA adducts *in vivo*. Thus, measurement of etheno and other exocyclic DNA adducts in epidemiological studies will be a useful tool to elucidate the role of dietary fat intake, oxidative stress and protective dietary antioxidants on endogenous DNA damage.

11. Conclusion.

The development of highly sensitive and specific analytical techniques, in particular mass spectrometric measurements, combined with immunoaffinity purification has given new insights into mechanisms and etiological agents that seem to play a role in human carcinogenesis. Future progress will depend on the systematic application of these high-resolution techniques in human biomonitoring and molecular epidemiology studies.

Acknowledgements: In part supported by EU-contract EV5V-CT940409. The skilled secretarial assistance of Mrs. G. Bielefeldt is gratefully acknowledged.

References

1. Bartsch, H. (1996) DNA adducts in human carcinogenesis: etiological relevance and structure-activity relationship. *Mutat. Res.*, in press.
2. Alexandrov, K., Rojas, M., Geneste, O., Castegnaro, M., Camus, A.M., Petruzzelli, S., Giuntini, C. and Bartsch, H. (1992) An improved fluorometric assay for dosimetry of benzo[*a*]pyrene diol-epoxide-DNA adducts in smokers' lung: comparisons with total bulky adducts and AHH activity, *Cancer Res.*, **52**, 6248-6253.
3. Rojas, M., Alexandrov, K., van Schooten, F.J., Hillebrand, M., Kriek, E. and Bartsch, H. (1994) Validation of a new fluorometric assay for benzo[*a*]pyrene diolepoxide-DNA adducts in human white blood cells: comparisons with ^{32}P-postlabelling and ELISA. *Carcinogenesis*, *15*, 557-560.
4. Anttila, S., Hietanen, E., Vainio, H., Camus, A.-M., Gelboin, H.V., Park, S.S., Heikkilä, L., Karjalainen, A. and Bartsch, H. (1991) Smoking and peripheral type of cancer are related to high levels of pulmonary cytochrome P450-1A1 in lung cancer patients, *Int. J. Cancer*, **47**, 681-685.
5. Hirvonen, A., Husgafvel-Pursiainen, K., Anttila, S. and Vainio, H. (1993) The GSTM1 null genotype as a potential risk modifier for squamous cell carcinoma of the lung. *Carcinogenesis*, **14**, 1479-1481.
6. Rojas, M., Alexandrov, K., Auburtin, G., Wastiaux-Denamur, A., Mayer, L., Mahieu, B., Sebastien, P. and Bartsch, H. (1995) *Anti*-benzo[*a*]pyrene diolepoxide-DNA adduct levels in peripheral mononuclear cells from coke oven workers and the enhancing effect of smoking. *Carcinogenesis*, **16**, 1373-1376.
7. Izzotti, A., De Flora, S., Petrilli, G.L., Gallagher, A., Rojas, M., Alexandrov, K., Bartsch, H. and Lewtas, J. (1995) Cancer biomarkers in human atherosclerotic lesions: detection of DNA adducts. *Cancer Epidemiol. Biomarkers Prev.*, **4**, 105-110.
8. Shuker, D.E.G. and Farmer, P.B. (1992) Relevance of Urinary DNA Adducts as Markers of Carcinogen Exposure, *Chem. Res. Toxicol.*, **5**, 450-460.
9. Shuker, D.E.G., Prevost, V., Friesen, M.D. and Bartsch, H. (1993) Noninvasive methods for measuring DNA alkylation in experimental animals and humans. *Environ. Health Perspect.*, **101** (Suppl. 3), 151-153.
10. Shuker, D.E.G. and Bartsch, H. (1994) Detection of human exposure to carcinogens by measurement of alkyl-DNA adducts using immunoaffinity clean-up in combination with gas chromatography-mass spectrometry and other methods of quantitation, *Mutat. Res.*, **313**, 263-268.
11. Eberle, G., Barbin, A., Laib, R.J., Ciroussel, F., Thomale, J., Bartsch, H. and Rajewsky, M.F. (1989) 1,N^6-etheno-2'-deoxyadenosine and 3,N^4-etheno-2'-deoxycytidine detected by monoclonal antibodies in lung and liver DNA of rats exposed to vinyl chloride, *Carcinogenesis*, **10**, 209-212.
12. Prevost, V., Shuker, D.E.G., Friesen, M.D., Eberle, G. Rajewsky, M.F. and Bartsch, H. (1993) Immunoaffinity purification and gas chromatography-mass spectrometric quantification of 3-alkyladenines in urine: metabolism studies and basal excretion levels in man. *Carcinogenesis*, **14**, 199-204.
13. Friesen, M.D., Garren, L., Prevost, V. and Shuker, D.E.G. (1991) Isolation of 3-methyladenine using immunoaffinity columns prior to determination by low-resolution gas chromatography-mass spectrometry, *Chem. Res. Toxicol.*, **4**, 102-106.
14. Prevost, V. and Shuker, D.E.G. (1996) Cigarette smoking and urinary 3-alkyladenine excretion in man. *Chem. Res. Toxicol.*, **9**, 439-444.
15. Kang, H., Konishi, C., Bartsch, H., Kuroki, T. and Huh, N. (1995b) Detection of smoking-related O^4-ethylthymine in human lung tissues. *Proc. Am. Assoc. Cancer Res.*, Abstract No. 810.
16. Bartsch, H., Barbin, A., Marion, M.-J., Nair, J. and Guichard, Y. (1994) Formation, detection and role in carcinogenesis of ethenobases in DNA, *Drug Metab. Rev.*, **26**, 349-371.
17. El Ghissassi, F., Barbin, A., Nair, J. and Bartsch, H. (1995) Formation of 1,N^6-ethenoadenine and 3,N^4-ethenocytosine by lipid peroxidation products and nucleic acid bases, *Chem. Res. Toxicol.*, **8**, 278-283.
18. Nair, J., Barbin, A., Guichard, Y. and Bartsch, H. (1995) 1,N^6-Ethenodeoxyadenosine and 3,N^4-ethenodeoxycytidine in liver DNA from humans and untreated rodents detected by immunoaffinity/^{32}P-postlabelling, *Carcinogenesis*, **16**, 613-617.
19. Nair, J., Sone, H., Nagao, M., Barbin, A., Bartsch, H. (1996) Copper-dependent formation of miscoding etheno-DNA adducts in the liver of Long Evans Cinnamon (LEC) rats developing hereditary hepatitis and hepatocellular carcinoma. *Cancer Res.*, **56**, 1267-1271.
20. Nair, J., Carmichael, P.L., Fernando, R.C., Phillips, D.H., Bartsch. H. (1996) Elevated 1,N^6-ethenodeoxyadenosine (εdA) and 3,N^4-ethenodeoxycytidine (εdC) DNA adducts in liver of patients with Wilson's disease and primary hemochromatosis. *Proc. Am. Ass. Cancer Res.*, Abstract # 813.

21. Nair, J., Vaca, C.E., Fang, J.-L., Velic, I., Mutanen, M., Bartsch, H. (1996) Effect of dietary fatty acid composition on the formation of etheno- and malonaldehyde-DNA base adducts in human white blood cells. Abstract for the Meeting 'DNA Adducts and Mutations in Human Biomonitoring', Stockholm, Sweden, June 1996.

MALDI MS OF OLIGO AND POLYSACCHARIDES

D. GAROZZO
Istituto per la Chimica e la Tecnologia dei Materiali Polimerici
Viale A. Doria 6, 95125 Catania, Italy

1. Oligosaccharides

1.1 IDENTIFICATION OF OLIGOSACCHARIDES BY POST SOURCE DECAY

Most problems in carbohydrate analysis arise from investigation of biological systems, so that we need highly sensitive methods of analysis. In the last two years, some papers on PSD MALDI MS of oligosaccharides have been published [1-4] and it appears that PSD may became the method of choice to distinguish isomeric oligosaccharides. In figure 1 the PSD spectra of the four isomeric lacto-N-fucopentaoses (LNFP) shown in table I are reported. The spectra are quite different allowing easily the differentiation, though the structures in table 1 are very similar.

TABLE I. Structures of the investigated oligosaccharides

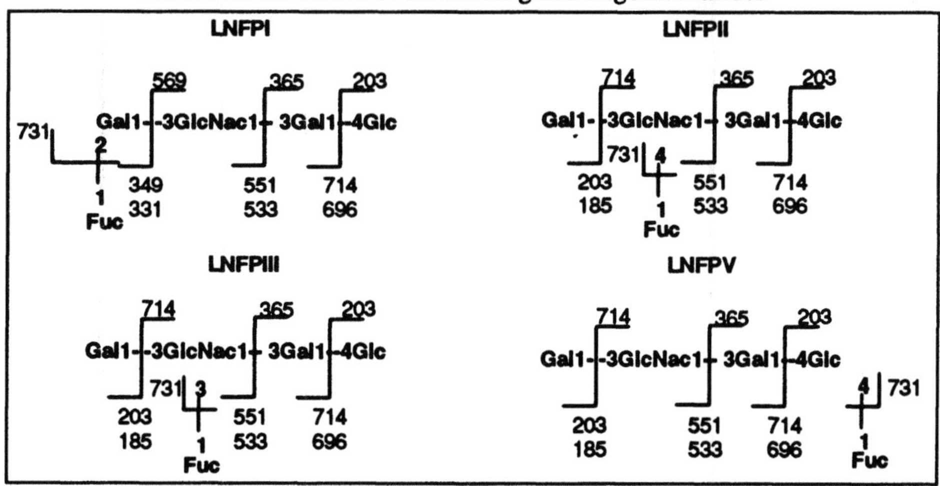

Figure 1a shows the PSD MALDI mass spectrum of LNFP I. Besides to the MNa^+ molecular ion at m/z 877, intense peaks corresponding to ions formed by the cleavages of all glycosidic linkages are present at m/z 731, 568, 365, 365, and 203 (Y

R. M. Caprioli et al. (eds.), Selected Topics and Mass Spectrometry in the Biomolecular Sciences, 477–499.

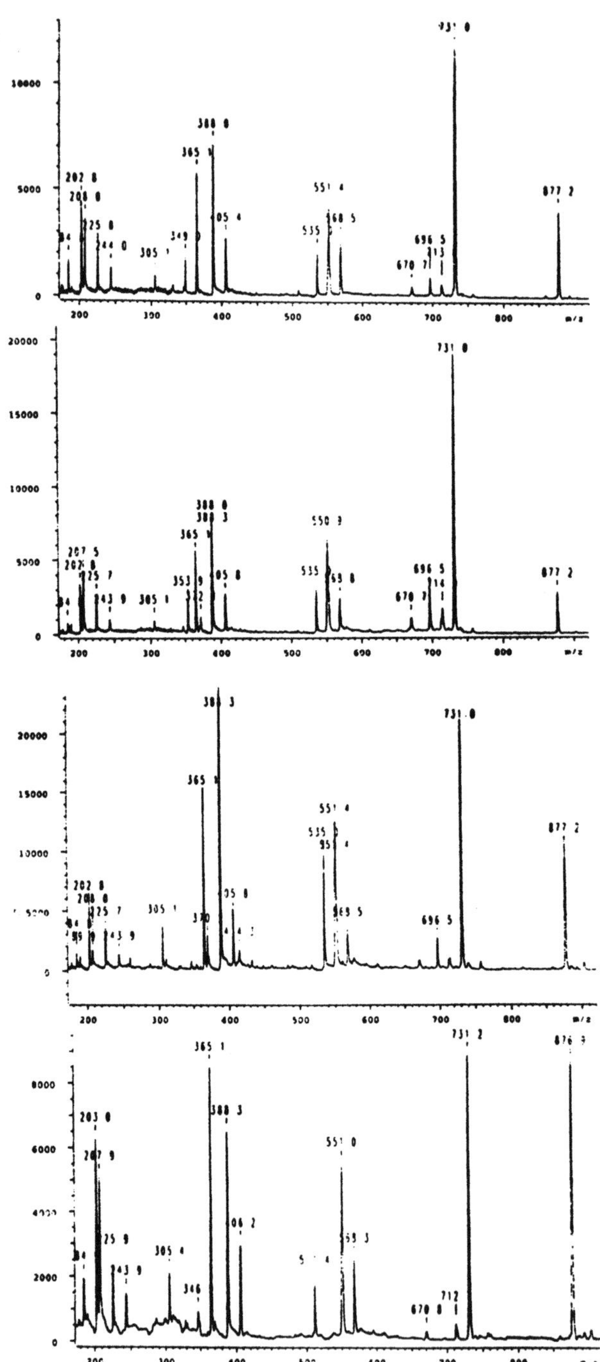

Figures 1 a, b, c and d Post source decay mass spectra of the pentasaccharides shown in table I.

series Domon and Costello nomenclature [5]), 696, 534, 331 (B series), 714, 552, 349 (C series). Ions produced from the cleavages of two glycosidic linkages (internal fragment) are also present in the spectrum at m/z 406, 388, 244, 226 and 208. These ions are formed by two Y-type cleavages or by a Y- and C-type cleavages. Very weak signals originating from a ring cleavage are present at m/z 305, 508 and 671, arising from a loss of 60 mass units from Y ions. The presence of the Z series ions at m/z 713, 551, 347, and 185 is uncertain, because peak at m/z 347 is missing and peaks at m/z 713, 551 and 185 present in the spectrum of figure 1 can originate from other fragmentation.

The PSD spectra of LNFPII and LNFP III are shown in figures 1b and 1c. These two compounds have the same sequence and one might expect identical spectra, but this is not the case. Major peaks arise from ions corresponding to Y,(m/z 731,714,365,and 203), C (m/z 714, 552, 203) and B (m/z 696, 535, 185) series, ions at m/z 568,406 and 388 arise from internal fragments. The peak at m/z 349 present in the spectrum of LNFPI is missing here as expected, since the sequence Fuc-Gal is not present.. Differences between the two spectra are the presence of a peak at m/z 354 in the spectrum of LNFP II and the peak at m/z 370 in the spectrum of LNFP III. They might arise from a loss of the substituent linked in 3-O from the B_3 ion.

Figure 1d shows the PSD spectrum of LNFP V. This spectrum is quite similar to that of figure 1a, and most of the considerations made for the spectrum of LNFPI are valid here except that the peak at m/z 568 is now an internal fragment. Major differences are that peaks at m/z 349 and 331 are missing and a new peak at m/z 511 is now present. These facts are easily explained; peaks at m/z 349 and 331 in the spectrum of figure 1a arise from the sequence Fuc-Hex (C_2 and B_2 ions) that is missing here and the peak at m/z 511 originates from the sequence Hex-Hex-Fuc. Peaks from ring fragmentation give very low signals at m/z 671 and 305, and ions from the Z series appear to be absent.

Although PSD is an excellent technique to distinguish isomeric oligosaccharides with 1 picomole of substance or less., information on the sequence cannot be easily obtained because the presence of "internal fragments" complicates interpretation of the spectra. The comparison of the fab spectra of a derivatized oligosaccharide with the PSD maldi is, in most cases, a resolutive method to obtain the sequence. In figure 2 is shown the fab mass spectrum of a reduced and peracethylated heptasaccharide with the monosaccharidic composition shown in table II. In this spectrum peaks corresponding to B ions at m/z 259, 303, 331, 546 and 619 are present; the information gained is shown in table III.

TABLE II. Monosaccharidic composition of the heptasaccharide analysed

Gal	1
Man	1
GlcNAc	1
GalA	2
Ara	1
Fuc	1

Although this data is quite useful, the complete sequence of the oligosaccharide cannot be obtained. In figure 3 the PSD maldi spectrum of the underivatized oligosaccharide is shown. From the two spectra now it is very easy to deduce the sequence shown in table IV.

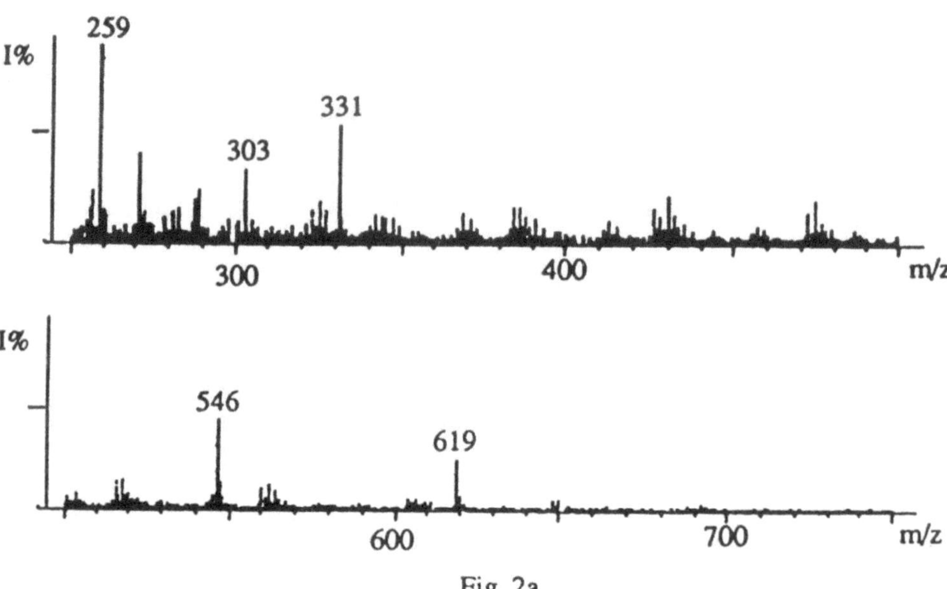

Fig. 2a

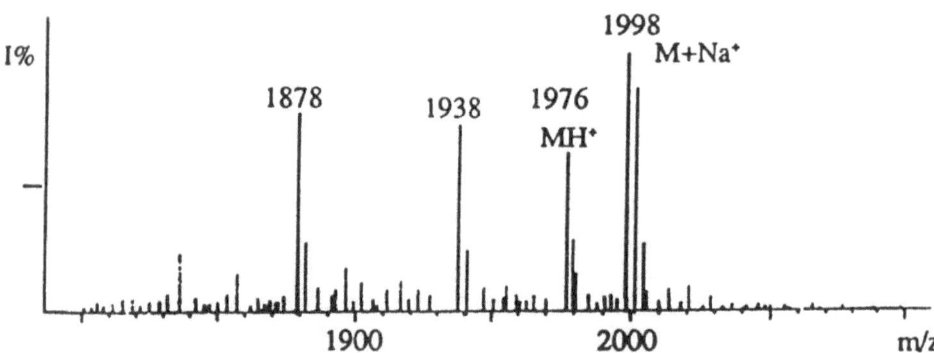

Figure 2. Fab mass spectrum of the heptasaccharide shown in table IV

TABLE III. Interpretation of B ions from spectrum in figure 2

m/z	structure
259	Ara
303	GalA
331	Hex
546	Ara-GlcNac
619	Hex-Hex

TABLE IV. Sequence of the heptasaccharide analysed

```
Ara--GlcNac--GalA--Fuc
         |        |
       GalA      Hex
                  |
                 Hex
```

1.2. QUANTITATIVE DETERMINATION

Cyclic oligosaccharides are a well-known family of carbohydrates. The interest in these compounds is mainly due to their ability to form inclusion complexes in water with molecules exhibiting hydrophobic character.

There is a need to determine the degree of polymerization, molecular weight distribution and purity of cyclic oligosaccharides by reliable analytical method. Fast atom bombardment mass spectrometry (fab ms.) and ^{13}C-n.m.r. spectroscopy have been used for the analysis of these compounds, and have shown that they are unbranched, cyclic molecules [6]. Despite of these excellent results, some analytical problems persist. In the f.a.b. mass spectra of a mixture of cyclic sophoraoses peaks corresponding to the higher molecular weight compounds were missing, whereas permethylation was shown to produce some degradation [6].

Matrix assisted laser desorption mass spectrometry (MALDI MS) is a recent developed analytical technique [7,8] which is able to produce intact molecular ions from polar compounds such as proteins [9], oligosaccharides [10], oligonucleotides [11], gangliosides [12], and polymers [13]. MALDI MS can be used for direct analysis of complex mixture such as enzymatic digests [14] or polyesters methanolyzed [15] without requiring purification of the reaction mixture. The advantages of this method are easy preparation of samples, short times of analysis and very high sensitivity.

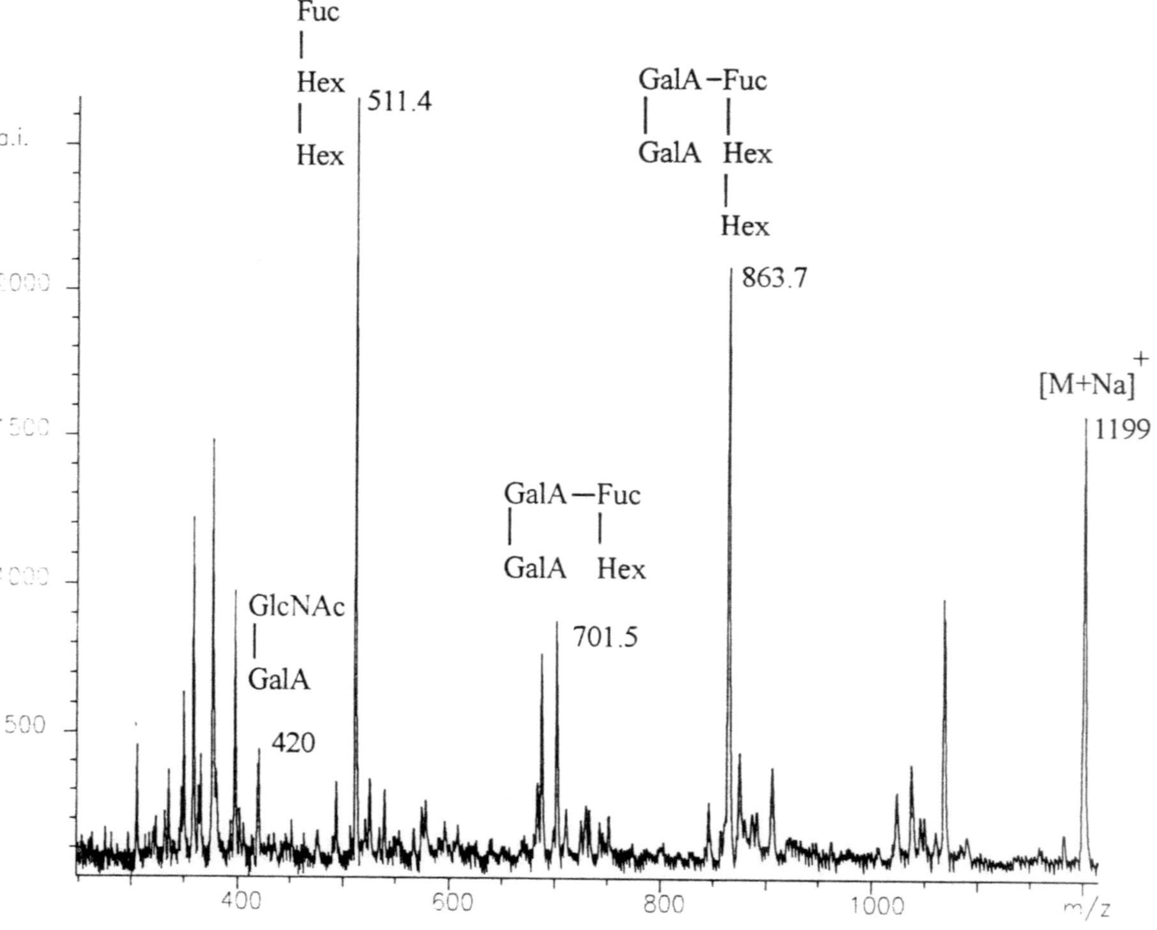

Figure 3. Post source decay mass spectrum of the heptasaccharide shown in table IV

We report here on the use of MALDI MS for the determination of the molecular weight distribution of a mixture of cyclic sophoraoses. Comparison with the molecular weight distribution obtained by HPLC separation of the same cyclosophoraoses sample was also performed and it has been found that the two sets of measurements match pretty well (table V).

The MALDI mass spectrum, obtained with the BRUKER Reflex instrument using DHB as matrix, of the cyclic β-(1-2) linked glucans from Rhizobium trifolii containing a number of monomers from 17 to, at least, 26 [17] is reported in figure 4. This spectrum contains peaks up to mass 4400 corresponding to the $(M+Na)^+$ pseudo molecular ions from 17^{th}mers to 27^{th}mers at m/z 2780 + n162. Besides these peaks, very weak signals corresponding to the protonated molecular ions are present. Resolution achieved by using the Reflector is able to resolve these peaks from the sodiated ions. No fragment peaks are present, confirming that MALDI is a very soft ionization technique. Addition of potassium chloride produces the spectrum shown in figure 5. In this spectrum besides the peaks observed in the spectrum in figure 4, peaks corresponding to $[M+K]^+$ are present with a shift of 16 masses. Capability of cyclic sophoraoses to form cluster ions with sodium and potassium was observed also in fab ms. [6].

These spectra show that MALDI MS is an excellent technique for the qualitative analysis of cyclic glucans in mixture. In the f.a.b. mass spectrum of the same sample [17] only peaks from 17^{th}mer to 22^{th}mer were observed.

Because in the mass spectrum reported in figure 4 only pseudo molecular ion signals are present, we have compared the quantitative distribution obtained by MALDI MS with that obtained by H.P.L.C.-RI and the two sets of data are in good agreement (table V). This is somehow surprising because it has been reported [18,19] that the quantitative response of MALDI MS is not linear. In particular, Harvey studied a group of oligosaccharides of structure $(Man)_n$ $(GlcNac)_2$ (n=5-9) and he showed that the signal strength decreases with increasing molecular weight [18].

Quantitative response in MALDI MS depends on several factors; the most important are the cocrystallization of analyte with matrix, the laser fluence and the reproducibility.

In order to investigate MALDI MS reproducibility of these compounds, we have sent samples of cyclic sophoraoses to MS manufacturers to perform spectra with different instruments and matrices. M.W. distributions obtained with the Fisons TofSpec instrument (DHB matrix) and using the Finningan Lasermat (a-Cyano-4-hydroxycinnamic acid matrix) are reported in table 1. Although peaks corresponding to 26_{th}mer and 27_{th}mer are not well distinguishable from the noise, the comparison between the mass spectra obtained shows that MALDI MS of these compounds is well reproducible. Furthermore, intensities obtained by MALDI MS are in good agreement with those obtained by HPLC-R.I., thus MALDI MS can be used for the quantitative analysis of these compounds.

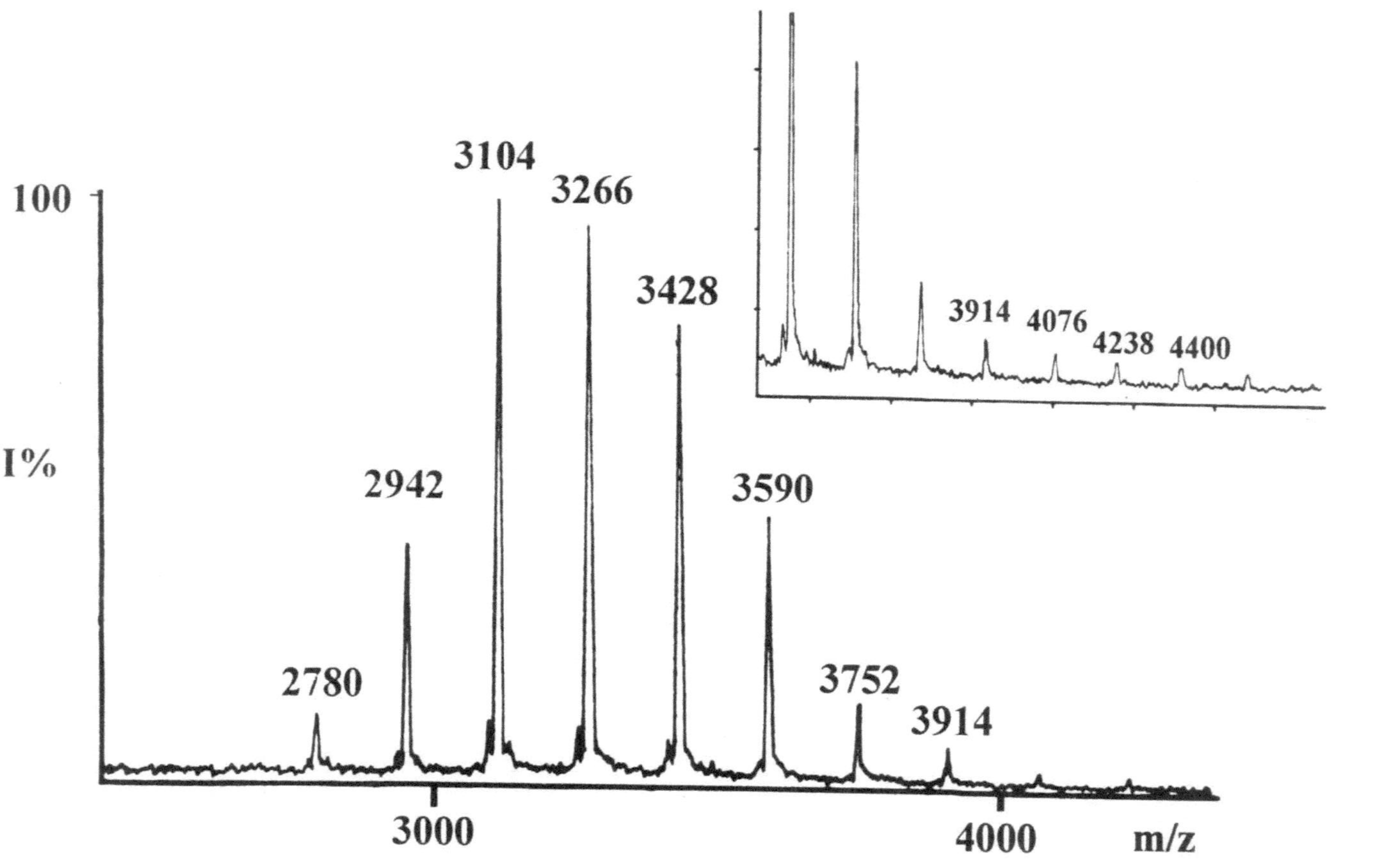

Figure 4. Maldi mass spectrum of cyclosophoraoses mixture accumulated from 50 single laser shots, Bruker maldi tof operating at 20 KV, DHB matrix, reflector 21 KV, positive ionization.

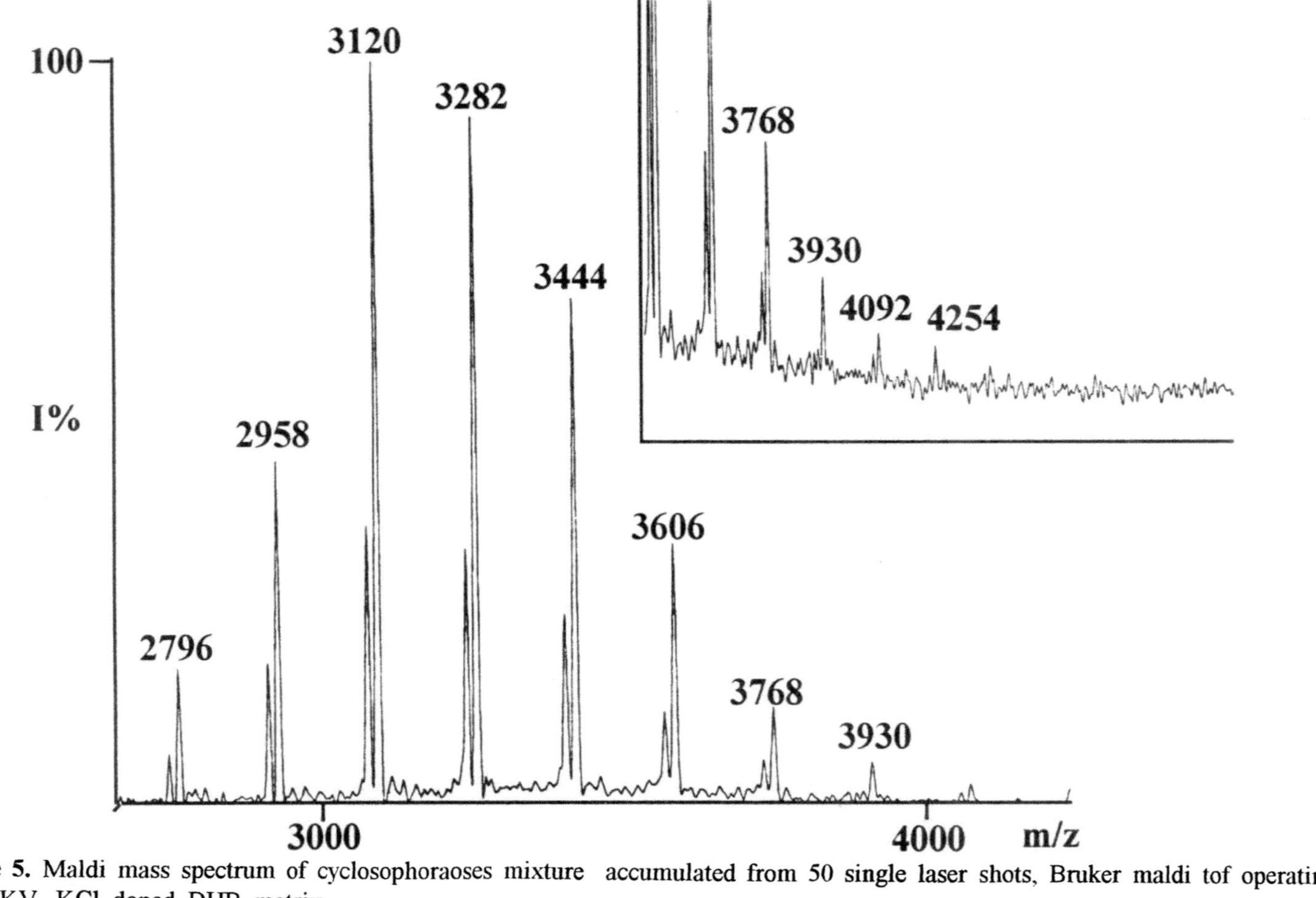

Figure 5. Maldi mass spectrum of cyclosophoraoses mixture accumulated from 50 single laser shots, Bruker maldi tof operating at 20 KV, KCl doped DHB matrix,

TABLE V. Comparison of the molecular weight distribution of the cyclic glucans obtained by different techniques and instruments (see text).

N° of repeating units	Mass (MNa^+)	HPLC[a] (RI)	MALDI[a] Bruker DHB matrix	MALDI[a] Finningan CHC matrix	MALDI[a] VG DHB matrix
17	2780	13	13	15	15
18	2942	40	41	40	42
19	3104	93	100	100	100
20	3266	100	95	98	99
21	3428	78	79	77	72
22	3590	48	47	45	48
23	3752	16	15	15	18
24	3914	8	8	8	10
25	4076	4	4	4	3
26	4238	3	3	-	3
27	4400	2	2	-	-

a) Relative abundance measured by the peak height

This result is not necessarily in disagreement with that of Harvey [18]. In fact, in that study oligosaccharides with m.w. from 1192 to 1840 (DP from 7 to 11) were examined, whilst in the present study the m.w. varies from 2780 to 4400 (DP from 17 to 25). Thus, in our case, the sample has a higher mass value. Although further evidence has to be collected, it is reasonable to assume that the mass discrimination observed by Harvey has leveled off in our case.

The studies reported here demonstrate the following: 1) The sensitivity of MALDI MS for molecular weight determination of native cyclosophoraoses is at least two orders of magnitude better than that of fab ms, 2) No fragmentation is observed, 3) The MALDI mass spectra of native β (1-2)cyclic glucans show mass discrimination and can be used for quantitative determination and m.w. distribution.

2. Polysaccharides

Matrix assisted laser desorption ionization mass spectrometry (MALDI MS) introduced in 1988 [20,21] is able to produce intact molecular ions from high polar compounds such as proteins [22], oligosaccharides [23], oligonucleotides [24] and synthetic polymers [25]. MALDI MS can be used for direct analysis of complex mixture such as enzymatic digests [26] or methanolyzed polyesters [27] without requiring any purification of the reaction mixture.

For polysaccharides, several studies showing good and sensitive MALDI mass spectra of compounds like dextrans of molecular weight (MW) 1,500 and 5,000 [23], complex oligosaccharides [24], mixtures of cyclic β 1-2 glucans [28], gangliosides [29], inulin [30] and heparin derived oligosaccharides [13] were reported, but the mass range has, as yet, only been extended to about 15,000 daltons [32]. This situation is quite different from that encountered in the MALDI MS analysis of, for instance, proteins, where molecular masses up to 300,000 daltons have been measured [32], and no explanation has been suggested for this behaviour.

Dextrans are important microbial polysaccharides, their largest outlet being the pharmaceutical and fine chemical industries [33]. The MW and the molecular weight distribution (MWD) of these products greatly influences their properties and use, and fast, absolute methods for their determination would be certainly welcome. In this respect, MALDI MS looks very promising, provided that the limit of 15,000 daltons be overrun.

The MALDI mass spectra of dextrans with MW 1500 and 5000, showing peaks up to more than 10.000 daltons, were reported [23,24 32] and the oligomeric distribution obtained is in excellent agreement with that obtained by classical methods. In spite of these good results, higher molecular weight dextrans give very bad MALDI mass spectra: a series of peaks separated by 162 amu, the incremental mass for a glucose unit, is discernible up to m/z 10,000 and beyond. However, the intensity of these signals in relation to their m/z values is far from what is to be expected from the molecular weight of the sample. As the higher MW dextrans 25,000, 50,000, and 80,000 were analyzed, larger and larger discrepancies were found between the MALDI mean numerical MW (**Mn**) and those reported for the samples, irrespectively of the laser fluency and matrix to analyte ratio used. This apparent error in estimating the molecular weight of dextrans made also senseless to derive a polydispersity index (D) measure from those spectra.

In an attempt to overcome these problems, we subjected the dextran samples to permethylation. This derivatization procedure is well established in FAB-MS to enhance sensitivity for oligosaccharides in the range 2,000-10,000 daltons [36].

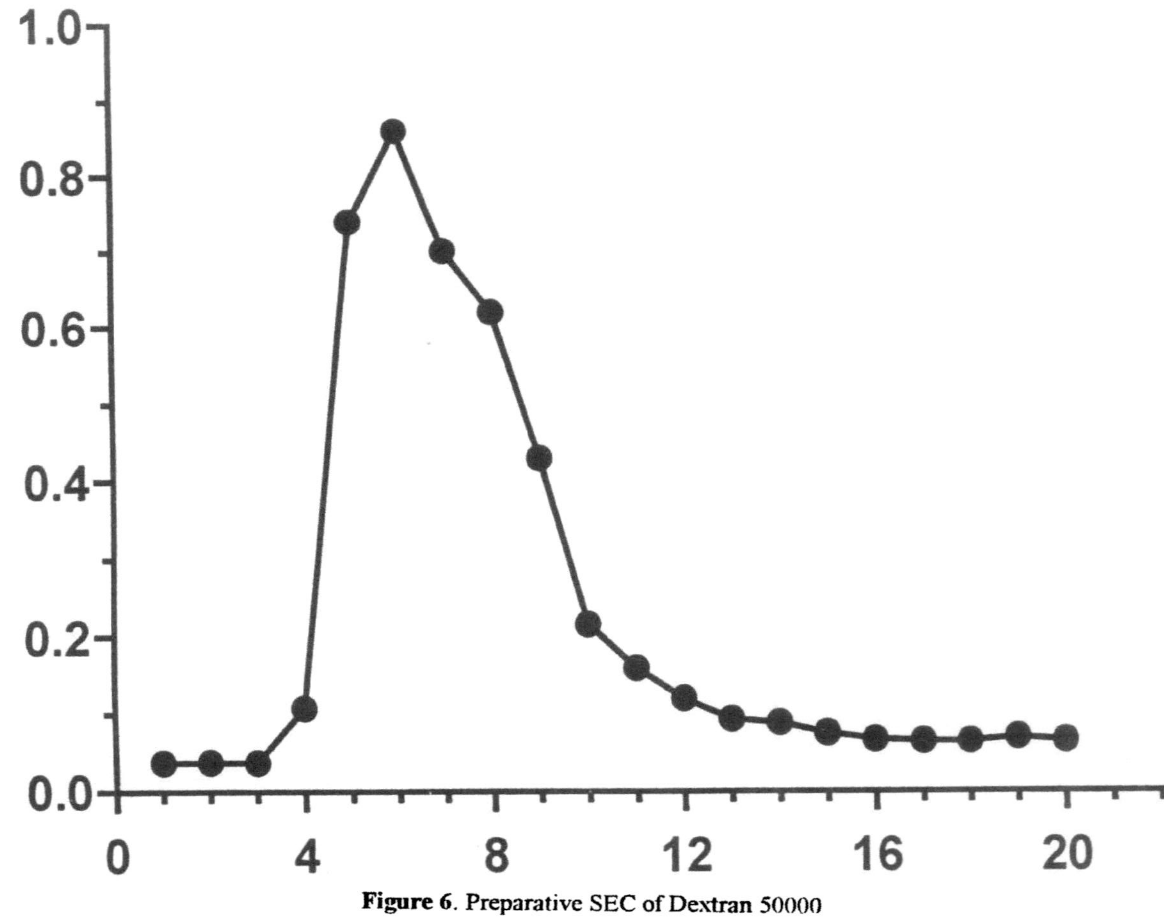

Figure 6. Preparative SEC of Dextran 50000

Recently, Juhasz and Costello [29] applied it to the MALDI analysis of gangliosides, and found an increase in sensitivity of two order of magnitude. The MALDI mass spectrum of a sample of permethylated dextran 50,000(not reported here) shows series of peaks corresponding to permethylated oligosaccharides, but peaks above mass 15,000 are missing, and the **Mn** measured from the spectrum is about 4,000, while the non-methylated polymer has a **Mn** of 35,000. A tentative explanation for the failure of MALDI-MS of both free and permethylated polysaccharides of adequately represent the molecular weight distribution of these samples may be that the lighter molecules are preferentially desorbed and ionized in the MALDI process, suppressing the desorption and ionization of larger molecules.

In order to check out this hypothesis, we decided to obtain samples with narrower MWD and to fractionate the dextrans by SEC on Sephadex columns. The eluate from the SEC columns was collected and fractions assayed for hexose by the anthrone colorimeritric method. The SEC separation of dextran 50,000 on a low pressure Sephadex G 75 column is shown in figure 6. Fractions whose **Mn** was determined by MALDI MS are reported in Table VI, and the spectra relative to fractions 6, 7 and 8 are shown in figure 7(a), (b) and (c), respectively. In these spectra peaks at mass 50,000, 42,000 and 34,000 are present in the order, together with minor peaks corresponding to the doubly charged molecular ions and to the singly charged dimers. From the spectra **Mn** and the average ponderal molecular weight (**Mw**) of each fraction, and consequently D (**Mw**/**Mn**) can be calculated. For the samples in figure 7, D is less than 1.1.

TABLE VI. MALDI data for fractions from dextran 50 000

Fraction	M_n
6	50 000
7	40 000
8	34 000
9	29 000
10	26 000
11	24 000
12	20 000
13	18 000
14	17 000
20	3 400

The same approach was used for dextran 80,000. The SEC separation on Sephadex G 150 is shown in figure8, and Table VII summarizes the results of MALDI analysis of some fractions. MALDI mass spectra of fractions 14, 17 and 18 are shown in figure 9(a), (b) and (c), respectively. From these spectra **Mn** of 90,000, 51,000 and

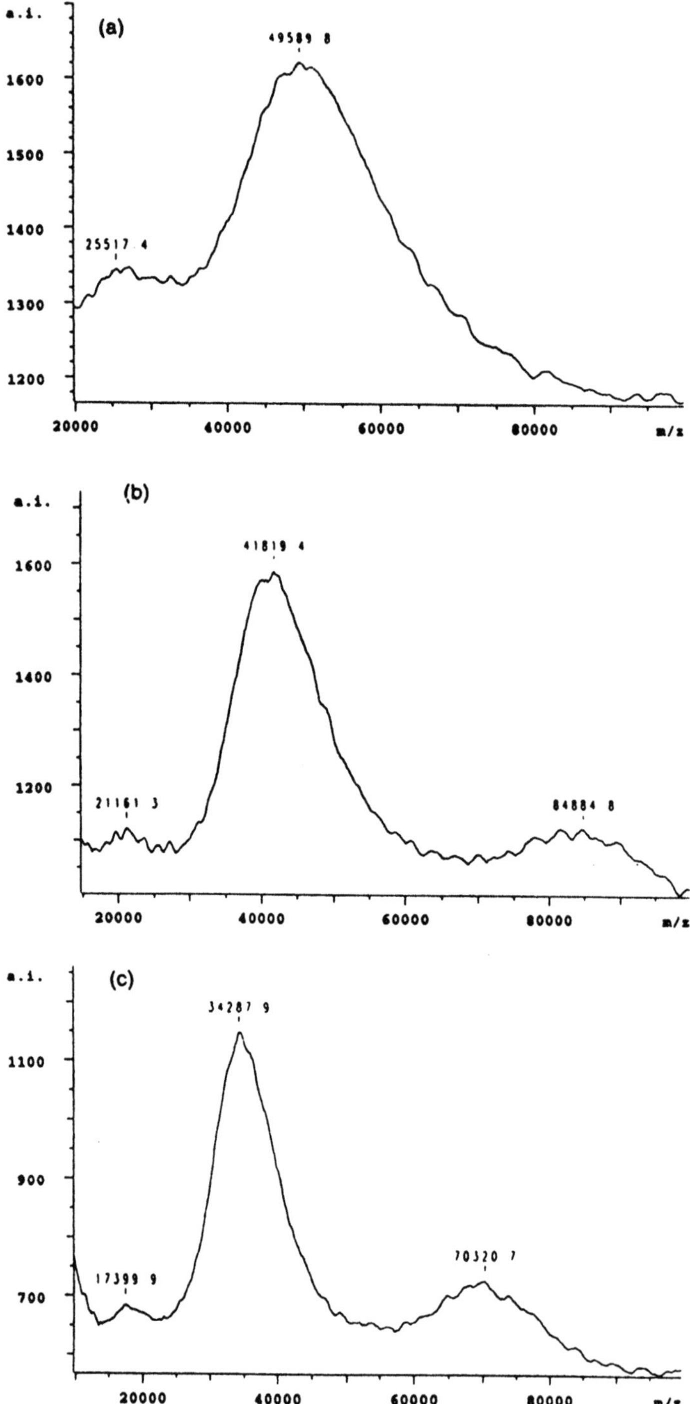

Figures 7 a, b and c. MALDI mass spectra of fractions 6,7 and 8 of Dextran 50000

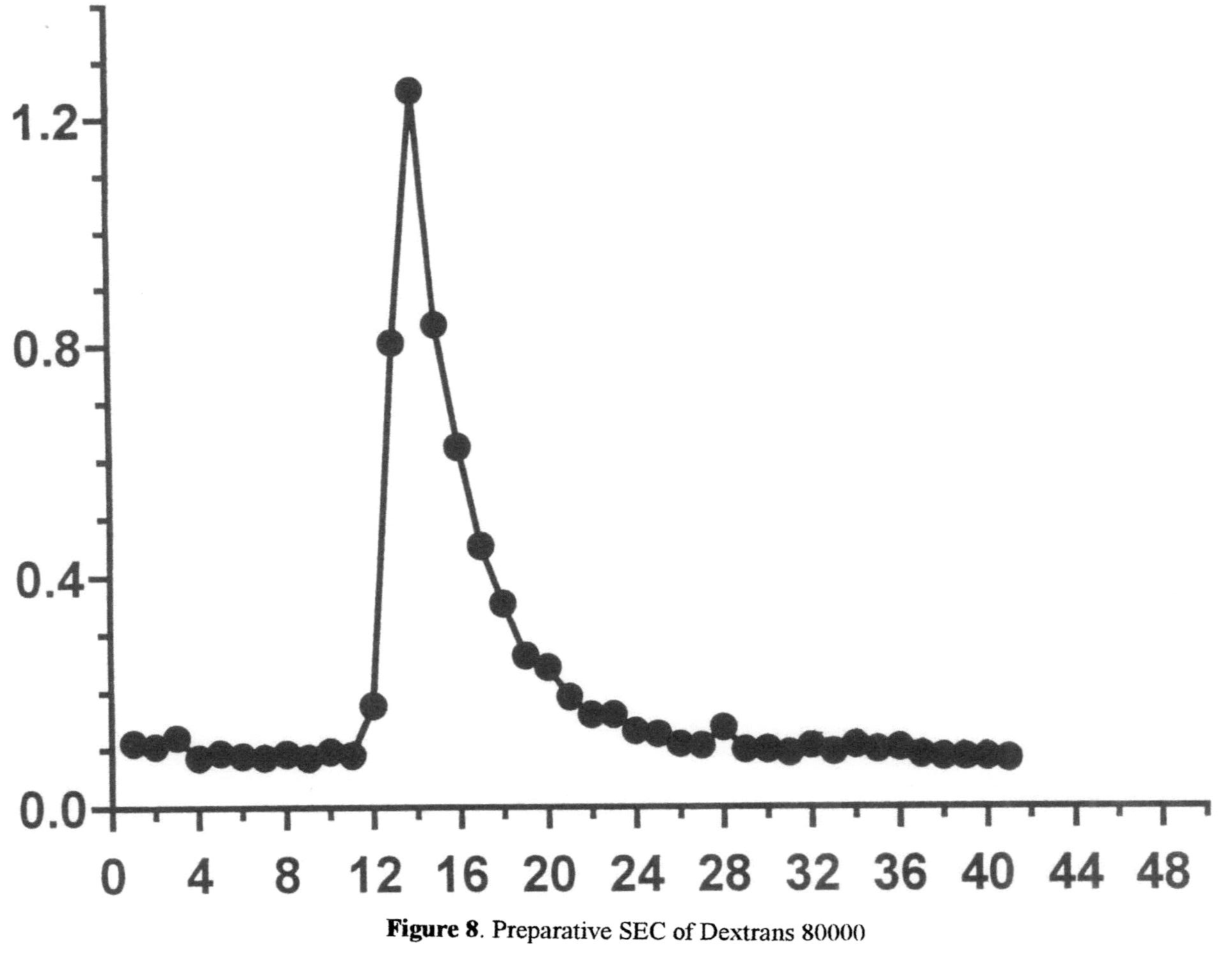

Figure 8. Preparative SEC of Dextrans 80000

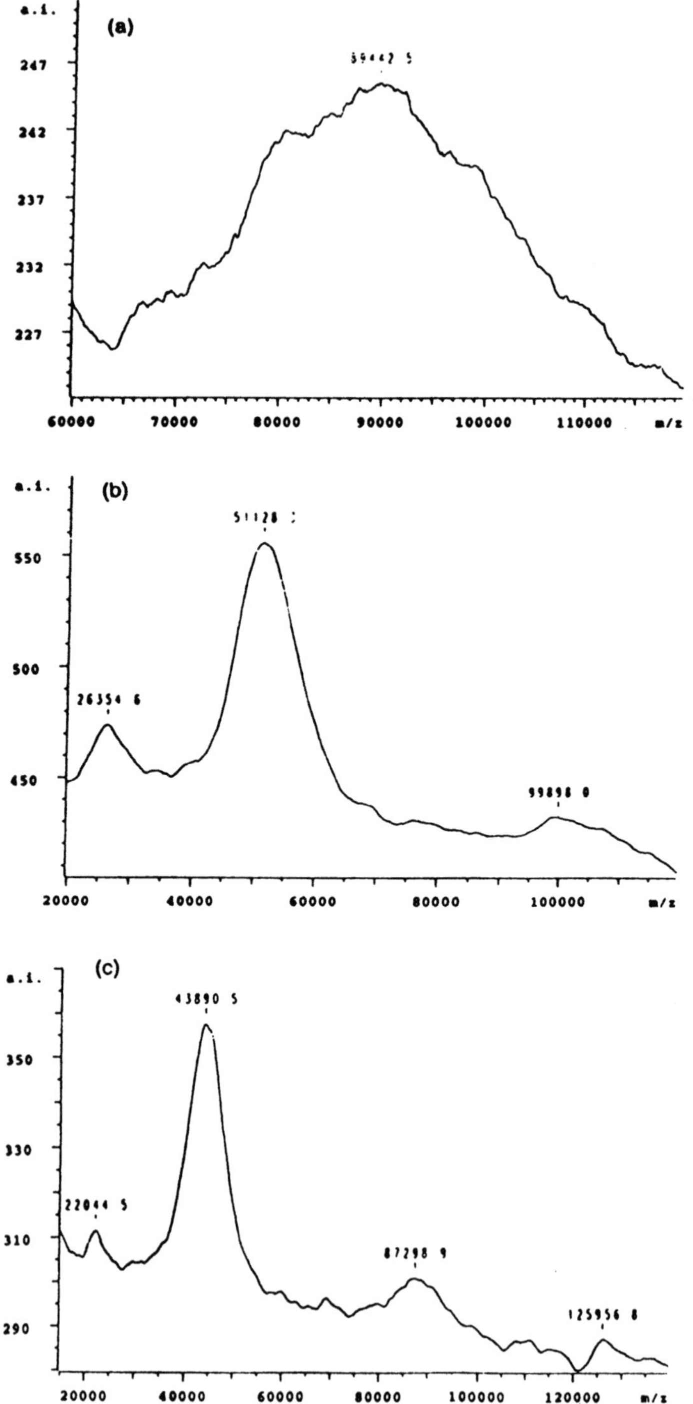

Figures 9 a, b and c. MALDI mass spectra of fractions 14, 17 and 18 of Dextran 80000

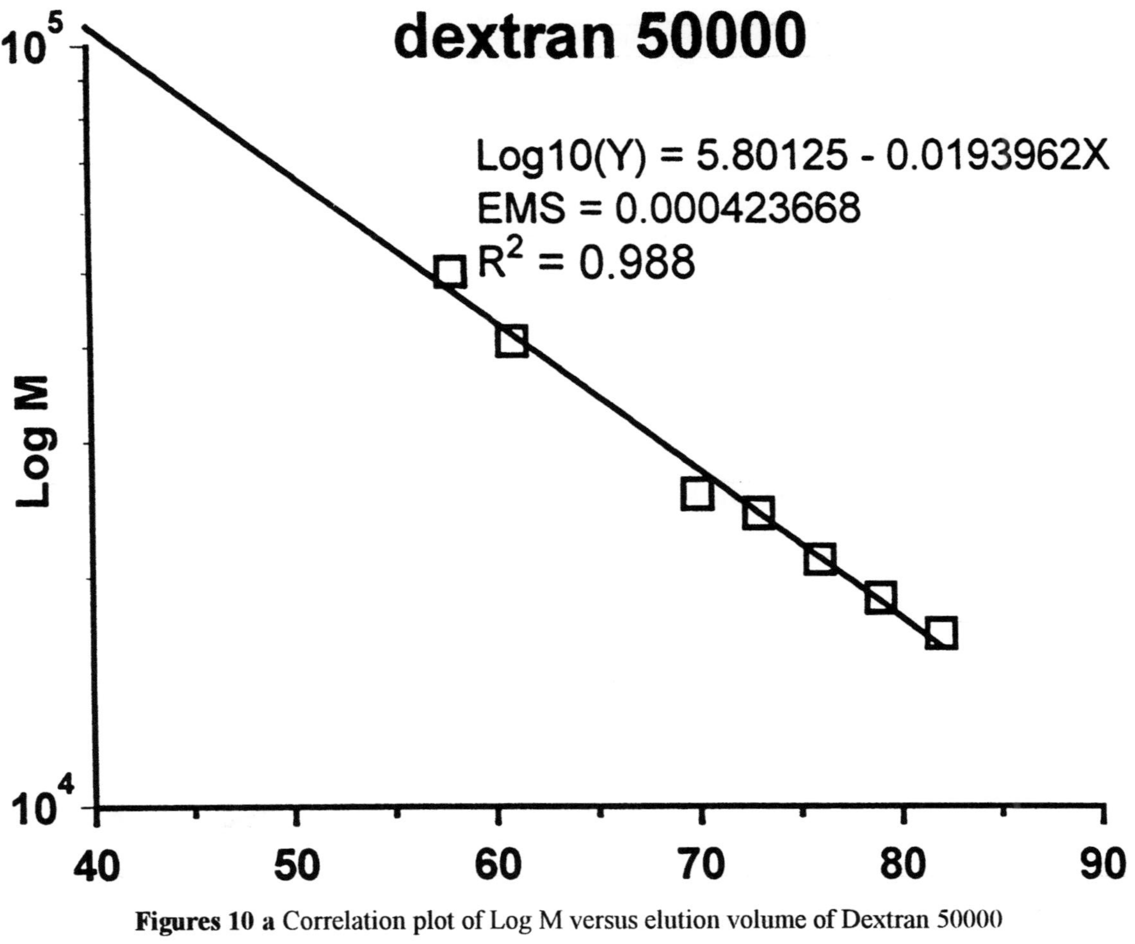

Figures 10 a Correlation plot of Log M versus elution volume of Dextran 50000

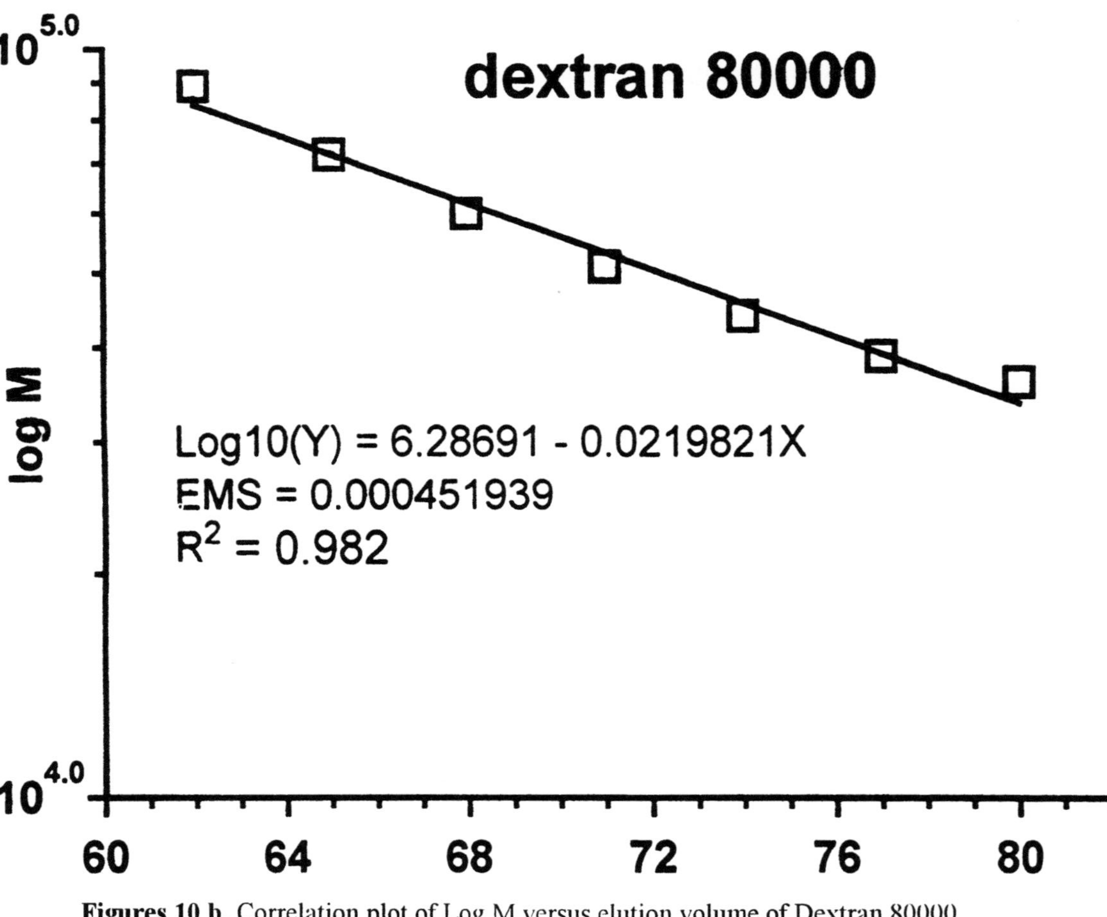

Figures 10 b. Correlation plot of Log M versus elution volume of Dextran 80000

44,000 with D < 1.1 are found, together with minor peaks corresponding to the doubly charged molecular ions and to the singly charged dimers.

TABLE VII. MALDI data for fractions from dextran 80 000

Fraction	M_n
14	89 000
15	72 000
16	60 000
17	51 000
18	44 000
19	39 000
20	36 000

These results represent a 10-fold increase in the mass range amenable to MALDI MS of dextrans, and support the hypothesis that lower MW species suppress the desorption-ionization of larger molecules.

In order to verify the accuracy of these measurements, we analyzed by high performance SEC the Sephadex G 75 fraction 6 of dextran 50,000 and the Sephadex G 150 fraction 14 of dextran 80,000 on TSK high pressure columns calibrated with dextran MW standards, using diffrative refractive index (DRI) and LALLS detectors. The MW data obtained from the three types of detectors (MS, DRI and LALLS) are compared in Table VIII. The data matches pretty well with each other, confirming that the **Mn** measured by MALDI MS after SEC separation are exact.

TABLE VIII. Comparison of M_n values obtained by MALDI MS, HPSEC and LALLS

Sample	M_n MALDI	M_n HPSEC	M_n LALLS
Dextran 50.000 fraction 6	50 000	51 000	56 000
Dextran 80 000 fraction14	90 000	92 000	94 000

Finally, it is known that in SEC analysis a linear correlation exists between the elution volume and the logarithm of **Mw**. In figure 10 we report the plots of this correlation for dextran 50,000 on Sephadex G 75 and of dextran 80,000 on Sephadex G 150 using the data of Tables VI and VII. The linearity of these relationships proves once again the correctness of the MALDI measurements.

It must be concluded that it is possible to obtain MALDI mass spectra of polysaccharides with **Mn** almost up to 100,000 daltons when D < 1.1, whereas no reliable spectra are obtained from polydisperse samples. These findings appear to be a general fact; if we look at the flourishing literature on MALDI MS of synthetic polymers, all the high MW samples analysed are SEC standards or polymers obtained by anionic polymerisation, known to produce low D samples. The same consideration applies to proteins that are essentially monodisperse polymers even if heterogeneity, cation and matrix molecule attachment, or varying degree of glycosilation are present.

In conclusion, when analysing by MALDI MS polydisperse samples, a previous **Mw** fractionation should be performed, as that obtained by SEC.

Acknowledgments

Partial financial support from the Italian Ministry for University and for Scientific and Technological Research (MURST) and from the National Council of Research (CNR) is gratefully acknowledged.

References

1) B. Spengler, D. Kirsch, R. Kaufmann, J. Lemoine *J. Mass Spectrom.* **30**, 782 (1995)

2) D.J. Harvey, P.M. Rudd, R.H. Bateman, R.S. Bordoli, K. Howes, J.B. Hoyes and R. G. Vickers *Org. Mass Spectrom.* **29**, 753 (1994)

3) D. J. Harvey *J. Chromatogr.* **720**, 429 (1996)

4) G. Talbo, M. Mann *Rapid Commun. Mass Spectrom.* **10**, 100 (1996)

5) B. Domon, C.E. Costello *Glycoconnj* **5**, 397 (1988)

6) A. Dell, W.S. York, M. McNeal, A.G. Darvill, P. Albersheim *Carb. Res.* **117**, 185 (1983).

7) M. Karas, D. Bachmann, U. Bahr, F. Hillenkamp *Int. J. Mass Spectrom. IOn Proc.* **78**, 53 (1988).

8) K. Tanaka, H. Waki, Y. Ido, S. Akita, Y. Yoshida, *Rapid Commun. Mass Spectrom.* **8**, 151 (1988).

9) T-W. D. Chan, A. W. Colburn, P. J. Derrick *Org. Mass Spectrom.* **27**, 53 (1992).

10) B. Stahl, M. Steup, M. Karas, F. Hillenkamp *Anal. Chem.* **63**, 1463 (1991).

11) A. Overberg, A. Hassenburger, F. Hillenkamp in *Mass Spectrometry in Biological Sciences: A Tutorial;* M.L. Gross Ed., Kluwer Academic Publishers: Dordrecht, p. 181 (1992).

12) P. Juhasz, C.E. Costello *J. Am. Soc. Mass Spectrom.* **3**, 785 (1992).

13) C. F. James, C. L. Wilkins *J. Am. Chem. Soc.* **110**, 2687 (1988).

14) M. Chevrier, R.J. Cotter *Rapid Comm. Mass Spectrom.* **12**, 611 (1991).

15) R. Abate, A. Ballistreri, G. Montaudo, D. Garozzo, G. Impallomeni, G. Critchley, K. Tanaka *Rapid Comm. Mass Spectrom.* 7, 1033 (1993).

16) M.W. Breedveld, L.P.T.M. Zevenhuizen, A.J.B. Zehnder, *Appl. Envinromental Microbiol.* **56**, 2080 (1990).

17) R. Rizzo, V. Crescenzi, F. Gasparrini, G. Gargaro, D. Misiti, A.L. Segre, L.P.T.M. Zevenhuizen, R.H. Fokkens in *Biomedical and Biotechnological Advances in industrial Polysaccharides;* V. Crescenzi Ed.; Gordon and Breach Science Publishers (1989) p. 485.

18) D. J. Harvey *Rapid Comm. Mass Spectrom.* **7**, 614 (1993).

19) K. Tang, S. L. Allman, R. B. Jones, C. H. Chen *Anal. Chem.* **65**, 2164 (1993).

20) M. Karas, D. Bachmann, U. Bahr, F. Hillenkamp *Int. J. Mass Spectrom. Ion Proc.* **78**, 53 (1988).

21) K. Tanaka, H. Waki, Y. Ido, S. Akita, Y. Yoshida, *Rapid Commun. Mass Spectrom.* **8**, 151 (1988).

22) T-W. D. Chan, A. W. Colburn, P. J. Derrick *Org. Mass Spectrom.* **27**, 53 (1992).

23) B. Stahl, M. Steup, M. Karas, F. Hillenkamp *Anal. Chem.* **63**, 1463 (1991).

24) A. Overberg, A. Hassenburger, F. Hillenkamp in *Mass Spectrometry in Biological Sciences: A Tutorial;* M.L. Gross Ed., Kluwer Academic Publishers: Dordrecht, p. 181 (1992).

25) C. F. James, C. L. Wilkins *J. Am. Chem. Soc.* **110**, 2687 (1988).

26) M. Chevrier, R.J. Cotter *Rapid Comm. Mass Spectrom.* **12**, 611 (1991).

27) R. Abate, A. Ballistreri, G. Montaudo, D. Garozzo, G. Impallomeni, G. Critchley, K. Tanaka *Rapid Comm. Mass Spectrom.* **7**, 1033 (1993).

28) D. Garozzo, E. Spina, L. Sturiale, G. Montaudo, R. Rizzo, *Rapid Comm. Mass Spectrom.* **8**, 358, (1994)

29) P. Juhasz, C.E. Costello *J. Am. Soc. Mass Spectrom.* **3**, 785 (1992).

30) J.O. Metzger, R. Woisch, W. Tuszynski, R. Angermann *Fresenius J. Anal. Chem.* **349**, 473 (1994)

31) P. Juhasz, K. Biemann in *Proceedings of the 42nd ASMS Conference on Mass Spectrometry and Allied Topics*, p. 890

32) U. Bahr, M. Karas, F. Hillenkamp *Fresenius J. Anal. Chem.* **348**, 783 (1994)

33) N.K. Matheson, B.V. Mccleary in *The Polysaccharides*, G. O. Aspinall Ed.,

Academic Press (Orlando, Fl) vol. 3, p. 10

34) S. Hakomori, *J. Biochem.*, (Tokyo), **55**, 205 (1964)

35) W. York, A.G. Darvill, M. McNeill, T.T. Stevenson, P. Albersheim, *Methods in Enzymology*, Academic Press Inc., **Vol. 118**, 3 (1985).

36)A. Dell, J.E. Thomas-Oates in *Analysis of Carbohydrates by GLC and MS*, C.J. Biermann and G.D. McGinnis eds, CRC Press Inc. p. 217 (1989).

MASS SPECTROMETRY OF ISOEICOSANOIDS AS NOVEL LIPID SUBSTANCES

ROBERT C. MURPHY and RUSSELL J. WAUGH
National Jewish Center for Immunology and Respiratory Medicine
1400 Jackson Street, Denver, CO 80206 USA

1. Introduction

Mass spectrometry has played an important role in both the qualitative and quantitative analysis of lipids for almost 50 years. The pioneering work of Ryhage and Stenhagen [1] in the late 1950s as well as that of McLafferty [2] during the same time period, led to a detailed understanding of the ion chemistry involved in the electron ionization of fatty acid esters. Experience and confidence gained with the successful application of mass spectrometry to the analysis of these rather simple lipids led to the investigation of more complex lipid compounds including prostaglandins [3], thromboxanes [4], and more recently, leukotrienes [5]. Djerassi and co-workers [6] studied in detail the behavior of steroids which added a great deal of basic information to our current concepts of unimolecular ion decompositions. The analysis of other lipid classes such as vitamins [7], bile acids [8], and triglycerides [9] was made possible with the development of micro-derivatization protocols to yield volatile compounds suitable for gas chromatographic/mass spectrometric analysis.

The analysis of more complex, highly polar lipids such as phospholipids [10], glycolipids [11], and peptidolipids [12] was not possible until the development of desorption and spray ionization techniques which have only emerged within the last 10-15 years. These rather soft ionization techniques yield little structural information, but produce rather abundant ions corresponding to the molecular weight of the compound. The development of the tandem mass spectrometer to collisionally activate these molecular ions added the requisite capability of decomposing the molecule following ionization to yield structurally significant product ions. A review of the application of mass spectrometry to the analysis of lipids from simple to complex is beyond the scope of this chapter; therefore, an example of the use of the various ionization techniques applied to the analysis of a recently discovered class of lipid molecules will be described to illustrate the applicability of mass spectrometric techniques to the study of complex lipids. The implication of these structural discoveries will also be discussed. Further information about the mass spectrometry of lipids can be obtained in recent reviews [13,14] and a monograph [15].

2. Eicosanoids

The enzyme catalyzed oxidation of arachidonic acid leads to several families of metabolites, termed eicosanoids, which have profound biological activity and are known to mediate many intracellular events within tissues and organs [16]. These products include prostaglandins and thromboxanes which are derived from the enzyme PGH synthase; also called cyclooxygenase, and the leukotrienes which are derived from the 5-lipoxygenase pathway of arachidonic acid metabolism. The enzymes involved in the oxidation of arachidonic

R. M. Caprioli et al. (eds.), Selected Topics and Mass Spectrometry in the Biomolecular Sciences, 501–515.

acid catalyze what is thought to initiate the free radical addition of oxygen to arachidonic acid [17]. Since the formation of free radical species within biological systems is known to take place under many conditions, it was not surprising that molecules with similar structures to that of the enzymatically-derived prostaglandins, thromboxanes, and leukotrienes have been identified as products of the free radical oxidation of arachidonic acid. One unique aspect of the enzymatic formation of these eicosanoids is the requirement of free arachidonic acid as substrate. In contrast, free radical derived eicosanoids are initially formed within the phospholipid bilayer and exist as complex glycerophosphocholine lipids [18]. The reason for this is that most arachidonic acid within cellular systems is found esterified to the *sn*-2[1] position of glycerophospholipids in the lipid bilayer and does not exist as the free carboxylic acid. Thus, the random nature of the free radical process leads to the initial oxidation of those glycerophosphocholine lipids containing arachidonic acid.

In the past, the study of lipid peroxidation has been largely investigated through the measurement of stable molecules of small size such as pentane, malonyldialdehyde and 4-hydroxynonenal [19]. Recently, more complex structures of free radical derived eicosanoids have been suggested to be initial oxidation products and more useful as markers of lipid peroxidation [20]. Perhaps even more striking was the discovery that several of these eicosanoids possess biological activity and as such may mediate some of the events of free radical induced injury of tissues within the biological system.

The first such eicosanoid discovered was termed F_2-isoprostane because it was isomeric to prostaglandin $F_{2\alpha}$ derived from cyclooxygenase [21]. The mechanism of formation of this molecule has been proposed to be similar to that of the enzyme catalyzed reaction, however, the fact that it is a free radical process means stereochemical control of the reaction is lost. Thus, the initial oxidation of arachidonic acid by a free radical process can result in the formation of four separate regioisomeric families of F_2-isoprostanes with each regioisomeric family containing eight different isomeric compounds differing at the asymmetric centers. This results in a theoretical total of 64 different compounds produced by the free radical process as outlined in Scheme 1 for the formation of F_2-isoprostanes [21].

3. Mass Spectrometry of F_2-isoprostane Containing Phospholipids

The technique of fast atom bombardment mass spectrometry altered completely the strategies used to structurally characterize glycerophosphocholine lipids. Prior to this desorption ionization technique, it was difficult if not impossible to obtain information of the intact molecule because of the very polar nature of the phosphate ester polar head group. While field

---

[1] The IUPAC-IUB recommended nomenclature of lipids can be found in Biochem. J. *178*:21-35 (1978). The stereospecific numbering system (*sn*) for glycerol is used to assign the positions of substituents as *sn*-1, *sn*-2 or *sn*-3 for the three different carbon atoms of glycerol.

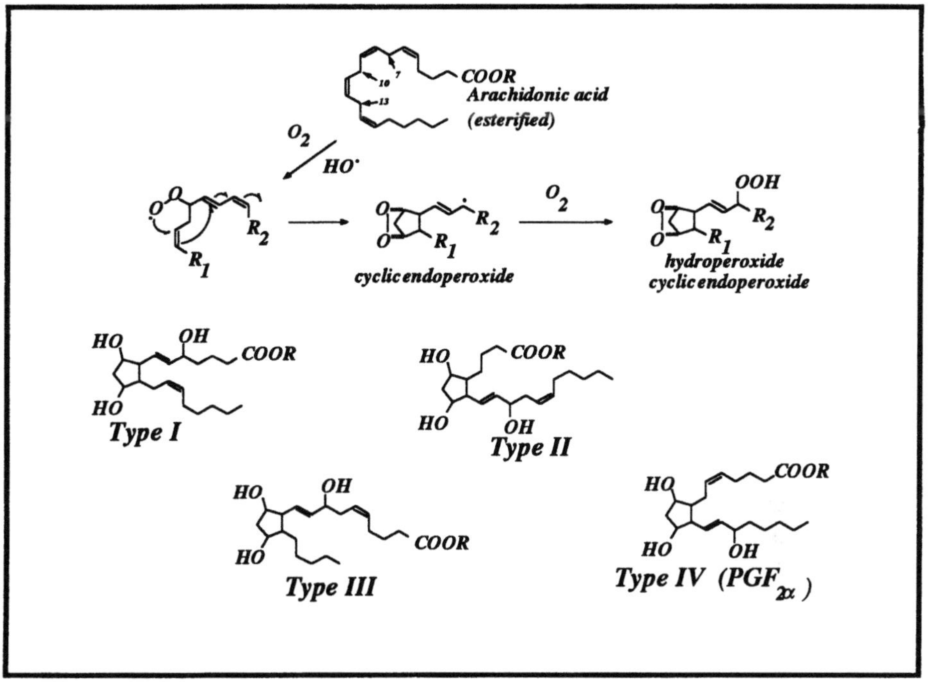

Scheme 1 Mechanism of free radical oxidation of arachidonic acid leading to isoprostanes. With permission, American Society for Mass Spectrometry (37).

desorption mass spectrometry offered some promise to the analysis of these compounds [22], it was not until the development of fast atom bombardment that ions of the intact chemical structure of phospholipids could be obtained routinely with very small quantities of isolated phospholipid. Very abundant positive ions could be obtained for the quaternary ammonium glycerophosphocholine lipids as well as glycerophosphoethanolamine and glycerophosphoserine phospholipids [15]. Negative ions were abundant for the acidic phospholipids such as glycerophosphoinositol and its polyphosphorylated analogs [23]; however it was still rather difficult to analyze larger lipids such as cardiolipin by FAB [24]. Nevertheless, it was possible to determine unambiguously both the polar head group and the fatty acyl composition of such glycerophospholipids. For example, major ions following collision induced decomposition of the $[M+H]^+$ ions could be used to distinguish glycerophosphocholines (*m/z* 184) [10,25], glycerophosphoethanolamine $[M+H-141]^+$, and glycerophosphoserine $[M+H-185]^+$ [10]. Collision induced decomposition of $[M-H]^-$ ions from glycerophosphoinositol yielded a characteristic set of complex ions at *m/z* 241, 259, and 299 [23].

Most striking was the facile determination of molecular species of glycerophospholipids for the first time. This was a consequence of the ability to study the abundant negative ions produced by all of the glycerophospholipids including glycerophosphocholine lipids which lose a methyl group from the quaternary ammonion ion choline residue to yield a negative ion [M-

15]⁻ [10]. The carboxylate anions from the esterified fatty acyl groups appeared as the most abundant ions at low mass following collisional activation of the molecular anion species. The carboxylate anions could be used to assign not only the nature of both fatty acyl groups at *sn*-1 and *sn*-2, but also their distribution as being precisely at *sn*-1 or at *sn*-2. An example is shown in Figure 1 for the fast atom bombardment mass spectrum of 1-octadecanoyl-2-arachidonoyl-glycerophosphocholine following collisional activation of the negative ion [M-15]⁻ [25]. The abundant carboxylate anion at *m/z* 283 established the presence of hexadecanoic acid in this phospholipid and *m/z* 303 the presence of arachidonic acid. The relative ratios of *m/z* 283 to 303 in the spectrum allowed a first approximation of the position of the arachidonate at *sn*-2 with *m/z* 303 being more abundant than *m/z* 283. A more precise determination requires the analysis of a different ion [M-86]⁻ which undergoes collision induced decomposition to lose the *sn*-2 position as a free carboxylic acid in a more facile manner than the *sn*-1 carboxylic acid loss. The ratio of these two ions was found to reflect the position of the fatty acyl group at either *sn*-1 or *sn*-2 in an all molecular species that have been studied to date [28].

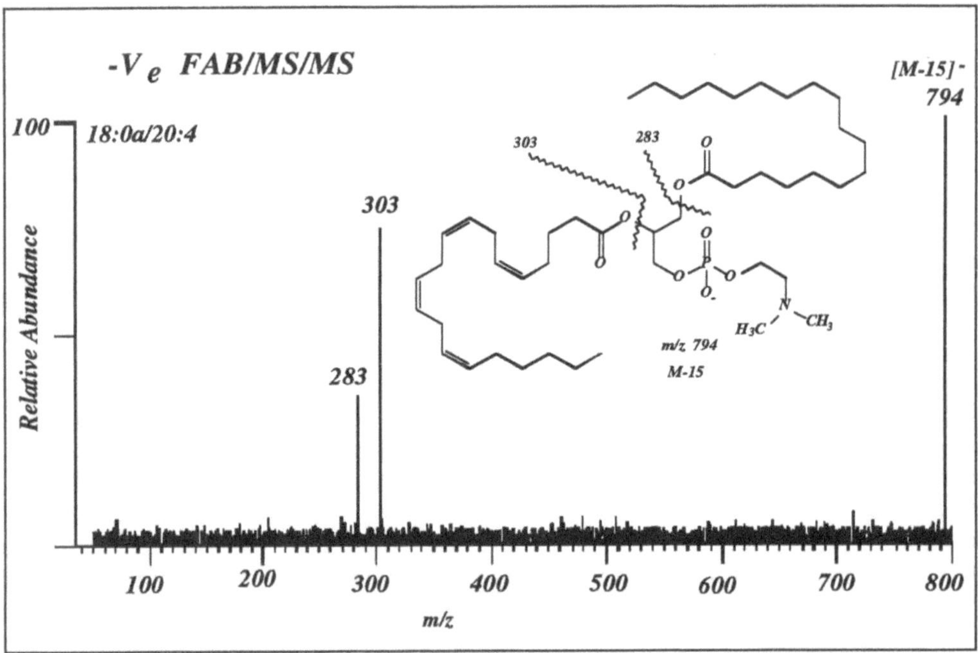

Figure 1 CID MS/MS of $[M\text{-}CH_3]^-$ produced by fast atom bombardment mass spectrometry of 1-octadecanoyl-2-arachidonoyl-glycerophosphatidylcholine.

Fast atom bombardment mass spectrometry also permitted the study of lipids extracted from rat livers after the administration of carbon tetrachloride to induce a cytochrome P-450 dependent lipid peroxidation *in vivo* [18]. The phospholipids were isolated and purified by HPLC and then analyzed by fast atom bombardment mass spectrometry. Those glycerophosphocholine lipids which contain the F_2-isoprostane moiety esterified to the glycerophosphocholine backbone were determined using precursor ion scans for *m/z* 353 which corresponds to the carboxylate anion of F_2-isoprostane ($C_{20}H_{33}O_5^-$). The mass spectrum in Figure 2 shows abundant ions at *m/z* 844, 816, 799, 773, 771, and 745 which were collisionally activated to yield this ion. Since the fast atom bombardment ionization of glycerophosphocholines leads to a triad of ions corresponding to $[M\text{-}15]^-$, $[M\text{-}60]^-$, and $[M\text{-}86]^-$, two compounds were identified with this triad of ions consistent with F_2-isoprostanes esterified to the glycerol backbone. The most abundant of this series corresponded to *m/z* 844 $[M\text{-}CH_3]^-$ and collision induced decomposition yielded the product ion spectrum shown in Figure 2B. Major ions at *m/z* 283 and 353 resulted corresponding to octadecanoic acid and the F_2-isoprostane carboxylate anion, respectively. The ratio of these ion abundances suggested that the F_2-isoprostane carboxylate anion was esterified to the *sn*-2 position. Precise determination of this was carried out by collision induced decomposition of the $[M\text{-}86]^-$ ion at *m/z* 773 with a more abundant loss of the carboxylic acid with a molecular weight of 354 (*m/z* 419) to that of the loss of the carboxylic acid with a molecular weight of 284 observed at *m/z* 489 [28]. The other abundant isoprostane phosphatidylcholine species corresponded to the 1-hexadecanoyl-2-F_2-isoprostane glycerophosphocholine.

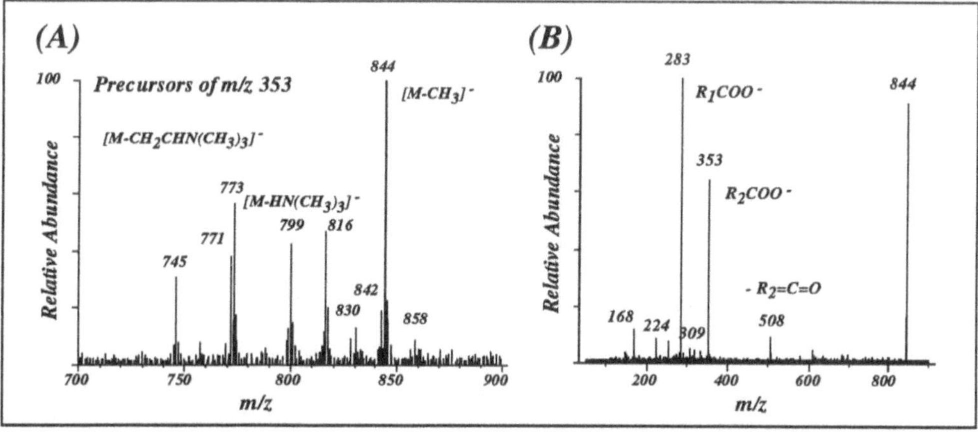

Figure 2 (A) Precursor ion scan for *m/z* 353 ($C_{20}H_{33}O_5^-$) from negative ions produced by FAB of GPC fraction from rat liver lipid extract following administration of CCl_4 to the rat. FAB of GPC produces high mass negative ions corresponding to losses of parts of the choline moiety. Ions at *m/z* 844, 799, and 773 correspond to the $[M\text{-}CH_3]^-$, $[M\text{-}HN(CH_3)_3]^-$, and $[M\text{-}CH_2CHN(CH_3)_3]^-$ ions from a major GPC molecular species that fragment to produce an ion with *m/z* 353. Ions at *m/z* 816, 771, and 745 correspond to these same ions for another GPC molecular species. (B) CID product ion spectrum of *m/z* 844, the $[M\text{-}CH_3]^-$ ion from a major precursor of *m/z* 353. With permission, J. Lipid Research (18).

Further confirmation of the nature of the F_2-isoprostane phospholipid (Scheme 2) was obtained by collision induced decomposition of the carboxylate anion at *m/z* 353, which was obtained during the fast atom bombardment process and observed as an ion source derived species. The collision induced decomposition of $PGF_{2\alpha}$ had been previously studied [29] and found to be somewhat unique in that it underwent a facile loss of C_2H_4O from *m/z* 353 to yield an ion at *m/z* 309. This loss of 44 Da was characteristic for several compounds containing a

1,3- diol in a cyclic structure. The collision induced decomposition of *m/z* 353 obtained from the analysis of the F_2-isoprostane glycerophosphocholine lipids yielded a mass spectrum qualitatively very similar to that of authentic $PGF_{2\alpha}$ as shown in Figure 3. The abundant ion corresponding to the loss of 44 Da was consistent with the presence of 1,3-diol cyclopentane structure in this carboxylate anion.

Scheme 2 Isoprostane phosphatidylcholine

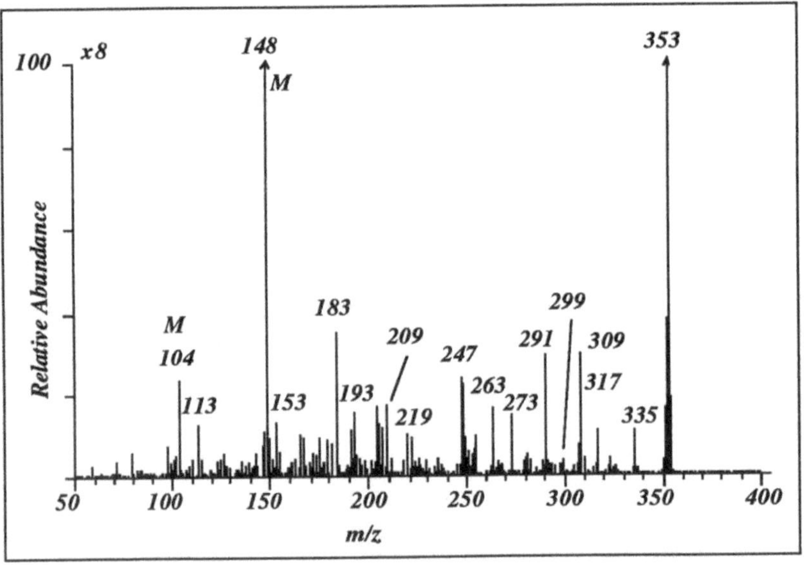

Figure 3 CID product ion spectrum of *m/z* 353 produced by FAB of isoprostane-GPC fraction. Ions at *m/z* 335, 317, 309, 291, and 273 are indicative of the presence of a F_2-ring and three hydroxyl moieties. Several matrix fragment ions (marked M) are present due to the overlap of *m/z* 353 with matrix ions. Relative abundances are magnified 8-fold. With permission, Journal of Lipid Research (18).

Electrospray ionization mass spectrometry has been found to be an even more sensitive technique for the generation of structurally significant positive and negative ions from intact glycerophospholipids [30-33]. Very little differences have been observed for electrospray ionization in the types of ions formed for the different phospholipid classes compared to fast atom bombardment ionization. Furthermore, the collision induced decomposition products of the abundant molecular ion species have been identical. For glycerophosphocholine lipids there do appear to be abundant adduct anions such as $[M+Cl]^-$ and $[M+acetate]^-$ anions when trace chloride and acetate containing buffers are employed in the spray solution [32]. These adduct anions decompose to the $[M-15]^-$ anion typically observed in the fast atom bombardment experiment. Collision induced decomposition of either positive or negative ions provide a wealth of structural information from these ions obtained by electrospray ionization identical to that described above. Additionally, electrospray ionization enables the on-line connection of a liquid chromatograph to the mass spectrometer ion source. This degree of freedom of online LC/MS and LC/MS/MS has been an exceedingly powerful technique for phospholipid analysis in terms of sensitivity and complex mixture analysis.

4. F_2-isoprostanes

In 1988, Roberts and co-workers [34] were studying the metabolism of prostaglandin D_2 in human subjects and reported the appearance of a family of isomeric prostaglandin $F_{2\alpha}$ compounds that were thought to be metabolites. Subsequent to this, they found that these $PGF_{2\alpha}$ compounds increased in abundance in plasma or urine after storage at -20° C for one month [34]. This suggested such molecules resulted from *in vitro* oxidation of $PGF_{2\alpha}$ which led to the realization that such compounds isomeric to $PGF_{2\alpha}$ could, in fact, be formed *in vitro* by free radical oxidation of arachidonic acid. The detection of these compounds was carried out using mass spectrometry and negative ion chemical ionization of the pentafluorobenzyl ester trimethylsilyl ether derivative (Figure 4A). The electron ionization mass spectrum of one of the F_2-isoprostanes, *viz.* 8-epi-$PGF_{2\alpha}$, is shown in Figure 4B using electron ionization conditions (positive ions). A wealth of structural information is available in this mass spectrum as seen from the numerous fragment ions and specific structural details can be gleaned from this mass spectrum; however, it is somewhat difficult to determine the exact location of double bonds. The position of the side chain hydroxyl group is indicated by the significant ion at *m/z* 173.

The typical analysis of free radical derived F_2-isoprostanes isolated from biological matrices such as urine, plasma or liver tissue following saponification reveals a family of closely related compounds eluting from the gas chromatograph. A typical GC/MS trace using negative ion chemical ionization of the pentafluorobenzyl ester trimethylsilyl ether derivative (PFB,TMS) is shown in Figure 5. Measurement of the abundant carboxylate anion (*m/z* 569) is the basis of the quantitative assay for F_2-isoprostanes by negative ion CI mass spectrometry using either D_4-$PGF_{2\alpha}$ or D_4-8-epi-$PGF_{2\alpha}$ as internal standard. As can be seen in Figure 5, a number of isoprostane components are present in this sample with the internal standard D_4-$PGF_{2\alpha}$ eluting as shown.

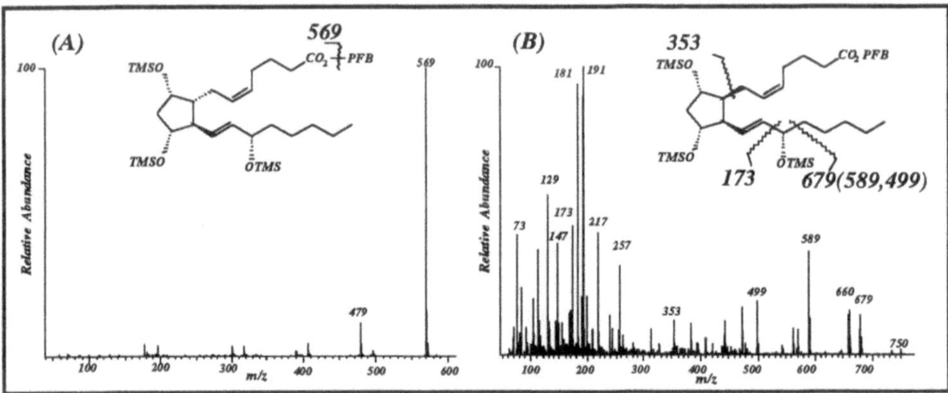

Figure 4 (A) Negative ion chemical ionization mass spectrum of the pentafluorobenzyl ester trimethylsilyl ether derivative of $PGF_{2\alpha}$. (B) Electron ionization mass spectrum of the pentafluorobenzyl ester trimethylsilyl ether derivative of 8-epi-$PGF_{2\alpha}$.

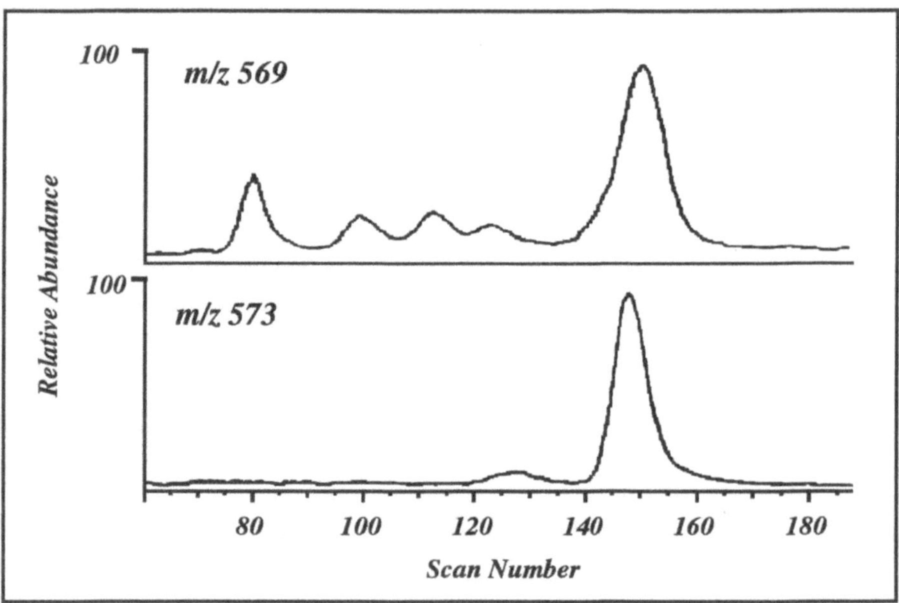

Figure 5 GC/MS trace of an extracted urine sample (3 mL) using negative ion chemical ionization of the pentafluorobenzyl ester trimethylsilyl ether derivative and selected ion monitoring for *m/z* 569 (F_2-isoprostanes) and *m/z* 573 (D_4-$PGF_{2\alpha}$ as internal standard, 50 ng).

Mass spectrometry has also been used to identify D_2/E_2-isoprostane formed in the liver of rats treated with carbon tetrachloride to induce lipid peroxidation [35]. Negative ion chemical ionization mass spectrometry was used to detect the elution of the derivatized D_2/E_2-isoprostane (*m/z* 524) in the GC/MS analysis and *m/z* 528 to indicate the elution of the D_4-PGE_2 internal standard. For these experiments, the methoxime PFB,TMS derivative was made which resulted in two peaks (*syn* and *anti* isomers) of the O-methyloxime derivative eluting from the gas chromatograph. As was the case for the F_2-isoprostane analysis by negative ion chemical ionization GC/MS, multiple peaks were evident in revealing the free radical nature of formation of these molecules.

Direct analysis of oxidized phosphocholine lipids purified by HPLC was also carried out [35]. Collision induced decomposition of a candidate $[M-15]^-$ anion at *m/z* 834 revealed the presence of an anion at *m/z* 351 which was the expected mass for a PGD_2/E_2 carboxylate anion. Collision induced decomposition of *m/z* 351 yielded product ions that were very similar to the decomposition products of cyclooxygenase derived PGD_2 and PGE_2 [29]. In separate studies, one of the E_2 isoprostanes (8-iso-PGE_2) was found to be a potent renal vasoconstrictor in the rat. This biological activity could be reduced by a thromboxane receptor antagonist suggesting that 8-epi-PGE_2 exerted its effect through the TxA_2/PGH_2 receptor.

Mass spectrometry remains the standard reference method for the determination of F_2-isoprostanes and the more recently discovered E_2/D_2-isoprostanes [35]. Other assays have been developed using specific antibodies [36], however, little is known concerning the cross reactivity of different regioisomeric F_2-isoprostanes with the specific antibodies raised against 8-epi-$PGF_{2\alpha}$.

5. F_2-isoprostane Regioisomers

One of the questions which has been asked is whether or not specific regioisomers of F_2-isoprostanes are formed *in vivo* or *in vitro* by free radical oxidation of arachidonic acid. In order to address this question, it has been necessary to develop methods by which one can distinguish the different regioisomeric compounds from each other. F_2-isoprostanes exist as four different regioisomeric families. For example, cyclooxygenase produces only Type IV regioisomeric family types (Scheme 1). The four F_2-isoprostane regioisomeric families are structurally similar in that all contain the 1,3-diol in a cyclopentane ring; however, they differ as to the length of the methyl terminus alkyl chain and the position of the third hydroxyl group not present in the cyclopentane ring. These structural features have provided a means by which mass spectrometry can be used to distinguish the four regioisomeric family subtypes from each other. Electron ionization mass spectrometry provides sufficient information to locate the position of the hydroxyl moiety on the alkyl chain whose position differs in each of the subtypes. In order to maximize the information concerning the position of this hydroxyl group, it was found that reduction of all the double bonds lead to an enhancement of electron ionization induced fragmentation adjacent to the trimethylsilyl ether group known as α-cleavage [37]. As shown in Figure 6, the electron ionization mass spectra of the methyl ester trimethylsilyl ether derivative of the reduced F_2-isoprostanes have been obtained for each of the regioisomeric subtypes. The diagnostic ions for Type I appeared at *m/z* 203 and 397 as abundant ions, Type II at *m/z* 215 and 385, Type III at *m/z* 245 and 355, and Type IV at *m/z* 173 and 427. These ions establish the position of the side chain hydroxyl group to be at C-5 (Type I), C-12 (Type II), C-8 (Type III), and C-15 (Type IV, $PGF_{2\alpha}$), respectively. Additional ions corresponding to cleavage adjacent to the cyclopentane ring leads to the formation of other additional unique ions for each regioisomeric subtype. The combination of HPLC separation and capillary gas chromatographic separation separates many of the isomeric F_2-isoprostanes, but even with these two powerful chromatographic techniques used sequentially, it has not been possible to completely separate all of the stereoisomers.

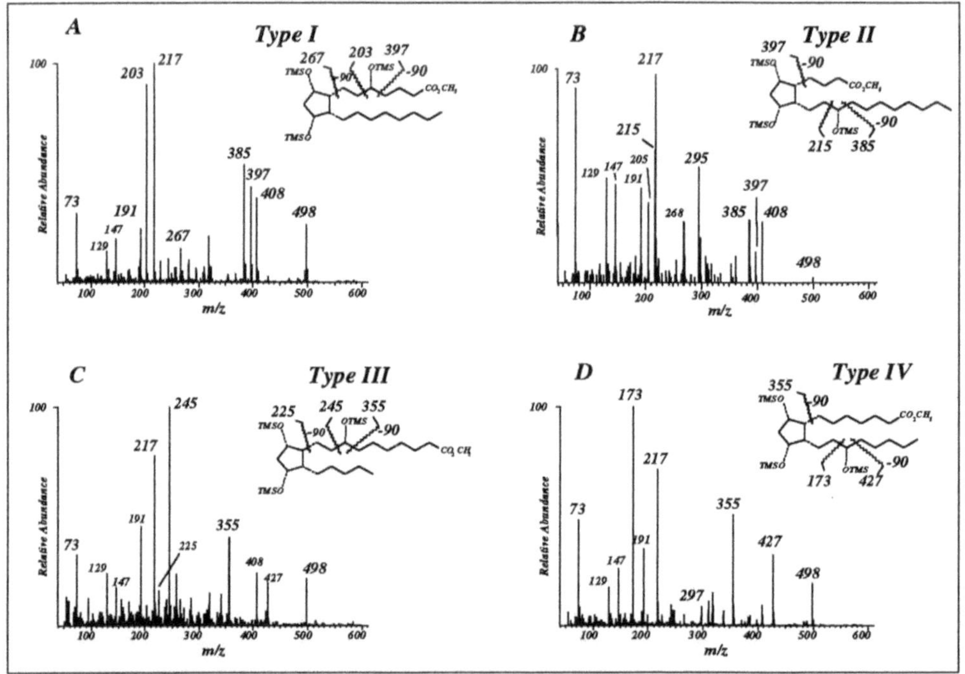

Figure 6 Electron ionization mass spectrum of reduced methyl ester trimethylsilyl ether derivative of F_2-isoprostane regioisomeric families obtained by free radical oxidation (modified Fenton reaction) of arachidonic acid. (A) Type I F_2-isoprostane component that displays abundant α-cleavage ions at *m/z* 203 and 397. (B) Type II F_2-isoprostane that displays abundant α-cleavage ions at *m/z* 215 and 385. (C) Type III F_2-isoprostane that displays abundant α-cleavage ions at *m/z* 245 and 355. (D) Type IV F_2-isoprostane that displays abundant α-cleavage ions at *m/z* 173 and 427. Dashed lines in the chemical structures are used to indicate unknown stereochemistry. With permission, American Society for Mass Spectrometry (37).

The direct connection of the liquid chromatograph to the mass spectrometer has also been used to develop a procedure to identify the F_2-isoprostanes eluting directly from the HPLC [37]. This tandem mass spectrometric technique is based upon the unique loss of 44 Da (C_2H_4O) from the carboxylate anion at *m/z* 353 of the underivatized F_2-isoprostane. This transition (*m/z* 353 → 309) can be used in a multiple reaction monitoring experiment to directly detect the elution of F_2-isoprostanes during an HPLC separation as shown in Figure 7. The addition of D_4-8-epi-PG$F_{2\alpha}$ and D_4-PG$F_{2\alpha}$ uniquely marks the elution of these specific eicosanoid isomers using the transition *m/z* 357 → 313 and enables either molecule to be used as an internal standard for quantitative analysis. This LC/MS/MS technique is highly sensitive and picomole quantities of F_2-isoprostanes can be detected.

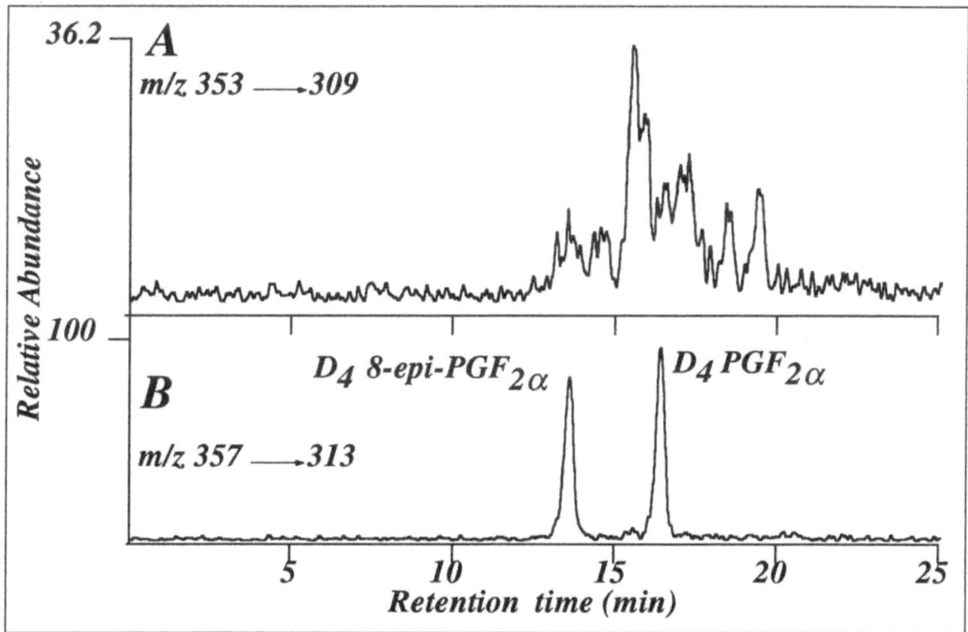

Figure 7 (A) Direct liquid chromatography-tandem mass spectrometry analysis obtained by electrospray ionization for the loss of 44 u from the carboxylate anion of F_2-isoprostanes (*m/z* 353). The F_2-isoprostanes were obtained by free radical oxidation (modified Fenton reaction) of arachidonic acid. (B) A loss of 44 u from *m/z* 357 corresponds to the deuterium labeled internal standards, [$^2H_4$]8-epi-PG$F_{2\alpha}$ and [$^2H_4$]PG$F_{2\alpha}$ as indicated. with permission, American Society for Mass Spectrometry (37).

Analysis of the regioisomeric families of F_2-isoprostanes generated by free radical reactions *in vitro* or *in vivo* has revealed a rather even distribution of each of the family subtypes. No unique formation of any one F_2-isoprostane family or stereoisomer has been observed; for example, there has been no enhanced formation of 8-epi-PG$F_{2\alpha}$ observed in the *in vitro* oxidation of arachidonic acid or in the *in vivo* formation of F_2-isoprostanes following carbon tetrachloride administration to rats (unpublished data). Thus, the measurement of any individual F_2-isoprostane would be an adequate marker for the formation of these unique molecules.

6. B_4-isoleukotrienes

Free radical oxidation of phospholipids containing arachidonic acid *in vitro* was found to yield numerous other products including the above mentioned F_2-isoprostanes, E_2/D_2-isoprostanes (*vide supra*). In addition, components have been found which contain a conjugated triene moiety as detected by UV absorption between 255 and 285 nm to indicate the elution of these compounds during reverse phase HPLC analysis [38]. Tandem mass spectrometry of those HPLC fractions, which displayed the characteristic UV chromophore of 270 nm absorption maximum with shoulders at 260 and 280 nm, revealed phospholipids which had a carboxylate anion at *m/z* 335. Such a carboxylate anion would correspond to the addition of one molecule of oxygen to arachidonic acid. Following saponification of these target phospholipids, subsequent reverse phase HPLC separation revealed several UV-absorbing components which had electrospray MS/MS spectra consistent with 5,12-dihydroxyeicosatetraenoic acid. This was based on the very abundant decomposition of *m/z* 335 to *m/z* 195 which has been studied in some detail to involve the decomposition shown in Scheme 3 [39]. Electron ionization mass spectra of the PFB,TMS derivative following saturation of the double bonds by catalytic reduction were identical to synthetic 5,12-dihydroxyeicosatetraenoic acid. Specifically, abundant ions at *m/z* 555, 401, 369, and 215 which corresponded to the characteristic α-cleavage ions adjacent to the trimethylsilyl ether derivative for the 5,12-dihydroxyeicosatetraenoic acids were abundant. Of further interest was the finding that certain B_4-isoleukotriene isomers were quite potent in eliciting an increase in intracellular free calcium levels in the human polymorphonuclear leukocyte [38]. The dose response curve for this effect yielded an EC_{50} value of approximately 30 nM which is 30-100 fold less potent than LTB_4, but nonetheless, quite potent in its own right as an activator of the human neutrophil.

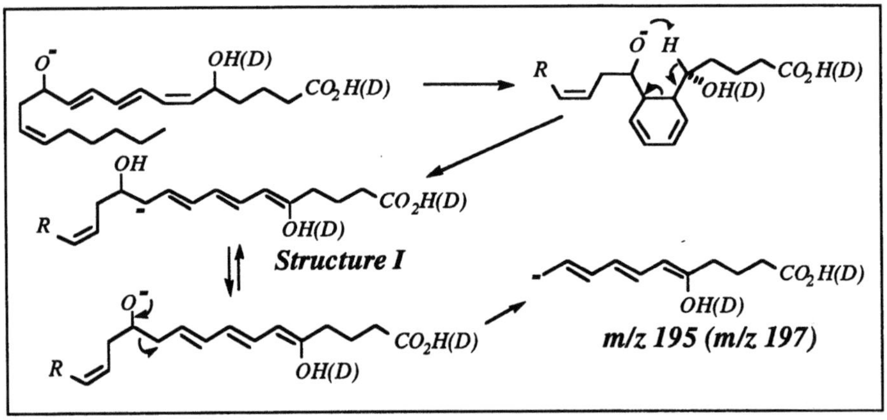

Scheme 3 Proposed mechanism for the abundant ion at *m/z* 195 indicating the presence of 5,12-dihydroxyeicosatetraenoic acids. With permission, American Society for Mass Spectrometry (39).

7. Lipid Mediators in Evolution

The mass spectrometric identification of the free radical products of arachidonic acid metabolism described above has unveiled a family of interesting and structurally complex molecules derived from the reaction of molecular oxygen with the polyunsaturated fatty acid, arachidonic acid. In addition to the F_2-isoprostanes, E_2/D_2-isoprostanes, and B_4-isoleukotrienes, other biologically active metabolites derived from the oxidation of glycerophospholipids have been discovered. These include oxidatively fragmented phospholipids derived either through oxidation of phospholipids *in vitro* [40] or from the oxidation of LDL particles *in vivo* [41]. Such oxidized phospholipids are known to exert their biological activity through the platelet activating factor receptor [40] as well as being inactivated by the platelet activating factor hydrolyzing enzyme, PAF acetyl hydrolase [42]. While the structure of these oxidized phospholipids has not been fully elucidated, it is known that they are not PAF itself which is 1-hexadecyl-2-acetyl glycerophosphocholine [43].

A new picture is now emerging from the structure elucidation of free radical derived oxidized metabolites of glycerophospholipids containing arachidonic acid and the discovery of biological activity of certain isomers mediated by cell membrane G-protein linked receptor proteins. This picture concerns the development of the complex machinery involved in lipid mediator biochemistry and the question - which came first, the lipid mediators or the receptors that recognize lipid mediators? One possibility is that the free radical generation of oxidized arachidonic acid compounds may have emerged as an early signal of lipid peroxidation causing damage to those earliest evolving cells. For those cells which had proteins that could recognize these oxidized products and induce specific biochemical responses, oxidative stress could be detected at the cellular level. In such a scenario, this would serve as a selection process to enhance survival of those cells which could effectively respond to the oxidative stress. During the evolutionary time scale, it is possible that both the recognition apparatus (protein receptors) for such oxidized molecules, as well as the biochemistry which led to the formation of such molecules. Specifically, the formation of eicosanoids under enzymatic control, could be refined so that only one unique isomer would emerge from the oxidation of arachidonic acid. Mass spectrometry has played a central role in structure elucidation of the large number of products formed by free radical oxidation of arachidonic acid and perhaps has given us a glimpse into the development of the complex and highly regulated events occurring within cells leading to the production of thromboxanes, prostaglandins, and leukotrienes as lipid mediators.

Acknowledgment: This work was supported in part by a grant from the National Institutes of Health (HL34303).

8. References

1. Ryhage, R. and Stenhagen, E. (1959) Mass spectrometric studies. I. Saturated normal long-chain methyl esters, *Arkiv Kemi* **13**, 523-533.
2. McLafferty, F.W. (1959) Mass spectrometric analysis: Molecular rearrangements, *Anal. Chem.* **31**, 82-87.
3. Green, K. (1969) Gas chromatography-mass spectrometry of O-methyloxime derivatives of prostaglandins, *Chem. Phys. Lipids* **3**, 254-272.

4. Hamberg, M., Svensson, J. and Samuelsson, B. (1975) Thromboxanes: A new group of biologically active compounds derived from prostaglandin endoperoxides, *Proc. Natl. Acad. Sci. USA* **72**, 2994-2998.
5. Murphy, R.C. (1995) Lipid mediators, leukotrienes, and mass spectrometry, *J. Mass Spectrom.* **30**, 5-16.
6. Budzikiewicz, H., Djerassi, C. and Williams, D.H. (1967) *Mass Spectrometry of Organic Compounds*, Holden-Day, San Francisco, pp. 3-17.
7. Hill, R.E. and Whelan, D.T. (1984) Mass spectrometry and clinical chemistry, *Clinica Chemica Acta* **139**, 231-294.
8. Gilpin, R.K. and Pachla, L.A. (1985) Pharmaceuticals and related drugs, *Anal. Chem.* **57**, 29R-46R.
9. Hites, R.A. (1970) Quantitative analysis of triglyceride mixtures by mass spectrometry, *Anal. Chem.* **42**, 1736-1740.
10. Murphy, R.C. and Harrison, K.A. (1994) Fast atom bombardment mass spectrometry of phospholipids. *Mass Spectrom. Rev.* **13**, 57-75.
11. Peter-Katalinić, J. And Egge, H. (1990) Desorption mass spectrometry of glycosphingolipids, *Methods in Enzymology* **193**, 713-733.
12. Morris, H.R. (1980) Biomolecular structure determination by mass spectrometry, *Nature* **286**, 447-452.
13. Rohwedder, W.K. (1985) Mass spectrometry of lipids labeled with stable isotopes. *Progress in Lipid Research.* **24**, 1-18.
14. Myher, J.J. and Kuksis, A. (1995) General strategies in chromatographic analysis of lipids, *J. Chromatogr.* **671**, 3-33.
15. Murphy, R.C. (1993) *The Handbook of Lipid Research: Mass Spectrometry of Lipids.* Plenum Press, New York.
16. Smith, W.L., Marnett, L.J. and DeWitt, D.L. (1991) Prostaglandin and thromboxane biosynthesis, *Pharmacol. Ther.* **49**, 153-179.
17. Smith, W.L. (1992) Prostanoid biosynthesis and mechanisms of action, *Am. J. Physiol.* **263**, F181-F191.
18. Kayganich, K.A., Rose, D.M., Murphy, R.C., Morrow, J.D. and Roberts, L.J. II (1993) Collision induced dissociation of isoprostane-containing phospholipids, *J. Lipid Res.* **34**, 1229-1235.
19. Halliwell, B. and Gutteride, J.M.C. (1989) *Free Radicals in Biology and Medicine*, Oxford University Press, United Kingdom.
20. Morrow, J.D., Hill, K.E., Burk, R.F., Nammour, T.M., Badr, K.F. and Roberts, L.J. II (1990) A series of prostaglandin F_2-like compounds are produced *in vivo* in humans by a non-cyclooxygenase, free radical-catalyzed mechanism, *Proc. Natl. Acad. Sci. USA* **87**, 9383-9387.
21. Morrow, J.D., Awad, J.A., Kato, T., Takahashi, K., Badr, K.F., Roberts, L.J.I. and Burk, R.F. (1994) Formation of novel noncyclooxygenase-derived prostanoids (F_2-isoprostanes) in carbon tetrachloride hepatotoxicity: An animal model of lipid peroxidation, *J. Clin Invest.* **90**, 2502-2507.
22. Wood, G.W. and Perkins, S.E. (1982) Purification technique for enhancement of field desorption mass spectra of phospholipids, *Anal. Biochem.* **122**, 368-371.
23. Sherman, W.R., Ackerman, K.E., Bateman, R.H., Green, B.N. and Lewis, I. (1985) Mass-analyzed ion kinetic energy spectra and B_1E-B_2 triple sector mass spectrometric analysis of phosphoinositides by fast atom bombardment, *Biomed. Mass Spectrom.* **12**, 409-413.
24. Jensen, N.J., Tomer, K.B. and Gross, M.L. (1987) FAB MS/MS for phosphatidylinositol, -glycerol, -ethanolamine and other complex phospholipids, *Lipids* 22:480-489.
25. Kayganich, K. and Murphy, R.C. (1991) Molecular species analysis of arachidonate containing glycerophosphocholines by tandem mass spectrometry, *J. Am. Soc. Mass Spectrom.* **2**, 45-54.
26. Kayganich, K. and Murphy, R.C. (1992) Fast atom bombardment tandem mass spectrometric identification of diacyl, alkyl-acyl, and alk-1-enyl acyl molecular species of glycerophosphoethanolamine in human neutrophils, *Anal. Chem.* **64**, 2965-2971.
27. Münster, H. and Budzikiewicz, H. (1988) Structural and mixture analysis of glycerophosphoric acid derivatives by fast atom bombardment tandem mass spectrometry, *Biol. Chem. Hoppe-Seyler* **369**, 303-308.
28. Huang, Z.H., Gage, D.A. and Sweeley, C.C. (1992) Characterization of diacylglycerophosphocholine molecular species by FAB-CAD-MS/MS: A general method not sensitive to the nature of the fatty acyl groups, *J. Am. Soc. Mass Spectrom.* **3**, 71-78.
29. Zirrolli, J.A., Davoli, E., Bettazzoli, L., Gross, M.L. and Murphy, R.C. (1990) Fast atom bombardment and collision-induced dissociation of prostaglandins and thromboxanes: Some examples of charge remote fragmentation, *J. Am. Soc. Mass Spectrom.* **1**, 325-335.
30. Smith, P.B.W., Snyder, A.P. and Harden, C.S. (1995) Characterization of bacterial phospholipids by electrospray ionization tandem mass spectrometry, *Anal. Chem.* **67**, 1824-1830.

31. Kerwin, J.L., Tuininga, A.R. and Ericsson, L.H. (1994) Identification of molecular species of glycerophospholipids and sphingomyelin using electrospray mass spectrometry, *J. Lipid Res.* **35**, 1102-1114.
32. Harrison, K.A. and Murphy, R.C. (1995) Negative electrospray ionization of glycerophosphocholine lipids: Formation of [M-15]⁻ ions occurs via collisional decomposition of adduct anions, *J. Mass Spectrom.* **30**, 1772-1773.
33. Han, X. and Gross, R.W. (1994) Electrospray ionization mass spectroscopic analysis of human erythrocyte plasma membrane phospholipids, *Proc. Natl. Acad. Sci. USA* **91**, 10635-10639.
34. Wendelborn, D.F., Seibert, K. and Roberts, L.J. II (1988) Isomeric prostaglandin F_2 compounds arising from prostaglandin D_2: A family of eicosanoids produced *in vivo* in humans, *Proc. Natl. Acad. Sci. USA* **85**, 304-308.
35. Morrow, J.D., Minton, T.A., Mukundan, C.R., Campbell, M.D., Zackert, W.E., Daniel, V.C., Badr, K.F., Blair, I.A. and Roberts, L.J.II (1994) Free radical-induced generation of isoprostanes *in vivo*: Evidence for the formation of D-ring and E-ring isoprostanes. *J. Biol. Chem.* **269**, 4317-4326.
36. Wang, Z., Ciabattoni, G., Créminon, C., Lawson, J., Fitzgerald, G.A., Patrono, C. and Maclouf, J. (1995) Immunological characterization of urinary 8-epi-prostaglandin $F_{2\alpha}$ excretion in man, *J. Pharm. Exp. Ther.* **275**, 94-100.
37. Waugh, R.J. and Murphy, R.C. (1996) Mass spectrometric analysis of four regioisomers of F_2-isoprostanes formed by free radical oxidation of arachidonic acid, *J. Am. Soc. Mass Spectrom.* **7**, 490-499.
38. Harrison, K.A. and Murphy, R.C. (1995) Isoleukotrienes: Biologically active free radical products of lipid peroxidation, *J. Biol. Chem.* **270**, 17273-17278.
39. Wheelan, P., Zirrolli, J.A. and Murphy, R.C. (1996) Negative ion electrospray tandem mass spectrometric structural characterization of leukotriene B_4 (LTB_4) and LTB_4-derived metabolites. *J. Am. Soc. Mass Spectrom.* **7**, 129-139.
40. Smiley, P.L., Stremler, K.E., Prescott, S.M., Zimmerman, G.A. and McIntyre, T.M. (1991) Oxidatively fragmented phosphatidylcholines activate human neutrophils through the receptor for platelet-activating factor, *J. Biol. Chem.* **266**, 11104-11110.
41. Lehr, H.-A., Menger, M.D. and Messmer, K. (1993) Oxidized LDL-induced leukocyte/endothelium interaction *in vivo* involves the receptor for platelet-activating factor, *Arteriosclerosis Thrombosis* **13**, 1013-1018.
42. Stremler, K.E., Stafforini, D.M., Prescott, S.M. and McIntyre, T.M (1991) Human plasma platelet-activating factor acetylhydrolase: Oxidatively fragmented phospholipids as substrates, *J. Biol. Chem.* **266**, 11095-11103.
43. Summers, J.B. and Albert, D.H. (1995) *Advances in Pharmacology*, Academic Press, New York, pp. 67-168.

THE ROLE OF MASS SPECTROMETRY IN THE CHARACTERIZATION OF PLANT METABOLITES FROM CRUDE EXTRACTS

ANTONIO DE NINO, ANNA NAPOLI and GIOVANNI SINDONA
Dipartimento di Chimica
Università della Calabria
I-87030 ARCAVACATA DI RENDE

ABSTRACT. A brief description of the capabilities of the modern methodologies of mass spectrometry in the plant kingdom is presented. The structure determination of some phytohormones, antibiotics and secondary plant metabolites extracted from plant is discussed.

1. Introduction

Hundreds of biologically active compounds are provided by the plant kingdom and many of them have shown important pharmacological activity. The complexity of crude plant extracts, whose components may differ considerably in their chemical structure, poses a formidable task when the characterization of each particular species present in the mixture has to be carried out by any of the available analytical methods. Several consecutive steps of chromatographic purification are routinely performed in order to isolate pure compounds. This procedure is often accompanied by conventional derivatization steps to enhance the lipophilicity of those hydrophilic compounds present in the crude extracts usually obtained with highly polar solvents [1].
Mass spectrometry is nowadays perfectly coupled with chromatographic separation methods and, with respect to other analytical methodologies, has the advantage of a specificity which allows structure determination of analytes in very small quantities. Moreover, when the charged species, produced during the ionization of the analytes are submitted to a multistage ion analysis the structure elucidation can be performed through well established mass spectrometry/mass spectrometry (MS/MS) methods [2]. The latter allow a straightforward correlation between the reacting ions and the products of their gas-phase fragmentation with a good to excellent signal-to-noise ratio without any need of previous separation of complex mixtures.

R. M. Caprioli et al. (eds.), Selected Topics and Mass Spectrometry in the Biomolecular Sciences, 517–531.

2. Plant diseases

Like any living species, plants are exposed to the action of phytopathogenic agents which may affect their growth. Bacteria from the *Pseudomonas* family are the casual agent of some widespread typical diseases [3,4]. *Pseudomonas syringae* causes the canker of Mediterranean olive trees as well as that of cherry and other stone fruit trees grown in the Pacific Northwest area. *Pseudomonas amygdali* is responsible for the hyperplastic canker of almonds. *Fusarium* is a field fungus that can attack a wide variety of plant kinds, particularly cereals, causing root, stem and ear rots [5]. Under certain environmental conditions the *fusarium* mould may produce mycotoxyns that can affect animals and , when contaminated plants are fed to the more tolerant livestock species, can accumulate in edible products, i.e. meat, milk, and eggs [6]. The number of available MS methods enables the selection of a technique which is suited for the specific problem to be solved.
Most of the recent applications in plant analysis are based on the use of well established liquid chromatography/mass spectrometry (LC/MS) techniques [7, 8]. In the following sections the role of MS/MS methods in plant analysis will be exploited with reference to some specific applications.

2. 1. *CHARACTERIZATION OF HORMONES OF THE CYTOKININ FAMILY*

Cytokinins, phytohormones isolated from culture filtrates of *Pseudomonas* bacteria, have the structure of modified adenines (**1-4**) and of their corresponding ribo (**5-7**) and deoxyribosides (**8**).

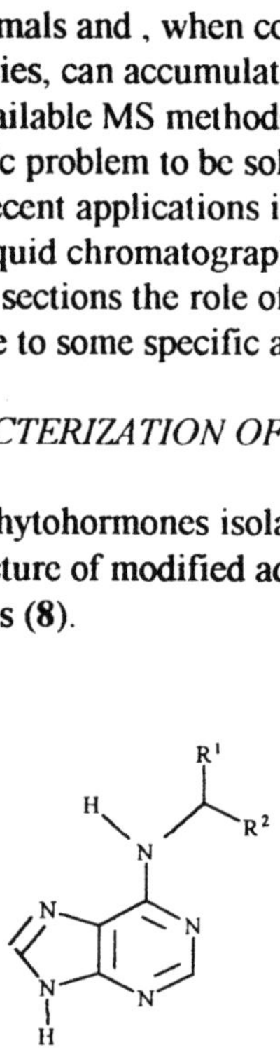

R^1 H N R^2 N N N N H

1-4

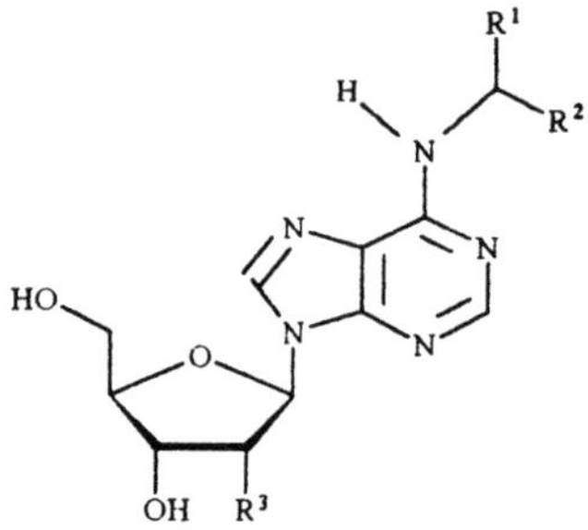

R^1 H N R^2 N N N N HO O OH R^3

5-8

1; R^1 = H, R^2 = 1-(2-methyl-1-propen-1-yl)
2; R^1 = H, R^2 = 1-(3-hydroxy-2-methyl-1-propen-1-yl)
3; R^1 = CH_3, R^2 = 1-(3-hydroxy-2-methyl-1-propen-1-yl)
4; R^1 = H, R^2 = 1-(3-hydroxy-2-methyl-1-prop-1-yl)
5; R^1 = H, R^2 = 1-(2-methyl-1-propen-1-yl), R^3 = OH
6; R^1 = H, R^2 = 1-(3-hydroxy-2-methyl-1-propen-1-yl), R^3 = OH
7; R^1 = CH_3, R^2 = 1-(3-hydroxy-2-methyl-1-propen-1-yl), R^3 = OH
8; R^1 = H, R^2 = 1-(3-hydroxy-2-methyl-1-propen-1-yl), R^3 = H

These highly polar and labile compounds are not amenable to those mass spectrometric techniques which require the evaporation of the analyte. Electron ionization (EI) [9, 10], and chemical ionization (CI) [11, 12], can be in fact successfully applied only on their trimethyl silyl and t-butyldimethylsilyl derivatives. The direct analysis of these fragile molecules can be performed by fast atom bombardment (FAB, 13) [14] or Ion spray ionization methods (ISI, 15) [16]. The importance of desorption methods is directly related to the possibility of detecting these molecules directly in the crude filtrates or likely in the plant tissues themselves. The gas phase chemistry of purine and pyrimidine nucleobases as well as that of their ribo and deoxyribosides has been extensively studied by FAB [17, 18], and has been also applied in the characterization of natural and chemically modified nucleosides [19, 20]. The drawbacks arising from the presence of matrix background peaks are almost inevitable when FAB is used for low molecular weight compounds, such as those of natural cytokinins, at sub-nanomolar scale, however, the complementary use of MS/MS techniques allowes an efficient characterization of all cytokinins **1-8** with a good signal-to-noise ratio [14]. Zeatins (**2-4**) and isopentenyladenine (**1**) are cytokinins formally derived by alkylation of adenine at the N-6 position. Their FAB positive (+) spectra are characterized by abundant $[M+H]^+$ ions and diagnostic fragments. A very useful peak at *m/z* 136, present in all the spectra, indicates that the zeatins have a common adenine "core" which preferentially eliminates the side chain groups as neutrals (scheme 1)

R¹ H R² N N N N N H +H⁺ **1-4** → NH₂ N N N N H +H⁺ m/z 136

scheme 1

The perfect match of the collision induced dissociation (CID) spectra of the species at *m/z* 136 with that of an authentic sample of adenine, in typical MS/MS experiments, enables a straightforward structural assignment of the common "core" of zeatins. This approach is suitable therefore for the identification of the purine nucleus and provide the first structural element for the characterization of the analytes. In the same conditions the MS/MS spectra of the $[M+H]^+$ species obtained from **1-4** provide the finger-printing of the zeatin derivatives so far examined and can be used for the characterization of their ribo and deoxyribosides **5-8**. The latter, in fact, follow the well established rules of gas-phase fragmentation of protonated nucleosides [17, 18] (scheme 2).

scheme 2

The aglycone of each cytokinin **5-8** is, therefore, released as a charged species and is present in their FAB (+) spectra at *m/z* 204, 220, 220 and 234, with 62, 84, 50 and 37% relative abundances, respectively. The structure determination of the protonated purines released from **5-8** was, therefore, carried out by MS/MS comparing their CID spectra with those of the appropriate zeatin aglycones **1-3**. A perfect match was always obtained as can be seen from figure 1 where the CID spectra of *m/z* 220 from **6** and **2** are reported.

The know-how obtained in nearly three decades in the nucleic acid field [17, 18] has therefore been used in a straightforward manner in the characterization of cytokinins from plants: this is probably one of the major achievements of MS/MS methods, i.e. their reproducibility and reliability.

The application of ESI/MS/MS, recently proposed [16], is of considerable interest since it could allow the direct identification of this plant tumoral marker directly from crude extracts.

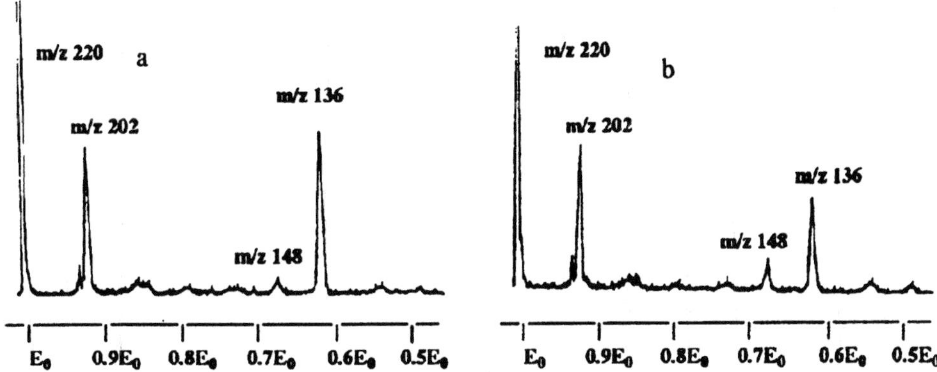

Figure1. CID spectra of the species at *m/z* 220 from cytokinins (a) **6** and (b) **2**

Interestingly, a natural defense mechanism against *Pseudomonas* has been proposed [21]. It has been found that hydroxytyrosol, present in the waste waters of olive oil mills and conjugated in olive leaf and fruits [22] (see 2.3.1.) is toxic to these bacteria. It might be possible in the future to recycle those natural polyphenols present in the vegetation waters (wastes from the olive oil industry) in an integrated pest management program. FAB, once more, was extensively used in the evaluation of the phenol contents of vegetable waters due to the polar character of the analytes and to their thermal fragility [23, 24].

2.2 *BIOLOGICALLY ACTIVE PYRONES FROM PLANTS*

Certain *Fusarium* species, whose pathogenic action was briefly described above, produce mycotoxyn, phytotoxins and antibiotics [25].

HO OH OH OH OH O O O R

9; R = OH; **10**; R = H

Two untifungal metabolites have been isolated from a strain of *fusarium semitectum* grown on rice kernels which are characterized by a common α-pyrone "core"[26]. The biological activity of compounds containing the pyrone structure produced by microorganisms is well documented [27], their isolation from crude filtrates and their structural characterization requires the synergetic use of all the possible analytical

methods. Mass spectrometry, however, is of unique importance since structural data can be obtained also from crude compounds by means of MS/MS techniques. The procedure adopted for the isolation of the two metabolites, called fusapyrone (9) and deoxyfusapyrone (10), which showed antifungal activity was complicated by the polar nature of the two molecules which had been isolated after complete acetylation of the fraction which gave a positive response to biological assays.
FAB (-) provided $[M-H]^-$ at *m/z* 605 (9, 100%) and *m/z* 589 (10, 100%). FAB(+) spectra from a nitrobenzyl alcohol matrix showed abundant protonated and sodiated molecular ions thus enabling the identification of the molecular weight of the two pyrones. The peracetylated derivatives (11, 12) also gave useful information confirming the presence of the exact number of free alcoholic functionalities.
The peculiarity of the fragmentation pathway taken by the two species is represented by the disconnection of the C_{13}-C_{14} bond (scheme 3) which shows that there is a gas phase process which does not depend on the location of charge.

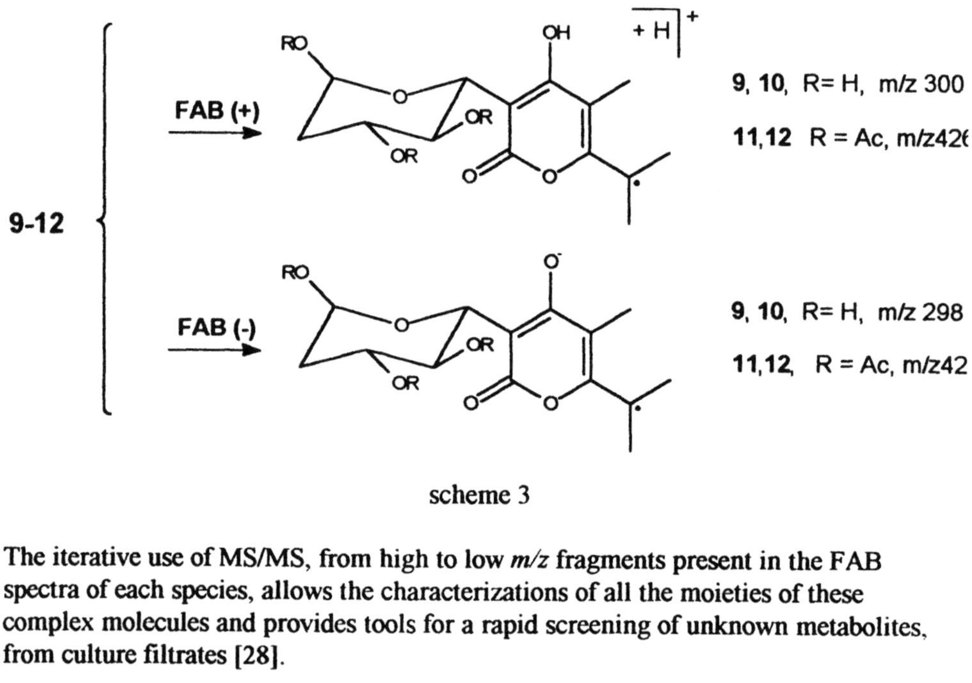

9-12

FAB (+) → [+ H]+ 9, 10, R= H, m/z 300; 11,12 R = Ac, m/z426

FAB (-) → 9, 10, R= H, m/z 298; 11,12, R = Ac, m/z42

scheme 3

The iterative use of MS/MS, from high to low *m/z* fragments present in the FAB spectra of each species, allows the characterizations of all the moieties of these complex molecules and provides tools for a rapid screening of unknown metabolites, from culture filtrates [28].

2.3 *IRIDOIDS AND SECOIRIDOIDS*

Among the great variety of differently structured metabolites present in plants, the terpenes deserve particular interest because they are widespread and possess a great variety of biological activity. The family of terpenes includes molecules where the isoprenic unit indicated as C_5, is present in the molecular structure.

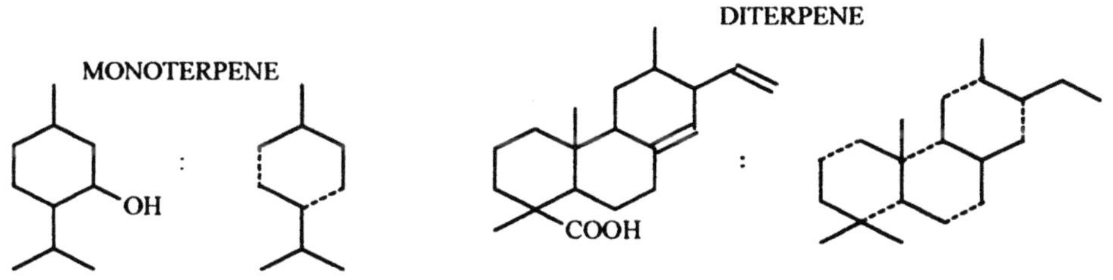

DITERPENE

MONOTERPENE

OH

COOH

MENTHOL

PIMARIC ACID

The term terpene or monoterpene is deserved to a ten carbon atom molecule, two C_5 units, regardless of the chemical functionality present on it, whereas a diterpene possess a total of twenty carbon atoms. Accordingly higher and lower homologues are easily defined.

The iridoids represent a sub-group of terpenes which include hundreds of natural monoterpenes, characterized by the presence of a cyclopentane ring. In many cases the position 1 and 3 are bound through an hemiacetal bridge and form a α-hydroxytetrahydropyran or dihydropyran structure fused on the cyclopentan ring. The complete structure of the molecules includes a double bond at C_3.

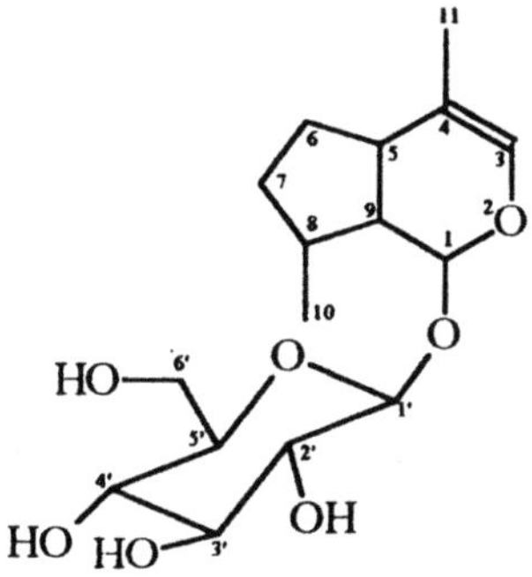

13

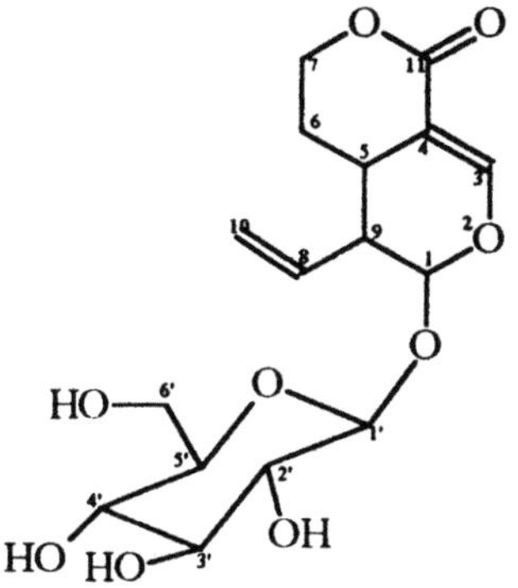

14

They are usually isolated from plant tissues (leaf, seeds, etc..) as glycosides where the aglycone is bound to the sugar through its C_1 carbon atom such as iridene glycoside (**13**) and sweroside (**14**).

The secoiridoids can be derived from the iridoids by a formal C_7—C_8 bond breakage of the cyclopentan ring. A secoiridoid glucoside, of typically bitter taste [29, 30], present both in leaves and the fruit of olive trees is Oleuropein (**15**). Oleuropein belongs to the pool of natural polyphenols which are supposedly active, among other, things as antioxidant, hypoglycaemic and anticholesteric agents. Several other phenolic compounds present in lower amounts in olive leaves such as demethyloleuoropein, ligstroside, verbascoside, oleoside dimethyl ester, nonglycosidic secoiridoids and oleuroside, have been isolated and tested for their potential therapeutic use.

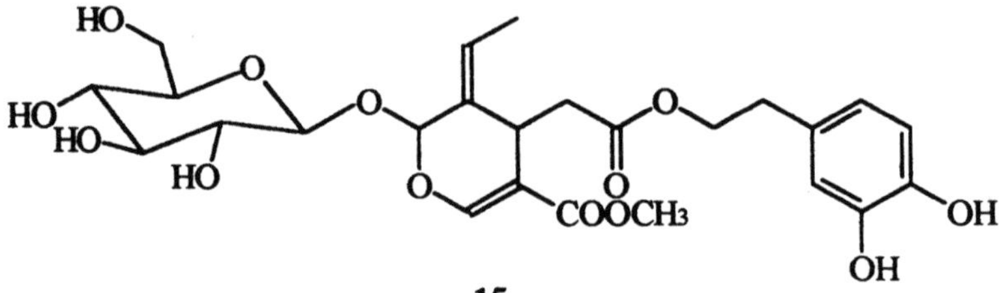

HO HO HO HO O O O O O COOCH3 OH OH

15

2.3.1-*MASS SPECTROMETRY OF IRIDOIDS FROM PLANT EXTRACTS*

Quite recently mass spectrometry has been applied with a good degree of success to the analysis of crude plant extracts. Gas-chromatography /mass spectrometry (GC/MS) is an analytical method suitable for the identification of peracetylated, permethylated and trimethylsylilated derivatives of iridoid and secoiridoid glycosides [31]. However, the information available has always been scarce and limited to the structure of the sugar moiety. The E.I. spectrum of peracetylated **15** showed a nearly absent molecular ion and some diagnostic fragments [32]. Field Desorption Mass Spectrometry [33] provided information on the molecular weight of the analyte through the appearance of $[M+H]^+$, $[M+Na]^+$ and $[M+K]^+$ ions [32]. Earlier applications of electron ionization (EI) methods on underivatised iridoids had led to the identification of some common pathways for the unimolecular dissociations of the aglycone (scheme 4) [34]. In EI mode, in fact, the glycoside did not survive the ionization region intact and, in the absence of molecular ions, abundant fragments corresponding to the ionized aglycones were always present. The latter undergo a series of consecutive neutral losses which are consistent with the existence of a cyclopentane ring.

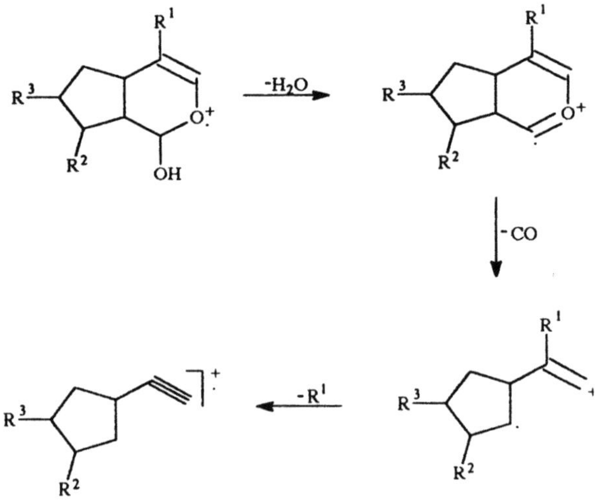

R^1 R^3 O^+ R^2 OH $-H_2O$ R^1 R^3 O^+ R^2 -CO R^1 R^3 R^2 $-R^1$ R^3 R^2

scheme 4

The presence of the dihydropyran ring was ascertained by the formation of an aromatic pyrilium ion, whose relative abundance was correlated with the nature of the

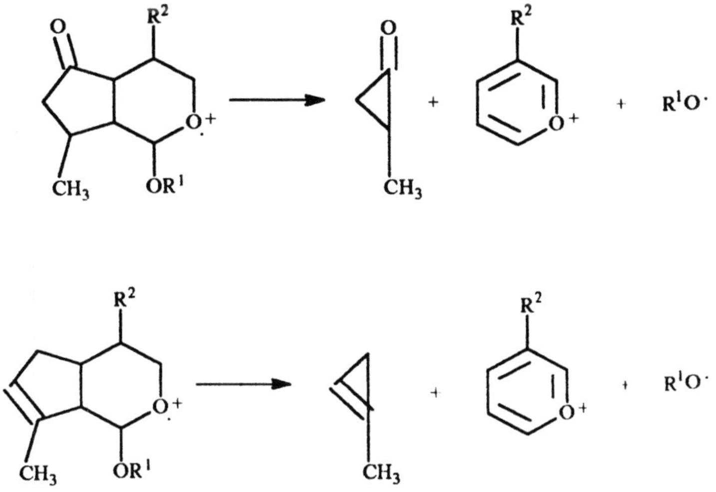

scheme 5

substituent in position 6 of the cyclopentane ring, the maximum intensity was obtained when a keto function was present (scheme 5).
The use of desorption chemical ionization (DCI) has allowed for the first time a proper identification of intact iridoid glycosides [35]. In this ionization mode, the molecular mass of the analytes can be easily determined by the presence in the spectra of the $[M+NH_4]^+$ and $[M+H]^+$ ions.The fragments due to sugar and aglycon moieties provide some general rules for theinterpretation of the spectra of iridoids [36]. The most abundant ionic species found in the spectra of four iridoid and seven secoiridoid glucosides of the Loganine and Gentiopicroside type, respectively, fit with the fragmentation path reported in scheme 6.

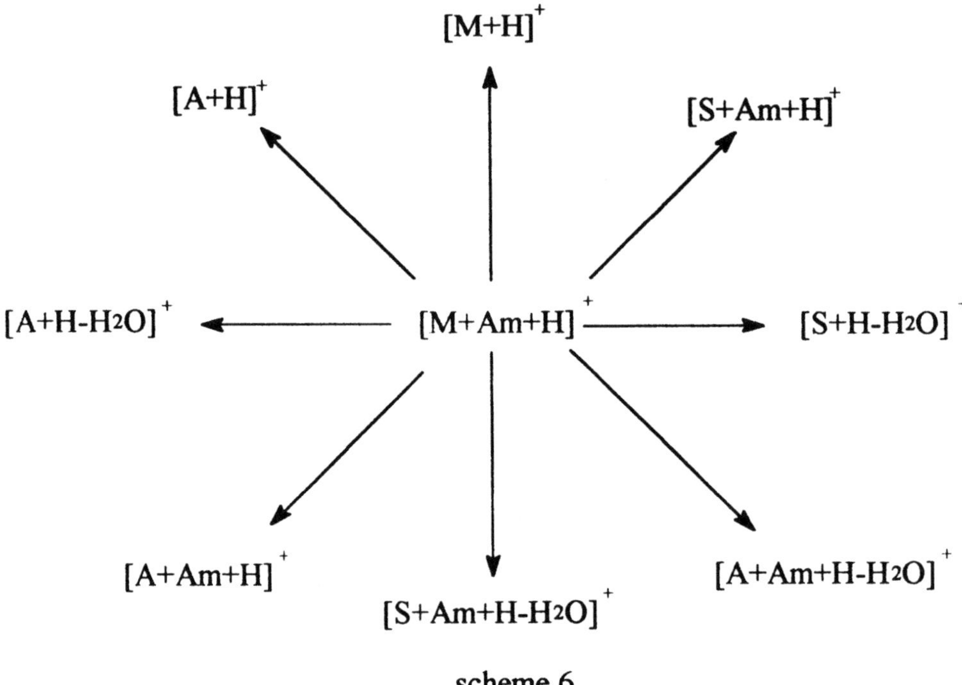

scheme 6

FAB has been used fo the identification of the molecular mass from $[M+Na]^+$ $[M+K]^+$ and $[M+H]^+$ ions[23, 24, 37-39], however a unique role in the analysis of crude plant extracts is played by LC/MS methods [1]. An exhaustive study was recently published by Hostettmann et al on the many approaches available for structure elucidation of secondary plant metabolites in crude extracts [8]. All the various mass spectrometric techniques were compared and the complementary nature of thermospray, continuous flow FAB, electrospray and atmospheric pressure chemical ionization was discussed. A significant number of iridoids are present in plants of the Oleaceae family, including olive trees, which are widespread in the Mediterranean area and are of significant importance for human nutrition. Phenolic glycosides are microelements of olive trees that contribute to certain plant properties, such us the colour of olive fruit [40], the nutritional value and stability of the oil [41], and their resistance to microorganisms [42]. The complexity of the chromatographic profiles of the polar fraction of the olive leaf extracts clearly indicate that many phenolic derivatives could be present in the natural matrix. Moreover, it cannot be excluded that both extraction procedures and sample treatments alter the structure of some of the secondary metabolites originally present in the plant tissues. (fig.2)

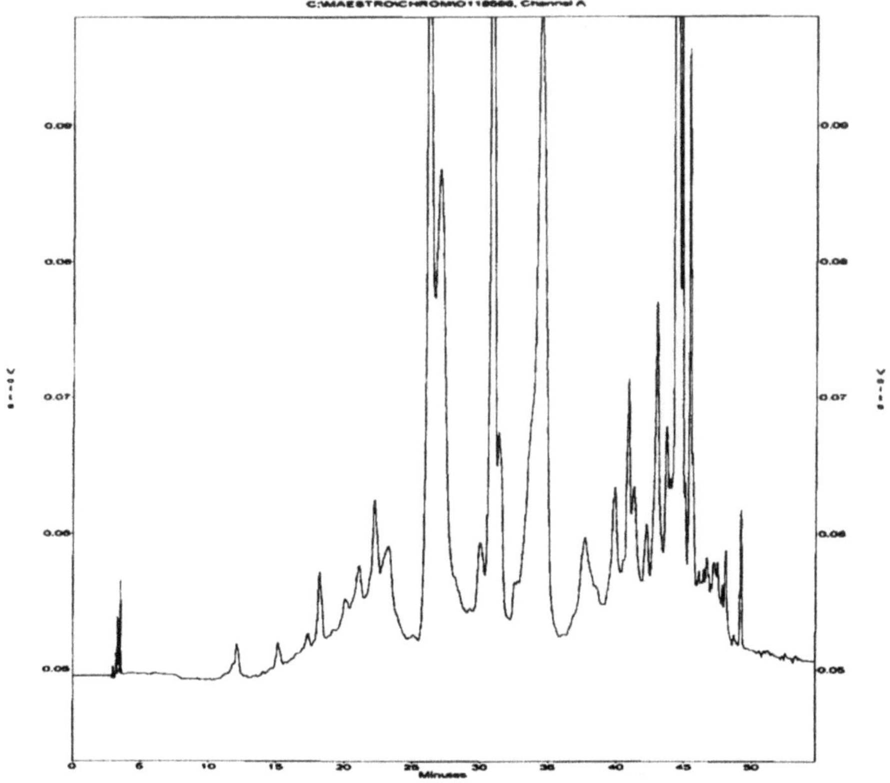

Figure2. Methanolic extract of "Carolea" olive leaf. HPLC, C_{18} i.d. 4.6 mm, length 25 cm, particle size 5μm, U.V. detector.

Ionspray[15] was used to nebulize and ionize the HPLC eluate in the atmospheric pressure source of the instrument, where the ionization of the analytes is driven by the field at the needle tip that disperses the emerging charged droplets. The extracts of young olive leaves from *Olea europaea* L. cv Cassanese, produced the reconstructed chromatogram reported in figure 3 [22].

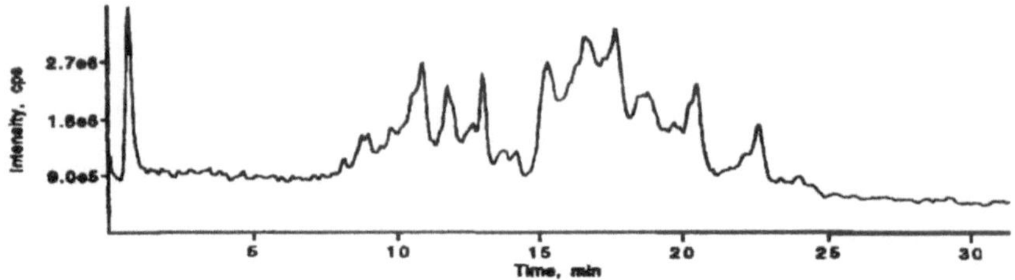

Figure 3. Methanolic extract of "Carolea" olive leaf. Reconstructed ion chromatogram. ISI-MS, C_{18} i.d. 4.6 mm, lenght 3 cm, particle size 3μm.

The presence of Oleuropein was very easily detected by matching the MS/MS spectrum of the $[M+NH_4]^+$ ion at *m/z* 558 (fig. 4a) in the extracted ion chromatogram with that of an authentic sample (fig. 4b) analyzed in the same experimental conditions (fig.4b).

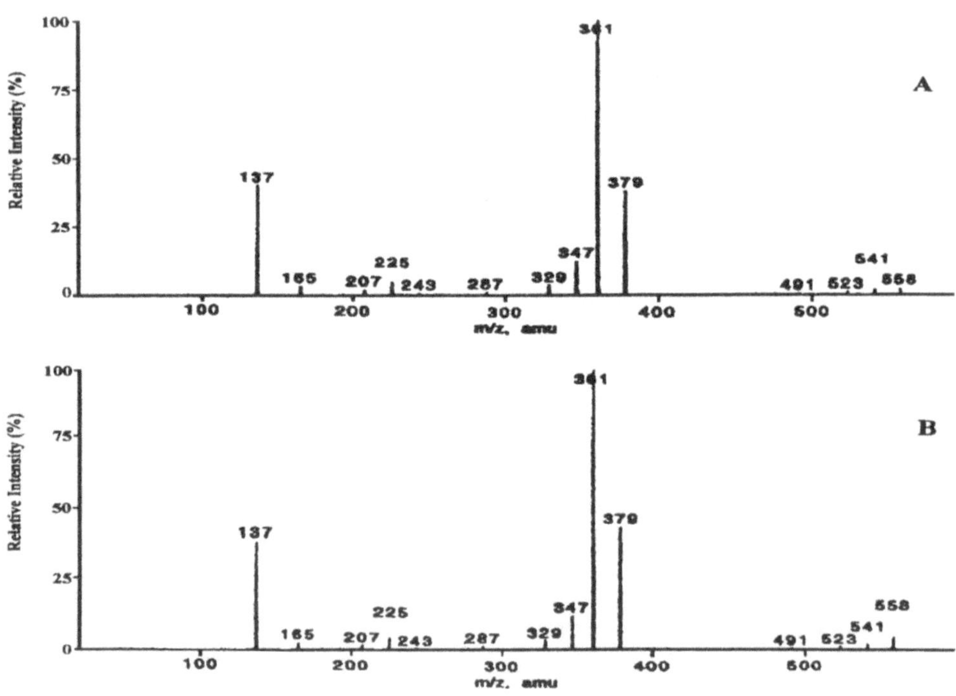

Figura 4. ISI-MS/MS spectra of $[M+NH_4]^+$ ion. (A) from ion chromatogram of fig. 3 and (B) from a standard sample.

The species at *m/z* 379 and 361 correspond to the Z^+ and Y^+ fragments of a glycosyl bond in protonated carbohydrates (scheme 7) [43] and are of foremost importance in structure determination. It was proposed [22] that the $[M+NH_4]^+$ species of oleuropein react, in the gas phase, through the elimination of neutral ammonia to give a transient $[M+H]^+$ which posses enough internal energy to undergo further fragmentation. The fragments at *m/z* 243 and 225, among others, are likely due to a McLafferty type rearrangement of the ester function involving the hydroxytyrosol moiety of Y^+ and Z^+ precursors, respectively. The species at *m/z* 137 is a very significant fragment since it allows the characterization of an important part of the aglycone, i.e. the phenol moiety. It may be formed as suggested in scheme 7 or directly from ionized **15** or from other ionic intermediates such as *m/z* 379. The relative abundance of *m/z* 137 reflects its highly stabilized benzyl cation structure to which the originally formed ion may rearrange. In conclusion the ISI MS/MS spectrum of oleuropein provides evidence of (i) the monosaccharide unit (Y^+ and Z^+ fragments), (ii) the aglycone moiety (*m/z* 243, 225 and 207, 165) and of (iii) the phenolic constituent (*m/z* 137). These ions can be considered as "markers" of the phenolic components of olive leaves which would allow their identification in complex mixtures by mass spectrometry.

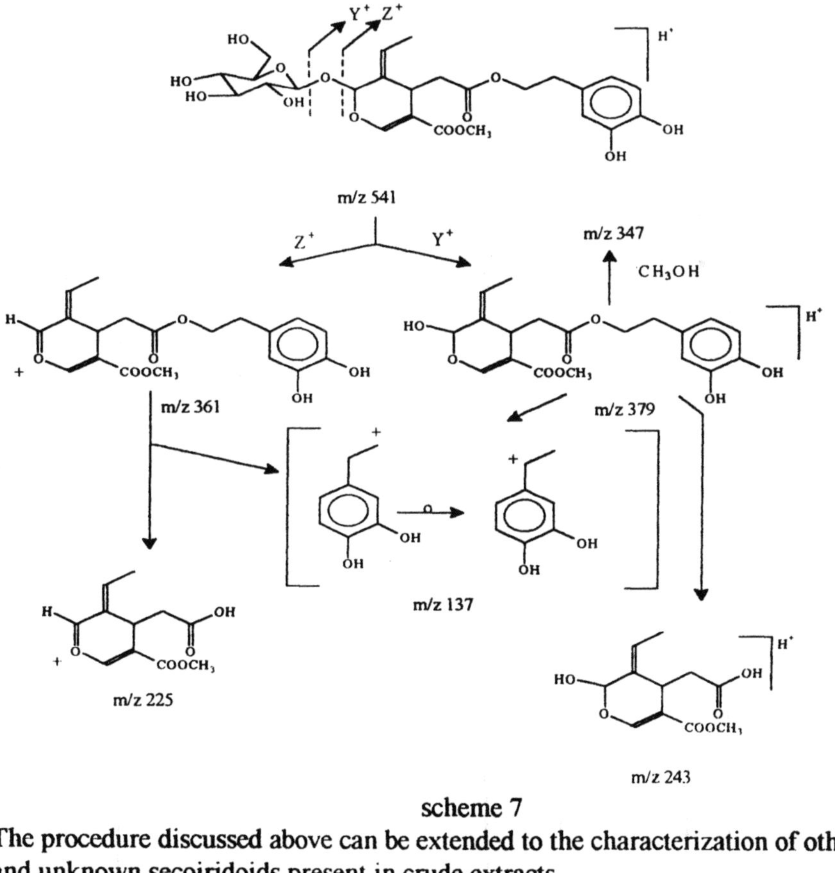

scheme 7

The procedure discussed above can be extended to the characterization of other known and unknown secoiridoids present in crude extracts.

3. References

1. Wolfender and K. Hostettmann, *Spectroscopy News*, **8**, 7 (1996)
2. (a) O. V. Nemirovskiy, J. K. Gooden, R. Ramanathan and M. L. Gross, this book; (b) K. L. Busch, G. L. Glish and S. A. McLuckey, *Mass Spectrometry/Mass Spectrometry: Techniques and Applications of Tandem Mass Spectrometry*, VCH, New York (1988)
3. A. Evidente, G. Surico, N. S. Iacobellis and G. Randazzo, *Phytochemistry*, **25**, 525 (1986)
4. A. Evidente, N. S. Iacobellis, R. Vellone, A. Sisto and G. Surico, *Phytochemistry*, **28**, 2603 (1989)
5. P. E. Nelson, T. A. Toussoun, R. J. Cook, *Fusarium-Diseases, Biology and Taxonomomy*, The Pennsylvania State University Press, University Park (1981)
6. W. F. O. Marasas, P. E. Nelson, T. A. Toussoun, *Toxigenic Fusarium Species-Identity and Mycotoxicology*. The Pennsylvania State University Press, University Park (1984)
7. C. R. Blakley and M. L. Vestal, *Anal. Chem.*, **55**, 750 (1983)
8. J-L. Wolfender, S. Rodriguez, K. Hostettmann and W. Wagner-Redeker, *J Mass Spectrom. & Rapid. Commun. Mass Spectrom.*, **S35** (1995)
9. J. Badenoch-Jones, D. S. Letham, C. W. Parker and B. G. Rolfe, *Plant Physiol.*, **75**, 1117 (1984)
10. C. H. Hocart, J. Badenoch-Jones, C. W. Parker, D. S. Letham and R. E. Summons, *J. Plant Growth Regul.*, **7**, 179 (1988)
11. C. H. Hocart and U. P. Schlunegger, *Rapid Commun. Mass Spectrom.*, **3**, 36 (1989)
12. C. H. Hocart and U. P. Schlunegger, *Rapid Commun. Mass Spectrom.*, **3**, 249 (1989)
13. M. Barber, R. S. Bordoli, R. D. Sedgwick and A.N. Tyler, *J. Chem. Soc. Chem. Commun.*, 325 (1981)
14. F. Greco, G. Sindona, N. Uccella and A. Evidente, *Org. Mass Spectrom.*, **27**, 750 (1992)
15. a) A. P. Bruins, J. R. Covey and J. T. Henion, *Anal. Chem.*, **59**, 2642 (1987); b) A. Raffaelli, this Book
16. T. Bartok, G. Borcsok, R. Komoroczy and F. Sagi, *J. Mass Spectrom.*, **31**, 756 (1996)
17. P. F. Crain, *Mass Spec. Review*, **8**, 505 (1990)
18. G. Sindona, in *Mass Spectrometry in the Biological Sciences: a Tutorial*, ed. by M. L. Gross, pp. 383-405, Kluwer, Dordrecht (NL) (1992)
19. A. Liguori, F. Greco, G. Sindona and N. Uccella, *Org. Mass Spectrom.*, **25**, 459 (1990)
20. A. Liguori, G. Sindona and N. Uccella, *Nucleosides Nucleotides*, **9**, 373 (1990)
21. R. Capasso, A. Evidente, L. Schivo, G. Orru, M. A. Marcialis and G. Cristininzio, *J. Appl. Bacteriol.* **79**, 393 (1995)

22. De Nino, N. Lombardo, E. Perri, A. Procopio, A. Raffaelli and G. Sindona, *J. Mass Spectrom.*, **32**, 000 (1997)
23. R. Capasso, G. Cristininzio, A. Evidente, F. Scognamiglio, *Phytochemistry*, **31**, 4125 (1992)
24. R. Capasso, A. Evidente, C. Visca, L. Gianfreda,M. Maremonti and G. Greco Jr., *Appl. Biochem. Biotech.*, **60**, 183 (1996).
25. R. F. Vesonder, P. Golinski, in *Fusarium-Mycotoxyns, Taxonomy and Pathogenecity*, ed. by J. Chelkowski, pp 1-39, Elsevier, Amsterdam (NL), (1989)
26. A. Evidente, L. Conti, C. Altomare, A. Bottalico, G. Sindona, A.L. Segre and A. Logrieco, *Natural Toxins*, **2**, 4 (1994)
27. J. M. Dickinson, *Nat. Prod. Report*, **10**, 71 (1993)
28. A. Napoli, G. Sindona, in preparation
29. L. Panizzi, M. L. Scarpati and G. Oriente, *Gazz. Chim. Ital.*, **90**, 1449 (1960)
30. H. Inouye, T. Nishioka, M. Kaniva, *Phytochemistry*, **14**, 304 (1975)
31. H. Inouyc, K. Uobe, M. Hiraf, Y. Masada, K. Hashimoto, *J. Chromatogr.*, **118**, 201 (1976)
32. A. Selva, U. Vettori, S. Popov, N. L. Marekov, *Boll. Chim. Farm.*, **117**, 77 (1978)
33. H. D. Beckey, *Principles of Field Ionization and Field Desorption Mass Spectrometry*, Pergamon Press, Oxford (1977)
34. T. Bentley, R. Johnstone and J. Grimshan, *J. Chem. Soc. (C)*, 2234 (1967)
35. K. Hostettmann, J. Doumas, M. Hardy, *Helv. Chim. Acta*, **64**, 28 (1981)
36. N. N. Mollova, N. V. Handajeva, S. S. Popov, *Org. Mass Spectrom.* **24**, 1001 (1989)
37. J. Garlia, A. J. Chulia, *Planta Medica*, 327 (186)
38. Y. Takeda, S. Tsukida andT. Fujita, *Phytochemistry*, **26**, 2303 (1987)
39. B. Gering, P. Junior and P. Wichtil, *Phytochemistry*, **26**, 3011 (1987)
40. M. Brenes, C. Romero, P. Garcia, A. Garriddo, *J. Sci. Food Agric.*, **67**, 35 (1995).
41. G.Montedoro, M. Servili, M. Baldioli, E. Miniati, *J. Agric. Food Chem.*, **40**, 1571-, (1992).
42. A. Heredia-Moreno, J. Fernàndez -Bolanos, R. Guillen, *Food Chem.*, **38**, 69-73, (1990).
43. Domon, C. E. Costello, *Glyconjugate J.*, **5**, 397-409, (1988).

ENVIRONMENTAL APPLICATIONS OF MASS SPECTROMETRY: TOXAPHENE ANALYSIS

F. I. ONUSKA and R. J. MAGUIRE
National Water Research Institute, Department of the Environment,
867 Lakeshore Road, Burlington, Ontario, Canada, L7R 4A6

1. Introduction

Several methods for the elucidation of unknown structures are available in organic chemistry. In environmental trace analysis mass spectrometry is the most suitable technique due to its specificity and unsurpassed sensitivity. However there are problems due to the fact that analyte concentrations in environmental samples are usually small, and there are many other contaminants and naturally-occurring chemicals in such samples. The identification and quantitative analysis of contaminants requires a large range of sampling, pretreatment, and selective and sensitive instrumental analytical procedures. An illustration of the various techniques available is provided using the pesticide toxaphene as an example. The intention is not to review the environmental chemistry and ecological effects of toxaphene exhaustively, but to give an overview and to discuss certain aspects of its analysis.

2. Structure, Properties and Production

Technical toxaphene (Chemical Abstracts Service Registry Number 8001-35-2) is not an individual chemical compound but a very complex mixture of chlorinated camphenes containing over 300 congeners. General structures are illustrated below [1] which show the numbering for the C positions in the bornane skeleton (A) and letters for Cl/H positions (B).

R. M. Caprioli et al. (eds.), Selected Topics and Mass Spectrometry in the Biomolecular Sciences, 533–558.

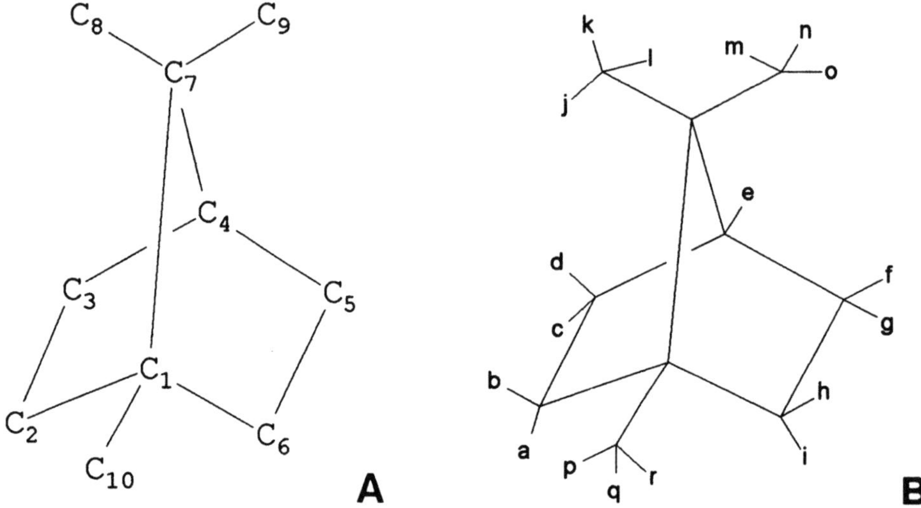

As Vetter [1] has indicated, the toxaphene skeleton has a plane of symmetry through carbons C_1, C_7 and C_4, and all positions from *a* to *i* are different with possibilities for either chlorination or no chlorination. However, the three positions of each methyl group at carbon atoms C_8, C_9, and C_{10} are identical. Consequently there are 32768 possible structural combinations, including the unchlorinated bornane molecule. Congeners with the same substituents at pairs a-i, b-h, c-g and d-f, as well as an identical number of chlorine atoms at carbons C_8 and C_9 are not optically active. Thus, only 512 of all possible chlorinated bornanes are optically inactive, while 16128 congeners are enantiomeric pairs. The example of a heptachlorobornane below shows chiral carbons (*) in the molecule. As can be seen, there are carbon atoms such as C_2, C_3, C_5 and C_6 that are asymmetric when they are substituted with one Cl and one H atom.

Structure Mirror Mirror-image

Different chlorine substitution at carbons C_8 and C_9 generates an asymmetric center at carbon C_7. Carbon atoms C_1 and C_4 are asymmetric when the carbon pairs C_2-C_6, C_3-C_5 and C_8-C_9 are not identically substituted. This means that a chlorinated bornane may have from 0 to 7 asymmetric carbon centers. The theoretical total number of congeners and enantiomers of toxaphene is summarized in Table 1.

Table 1. Total number of congeners and enantiomers of toxaphene (Reproduced from reference 1 with permission)

# of chlorines	Possible isomers	Optically inactive	Enantiomers	Enantiomeric pairs
0	1	1	0	1
1	12	2	10	5
2	69	7	62	31
3	256	12	244	122
4	696	22	674	337
5	1488	3	1456	728
6	2608	42	2566	1283
7	3840	52	3788	1894
8	4818	56	4762	2381
9	5192	60	5132	2566
10	4818	56	4762	2381
11	3840	52	3788	1894
12	2608	42	2566	1283
13	1488	32	1456	728
14	696	22	674	337
15	256	12	244	122
16	69	7	62	31
17	12	2	10	5
18	1	1	0	0
Σ	32768	512	32256	16128

Toxaphene was introduced in the U.S.A. by Hercules Co. as an insecticide to control a variety of pests, in particular on cotton. Because it was a very potent insecticide with a broad spectrum of activity, it was also used on vegetables, small grains, soybeans, and as an external veterinary insecticide on livestock [2]. It was estimated that 500,000 tonnes of toxaphene were manufactured in the U.S.A. in the period 1946-1982 [3]. It was formulated into emulsifiable concentrates, wettable powders and dusts. Toxaphene was manufactured under many different trade names, as shown in Table 2.

Table 2. Common names and trade names of toxaphene

Common Names	Trade Names	Trade Names
Toxaphene	Allotox	Melipax
Camphechlor	Chemphen	Penphene
Chlorinated camphene	Estonox	Toxakil
Polychlorocamphene	Geniphene	Toxiphen
Octachlorocamphene	Hercules 3956	Toxaphen

The most common synonym is camphechlor, which is widely used in Europe [4]. Some chemical properties of toxaphene are shown in Table 3.

Table 3. Chemical properties of toxaphene

Molecular formula	$C_{10}H_{10}Cl_8$
Molecular weight	414
Melting point	waxy solid melts between 65 to 90 °C
Density	1.65 at 20 °C
Vapor pressure	$6.7 \cdot 10^{-6}$ atm at 20 °C
Solubility	3 mg/L in water; soluble in most organic solvents
Corrosivity	corrosive to metals under moist conditions
Incompatibility	Fe; Al and alkaline formulations
Octanol-water partition coefficient	825

In Canada registration of most uses of toxaphene were suspended in 1980, and the registration of the last registered products were withdrawn by the registrant in 1983 [5]. In 1977 the U.S. Environmental Protection Agency issued a rebuttable presumption against registration of toxaphene, and in 1982 most uses of toxaphene were banned in the U.S.A. However, stocks were allowed to be used up until 1986. At the present time, all uses of toxaphene have been banned in many other countries. It appears that the last known producer of toxaphene in the world ceased operations in 1991 [6].

3. Environmental Distribution

Although toxaphene use has been eliminated in many industrialized countries, it is continually found in rain, soil, sediment and biota. In addition, because of its persistence and its long-range atmospheric transport, it is found in remote regions such as the Arctic, where it was never used.

3.1 AIR

A number of studies have been conducted to determine toxaphene in air. Toxaphene has been detected in the Arctic in air, sea water, vegetation and biota [7,8]. Toxaphene levels in air samples varied greatly at a South Carolina estuary, from 0.33 to 7.2 ng/m^3 air [9]. Mean toxaphene concentrations in air in the Mississippi delta were 1.89 μg/m^3 with a maximum concentration of 8.7 μg/m^3 in 1970 to 1972 and the mean value was 167 ng/m^3 for 1972 to 1974 [10].

3.2 PRECIPITATION

Toxaphene has been found in precipitation samples close to where it was used, e.g., in the southern U.S. cotton belt, and in remote areas where it was never used [8]. The mean toxaphene concentration in rain water was found to be 45 ng/kg in a South Carolina estuary [9]. Toxaphene has also been found in rain on Isle Royale in Lake Superior [11,12]. Snow also has been analyzed for toxaphene at several very remote locations. Sampling at the North Pole and north, central and south Finland [13] indicated no toxaphene in the period 1983-85.

3.3 WATER

There are few data on the occurrence of toxaphene in natural waters, but it has been found, for example, in water in Siskiwit Lake on Isle Royale in Lake Superior [12] and in the Arctic [7]. The International Joint Commission [14] has recommended a maximum concentration of toxaphene in water of 8 pg/L for the protection of aquatic life.

3.4 AQUATIC BIOTA

Toxaphene was used early on to eradicate rough fish and sea lamprey, but its high toxicity and lack of selectivity also led to the death of many desirable species, and to subsequent investigations of its aquatic toxic effects and bioaccumulation [15-17]. For example, adverse effects of toxaphene were noted in brook trout *(Salvelinus fontinalis)* fry at 39 ng/L, fathead minnow *(Pimephales promelas)* fry at 97 ng/L and channel catfish *(Ictalurus punctatus)* at about 299 ng/L. Residues of toxaphene in eggs, fry and adults were measured for the species exposed at six concentrations of toxaphene (0, 39, 68, 139, 288 and 502 ng/L). Residues increased with concentration in all cases, and bioconcentration factors were 7.6×10^5, $5.5\text{-}6.5 \times 10^5$, and $2.7\text{-}5.0 \times 10^5$, respectively, for the brook trout, fathead minnow, and channel catfish.

Whole fish samples from Tule Lake and Lower Klamath Refuges, part of an agricultural irrigation system in north-eastern California and Oregon, had toxaphene residues ranging from 0.1 to 8.0 mg/kg, while invertebrates collected at that time contained a maximum of 0.2 mg/kg in 1960 to 1961 [18].

3.5 SOIL, SEDIMENT AND CROPS

Because the major use of toxaphene was as an insecticide on crops, it was not surprising to find toxaphene in crops, the surrounding soil and in the sediments of adjacent rivers and lakes because of runoff. Soils and sediments are probably the ultimate sinks for toxaphene and its degradation products. In their review of toxaphene, Pollock and Kilgore [2] also examined the toxaphene residue issue and found much confusion as to the identification of residues. This problem of unequivocally identifying toxaphene has continued to the present time, and will be discussed further below.

Carey et al. [19] collected composite samples of soils and mature crops from 37 states in the U.S.A. Toxaphene was not found in 13 of the 29 crop groups surveyed. The arithmetic mean of positive detections ranged from 0.01 to 10.2 mg/kg on a dry weight basis over these crops with the highest value being found on cotton stalks for 70.5% of the samples. Non-cropland soil samples were collected for 11 other states. Toxaphene was found at a mean residue level of 0.07 mg/kg in 4.2% of cropland samples, and concentrations ranged as high as 11.72 mg/kg.

Compared to soils, there are relatively few sediment toxaphene data in the literature. Durant [20] found 32.8 mg/kg toxaphene in a single composited sediment core sample near a Georgia pesticide plant. Near the outfall from the plant levels approached 1900 mg/kg in the top sediment layer. In a southern Florida study, Mattraw [21] found only 3.2% of soil samples contained detectable toxaphene, and none was found in 146 water samples.

3.6 BIRDS, MAMMALS AND HUMANS

Birds, mammals and humans are near or at the top of food chains, and thus it is expected that in certain areas they would show evidence of accumulation of residues of toxaphene, or effects resulting from exposure to toxaphene and/or its metabolites [22].

Two studies [23,24] identified traces of toxaphene in Scandinavian individuals. In one study toxaphene was identified in a 10 kg composite human milk sample by high resolution gas chromatography/mass spectrometry (HRGC/MS). Both electron impact and negative ion chemical ionization mass spectrometric detection were employed with negative ionization being more useful for the quantitation of toxaphene in the presence of other organochlorine pesticides. It was noted that the reported concentrations could have been affected by differences between the chromatograms of the toxaphene analytical standard and that of the human milk sample, differences that might have been due to metabolism. Pyysalo and Antervo [24] reported that human milk contains 1-10 μg/kg of toxaphene on a whole milk basis. Lower chlorinated terpenes were not observed, indicating partial metabolism had occurred. Mussalo-Rauhamaa [25] found metabolized toxaphene in a number of human adipose tissues at concentrations similar to that of chlordane. Finnish adipose tissue toxaphene levels were reported to be in the order of 0.01-0.1 mg/kg.

4. Metabolism, degradation and persistence

There have been many studies of the metabolism, persistence and fate of toxaphene [*e.g.*, 26-28], using a variety of techniques. For example, a study of toxaphene metabolism in rats showed significant amount of metabolites of toxaphene in the feces [29].

Cairns *et al.* [30] examined metabolized toxaphene in dairy cow milk fat samples. The animals were fed toxaphene at the 5 mg/kg level. No toxaphene was evident in their milk prior to feeding and feed was carefully monitored. The electron capture detection chromatogram indicated substantial metabolism of toxaphene, as seen by the lack of resemblance to the chromatogram of the reference toxaphene fed to the cows. In electron-capture chemical ionization mass spectrometric detection in tandem with high resolution gas chromatography using methane as the reagent gas C_{10}-polychlorinated fragments at m/z 199,235, 271, 307, 343 and 411 dominated the elution profile as shown in Fig.1. Lower chlorinated polychlorinated bornanes became more pronounced in the milk samples.

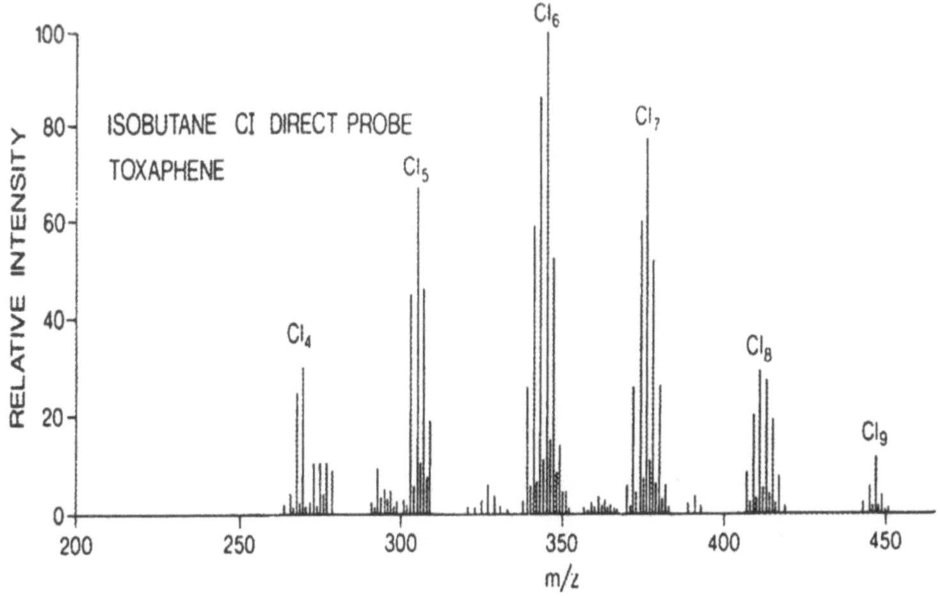

Figure 1. Chemical ionization mass spectrum of toxaphene.

For low level detection of toxaphene and its metabolites in milk careful selection of ions was recommended. The most abundant ions were observed at m/z 159, 199, 235, 271, 307, 343 and 377. The study was qualitative and did not attempt to identify metabolic residues. However, the multicomponent nature of toxaphene makes quantitation of toxaphene and its metabolites very difficult.

Williams and Bidleman [31] examined toxaphene degradation in sediments and found that breakdown did not occur if the iron redox system was not present. Anoxic salt marsh sediments stimulated the rapid breakdown of toxaphene. The breakdown products were not determined nor was their toxicity. Hydroxylation and/or dehydrochlorination were suggested to be important metabolic routes.

5. Sampling

5.1 AIR

Sampling for airborne toxaphene is done by passing a known volume of air at a predetermined rate through a suitable filter [32,33]. Trapping materials commonly used in high-volume air sampling for toxaphene are glass-fiber filters, polyurethane foam plugs [33], Tenax and XAD-4 macroreticular resin. The toxaphene is extracted from the resin with acetone and the extract is further cleaned up on a Florisil column prior to HRGC/MS analysis.

5.2 WATER

Water samples are generally collected in clean, solvent-rinsed glass, Teflon or stainless steel containers [*e.g.*, 32]. Sampling at various depths can be accomplished using Teflon tubing and a pump with Teflon or stainless steel bellows. The containers should be filled to the top to reduce volatilization of toxaphene into the headspace. The toxaphene must be extracted from the sample as soon as possible after collection to minimize losses due to volatilization or adsorption to container surfaces. The sample should be kept at 4 °C until processed. Sample extracts can be stored for months provided they are placed in properly sealed containers and are protected from photodegradation [*e.g.*, 34].

5.3 SEDIMENT

Wet sediment samples (at least 10-20 g), in clean solvent-rinsed jars, should be frozen until analysis. Samples may be taken by a variety of dredges and manual sampling methods. Samples can be freeze-dried before extraction, or dried with sodium sulfate prior to extraction. Details are provided in the literature [*e.g.*, 35].

5.4 BIOTA

Fish and other biota samples are wrapped in hexane-rinsed aluminum foil or placed in glass jars and frozen until analysis. Sample representativeness is important, since the biological variability between individual organisms is large [36]. In general, it is best to analyze a reasonably large number of individuals or composite a large number into a single sample. To achieve rapid freezing dry ice is recommended. Algae and plankton are collected from water by filtration through different mesh-size nets and small invertebrates such as mysids or shrimps can be sampled by dragging a 1-mm mesh metal screen on a sled [*e.g.*, 36].

6. General Analytical Procedures

In general, the determination of chemicals in environmental samples requires sample pretreatment before instrumental analysis. In order to increase the sensitivity of the analyses, extraction and cleanup procedures are required [*e.g.*, 37], and some common procedures are described below.

6.1 EXTRACTION, PRECONCENTRATION AND CLEANUP

A major problem in liquid-liquid extraction procedures is contamination with impurities from the glassware and particularly from the solvents. It is difficult and expensive to obtain solvents with a high purity. An alternative technique is that of solid-phase extraction whereby the pollutants that are adsorbed (*i.e.*, preconcentrated) on a suitable adsorbent are desorbed with a small amount of solvent. Another approach is to perform a liquid-liquid extraction using a Goulden continuous extractor which can provide up to 1000 fold preconcentration of analytes from water into methylene chloride [38]. This method ensures a high preconcentration rate and a minimum of contamination as a result of the closed system and the use of a small amount of methylene chloride. Extraction techniques and solvents vary depending upon the matrix being analyzed. Details can be found in review articles such as [39].

Soil and sediment samples can be extracted using a Polytron extractor [40]. The soil sample (1 g) is extracted with 50 mL CH_2Cl_2-acetone (1:1) in a 250 mL stainless-steel beaker. The sonicator is adjusted to pulsed operation and output control is set to full power and duty cycle to 50 %. The sample is sonicated for 3 min. The extraction is repeated twice with fresh solvents. Combined extracts are dried by passage through a Na_2SO_4 column. After filtration, 10 mL of isooctane is added to the filtrate, and the filtrate is evaporated to 2-3 mL at 40 °C. A Florisil column is prepared by placing a charge of activated Florisil in a chromatographic column (30 x 1 cm i.d.) over a 1-cm layer of anhydrous sodium sulfate. Five grams of 120-150 μm Florisil previously activated at 130 °C for 16 hrs is employed. Each column is pre-washed with 20 mL of n-hexane. When the solvent reaches the top of the Na_2SO_4 layer, the concentrated extract is quantitatively transferred onto the column and is allowed to drain onto the bed of Florisil. The column is eluted with 50 mL n-hexane-diethyl ether (94:6, v/v). The eluate is collected for further analysis. It contains aldrin, benzene hexachloride (BHC), chlordanes, DDD, heptachlor, DDE and DDT isomers, lindane, methoxychlor, mirex, PCB and toxaphene. The effluent is further treated with 5 mL of cold sulfuric acid - fuming nitric acid (1:1) for 15 minutes at room temperature to remove DDT and its metabolites. PCB are removed by silica gel column chromatography. The eluate volume is reduced by rotary evaporation and the volume of the resulting solution is adjusted to 1 mL with isooctane prior to analysis by HRGC with electron capture - negative chemical ion mass spectrometry (HRGC/EC-NCI-MS).

Biota samples can be extracted by techniques similar to those just described. Co-extracted lipids that interfere with the analysis are removed by gel permeation chromatography [*e.g.*, 41]. Excellent recoveries for toxaphene are obtained while simultaneously removing greater than 98 % of the lipids. Lipid removal is essential to the successful chromatographic cleanups, since adsorbents used for cleanup have a low capacity for lipids. Automated gel permeation cleanup systems are commercially available.

6.2 WALL-COATED OPEN TUBULAR COLUMN GAS CHROMATOGRAPHY

We will use the term *wall coated open tubular columns* (WCOT) to specify open tubular columns in which the liquid phase is deposited directly on the inner wall of the pretreated (deactivated) fused silica surface without the inclusion of any particles that might be considered solid support.

The degree of resolution and the time necessary for a separation is a function of six parameters, *i.e.*, column length, column internal diameter, film thickness, column temperature, carrier gas, and carrier gas velocity. The first three parameters are characteristic of the column and the last three parameters are operational parameters that can be easily changed, even during analysis. A thorough understanding of how these parameters influence resolution and time of analysis is invaluable, and such information can be found in monographs [*e.g.*, 42]. Their optimization, especially in tandem with mass spectrometric detection, is very important. For example, a large diameter column (0.32 mm) increases the phase ratio over that of a smaller diameter column (0.15 mm), and this is manifested in a shorter analysis time. Column temperature rather than column length is usually used to control the separation. A typical chromatogram of toxaphene is shown in Fig. 2.

The use of WCOT columns places special requirements on the gas chromatograph, and such parts as the injector, pneumatic controls, column oven and detectors must be carefully optimized.

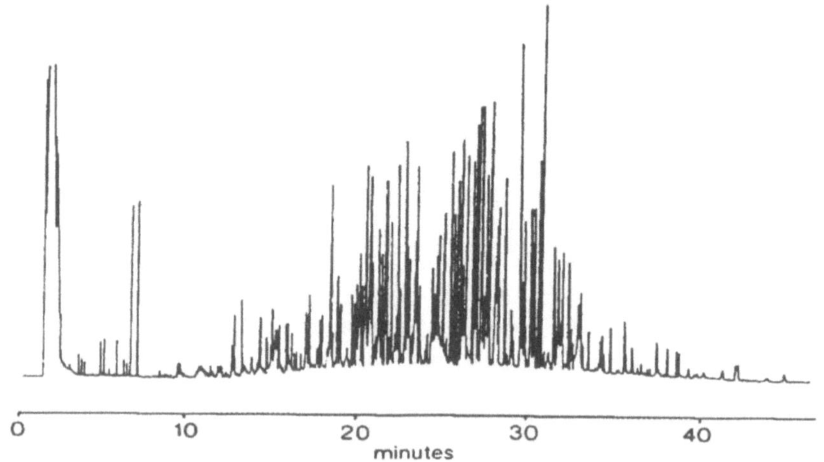

Figure 2. Open tubular column gas chromatogram of toxaphene (SE-52 20 m x 0.1 mm i.d.; d_f = 0.2 μm) using electron capture detection.

6.3. GAS CHROMATOGRAPHY - MASS SPECTROMETRY

The combination of a capillary column gas chromatograph with a mass spectrometer (HRGC/MS) is a powerful system for the identification and quantitation of low levels of xenobiotics in environmental samples. Advances in automation and computerized data handling in the past ten years have greatly aided the development of HRGC/MS techniques. Table 4 lists some of the common types of operations found in HRGC/MC. Details are discussed in other chapters in this book.

Table 4. Modes of operation of the mass spectrometer and data acquisition and reduction used in environmental applications

Mode of operation	Type of operation and /or acquisition
Ionization techniques	Electron impact
	Chemical ionization
	Field ionization
	Field desorption
Resolution setting	Low resolution
	Medium resolution
	High resolution
Scan	Full scan
	Single ion monitoring
	Multiple ion monitoring
Data acquisition/Data reduction	Strip chart recorder
	Multichannel analyzer
	Computer
Interfaces	Direct connection in GC/MS
	Thermospray in HPLC/MS
	Electrospray in HPLC/MS

Although qualitative analysis by HRGC/MS will always be important, a trend in environmental analyses has been toward quantitative analysis. Unlike qualitative studies which have as their aim the identification of unknown compounds by deducing fragmentation patterns, quantitative determination presupposes this knowledge, and the presence of a compound may often be deduced by the existence of a particular characteristic ion, molecular ion or fragment, particularly when this is coupled with appropriate retention time data from a chromatographic method. The amount of analyte present is deduced from the area response of that ion [43]. Two quantitation methods are in general use in HRGC/MS, one based on selected ion monitoring, the other one on a total integrated ion current, which may eliminate the need for one or more cleanup steps. Because instrumental parameters are so varied and it is virtually impossible to maintain them all at a constant level of performance over a reasonable period of time, it is in practice always necessary to use comparative methods with an added internal standard or a surrogate standard.

6.4. MASS SPECTROMETRIC ANALYSIS OF TOXAPHENE

6.4.1 *Fragmentation of toxaphene under electron impact ionization*

The electron impact ionization mass spectra of toxaphene congeners are dominated by characteristic fragments. It has been noted that a fragmentation characterizing the mass spectra of toxaphene congeners involves elimination of the HCl or chlorine of the toxaphene moiety. This fragmentation of the toxaphene gives rise to the cleavage of the chlorine-carbon linkage and the subsequent breakdown produces abundant ions derived from the toxaphene molecule. Thus these fragments due to the fragmentation of toxaphene molecule appear in every congener.

As stated by Stern *et al.* (44) the electron impact ionization mass spectra are complex but structurally informative fragmentation is observed. In particular, fragmentations involving elimination of $C_2H_2Cl_2$ indicate that two chlorine atoms are bound to the C_2-C_3 and C_5-C_6 carbons of a bornane moiety.

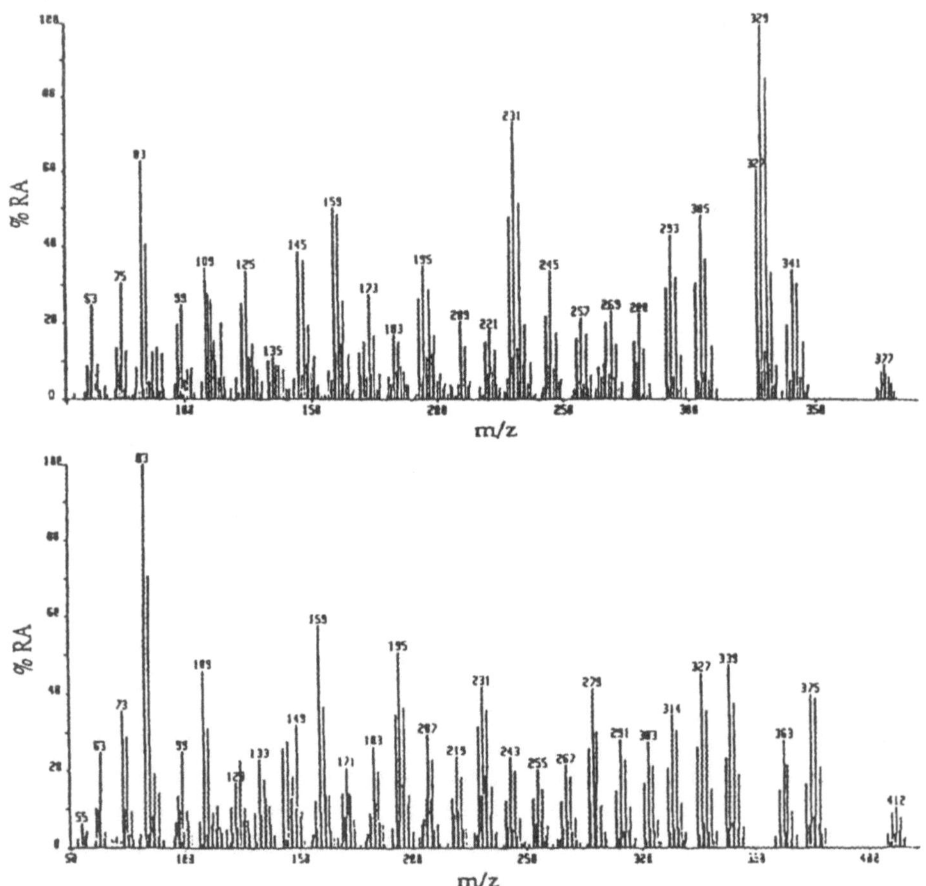

Figure 3. Electron impact ionization mass spectra of T2-(top) and T12 (bottom) at 70 eV (Reproduced from reference 44 with permission)

Furthermore, from the $M^{+\cdot}$ ion, elimination of HCl is enhanced with respect to $Cl^{\cdot}$, in Toxicants A, a mixture of two octachloro bornanes, 2,2,5-*endo,* 6-exo, 8,8,9,10- and 2,2,5-*endo,* 6-*exo,* 8,9,9,10-octachlorobornane) and toxicant B (2,2,5-*endo,* 6-

exo,8,9,10-heptachlorobornane), which are major components of commercial toxaphene. This finding suggests that one chlorine atom is bound to each of C_2, C_3, C_5 and C_6 carbon. Both compounds have $CHCl_2$ substituents [45].

Recently, two components of toxaphene were characterized as 2-*exo*, 3-*endo*-5-*exo*, 6-*endo*, 8,8,10,10-octachlorobornane and 2-*exo*,3-*endo*,5-*exo*, 6-*endo*, 8,8,9,10,10-nonachlorobornanene, hereafter referred to as T2 and T12 respectively [44]. The EI ionization mass spectra of T2 and T12 are shown in Figure 3.

For the analysis a Hewlett-Packard 5890 gas chromatograph was coupled to a VG-7070E HF double-focusing mass spectrometer of EB geometry. The analysis was done using a 60 m x 0.25 mm i.d. DB-5 WCOT column with helium as the carrier gas and the following operating conditions: injector temperature 220 °C; initial oven temperature 100 °C, held for 2 min., then programmed at 15 °C/min. to 265 °C with final hold for 20 min. The effluent was transferred to the ion source of the mass spectrometer. For electron impact ionization (EI), the electron beam energy was 70 eV, the electron emission current was 200 μA, the ion acceleration voltage was 6 kV, and the ion source temperature was 175 °C. Exact masses of ions were measured at M/ΔM 10,000 resolution using perfluorokerosene as a mass calibrant. Major ions resulting from the fragmentation of the toxaphene congeners were at m/z 83 , 109, 125, 159, 195, 231, 329, 377 and 412, for the majority of congeners present in the mixture. The relative intensity of the fragment ion at m/z 159 varies, but it is present in more than 70 percent of all peaks and for this reason is used as a principal ion for the selected ion monitoring (SIM) of toxaphene by EI [39]. It should be noted that polychlorinated biphenyls (PCB) and all of the organochlorine hydrocarbons (OC) insecticides do not show any interference at m/z 159.

6.4.2 *Fragmentation of toxaphene under chemical ionization*

In chemical ionization (CI) mass spectrometry, ionization of the molecule is effected by reaction between molecules of the analyte (M) and reactant ions (RH^+) as follows:

$$RH^+ + M \rightarrow MH^+ + R$$

This reaction results in initial formation of the protonated molecule MH^+ and represents by far the most common CI reaction. Most studies have used methane as the reactant gas. Chemical ionization mass spectra have also been generated using isobutane and ammonia as the ionizing reagent gases. At a source pressure of approximately 1 torr the predominant ion is the tert-butyl $(C_4H_9)^+$ ion formed by hydrogen abstraction from the isobutane molecule. The isobutane-induced mass spectrum is virtually monoionic and the reactant ion behaves as a mild Bronsted acid, protonating the sample molecules [46].

Examination of toxaphene by positive ion CI-MS with methane resulted in mass spectra characterized by multiple ion clusters as shown in Figure 4. These cluster ions represent isomeric series of different congeners containing variable number of hydrogen atoms for each chlorine substitution.

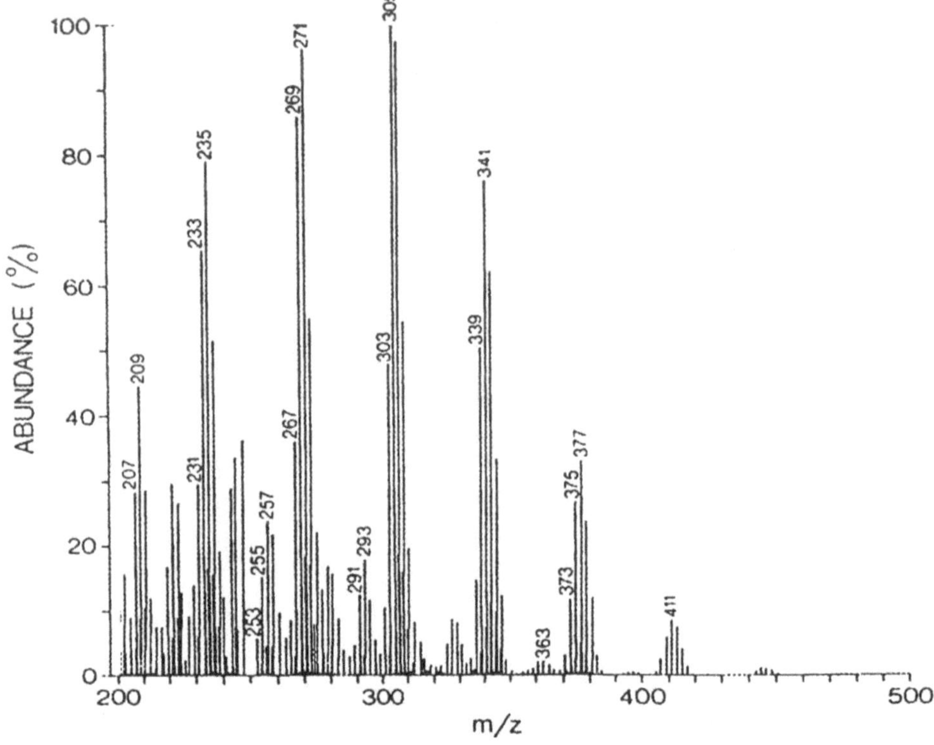

Figure 4. Positive methane CI mass spectra of toxaphene

The formula weights of several possible empirical formulas for increasing number of chlorine atoms of major homologues of chlorinated camphene having general formula $C_{10}H_{14-n}Cl_n$ are shown in Table 5.

Table 5. Formula weight for camphene and chlorinated bornanes

Name	Formula	Formula weight
Camphene	$C_{10}H_{18}$	138.140
Dichlorobornanes	$C_{10}H_{16}Cl_2$	206.062
Trichlorobornanes	$C_{10}H_{15}Cl_3$	240.023
Tetrachlorobornanes	$C_{10}H_{14}Cl_4$	273.984
Pentachlorobornanes	$C_{10}H_{13}Cl_5$	307.945
Hexachlorobornanes	$C_{10}H_{12}Cl_6$	341.906
Heptachlorobornanes	$C_{10}H_{11}Cl_7$	375.867
Octachlorobornanes	$C_{10}H_{10}Cl_8$	409.826
Nonachlorobornanes	$C_{10}H_9Cl_9$	443.789
Decachlorobornanes	$C_{10}H_8Cl_{10}$	477.750

6.4.3. Electron-capture negative ion chemical ionization mass spectra of toxaphene

This technique represents an extension of the nonreactive enhancement of negative ion mass spectra. Because of the electronegative chlorine atom, chlorinated compounds have a higher cross-section for the electron capture processes than their non-chlorinated analogs. In contrast to positive ion chemical ionization, the major route for negative ionization ion formation involves electron capture, which can be dissociative or non-dissociative in nature. It is known that the use of methane as the reactant gas results in large yields of thermal energy electrons, while the addition of a small amount of water results in the formation of OH^- as the major reactant ion. Using methane as the reactant ion, the major feature in mass spectra of toxaphene and other chlorinated compounds is the cluster of peaks due to chlorine attachment $(M+Cl)^-$. The negative CI mass spectrum of toxaphene is illustrated in Figure 5. Methane at 100 Pa was used as reactant gas.

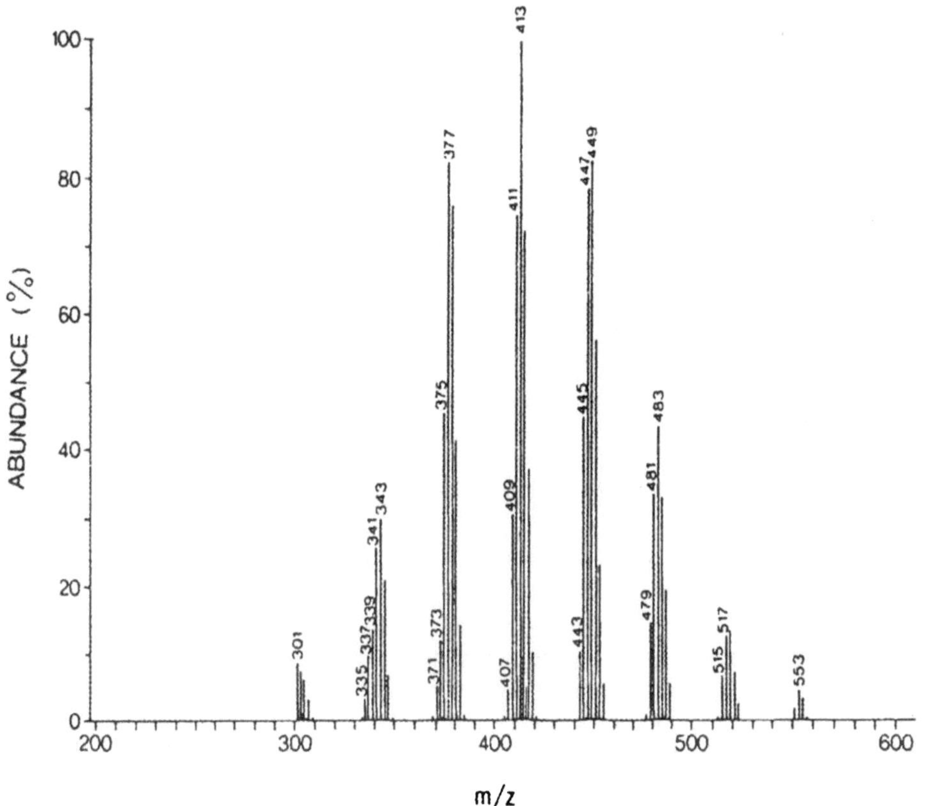

Figure 5. Electron capture negative ion chemical ionization mass spectrum of toxaphene.

Major ions in the EI-NCI-MS mode are at m/z 309, 311, 343, 345, 379, 381, 413, 415, 447, 449, 481,483. Ions at m/z 413, 415, 447 and 449 should be checked carefully because of potential interference from chlordanes.

The determination of individual toxaphene congeners is inherently difficult because of the complexity of the mixture. For example, in a comparison of toxaphene profiles from environmental residues with standard material, quantification using electron-capture negative ionization (ECNI) mass spectra *vs.* electron capture detection GC data showed a lower response by ECNI-MS analysis [47]. The authors claimed that the concentrations of toxaphene, when quantified by open tubular column GC-ECD were two to four times greater than determined by ECNI mass spectrometry. An alternative explanation is that the lower response produced by the bulk of toxaphene residue was a consequence of the "weathering" of the toxaphene mixture.

The mechanism in EI-NCI-MS fragmentation of toxaphene is still not resolved. The dominant reaction appears to be the formation of a stable negative ion which is directly dependent upon the electron attachment cross-section of the molecule, the number of degrees of freedom inherent in the molecule, and the electron affinity of the molecule [48]. The most stable ion produced by electron capture is its substituted chlorine anion. The molecular ion becomes the most stable one with increasing degree of chlorination on a camphene moiety. Once the molecule is fully chlorinated, the number of degrees of freedom decreases owing to steric hindrances caused by neighboring chlorine atoms. A decrease of molecular response could be a consequence of its stereo-configuration.

Differences among EI mass spectra of Toxicants A and B and T2 and T12 can be attributed to the different bridge and bridgehead moieties, i.e. CH_3, CH_2Cl or $CHCl_2$. There are also differences related to presence of the chlorines and hydrogen atoms attached to the ring carbons in position 2, 3, 5 and 6. In particular, while mass spectra of toxicant A and B exhibit elimination of $Cl^\cdot$ radical from molecular ions (see Fig. 3), the mass spectra of T2 and T12 show competitive losses of HCl and $Cl^\cdot$ from their molecular ions. The enhanced HCl elimination from the molecular ions of T2 and T12 with respect to toxicants A and B is probably due to the greater acidity of the *5-endo* and *6-exo* hydrogen atoms compared to the *5-exo* and *endo* hydrogen atoms of toxicants A and B.

The results and fragmentation pathways are illustrated in Figures 6 and 7 for the T2 and T12 molecules.

A notable difference between the mass spectra of T2 and T12 is that the $[M\text{-}CHCl_2]^+$ ion at m/z 327 is the most abundant ion for the T2 molecule, while the $[CHCl_2]^+$ ion at m/z 83 is the most abundant ion for T12. The EI mass spectrum of Toxicant A shows the most abundant ion at m/z 83, while the m/z 233 ion corresponding to $[M\text{-}HCl\text{-}Cl\text{-}Cl_2\text{-}HCl]^+$ is the dominant ion in the EI mass spectrum of Toxicant B.

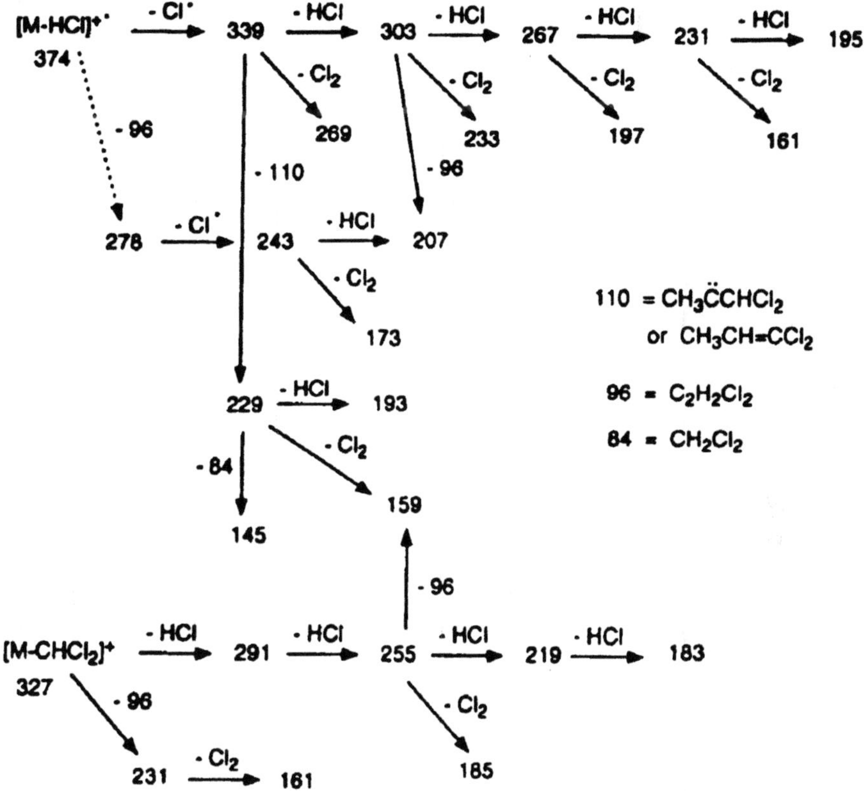

Figure 6. Fragmentation pathway obtained by linked-field scanning for the T2 congener (Reproduced from ref. 44 with permission).

6.5. QUANTITATIVE ANALYSIS OF TOXAPHENE

The trace analysis of complex mixtures of environmental contaminants by HRGC/MS presents a formidable challenge. The quality of such analyses depends upon the *resolution* of individual congeners in the mixture, the *selectivity* of a stationary phase of the open tubular column, the *sensitivity* of the detector, and the *availability of particular standards.* Following is a description of the effect of some modes of operation on the sensitivity of the analyses.

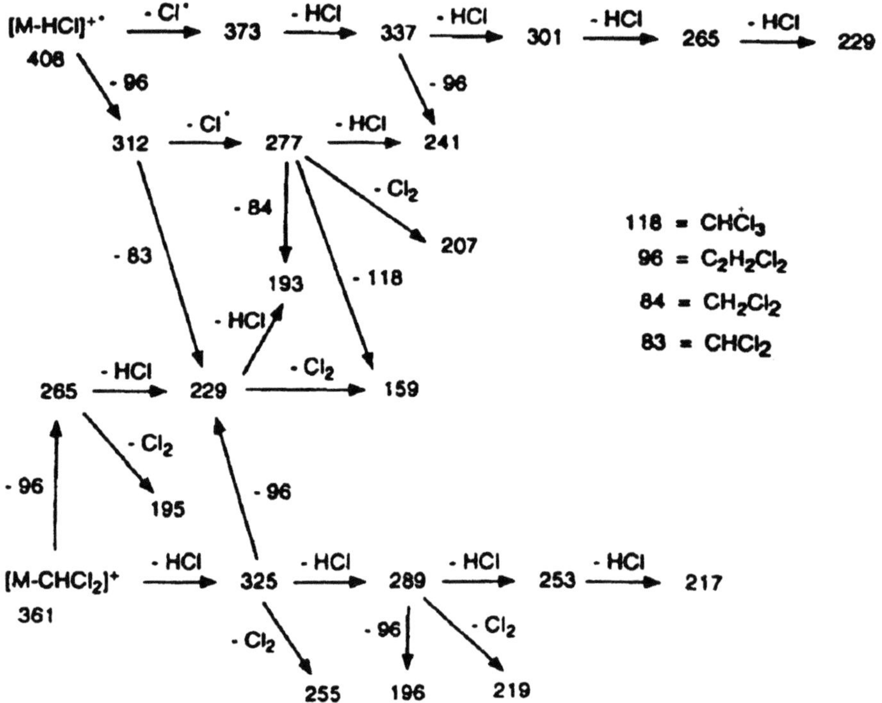

Figure 7. Fragmentation pathway obtained by linked-field scanning for the T12 congener (Reproduced from ref. 44 with permission).

6.5.1. *Selected ion monitoring*

The processes of switching the mass analyzer voltages to one or more preselected masses for the purpose of recording the changing mass-to-charge ratios at different dwell times are called *selected ion monitoring* (SIM) or *multiple ion monitoring* (MID). Such techniques provide much greater sensitivity than full-range scanning. SIM can be used for confirmation purposes when high resolution MS is applied by focusing the mass analyzer voltage on an ion characteristic of that compound. Because the mass analyzer remains permanently focused on the same mass value, instabilities associated with voltage switching are removed, and also the output from the electron multiplier current can be integrated over a long period of time. This technique represents the most sensitive quantitation of all mass spectrometric means of detection.

When more than one ion is monitored by the switching voltages of the analyzer, the sampling rate is determined by the dwell times, that is, the times spent integrating the ion current at selected masses. The optimum combination of dwell time, electron multiplier gain and frequency filter must be found empirically to obtain the best signal-to-noise ratio. These techniques for quantifying analytes of interest can be used in tandem with gas or liquid chromatography or electrophoretic separation methods.

6.5.2. *Direct probe quantitative analysis of toxaphene*

Most organic compounds can be introduced into the ion source of a mass spectrometer by means of a *direct insertion device.* This device allows the introduction of much larger volumes than would normally be accomplished by gas chromatographic introduction. It also allows the introduction of samples which have insufficient thermal stability to be injected into a gas chromatograph. Typically, the sample is introduced directly into the ion source on the cooled tip of the probe, and the end of the probe is heated quickly. An increased vapor pressure in a high vacuum of the ion source permits a mass spectrum to be obtained using this technique. This technique is very quick and is applicable to multicomponent mixtures. Amounts of toxaphene as low as 50 pg have been analyzed by direct probe introduction and the SIM-electron impact ionization mode. A sample output signal is shown in Figure 8.

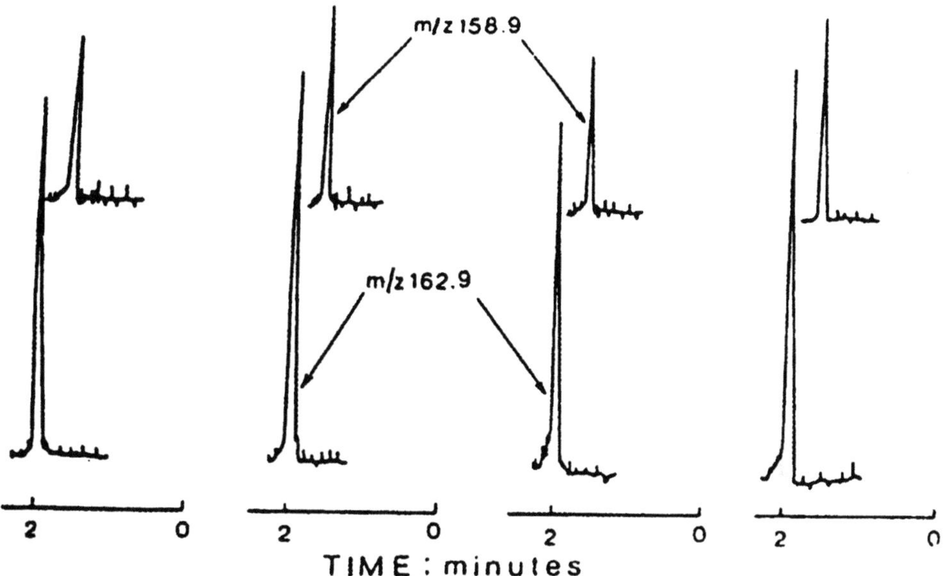

Figure 8. The recorder output for m/z 158.9 and 162.9 of 1 ng toxaphene and its replicate analyses.

As has been demonstrated in our laboratory, toxaphene can be determined by monitoring the fragment ions at m/z 158.9768; 160.9739 and 162.9709 using a high resolution mass spectrometer [49]. The ion at m/z 158.9 corresponds to a dichlorotropylonium moiety and is formed through a series of dechlorination steps under electron impact ionization. It can be monitored during a fast temperature programming of the probe tip (20 °C/s). During the subsequent evaporation of a sample, the preselected masses were monitored for 45 sec using a 2-pen recorder [50]. Positive responses in the SIM of its accurate mass at m/z 158.9768 and the isotope mass at 162.9709 at a resolution of 10,000 with a correct isotope ratio among chlorine atoms of 100:65:11 percent provides a clear confirmation of the presence of chlorinated bornanes.

As discussed Lau *et al.* [49] this technique has several advantages that can be summarized as follows:

- high resolution mass spectrometers are equipped with electron impact ionization sources;
- EI conditions are reproducible, responses are stable and it is easy to operate the system on a daily basis;
- the characteristic ions can be considered universal ones for all toxaphene congeners regardless of the degree of chlorination of the homologs in the toxaphene mixture;
- there are no significant interferences from the common environmental pollutants such as PCBs or organochlorine pesticides;
- relative response factors of individual pure congeners indicate that different congeners exhibit similar and reproducible responses;
- the resulting mass chromatograms resemble electron capture detector responses obtained by GC; since the m/z 158.9/160.9 ion ratio represents all toxaphene components, determination of **total toxaphene** can be obtained based on the comparison of the ratio of the area of unknown toxaphene to the area of an internal standard. this method allows the analysis of up to 40 samples in 8 hrs. compared to 6 to 8 samples by HRGC-MS.

The major disadvantage is that the identity of the original components is not known. However, it should be noted that only about 30 congeners have been positively identified in ECD chromatograms. An additional disadvantage is that because the high resolution mass spectrometer is an expensive investment, it is not routinely used in environmental analyses.

Lau *et al.* [49] have also noted that EI-NCI mass spectra of toxaphene do not always show an abundant $[M-Cl]^-$ ion from some congeners, and there may be very weak or even absent responses in SIM. In the past few years, advances have been made in the isolation and characterization of 27 individual chlorinated bornanes and camphene congeners by Parlar *et al.* [50]. Two mixtures are currently available and their ECD-chromatograms are illustrated in Fig. 9 and 10.

The direct probe technique can also be used in electron capture negative ion mass spectrometric analysis of toxaphene by monitoring m/z 343, 377 and 413 ions.

There is no doubt that toxaphene is a challenge to determine in environmental samples, even using specific techniques such as HRGC/EI-NCI-MS. Because of metabolism and weathering the residual composition of toxaphene in environmental samples is difficult to relate to the technical formulation of toxaphene used in calibration standards. This results in a large degree of uncertainty in measured concentrations of toxaphene, especially in fish and biota samples. Differences in the toxaphene concentrations obtained by HRGC/ECNI-MS using several quantification schemes can be employed. The main problem in the determination of toxaphene arises from the significant difference in peak profiles in environmental/biological samples and the lack of congener-specific analytical standards. Concentrations of toxaphene from the same sample vary depending on the method of detection employed [40,44].

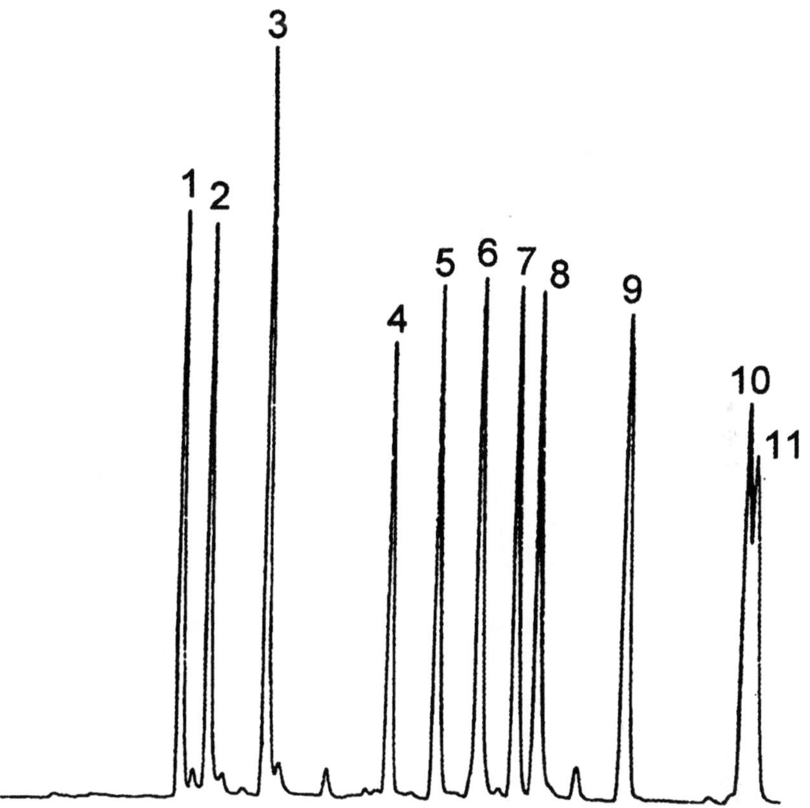

Figure 9. Mixture of chlorinated bornanes used for determination of toxaphene.
Mixture containing 666 pg/μL; HRGC chromatogram- DB-1 60 m x 0.32 mm i.d. column; d_f = 0.25 μm; nitrogen as a carrier gas at 2 mL/min. Temperature programming: 60 ° to 180 °C @ 10 °C/min, hold 1 min. programmed to 250 °C at 4 °C/min; ECD-detector.

1 2-exo, 3-endo, 5-exo, 6-endo, 8,8,10,10-octachlorobornane, (T2);
2 2,2,5-endo, 6-exo, 8,9,10-heptachloro-bornane;
3 octachlorobornane;
4 2-exo, 3-endo, 5-exo, 6-endo, 8,8,9,10,10-nonachlorobornane, (T12);
5 nonachlorobornane;
6 2,2,5-endo, 6-exo, 8,8,9,10,10-nonachlorobornane;
7 2,2,5,5,8,8,9,10,10-nonachlorobornane;
8 2-exo, 3-exo, 5-endo, 6-exo, 8,8,9,10,10-nonachlorobornane;
9 2,2,3-exo, 5-endo, 6-exo, 8,8,9,10,10-decachlorobornane;
10 2,2,5,5,6-exo, 8,8,9,10,10-decachlorobornane;
11 2,2,3-exo, 5,5,8,8,9,10,10-decachlorobornane.

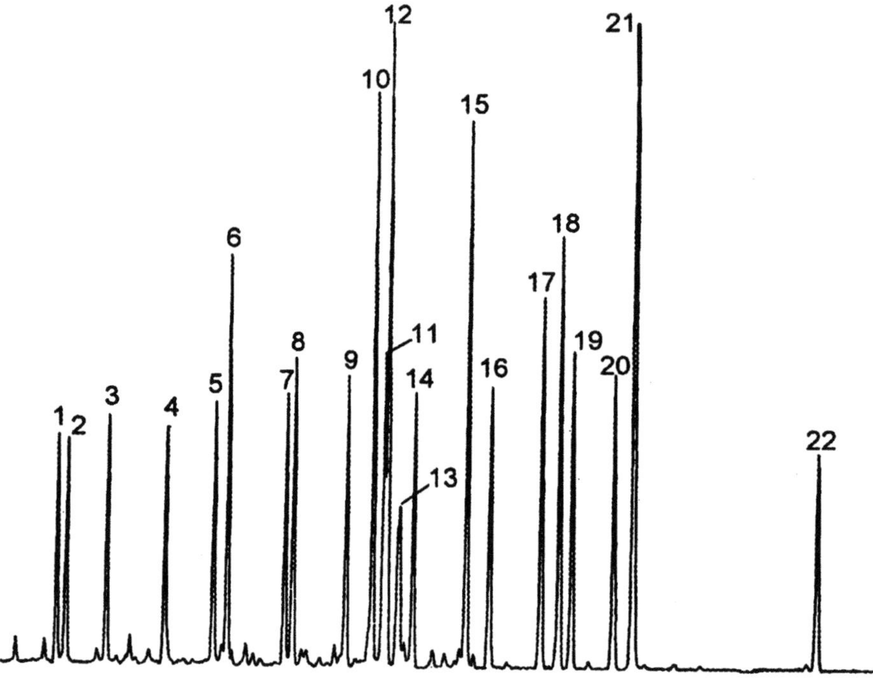

Figure 10. Standard mixture of 22 chlorinated bornane congeners for toxaphene analysis.

Identification:

1 2,2,3-exo,8b,9c,10a-hexachlorocamphene
2 2-exo,3-endo,8b,8c,9c,10a-hexachloro-camphene
3 2-exo,3-endo, 7a,8b,9c,10a- hexachlorocamphene
4 2,2,5,5,8c,10a,10b-heptachlorobornane
5 2,2,3-exo,8b,8c,9c,10a-heptachlorocamphene
6 2-exo,3-endo,5-exo,6-endo,8b,8c,10a,10b-octachlorobornane (T2)
7 2,2,3-exo,8a,8b,9b,9c,10a-octachlorocamphene
8 2,2,5-endo,6-exo,8b,9c,10a-hepta-chlorobornane (Tox B)
9 2,2,5,5,8b,8c,10a,10b-octachlorobornane
10 2,2,3-exo,5-endo-,6-exo, 8b, 9c, 10a-octachlorobornane
11 2-exo,3-endo,5-exo,6-endo,8b,9c,10a,10b-octachlorobornane
12 2-exo,3-endo,5-exo,8b,8c,9c,10a,10b-octachlorobornane; 12 2, 2, 5-endo, 6-exo, 8b, 8c, 9c, 10a- and 2, 2, 5-endo, 6-exo, 8c, 9b, 9c, 10a-octachlorobornane
13 2-exo, 5,5,8b,9c,10a,10b-octachlorobornane
14 2-exo, 3-endo, 5-exo, 6-endo, 8b,8c,9c,10a,10b-nonachlorobornane (ToxAc/T12)
15 2,2,5,5,8b, 9c,10a,10b-octachlorobornane
16 2,2,5-endo, 6-exo, 8c9b,9c,10a,10c-nonachlorobornane
17 2,2,3-exo, 5,5,8b,9c,10a,10b-nonachlorobornane
18 2,2,5-endo, 6-exo, 8b,8c,9c,10a,10c-nonachlorobornane
19 2,2,5,5,8b,8c,9c,10a, 10b-nonachlorobornane
20 2-exo, 3-exo, 5-endo, 6-exo, 8b,8c,9c,10a,10b-nonachlorobornane
21 2,2,5,5, 6-exo, 8b,8c,9c,10a,10b-decachlorobornane
22 octachloro derivative
24 nonachloro derivative
25 2,2,3-exo, 5-endo, 6-exo, 8,8,9,10,10-decachlorobornane
26 2,2,3-exo, 5,5,8,8,9,10,10-decachlorobornane

A homologue-specific method should be based on the availability of individual congener standards of chlorinated bornanes. Corresponding masses of $[M]^-$ or $[M-Cl]^-$ in the retention window of each homologue group can be summed and compared *vs.* responses of available homologues in the standard mixture and the commercially available toxaphene sample. An example is shown in Fig. 11.

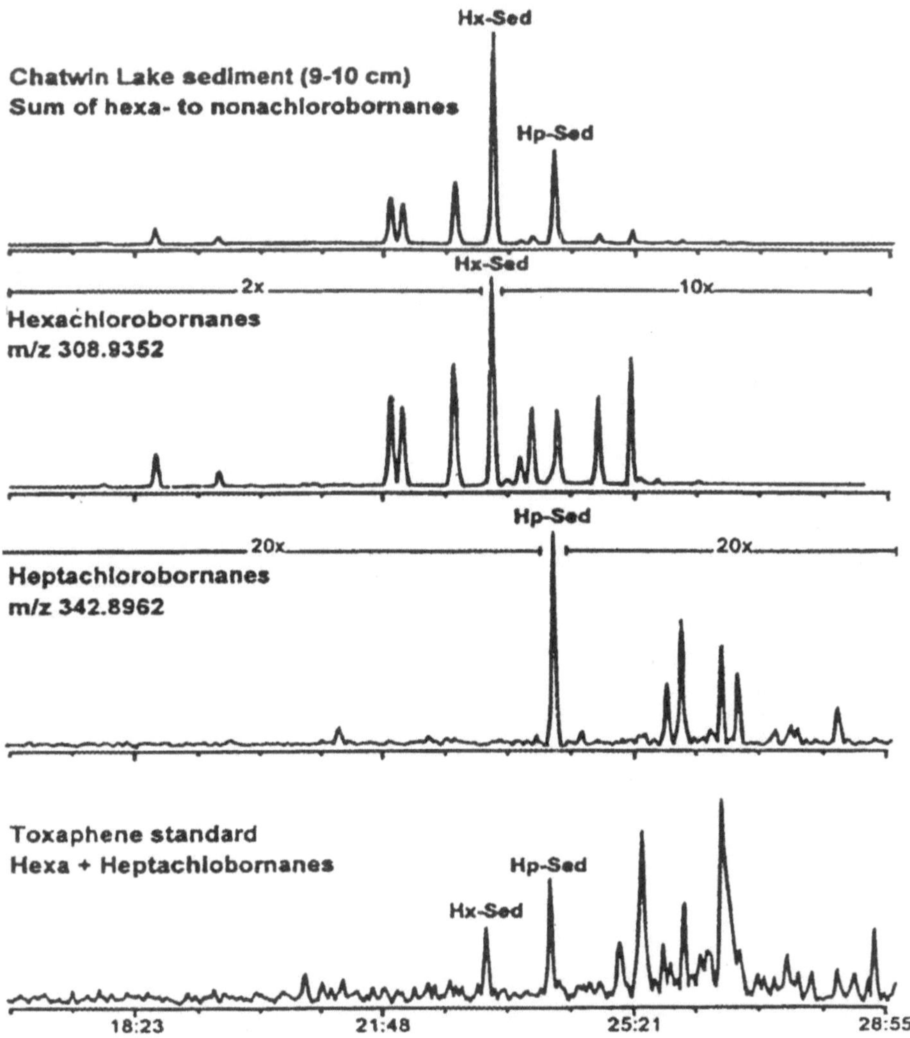

Figure 11. **HRGC-ECNIMS-SIMs for chlorinated bornanes in sediment extracts and in the toxaphene standard.** Reproduced from ref. 52 with permission of the publisher.

Although congener-specific methods are usually thought of as the best methods in the analysis of complex mixtures of closely-related congeners, they are time-consuming, very expensive and involve painstaking work. We are not aware of any laboratory that routinely analyzes all 209 PCB congeners, for example, and there are far more toxaphene congeners.

7. Quality Assurance

The ultimate objective of the toxaphene method validation process is to provide evidence that the method does what it is intended to do, reliably and reproducibly. Certified standard reference materials (SRMs) containing toxaphene in water, sediment, fish, and biota are available from the U.S. Environmental Protection Agency repository, Cincinnati, OH, U.S.A.

These analyzed standards should be used to ensure the validity of data generated in the laboratory. Reagent blanks, method blanks, method spike recoveries, surrogate sample matrix spikes and at least duplicate samples are mandatory.

8. Conclusion

Despite the fact that toxaphene has been banned in many countries, it continues to be found in many locations, including remote Arctic regions where it was never used. Such findings are the result of its persistence and mobility, *i.e.*, through long-range atmospheric transport. Toxaphene is highly toxic, but the toxicological significance of toxaphene residues in soils and sediments requires further research. The analysis of toxaphene is very complicated because of the large number of congeners in technical formulations, far more, for example, than the 209 PCB congeners. It appears that the development of congener-specific methods of analysis of toxaphene would be arduous and very expensive. In our opinion a more practical approach would be to determine total toxaphene residues by solid probe electron impact ionization and electron capture negative ion mass spectrometry. This technique is up to 10 times faster than gas chromatography with the same detector, and more precise.

9. References

1. W. Vetter, *Chemosphere*, **26**, 1079 (1993).
2 G. A. Pollock and W. W. Kilgore, *Residue Rev.*, **69**, 87 (1978).
3. L. V. Zyglado, Domestic usage of toxaphene. U.S.-EPA, Office of Pesticides Programs, Economic Analysis Branch, p. 9 (1982).
4. IPCS (International Programme on Chemical Safety), Camphechlor: Health and Safety Guide No. 40, World Health Organization, Stuttgart, Germany, (1990).
5. Pest Management Regulatory Agency, Department of Health, Hull, Québec J8Y 3Z5, Canada (personal communication), (1995).
6. Meeting Background Report 1995. *International Experts Meeting on Persistent Organic Pollutants: Towards Global Action.* Vancouver, Canada, June 4-8, 1995.
7. T. F. Bidleman, R. L. Falconer and M. D. Walla, *Sci. Total Environ.*, **160-161**, 55 (1995).
8. T. F. Bidleman, C. P. Rice and C. E. Onley, High molecular weight chlorinated hydrocarbons in the air and sea: rates and mechanisms of air/sea transfer, in *Marine Pollutant Transfer*, ed. by H. L. Windom and R. A. Duce, Lexington Books, Lexington, MA (1981).
9. H. W. Harder, E. C. Christensen, J. R. Mattews and T. F. Bidleman, *Estuaries*, **3**, 142 (1980).
10. R. D. Arthur, *Bull. Environ. Contam. Toxicol.*, **15**, 129 (1976).
11. W. R. Swain, *J. Great Lakes Res.*, **4**, 398 (1980).
12. C. P. Rice and M. S. Evans, Toxaphene in the Great lakes, in *Toxic contaminants in the Great Lakes, ed. by J. O.* Nriagu and M. S. Simmons, John Wiley & Sons/Interscience, New York, N.Y. (1984).
13. J. Paasivirta, M. Knuutila and R. Paukku, *Chemosphere*, **14**, 1741 (1985).
14. New and revised Great Lakes water quality objectives. Vol. III., International Joint Commission Canada-U.S.A., Windsor, Ont. (1977).
15. W. E. Gaylord, Treatment of East Bay, Alger County, Michigan with toxaphene for control of sea lampreys, in *Resource Publication II.*, U.S. Department of the Interior, Fish and Wildlife Service, Washington D.C., January 1966.
16. R. A. Hughes and G. F. Lee, *Environ. Sci. Technol.*, **7**, 934 (1973).
17. H. P. Nicholson, *Limnol. Oceanogr.*, **9**, 310 (1964).
18. J. O. Keith, *J. Appl. Ecol.* Suppl., **3**, 71 (1966).
19. A. E. Carey, J. A. Gowen, H. Tai, W. G. Mitchell and G. B. Wiersma, *Pestic. Monit. J.*, **10**, 137 (1978).
20. C. J. Durant, *Pestic. Monit. J.*, **6**, 94 (1972)
21. H. C. Mattraw, Jr., *Pestic. Monitor. J.*, **9**, 106 (1975).
22. G. F. Lee, R. A. Hughes and G. D. Veight, *Water, Air, Soil Pollut.*, **8**, 479 (1977).
23. R. Vaz and G. Blomkvist, *Chemosphere*, **14**, 223 (1985)
24. H. Pyysalo and K. Antervo, *Chemosphere*, **14**, 1723 (1985).
25. H. Mussalo-Rauhamaa, H. Pyysalo and R. Moilanen, *J. Toxicol. Environ. Health*, **13**, 689 (1984).
26. M. A. Saleh and J. E. Casida, *J. Agric. Food Chem.*, **26**, 583 (1976).
27. S. Sandborn, *Env. Entomol.*, **5**, 533 (1976).
28. L. C. Terriere, *J. Agric. Food Chem.*, **14**, 1041 (1966).
29. T. Oshawa, J. R. Knox, S. Khalifa and J. E. Casida, *J. Agric. Food Chem.*, **23**, 98 (1975).
30. T. Cairns, E. G. Siegmund and J. E. Froberg, *Biomed. Mass Spectrom.*, **8**, 569 (1981).
31. R. R. Williams and T. F. Bidleman, *J. Agric. Food Chem.*, **26**, 280 (1978).
32. T. F. Bidleman and C. E. Onley, *Science*, **183**, 516 (1974).
33. T. F. Bidleman and C. E. Onley, *Nature*, **257**, 475 (1975).
34. S. F. Stepan, J. F. Smith, U. Flego and J. Reukers, *Water Res.*, **12**, 447 (1978).
35. W. M. J. Strachan, Environmental distribution of toxic chemicals, in *Current Research* Environmental Contaminants Div., NWRI, Environment Canada, p. 42 (1984).
36. D. L. Stalling and J. N. Huckins, *EPA-602/3-76-076.* U.S.-Epa, Washington, DC, U.S.A. (1976).
37. F. I. Onuska, *J. High Resol. Gas Chromat. & Chromat. Commun.*, **7**, 660 (1984).
38. F. I. Onuska, *J. High Resol. Chromatogr.*, **12**, 4 (1989).
39. D. B. Sergeant and F. I. Onuska, Analysis of toxaphene in environmental samples, in *Analysis of Trace Organics in the Aquatic environment*, ed. by B. K. Afghan and A. S. Y. Chau, CRC-Press Inc., Boca Raton, Florida, U.S.A. (1989).
40. F. I. Onuska, K. A. Terry, A. Seech and M. Antonic, *J. Chromatogr. A*, **665**, 125 (1994).
41. K. R. Griffitt and J. C. Craun, *J. Assoc. Off. Anal. Chem.*, **57**, 168 (1974).

42. F. I. Onuska and F. W. Karasek, *Open Tubular Column Gas Chromatography in Environmental Sciences,* Plenum Press, New York (1984).
43. F. W. Karasek and R. E. Clement, *Basic Gas Chromatography-Mass Spectrometry,* Elsevier Publishing Co., Amsterdam, The Netherlands (1988).
44. G. A. Stern, D. C. G. Muir, J. B. Westmore and W. D. Buchannon, *Biol. Mass Spectrom.*, **22,** 19 (1993).
45. W. V. Turner, J. L. Engel and J. E. Casida, *J. Agric. Food Chem.*, **25,** 1394 (1977).
46. A. G. Harrison, *Chemical Ionization Mass Spectrometry,* CRC Press, Boca Raton, FL (1983).
47. T. F. Bidleman, U. Wideqvist, B. Jansson and R. Soderlund, *Atmospheric Environ.*, **21,** 641 (1987).
48. S. Erhardt-Zabik, J. T. Watson and M. J. Zabik, *Biomed. Environ. Mass Spectrom.*, **19,** 101 (1990).
49. B. Lau, D. Weber and P. Andrews, *Chemosphere,* **32,** 1021 (1996).
50. F. I. Onuska and K. A. Terry, *J. Chromatogr.*, **471,** 161 (1989).
51. H. Parlar, J. Burhenne and D. Hainzl, Isolation and identification of twenty different chlorinated terpenes found in environmental samples. *Workshop on the analytical and environmental chemistry of toxaphene,* sponsored by Dept. of Indian and Northern Affairs, Atmospheric Environ. Service, Dept. of Fisheries and Oceans, Burlington, Ont. Canada, February 4-6, 1993.
52. B. M. Miskimmin, D. C. G. Muir, D. W. Schindler, G. A. Stern and N. P. Grift, *Environ. Sci. Technol.*, **29,** 2490 (1995).

THE DETERMINATION OF POLAR COMPOUNDS IN THE AQUATIC ENVIRONMENT

MARC J.-F. SUTER
Department of Chemistry
Swiss Federal Institute for Environmental Science and Technology
EAWAG
CH-8600 Duebendorf
Switzerland

1. Introduction

A large amount of anthropogenic organic compounds are released into the aquatic environment through municipal and industrial discharges. Other important sources of contamination are e.g. dry and wet deposition from the atmosphere, run off from agriculturally used soils, urban surfaces, and leachates from landfills. As a result, a large number of organic pollutants can be found in surface and groundwaters that may in the end affect the quality of drinking water, or may have other adverse effects on the biosphere.
The effects of anthropogenic organic compounds on the environment vary, according to their fate and behavior in the biosphere and the quantities involved. For instance, branched alkylbenzene sulfonates (ABS), an anionic surfactant used in high quantities up to the late sixties in western Europe, the US and Japan, caused heavy foam production on rivers. This was due to the fact that these compounds are not easily degraded by biological processes in the natural environment or waste water treatment plants.
Other compounds, like ethylenediaminetetraacetate (EDTA), a chelating agent used in high quantities in industrial processes, have the potential to remobilize heavy metals in sediments of surface waters (rivers, lakes, oceans). In the case of EDTA, this danger arises from the fact that it is a persistent compound that is not degraded in waste water treatment plants. According to *Kari and Giger*, and *Karametaxas et al.* [1,2], the major elimination pathway, is through photodegradation of the Fe^{III}-EDTA complex.
Pesticides on the other hand are meant to prevent weed or fungi growth, and kill pest that would otherwise reduce crops. *Fent et al.* showed that similar problems are caused by underwater boat paints, so called anti-foulings that accumulate in harbor sediments and are highly toxic for aquatic species [3].
Other compounds have no serious effects on the ecosystem as such, but are transformed to highly problematic pollutants. A well known case, described by *Giger et al.*, are the nonylphenolpolyethoxylates (NP_nEO) that were widely used as nonionic surfactants [4]. In Europe they have been replaced in household detergents by other surfactants. How-

R. M. Caprioli et al. (eds.), Selected Topics and Mass Spectrometry in the Biomolecular Sciences, 559–573.

ever they are still in use in e.g. the US. Their major degradation product 4-nonylphenol has been shown by *Jobling and Sumpter* to be estrogenic to fish [5].
The impact of all these organic pollutants on ecosystems needs to be fully understood in order to prevent unwanted side effects, like accumulation of pollutants in the food chain, or reduction of biodiversity. For this reason, sophisticated analytical techniques are needed for enrichment, separation, identification and quantitation of these compounds in complex matrices.

2. Environmental Analytical Organic Chemistry

Environmental analytical chemistry deals with very low concentrations in the parts per million (ppm) and lower levels of ecotoxicologically important compounds, found in the biosphere, sewage sludge, soil, sediments and water samples, be that ground, drinking, surface water or samples from waste water treatment plants. This always implies identifying and quantifying problem analytes in complex matrices. Therefore, sample work up, separation and detection are major corner stones of environmental analytical chemistry. Also, sampling is a potential pitfall that, however, will not be discussed in this chapter.

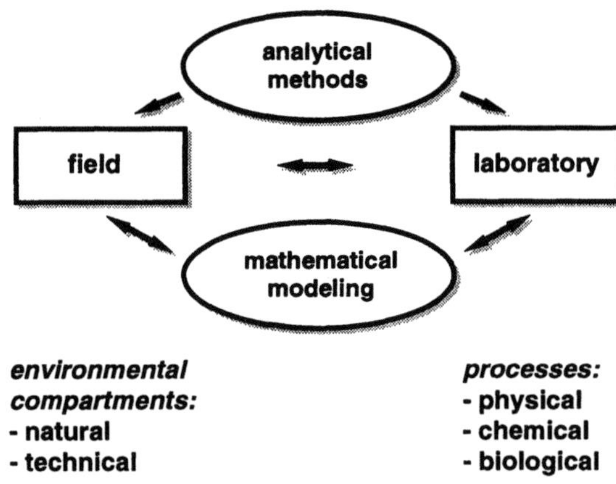

Figure 1: Concept for fate and behavior studies of organic micropollutants.

The strategy for fate and behavior studies of organic contaminants involves analytical methods development, field and laboratory studies and mathematical modeling, as schematically represented in figure 1. A chemical compound is exposed to physical, chemical and biological processes in natural or technical (drinking water production, waste water treatment) compartments that need to be accounted for in modeling its transport, transfer and transformation. This strategy is aimed at predicting a compounds ecotoxicological effect on the environment. For example, processes affecting organic pollutants in lakes can be described with a model presented by *Ulrich et al.* [6].

The analytical method used for identification and quantitation of environmental contaminants depends on their physical properties. The diagram shown in figure 2 lists volatility *versus* polarity. This sort of diagram helps in method development for the analysis of new compounds, especially regarding sample enrichment and separation. Gas chromatography (GC) is used for mixtures of lipophilic molecules with high volatility, while polar and non-volatile compounds are typically analyzed by high performance liquid chromatography (HPLC) or capillary electrophoresis (CE).

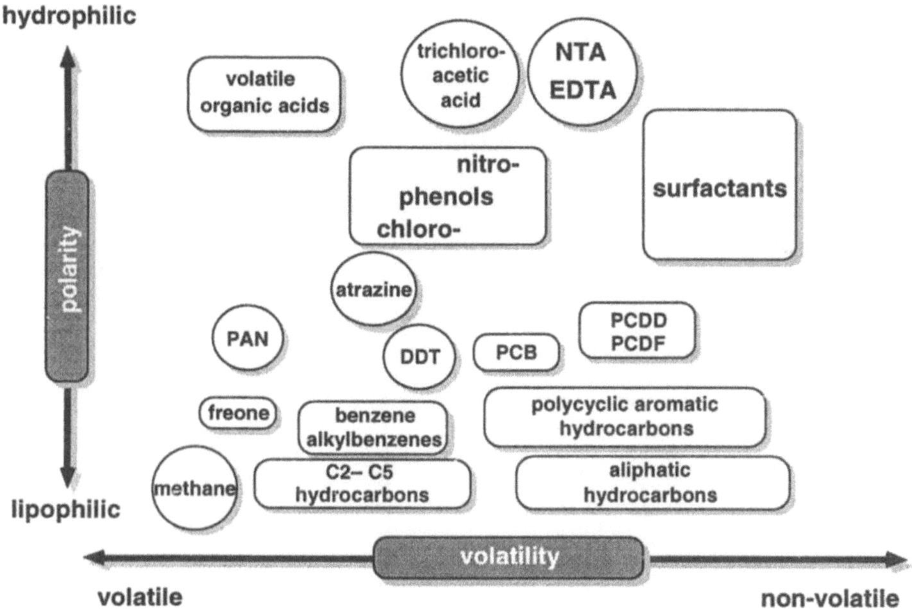

Figure 2: Volatility *versus* polarity diagram for some organic pollutants.

The classical approach for the determination of polar non-volatile compounds involves derivatization, to make them amenable for GC analysis (for more information on chemical derivatization see [7]). For instance, the chemical reaction shown in Figure 3, transforms an anionic linear alkylbenzene sulfonate (LAS) to its corresponding trifluoroethanol ester [8,9]. In this case, trifluoroethanol was chosen because of the good response in electron capture detection and the high volatility of the resulting fluorinated compound.

Today, mass spectrometry plays a key role as universal detector in environmental organic chemistry. The only prerequisite for GC/MS analysis is volatility of the analytes. Specificity in mixture analysis, however, requires that either the molecular ion, or typical fragment ions be present in the mass spectrum of the compound under investigation. If either is absent, chemical ionization (CI) has to be used for the production of ions, instead of electron ionization (EI). Very often, CI is preferred, when sensitivity is crucial, because fewer fragments are produced under CI conditions, which also means that the specific signal will have a higher signal-to-noise ratio.

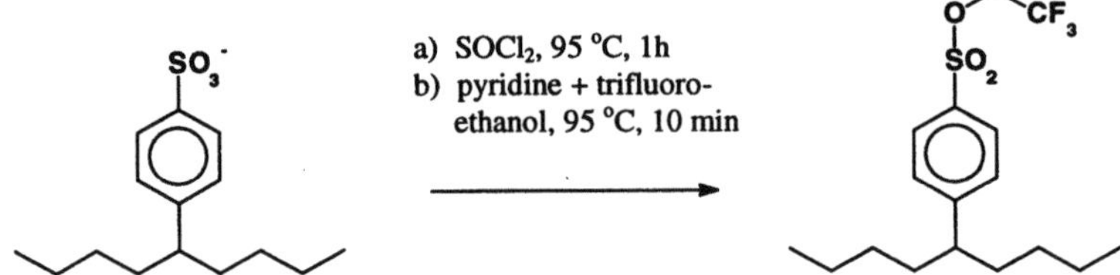

Figure 3: Derivatization of an anionic surface active compound. The transformation of the sulfonate group to a trifluoroethanol ester removes the negative charge and increases the volatility of the molecule.

With the advent of soft ionization techniques, like fast atom bombardment (FAB, *Barber et al.* [10]) and electrospray ionization (ESI, *Yamashita and Fenn* [11]), polar and charged compounds have become compatible with mass spectrometry, making derivatization unnecessary. Among others, these techniques allow the direct coupling with HPLC or CE.

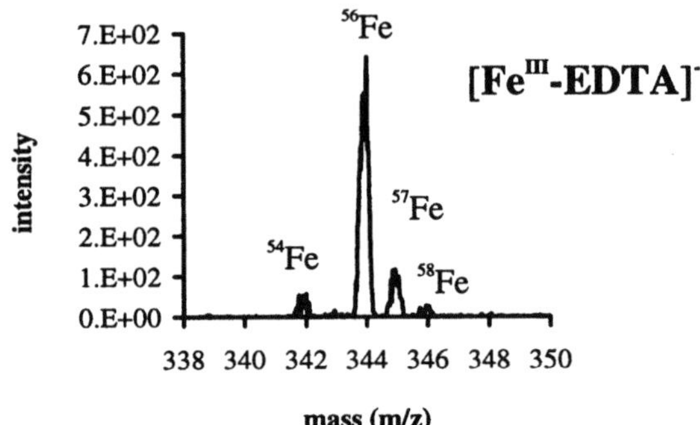

Figure 4: Negative ion electrospray mass spectrum of the metal complex Fe^{III}-EDTA.

Classical analytical techniques require derivatization of EDTA that prevents the identification and speciation of the intact complex [2]. However, this is now possible using ESI, because the energy transferred to the target compound in the desorption process is small enough to prevent dissociation of the complex. Figure 4 shows the negative ion electrospray mass spectrum of the metal complex Fe^{III}-EDTA, with the isotope pattern of iron clearly visible.

In a similar way, compounds like LAS can be analyzed directly using ESI, without chemical derivatization, thus effectively removing one potential source of errors of quantitation.

3. GC/MS and GC/MS/MS Differentiation of LAS and ABS

Branched alkylbenzene sulfonates were substituted by their linear analogs in the sixties because of their better degradability. An account of this voluntary replacement initiated by the industry is accessible in the sedimentary records of lakes in Switzerland. However, in samples that contain both ABS and LAS, it is very difficult to distinguish between the two compound classes. LAS is a mixture of several homologs of increasing alkyl chain length and positional isomers. In addition to that, ABS also shows a wide variety of branching in the alkyl chain. The complexity of the mixture is shown in figure 5 with LAS on the left and ABS on the right.

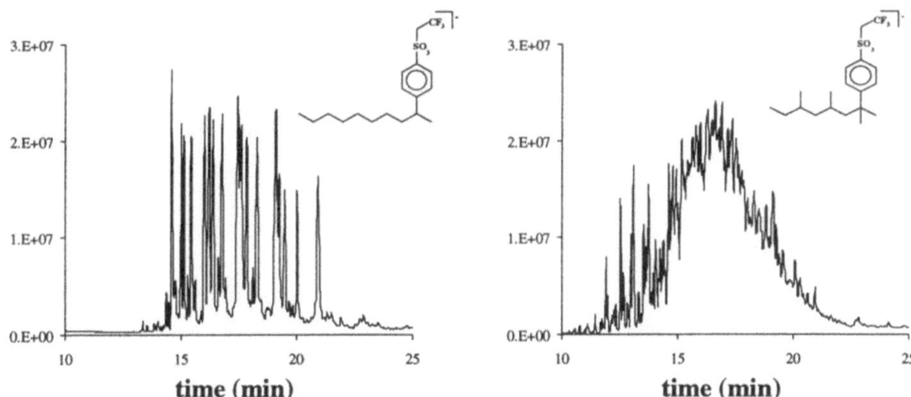

Figure 5: GC/MS total ion chromatogram for a commercial mixture of derivatized LAS (left) and ABS (right).

Without the mass selectivity of a mass spectrometer, it would be impossible to distinguish between the various homologs and their positional isomers. Figure 6 shows the reconstructed ion chromatograms of three homologs C_{10}-, C_{11}- and C_{12}-LAS. Separation of the positional isomers is obtained through their varying GC retention times. The isomers with the functional group towards the center of the alkyl chain will elute earliest with the GC conditions used [9]. For the C_{10}-homolog the 5-C_{10}-isomer, with the benzene sulfonate in position 5 of the alkyl chain, will be first, while 2-C_{10} will elute last. 1-C_{10} is not a product of the synthesis most commonly used. The different homologs can be identified by their characteristic masses. In figure 6 for instance, the intense signal m/z 281, which corresponds to the loss of [·O-CH_2-CF_3] from the molecular ion, was extracted from full scan data for the C_{10}-LAS. M/z 295 was recorded for C_{11}-LAS and m/z 309 for C_{12}-LAS.

Figure 7 shows a depth profile of the C_{14}-homolog of LAS in a sediment core. LAS concentrations were determined using GC/MS after derivatizing sediment extracts, following the procedure shown in figure 3.

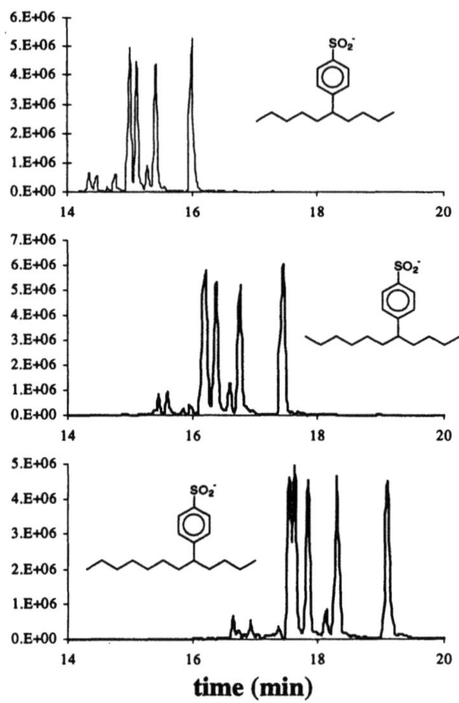

Figure 6: Reconstructed ion chromatograms of C_{10}- (top), C_{11}- (middle) and C_{12}-LAS (bottom) homologs and their respective positional isomers.

The dating of a sediment core is usually achieved by measuring Cs^{137} activity. The maxima found can be attributed to the fallout from nuclear weapons testing (1963), a leakage from a local nuclear power plant (1976), and the accident in Chernobyl (1986). The steady increase of C_{14}-LAS beginning in the early sixties indicates the substitution of ABS with LAS. The concentration reaches a maximum in 1972 and then diminishes again, due to the fact that longer chain homologs were later reduced in LAS formulations and an increasing number of waste water treatment plants were built.

The filled squares represent data obtained by mathematical modeling. *Reiser* [12,13] based the simulation of the fate and behavior of the 2-C_{14}-LAS in the sediment (lake Greifen, near Zürich, Switzerland) on mathematical models proposed by *Berner* [14]. He was using a Gaussian distribution for the period 1950-1972 and an exponential decay for 1972-1994, based on the fact that waste water treatment plants came online during that time. A correction term took into account the steadily increasing consumption of LAS. The processes affecting LAS were assumed to be diffusion and advection only, while biodegradation was expected to be negligible under the conditions present in the sediment. The simulation fitted the experimental data best, when diffusion and sorption were the main processes used in the model. In this way it could be shown that homologs with chain lengths > C_{14} were strongly sorbed to the sediment, with very little diffusion taking place.

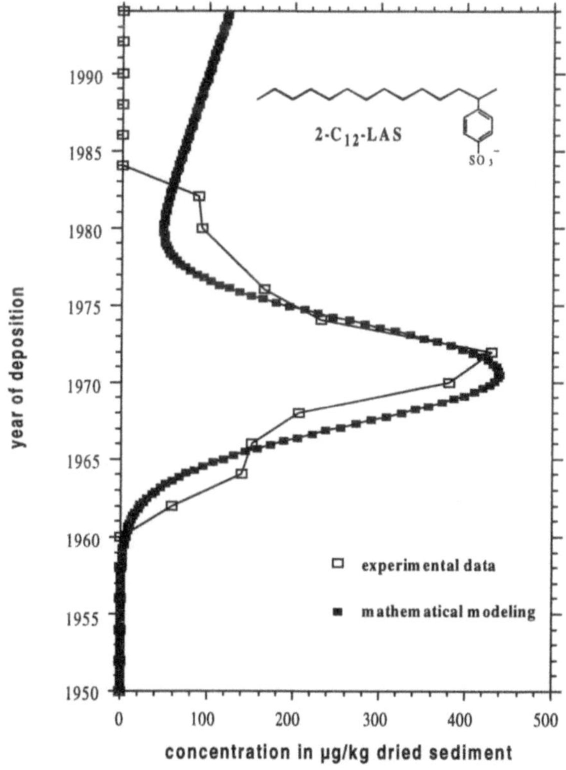

Figure 7: Depth profile of 2-C_{14}-LAS concentrations in a sediment core [12].

Because the gradual fading out of ABS overlapped with the increasing use of LAS, the lacustrine sediments contain zones, where both types of anionic surfactants can be detected. Yet, the identification of LAS in the presence of ABS, is not easy, due to the fact that the corresponding homologs have the same mass and to some extend overlapping GC retention times. However, *Lyon et al.* showed that LAS and ABS can be distinguished under FAB conditions, based on their different fragmentation pathways [15]. The same differentiation can be achieved under negative CI, following the fragmentation shown in figure 8 [9]. The LAS components predominantly form a fragment of mass m/z 167, while ABS form a fragment of mass m/z 181 due to their branching. When monitoring the formation of m/z 167 and m/z 181 during a GC run, ABS and LAS can be differentiated as shown in figure 9. The trace in figure 9 corresponds to the fragmentation of m/z 295, generated from derivatized C_{11}-LAS or -ABS. Similar chromatograms can be obtained for all homologs, when monitoring the fragmentation of $[M - OCH_2CF_3]^- \rightarrow$ m/z 181, 167.

Compared to figure 5, all positional isomers and all homologs can be identified using a mass spectrometer as GC detector (see figure 6). Furthermore, when applying MS/MS to the problem, even linear and branched alkylbenzene sulfonates can be differentiated as shown in figure 9.

m/z 394 m/z 295 m/z 167

m/z 394 m/z 295 m/z 181

Figure 8: Fragmentation pathways of LAS and ABS under negative CI.

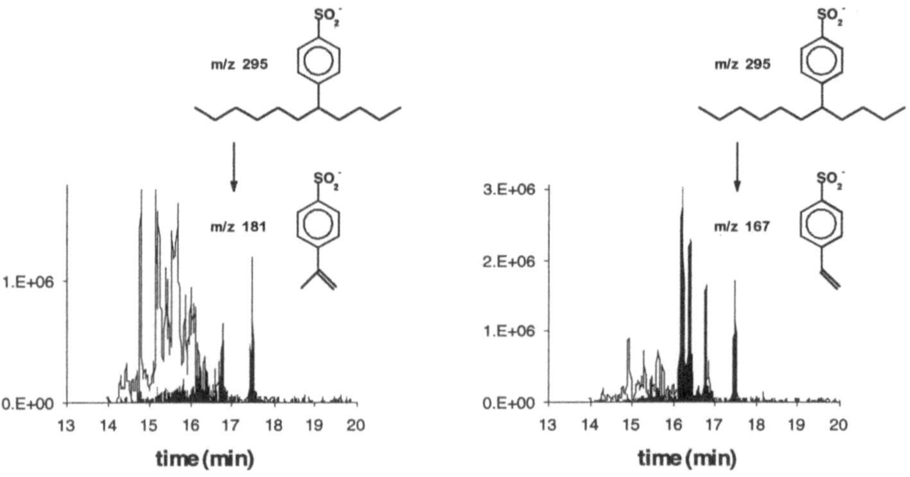

Figure 9: Superimposed reconstructed ion chromatograms from a fragment ion scan of mass 295 that corresponds to C_{11}-LAS or C_{11}-ABS after loss of [·O-CH_2-CF_3]. The left trace monitors the transition m/z 295 → 181, a fragmentation favored by ABS, and the right trace m/z 295 → 167, both recorded for C_{11}-LAS (black) and C_{11}-ABS (clear).

4. LC/MS Analysis of Naphthalene Sulfonates

Benzene and naphthalene sulfonates are used in a wide variety of industrial applications and are intermediates for many industrial products, such as azo-dyes, cement plasticisers, tanning agents, and others. Because they are ionic xenobiotics with a slow biodegradability, they are released into the environment through effluents of wastewater treatment plants and leachates of landfills [16].

The qualitative and quantitative determination of these chemicals, found at varying levels in environmental compartments is typically done using ion-pair HPLC with tetrabutylammonium (TBA) in combination with UV absorption or fluorescence detection. Both detection techniques, however, rely either on spectral libraries, or on known fluorescence wavelengths. In contrast, electrospray ionization mass spectrometry (ESI/MS) provides a more direct identification of unknowns, without the need of extensive libraries.

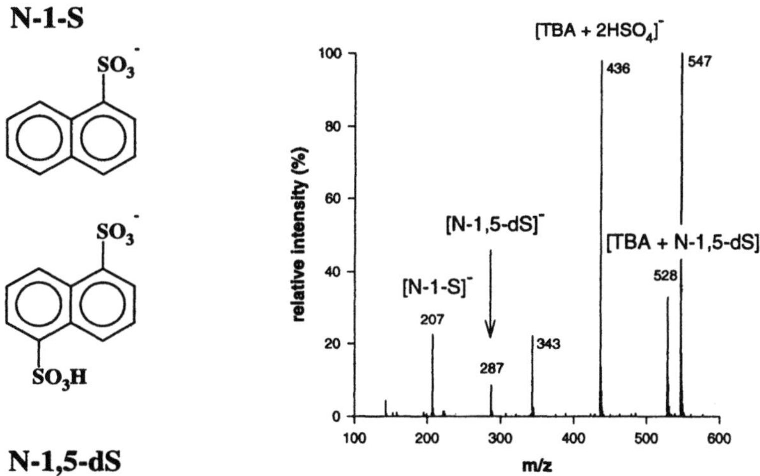

Figure 10: The use of TBA as ion-pair reagent causes important cluster formation in electrospray spectra. Naphthalene monosulfonate (N-1-S) appears at mass 207, and the disulfonate (N-1,5-dS) at mass 287. The signal at mass 528 corresponds to the naphthalene disulfonate-TBA cluster.

For that reason, reversed-phase capillary HPLC, using tetrabutylammonium (TBA) as ion-pair reagent for the separation of naphthalene disulfonates was directly coupled with ESI/MS [17,18]. Figure 11 shows the reconstructed chromatogram of a capillary HPLC/MS run of a standard mixture, containing 6 naphthalene sulfonate isomers. The ions monitored were m/z 207 and 528.

For quantitation purposes it is profitable to obtain linear response for areas measured at different concentrations. Non-linear response caused by competing cluster formation, especially when using cluster signals for quantitation (e.g. m/z 528 in figure 10) and when compounds coelute, is not a problem in the case of the napthalene disulfonates, as shown in figure 12.

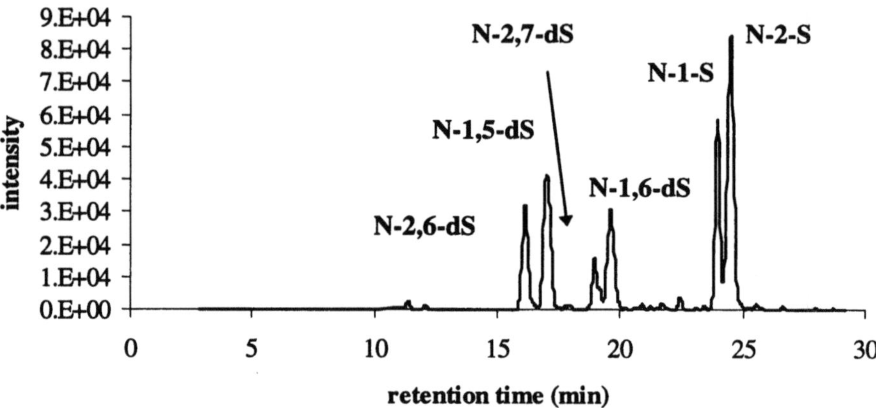

Figure 11: Capillary HPLC/MS reconstructed ion chromatogram of a standard mixture, containing 6 naphthalene sulfonate isomers.

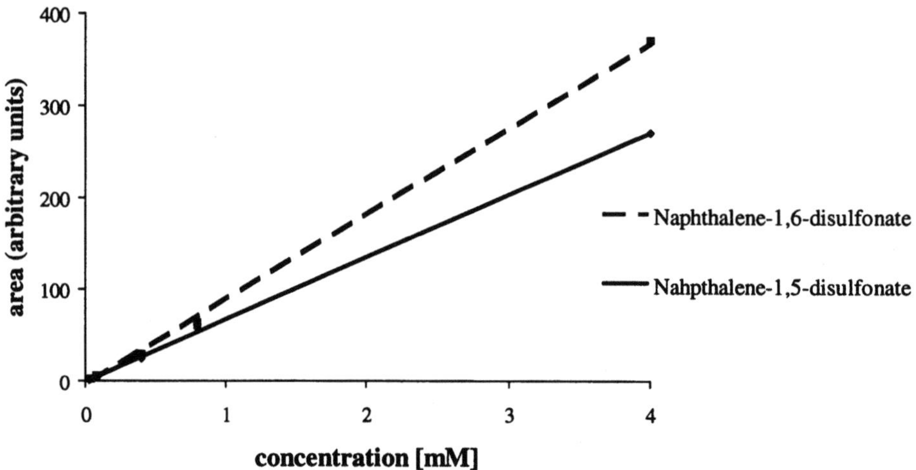

Figure 12: Calibration curve for naphthalene-1,5- and naphthalene-1,6-disulfonate. Both isomers show very good linear response over three orders of magnitude (N-1,5-dS: $y = 92.07\ x - 5.39$, $r = 0.999$; N-1,6-dS: $y = 68.36\ x - 0.42$, $r = 0.999$). The chromatograms were acquired in single-ion-monitoring mode, using the fragment ion [207 $M\text{-}SO_3]^-$ and a matrix cluster as lock-mass.

Another source of discrimination are differing interface settings for optimal transmission of different ions. For instance, singly and doubly charged ions can have other optimal settings for maximal signal intensity. However, this effect is essentially independent of the analyte concentration and thus would be the same over the whole concentration range. In a similar way, cluster formation can be disfavored, by setting the voltages in the gas expansion zone of the interface in such a way that collisions between ions and neutral gas molecules occur. This effect is often used for the generation of fragment ions in the electrospray interface (see for instance [19]). These problems can be avoided, by only monitoring signals of equally charged quasi-molecular ions, or cluster ions.

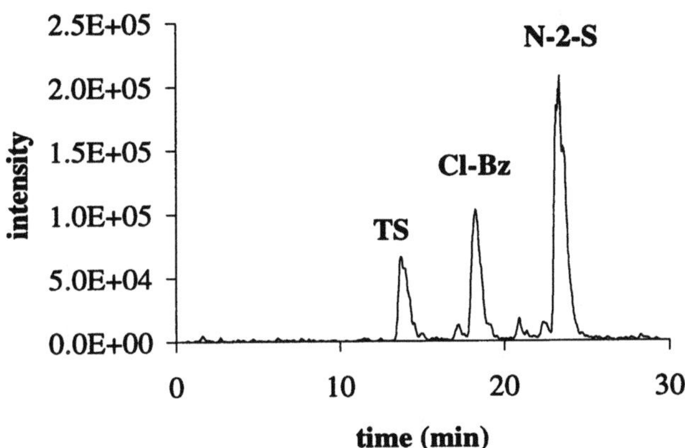

Figure 13: The microbore HPLC/MS chromatogram from a sample taken from a hazardous waste landfill leachate in Kölliken, Switzerland, shows the presence of toluene- (TS), chlorobenzene- (Cl-Bz) and naphthalene-2-sulfonate (N-2-S) (enrichment factor: 300).

Individual benzene and naphthalene sulfonates were separated by using microbore LC without ion-pair reagent and also monitored with ESI/MS. The chromatogram shown in figure 13 was acquired without the ion-pairing reagent TBA, using an acetic acid buffer and a 1mm reversed-phase column. It proves the presence of toluene-, chlorobenzene- and naphthalene-2-sulfonate in a sample taken from a hazardous waste landfill leachate. Using these analytical methods, the fate of sulfonates in landfill leachates and waste water treatment plants can be investigated.

5. Alkylpolyethoxylates (AEO_x)

Linear alkylpolyethoxylates are commonly used nonionic surfactants with the alkyl chain length varying from C_8 to C_{18} and an average number of ethoxy units from 5 to 15, depending on their origine [20]. Using classical GC/MS techniques, only multimers of up to 15 ethoxy units can be detected, due to the temperature limit imposed by the

GC interface. With high-temperature GC (up to 400 °C) e.g. up to 27 ethoxy units of a $C_{16}EO_x$ homolog can be analyzed [21].

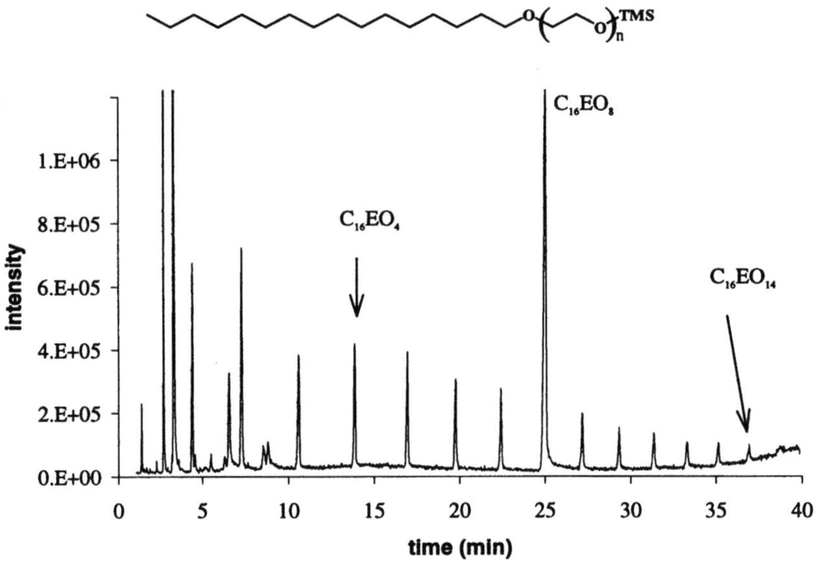

Figure 14: GC/MS chromatogram of t-butyldimethylsilyl derivatives of linear $C_{16}EO_X$, acquired on an instrument with a conventional interface using chemical ionization. The interface temperature was 310 °C and the GC temperature was run up to 300 °C.

Evans et al. have shown that quantitative determination of nonionic surfactants can be achieved, using thermospray LC/MS, another soft ionization technique [22]. *Lattimer et al.* have shown that alkali metal cations are an easy way of enhancing signal intensities in FAB/MS, especially when using Li^+ [23]. As in FAB/MS, electrospray sensitivity is best, when preformed ions are present in the analyte sample. If the analyte is nonionic, addition of cations for the formation of clusters (H^+ , Li^+ , Na^+ , etc.), or derivatization becomes necessary. Figure 15 shows the positive ion ESI spectrum of a commercial nonionic surfactant, dissolved in 50 mM LiCl in methanol. There are two series visible in the spectrum. The one starting at m/z 203 corresponds to $[C_{11}EO_x + Li]^+$, while the one shifted to higher masses is generated by $[C_{11}EO_x + Na]^+$. If the nonionic surfactant consists of ethoxymers of only one alcohol, then direct infusion into the ESI interface gives good results. If however, more than one alcohol are used for the formulation, then a separation step is necessary, in order to be able to quantify all homologs and ethoxymers [22].

Several pathways of aerobic biodegradation lead to fatty acids, carboxylated alkylpolyethyleneglycols or carboxylated polyethyleneglycols as shown by *Schöberl* and *Steber* [24,25], as do some anaerobic biodegradation pathways (*Wagener and Schink* [26]).

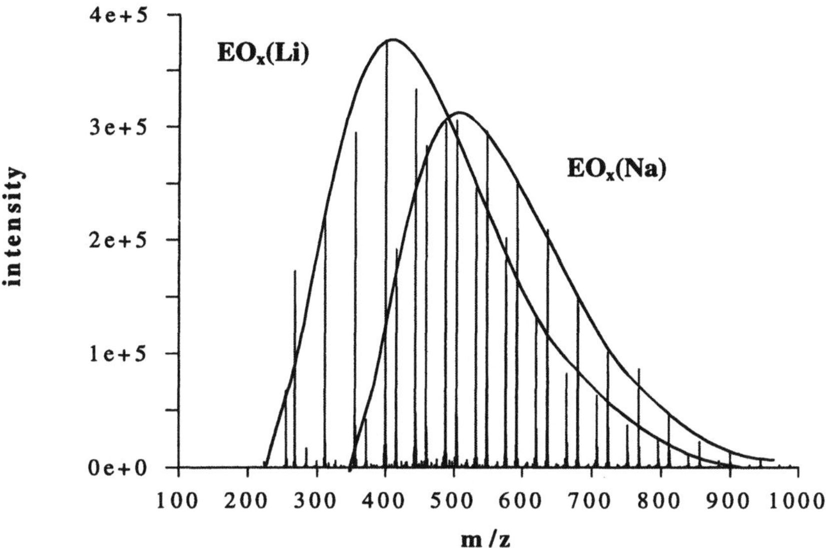

Figure 15: Positive ion ESI spectrum of a commercial mixture of lineare C_{11} ethoxymers. The sample was dissolved in a 1:1 mixture of methanol and water, containing 50 mM LiCl. The two envelopes correspond to the Li and Na cluster ions.

Figure 16 shows the negative ion ESI spectrum of a C_{12}- and C_{13}-EO_x mixture, with the terminal hydroxy group oxydized to the carboxylate. The lowest visible signal corresponds to the C_{12}- and C_{13}- carboxylic acid (m/z 199 and m/z 213), followed by increasing number of ethoxy units.

The case of the AEO_X shows the complexity of a fate and behavior study. Not only are the analytes composed of wild mixtures of isomers and homologs as is also the case with LAS and ABS (see chapter 3), but very often, biotransformation generates additional mixtures. In the case of AEO_X, this also means that some of the products are negatively charged, while the initial compounds are nonionic.

Using ESI, nonionic surfactants can be identified and quantified in environmental compartments.

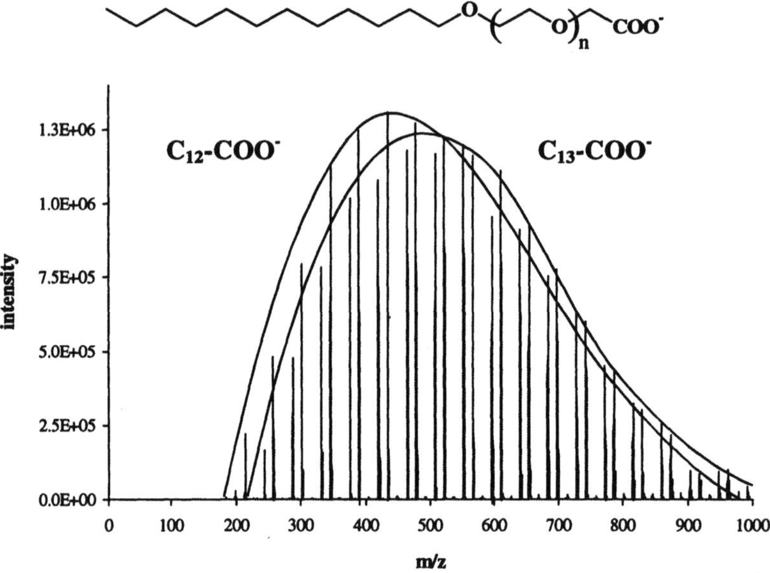

Figure 16: Negative ion ESI spectrum of a mixture of C_{12}- and C_{13}-EO_x, with the terminal hydroxy group oxydized to the carboxylate.

6. Conclusion

The determination of polar compounds in environmental compartments can be achieved by classical means, using GC/MS after a derivatization step, or directly with more recent soft ionization techniques like FAB, thermospray or electrospray ionization. GC/MS is a well established technique, used in a wide area of environmental analysis and has the advantage of reproducibility and huge spectral libraries at its disposition, especially if EI can be used as method of ionization. The soft ionization techniques have the advantage of producing simple spectra, very often with a predominant quasi-molecular ion, but have the disadvantage of non or little structural information, making MS/MS mandatory for structur elucidation. Furthermore, as in CI, competing gas-phase reactions can cause discrimination, and thus non linear response for quantitation. Matrix effects, like cluster formation, often make it difficult to identify mixtures of unknown composition. However, ESI is an easy way for screening environmental samples, because no derivatization is necessary and the small amount of energy deposited into the analyte allows determination of e.g. metal complexes as in the case of Fe^{III}-EDTA.

7. Acknowledgements

The author wishes to thank Walter Giger for figures 1 and 2 and René Reiser and Sonja Riediker for their contributions to chapter 3 and 4, respectively.

8. References

1. F.G. Kari, and W. Giger, *Environ. Sci. Technol.* **29**, 2814-2827 (1995).
2. G. Karametaxas, S.J. Hug, B. Sulzberger, *Environ. Sci. Technol.* **29**, 2992- 3000 (1995).
3. K. Fent, J. Hunn, M. Sturm, *Naturwissenschaften* **78**, 219-221 (1991).
4. W. Giger, P.H. Brunner, C. Schaffner, *Science* **225**, 623-625 (1984).
5. S. Jobling, and J.P. Sumpter, *Aquat. Toxicol.* **27**, 361-372 (1993).
6. M.M. Ulrich, D.M. Imboden, R.P. Schwarzenbach, *Environ. Software* **10**, 177-198 (1995).
7. K. Blau, and J. Halket, *Handbook of Derivatives for Chromatography*, 2nd edition, J. Wiley & Son, Chichester (1993).
8. M.L. Trehy, W.E. Gledhill, R.G. Orth, *Anal. Chem.* **62**, 2581-2586 (1990).
9. M.J.-F. Suter, R. Reiser, W. Giger, *J. Mass Spectrom.* **31**, 357-362 (1996).
10. M. Barber, R.S. Bordoli, R.D. Sedgewick, N.A. Tyler, *J. Chem. Soc., Chem. Commun.* 325-327 (1981).
11. M. Yamashita, J.B. Fenn, *J. Phys. Chem.* **88**, 4451-4459 (1984).
12. R. Reiser, Ph. D. Thesis, in preparation, ETH Zürich (1996).
13. R. Reiser, H. Toljander, W. Giger, *Proceedings of the 17th international Meeting on Organic Geochemistry, September 1995, San Sebastian, Spain,* 583-585 (1995).
14. R.A. Berner, *Early Diagenesis. A Theoretical Approach.* Pinceton Series in Geochemistry, Princeton University Press, Princeton, New Jersey (1980).
15. P.A. Lyon, W.L. Stebbings, F.W. Crow, K.B. Tomer, D.L. Lippstreu, M.L. Gross, *Anal. Chem.* **56**, 8-13 (1984).
16. B. Altenbach, W. Giger, *Anal. Chem.* **67**, 2325-2333 (1995).
17. S. Riediker, Ph. D. Thesis, in preparation, ETH Zürich (1996).
18. M.J.-F. Suter, S. Riediker, W. Giger, *Presented at the 44th Annual Conference on Mass Spectrometry and Allied Topics, Portland, OR* (1996).
19. R.D. Smith, J.A. Loo, C.J. Barinaga, C.G. Edmonds, H.R. Udseth, *J. Am. Soc. Mass Spectrom.* **1**, 53-65 (1990).
20. B. Fell, *Tenside Surf. Det.* **28**, 385-395 (1991).
21. D. Scheidegger, W. Giger, *Presented at the 13th International Symposium on Capillary chromatography, Riva del Garda, Italy* (1991).
22. K.A. Evans, S.T. Dubey, L. Kravetz, I. Dzidic, J. Gumulka, R. Mueller, J.R. Stork, *Anal. Chem.* **66** 699-705 (1994).
23. R.P. Lattimer, H. Muenster, H. Budzikiewicz, *Int. J. Mass Spectrom. Ion Processes* **90** 119-129 (1989).
24. P. Schöberl, *Tenside Deterg.* **19** 329-39 (1982).
25. J. Steber, P. Wierich, *Appl. Environ. Microbiol.* **49** 530-537 (1985).
26. S. Wagner, B. Schink, *Appl. Environ. Microbiol.* **54** 561-565 (1988).

SOME MASS SPECTROMETRY CONTRIBUTIONS IN THE STUDY OF CULTURAL HERITAGE

PASQUALE AGOZZINO
Dipartimento di Chimica e Tecnologie Farmaceutiche
Università di Palermo.
Via Archirafi, 32. 90123 Palermo - Italy

In the last decades the application of various scientific methods and technologies to study materials and handicrafts of historical, artistic or archaeological interest have dramatically increased. These studies contribute to successfully solve various problems of identification, dating, utilization of artistically or historically important objects.

In fact it is possible to obtain very important information on social, religious and familiar customs of the ancient peoples; furthermore the finds are related to the commercial trades between different, and sometimes far, geographical areas, and also to reciprocal influences.

In this way our knowledge on the ancient civilisations, besides getting information from historiographical sources, grows richer also by serious scientific research on the finds. Thus, the science of archaeology has made considerable advances helped by contribution of scientists in a number of other disciplines.

Analytical chemistry, biochemistry, DNA analysis, electron microscopy, isotope measurements and all the spectroscopic techniques played an important part and have been used to study archaeological remains and artistic objects with a precision not available until recently.

The Mass Spectrometry (MS) is a powerful tool also in research in this fascinating field thanks to its high analytical sensitivity and specificity.

Naturally many applications are in the study of organic materials of various origin and provenance. In archaeological field mummies, bones, food remains, seeds, textiles, bitumens, resins, waxes, natural dyes and various patinas in pottery and vitreous vessels are investigated.

Another interesting application of MS in the cultural heritage field is the identification possibility of paint vehicles and pigments in ancient masterpieces in order to study and also in view of a restoration project. The analytical problems related to identification of organic compounds used in painting are complex, mainly due to the empirical approaches employed by the Artists in their use and to the chemical variety of the binders (proteinaceous glues, drying oils, resins, waxes).

In these cases the Pyrolysis-MS (Py-MS) or the Pyrolysis-Gaschromatography-MS (Py-GC-MS) couplings are frequently used to obtain reproducible fingerprints.

However research in all these fields is rather difficult, because of the variety and intrinsic complexity of materials, because of the very small quantities of samples, sometimes at trace level, and at last the need of extrapolation on account of long ageing, and possible environmental contamination.

R. M. Caprioli et al. (eds.), Selected Topics and Mass Spectrometry in the Biomolecular Sciences, 575–579.

Ever since 1976 J. Condamin and F.Formenti, at completion of previous chromatographic works, used the GC-MS technique in order to detect olive oil traces in sherds of transport anphoras, prevalently Dressel 20 type, and in Gallo-Roman oil lamps. Fatty acids analysis confirmed the presence of olive oil, besides a better conservation of oil components in submarine remains was found. In fact it has been later demonstrated that ceramics are the major loci for the survival of residual organic matter absorbed into the porous microstructure of the fired clay walls. Other GC-MS analysis of pitches used to line the inner of wine transport anphoras show the presence of diterpene resin acids.

A GC-MS work carried out in 1979 by A.Malorni & Coll. on a resinous matter in an anphora find in Oplonti (old town buried by Vesuvius eruption in 79 A.D.) identified abietic and dehydroabietic acids: triciclyc diterpenes of coniferous resins. Pine pitches are also present in anphoras found in the Carthage harbour.

An accurate MS study is carried out (C.W. Beck & C. Borromeo) on various pitches found in the Hellenistic Kyrenia wreck by separation of acidic and neutral fractions. Significant variations in the samples composition indicate differences among original pitches and may permit inferences about their botanical source and their manifacturing techniques. The occurrence of a large amount of resin acid methyl esters shows that the pitches are made by pyrolysis of wood; this process forms methanol (wood spirit) and acidic phenolic products: favourable conditions for esterification of acidic terpenes. A similar procedure, in order to obtain resinous products, is described in the "*Naturalis Historia*" by the Roman writer Pliny the Elder. In addition the presence, in a sample from the strake coating of the ship, of substantial amount of longchain normal alkanes indicates addition to the pine pitches of petroleum product (bitumen).

By GC-MS, Hairfield (1990) identified several compounds typical of resin from *Pistacia* genus tree (terebinth) in the Ulu Burun wreck of the late Bronze Age (about the fourteenth century B.C.) .

In 1992, using a FAB source and the MS-MS technique, researchers of the University of Illinois (M.L.Proefke & Coll.) studied the embalming materials in an Egyptian mummy of Roman period (first-second century A.D.). Embalming matters are complex mixtures of resins, hidrocarbons and other natural compounds; the composition indicate the presence of Judaea bitumen from the Dead Sea confirmed by recovery of Ni, Mo and V as microelements. The presence of this ancient source of carbon in these samples explains the inconsistent date assigned to the mummy by ^{14}C analysis.

German researchers (S. Balabanova & Coll.) in the same year reported, in a GC-MS study performed on nine Egyptian mummies the surprising discovery of nicotine, cocaine and hashish traces. Further works confirmed the Judaea bitumen occurrence in embalming matter.

High Performance Liquid Chromatography - MS coupling (HPLC-MS) enabled Japanese researchers (R. Yamaoka & Coll.) to identify the Alizarin in a red dye on Egyptian textiles of Greek-Roman period.

By means of a single ion monitoring technique (SIM) French researchers (P. Richardin & Coll.) discovered and confirmed very small leather traces in old objects, detecting a 4-hydroxy-proline derivative, typical aminoacid of the leather collagen.

The GC-MS coupling is valued in order to identify organic remains in four little Corinthian "figure vases" (so called *askoi,* three in the shape of a hare and one in the form of a bird with a Gorgon' head) made in the seventh and sixth centuries BC. The vase contents were extracted by non-destructive method, not changing the integrity of art object, by suitable solvents. Several of the large number of compound present were significant with regard to identity of the original contents: the terpenoid ether manoyl oxide is indicative of some oleoresin of pine, cypress or juniper; alkanes C_{14} - C_{17} are typical of flowers; linalool oxide and canphor are indicative of plant material; cedrol and cedrene (only in Gorgon vase) are characteristic of cedar wood oil; finally cholesterol presence relates to animal fats. The figure vases thus are employed as perfume bottles and maybe the shapes relate to a particular scent.

Samples of resinous substances from the Manila galleon and other shipwrecks in waters off Thailand and Marianas Islands have been identified by American researchers (R. Gianno & Coll.) using GC-MS, as derived from a *Styrax benzoin* tree; this material has been considered an important incense and medicinal resin since ancient times it was traded in China and exported in Southeast Asia. Other samples from the ships are triterpenoid resins, possibly from dipterocarp trees, used as caulking between ship timbers.

New knowledges are now achieved on the everyday life and diet of ancient peoples. Some researches carried out by GC-MS technique (Evershed & Coll.) on pottery sherds enable the archaeochemist to address a range of issues involving trade, cooking, processing and storage of foods.

Lipidic residues, in particular, are often a direct reflection of the original contents and usage of ceramic vessel (*vases, pans, pots, containers and other basic utensils of various shape and size*). Lipids are well preserved in favourable environments and have potential for use of biogenic organic matter on chemiotaxonomic and biosistemic ground. An additional work of Heron, Evershed and Goad showed that contamination arising from migration of soil components, or compositional alteration through microbial activity are minimal for lipids absorbed in potsherds. For example, the repeated presence of the long-chain hydrocarbon nonacosane (C 29) and their derivatives nonacosan-15-ol and nonacosan-15-one in characteristic proportions indicate the large use of cabbage in England during mediaeval and Saxon times, because the pattern of components is typical of the epicuticolar leaf wax of *Brassica oleracea.*

The presence of cholesterol provides a strong evidence of animal fats and traces of beeswax in some potsherd indicate a beekeepers works and a liking for honey.

The identification of organic residues provide direct evidences of some ancient activities; for example: the use of an adhesive matter, obtained from birch bark, to repair a Roman period jar is evidenced .

Austrian researchers identified by GC-MS a birch resin as an adhesive used for fixing the arrow points in Copper Age.

Pyrolysis-GC-MS studies, were principally applied in paint analysis, but recently are extended in other fields: food remains in archaeological pottery, amber samples from archaeological objects (sometimes to unmasking amber imitations) and organic patinas on statues exposed in urban environment (G.Chiavari & Coll., University of Bologna).

The secondary ions mass spectrometrry (SIMS) resulted a sensitive and reliable technique in determination of trace level elements in old vitreous surfaces (S. Daolio & Coll.), this is useful to characterize the origin, the manufacture, the age and the decay of vitreous old objects.

In 1995 Jens Glastrup (Danish National Museum Laboratory) studied the paint layers on the wooden planks of a Phoenician shipwreck dated 400 B.C. using GC-MS. The results indicates that te painting medium was esparto wax mixed with resin, and thah an encaustic technique was used (an old painting technique using the pigments dispersed in a melt binding medium).

A research group in University of Palermo is also working on ancient finds (potsherds, pitches, waxes, waterlogged woods) using MS and other chemical and instrumental techniques. Recently (work in progress) we studied by GC-MS a glass ampoule of Greek period from center of Sicily, finding some fatty acids and pine resin components. The first results suggest that the ampoule was used to contain an ointment .

In conclusion of this short and non exhaustive revue on the use of mass spectrometry and ancillary techniques to study and solve some important problems of knowledge and conservation, we can foresee an increase of MS application in the field of cultural heritage studies.

References

1. Condamin, J., Formenti, F., Metais, M.O., Michel, M., Blond, P. (1976) The application of GC to the tracing oil in ancient anphorae, ***Archaeometry*, 18**, 195-201.
2. Forbes, R.J. (1965) ***Studies in ancient technologies***, vol.3. E.J. Brill, Leiden.
3. Enzell C.R. and Wahlberg J. (1969) Mass spectrometric studies of Diterpenes. *Acta Chem. Scand.* **23**, 871-91
4. Mills, J.S. and White, R. (1977) Natural resin of art and archaeology their sources, chemistry and identification. ***Studies in Conservation*, 22**, 12-31
5. Addeo F., Barlotti L., Boffa G., A.DiLuccia, A.Malorni, G.Piccioli. (1979) Costituenti acidi di una oleoresina rinvenuta in anfore vinarie durante gli scavi archeologici di Oplonti, *Ann. Fac. Sci. Agr. Univ. Napoli*, **13**, 144-48.
6. Mills J.S. and White R. (1987) *The organic chemistry of museum objects*, Butterworths, London
7. Hairfield, H.H. Jr., Hairfield, E.M. (1990) Identification of a late bronze age resin. *Anal. Chem.* **62**, 41-45 A.
8. Evershed, R.P., Heron, C., Goad,L.J. (1990). Analysis of organic residues of archaeological interest by high temperature GC and GC-MS. *Analyst*, **115**, 1339-42.
9. Beck, C.W., Borromeo, C. (1990) Ancient pine pitch: Technological perspectives from a Hellenistic shipwreck. ***MASCA Res. Pap. Sci. Archaeol.*, 7**, 51-58.
10. Gerhardt, K.O., Searles, S., Biers,W.R. (1990) Corinthian figure vases: non-destructive exraction and GC-MS. ***MASCA Res. Pap. Sci. Archaeol.*, 7**, 41-49.
11. Gianno, R., Erhardt, D., von Endt, D.W., Hopwood, W, Baker, M.T. (1990) Archaeological resins from shipwrecks off the coast of Saipan and Thailand. ***MASCA Res. Pap. Sci. Archaeol.*, 7**, 59-67
12. Heron, C., Evershed, R.P., Goad,L.J. (1991) Effects of migration of soil lipids on organic residues associates with buried potsherds. ***J. Archaeol. Sci.*, 18**, 641-59.
13. Proefke,M.L., Rinehart, K.L.. (1992) Analysis of an Egyptian mummy resin by MS. , ***J. Am. Soc. Mass Spectrom.*, 3**, 582-89.
14. Evershed, R.P. (1993) Archaeology and analysis: GC-MS studies on pottery

fragment shed new light on the past, *Spectroscopy Europe* **5/4**, 21-26.

15. Evans, K., Heron, C. (1993) Glue, Disinfectant and Chewing Gum: Natural product Chemistry in Archaeology, *Chemistry &Industry*, 446-49.

16. Yamaoka R., Shibayama N., Yamada T., Sato M. The identification of dyes of Greco-Roman period fabric in Egypt using LC/MS . (1989) *Shisuryo Bunseki* , **37**, 249-53.

17. Chiavari G., Ferretti S., Galletti G., Mazzeo R. (1991) Analytical Pyrolysis as a tool for the characterization of organic substances in artistic an archaeological objects. *J. Anal. Appl. Pyrolysis.* **20**, 253-61.

18. Richardin P., Bonassies S., Chahine C. Detection of trace quantities of 4-Hidroxiproline using mass specrometry in archaeological leathers. (1991) *Leder,* **42**, 201- 4.

19. Balabanova S., Parsche F., Prisig W. (1992) First identification of drugs in Egyptian mummies. *Naturwissenshaften* **79**, 358.

20. Daolio S., Facchin B., Pagura C., Zecchin S., Verità M., Hreglich S. (1995) Metodologie SIMS nella caratterizzazione di vetri antichi; *Atti "MASSA 95"* p. 83; Palermo, april 25-29 1995.

21. Glastrup J. (1995) A note on the analysis of the binding medium from a Phoenician shipwreck. *Studies in Conservation*; **40**, 65- 68 .

INDEX

—D—

—E—

—F—

—G—

MIX
Papier aus verantwortungsvollen Quellen
Paper from responsible sources
FSC® C105338

If you have any concerns about our products,
you can contact us on
ProductSafety@springernature.com

In case Publisher is established outside the EU,
the EU authorized representative is:
Springer Nature Customer Service Center GmbH
Europaplatz 3, 69115 Heidelberg, Germany

Printed by Libri Plureos GmbH
in Hamburg, Germany